ADVANCED BRAIN NEUROIMAGING TOPICS IN HEALTH AND DISEASE – METHODS AND APPLICATIONS

Edited by **T. Dorina Papageorgiou,**
George I. Christopoulos
and **Stelios M. Smirnakis**

Advanced Brain Neuroimaging Topics in Health and Disease - Methods and Applications

http://dx.doi.org/10.5772/58256

Edited by T. Dorina Papageorgiou, George I. Christopoulos and Stelios M. Smirnakis

Contributors

Nasser H. Kashou, Christopher W. Tyler, Lora T. Likova, Luis Hernandez-Garcia, Hesamoddin Jahanian, Gisela E. Hagberg, Elisa Tuzzi, Yeji Han, Sung Suk Oh, Joong Koo Kang, HyunWook Park, Reza Farivar, Wim Vanduffel, Wei Gao, Hongtu Zhu, Kelly S. Giovanello, J. Keith Smith, Dinggang Shen, John H. Gilmore, Weili Lin, John Schlerf, Tobias Wiestler, Timothy Verstynen, Jörn Diedrichsen, Hiroki Yamamoto, Natasha Sigala, Tessa M. van Leeuwen, Karl Magnus Petersson, Oliver Langner, Mark Rijpkema, Peter Hagoort, Alyssa A. Brewer, Brian Barton, T. Dorina Papageorgiou, Amalia Papanikolaou, Stelios M. Smirnakis, Deborah A. Hall, Cornelis P. Lanting, Douglas E. H. Hartley, Aymeric Guillot, Franck Di Rienzo, Christian Collet, Aaron J. Suminski, Robert A. Scheidt, Victoria A. L. Mosher, Einat Liebenthal, Bradley G. Goodyear, Julia Kivistö, Hilkka Soininen, Maija Pihlajamäki, Long Jiang Zhang, Guang Ming Lu, Hui Mao, Vincent Lai, Kirsten G. Volz, Gerd Gigerenze, Ann H. Harvey, Ulrich Kirk, P. Read Montague, Richard S. Pond, Jr., Stephanie B. Richman, David S. Chester, C. Nathan DeWall, D.Y. Phua, G.I. Christopoulos

Published by InTech

Janeza Trdine 9, 51000 Rijeka, Croatia

Publishing Process Manager Jana Sertic

Technical Editor InTech DTP team

Cover InTech Design team

First published May, 2014

Printed in Croatia

Additional hard copies can be obtained from orders@intechopen.com

Advanced Brain Neuroimaging Topics in Health and Disease - Methods and Applications, Edited by T. Dorina Papageorgiou, George I. Christopoulos and Stelios M. Smirnakis

p. cm.

ISBN 978-953-51-1203-7

For EU product safety concerns:
IN TECH d.o.o., Prolaz Marije Krucifikse Kozulić 3, 51000 Rijeka, Croatia,
info@intechopen.com or visit our website at intechopen.com.

Meet the editors

Dr. Papageorgiou obtained a B.A. in Psychology and Sociology (University of Georgia), a M.H.Sc. in Psychiatric Epidemiology (Johns Hopkins University), and a Ph.D. in the Biological Sciences with a focus on the brain neuroimaging of opioids (University of Texas - M.D. Anderson Cancer Center; MDACC). She continued with three postdoctoral fellowships: (i) neuroimaging of pain in cancer (MDACC); (ii) real-time fMRI neurofeedback of speech (Baylor College of Medicine - BCM); and (iii) real-time fMRI neuro-feedback of cortical blindness (BCM). Her research interests focus on the mechanisms of cortical plasticity, and neuro-rehabilitation of cortical blindness, speech impairment and, chronic pain syndromes, as a result of neurological disorders, traumatic brain injury or, cancer-related symptoms using targeted/individualized real-time fMRI neurofeedback methods.

George Christopoulos holds a Ph.D. in Cognitive Neuroscience (University of Cambridge) and has extensive post-doctoral research experience (Cambridge, Virginia Tech and Baylor College of Medicine). (By the way, he is the guy on the left in the picture). Dr. Christopoulos' research combines game theory, behavioural economics, reinforcement learning and cognitive neuroscience in an effort to uncover the mechanisms of human decision making. His target is to help people make better decisions. Dr. Christopoulos' enjoys multi-disciplinary work and his research has a mutual attraction with many disciplines, including culture science, organizational behaviour, psychopathology and social psychology. Dr. Christopoulos is currently an Assistant Professor at Nanyang Business School, Nanyang Technological University (Singapore) and the Research Manager of the Culture Science Institute.

Dr. Stelios Smirnakis was born in Iraklion, Crete, Greece. He obtained a BA in Physics and the MD/PhD in Medicine and Physics at Harvard University. After training in Neurology at the Partners Neurology Program, he specialized in Vascular Neurology and Critical Care. He served as a Research Scientist at the Max Planck Institute for Biological Cybernetics (Dept. Logothetis) in Tuebingen, Germany, working on a primate fMRI model of visual cortex plasticity after injury, then as Assistant Professor of Neurology at Harvard University, before moving to Baylor College of Medicine in March 2008 as an Assistant Professor of Neurology & Neuroscience. His research interests focus on studying cortical circuit function in health and disease using a range of methods including functional magnetic resonance imaging, two-photon imaging and classical electrophysiology.

Contents

Preface

"New directions in science are launched by new tools much more often than by new concepts. The effect of a concept-driven revolution is to explain old things in new ways. The effect of a tool-driven revolution is to discover new things that have to be explained."

Freeman Dyson, "Imagined Worlds", Harvard University Press, 1997.

The purpose of this book is to compile a sample of recent developments in brain functional magnetic resonance imaging (fMRI). In the last two decades, we witnessed several advances in neuroimaging, each opening new vistas in the exploration of the workings of the human brain. Improvements in brain mapping methods, in particular, now afford us an unprecedented opportunity for studying the structure and function of neural networks, informing our understanding of sensory perception, cognitive processing and motor function.

The non-invasive technology of fMRI has revolutionized our understanding of brain function over the last twenty years. fMRI measures the Blood-Oxygen-Level-Dependent (BOLD) signal, which is a function of the hemodynamic and metabolic response to neural activity. This surrogate measure of neural activity can then be used to study brain function in a number of settings. A host of different experimental paradigms have been used to study neural processing, and it is well beyond the scope of this book to do them justice. We therefore have selected a few examples that illustrate promising areas of recent development. The application of novel technologies discussed will allow us to not only elucidate the cortical and subcortical circuits' mechanisms but also achieve more accurate diagnoses by inducing plasticity or reorganization with the goal to restore injured or diseased brain activity patterns to normal patterns of activity. These advances will foster the development of broader biomedical applications, and will produce a host of associated economic benefits. In addition, introductory chapters of this book (1, 6, and 25) provide basic information to non-experts, such as clinicians and social scientists in the field of brain neuroimaging. The list of chapters is not exhaustive, and we have certainly omitted much more worthy material than it was possible to include.

The book has two sections. The *Methods* section (Section 1), which contains a practical guide to fMRI experiments (chapter 1), and then focuses on select technological advances, such as the neurometabolic underpinnings of BOLD signal dynamics (chapter 2), the promise and limitations of perfusion-based fMRI (chapter 3), the decoding of the phase signal to acquire additional information (chapter 4), the combination of fMRI and EEG measurements and (chapter 5), the ability to combine invasive recording and micro-stimulation methods with fMRI for studying the primate brain (chapter 6), the resting state in the developing brain (chapter 7), as well as maps of cerebellar function (chapter 8).

The *Applications* section (Section 2) presents examples of how recent fMRI methodology can be used to address specific questions about brain function. We discuss latest applications to the visual system (1st subsection: *"Advances in Visual and Auditory Cortical Network Applications"*), such as the development and application of probability- and entropy-based atlases to quantify the inconsistency in localization of human visual areas (chapter 9), the study of perceptual and visual categorization of stimuli using fMRI (chapter 10), the color specificity in the human V4 complex (chapter 11), the developmental plasticity of the human visual system (chapter 12), cross-modal plasticity in blind patients (chapter 13), as well as the potential of real-time fMRI neuro-feedback strategies for enhancing cortical visual rehabilitation (chapter 14). In this 1st subsection, we also discuss the auditory system's plasticity and the framework of theoretical and methodological challenges associated with fMRI auditory system studies (chapter 15).

The 2nd subsection of the *Applications* section, *"Advances in Motor Cortical Network Applications"* presents applications of the motor system, including studies of motor imagery as a tool to enhance cognitive and motor performance (chapter 16), the characterization of feedback regulation of limb position using fMRI (chapter 17), and fMRI predictors of surgical outcome (chapter 18).

The 3rd subsection of the *Applications* section, *"Brain Neuroimaging Applications in Disease Processes"*, discusses fMRI findings on the pathophysiology underlying Alzheimer's disease (chapter 19), and hepatic encephalopathy (chapter 20), and examines the use of diffusion- and perfusion-weighted imaging in acute ischemic stroke (chapter 21).

The 4th and final subsection of the *Applications* section, *"Neuroimaging of Decision Making and Social Valuation"*, offers a comprehensive review of social neuroscience tasks, aiming to introduce social scientists to fMRI (chapter 22), presents the cortical and subcortical underpinnings of decision making and valuation mechanisms (chapter 23), demonstrates the importance of prediction (as compared to fitting) of behavior in human decision making (chapter 24), and finally elucidates the biological and psychological mechanisms underlying social (emotional) pain (chapter 25).

The brain is the most complex computational device we know, consisting of highly interacting and sometimes redundant networks of areas, supporting specific brain functions. The rules by which these areas organize themselves to perform specific computations have only now started to be uncovered. Advances in neuroimaging already play a critical role in our effort to understand the functional anatomy of distributed cortical circuits and to define how these circuits malfunction in various disease states. The next few years will bring an exponential growth in understanding basic brain processes, largely as a result of further advances in neuroimaging methods and applications.

On April 2nd, 2013, the President of the United States of America, Barack Obama unveiled an initiative to map the brain with these words: *"We have a chance to improve the lives of not just millions, but billions of people on this planet through the research that's done in this BRAIN Initiative alone."* On the other side of the Atlantic, the European Union has heavily funded the "Human Brain Project" led by Henry Markram at the École Polytechnique Fédérale de Lausanne (EPFL) in Switzerland. Markram and his team aim to discover *"…profound insights into what makes us human, develop new treatments for brain diseases and build revolutionary new computing technologies."* Here we present some recent contributions towards these goals, focusing on the field of human brain neuroimaging and its applications on health and disease.

We would like to express our sincere appreciation to all the scientists that contributed to this book, as well as to InTech publishing company for their valuable and constant assistance.

Finally, we would like to whole-heartedly thank the McNair Medical Institute (MMI), the Board of the McNair Medical Institute, and in particular Mr. and Mrs. Robert McNair, the Medical and Scientific Director, neurosurgeon, Dr. Charles Neblett, MMI Board member Ms. Ruth Smith, as well as the MMI Executive Director, Ms. Joanie Haley, whose support made this book possible. We would also like to thank the Ministry of Education of Singapore (AcRF, Tier 1, RG 1/11 M4010946.010) for partially supporting the second editor.

T. Dorina Papageorgiou, Ph.D., M.H.Sc.
Assistant Professor of Neurology
McNair Medical Institute
Baylor College of Medicine, U.S.A.

George I. Christopoulos, Ph.D., M.Sc.
Assistant Professor NBS, Nanyang Technological University, Singapore
Research Director, Culture Science Institute, Singapore

Stelios M. Smirnakis, M.D., Ph.D.
Assistant Professor of Neurology and Neuroscience
Baylor College of Medicine, U.S.A.

Section 1

Development of Advanced Brain Neuroimaging Methods

Chapter 1

A Practical Guide to an fMRI Experiment

Nasser H Kashou

Additional information is available at the end of the chapter

1. Introduction

Functional Magnetic Resonance Imaging (fMRI) has been around for two decades now and research in this field has been exponentially rising. Many of this research has been dominated by basic science. Recent trends have brought the clinical realm into play in which valuable contribution can still be made. Allowing the clinician to understand the basic concepts behind an fMRI experiment is crucial to further developing and evaluating functional paradigms and research. Part of designing an fMRI experiment is understanding the physics and how fine tuning scanning parameters affect the image quality which in turn affect the findings of an fMRI study. The other part is understanding the physiology behind the acquired signal as well as the anatomy of the brain. To appreciate the complexity of the fMRI process see [1, 2].

In this chapter we present a practical guide to the novice on what important aspects needed to perform an efficient fMRI experiment from idea formulation to understanding the possible limitations of the results. In doing so, the basic concepts of fMRI beginning with image resolution and physics will be discussed along with advice on possible pearls and pitfalls of this process. Points covered will include paradigm design, scanning protocol, and limitations.

2. Basics physics

How is an image acquired in MRI? In this section a brief overview of the physics and steps needed is introduced. The main components to acquire an image in MRI are a magnet, three gradients and a radio frequency (RF) coil. In contrast to other imaging modalities, the magnet is always on, very strong and can range anywhere from 1.5 to 8 Tesla (higher for animal systems). Currently 1.5 and 3 Tesla are the standard magnets for clinical MRI. To get a grasp on the strength of the magnet, consider that the earth's magnetic field is equal to 0.5 Gauss where 10,000 Gauss is equal to 1 Tesla. This means if working with a 1.5 or 3 Tesla system then the magnet is 30,000 and 60,000 times stronger than the earth's magnetic field. This fact is probably the most important thing to know about MRI scanners, namely because of safety. Any ferrous material should never be brought near the scanner room. The magnet itself cannot provide images without having the two other components; the RF coil and three gradients (in x, y, and x planes). The RF coil is used to send a pulse at the same precessing frequency of the hydrogen nuclei (see below). In some systems the RF coil both transmits and receives. The gradients are turned on and off to cause a slight gradient increase in the magnetic field. All together these components initiate the process of acquiring viewable images. They also form the physical basis of scanning sequences.

But what is being measured? The interaction of protons in the magnetic field,i.e. nuclear magnetic resonance (NMR). Specifically, the hydrogen nuclei from water molecules. In general they have spins, like tops with orientation and a frequency ($\gamma = 42.56 MHzT^{-1}$). γ is a proportionality constant called the gyromagnetic ratio. These protons are randomly pointing in different directions in our bodies, Fig. 1.

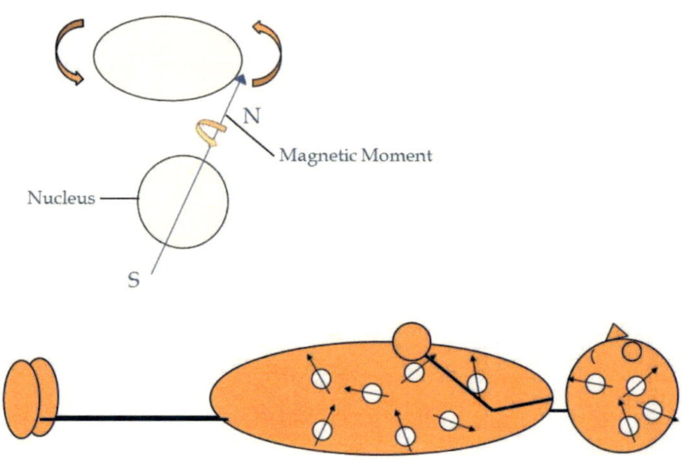

Figure 1. The human body is mainly water thus the hydrogen nuclei are exploited to extract an image in MRI. The nuclei (precess at 42.58 MHz/Tesla) have a spin with orientation and frequency and a magnetic moment, direction it points. These directions are random and thus no net magnetization.

Once we enter an MRI scanner there will be a net magnetization in the direction of the B_o field caused by the magnet. The larmor equation describes this interaction.

$$\omega_o = \gamma B_o \tag{1}$$

where ω_o is the angular frequency of precession of protons in an external magnetic field, and B_o is the strength of the external magnetic field.

The following steps illustrated in Figs. 2 and 3 are needed for the entire process. First the participant, after safety screening of any ferrous objects, is laid down and entered into the scanner. At this point there will be a net magnetization in the direction of the magnetic field (B_o) as a result of some protons aligning with the magnetic field. In a 1.5 Tesla scanner, 1 out of 100,000 hydrogen nuclei aligns itself with B_o [3]. Next some calibration and fine tuning is performed, called shimming. Then an RF pulse (typically 90°, known as flip angle, ϕ) is sent at the appropriate frequency based on the larmor equation. This knocks the particular protons of interest over by 90° as they absorb the energy however they will try to realign with the field and emit energy during this process. After turning off the RF pulse and as the protons return to their original orientations, they emit energy in the form of radio waves. Anytime between this process of realigning with B_0 the RF coil can measure the radio waves being emitted. This gives a signal from one location so the gradients are needed to encode space, Fig. 3. One gradient will phase encode while the second will frequency encode and the final one will slice encode. How this is done is dependent on the imaging parameters and pulse sequences (will not be discussed in this chapter) used. The use of different values of echo time (TE) and repetition time (TR) will determine the contrast or type of image acquired. Simply,TR is the time between transmitted RF pulses and TE is the time to listen to the signal. Varying these two parameters determines the contrast of an anatomical MRI allowing for a T1-weighted, T2-weighted or proton density (PD) image. In a T1-weighted image fat has a high signal (bright) and the cerebral spinal fluid (CSF) has a low signal (dark). In a T2-weighted image fat has low signal (dark) and CSF has high signal (bright). In a PD image the contrast is between that of T1 and T2. They are called T1 and T2 weighted because they are based on the T1 and T2 relaxation times. T1 relaxation is a measure of how quickly the protons realign with B_0 (returning to equilibrium) and T2 relaxation is a measure of how quickly the protons interact with each other (dephase).

Remember that radio waves have to be the right frequency to excite the protons. This frequency is proportional to the strength of the magnetic field. Turning on the gradients slightly increase the magnetic field allowing for manipulation of the frequencies that will affect protons in different parts of space. The slice encode process determines the slice thickness (ST) per user specification. This is repeated until the desired dataset is acquired.

The final step is to convert the acquired frequencies to image space/domain. This is done by using the inverse fast fourier transform (IFFT), Fig. 4. This is a quick overview of generating an image which is necessary to understand in order to properly design an efficient fMRI experiment.

3. Image quality

A key aspect and development in this field is the image quality. MRI image quality is dependent on many factors, some of which are TE, TR, the number of signal averages (NA), and resolution. There are tradeoffs among all these factors. Increasing TE allows for more T2 weighting, but decreases the signal to noise ratio (SNR). A long TR allows for more slices to be acquired, while decreasing T1 contrast and increasing scanning time. Thus, to decrease the scanning time, TR can be decreased which allows for more T1 contrast, but the SNR is lowered and fewer slices can be acquired. SNR increases as NA is increased, but only by a factor of $\sqrt{NA}$. Scan time, in turn, is directly proportional to NA. Resolution is a function of the number of phase encode (PE) steps, the number of frequency encode (FE) steps (matrix size), field of view (FOV), and slice thickness (ST). Thinner slices reduce partial volume effects (PVE) and increase resolution, but SNR and anatomic coverage are reduced proportionally. In-plane resolution is increased as FE steps and/or PE steps are increased, but SNR decreases. Scan time increases with the number of PE steps. However, changing the FOV or increasing the number of FE steps does not affect scan time. The smaller the FOV, the higher the resolution and the smaller the voxel size [4].

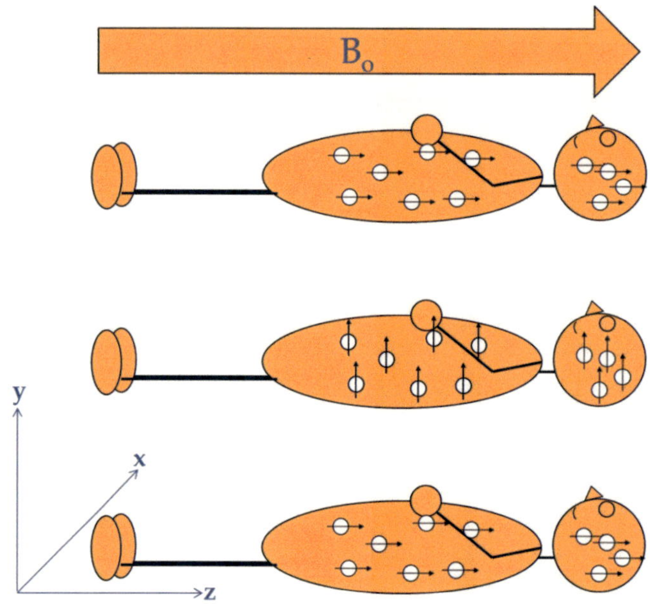

Figure 2. Illustration of the interaction between protons, the magnetic field and the RF pulse. In the magnet there is a magnetic field (B_o) 60,000 times stronger (for 3 Tesla) than the earth's magnetic field. When someone enters the bore of the magnet some of the hydrogen nuclei align with Bo causing a net magnetization (top). An RF pulse is sent at the same precessing frequency of the hydrogen nuclei. This tips the nuclei by 90°. (middle). The B_o field will force them to begin to realign, as this is happening the signal is being measured by the RF coil (bottom).

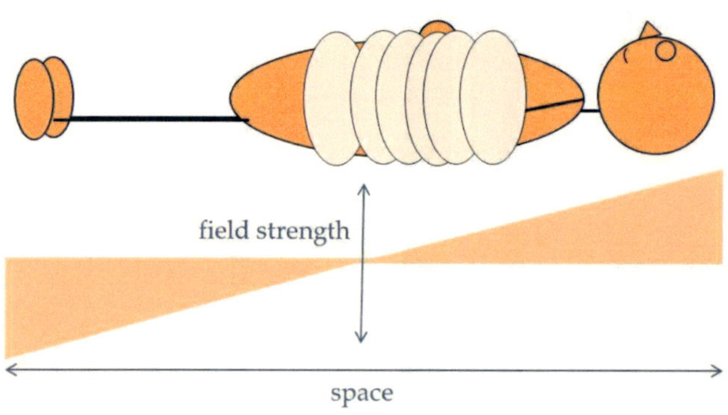

Figure 3. Using the gradient coils adds the ability to spatially encode the signals from previous process (see Fig. 2) by slightly modifying the magnetic field, thus a 2D/3D dataset in k-space can be mapped. The lower the magnetic field targets lower frequencies and higher field higher frequencies.

Note a common mistake is that displayed resolution is sometimes used interchangeably with nominal resolution. The nominal spatial resolution is the smallest tissue volume size that can be represented on an image. It is defined as the prescribed FOV in the frequency- and phase-encoding directions divided by the number of frequency- and phase-encoding points ($[\Delta x = FOV_x/N_x]$ and $[\Delta y = FOV_y/N_y]$), respectively [5]. For instance, on older scanner platforms this arises when a matrix of 96x96 is set. As a result the scanner automatically upsamples by zero padding the raw data to 128x128, and thus the displayed resolution

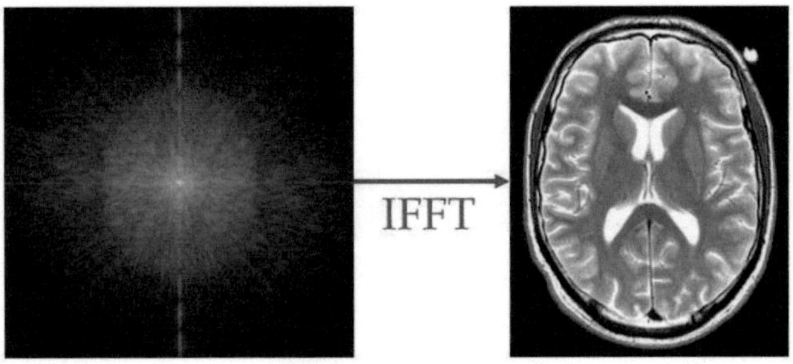

Figure 4. K-space data generated and collected by the gradients and RF coil is converted to the image domain via the inverse fast fourier transform (IFFT) as the final process to generating an MRI/fMRI image.

is $FOV_x/128$ and $FOV_y/128$ instead of the nominal (acquired) resolution of $FOV_x/96$ and $FOV_y/96$. This is because in the past these scanners interpolated anything that was not a factor of two from 64x64, e.g. 96x96 so the displayed resolution would not be the same as the nominal.

It is immediately apparent that from these acquisition parameters many things can either make for an efficient or inefficient dataset before beginning the fMRI task paradigm. In MRI, higher image resolution, i.e. smaller pixel/voxel sizes, is directly proportional to the magnet strength. Thus, going from 1.5 to 3 Tesla, the resolution can be doubled while using the same imaging parameters with the added advantage of shorter acquisition time. But with increase in resolution or signal there is also an increase in the noise. Note, for some applications (i.e. infant imaging) it may be advantageous to image at 1.5 Tesla.

In fMRI it is standard to achieve isotropic in plane resolution. For instance to attain a 3x3 pixel size then a matrix of 64x64 with a field of view (FOV) of 192mm can be set in the scanner. The voxel is determined by the additional parameter of slice thickness which is dependent on the third (slice encode) gradient as was previously alluded to, Fig. 5.

3.0.1. SNR

The signal to noise ratio (SNR) can be adjusted and optimized by choosing the proper imaging parameters. Note, there will be a tradeoff between resolution and SNR. For example if an increase from the 3x3x5 voxel resolution is desired then simply adjusting the matrix size to 128x128 while keeping all other parameters the same would yield a voxel of 1.5x1.5x5. Here the in plane resolution is doubled however the signal also suffers more noise. Similarly if a smaller slice thickness is chosen then SNR decreases and chances of partial volume effects increases. Next an explanation of the differences between MRI and fMRI is presented.

3.1. MRI vs fMRI

For both MRI and fMRI the process explained above applies. The difference is in the acquisition parameters and pulse sequences used. The most obvious difference is in the resolution, Fig. 6. MRI is denoted as the anatomical high resolution ($<$ 1 mm in-plane) image. In general this is one anatomical (T1) dataset with three spatial dimensions. Whereas fMRI is a set of low resolution (3 mm in plane) datasets with the addition of a time dimension, 4D. The difference in resolution is based on the imaging sequence used to acquire the data.

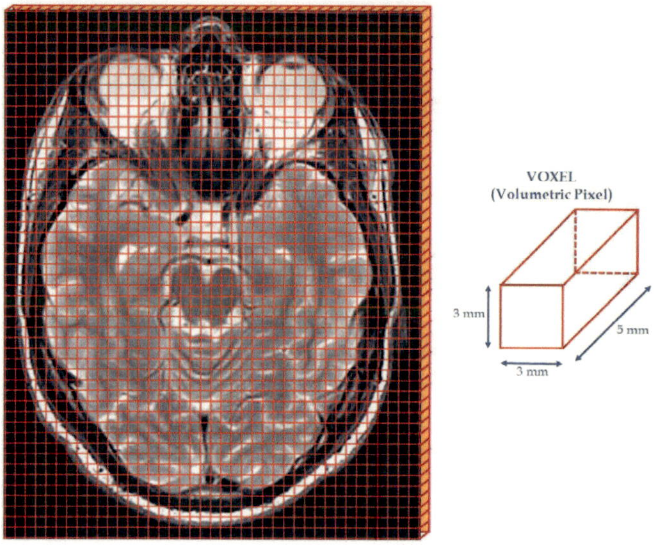

Figure 5. Image resolution is determined by the number of pixels in the x,y plane know as matrix size. This is dependent on how the gradients are phase and frequency encoded. In this case x=40, y=50. Assuming a field of view of 120 mm yields an in plane resolution of (120/40, 120/50) or (3, 2.4). Typically fMRI resolution at 3 Tesla is 3x3x5 with a FOV of 192 mm and matrix size of 64. The slice thickness is determined by a third gradient while performing the slice encoding phase.

In MRI about 5 minutes are used to scan the entire brain which allows for a very fine grid whereas in fMRI more than 100 volumes are acquired in the same amount of time. So, rather than 5 minutes to acquire one brain (MRI), only 2 seconds are allotted to acquire a brain (fMRI). When scanning for an anatomical image the participant simply lays in the scanner. But for an fMRI there is a specific task that needs to be repeated by the participant over the span of the 5 minutes. In doing so, the functional signal can be derived and analyzed. A description on preparing this task is given in Sections 5 and 6.

3.1.1. Artifacts

The main artifacts in fMRI are susceptibility due to structures such as the ear and nasal canals because of the air tissue interface. These artifacts cause a signal loss in the auditory and frontal regions respectively. If an experiment is interested in frontal lobe activation then the imaging parameters need to be optimized to minimize this. For a 3 Tesla system, being conservative and using 64x64 helps alleviate this.Other common artifacts are caused by retainers and braces which also cause a signal loss. This should be kept in mind when recruiting participants for a study. With a basic understanding of MR imaging an explanation of the origins of the functional signal can now follow.

4. BOLD

In the brain different regions are specialized to perform different sensory, motor and cognitive functions. Over the last two decades fMRI has been developed as a technique for mapping brain activation in these regions and has found widespread interest in basic and clinical research aimed at better understanding brain function. The fMRI technique is

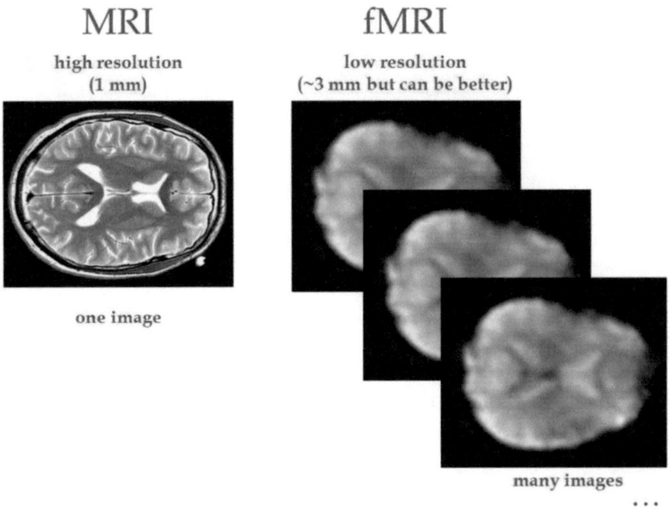

Figure 6. Illustrated is a visual comparison of the difference between MRI and fMRI. MRI is a set of high resolution slices that make up one 3D dataset while fMRI is a series of low resolution slices that make up many volumes, a 4D dataset (volumes + time). Note, images not to scale.

based on the detection of local perturbations of the deoxyhemoglobin concentration in the vicinity of neuronal activity. Neuronal activity at the synaptic level results in both an increase in oxygen consumption by the active cortex and an even greater increase of blood flow to the site. Because oxygen delivery exceeds oxygen utilization, the net effect is a local decrease in deoxyhemoglobin concentration near the activation site. The decrease in deoxyhemoglobin concentration at the site of neuronal activity causes a local increase the magnetic resonance signal, Fig. 7. This effect has been termed the Blood Oxygenation Level Dependent (BOLD) contrast mechanism [6]. This is possible because of the magnetic properties of hemoglobin which contains four Fe2+ ions. Specifically, deoxygenated blood is paramagnetic, meaning it has a small additive intrinsic magnetic field and oxygenated blood is diamagnetic meaning it tends to oppose external magnetic field. The ratio of deoxygenated to oxygenated blood changes when a particular task is performed as a result of the neurons firing which cause an increase in both blood flow into the neurons and oxygen consumption level to increase in that particular region of the brain. However, the blood flow increase is larger than proportional oxygen consumption. The result of this brief stimulus and in turn neural activation is the hemodynamic response function (HRF). The HRF has a characteristic shape with an initial dip immediately following the stimulus then an increase and finally an undershoot but can vary between participants [7]. Understanding the behavior of the HRF can help in designing an efficient fMRI paradigm. In summary BOLD fMRI capitalizes on the difference between two conditions; an active condition in which stimulus related neural activity is generated and a passive condition in which stimulus related neural activity is absent or kept to a minimum, Fig. 8.

5. Paradigms

There are several types of fMRI design paradigms: block, event-related (widely spaced and rapid) and mixed (block and event related). The development of the event related studies

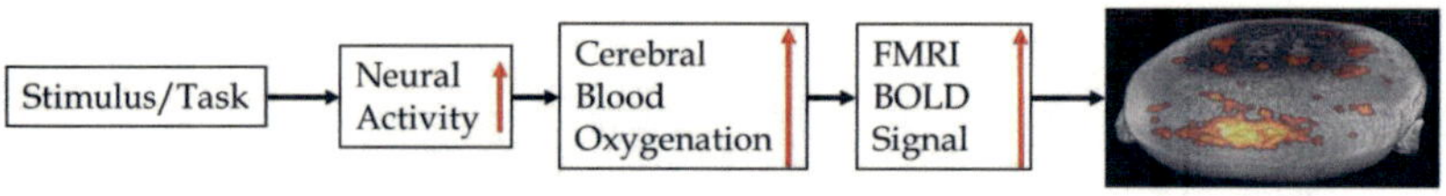

Figure 7. A simplified flow of events that lead to the BOLD fMRI signal. A specific stimulus/task causes an increase in neural activity which triggers a complex chain of changes to cause an increase in cerebral blood oxygenation. This complex series of events includes an increase in cerebral blood flow (CBF), an increase in cerebral metabolic rate for oxygen which in turn causes the cerebral blood volume (CBV) to increase. These events cause a decrease the local deoxyhemoglobin (HHb) content. This then allows for the detection of the signal which after post processing can be overlaid on an anatomical MRI.

[8–19] came several years after the advent of BOLD fMRI. Choosing the proper TR, TE FOV and matrix size values are all important and are dependent on the problem or question that is being investigated as will be discussed in section 6, but of equal importance is the type of of paradigm used. This in fact is intertwined with the imaging parameters. The importance of choosing a suitable TR to and interstimulus interval (ISI) ratio has been known early on, namely because BOLD is not necessarily steady-state rather transient signal [20]. For block designs, this is fairly straight forward. If using an event related paradigm then caution should be taken in choosing the proper TR. Software such as optseq (http://surfer.nmr.mgh.harvard.edu/optseq/) allows for a good randomization of stimuli for rapid event related designs based on specific imaging parameter. This ensures a robust randomized design.

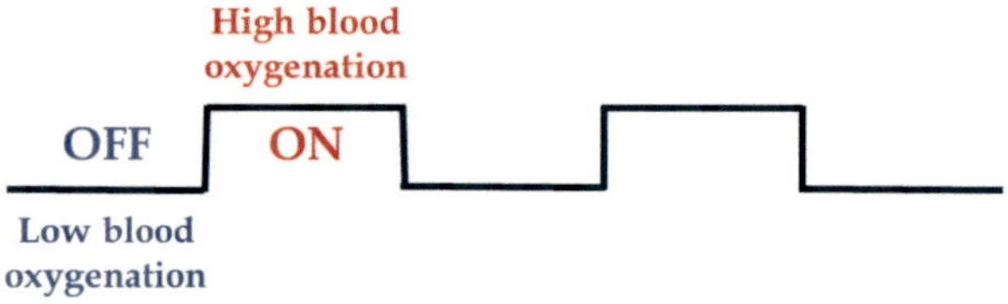

Figure 8. A typical block design should consist of two states, OFF (no task) and ON (task). The ratio of blood oxygenation is measured to determine the pixel with the fMRI BOLD signal change.

Block designs commonly consist of two states, Fig. 8, A (rest) and B (task), however in some situations where time or a factor like budget is an issue and a third or even a fourth state is added. Having four states complicates the design and strategies should be taken to design efficiently. The minimum block time and ISI can be around 20 seconds with an absolute minimum of 6 seconds and the maximum at 60 seconds. In two states the conditions would alternate a suggested minimum of three times, e.g. ABABABA. It is good practice to always end on a rest state. For three conditions there are several combinations for presentation, e.g. CACBCACBC; CABABABAC; CABCABCABC where C can be the rest condition and A and B are task 1 and task 2 respectively. The disadavantage of this paradigm is that the participant may start to predict or anticipate the task. In contrast event related can be more easily randomized because of the small stimulus duration. But how long should the stimulus last? Several groups have noted different durations that can still be detected by fMRI such as 2 seconds [21], 0.5 seconds [22] and 34 milliseconds [23]. This temporal resolution is a clear advantage in event related over block designs, however there is a tradeoff of SNR in the fMRI signal. It has been shown that going from a block to a widely spaced event related design decreases the SNR by approximately 33% [24] and about 17% [25] from a widely spaced to a rapid event related pardigm. In general block designs have larger BOLD signal change

[26, 27] and better statistical power [28]. An optimal ISI for a fixed stimulus duration of less than 2 seconds is about 12 seconds [24]. Additionally by randomizing the ISI the statistical power increases and allows for reducing the ISI [24, 29]. A third possibility is using a mixed design which is a combination of event-related and block design. The problem with this is the statistical processing becomes even more complicated and if not fully understood can lead to errors in data processing, analysis and interpretation. It is best to keep the paradigm as simple as possible. Note, for clinical use the majority if not all the fMRI scans will be block however this may change quickly as the experimental design aspects of fMRI continue to advance. Similar to choosing the imaging parameters, determining which paradigm to use will depend on the goal of the experiment.

6. Preparing an experiment

With the basics of MR physics and fMRI paradigms presented a more informed decision can be made on the experiment. In begining the journey into an fMRI experiment some basic questions need to be asked. Why are we doing this experiment, is generally our hypothesis. What are we looking for? For example, a specific behavior we are interested in. Where? This involves knowing the neuroanatomy. How? This is the type of fMRI design. The best way to explain this is to walk through an example. Why? We hypothesize that eye movements will cause activation. What? Moving the eyes. Where? In the cortex. How? Tell participant to move eyes. Initially this may look like a good set of answers but they are not. What is wrong with them? They are too general and leave room error, confounds, and reproducibility will be difficult as nothing was mentioned about the paradigm or scanning protocol. Several groups can follow the same logic and come up with different activation maps even if the scanning protocol was mentioned. For instance, full brain coverage axial scan with TR=3 seconds, TE=35ms, $\phi = 90^o$, matrix size=128x128, FOV=256mm ST=8mm. This would yield a voxel resolution of 2x2x8mm, SNR=75%, and allow for about 45 slices to be acquired. If full brain coverage was necessary then these parameters would be good, however from the vague answers this can not be determined. Also, note scanning time is still missing and no mention of the number of volumes to be acquired.

Let us now try to reword the answers to come up with a better starting point. Why? We hypothesize that eye movements will activate the visual cortex. What? Specifically moving the eyes to from left to right in a saccadic fashion. Where? Want to see activation differences between V1 and V2 in visual cortex. How? Will have a block paradigm of 30 seconds fixation followed by 30 seconds of guided saccadic eye movements. A backprojection system will be used to display the visual stimulus that will cue the participant to fixate on a white dot in the center of the screen. The dot will then automatically proceed to alternate back and forth between the center and the left side of the screen, note this is flipped because participant is looking through a mirror. A common mistake made by novice users. In regards to any stimulus that involves the visual cortex, it is good practice to either turn the lights off or keep on for all participants. Again, the number of variables and confounds should be at a minimum. It is always recommended to test the paradigm in the scanner before recruiting participants in order to debug and pick out issues like this. For statistical power want to repeat this on-off cycle six times. Using the same imaging protocol just mentioned allows for 120 volumes to be acquired in 6 minutes and 21 seconds. The 21 seconds are used for dummy scans to allow the scanner to warm up. For a particular participant whole brain coverage may only require 20 slices thus total slices acquired would be 120*20=2400. By providing a more informed set of answers and paradigm the study can be changed from an inefficient fishing expedition to an efficient specific study. Is this the most optimized design? Probably not but it is a better starting point. Since the interest is in the known regions of the visual cortex then whole brain coverage may not be necessary. That can buy more time, SNR, statistical power and/or image resolution.

Answering the questions differently will affect the way we want to scan the brain. If the following questions are asked then options from Fig. 9 will change. What brain areas are active during visual imagery? This is a very general question and may require an axial full brain coverage scan. Is the primary visual cortex (area 17) active during visual imagery? This is more specfic and allows for a more localized scan in the visual cortex. What areas of the brain are active during working memory? Again, this is very general. Is the prefrontal cortex active during working memory? Here the frontal cortex is targeted hence no need to scan the visual cortex.

The conventional scanning plane has been the axial however there is no rule against scanning in other planes or at an oblique angle. In fact, there are situations where going away from convention is advantageous. Depending on the answers given to the list of questions a very specific orientation and slices desired can be determined.

For example in Fig.9 it can be seen that it may not be necessary or the most efficient use of time and resolution to choose the axial or coronal planes even for visual cortex activation. The oblique plane scan at the visual cortex may be the best choice. Also if the other scans were chosen it would not have been necessary to acquire slices from the entire brain rather enough to cover the visual cortex. This saves acquisition time which can be invested in acquiring more volumes and increasing the resolution. A priori knowledge of the problem to be studied can go a long way in optimizing the fMRI experiment. Indeed it is advantageous to do full brain coverage if that is what the problem entails however it is not a good idea to go on a fishing expedition for activation sites. These are some aspects of several questions to ask before preparing an experiment. If thought out thoroughly they would minimize errors and confounding variables and in turn optimize the fMRI design and experience.

7. Collecting data

Once the questions to the problem are answered, the parameters are entered into the console of the MR scanner. This will then give the time required, number of slices, number of volumes, etc. to be acquired based on the matrix size, resolution and FOV chosen. It is recommended that the participant is made familiar with the task outside the scanner to minimize confusion and error. After entering the scanner the anatomical high resolution MR images are acquired followed by the fMRI scan. The latter is where the participant is instructed to simulataneously perform a task. This process can take an average of an hour. It is not recommended to go over this time too much mainly because of the attention and fatigue factor on the participant's end. If prepared and thought out properly the hour should be sufficient to collect the MRI and two or three fMRI sessions. At the end of the scanning session the images can be copied onto a DVD in DICOM (http://medical.nema.org/) format and taken to a local workstation for analysis. In the case of strictly clinical fMRI, for example, presurgical planning support, the scanners have a built in analysis tool that can output the results of activation on the console itself and these results can be pushed onto a PACS system. In this case the paradigm timing and design need to be entered into the console before the scanning begins. For research purposes using the built in system is not recommended. One reason being that the console can only process a single participant meaning no group analysis

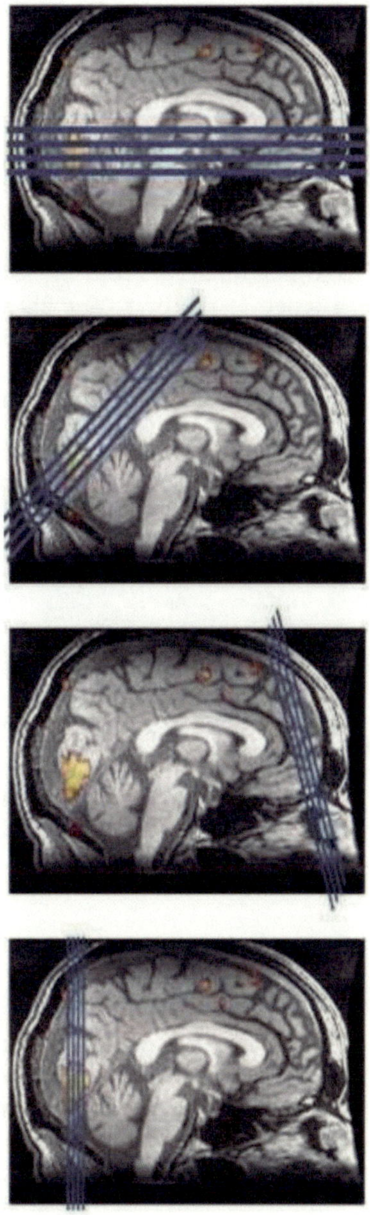

Figure 9. Illustration of four possible orientations overlaid on top fMRI results from an oculomotor paradigm coregistered withan anatomical MRI. In this case, all orientations except the third from top (oblique orientation at frontal cortex) would havecaptured the visual cortex activation.

can be performed. To date vendors have not added the option to do group analysis. This could be because the main users are clinicians whose priamry focus is treat one patient at a time. Section 8.2 lists possible software packages that can be used instead. For clinicians, it is also suggested that they use these packages for a more robust analysis especially for case studies and for confirming the results in situations of more critical care such as surgical resections.

8. Preparing data

After fMRI data is acquired, motion correction and filtering may be required. Each slice has to be aligned with the next within the volume and then registration is performed between volumes using the first as the fixed (stationary). From the saccadic eye movement example this means 19 slices need to be aligned to the first. The 2D slice alignment is repeated for all 120 volumes independently. Afterwards the slice aligned volumes are registered to each other in 3D. So 119 volumes are registered to this first (fixed volume). This is repeated for each participant dataset. For group comparisons the processed 4D datasets then have to be registered to a common space.

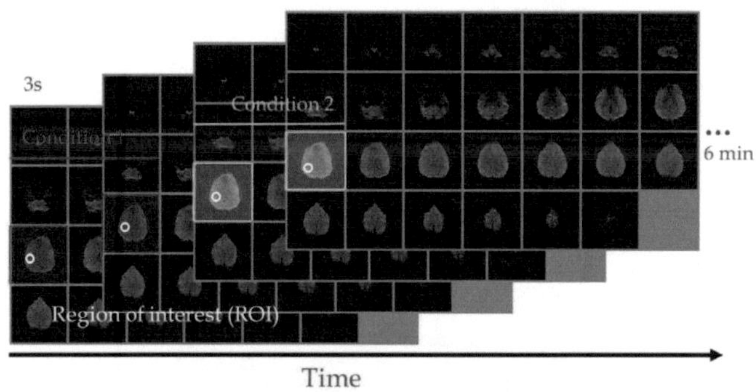

Figure 10. The signal intensity of a spatial region of interest (ROI) or a voxel can be traced over time, in this case every 3 secs for 6 minutes between two conditions. This intensity is fitted to the ON-OFF paradigm design (see Fig. 8) by using a general linear model regression analysis to determine if there is significant activation.

8.1. Statistics

Once all the data sets are aligned, time courses can be plotted for each voxel, Fig. 10. In the case of a block paradigm during condition 1 (fixation), it is expected that signal is low from a particular pixel and high for condition 2 (visually guided saccades). Regression analysis is performed on the voxels and the ones that follow this trend or fit the boxcar (from Fig. 8) can be identified as significantly activated regions of the brain responsible for the associated task or stimulus chosen. The statistics get more complicated for event related designs. If each block last 30 seconds with TR=3 seconds that means the intensity of one voxel from each volume will give result in a time series of 120 points with time as the x-axis and fMRI signal as the y-axis, again with a total time of 6 minutes (3*120=360 seconds). The mean and standard deviations for the duration between the two conditions is then calculated Fig. 11. The voxels that significantly follow this trend are then denoted by an orange color map and overlaid onto the original fMRI dataset. Depending on the resolution of the dataset this can be performed on more than 100,000 voxels, 128x128x20 = 327,680 voxels from the above example. Because of this number many voxels will be activated by chance (for $P<0.01$, 0.01*327,680 = 3,276) so bonferroni corrections are used. Rather than use $P<0.01$, $P<0.0000001$ is used (0.01/100,000 = 0.0000001). An alternative analysis that is commonly used is cluster grouping rather than individual voxels. In this way if a neighborhood of voxels are all following the paradigm then the chances of random activation can be eliminated.

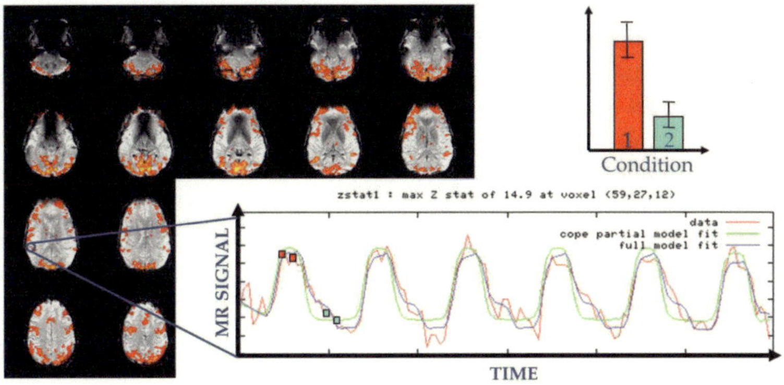

Figure 11. The orange pixels are the significant activation results of the general linear model (GLM) overlaid on top of the fMRIs. Looking at the ROI the raw signal is found to fit the model well for the two conditions.

8.2. Software packages

There are several software packages available for fMRI analysis. FSL (http://www.fmrib.ox.ac.uk/fsl/), AFNI (afni.nimh.nih.gov/afni), and SPM are all free with the exception that the latter requires a Matlab (http://www.mathworks.com/) license. FSL and SPM are available for all operating systems. AFNI is not available for Windows. More groups seem to be using a combination of these packages. For instance, using FSL for pre processing and SPM for post processing while others like to use AFNI for evaluating the susceptibility effects on the data to test efficacy of scanning protocol before beginning the experiment. The packages are all good in their own way and it comes down to user preference.

9. Pitfalls and limitations

Throughout the entire process of an fMRI experiment we need to be aware of our limitations. Are we using a 1.5, 3, or 7 Tesla magnet? The same exact design and protocol can be run on all three magnet strengths and give different results. All of these have their individual pearls and pitfalls and we should be able to identify them. The stimulus can easily be confounded if not designed efficiently. If interested in visual perception and using a visual stimulus then care needs to be taken to ensure that the lighting in the room, the stimulus size, contrast, duration, spatial and temporal frequency and distance to the participant stays fixed. If eye movements are of primary concern then investing in an MR compatible eye tracker would be necessary in order to confirm that indeed the participants are performing the task. In addition fMRI is susceptible to motion thus the head has to be held still and padding used. If motion is present then post processing algorithms can be utilized to correct them however only if they are minor. Otherwise the data will not be useful. It is crucial that the participants understand all the instructions. In an eye movement study, using the same visual stimulus but changing the instruction slightly found that significantly different regions were activated and recorded [30]. Participant comfort is a factor; if they are in pain then that will reflect on the results. At the same time if they are too comfortable then they will fall asleep. In designing the task paradigm things to consider are the timings, e.g. 15 vs 30 seconds. The latter may be to taxing on the eyes. It may advantageous to use a two state block design over a three state to alleviate participant confusion and simplfying the statistical analysis process. Reducing the time to 15 seconds and using only two states allow for more cycles to be repeated and give more power for statistical averaging. Alternatively, 15 seconds can be used to add a third state yet keep the total scan time of 6 minutes and introducing pseudo randomized cycles, however that would increase the complexity. Other limitations include both anatomical and physiological components. The size and location of the neuroanatomical site can limit what can be derived also the vascular response time may vary from approximately 2 ms to 6 s. The time of day certainly makes a difference on the fMRI signal so this should be kept in mind. Also, the amount of familiarity the participant has with the task will affect the intensity of the signal. Switching coils, or scanners in a middle of a study will definitely confound the data. This is of a bigger issue for multicenter trials, even if the same MR vendor is being used. All these factors can be minimized before the study begins and the data is collected.

After the data is collected, other issues arise. What is the minimum signficance level that will be accepted, p=0.05 vs p=0.01? What about gaussian smoothing the data, none, 5mm, 10mm? This depends on the SNR of the dataset. By smoothing the data activation clusters will spread and localization limited. Also, more regions will be activated. If fMRI voxel resolution is 2x2x2 mm and a smoothing kernel of 10mm is used then that decreased the resolution by a factor of five (10x10x10 mm). That in way defeats the purpose of sacrificing other parameters to achieve the 2mm^3 voxel size. The degrees of freedom in the registration process need to be understood as well. Simply choosing 3, 6, 7 or 12 can cause alignment errors and erroneous activation maps. This becomes a bigger issue if partial brain images were acquired for fMRI and now they are trying to be aligned with a whole brain high resolution MRI. Eventhough the software packages above have built in tools for these steps the user has to be cognizant of the potential pitfalls. Remember there is to magic button to push and everything will come out perfect. The registration algorithms vary and the types of interpolation will introduce more blurring on top of the gaussian smoothing chosen. Many users think that because the fMRI activation maps are overlaid ontop of a high resolution anatomical MRI, that they have the same accuracy and resolution and do not appreciate the amount of approximations involved in the process. Overall an fMRI study needs much thought. There are many variables and confounds involved in setting up an fMRI design and this section touched on the importance of being aware of the limitations and keeping these variables as constant as possible when clinical, longitudinal or group studies are performed.

10. Conclusion

In designing an fMRI experiment there has to be a realization of the fact that there is no real (one) gold standard or solution. So rather than asking what is the solution? We should be asking, what can we control? Here are some concluding tips in answering this question. Before beginning anything we need to fully understand the experimental goals. Inevitably tweaking the fMRI imaging parameters need to performed. As a result the paradigm will also be adjusted accordingly, especially in the case of event related designs. Here are some design questions to answer: Design type? Blocked, Event-Related, Mixed? How many subjects? How much data to collect for each subject? How many stimulus conditions? How many repetitions for each condition? When should each stimulus be presented? TR? Number of slices? Getting in the habit of asking many questions before starting a study is key. Overall, the best solution is experience.

Author details

Nasser H Kashou

The Children's Radiological Institute, Nationwide Children's Hospital, USA

References

[1] Robert L Savoy. Experimental design in brain activation mri: cautionary tales. *Brain Res Bull*, 67(5):361–367, Nov 2005.

[2] Edson Amaro and Gareth J Barker. Study design in fmri: basic principles. *Brain Cogn*, 60(3):220–232, Apr 2006.

[3] R. L. Savoy. History and future directions of human brain mapping and functional neuroimaging. *Acta Psychol (Amst)*, 107(1-3):9–42, Apr 2001.

[4] V. D. KÃűchli D. Weishaupt and B. Marincek. *How Does MRI Work? An Introduction to the Physics and Function of Magnetic Resonance Imaging*. Springer, 2006.

[5] G. S. Slavin. Spatial and temporal resolution in cardiovascular mr imaging: Review and recommendations. *Radiology*, 234:330–338, 2005.

[6] S. Ogawa, T. M. Lee, A. R. Kay, and D. W. Tank. Brain magnetic resonance imaging with contrast dependent on blood oxygenation. *Proc Natl Acad Sci U S A*, 87(24):9868–9872, Dec 1990.

[7] G. K. Aguirre, E. Zarahn, and M. D'esposito. The variability of human, bold hemodynamic responses. *Neuroimage*, 8(4):360–369, Nov 1998.

[8] R. L. Buckner, P. A. Bandettini, K. M. O'Craven, R. L. Savoy, S. E. Petersen, M. E. Raichle, and B. R. Rosen. Detection of cortical activation during averaged single trials of a cognitive task using functional magnetic resonance imaging. *Proc Natl Acad Sci U S A*, 93(25):14878–14883, Dec 1996.

[9] R. L. Buckner, J. Goodman, M. Burock, M. Rotte, W. Koutstaal, D. Schacter, B. Rosen, and A. M. Dale. Functional-anatomic correlates of object priming in humans revealed by rapid presentation event-related fmri. *Neuron*, 20(2):285–296, Feb 1998.

[10] M. A. Burock, R. L. Buckner, M. G. Woldorff, B. R. Rosen, and A. M. Dale. Randomized event-related experimental designs allow for extremely rapid presentation rates using functional mri. *Neuroreport*, 9(16):3735–3739, Nov 1998.

[11] V. P. Clark, J. M. Maisog, and J. V. Haxby. fmri study of face perception and memory using random stimulus sequences. *J Neurophysiol*, 79(6):3257–3265, Jun 1998.

[12] A. M. Dale and R. L. Buckner. Selective averaging of rapidly presented individual trials using fmri. *Hum Brain Mapp*, 5(5):329–340, 1997.

[13] M. D'Esposito, E. Zarahn, and G. K. Aguirre. Event-related functional mri: implications for cognitive psychology. *Psychol Bull*, 125(1):155–164, Jan 1999.

[14] K. J. Friston, P. Fletcher, O. Josephs, A. Holmes, M. D. Rugg, and R. Turner. Event-related fmri: characterizing differential responses. *Neuroimage*, 7(1):30–40, Jan 1998.

[15] O. Josephs, R. Turner, and K. Friston. Event-related f mri. *Hum Brain Mapp*, 5(4):243–248, 1997.

[16] B. R. Rosen, R. L. Buckner, and A. M. Dale. Event-related functional mri: past, present, and future. *Proc Natl Acad Sci U S A*, 95(3):773–780, Feb 1998.

[17] A. D. Wagner, D. L. Schacter, M. Rotte, W. Koutstaal, A. Maril, A. M. Dale, B. R. Rosen, and R. L. Buckner. Building memories: remembering and forgetting of verbal experiences as predicted by brain activity. *Science*, 281(5380):1188–1191, Aug 1998.

[18] E. Wiener, L. R. Schad, K. T. Baudendistel, M. Essig, E. MÃijller, and W. J. Lorenz. Functional mr imaging of visual and motor cortex stimulation at high temporal resolution using a flash technique on a standard 1.5 tesla scanner. *Magn Reson Imaging*, 14(5):477–483, 1996.

[19] E. Zarahn, G. Aguirre, and M. D'Esposito. A trial-based experimental design for fmri. *Neuroimage*, 6(2):122–138, Aug 1997.

[20] C. J. Price, D. J. Veltman, J. Ashburner, O. Josephs, and K. J. Friston. The critical relationship between the timing of stimulus presentation and data acquisition in blocked designs with fmri. *Neuroimage*, 10(1):36–44, Jul 1999.

[21] A. M. Blamire, S. Ogawa, K. Ugurbil, D. Rothman, G. McCarthy, J. M. Ellermann, F. Hyder, Z. Rattner, and R. G. Shulman. Dynamic mapping of the human visual cortex by high-speed magnetic resonance imaging. *Proc Natl Acad Sci U S A*, 89(22):11069–11073, Nov 1992.

[22] P. A. Bandettini, A. Jesmanowicz, E. C. Wong, and J. S. Hyde. Processing strategies for time-course data sets in functional mri of the human brain. *Magn Reson Med*, 30(2):161–173, Aug 1993.

[23] R. L. Savoy, P. A. Bandettini, K. M. OâĂŹCraven, K. K. Kwong, T. L. Davis, J. R. Baker, R. M. Weisskoff, and B. R. Rosen. Pushing the temporal resolution of fmri: Studies of very brief visual stimuli, onset variability and asynchrony, and stimulus correlated changes in noise. In *Proc. Soc. Magn. Reson. Med. Third Sci. Meeting Exhib.*, volume 2, page 450, 1995.

[24] P. A. Bandettini and R. W. Cox. Event-related fmri contrast when using constant interstimulus interval: theory and experiment. *Magn Reson Med*, 43(4):540–548, Apr 2000.

[25] F. M. Miezin, L. Maccotta, J. M. Ollinger, S. E. Petersen, and R. L. Buckner. Characterizing the hemodynamic response: effects of presentation rate, sampling procedure, and the possibility of ordering brain activity based on relative timing. *Neuroimage*, 11(6 Pt 1):735–759, Jun 2000.

[26] G. H. Glover. Deconvolution of impulse response in event-related bold fmri. *Neuroimage*, 9(4):416–429, Apr 1999.

[27] R. B. Buxton, E. C. Wong, and L. R. Frank. Dynamics of blood flow and oxygenation changes during brain activation: the balloon model. *Magn Reson Med*, 39(6):855–864, Jun 1998.

[28] K. J. Friston, A. P. Holmes, C. J. Price, C. BÃijchel, and K. J. Worsley. Multisubject fmri studies and conjunction analyses. *Neuroimage*, 10(4):385–396, Oct 1999.

[29] A. M. Dale. Optimal experimental design for event-related fmri. *Hum Brain Mapp*, 8(2-3):109–114, 1999.

[30] Nasser H Kashou, Lawrence E Leguire, Cynthia J Roberts, Nick Fogt, Mark A Smith, and Gary L Rogers. Instruction dependent activation during optokinetic nystagmus (okn) stimulation: An fmri study at 3t. *Brain Res*, 1336:10–21, Jun 2010.

Chapter 2

The Neurometabolic Underpinnings of fMRI BOLD Dynamics

Christopher W. Tyler and Lora T. Likova

Additional information is available at the end of the chapter

http://dx.doi.org/10.5772/31488

1. Introduction

The primary indicator of brain activity in the fMRI technique is known as the blood oxygenation level dependent (BOLD) signal, which derives from the hemodynamic response of the local blood vessels. Although much effort has gone into the analysis of the temporal dynamics of this signal as a reflection of the underlying neural activity that elicits it, the link between neural activity and the hemodynamic response is not direct, but involves a chain of metabolic processes mediated by the astrocytic glial cells in the cerebral cortex. This review focuses on the role of the metabolic pathways mediated by the glial cell in coupling the neural activity to the BOLD responses in the blood vessels, and the consequent implications for the processes governing the fMRI BOLD dynamics.

2. The chain of metabolic processing

In general terms, the stimulus impinging on the subject generates a sequence of neural responses starting with the transduction into a neural signal within the sensory receptors, which then propagates to the brain and activates various populations of neurons within the voxels being analyzed by the fMRI technique (Logothetis, 2003). For instance, the neural signals arriving from the visual pathway generate synaptic activation of the populations of neurons in the occipital cortex, which generates an energetic demand for the restoration of the neurotransmitter molecules carrying the activation signals across the synapses. The chain of cortical processing progresses from the neural events through the metabolic demand mediated by the glucose/lactate cycle in neighboring astrocyte glial cells to the processes of oxygen delivery by the adjacent arterioles and capillaries that is detected by the

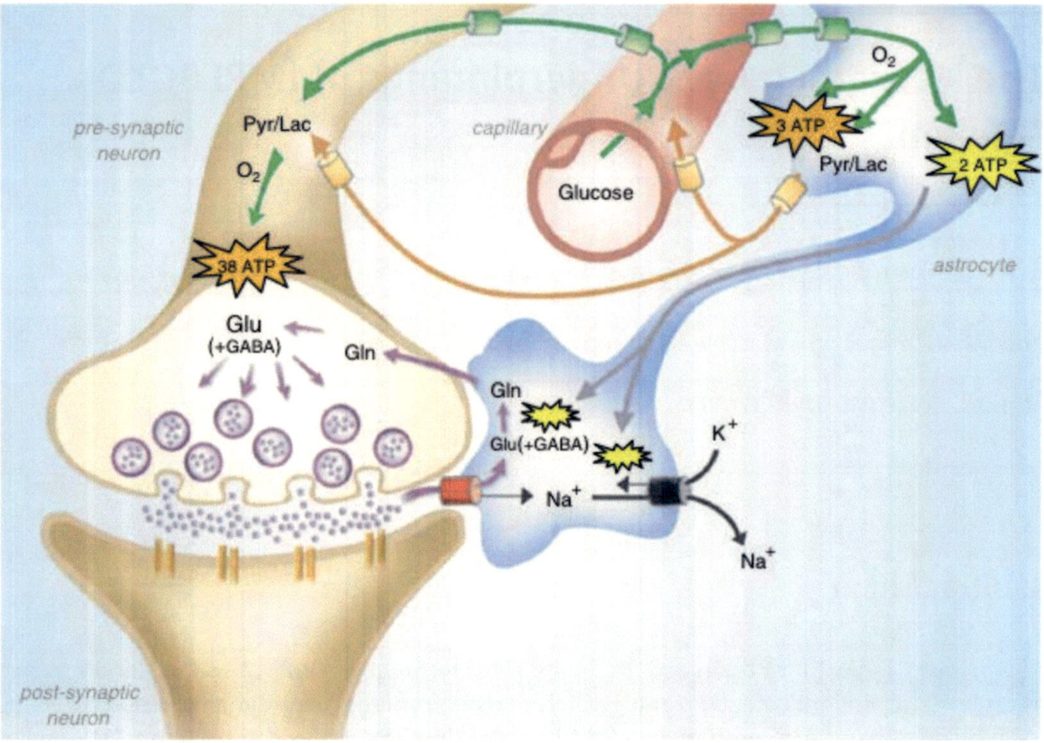

Figure 1. The astrocytes as the substrate for the neurovascular coupling of the neural metabolism. (From Hyder et al., 2006, with permission).

imaging methodology. The first element in this metabolic chain is the local astrocyte processes, which provide glucose to the neuron and replenish its supply by an ATP metabolism, which is fueled by oxygen diffusing through the astrocyte cytoplasm from their endfeet connections to neighboring blood vessels (Magistretti & Pellerin, 1999; Magistretti, 2009). The integrated metabolic demands are met primarily by the astrocytes, which integrate the required energy consumption over time and space and make a complementary metabolic demand on the adjacent vasculature.

In detail, the pathways are complex, involving lactate, glucose and glycogen metabolic mechanisms, mediated by both intracellular astrocyte and supplementary extracellular transport (Dienel & Cruz, 2008; Ghandi et al., 2009) but the connection between this local glucose metabolism close to the synapse and the oxygen-based hemodynamics in the blood vessel remains unclear. Three hypotheses for this neuro-hemodynamic coupling may be adduced, although all thee remain to be tested:

- aerobic glucose metabolism. This is the concept of a direct coupling between the neural glucose metabolism and the vascular oxygen supply, in which the neural metabolism is supported by oxygen transported from the blood vessels either within the encapsulating astrocytes or through the extracellular cytoplasm (having passed through the astrocytic sheath) to reach the site of the neural synapses and provide the oxidative metabolism of the glucose to reconstitute the ATP used in the neural response. Studies in rat cortex have demonstrated a linear relationship between neural activity, glutamatergic neurotransmitter flux and the cerebral rate of oxygen metabolism (Hyder et al., 2002; Smith et al., 2002). Since cells are predominantly linear summators of the excitatory and inhibitory transmitter release across their synaptic population, the energetic demand driving the BOLD signal is most closely coupled to the net transmitter signal impinging on the cells, and hence to intracellular potential in the cells. The problem with this hypotheses is that it evokes a process with a time constant of the order of minutes (Margaria et al., 1933), which is too slow to account for the 5 s time constant of the BOLD response, despite its high efficiency.

- anaerobic glucose metabolism. This is the concept that the entire neural metabolic process is based on anaerobic glucose delivery from the blood vessel, and that the variation in the oxygenation fraction of the hemoglobin is an epiphenomenon. On this hypothesis, the metabolic demand the site of the neural synapse generates a signal in the astrocytes to release nitric oxide in the filopodia (endfeet) wrapping the arterioles, stimulating an increase in the arteriolar volume with a consequent increase in the proportion of oxyhemoglobin in that region of tissue. The ~5 s time constant of decay of the nitric oxide in the presence of hemoglobin (Hakim et al., 1996) meshes well with that of the BOLD response dynamics. The nitric oxide release signal is presumably mediated by a calcium signal traveling 'antidromically' through the astrocyte (Bezzi et al., 1998; Koehler et al., 2006), accounting for the 1-2 s onset delay in the onset of the BOLD signal.

- anaerobic stimulation of the combined metabolic pathway. This is a mixed concept in which the anaerobic metabolic demand from the neural glucose metabolism stimulates the NO-mediated arteriolar dilation with the consequent increase in both glucose and oxygen transport into the astrocytes (and extracellular cytoplasm). On this concept, the slow diffusion of the oxygen and glucose molecules to the synaptic sites is irrelevant to the BOLD response time course. The critical factors is that the metabolic demand generated by the neural glycolysis is fast enough to elicit an NO signal to the arterioles that, together with its decay time constant, again generates an arteriolar volume time course compatible with the measured BOLD dynamics. On this interpretation, the question of how long the transported products take to reach their metabolic targets, to provide the needed aerobic recovery from the anaerobic depletion, is invisible to the BOLD signal probe.

Thus, the most likely basis of the metabolic demand driving the cortical BOLD signal is the energetic load deriving from the total conductance changes in the postsynaptic membrane generated by a range of processes consequent on transmitter release at the synaptic inputs to each neuron. The summed metabolic demand in the nexus of active cortical neurons adjacent to a capillary forms the drive for the metabolic response in that region of cortex. Thus, the transmitter release is tightly coupled to the activation of the post-synaptic receptors on the recipient cell membrane and consequently to the energetic demands of the membrane receptor activation and to a lesser extent to the subsequent recycling of the transmitter molecules. The majority of these energetic demands are met by the conversion of glutamate to glutamine in the neighboring astrocytes (Sibson et al., 1998, 2001; Rothman & Schulman,

1998). The glutamine is then taken up by the neurons for reconversion to glutamate for use as a transmitter, releasing energy within the neuron in the process.

3. The time course of astrocytic responses

The first element in this chain is the astrocytes surrounding the neuron, which provide glucose to the neuron and replenish its supply by ATP metabolism fueled by oxygen from neighboring blood vessels. It may be emphasized that the astrocyte metabolic processes are as slow, relative to the intracellular signal dynamics, as are the processes of hemodynamic oxygen supply. The time constant of the astrocyte responses at the cell body is known to be of the order of several seconds (Kelly & van Essen, 1974; Filosa, Bonev & Nelson, 2004; Metea & Newman, 2006), and it is clear that these slow responses must mediate the hemodynamic component.The hemodynamic processes provide the requisite oxygen exchange to replenish the energy depletion in the astrocytes and other intermediary processes. The hemodynamic response of the blood vessels to increase their oxygen supply in response to the neural metabolic demand G(t) is mediated by coupling with the astrocyte endfeet. The fMRI BOLD analysis provides an estimate of the ratio of oxygenated to deoxygenated hemoglobin in the blood complement of a given voxel. The post-neural processing stages are often modeled as a linear hemodynamic response kernel convolved with the presumed neural signal. However, this approach overlooks the key role of the pre-hemodynamic processes of the glial and other intermediaries. To reflect the contributions of these intermediary processes, we will term it a 'metabolic response kernel' (MRK) incorporating both the glial and hemodynamic components of the metabolic recovery processes. The metabolic demand has been shown to have a short time course, on the order of the membrane response to a single presynaptic spike. The time constant of the metabolite diffusion through the astrocyte cytoplasm is known to be of the order of several seconds (Filosa, Bonev and Nelson, 2004; Metea and Newman, 2006), so it is clear that there must be a substantial pre-hemodynamic component from these slow responses.

This analysis is borne out by an examination of the time courses of the astrocyte signals. These represent some of the shortest time courses that could be found in a search of the literature on astrocyte signals. These studies make it clear that the calcium signal represents the final control signal for the vasodilation that initiates the sequence of hemodynamic processes that generate the BOLD signal. This calcium signal again has the about the same time constant as the other processes, peaking at times of approximately 5 sec after the stimulus onset (Kelly & Van Essen, 1974; Filosa, Bonev & Nelson, 2004; Metea & Newman, 2006), which implies that the astrocyte response to the neural signal is a slow process with a time constant of the same order as that of the typical BOLD signal.

4. The form of the metabolic response kernel

The time domain approach of present analysis allows the extraction of the maximum possible information about the temporal evolution and processing nonlinearities of the neural signals underlying particular BOLD activation profiles. Our analysis reveals that much information about the neural response properties are reflected in the BOLD signals, even if the time-resolution is insufficient to reproduce the exact neural signal. Examples of such 'transparency' of response properties are particularly clear in the case that the full metabolic response kernel is monophasic.

This monophasic assumption was tested for rat cortex by both Martindale et al. (2003) and de Zwart et al. (2005), who showed that the empirical dispersion of the BOLD response generated increasing delay with distance from the activation site, but were always well-fit by a monophasic model of the BOLD impulse response. Similarly direct measurements of cerebral blood flow and the concentrations of oxygenated and deoxygenated hemoglobin in the human brain (Hoge et al., 2005) reveal only a monophasic temporal waveform for each of these contributors to the BOLD response. These results are all compatible with the inference of a dominant monophasic positive BOLD response in cat LGN and cortex reviewed above (Thompson et al., 2003, 2004, 2005). This form also seems to be a fair approximation in the case of human fMRI because the canonical response kernel (commonly known as the HRF) provided in the SPM-5 software package, although biphasic, has a negative lobe of less than 10% of the amplitude of the positive lobe. It is therefore only a minor modification of this standardized kernel to assume that it has no negative lobe, which is the assumption made for the following analysis. (As will be explained below, the residual biphasic component can be readily attributed to *neural* rather than hemodynamic rebound signals.) Armed with this assumption, we show how several properties of the neural signal are reflected in the recorded BOLD waveform.

5. Implications for the variety of BOLD response waveforms

Armed with this assumption, we show how several properties of the neural signal are reflected in the recorded BOLD waveform. This demonstration assumes a linear relationship between the neural response and the BOLD waveform, in order to make its properties clear before introducing the nonlinear aspects of the analysis.

- **Polarity**. The BOLD response polarity will be an accurate reflection of the polarity of the neural response for any monophasic neural response.
- **Latency**. Any delay in the neural response will also be reflected in the consequent BOLD response. Of course, the metabolic processing sequence may introduce additional delays, but neural delays such as response reaction times or perceptual ambiguity delays should be accurately reflected in the BOLD waveform once the inherent delays of the MRK are taken into account.

- **Transience.** It is an established property of the BOLD response that it is largely sustained for an appropriately sustained neural response (Birn, Saad & Bandettini, 2001; Boynton et al., 1996; Glover, 1999; Logothetis, 2002, 2003; Shmuel et al., 2006). Hence, transience of the BOLD response for a sustained stimulus implies a transience of the underlying neural response. For example, the onset of a sustained light is known to generate transient neural responses in most cortical neurons responding to it. Such stimulation will generate a transient BOLD response even though the stimulation and photoreceptor response are sustained.

- **Number of phases.** For a monophasic MRK (HRF), the BOLD waveform will have the same number of phases as the input stimulus, if the neural input is balanced for positive and negative lobes. Thus, the fact that the measured BOLD waveform is typically biphasic does not imply that the MRK is necessarily biphasic. The negative lobe may derive from a biphasic neural response to a stimulus rather than to blood dynamics.

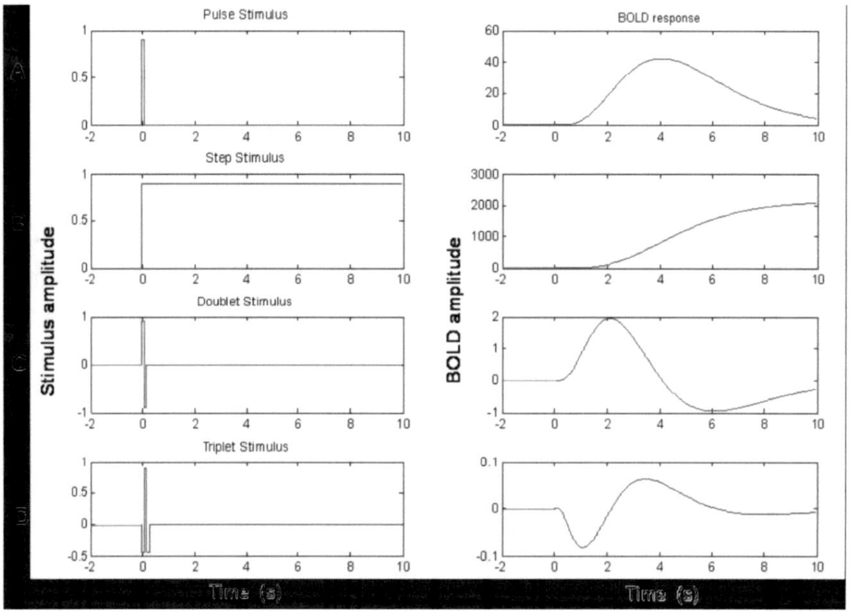

Figure 2. Left panels: (A) neural impulse response, (B) step response, (C) balanced doublet and (D) balanced triplet response. Right panels present convolution of each of these responces with the MRk shown at right. Note that differences in neural response characteristics (left panels) at the tim scale of 100 *ms* generate profound changes in the simulated BOLD waveforms (right panels) on a much longer timescale, which in turn are diagnostic of the differences in the neutral signals.

Some of these properties of the BOLD response are illustrated in the simulation of Fig. 2, for which the MRK is assumed to be the gamma function shown in Fig. 2A, right, as the basis function for the formal analysis. Gamma bases are statistical descriptors of the occurrence of discrete particles (Stacey, 1962; Farwell & Prentice, 1977) that have a simple analog implementation that is a cascade of identical lowpass filters (De Vries & Principe, 1992; Celebi &

http://dx.doi.org/10.5772/31488

Principe, 1995; Chen, 2006). In terms of molecular diffusion processes within neurons, the gamma function represents an optimal description of the cooperative process of the arrival of effector molecules at the channels controlling current flow through the cellular membrane (Shao, 1997). We derive the normal equation for the optimum value of the time scale parameter and decouple it from that of the basis weights. Using statistical signal processing tools, we further develop a numerical method for estimating the optimum time scale.

6. Direct forward modeling approaches

A few previous studies have made estimates of the effects of different neural models on the form of the BOLD response dynamics. For example, Mechelli et al. (2001) report a simulation study of the estimated regional cerebral blood flow (rCBF) and blood-oxygenation-dependent (BOLD) signals as a function of the duration, onset asynchrony and relative amplitudes of two brief stimuli. They included a basic model of neuronal dynamics and varied one parameter of this model – the amplitude of a slow late transient – to show its effect on the simulated BOLD responses. This exercise constitutes an unvalidated forward model of the effect of only one parameter of a simulated neural response. However, what they offer as the analysis of the effects on the BOLD waveform is described (incomprehensibly) as "the BOLD parameter estimates", which is actually a single-valued function with no specification of which parameter(s) is/are being estimated. Since no comparison is made with empirical data or their noise limitations, the fact that this study includes a proposed model for the neural dynamics does not qualify it as a validated procedure for estimating the neural population dynamics underlying the local BOLD signals (which is the goal of our study).

Buxton et al. (2004) extend their balloon model of the hemodynamic response leading to the BOLD signal by proposing a model of the neural response to account for the temporal nonlinearity in BOLD responses as a function of duration. This neural response incorporates a slow subtractive inhibitory component to the net neural signal, which has the effect of producing a neural response consisting of an initial transient followed by a sustained plateau. Model responses for three kinds of stimuli – a single short pulse, two short pulses and one long pulse, are offered as a demonstration of the properties of this model. As with Mechelli et al. (2001), no attempt is made to compare the model outputs with actual BOLD recordings, so the Buxton et al. (2004) study again does not qualify as a validated procedure for estimating the neural population dynamics underlying the local BOLD signals.

Both Mechelli et al. (2001) and Buxton et al. (2004) include parameters intended to account for the temporal nonlinearity of short duration responses (which do not fall linearly as response duration is reduced; Boynton et al., 1996; Birn, Saad & Bandettini, 2001). Both studies demonstrate the required lack of reduction in a single example of a short-duration response, but neither study provides a validation of either the waveform or the amplitude response function relative to empirical BOLD data. In principle, either model could provide a platform for such validation or the further estimation of the dynamics of the underlying neural population response, but they neither do so nor suggest procedures by which such estimation could be achieved.

7. Analytic model of the neural-BOLD coupling

To start the analysis, a specific model structure of the processes leading to the BOLD paramagnetic signal of fMRI recordings incorporates aspects of the biophysical processes that are not present in the linear convolution analyses of Friston et al. (1997, 1998, 2000), although somewhat condensed in comparison with the recent biophysical/metabolic derivations of Mechelli et al. (2001), Buxton et al., (2004) or Sotero & Trujillo-Bareto (2007, 2008).

The metabolic demand through the biophysical chain to the measured BOLD signal has two sources of nonlinearities. The major source of the BOLD waveform measured by functional Magnetic Resonance Imaging (fMRI) is the summed local field potential in the cortical region contributing to the signal for each voxel. The metabolic demand generated by the underlying neural activity is summed in parallel across the local neural sources, and is thus nonlinear with respect to the linear systems prediction for a serial system. The coupling itself also exhibits saturation nonlinearities when pushed beyond the region of small-signal linearity.

The neural responses within each voxel are treated as generated by sets of homogeneous populations with similar signal waveforms Ni(t) within each population (Fig. 3). Each neural response then generates a local metabolic demand Mi(t) that may have a nonlinear relationship to the neural signal waveform (Chatton, Pellerin & Magistretti, 2003). The MRK will be convolved with a non-linear transform of the presumed neural signal to provide an estimate of the neural metabolic demand that is being met by the combined glial and hemodynamic metabolic response. The fMRI analysis also has a finite dynamic response time, but it will be treated as incorporated in the MRK of the glial/hemodynamic response.

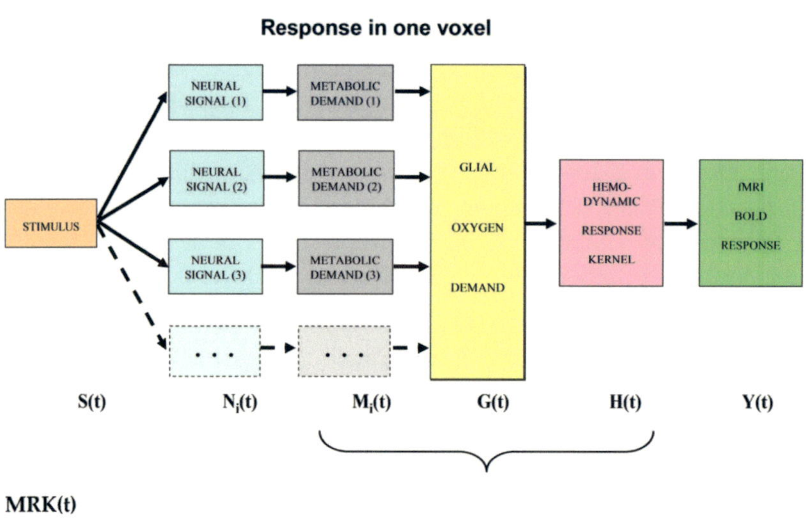

Figure 3. Block diagram of the main processing stages that lead up to the BOLD signal. The *i* subscript indicates that the stage incorporates multiple components within the voxel. See text for details.

These nonlinearities provide the opportunity to develop an analytic model of the neural signal leading to the BOLD activation. The model assumes the presence of neuronal subpopulations having response dynamics with various decay time constants in response to the stimulus. Pooling among the subpopulation responses can then explain the multiple decay characteristics of the recorded local field potentials (LFPs). Subsequent convolution with a characteristic metabolic response kernel then generates the predicted response Y(t) for the BOLD waveform for the region of cortex generating the LFP signal, accounting for concurrently recorded BOLD waveforms.

The neural responses within each voxel are modeled as sets of homogeneous populations Ni(t) with similar signal waveforms within each population. Each neural population response then generates a local metabolic demand Mi(t) that may have a nonlinear relationship to the neural signal waveform. The integrated metabolic demands are met primarily by the astrocytes, which integrate the required energy consumption over time and space and make a complementary metabolic demand G(t) on the adjacent vasculature. The hemodynamic processes H(t) replenish the energy depletion in the astrocytes, leading to the paramagnetic response that generates the BOLD signal of the differential precession rates of the oxygenated/deoxygenated hemoglobin moieties.

8. Implications for the analysis of fMRI BOLD signals

A proper understanding of the metabolic pathway underlying the fMRI BOLD signal is a necessary precursor to an analytic capability for neural signal estimation from the BOLD waveforms recorded throughout the cortex. Any such attempt must be based on a model of the known neural population dynamics underlying the BOLD metabolic signal generation, which may be progressively refined as more information becomes available about both neural response characteristics and the metabolic cascade. Given adequate signal/noise ratio, it is possible to develop approaches that overcome the temporal limitations of BOLD signal and are able to reveal relevant properties of the underlying neural signals. This analysis can provide a direct linkage between the live assessment of the functioning brain and the direct neurophysiological recordings in other species.

FMRI analysis techniques for estimating the BOLD signal typically employ Generalized Linear Model, which incorporates the convolution approach to the estimation of the underlying neural signal. Convolution is based on the assumption of a unitary BOLD waveform kernel that generates the straightforward prediction of the BOLD response waveform for any stimulus type or duration in any brain area. In fact, however, major deviations from a standard BOLD waveform may be found, even in the same cortical regions, for variations in stimulus conditions. D'Avossa, Shulman & Corbetta (2003), for example, reported strong differences in waveform when the response to the motion or color of a cue/stimulus pairing was modulated by attention. Such local waveform differences must derive from differences in the neural signals driving the BOLD activation, since the metabolic and hemodynamic processes that mediate the paramagnetic signals should be invariant within a given cortical region.

The BOLD signal measured by fMRI is often characterized as having high spatial resolution (up to a quarter of a million voxels), but low temporal resolution (0.5 – 5 s) relative to other methods for mapping human brain function (such as high-density EEG analysis). Estimation of single-parameters of the waveform, such as response delay alone, can improve the temporal resolution for the neural signal delays to 100 ms or better for narrowly targeted brain regions (Menon, Luknowsky & Gati, 1998; Henson et al., 2002) but this requires the assumption that the BOLD signal has a unitary waveform, which is often not the case (Aguirre et al., 1998; D'Avossa, Shulman & Corbetta, 2003; Handwerker, Ollinger & D'Esposito, 2004; Likova & Tyler, 2008; Tyler & Likova, 2009). Even minor deviations from a stable waveform violate the assumptions of such single-parameter analysis and invalidate the delay measure.

To evaluate the neural contribution to the differential BOLD response waveforms within the same cortical regions, Likova & Tyler (2008) developed an "instantaneous stimulus paradigm" to evoke BOLD responses to instantaneous stimulus transitions. It is typically assumed that, despite their perceptual differences, such "instantaneous" stimuli would all generate the same BOLD waveform (effectively equivalent to the MRK) for all different stimulus types. Thus, revealing any significant deviation from that prediction in the BOLD waveform elicited in one and the same cortical area can contribute to revealing the specifics of the underlying neural processing and enhance the understanding of the recurrent network of extended perceptual responses to complex stimulus configurations. Indeed, the instantaneous stimulus paradigm generated a wide variety of BOLD waveforms. Moreover, the data show dramatic differences in the BOLD waveforms properties (e.g., latency, sign, amplitude and width) even within the same brain areas as a function of the stimulus type. As there can be no vascular heterogeneity within the same area, these dramatic waveform variations must be attributed to differences in the underlying neural dynamics, not to spatial variations in the MRK. These results further imply that fMRI signals contain much more information about the neural processing than is commonly appreciated, and thus have the potential to capture them through an appropriate approach.

9. Nonlinear dynamic forward optimization

However, until recently there has been no method of transcending the BOLD temporal limitations in order to estimate the dynamics of the neural signals underlying the measured fMRI waveforms. Tyler & Likova (2009, 2011) therefore proposed a Dynamic Nonlinear Forward Optimization (NDFO) approach for the time-resolved estimation of the neural signals underlying the particular characteristics of the temporal BOLD waveforms for a particular stimulus processed by a particular cortical region. The philosophy of this approach is to utilize the information available from neurophysiological studies of the neural population dynamics and biochemical studies of the metabolic pathway coupling to the measurable blood response to provide Bayesian priors as to the likely temporal structure of the component neural signals. This method provides a compact account of the measured waveform with the minimal number of neural predictors, based on prior knowledge of the expected temporal properties of neural signals and of their consequent metabolic demand. The result is the ability to specify the prevalence of each element in the otherwise inaccessible star. In the case of the neural signals, the goal is to estimate the amplitude and time course of each of the neural components whose metabolic effects, when summed, account for the measured BOLD waveform for a particular stimulation condition and cortical region.

Rather than simply characterizing the behavior of the BOLD waveform (Birn, Saad & Bandettini, 2001; d'Avossa, Shulman & Corbetta, 2003; Fox et al., 2005; Grotz et al., 2009) or attempting to infer the potentially complex properties of the underlying neural mechanisms from the form of the BOLD response by deconvolution (Glover, 1999; Logothetis, 2002, 2003a,b; Logothetis & Wandell, 2004), the concept of forward modeling is to incorporate as much knowledge as possible about the likely neural substrate and optimize the remaining details to best fit the BOLD waveform. Here the predictors are non-linear because there is a nonlinear relationship between the neural responses and the metabolic demand that they generate, but the summative property of the paramagnetic signals throughout a voxel implies that we can assume that the component metabolic demands sum linearly together (Fig. 3). The dynamic forward modeling may incorporate a variety of possible nonlinearities into the structure of the model. The neural model that we investigate in the present version of the analysis is the sum of positive and a negative component based on delayed gamma functions convolved with the stimulus waveform

9.1. Analytic framework for the neural temporal response

The starting point for the NDFO modeling is a model of the neural signal, whose first effect in terms of the cascade of BOLD dynamics is to create a metabolic demand G(t) in the neighboring glial cells (see Fig. 3). Gamma probability density functions have the analytic form a. kt-1. e-t/τ/Γ(k), where t is the time dimension, a is a scaling parameter and k and τ are generic waveform parameters. (To equate the area to unity, the function is defined with a = 1/τk), but for the present purposes a is a free optimization parameter.) They may be termed simply "gamma functions" to emphasize their analytic rather than statistical properties. In engineering the same function is known as the n-pole filter function (with n = k), and is used to describe the dynamics of a wide range of processes. For the present application, the temporal gamma function is assumed to have integer powers of k and corresponds to the solution of differential equations with real (non-imaginary) roots. The gamma function has the analytic advantage over many other functions, such as the Gaussian, that it is by definition causal because it has the value of zero at t = 0, and is defined as zero for t < 0 (i.e., the full specification is y = kt-1. e-t/τ for t >= 0; y = 0 for t < 0). Its shape progresses from highly asymmetric around the peak for small k to approximately Gaussian and symmetric for large k (see Fig. 4). Fig. 4. Examples of delayed gamma function step responses with exponents of k = 2 (left) and 6 (right). Successive functions (colors) introduce a wide range of peak latencies for the neural signal estimates (with time constants increasing in factors of 2 and a fixed delay Δt of 40 ms).

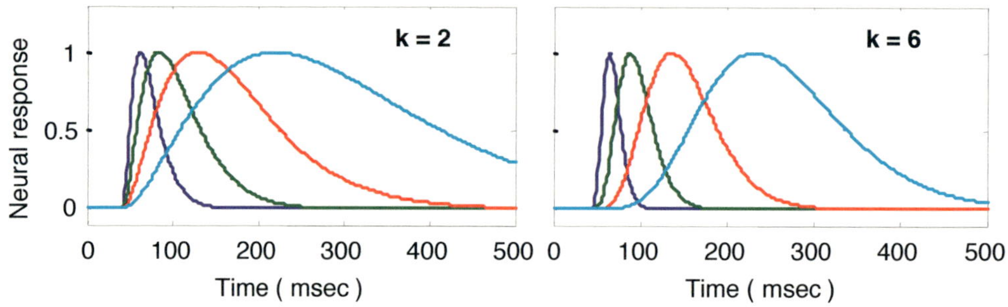

Figure 4. Examples of delayed gamma function step responses with exponents of k = 2 (left) and 6 (right). Successive functions (colors) introduce a wide range of peak latencies for the neural signal estimates (with time constants increasing in factors of 2 and a fixed delay Δt of 40 ms).

A key feature of this formalism is that the peak latency is determined by the time constant, τ, and is proportional to the width (at half height) of the response peak, which can be estimated to good accuracy by the methods described in the next section.

9.2. Neural model

A comprehensive model of the BOLD therefore requires an accurate model of the intracellular potential dynamics deriving from the sensory stimulation. Formally, we propose to use the full neural response model jointly specified in eqs 1 and 2:

$$N_p(t) = \left[|N(t)|_+ + \lambda \cdot |N(t)|_- \right] e^{-t/\eta} + \varepsilon(t) \tag{1}$$

where ε is the net source of additive noise and the function is governed by the parameter set p=$(a,\ k_\eta,\ \tau_\eta,\ b,\ \Delta t,\ \eta,\ \lambda)$

The neural signal NP(t) in eq 1 is the sum of positive and a negative component based on delayed gamma functions convolved with the stimulus waveform:

$$N(t) = a \cdot S(t) \otimes n(t - \Delta t) \tag{2}$$

where $n\,(t) = (1-t) \cdot\ t^{k_n - 1} e^{-t/\tau n} - \frac{b}{\tau_n} t^{k_n} e^{-t/\tau_n}$

(Note the convention that time series functions are capitalized, impulse response kernels are lower case, and vectors are bold-face.) The neural impulse response expression n(t) is set up so that its convolution with a step function is equivalent to the sum of a pure transient and a pure sustained component. In addition, the expression is specified with an additional transmission delay, Δt, that delays the response relative to the stimulus without affecting its waveshape with the parametrization specified in Table II, which defines the parameters in vector p of this equation. To illustrate the properties of the model, we analyze the effect of

http://dx.doi.org/10.5772/31488

varying the inhibitory ratio implied by the negative component weight b, and the offset/onset gain ratio λ.

a	scaling constant (a fitting parameter but not a waveshape parameter)
kn	integer exponent governing the rising phase
τn	time constant of falling phase in the neural response
Δt	transmission delay before response onset
b	sustained/transient ratio in the step response
η	time constant of nonlinear gain control in the neural response
λ	ratio of offset to onset gain in the neural response

Table 1. Nonlinear Forward Model Parameters

One issue that arises is how to measure the latency Δt of the delayed gamma functions of Fig. 4. A simple derivation can show that the peak latency tpeak of these responses is specified by the expression ($k\tau n + \Delta t$). Thus if k and Δt, both of which are well-determined from neurophysiological studies in monkey cortex, are set to means of their Bayesian priors on this basis, the peak latency tpeak is determined from the value of τ, which can be accurately derived from the model optimization.

9.3. Metabolic demand

Since little is known about the glial dynamics of transmitter recovery, we will pursue two options as to their effects. One option is that the metabolic demand driving the BOLD response derives from the transmitter recovery cycle following the activation by an axonal spike. Since axonal spikes represent only the positive aspect of the intracellular voltage and since 90% of cortical synapses are excitatory (Shank & Aprison, 1979; Wang & Floor, 1994), the signal transmitted from one cortical stage to the next may be treated as a half-wave rectified version of the dynamic neural signal. This prediction is shown as the blue curves in Fig. 5Aa (indexed in row/column notation), which is an overlay of the model estimates of the neural responses to stimulus pulses that double in duration from 8 ms to 16 s (eight doublings). For this example, the neural response has balanced excitation and inhibition, so even the prolonged pulses generate only an initial transient response, with the negative lobe at offset being thresholded out by the rectification. (Note that the local metabolic demand, Mi(t), has the same time course in this model as the transmitter recovery from which it derives. The energetic processes required for the recovery to the initial state, however, form an oxygen-based chain of glial metabolic response, G(t), that may have substantially slower time course at one or more stages.)

The other option is to consider the instantaneous metabolic demand of both excitatory and inhibitory cells, or both 'on' and 'off' cells, implying that the signal generating the metabolic demand is a full-wave rectified version of the intracellular voltage. This option is shown as the red curve in Fig. 5Aa, whose spikes represent the instantaneous metabolic demand for the offset transient for each of the pulse duration doublings (which were thresholded out in the blue curve).

9.4. Metabolic coupling

Having provided relevant variables for the linear and nonlinear components of the neural response dynamics, we may now consider the issue of the metabolic coupling of the neural signal to generate the BOLD response. This coupling has been extensively modeled over the past two decades, with the best-known example being the Buxton-Friston balloon model and the most elaborated version being by Sotero & Trujillo-Barreto (2007, 2008). However, these models are not well-validated by empirical human studies (because they contain too many interdependent variables to allow the independent assessment of each separately

10. Characteristics of the model

We may now evaluate the response to these two options for the nonlinearity of the metabolic demand through the biophysical chain of the metabolic processes to the measured BOLD signal (Table I). However, at present too little is known of the dynamics of transmitter recovery and/or the nonlinearities in the process to securely assign time constants to the astrocytic component relative to the hemodynamic component of the metabolic coupling. For the first pass, we will therefore treat the entire chain coupling the metabolic demand to the magnetic resonance signal in the traditional fashion, as a unitary linear kernel convolving the nonlinear neural signal. (As stated above, this kernel is often termed the hemodynamic response function HRF, but in view of its likely substantial astrocyte contribution, we give it the more general term of the metabolic response kernel MRK). Our main goal is to estimate the properties of the neural signal processing, and it will be seen that there is sufficient information to provide a rich analysis of these properties, and to account for the empirical nonlinearities of the BOLD signal, as long as the metabolic supply chain conforms to the linearity assumption. (As more information becomes available, nonlinear aspects of the metabolic coupling may readily be incorporated into the analysis we propose.)

The development of the peak amplitude in the BOLD temporal summation series is shown as the blue summation curve in Fig. 5Ad. The critical point of this plot is that the asymptotic corner of this summation curve occurs at 40 ms, which is the value of the time constant assumed for the neural signal in this example. Thus, the form of the BOLD amplitude summation series (Fig. 5, column d) provides a direct empirical estimate of the time constant of neural integration down to the millisecond range. There is no limit in principle to the temporal resolution that can be achieved by this methodology since it is estimated from the amplitude variation of the BOLD signal as a function of stimulus duration, not from its temporal aspects.

http://dx.doi.org/10.5772/31488

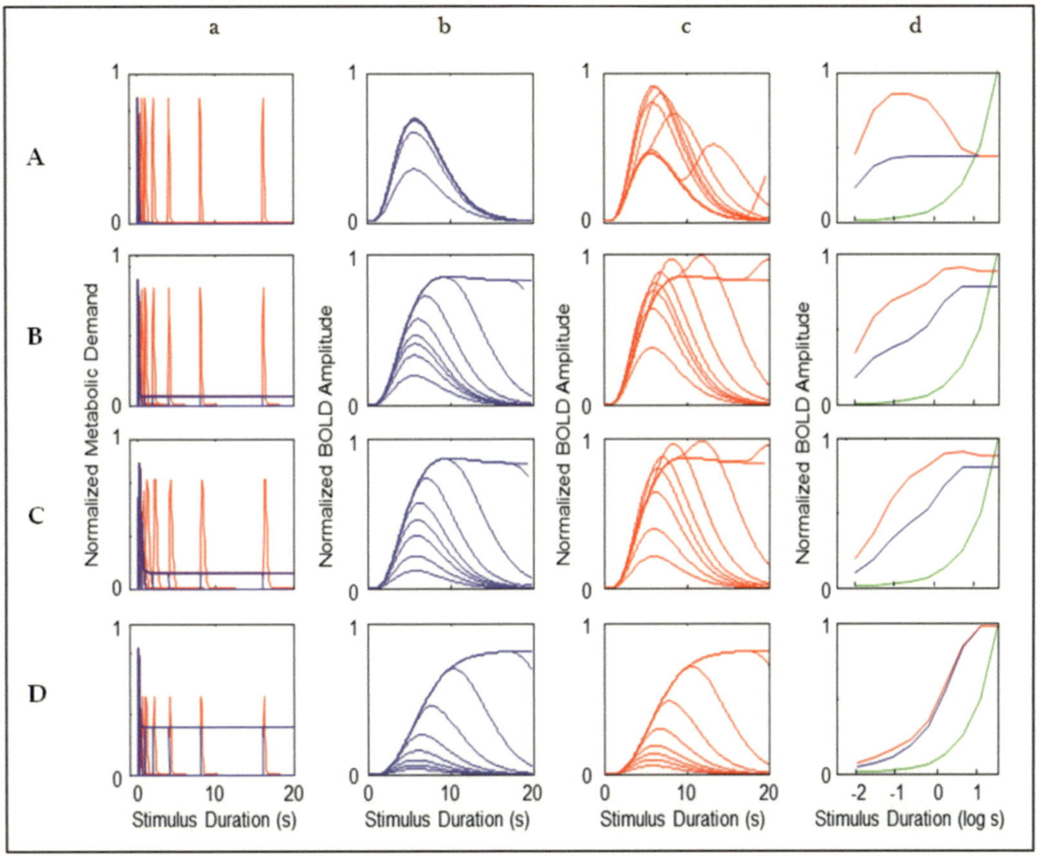

Figure 5. Simulations of four different types of BOLD response for monophasic metabolic demand signals (and a monophasic Metabolic Response Kernel, MRK). The rows represent the results for A: metabolic demands with a purely transient time course, B-D: responses with a mixed transient and sustained time course, with the sustained component at respectively 12%, 18% and 50% of the amplitude of the transient component (based on different ratios of neural excitation/inhibition). For each type, column (a) shows the assumed metabolic demand, column (b) plots the BOLD responses over duration for the half-wave-rectified model of metabolic demand, column (c) plots the BOLD responses over duration for a fully-rectified model, and column (d) plots the duration summation curves for peak amplitude (blue curve: half-rectified model, red curve: fully-rectified model, green curve: pure linear summation). Note the use of the logarithmic abscissa in column (d) to focus the analysis on the brief duration regime. The progression of the model BOLD responses with stimulus duration and the form of the summation curves are diagnostic of both the relative weighting between the sustained and transient components of the neural signal and the form of rectification feeding the metabolic demand.

The reduction in peak amplitude is captured in the red temporal summation curve of Fig. 5Ad, showing a reduction by a factor of two for long durations. (The green curve in Fig. 5Ad represents the values expected for fully proportional linear summation of the energy in the stimulus pulse; it is an accelerating curve due to the logarithmic abscissa.) The case of the full-rectification model of the metabolic demand is shown in Fig. 5Ac. Here the second neural response peak (i.e., that from the stimulus offset) plays a key role in varying the BOLD waveform, which first extends in time and then shows a two-peaked structure with reduced amplitude for long-duration stimuli.

Fig. 5Ba shows the half- and fully-rectified version of the metabolic demand to the same pulse duration series, where the neural inhibition is now assumed to be reduced in energy by 1.5% relative to the excitation. This small imbalance is magnified by the convolution with the sustained stimulus, and thus it results in a sustained component that is 12% of the amplitude of the initial transient (blue curve in Fig. 5Ba) and then into an almost fully sustained set of BOLD response functions (Fig. 5Bb, compare to Ab). Thus, the form of the BOLD response functions can be strongly diagnostic of even slight variations in the properties of brief neural signals. Moreover, the nature of the metabolic demand function (half- or fully-rectified) has a big impact on the form of the BOLD response, determining whether or not an offset peak occurs at the tail of the responses even when they are sustained (Fig. 5Bc). Such a peak has been reported in some studies but is not always evident. Thus it remains an empirical question to what extent rectification is representative of BOLD waveforms; intermediate forms of the rectification model are required to capture the empirical properties in detail.

Note that the amplitude series in Fig. 5Bb and c show bands of denser packing of the functions, where the amplitude changes were not spaced in proportion to the doublings of stimulus duration. Viewed in terms of the sequence of BOLD waveforms in Fig. 5Bb,c, the regions of dense packing form an intermediate "shelf" or partial asymptote in peak amplitude summation plots of Fig. 5Bd. It is again evident that the onset of this intermediate shelf in the summation curve corresponds to the integration time of the underlying neural signal, while the second asymptote at higher amplitude corresponds to the ~5 s integration time of the MRK (HRF). Accurate measurement of such summation functions can therefore provide discriminative characteristics that, when interpreted through the nonlinear model structure, can provide estimates of both the neural and the metabolic time constants in the neural-to-BOLD signal chain.

This point is emphasized by the response set in row C of Fig. 5, which probes the effect of varying the time constant of the neural transient. The key difference from the parameters used in row B of Fig. 5, is that the neural time constant was doubled from 50 ms to 100 ms (and the excitation/inhibition imbalance was also increased to 7% to maintain the same form of offset peak). It is evident that (i) the summation curve (Fig. 5Cd) takes a measurably different form, and that (ii) the accuracy of estimation of the neural time constant is limited not by the BOLD time constant but only by the variability of the BOLD amplitude measures. For example, this analysis shows that the neural time constant is estimable to within about 0.1

log units if the BOLD response functions can be measured to an achievable accuracy of about 10%.

The final case of duration summation analyses (Fig. 5Dd) shows the NDFO predictions of increasing the excitation/inhibition imbalance of the neural response to 50%, illustrative of a system that is predominantly sustained in nature. Under these conditions, the impact of the initial transient becomes essentially negligible, and the summation curves (Fig. 5Dd) become indistinguishable from proportional summation (i.e., they run parallel to the green curve). This manipulation illustrates that the power of the NDFO analysis depends on the neural processing being predominantly transient, and that the properties of the underlying neural mechanisms would not be accessible to this form of analysis in predominantly sustained systems. Luckily, however, the well-established deviation from proportionality for short-duration stimuli implies that the neural system is, in practice, predominantly transient and is therefore amenable to this form of NDFO analysis.

11. Neural response component analysis

In a study of the fMRI components, we have recorded (Tyler et al. 2008) the sets of BOLD waveforms generated by fields of dynamic noise-patterned stimuli in differenct eccentricity bands across the visual field (Fig. 6, uppermost row). Responses were recorded in a corresponding series of retinotopic regions of interest (ROIs) defined bilaterally on V1 on the basis of serparate retinotopic mapping stimuli. An example of the variety of cortical responses obtained in retinotopic area V1 combined for both hemispheres is shown in the upper panel of Fig. 6. The responses vary not just in amplitude across stimulus types, but markedly in the waveform of the responses. Forsome stimuli, the same area may show a double-peaked on-response, a rounded on-response, a negative on-response, or a classica boxcar response. The variation in response from single cortical regions cannot be explained by variations in the hemodynamics of the blood oxygenation and strongly implies a neural origin of the differences for different stimulus type

These results support once again the important point that BOLD waveforms within particular cortical regions may vary dramatically as a function of the stimulus type, despite the fact that the metabolic/hemodynamic response kernels must remain invariant in the region (d'Avossa et al., 2003; Likova & Tyler, 2008; Tyler & Likova, 2009). The only option for a differential MRK is if the underlying neural circuitry has differential cortical layout relative to the local arterioles controlled by the astrocytes, which seems a highly unlikely structural association (and one that, if it did exist, would produce minor stastical variations in the BOLD waveform, not the radical differences among many stimulus types, as seen here). Consequently, differences in the BOLD response waveform in this ROI should be interpreted as due to differences in the *neural* response waveforms to the *different* stimuli.

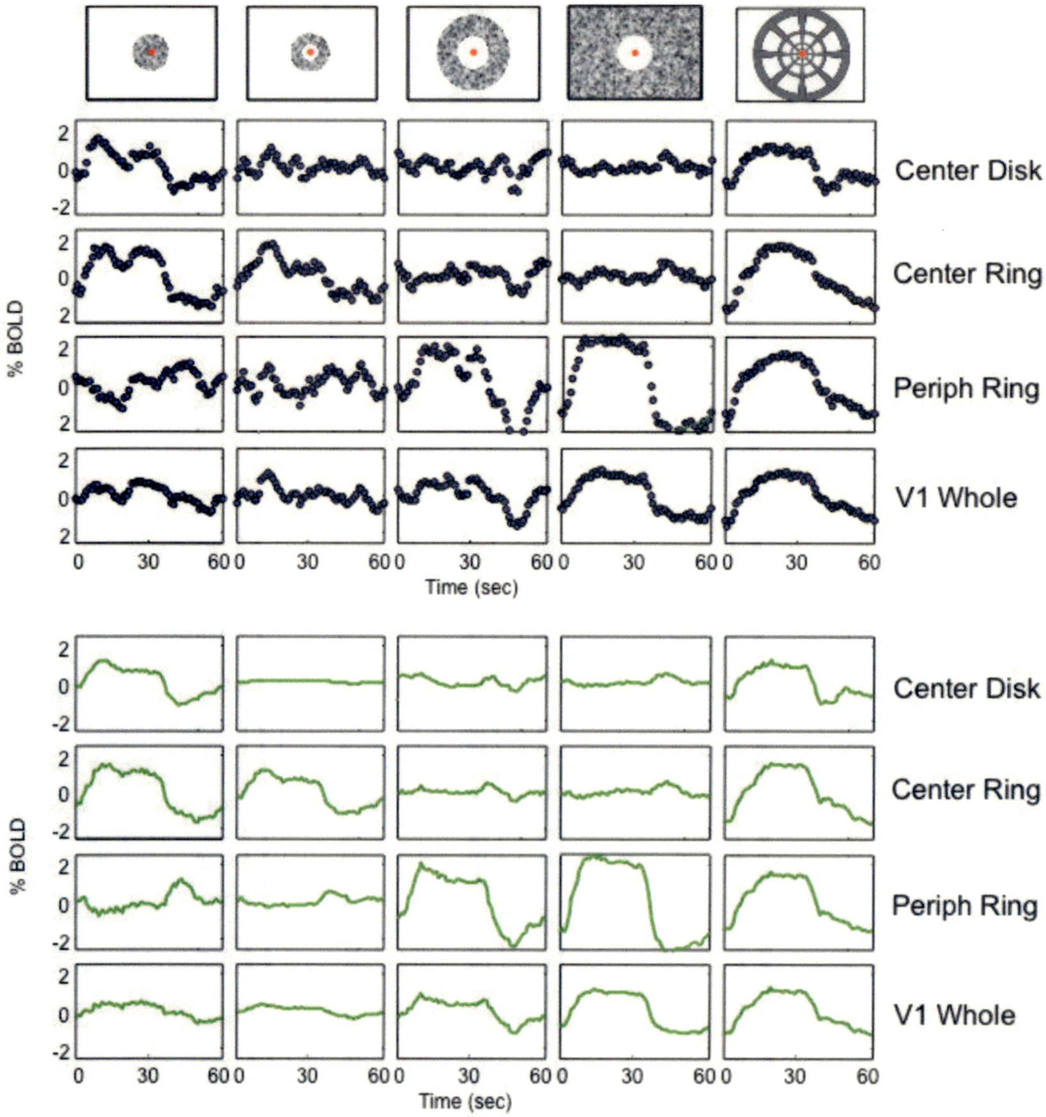

Figure 6. Upper plots: Variety of BOLD response waveforms obtained for the set of spatial noise stimuli shown along the top row in four matching ROIs deisgnated at right, in a block design of 30*s* on/ 30*s* off of fields of dynamic visual noise. Many of these responses differ substantially from the form of the typical GLM and from each other, particularly for the fine grid structure of the scaled grid in the last column. Lower plots: The independent components (IC) model does a good job of capturing this variety with the sum of the first three ICs, accounting for 92% of the variance in the empirical responses.

12. Multicomponent analysis of the neural signal contributions

Under the assumption that the neural signals in local regions of cortex consist of multiple components that are well-approximated by sets of delayed gamma-function components, they should be *dissociable* into the same vocabulary of multiple components as for the direct neural recordings through the NDFO procedure that takes into account the slow metabolic/ astrocytic dynamics following the neural activation signals. The goal is to resolve the signals into the same component vocabulary, i.e., delayed gamma functions, that approximate known components of the neural dynamics.

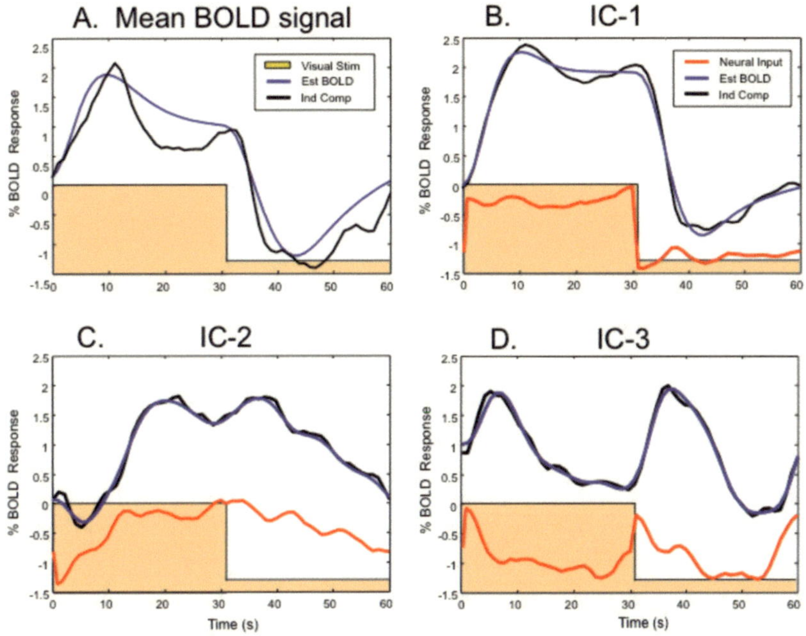

Figure 7. Dynamic forward modeling optimization of multiple neural components to the BOLD responses for cortical area V1. A. The average BOLD waveforms across the whole of V1 (black) with the fit for the Buxton-Friston balloon model. B-D. Plots of the optimized neural response estimates (red) for each of the three BOLD ICs (black curves), with the fit of the balloon model to each component (blue curves). Note the wide variety of temporal properties of the neural signals (red curves) selected by the optimization to account for the waveform differencesof the BOLD components. (from Tyler, Kontsevich & Ferree, 2008, with permission.)

The novel analysis we propose consists of (i) the simultaneous optimization of the set of neural components (ii) throught the Buxton-Friston balloon model, together with (iii) the rise and fall parameters of MRK. The results of the analysis for the V1 dataset of Fig. 6 s were sufficient to account for >95% of the BOLD variance overall (Fig. 7), which drew from the whole of the V1 ROI. Note that the mean BOLD waveform has an unusual double-peaked waveform, rising at the stimulus offset as well as the onset, and that the off-period (Fig. 7A). The IC analysis breaks this down into independent components that are differentially expressed across the individual voxels of V1 (some of this temporal waveform variety being captured in the specified ROIs within V1 shown in Fig. 6). The first IC (Fig. 7B) shows the classic form, following the predicted linear form for the balloon model (blue curve) accurately. The optimized neural signal for the fit to the actual wavform for the first IC is close to a boxcar function. The second IC has a very different form (Fig. 7B), with an early negative

BOLD peak at the stimulus onset and no corresponding rise at stimulus offset. The estimated neural signal has the same characteristic, although the negative peak has a much shorter latency. The third IC is even more non-linear, with approximately equal positive BOLD peaks at stimulus onset and offset as though responding to the stimulus events through a full-wave rectified nonlinearlity. Again, the estimated neural signal is a reflection of the same form of nonlinearity with much shorter latency.

13. Discussion

Our novel analysis of the neural signal underlying the BOLD response waveforms to a full temporal stimulus series constitutes a form of "temporal microscope" for the neural signals in the cortex generating the recorded BOLD waveforms. Thus, this study illustrates how the neural waveform parameters can be estimated from variety of stimulus effects. The estimated neural waveform for the present dataset in Fig. 5C has a strong brief transient, followed by a sustained response of about 1/10th of the peak transient response and a reduced, but rectified, offset response at the end of each stimulus. The transient should be presumed to match the corresponding neural response properties recorded in the cortex of other species. A useful survey of the dynamics of the unitary intracellular responses to single synaptic action potentials is provided by Thomson & Lamy (2007). They tabulate data from a large number of studies of the excitatory post-synaptic potential (EPSP) transient with a typical full width at half height of ~50 ms. We are not familiar with comparable meta-analyses for monkey cortical responses, but this value is clearly in line with the published examples of V1 transients from this species (such as those of Hegdé & van Essen, 2004).

The final analysis describes a search for independent response components in occipital are V1 in a relatively standard block-design stimulus paradigm incorporating a variety of spatial stimulation patterns (Tyler, Konstevich & Ferree, 2008). This analysis began with the assumptions of the balloon model of vascular hemodynamics, but the results revealed dramatic response nonlinearities, as expressed through an independent components analysis of the response variation across the cortical space of V1 and the stimulus variety (Figs. 6,7). Not shown in these figures is the way in which these nonlinear components were expressed across the cortical space, which implied that they were functional response components dependent on the relation between the stimulated and unstimulated regions, not structural differences in the vasculature. Thus, although the balloon model was an assumption of the analysis, the results raised the question of whether its assumed nonlinearities are hemodynamic in origin, given this result of profound functional nonlinearities across the space of V1, which are a fortiori of neural origin.

The neural signal components estimated to underlie the three primary ICs of Fig. 7 provide insight into the nature of the neural nonlinearities involved. Unlike the stable (linear) boxcar of Fig. 7B, the other two estimated neural waveforms have initial transients, one representing a half-wave rectification and the other a full-wave rectification. In neural signals, such nonlinearities are extremely well-known, and indeed are characteristic of neural processing in general (as "on" responses and "on-off" responses, respectively), while it is a stretch to conceptualize how they could arise from the hemodynamics per se. A more global mechanism that has such transient, rectified character is the top-down attentional mechanism, where activation may be drawn to regions of recent stimulus change in a transient fashion. These data are not sufficient to distinguish between the top-down and bottom-up hypotheses for the neural signals underlying the nonlinear BOLD responses revealed by this paradigm, but they make it clear that a rich array of neural nonlinearities should be expected to contribute to the fMRI signals recorded from the brain, which should allow the technique to answer a correspondingly abundant set of questions about the neural mechanisms involved in these response dynamics.

As knowledge evolves, more complex models of the neural/metabolic coupling and the hemodynamic response could be easily incorporated in our model. These all represent Bayesian information that, if well established, can be used to refine the model structure and enhance the fitting process when available. However, our understanding of the literature is that a) the first-order specification of the neural/BOLD coupling is well approximated by a linear response kernel, and b) that estimates of the second-order effects are contaminated by the assumption that the modeling has been purely hemodynamic, and has not taken into account potential and actual nonlinearities in the neural signals at the time scale of the BOLD signal. Thus, despite the best efforts of the proponents of elaborated hemodynamic modeling, there is no secure information about the nonlinearities of this process for the human brain in vivo.

Consequently, the present paper has tried to redress the balance by considering the effects plausible nonlinearities in the neural population responses to standard types of stimulus presentation. For clarity in this enterprise, we have assumed that the metabolic response is both linear and monophasic, illustrating a variety of BOLD response properties that could arise from such neural nonlinearities. We are not claiming to have proven that these BOLD properties are entirely determined at the neural level that, conversely, claims that they are purely properties of the vascular hemodynamics per se are suspect until they are replicated in paradigms that remove the neural component of the system. One such approach would be stimulation of the hemodynamic response by direct infusion of nitric oxide (NO) in the vicinity of the blood vessels. This experiment has apparently not been attempted, although suppression of the nitric oxide with application of the NO synthase inhibitor 7-nitroindazole completely abolishes the BOLD response while only marginally affecting the local field potentials (Burke & Bührle, 2006), establishing the role of NO in neurovascular coupling.

In particular, the nonlinearity of the transient responses at the offset of the neural response could be positive (rectifying) rather than negative (linear), and in either case show could show reduced amplitude relative to the onset response (adaptive gain control). In this example of the NDFO analysis, the offset responses are much smaller than the positive responses at stimulus onset. This analysis demonstrates that estimation of the neural response dynamics for each stimulus type is therefore not a Herculean task requiring extreme levels of signal/noise ratio, but is well within the capability of one session of normal fMRI methodology using however the appropriate combination of experimental design and theoretical analysis.

We note that a technique with a philosophy similar to the present approach has been successfully applied to the estimation of net spatial receptive field structure of small cortical regions by Dumoulin & Wandell (2008), although they used a linear rather than the full nonlinear model of the sequence of processes. Their spatial estimates were based on a model of the temporal signal to be expected as a stimulus swept across each defined point on the retina. Like us, they take the temporal stimulus waveform, convolve it with a spatiotemporal model of the response of the underlying neural population and then with a model of the hemodynamic response function to provide a basic forward model of the temporal BOLD response that is optimized to the measured BOLD response at each cortical location. Our approach takes the temporal analysis several steps further towards biological plausibility, and focuses on the temporal rather than spatial aspect of the neural population response.

14. Conclusion

The conceptualizations and techniques introduced in this paper provide an analytic capability to resolving the timing and neural signal estimation underlying the BOLD waveforms recorded throughout the cortex. Any such attempt must be based on a model of the known neural dynamics of the neural populations underlying the BOLD metabolic signal generation, which may be progressively refined as more information becomes available about both neural response characteristics and the metabolic cascade. Given adequate signal/noise ratio, the present analysis shows that it is possible to develop approaches that overcome the temporal limitations of BOLD signal and are able to reveal the relevant properties of the underlying neural signals down to their native temporal resolution. In combination, these approaches represent a notable advance in the capabilities of the fMRI technology, providing a direct linkage between the live assessment of the functioning brain and the direct neurophysiological recordings in other species.

Acknowledgements

Supported by NSF SLC grant 0846229, AFOSR grant #FA9550-09-1-0678 and CDMRP grant XWH-11-2-0066.

Author details

Christopher W. Tyler and Lora T. Likova

Smith-Kettlewell Eye Research Institute, USA

References

[1] Aguirre GK, Zarahn E, D'Esposito M (1998) The variability of human BOLD hemodynamic responses. NeuroImage, 8:360-9.

[2] Barth M, Norris DG (2007) Very high-resolution three-dimensional functional MRI of the human visual cortex with elimination of large venous vessels. NMR Biomed. 20:477-84.

[3] Bezzi P, Carmignoto G, Pasti L, Vesce S, Rossi D, Rizzini BL, Pozzan T, Volterra A (1998) Prostaglandins stimulate calcium-dependent glutamate release in astrocytes. Nature 391:281–285.

[4] Birn RM, Saad ZS, Bandettini PA (2001) Spatial heterogeneity of the nonlinear dynamics in the FMRI BOLD response. NeuroImage, 14: 817-26.

[5] Boynton GM, Engel SA, Glover GH, Heeger DJ (1996) Linear systems analysis of functional magnetic resonance imaging in human V1. J Neurosci 16: 4207-21.

[6] Buracas GT, Boynton GM (2002) Efficient design of event-related fMRI experiments using m-sequences. NeuroImage, 16: 801-13.

[7] Burke M, Bührle C (2006) BOLD response during uncoupling of neuronal activity and CBF. Neuroimage. 32:1-8.

[8] Celebi S, Principe JC (1995) Parametric least squares approximation using gamma bases. IEEE Transactions on Signal Processing, 43:781 – 784

[9] Chatton JY, Pellerin L, Magistretti PJ (2003) GABA uptake into astrocytes is not associated with significant metabolic cost: implications for brain imaging of inhibitory transmission. Proc Natl Acad Sci U S A. 100:12456-61.

[10] Chen S (2006) Signal processing. Euro. Trans. Telecomm, 17:99-110.

[11] d'Avossa G, Shulman GL, Corbetta M (2003) Identification of cerebral networks by classification of the shape of BOLD responses. J Neurophysiol, 90:360-71.

[12] De Vries B, Principe JC (1992) The Gamma Model: A new neural model for temporal processing. Neural Networks, 5:565-576.

[13] de Zwart JA, Silva AC, van Gelderen P, Kellman P, Fukunaga M, Chu R, Koretsky AP, Frank JA, Duyn JH. (2005) Temporal dynamics of the BOLD fMRI impulse response. Neuroimage 24:667-77.

[14] Dienel GA, Cruz NF (2008) Imaging brain activation: Simple pictures of complex biology. Ann NY Acad Sci 1147, 139-170.

[15] Efron B, Tibshirani RJ (1993) An Introduction to the Bootstrap. New York: Chapman and Hall.

[16] Farwell VT, Prentice RL (1977) A study of distributional shape in life testing, Technometrics 19:69–75.

[17] Filosa JA, Bonev AD, Nelson MT (2004) Calcium dynamics in cortical astrocytes and arterioles during neurovascular coupling. Circ Res. 95:73-81.

[18] Fox MD, Snyder AZ, Barch DM, Gusnard DA, Raichle ME (2005) Transient BOLD responses at block transitions. Neuroimage. 28:956-66.

[19] Friston KJ (1997) Transients, metastability, and neuronal dynamics. Neuroimage 5:164-71

[20] Friston KJ, Fletcher P, Josephs O, Holmes A, Rugg MD, Turner R (1998) Event-related fMRI: characterizing differential responses. Neuroimage. 7:30-40.

[21] Friston KJ, Mechelli A, Turner R, Price CJ (2000) Nonlinear responses in fMRI: the Balloon model, Volterra kernels, and other hemodynamics. Neuroimage. 12:466-77.

[22] Glover G (1999) Deconvolution of impulse response in event-related BOLD fMRI. Neuroimage. 9:416-29.

[23] Grotz T, Zahneisen B, Ella A, Zaitsev M, Hennig J (2009) Fast functional brain imaging using constrained reconstruction based on regularization using arbitrary projections. Magn Reson Med. 62:394-405.

[24] Handwerker DA, Ollinger JM, D'Esposito M (2004) Variation of BOLD hemodynamic responses across subjects and brain regions and their effects on statistical analyses. NeuroImage. 21:1639-51.

[25] Hakim TS, Sugimori K, Camporesi EM, Anderson G (1996) Half-life of nitric oxide in aqueous solutions with and without haemoglobin. Physiol. Meas. 17: 267

[26] Hansen KA, David SV, Gallant JL (2004) Parametric reverse correlation reveals spatial linearity of retinotopic human V1 BOLD response. NeuroImage. 23: 233-41

[27] Hegdé J, Van Essen DC (2004) Temporal dynamics of shape analysis in macaque visual area V2. J Neurophysiol. 92:3030-42.

[28] Henson RN, Price CJ, Rugg MD, Turner R, Friston KJ (2002) Detecting latency differences in event-related BOLD responses: application to words versus nonwords and initial versus repeated face presentations. NeuroImage, 15:83-97.

[29] Hoge RD, Franceschini MA, Covolan RJ, Huppert T, Mandeville JB, Boas DA (2005) Simultaneous recording of task-induced changes in blood oxygenation, volume, and flow using diffuse optical imaging and arterial spin-labeling MRI. Neuroimage. 25:701-7.

[30] Hyder F, Rothman DL, Shulman RG (2002) Total neuroenergetics support localized brain activity: implications for the interpretation of fMRI. Proc Natl Acad Sci USA 99:10771–10776.

[31] Hyder F, Patel AB, Gjedde A, Rothman DL, Behar KL, Shulman RG (2006) Neuronal-glial glucose oxidation and glutamatergic-GABAergic function. J Cereb Blood Flow Metab. 865-77.

[32] Kelly JP, Van Essen DC (1974) Cell structure and function in the visual cortex of the cat. J Physiol. 238:515-47.

[33] Koehler RC, Gebremedhin D, Harder DR (2006) Role of astrocytes in cerebrovascular regulation J Appl Physiol. 100: 307–317.

[34] Likova LT, Tyler CW (2008) Occipital network for figure/ground organization. Experimental Brain Research, 189:257-67.

[35] Lingnau A, Ashida H, Wall MB, Smith AT (2009) Speed encoding in human visual cortex revealed by fMRI adaptation. J Vis. 9:3.1-14.

[36] Logothetis NK (2002) The neural basis of the BOLD fMRI signal. Phil. Trans. R. Soc. Lond. 357:1003-1037.

[37] Logothetis NK (2003) The underpinnings of the BOLD functional magnetic resonance imaging signal. J Neurosci 23:3963-3971

[38] Logothetis NK, Wandell BA (2004) Interpreting the BOLD signal. Annual Review Physiology 66:735-769

[39] Magistretti PJ, Pellerin L (1999) Cellular mechanisms of brain energy metabolism and their relevance to functional brain imaging. Philos Trans R Soc Lond B Biol Sci. 354:1155-63.

[40] Magistretti PJ (2009) Role of glutamate in neuron-glia metabolic coupling. Am J Clin Nutr. 90:875S-880S.

[41] Margaria, R., Edwards, H.T., Dill, D.B. (1933) The possible mechanism of contracting and paying the oxygen debt and the role of lactic acid in muscular contraction. Am. J. Physiol. 106, 689–714.

[42] Martindale J, Mayhew J, Berwick J, Jones M, Martin C, Johnston D, Redgrave P, Zheng Y (2003) The hemodynamic impulse response to a single neural event. J Cereb Blood Flow Metab. 23:546–555;

[43] Menon RS, Luknowsky DC, Gati JS (1998) Mental chronometry using latency-resolved functional MRI. Proc Natl Acad Sci U S A, 95:10902-7.

[44] Metea MR, Newman EA (2006) Glial cells dilate and constrict blood vessels: a mechanism of neurovascular coupling. J Neurosci, 26:2862-70.

[45] Meyer EP, Ulmann-Schuler A, Staufenbiel M, Krucker T (2008) Altered morphology and 3D architecture of brain vasculature in a mouse model for Alzheimer's disease. Proc Natl Acad Sci U S A. 105:3587-92.

[46] Mulligan SJ, MacVicar BA (2004) Calcium transients in astrocyte endfeet cause cerebrovascular constrictions. Nature 431:195-9.

[47] Prentice RL (1974) A log-gamma model and its maximum likelihood estimation. Biometrika 61:539–544.

[48] Shank R, Aprison M (1979) Biochemical aspects of the neurotransmitter function of glutamate. In: Glutamic Acid: Advances in Biochemistry and Physiology. New York: Raven, p. 139–150.

[49] Shao XM (1997) Parametric survival analysis for gating kinetics of single potassium channels. Brain Research, 770: 96-104

[50] Shmuel A, AugathM, Oeltermann A, Logothetis NK (2006) Negative functional MRI response correlates with decreases in neuronal activity in monkey visual area V1. Nature Neurosci, 9: 569-577.

[51] Shulman RG, Rothman DL (1998) Interpreting functional imaging studies in terms of neurotransmitter cycling. Proc. Natl Acad. Sci. USA 95, 11993–11998.

[52] Sibson NR, Dhankhar A Mason GF, Rothman DL, Behar KL, Shulman RG (1998) Stoichiometric coupling of brain glucose metabolism and glutamatergic neuronal activity. Proc. Natl Acad. Sci. USA 95, 316–321.

[53] Sibson NR, Mason GF, Shen J, Cline GW, Herskovits AZ, Wall JE, Rothma DL, Shulman RG (2001) In vivo 13C NMR measurement of neurotransmitter glutamate cycling, anaplerosis, TCA cycle flux in rat brain during [2–13C]glucose infusion. J Neurochem 76:975–989.

[54] Smith AJ, Blumenfeld H, Behar KL, Rothman DL, Shulman RG, Hyder F (2002) Cerebral energetics and spiking frequency: the neurophysiological basis of fMRI. Proc Natl Acad Sci USA 99:10765–10770.

[55] Sotero RC, Trujillo-Barreto NJ (2007) Modelling the role of excitatory and inhibitory neuronal activity in the generation of the BOLD signal. Neuroimage. 35:149-65.

[56] Sotero RC, Trujillo-Barreto NJ (2008) Biophysical model for integrating neuronal activity, EEG, fMRI and metabolism. Neuroimage. 39:290-309.

[57] Stacy EW (1962) A generalization of the gamma distribution, Ann. Math. Stat. 33, 1187–1192.

[58] Tyler CW, Kontsevich LL, Ferree TC (2008) Independent components in stimulus-related BOLD signals and estimation of the underlying neural responses. Brain Research, 1229, 72-89.

[59] Tyler CW, Likova LT (2009) Neural signal estimation through time-resolved functional imaging. In: Brain Mapping Research Progress, Girard IC, Andre JS (Ed), Nova Scientific Publishers. 1-31

[60] Tyler CW, Likova LT (2011) Estimating neural signal dynamics in the human brain. Frontiers in Systems Neuroscience 5:33.

[61] Wang Y, Floor E (1994) Dynamic storage of glutamate in rat brain synaptic vesicles. Neurosci Lett, 180:175–178.

Chapter 3

Perfusion Based Functional MRI

Luis Hernandez-Garcia and Hesamoddin Jahanian

Additional information is available at the end of the chapter

http://dx.doi.org/10.5772/28969

1. Introduction

In this chapter, we will discuss the perfusion imaging for functional MRI experiments as an alternative and complementary technique to the conventional Blood Oxygenation Level Dependent (BOLD) effect. We begin by discussing the motivation behind the development of perfusion based functional MRI techniques. In order to do so, we briefly review the underlying relationship between the brain's vasculature and neuronal activity. Increases in neuronal activity result in metabolic increases, which in turn elicit a local, complex, vasodilatory response. With this physiology in mind, we will discuss the advantages and disadvantages of perfusion MRI and BOLD imaging, and which situations are appropriate for each technique. We will discuss how the types of artifacts and noise characteristics present in each technique determine its applicability.

We will then survey a few examples of ASL based fMRI studies. These typically include clinical studies of pathological conditions, longitudinal studies of cognitive function and studies requiring sustained periods of the condition or state of interest (i.e., greater than 30 seconds).

We then proceed to survey the wide array of current techniques available to image cerebral perfusion, including non-MRI techniques. We will describe in broad strokes the basics of PET and SPECT imaging and the use of paramagnetic tracer injections in MRI to map cerebral perfusion. However, we will narrow our focus down to arterial spin labeling (ASL) techniques, which use the water in the arteries as a tracer by labeling it with radiofrequency pulses. Because of ASL's ability to image perfusion dynamically as well as quantitatively, this technique will be the main focus of this chapter.

We will discuss the technical aspects of ASL techniques by first describing the physics of the labeling process and the different labeling schemes available Here, we will discuss the advantages and disadvantages of each of these strategies in greater detail.

Next, we will discuss the requirements for image acquisition following the labeling period. We will give a brief survey of different image acquisition schemes that are typically used in the ASL literature. While slower non-echo planar techniques can be used for perfusion measurements at rest, functional imaging applications typically require faster, single-shot, echo-planar techniques. 3D acquisition techniques are also gaining prominence in ASL imaging because of signal-to-noise advantages, but they are often accompanied by blurring artifacts if not implemented properly.

We will then proceed to discuss the technical considerations for the analysis of ASL based functional MRI. We divide this section into the *detection* of perfusion changes due to brain activity and the *quantification* of perfusion from the ASL signal.

Linear regression techniques are typically used for detection of activation in ASL data, just as with BOLD data. However, ASL techniques typically rely on the subtraction of images acquired following a labeling pulse from control images that are free of label, in order to extract the perfusion information. As we will discuss in greater detail, there are several subtraction schemes that can be used to isolate the perfusion signal from the raw data. Each of these schemes has an effect on the content and properties of the resulting perfusion weighted signals. We will describe methods to leverage those statistical properties in order to maximize the statistical power of the analysis.

We then address the question of quantifying perfusion from the ASL images. We will discuss the tracer kinetics theory necessary for quantification, and we discuss strategies for quantifying perfusion from time-series data where the perfusion level is changing due to a known stimulation paradigm. In particular, we will discuss how the parameter estimates obtained in the same linear regression procedure used for signal detection can be used to obtain quantitative measures of perfusion both at baseline and perfusion increases due to neuronal activation.

2. Why perfusion FMRI?

Blood Oxygenation Level Dependent (BOLD) contrast fMRI is currently the dominant technique for functional imaging and has yielded a wealth of information about brain function. The observed BOLD contrast arises from a combination of several indirect phenomena that correlate with neuronal brain activation, such as increases in blood volume and perfusion and a decrease in the concentration of deoxy-hemoglobin, that causes the observed BOLD signal intensity to increase in the active area. However, this signal increase is a non-linear function of many physiological parameters as well as the scanner's own characteristics (Boynton, Engel et al. 1996; Buxton and Frank 1997; Buxton, Wong et al. 1998; Vazquez and Noll 1998). As a consequence, BOLD imaging results are typically reported as unitless statistical significance maps without a clear, quantitative, physiological interpretation.

One issue plaguing the BOLD effect technique is that, since the BOLD effect is based on sensitivity to local changes in magnetic susceptibility, artifacts due to susceptibility gradients are also greatly exacerbated. These artifacts are especially problematic in areas of the brain that lie near air spaces, such as the roof of the mouth, nose and ear canals, as well as the sinuses (Bandettini, Wong et al. 1992; Kwong, Belliveau et al. 1992).

The structure of the drift noise within a session has been shown to be autoregressive (Lund, Madsen et al. 2006) and difficult to remove or model. Furthermore, the height and shape of the BOLD response depend on the baseline perfusion and metabolic levels, and thus effects of interest can be confounded by the conditions of the experimental resting state (Cohen, Ugurbil et al. 2002; Mulderink, Gitelman et al. 2002). Thus, noise properties and drift of the BOLD signal make studies with long activation periods nearly impossible, as the effects of interest are confounded with the slow scanner drifts (Smith, Lewis et al. 1999). The same reasons prevent BOLD FMRI studies from answering questions about baseline conditions. For example, what is the effect of a given drug, therapy, or training regimen on the baseline activity of a brain structure of interest? How do the neural substrates of a specific cognitive function change with age?

Hence, there is a clear need for development of alternative and/or complementary *quantitative* methods to BOLD imaging for FMRI. As technology develops, we expect a paradigm shift toward quantitative methods that offer meaningful physiological interpretation and comparison across laboratories.

This is where perfusion comes in. Perfusion is a readily quantifiable physiological parameter, and is easier to relate to neuronal metabolism than the BOLD response. Furthermore, animal studies conducted at high field, and high spatial resolution have indicated that neuronal activity produces perfusion (Cerebral Blood Flow - CBF) changes that are more localized to the parenchyma than the BOLD effect, consistent with the notion that the BOLD effect is more weighted toward draining veins and away from the active tissue (Duong, Yacoub et al. 2002; Pfeuffer, Adriany et al. 2002; Nencka and Rowe 2007; Olafsson, Noll et al. 2008). Thus, a fast, repeatable technique to measure cerebral perfusion *directly* would be very desirable and complementary to BOLD imaging.

We note that recent blood volume imaging techniques, such as VASO (Lu, Law et al. 2005) and AVIS (Vazquez, Lee et al. 2006), hold promise for quantitative FMRI, but they are severely hampered in terms of signal to noise ration (SNR), temporal resolution, and multi-slice imaging capability in their present state. The corresponding models for quantification require many assumptions and/or multiple measurements per time point. (Gu, Lu et al. 2006; Changwei, Kai-Hsiang et al. 2008; Jin and Kim 2008 ; Christopher, Ronald et al. 2009). Other methods that indirectly measure the rate of oxygen consumption by the brain are also available, but they require ASL data and calibration studies involving hypercapnia experiments (Davis, Kwong et al. 1998; Hoge, Atkinson et al. 1999). Thus, at this stage, they are not practical for dynamic quantitative functional imaging and will not be the focus of this chapter.

3. A survey of quantitative perfusion imaging methods

Blood flow through any organ has been used as an important measure of its health and functionality for a long time (Kety and Schmidt 1945; Lassen and Perl 1979). As such, many techniques have been developed to measure blood flow based on tracer injections. The central idea behind these is always the same: some substance that can be detected easily (i.e. - radioactivity or fluorescence) gets injected somewhere upstream of the organ of interest, either an artery feeding of the tissue, a vein, or directly into the left ventricle of the heart. Then the concentration of the tracer is measured with the appropriate detector as it travels through the tissue of interest. Perfusion can then be calculated from the uptake curve of the tracer in the tissue.

With the advent of new imaging techniques, the detection process was incorporated into the imaging process, as in autoradiography, PET, or SPECT scanners. These imaging modalities use radioactive tracers to generate the raw signals from which the images are reconstructed. Hence, it was a straightforward leap to use the timing information in order to quantify perfusion through the existing tracer kinetic models.

In the case of MRI, perfusion can also be obtained by imaging the passage of a tracer through the tissue. Instead of being radioactive, though, MRI tracers are molecules that change the relaxation rates of blood and tissue – typically gadolinium compounds, such as Gd-DOTA or Gd-DTPA. MRI has the added benefit that it allows for fairly rapid imaging (1-2 seconds for a whole brain volume) so one can sample the wash-in and wash-out of the tracer through the tissue. One can also simultaneously measure the amount of dispersion of the bolus that occurs during the transit from the injection site to the region of interest. Knowledge of this dispersion provides a more accurate estimate of the tissue's perfusion rate. Unfortunately, one can only do one contrast injection at a time, as the tracer takes many hours to clear the body, so it is not a very useful technique for FMRI.

Thus, our focus is on arterial spin labeling (ASL) techniques. The principle behind ASL imaging is conceptually simple. As with other techniques, the measurement consists of measuring the concentration of a tracer as it passes through a tissue of interest. We will discuss the details in more depth later in this chapter, but for now, note that ASL deviates from the usual tracer strategy in two ways. First and foremost, no tracer is injected. Instead, the tracer is "created" inside the arteries feeding the organ by radio frequency (RF) electromagnetic pulses. These pulses are generated by standard MRI coils and invert the magnetization state of the water protons in the blood. After allowing a short period of time for inflow of the tracer into the tissue of interest, an image is collected. Just like in previous techniques, perfusion can be quantified by measuring the signal change due to the tracer – typically by subtracting the image with the tracer from a control image without the tracer. The second important difference is that the tracer used in ASL decays very quickly, since the longitudinal relaxation rate of arterial blood is in the order of 1600 ms. As a result, there is a significant penalty in the signal-to-noise ratio (SNR) but, on the bright side, the experiment can be repeated immediately and as many times as desired. One can think of ASL as a tracer (also referred to as "label" or "tag") that is

http://dx.doi.org/10.5772/28969

completely non-toxic, rapidly decaying and it can be selectively injected into any artery (Detre, Leigh et al. 1992; Williams, Detre et al. 1992).

4. ASL and functional MRI

Although ASL is still limited by SNR and temporal resolution relative to BOLD imaging, some features of ASL make it a preferable option in many situations. The most important feature of ASL is the ability to quantify perfusion from the signal difference. This poses a significant advantage over BOLD contrast in that it allows the study of absolute baseline activity (i.e. – resting state) without comparison to an active state. This type of measurement is particularly desirable for studies concerned with pathological states, and/or testing the effects and specificity of different drugs. This ability to quantify perfusion is extremely useful in longitudinal studies over scanning sessions, whether involving activation or limited to the resting state. For example, we have used quantitative ASL to study the long-term effects of working memory training on brain activity, both at rest and during the performance of working memory tasks (Jaeggi, Studer et al. 2009).

Another interesting feature is that although the residual variance of an ASL time course within a single run is higher than with BOLD, it has been shown that the variance of non-quantitative ASL studies (i.e., relative CBF) across scanning sessions is dramatically less than that of BOLD, allowing studies that span days or even months (Aguirre, Detre et al. 2002; Wang, Aguirre et al. 2003; Wang, Aguirre et al. 2003). Furthermore, the variance across subjects is also lower in ASL than in BOLD, thus requiring fewer subjects to be scanned per experiment (Tjandra, Brooks et al. 2005).

Arterial spin labeling techniques allow greater flexibility in the image acquisition, especially when examining slow activation paradigms. Hence, they do not require T_2* weighting and one can use standard Spin Echo imaging to collect the image data, dramatically reducing susceptibility artifacts. Functional imaging studies of the inferior brain structures, such as the basal ganglia, and the orbito-frontal cortex would benefit greatly from such techniques.

5. Some examples of ASL studies

Due to above mentioned properties of functional arterial spin labeling (FASL), it is particularly useful for the study of gradual changes in neural activity, longitudinal and multi-site studies. (N.B. in this chapter, we use the term FASL to indicate ASL time series used for FMRI purposes, but there is no technical difference between FASL and ASL). FASL has been used in several basic and cognitive neuroscience applications. It has been used in many studies to investigate visual, motor and language functions at 1.5T and 3T (Aguirre, Detre et al. 2002; Wang, Aguirre et al. 2003; Kemeny, Ye et al. 2005; Tjandra, Brooks et al. 2005; Leontiev and Buxton 2007; Ances, Leontiev et al. 2008; Chen, Wieckowska et al. 2008; Raoult, Petr et al. 2011) and has been compared to BOLD fMRI (Aguirre, Detre et al. 2002; Chen, Wieckowska et al. 2008; Raoult,

Petr et al. 2011) (Kemeny, Ye et al. 2005) (Tjandra, Brooks et al. 2005) (Hermes, Hagemann et al. 2007) indicating that the intra-individual reproducibility of FASL in terms of the area of activation and activation quantification is comparable to that of BOLD fMRI. FASL however, detects smaller activation volumes than BOLD fMRI but the areas had a high degree of co-localization between subjects. FASL also shows higher specificity compared to BOLD fMRI while maintaining high sensitivity in activation detection in the activated area (Raoult, Petr et al. 2011).

Due to the drift in the BOLD signal, study of slow changes in neural activity using BOLD is quit challenging. FASL is a suitable tool for these studies as well. Motor learning is an example of the gradual changes in the neural activity over time, which cannot be easily assessed using BOLD fMRI. In (Olson, Rao et al. 2006) Olson et al. used FASL to study continuous, gradual changes in neural activity during motor learning. Subjects were required to use four fingers to press keys as quickly and as accurately as possible in response to the presentation of visual cues. Olson et al. reported that subjects performing this task demonstrated a reduction of neural activity in response to motor execution after training as compared to the start of training. Because the change in performance is slow and continuous, it is assumed for this study that the neural correlate of performance improvement during training is a gradual reduction in regional activity. The authors used FASL to detect these neural changes and reported reliable correlations between performance improvements and decreases in blood flow in premotor cortex and the inferior parietal lobe. This study suggests that FASL is a suitable tool for the study of the slow changes in the neural activity resulting from different cognitive tasks.

FASL has also been used in higher level cognitive activation studies. There is an increasing body of evidence pointing to a neurobiological basis of personality. Characterizing the biological bases of personality dimensions is important to explaining individual differences in brain activity associated with more dynamic changes in experience (e.g., a psychotic episode) and cognition (e.g., activation paradigms in functional MRI) (O'Gorman, Kumari et al. 2006). In (O'Gorman, Kumari et al. 2006) O' Gorman et al. explored the biological bases of the major dimensions of models of personality using FASL. In this study, the correlation between personality factors with regional brain function was successfully investigated using FASL. The authors reported associations between perfusion in brain regions (including basal ganglia, thalamus, inferior frontal gyrus, cerebellum and cuneus) and different personality dimensions. These results suggest that variations in perfusion in certain brain regions correlated with variations in personality dimensions may reflect variations in brain function (O'Gorman, Kumari et al. 2006).

FASL provides reliable quantification of absolute cerebral blood flow (CBF) along with excellent reproducibility over long time periods, making ASL a sensitive technique for reliable visualization of brain function during the resting state as well as during task performance. This property allowed Rao et al. (Rao, Gillihan et al. 2007) to employ ASL to investigate the effect of genetic variation of the human serotonin transporter gene on resting brain function of healthy individuals. They studied the alteration of resting brain function in emotion-related regions (including the amygdala and ventromedial prefrontal cortex) in healthy individuals caused by the 5-HTTLPR genotype. Valid and reliable inferences of resting activity for applications such as this study could not be derived from BOLD fMRI because it only measures *relative* changes in neural activity, and the changes in signal intensity from scanner drift are much larger than the effects of interest.

Also, using ASL, authors (Rao, Gillihan et al. 2007) demonstrated an association of 5-HTTLPR genotype with resting amygdala and ventromedial prefrontal cortex function in the healthy human brain. Such alterations suggest a broad role of the 5-HTT gene in brain function that may be associated with the genetic susceptibility for mood disorders such as depression.

Neuronal modulations found using functional ASL (FASL), such as those reported in (O'Gorman, Kumari et al. 2006) and (Rao, Gillihan et al. 2007), may also be important in interpretation of the different manifestations of BOLD response (which typically rely on the modulation of cerebral hemodynamics for detection of task-induced activation) to a particular stimuli in different groups.

The presence of low frequency signal drift in the BOLD signal impedes using a long task block. This problem makes it difficult to investigate slow processes such as learning, emotion, sustained attention and behavioral states in healthy and clinical populations. Because of its efficacy in longitudinal designs and low frequency paradigms, FASL is a good candidate for these studies. Kim et al. (Kim, Whyte et al. 2006) successfully utilized FASL to study an uninterrupted 6-min continuous performance of the two high-level cognitive tasks (visual sustained attention and verbal working memory) which prior to FASL was only feasible for suboptimal short data acquisition blocks (40-90s). Understanding the neural correlates of these cognition function is very important because sustained attention is also implicated in various clinical disorders including attention deficit hyperactivity disorder, traumatic brain injury, and Alzheimer's disease (Kim, Whyte et al. 2006).

In another study, Rao et al. (Rao, Wang et al. 2007) used FASL and BOLD to measure brain activation patterns associated with natural vision while subjects were freely viewing a cartoon movie. Rao et al. reported that cerebral blood flow increases in multiple visual pathway areas and frontal areas, and decreases in ventromedial frontal cortex and superior temporal cortex during movie viewing compared to resting states. Concurrent BOLD contrast revealed similar but weaker activation and deactivation patterns. This study demonstrates the feasibility of using FASL for imaging both sustained and dynamic effects in neural activation during natural and ecologically valid situations (Rao, Wang et al. 2007).

With excellent reproducibility over long-term time periods, FASL is ideal for imaging a sustained behavioral state, such as stress. Wang et al (Wang, Rao et al. 2005) used FASL technique to measure perfusion changes associated with mild to moderate stress. The authors demonstrated that a positive correlation exists between the change in perfusion induced by the stress task and subjective stress rating in the ventral right prefrontal cortex and left insula/ putamen area. This study provides neuroimaging evidence that psychological stress induces negative emotion and vigilance and that the ventral right prefrontal cortex plays a key role in the central stress response.

Pharmacological functional (phMRI) studies are making a significant contribution to our understanding of drug effects on brain systems. Since noise spectrum of arterial spin labeling signal contains relatively less power at low frequencies, it is particularly useful for phMRI studies in which a change in brain activity is expected over the course of minutes, hours, or days to asses low frequency between- and within-session drug-induced changes (Wise and Tracey 2006) (16). As an example, Gollub et al. (Gollub, Breiter et al. 1998) used FASL to measure neuronal activation during visual stimulation before and after cocaine and saline infusions. The authors used FASL to determine whether acute intravenous cocaine use would change global cerebral blood flow and demonstrated that cortical gray matter cerebral blood flow was unchanged after saline infusion but decreased after cocaine infusion.

These studies demonstrate that FASL can be successfully utilized for the investigation of the cerebral blood flow changes associated with development of human brain, personality, high level cognitive operations, the behavioral states such as attention, natural vision, psychological stress and assessing low frequency drug effects. Increased applications of FASL to the investigation of cognitively impaired populations are expected to follow.

5. ASL methods

Now that we've seen the bigger picture, let's examine the details of carrying out the acquisition and analysis of ASL data. Any ASL sequence is basically made up of a labeling period, during which the inversion label is created, some delay time to allow the label to reach the tissue of interest, and then an imaging pulse sequence where the image is acquired. There may be additional pulses to shape the input function before the acquisition part, and there may be multiple coils involved in some variants, but in general, the structure remains the same. Thus, it is convenient to discuss labeling and imaging schemes independently, as they can be combined in multiple ways.

6. Labeling schemes

There are a number of MRI pulse sequence strategies that can be used to obtain ASL perfusion images. The first class of arterial spin labeling pulse sequences is "Continuous arterial spin labeling" (CASL). In the original continuous ASL formulation, low-power, long pulses (in the order of a second or two) are applied at a plane inferior to the brain. Typically this is below the circle of Willis where the arterial supply to the brain is perpendicular to the labeling plane. As usual, in order to achieve slice selection, the long RF pulses are applied at some frequency offset and in the presence of a slice-selective gradient. These pulses have very different effects on the stationary and the moving spins. While the stationary spins at the labeling plane become saturated and lose phase coherence (essentially destroying their magnetization), the spins that are moving through the labeling plane experience a gradual change of their resonant frequency (recall that the resonant frequency is directly proportional to the magnetic field). The difference between the resonance frequency of the protons and the transmission frequency changes from positive through negative as the spins move through the inversion plane and is zero at the center of it.

http://dx.doi.org/10.5772/28969

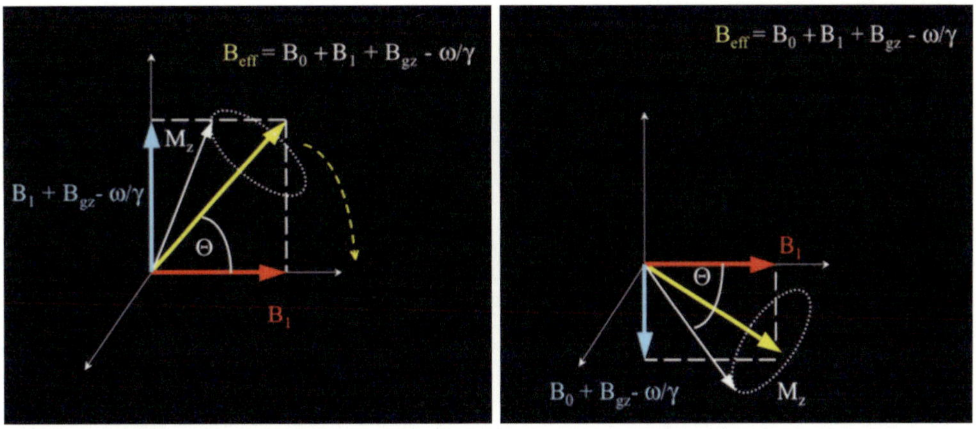

Figure 1. The Adiabatic Inversion Process - 1A – In the rotating frame of reference the magnetic fields due to the scanner's main magnetic field (B0), the RF coil's field (B1), the slice select gradient (Bz) and the opposing magnetization field add up to an "effective magnetic field" (Beff). The arterial water's magnetization vector precesses around Beff. 1B- As the spins move, the contribution from the slice select changes from positive to negative. As this happens, the magnetization vector continues to precess around the effective field, resulting in the inversion of the magnetization vector.

Because of this effective "frequency sweep", the net magnetic field generated by the combination of the main magnetic field (B0), the applied RF field (B1eff) and the magnetic field resulting from the spin's precession itself (ω/γ) experiences a rotation from pointing along the direction of the main magnetic field to pointing against it (see figure 1). If this rotation is slow relative to the precession frequency, the effect of the constant RF pulse on these moving spins is to create a "spin-locked" state. This means that the flowing proton's magnetization processes around the effective magnetic field and follows it as it rotated from pointing up to pointing down.

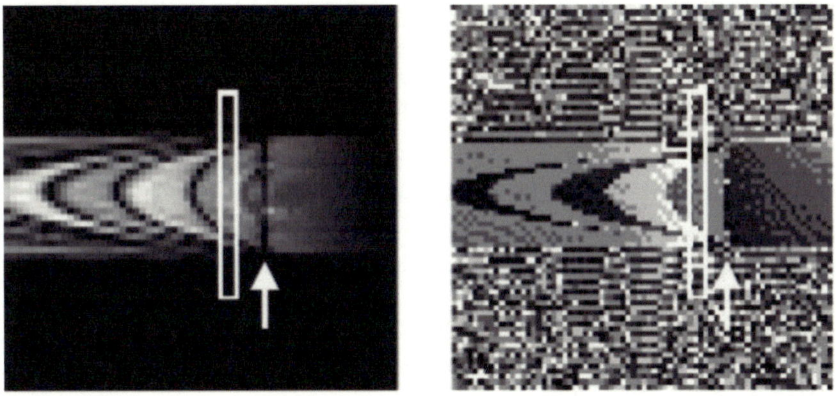

Figure 2. This image shows the magnitude (left) and phase (right) of an image of water flowing through a tube while experiencing adiabatic inversion. The banding pattern reflects periods where the pulses were turned on and off. Figure reproduced from (Hernandez-Garcia, Lewis et al. 2007).

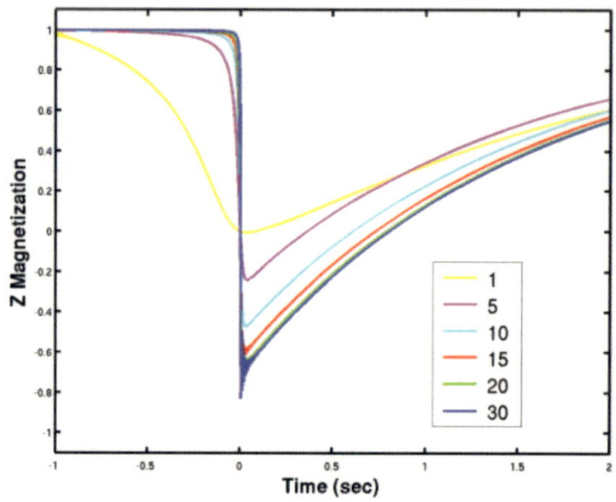

Figure 3. The longitudinal magnetization of the arterial water experiences adiabatic inversion at the labeling plane, then slowly decays back to its relaxed state. This phenomenon depends on a number of factors, such as the velocity of the spins, and the size of the slice select gradient and the RF fields used in the pulse sequence.

As the protons move through the inversion plane, the "effective" field (B_{eff}) to which they are locked rotates from pointing along the direction of the main magnetic field to pointing against it. The end result is that when the flowing spins leave the plane, their magnetization is inverted (figure 2 shows a practical demonstration of this phenomenon). This phenomenon is referred to as "flow driven adiabatic inversion" and is the basis for continuous arterial spin labeling techniques. It is important to note that immediately after inversion, the spins experience T1 relaxation and the label effectively decays in a matter of a couple of seconds, as depicted in figure 3.

The main caveat of continuous ASL is that the long inversion pulses applied at the neck for labeling purposes also produce a significant amount of magnetization transfer (MT) (Detre, Leigh et al. 1992; Williams, Detre et al. 1992) in the spins at the tissue of interest. Briefly, magnetization transfer consists of signal loss during acquisition when it is preceded by RF pulses, even if those are not applied on resonance. As a result of MT, subtracting a tagged image from a control image would indicate not only the perfusion effects but also the degree of magnetization transfer, which is of no interest to the investigator.

This would constitute a major obstacle, but a number of solutions have been devised. The original solution to this problem was to apply an MT preparation pulse with identical properties to those of the labeling pulse before collecting the control image. This preparation pulse was identical to labeling pulse but with a reversed slice select gradient such that the inversion pulse would be centered above the head. In that case, there is only magnetization transfer but no arterial spin labeling. While this works in principle, it is difficult to collect more than one slice at a time while appropriately compensated magnetization transfer symmetrically for all the slices. Another solution was to simply use a separate RF coil for labeling whose field pattern will not reach the brain (Zhang, Silva et al. 1995) (Talagala, Ye et al. 2004) (Hernandez-Garcia, Lee et al. 2004). While this approach works quite well and relatively simple to implement in principle, in practice it requires additional hardware that can be synchronized with the main MRI scanner's pulse sequence. It is also generally uncomfortable for the subjects to wear the additional neck coil and it can pose some logistical pulses.

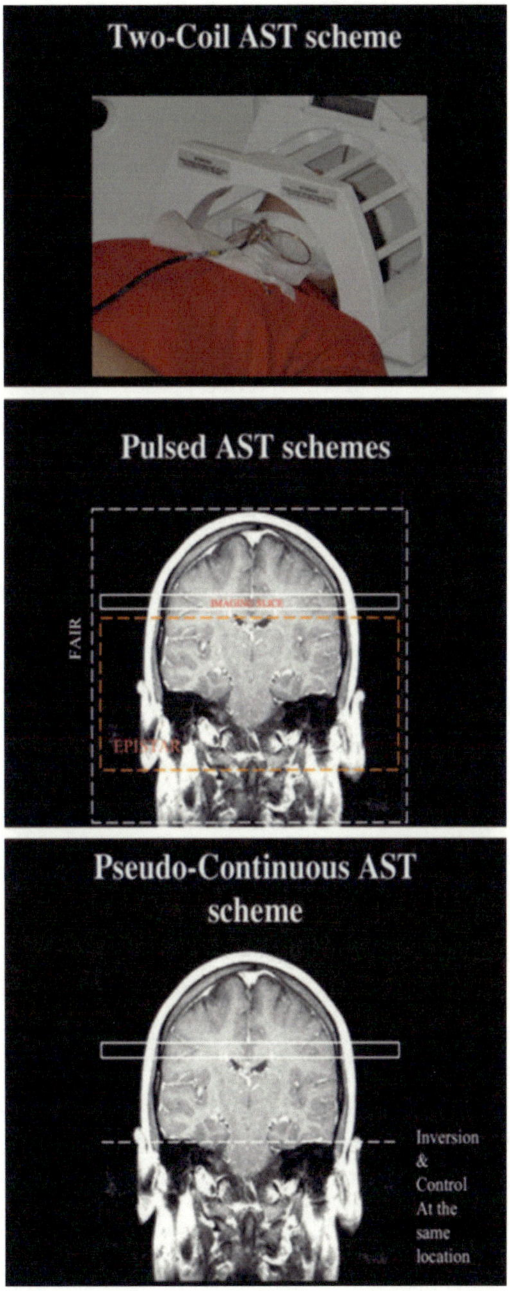

Figure 4. Labeling Schemes - Schematics indicating the relative locations of the inversion types of popular ASL schemes

Several other strategies to overcome the magnetization transfer problem have been tried to overcome this problem, such as amplitude modulated CASL (Wang, Zhang et al. 2005), but the most promising one is to break up the continuous labeling pulse into a train of pulses applied in rapid succession to achieve the same adiabatic inversion effect. In the control case, however, the same pulse train is applied but the phase of every other pulse is shifted by 180 degrees, such that every other pulse "un-does" the flip of the previous pulse. The end result is that both the control and the labeled image receive the same amount of magnetization transfer, the flowing blood's magnetization gets inverted prior to collecting the labeled image but it remains un-inverted during the collection of the control image. This technique, known as pseudo-continuous arterial spin labeling (PCASL) has greatly facilitated the use of continuous ASL because it addresses the magnetization transfer issue and is relatively easy to implement (Garcia, de Bazelaire et al. 2005; Wang, Zhang et al. 2005; Fernandez-Seara, Wang et al. 2007; Wu, Fernandez-Seara et al. 2007). The caveat is that PCASL is sensitive to magnetic field inhomogeneity (off-resonance) at the inversion plane. This effect can severely affect the efficiency of the labeling process, crippling the technique, and is exacerbated at higher magnetic fields. Fortunately, the loss of inversion efficiency can be recovered by adjusting the phase of the inversion pulses and introducing compensation gradients (Jahanian, Noll et al. 2011).

A second class of arterial spin labeling pulse sequences is "pulsed ASL" (PASL). This strategy is to label a slab containing the arterial supply to the organ of interest with a more standard, short RF inversion pulse and then allow the inverted spins to flow from that slab into the tissue of interest. The difference in signal intensity between the two images is provided by the un-inverted spins that flowed into the imaging slices during the inversion delay. This signal intensity difference can then be readily quantified and translated into perfusion images. This process mimics the injection of a bolus of tracer, rather than a constant infusion, as in the case of continuous ASL.

The simplest form of pulsed ASL is the EPISTAR technique (Edelman and Chen 1998), in which the inversion pulse (typically a hyperbolic secant) applied just below the region of interest followed by a delay to allow the inverted blood to flow into the tissue before collecting the image. Hyperbolic secant pulses are relatively short adiabatic pulses and exhibit no significant magnetization transfer effects, so there is no need for pre-compensation pulses before collecting the reference image. In the FAIR technique (the original pulsed ASL technique) the control image is preceded by an inversion pulse applied over the entire brain, followed by a short delay (Kim 1995; Kim and Tsekos 1997). The tagged image is collected following a similar hyperbolic secant inversion pulse, except that it is applied only over the slices of interest, followed again by the same short delay to allow for inflow of blood into the imaging region. There are many variants on this theme, but perhaps the most popular pulsed ASL technique is QUIPSS which uses additional pulses to saturate the trailing edges of the label, thus producing a clearly defined bolus (Wong, Buxton et al. 1997; Wong, Buxton et al. 1998).

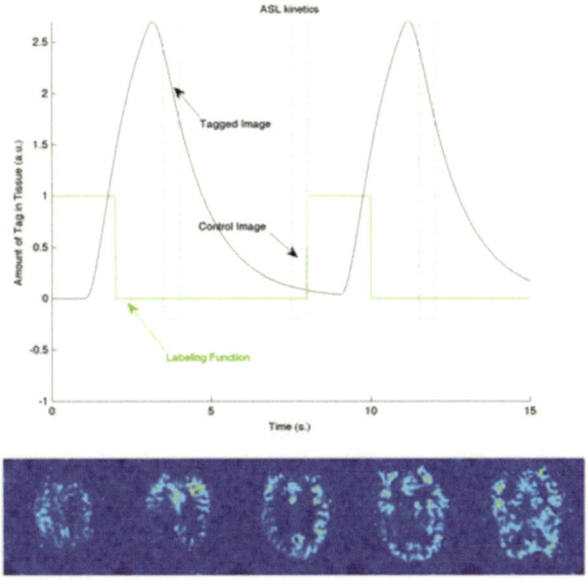

Figure 5. A - The uptake and release of the inversion label is relatively short, so there is only a limited amount of time to collect the images without a significant signal change. While including long post inversion delays reduce the sensitivity to the difference in slice timing, there is still only a limited amount of time to collect the images. If the process takes too long, different slices will have different SNR and sensitivity to perfusion changes from activation. B – This image, originally consisting of 24 slices, shows a noticeable change in SNR between the top and bottom slices. Here we only show slices 6, 9, 12, 15.

Whereas the pulsed approach only requires 1-2 seconds for the label to reach its maximum concentration in the tissue the continuous approach requires approximately 3 seconds before a steady state concentration of tag is reached. However, because of the longer inversion times of the continuous labeling scheme, the amount of tag to be detected is much larger and thus, the SNR of the method is also larger (Buxton, Frank et al. 1998; Wong, Buxton et al. 1998). Thus, the tradeoff between them is primarily one of speed versus SNR.

7. Image acquisition considerations in ASL

Image collection following the spin labeling process can be done in a number of ways, but it is important to take into account the goals of the specific application. As usual in MRI, there is a trade-off between acquisition speed, SNR and resolution.

In clinical and resting state applications, typical ASL scans use more standard image acquisition pulse sequences since the speed consideration is relaxed. Fast spin-echo acquisitions are a popular method because of their insensitivity to off-resonance artifacts and relatively high SNR and speed. For example (Fernandez-Seara, Wang et al. 2005; Fernandez-Seara, Wang et al. 2007) used a 3D GRASE acquisition sequence in conjunction with continuous ASL in order to boost the SNR of the acquired images in a functional imaging study of the medial temporal lobe.

Functional MRI requires image acquisition speed in order to capture the hemodynamic changes that take place during mental activity adequately. Fast imaging techniques that take advantage of echo-planar and echo-volumar acquisition and parallel imaging allow us to image the whole brain in a matter of a second or two with high spatial resolution. Consider the figure below, which indicates the concentration of tag in the tissue as it is taken up. Unfortunately, the rates at which the flows in and out of the tissue while decaying is such that we are constrained to roughly a half a second to acquire the whole volume.

If one slice is collected at the beginning of the uptake and another toward the end, the perfusion contrast will be vastly different. While one can account for it by including this delay in the quantification scheme, the SNR of the slices can still be significantly different.

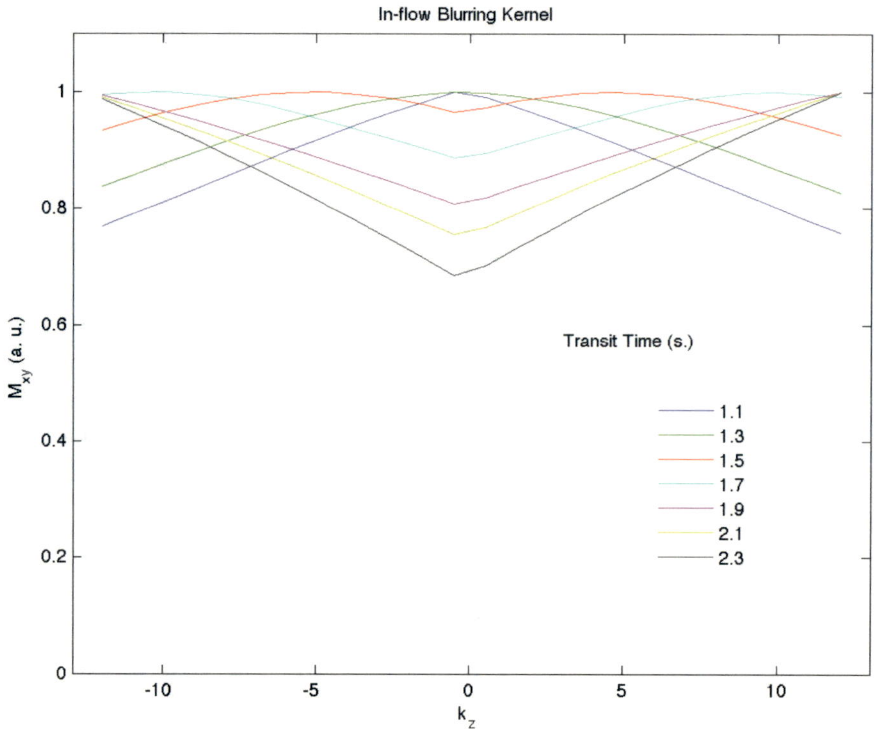

Figure 6. Depending on the transit time of the blood from the labeling plane to the imaging slab, the uptake of the label can induce a mild distortion along the slab-select location (or kz). Interestingly, this effect can be characterized by a point spread function that acts as either a high-pass or low-pass filter along the kz axis.

This acquisition time window means that there is a significant constraint in the number of slices that one can get after a single labeling period. For instance, using a fast spiral acquisition typically allows for acquisition of a single slice in at least 25 ms. At this rate one can acquire

20 slices in half a second, which is adequate but requires thick slices of about 6 mm thickness, assuming a 12 cm brain.

In this regard, there is recent interest in the development of 3D echo-planar and echo-volumar acquisition schemes. 3D imaging schemes can ensure the same acquisition delay for the entire imaging volume relative to the labeling period. The SNR of 3D imaging can be greater than 2D multislice because a larger volume contributes to each echo. For example, Gunther et al (Gunther, Oshio et al. 2005) proposed a multi-echo 3D acquisition that allows collection of a k-space plane at each echo. Unfortunately, this acquisition scheme imposes a T2 weighting on each plane of k-space depending on the acquisition order. The result is typically a severe blur of the volume along the z-direction. One can think of this as a blurring filter applied along the slice direction, and the shape of that filter's point spread function (PSF) is that of the T2 decay.

Another 3D alternative is to use multi-shot readouts as in (Talagala, Ye et al. 2004) but these typically require the use of low flip angles to preserve the magnetization throughout all the kz encoding steps. Low flip angles however result in reduction of the available signal. Gai et al (Gai, Talagala et al. 2011) have experimented with an increasing flip-angle schedule starting at 10 degrees that produces a constant signal along kz according to the Ernst formula. This approach produces a flat PSF, but at the cost of SNR. In recent investigation, we have found that a flip angle schedule that starts at 15 degrees and increases with a simple third order polynomial is a good compromise between SNR and blurring along the slice direction. The figure below illustrates this issue by calculating the PSF along the z-direction of different order polynomial flip angle schedules.

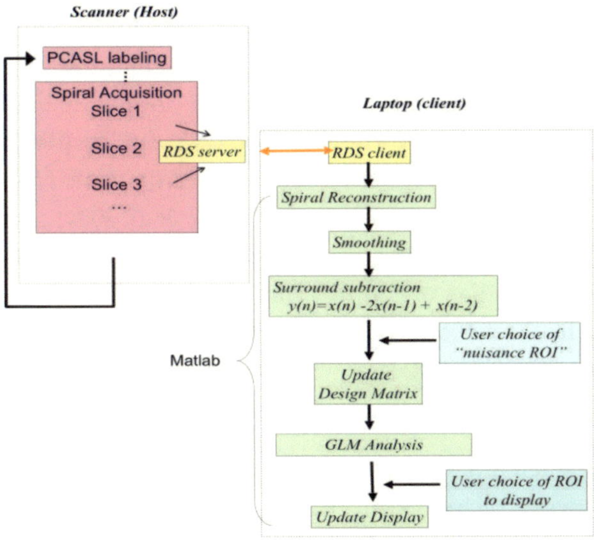

Figure 7. Real Time imaging with ASL requires coordination of the scanner's acquisition with an external computer that "catches" the data as it is generated by the scanner and carries out the analysis within the interval between two images. (Figure from (Hernandez-Garcia, Jahanian et al. 2011))

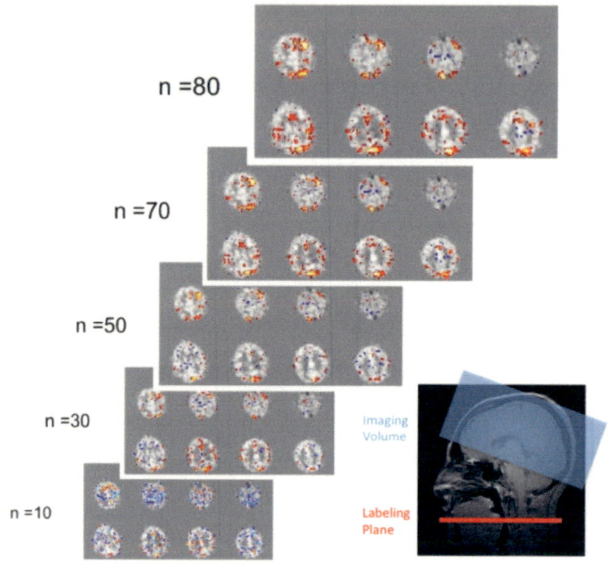

Figure 8. This figure shows an updated view of the real time FASL acquisition during a visuo-motor stimulation paradigm. As one can see, the activation maps evolve over time and become better defined as more data become available. (Figure from (Hernandez-Garcia, Jahanian et al. 2011))

Real-time functional magnetic resonance imaging (rtfMRI) is an exciting extension to conventional fMRI techniques that enables the user to analyze fMRI data as it is being collected. The requisite is that the reconstruction and analysis must be carried out before acquisition of the next image in the time series (about two seconds for standard FMRI). Thus, in rtfMRI the results are immediately available as the subject is being scanned, and can be used to reveal and/or guide the subject's cognitive processes. Thus, by collecting the data in real time, the investigator can fine tune the design's parameters to suit each specific subject (deCharms 2008). One can also design interactive paradigms based on the subject's dynamic functional activity (fMRI biofeedback or brain-computer interface (Yoo, Fairneny et al. 2004)). Furthermore, it allows the investigator to determine the subject's compliance during the experiment. Examples of real time FMRI also include studies of the modulation of motor-area cortical activation and emotional processing by the subjects themselves (Yoo and Jolesz 2002; Posse, Fitzgerald et al. 2003; deCharms, Christoff et al. 2004; Phan, Fitzgerald et al. 2004; Caria, Veit et al. 2007).

A number of processing strategies have been developed in order to carry out real time FMRI analysis, such as cumulative correlation (Cox, Jesmanowicz et al. 1995), sliding-window correlations with reference vector optimization (Gembris, Taylor et al. 2000), online general linear model [GLM] analysis (Nakai, Bagarinao et al. 2006). There are also a number of combined methods to collect behavioral, physiological and MRI data while performing near real-time statistical analysis (Voyvodic 1999). The recent advances in computational speed have made it relatively easy to implement any of these analysis methods.

http://dx.doi.org/10.5772/28969

8. Quantifying perfusion from the ASL signal

In the original implementation of continuous ASL, quantification was done by writing the well-known Bloch equations that describe the longitudinal magnetization with one modification: the authors included terms to account for the inflow of arterial blood and outflow of venous blood. (Recall that the Bloch equations describe the behavior of a magnetization vector in the presence of magnetic fields such as the ones present in an MRI pulse sequence Please see textbooks like Haacke's (Haacke 1999) or Bernstein (Bernstein, King et al. 2004) for an introduction). Like this $\frac{dM_{tissue}(t)}{dt} = f \cdot M_{art}(t) - \frac{f}{\lambda} \cdot M_{tissue}(t) - \frac{M_{tissue}(t)}{T1}$

Here, $M_{tissue}(t)$ would be the longitudinal magnetization of the tissue protons. $M_{art}(t)$ is the incoming arterial magnetization. In the control image it would be positive, but in the labeled image, it would be negative. Note that this must be adjusted to reflect the efficiency of the inversion efficiency (α) and its decay during transit time from the labeling plane to the tissue (Δ). Hence, $M_{art}(t) = (2\alpha e^{-(\Delta/T1art)} M_{art}(0))$.

T_{1art} and T_1 are the longitudinal relaxation rates of the arterial blood and the tissue. The parameter λ is blood brain partition coefficient and it relates the concentrations of label in blood and tissue at equilibrium. The key parameter is f, the perfusion rate that describes the rate at which water moves in and out of the tissue. By writing the steady state solutions for this equation in the case of control and tagged images, one can solve for f.

An equivalent but perhaps more generalizable approach was presented by Buxton et al (Buxton, Frank et al. 1998) in which they treated the ASL experiment as a standard tracer kinetics experiment. In this formulation, the concentration of "label" in the tissue, C_{tissue} is given by the subtraction of control and tagged images, and can be modeled as a linear system. In that case, the observed signal difference between control and tagged images is the convolution of an input function with a system function.

$$C_{tissue}(t) = f \cdot M_{art}(t) * [r(t)m(t)]$$

In this case, the input function is a fraction of the arterial blood (determined by the perfusion rate, f). The system function is a combination of the tissue "retention" function, e^{-ft}, which describes how the label clears away with an exponential decay, and also the T_1 decay function of the tag in the tissue, e^{-t/T_1}.

In subsequent work the pulse sequence's timing parameters, including the TR, duration of the tagging period (ω), post inversion delay (τ) and transit time (δ) were taken into account to yield the following equation relating the subtraction image (ΔM) to the perfusion rate. If the post inversion delay is longer than the transit time, the perfusion rate can be calculated by

$$f = \frac{\lambda \cdot \Delta M \cdot R_{1\,app}}{\left(\frac{M_{control}}{1-e^{-TR \cdot R_1}}\right) \cdot 2 \cdot \alpha \cdot e^{-\delta \cdot R_{1a}} \left(e^{(\delta-\omega) \cdot R_{1app}} - e^{(\delta-\tau-\omega) \cdot R_{1app}}\right)}.$$

Note that R_1 is simply the reciprocal of the T_1 relaxation rate (or $1/T_1$) used for convenience. In keeping with the notation of the authors, R_{1a} refers to the relaxation of arterial blood and R_{1app} refers to the apparent relaxation rate of the tissue.

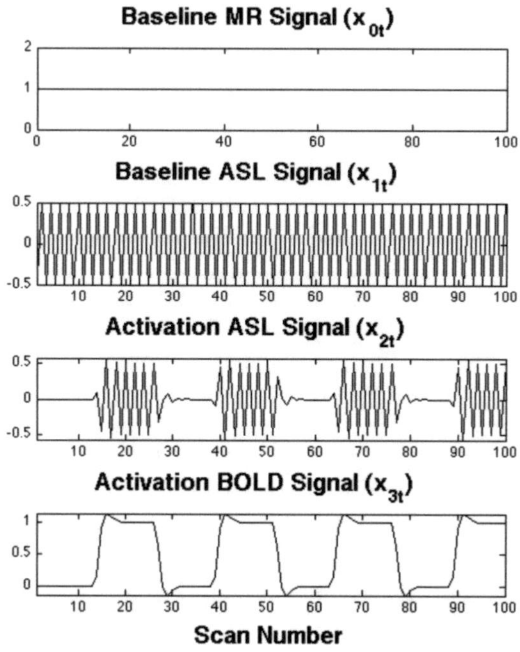

Figure 9. The regressors in a design matrix for a simple functional ASL experiment (only two conditions) include both the usual baseline and BOLD effects (regressors 0 and 2). In order to capture the variance due to resting and activation blood flow, we must include regressor 1 and 3, which are modulated by the presence of the arterial label. Reproduced from (Hernandez-Garcia, Jahanian et al. 2010)

On a practical note, R_{1a} and R_{1app} may vary from subject to subject when dealing with clinical populations or across age groups, so in those cases, it is important to make those measurements separately and use the measured values in the calculation. In the healthy subjects typically used for FMRI research, this is not so much of a factor and literature values are typically employed for the relaxation rates.

9. Quantification of perfusion in FASL data

While the above framework is useful for situations where perfusion is constant, it's more challenging to quantify dynamic perfusion changes as in the case of FMRI experiments. The very slow sampling rate of ASL makes it difficult to get perfusion values during the transitions between active and resting states, so investigators are often forced to settle for collecting only a few perfusion measurements obtained at the baseline and a few measurements at the plateau of the activation, and none from the transition. Unfortunately, this reduces the number of samples available for quantification, which can be fatal in a low SNR technique like ASL. Furthermore, it precludes event-related fMRI, where there is not much of a plateau of activation.

In order to overcome this problem, one can formulate a general linear model (GLM) that specifies BOLD and ASL effects. One can then estimate the coefficients of this model and translate them into meaningful, quantitative measures of perfusion by using standard tracer kinetic models. This concept is based on the simple realization that the difference images used in ASL quantitative models have a direct relationship to the coefficients (or amplitudes) of the regressors or the general linear model. Since estimation of a general linear model is the standard way of analyzing FMRI data by the bulk of the functional imaging community, this approach constitutes a natural next step in the analysis.

For example, let us consider an ASL FMRI experiment with a baseline condition and a single activation condition with tag and control ASL images acquired in each. One such experiment can be characterized by a linear model, as previously described (Mumford, Hernandez-Garcia et al. 2006). Let y_t be the time course (a vector) of image intensity at a particular voxel obtained from an ASL experiment. In Fig. 9 are the regressors of a simple ASL design matrix representing the linear model

$$y_t = \beta_0 + \beta_1 x_{1t} + \beta_2 x_{2t} + \beta_3 x_{3t} + \varepsilon_t$$

for time t=1, …, n. The first regressor, the baseline vector $x_{0t,}$ =1 and its coefficient parameter, the scalar β_0, indicate the baseline signal, and can be interpreted as a measure of spin density. The second regressor x_{1t} describes the baseline difference between control and tagged images ($\Delta\Delta M$) and thus its amplitude, β_1, is indicative of baseline perfusion. The third regressor x_{2t} with its regression coefficient β_2 describes differences between control and tagged images (ΔM) due to activation, and the fourth regressor x_{3t} with its regression coefficient β_3 describes the BOLD effect changes. By realizing that the amplitude of the oscillation (β_2) induced by the ASL scheme corresponds to the difference between control and tagged images, perfusion can be computed dynamically by translating the coefficient parameter estimates of the oscillations into the appropriate control-tag differences.

Quantification of perfusion effects in continuous ASL data without background suppression can be done directly by adapting the same kinetic model by Wang et al (Wang, Alsop et al. 2002) that we discussed in the previous section. The ΔM parameters are replaced with the GLM parameter estimates as follows

$$f_{effect,t} = \frac{\lambda \cdot R_{1app} \cdot \hat{\beta}_{effect}}{\left(\frac{\hat{\beta}_0}{1-e^{-TR/T_1}}\right) \cdot 2 \cdot \alpha \cdot e^{-\delta \cdot R_{1a}} \cdot \left(e^{(\delta-\omega) \cdot R_{1app}} - e^{(\delta-\tau-\omega) \cdot R_{1app}}\right)} \cdot x_{effect,t}$$

where $\hat{f}_{effect,t}$ is the estimated perfusion change due to the effect of interest, α is the inversion efficiency, $x_{effect,t}$ is the regressor for the effect at time t, $\hat{\beta}_{effect,t}$ is the coefficient parameter estimate of the regressor representing the effect of interest (for example, the amplitude of the perfusion changes due to activation in the above model are captured by β_2), $\hat{\beta}_0$ is the baseline state signal (T_1-weighted spin density), λ is the blood brain partition coefficient, R_1, R_{1a}, R_{1app}

are the longitudinal relaxation rates of arterial blood, tissue, and tissue in the presence of perfusion. The term δ is the arterial transit time, while TR, w, and τ are repetition time, post labeling delay, and labeling duration.

As previously noted (Wang, Alsop et al. 2002), when long post-inversion delays (w) are used, the equation becomes insensitive to changes in arterial transit time. Note also that the sign of $\hat{\beta}_{effect}$ can be positive or negative depending on the acquisition order of the control and tagged pairs.

But, what about the error on those parameter estimates? Both Ordinary and Generalized Least Squares estimation yield estimates of the linear model's parameters as well as their variances. Thus, the same relationship can be used to derive the standard deviation of the estimates in perfusion units. In order to calculate the variance of the estimated *perfusion* effects, we propagate the errors of the relevant coefficient parameter estimates through the kinetic model.

Propagation of error can be calculated in a straightforward way from the partial derivatives of the model relative to the coefficient parameters of interest (Bevington and Robinson 2003). We refer the readers to the article by Hernandez-Garcia et al (Hernandez-Garcia, Jahanian et al. 2010) for the derivation.

We have seen that instead of calculating a perfusion time series one image at a time, we can make a linear model of what that perfusion time series should be, and then estimate the corresponding coefficients and their variance. If the model is accurate, this yields much better estimates of the perfusion time series. To illustrate this point, figure 10 shows the perfusion time courses at a given voxel obtained from the traditional method versus the GLM method. Since the residual variance estimates are reduced, the activation maps and residual variances are greatly improved, as illustrated in figure 11.

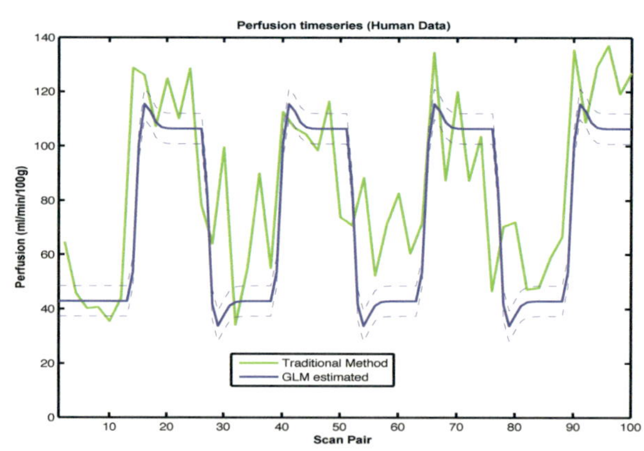

Figure 10. The green line shows the individual perfusion time series calculated at each individual time point. The propagation of errors through the equation makes the result noisy. However, by using all the data at once to estimate the coefficients of a linear model, one can also obtain such a time course if the activation paradigm is known. Reproduced from (Hernandez-Garcia, Jahanian et al. 2010)

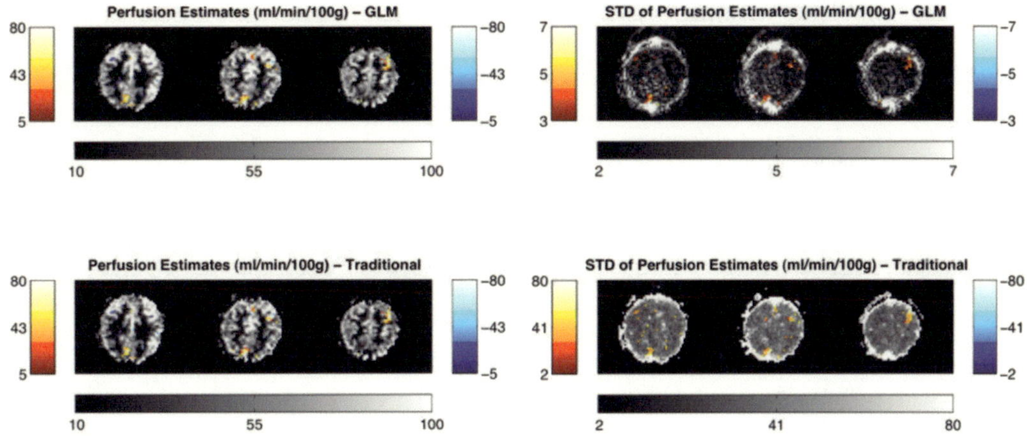

Figure 11. By translating the coefficients of a GLM into perfusion estimates, the residual variance is greatly reduced in comparison to calculating the perfusion time series first and then estimating a general linear model. The result is more accurate activation maps. As with all linear models, one must know the activation paradigm a priori. Reproduced from (Hernandez-Garcia, Jahanian et al. 2010)

10. About the subtraction of the ASL time series

As we have described earlier, the perfusion information in the ASL images is contained in the subtraction of tagged images from control images, which are previously acquired in an alternating order. While differencing the image time course reduces the degree of autocorrelation in ASL data (Aguirre, Detre et al. 2002 ; Liu and Wong 2005) and is widely used, this topic deserves some attention, as there are multiple options available with their corresponding side effects.

In its simplest form, subtraction is done by "pairwise differencing" in which every two images is used to obtain a single perfusion image. As a result, the raw data are no longer autocorrelated, but they are also sub-sampled severely and, consequently, aliasing can occur. By this sub-sampling one also gives up almost half of the degrees of freedom in the analysis, so it translates into a major loss of power. There are other ways to difference the data, however. For example, one can make a "running subtraction" in which one would subtract from every image the previous one, but reverse the sign of the subtraction at every image. Similarly, one would subtract from every control image the average of the two neighboring tagged images (i.e., "surround subtraction"). In this case, only two degrees of freedom are lost, and the data are still whitened. The caveat is that all these differencing processes have a smoothing effect on the time course so fast changes in perfusion are dampened, and we may not be able to observe them. The "sinc subtraction" scheme consists of interpolating the value of the control images at the time of acquisition of the tagged images and, likewise, interpolating the tagged images to obtain their value at the time of the control images. The two up-sampled time courses are then subtracted from each other.

pairwise

$$y_2[m] = y_c[m] - y_t[m] \Leftrightarrow D_2 = \begin{pmatrix} 1 & -1 & & & & 0 \\ & & 1 & -1 & & \\ & & & & \ddots & \\ 0 & & & & 1 & -1 \end{pmatrix}$$

running

$$y_3[n] = (-1)^n \cdot \left(y[n] - y[n+1]\right) \Leftrightarrow D_3 = \begin{pmatrix} 1 & -1 & & & 0 \\ & -1 & 1 & & \\ & & 1 & -1 & \\ & & & \ddots & \ddots \\ 0 & & & -1 & 1 \end{pmatrix}$$

surround

$$y_4[n] = (-1)^n \cdot \left(2y[n] - y[n+1] - y[n+2]\right) \Leftrightarrow \begin{pmatrix} 1 & -2 & 1 & & & \\ & -1 & 2 & -1 & & \\ & & 1 & -2 & 1 & \\ & & & \ddots & \ddots & \ddots \\ & & & -1 & 2 & -1 \end{pmatrix}$$

Figure 12. Examples of differencing matrices that can be used for functional ASL.Whether we use an ordinary least squares (OLS) approach on differenced data or a generalized least squares approach (GLS) on raw data can make a significant difference in the statistical power of the analysis. It's not important for blocked designs, but in the case of event related designs, whose frequency content is relatively higher, it makes a significant difference, whether the inter stimulus interval (ISI) is randomized or not. Reproduced from (Mumford et al. 2006)

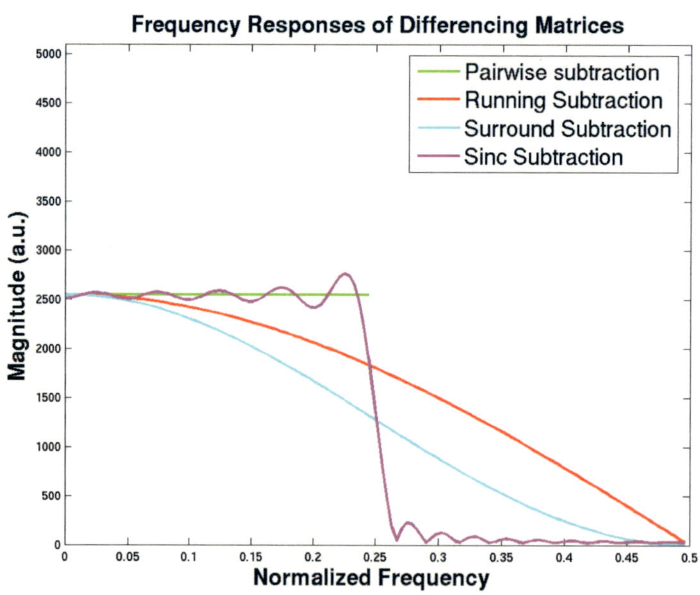

Figure 13. The effect of the differencing process on the frequency content of the ASL signal. Reproduced from (Mumford, Hernandez-Garcia et al. 2006).

http://dx.doi.org/10.5772/28969

From a signal processing point of view it is useful to note that the differencing schemes can be thought of as multiplying the time course by a "differencing matrix". Figure 12 shows some of the differencing matrices that are commonly used. This differencing process, in addition to isolating the perfusion weighting, also behaves like a filter applied to the data and frequency responses can be computed accordingly. For example, figure 13 shows the filtering effect on the frequency content by several differencing "filters".

Those frequency responses can be derived analytically for a given input y[n] whose discrete Fourier transform is given by $Y(e^{jw})$, yielding the equations in Table 1.

Note that pairwise subtraction and sinc subtraction do not have a straightforward linear response since they involve down-sampling the data. In both of those cases, the down-sampling process causes the top half of the frequency spectrum to alias into the bottom half before the filtering process. In terms of their frequency response, they are very similar except for the imperfections of the sinc kernel used in the implementation. In terms of detection and statistics, sinc subtraction preserves greater degrees of freedom than pairwise subtraction, which reduces the number of time points by half.

Type of Differencing		**Fourier Domain Expression**
D_1	No subtraction	$Y_1(e^{j\omega}) = Y(e^{j\omega})$
D_2	Pairwise subtraction	$Y_2(e^{j\omega}) = 0.5\cdot[Y(e^{j\omega/2})(1-e^{j\omega/2}) + Y(e^{j(\omega+2\pi)/2})(1-e^{j(\omega+2\pi)/2})]$
D_3	Running subtraction	$Y_3(e^{j\omega}) = Y(e^{j(\omega+\pi)})(1-e^{j(\omega+\pi)})$
D_4	Surround subtraction	$Y_4(e^{j\omega}) = Y(e^{j(\omega+\pi)})(-1+2e^{j(\omega+\pi)}-e^{2j(\omega+\pi)})$

Table 1. Frequency Responses of the differencing schemes

With that in mind, it is important to take the differencing matrix into account when constructing a general linear model for analysis. In the previous section we constructed a model that contained both BOLD effects and ASL effects (see figure 9). Let X represent the corresponding design matrix. We can then re-write the general linear model equation with an additional Differencing matrix, like this:

$$DY = DX\beta + D\varepsilon.$$

It is crucial to note that whatever we do to the data, we must also do to the model if that model is to be as accurate as possible. Hence we difference the data, we must also difference the model, including the noise term. If no differencing is applied, the matrix D can be the identity.

At this point, the objective is the same as in the standard GLM analysis: estimate the model's coefficients (β) and determine whether they are statistically significant. Our choices are to pick a differencing matrix, D, and solve by ordinary least squares (OLS), or do nothing to the data (use D = I) and solve be generalized least squares (GLS).

As it turns out, this decision largely depends on the type of experimental paradigm and on the design matrix. Figure 14 shows a comparison between the statistical power obtained from analyzing ASL data with different amount of noise the two strategies: differencing and not differencing. In one case the data are analyzed with a Generalized Least Squares (GLS) model and not differenced. In the other case, the data are pairwise differenced and analyzed with the usual ordinary least squares (OLS).

Specifically, it shows the statistical power for no differencing with GLS and that of pairwise differencing with OLS for the 3 study designs over SNR values ranging between 0.25 and 2. The dotted lines on the random event related figure indicate ± 2 standard deviations of the average power over the 100 simulations. As expected, the power is similar between the two methods for the block design, but the "no differencing GLS" model is shown to have larger power for the event related study designs. The random event related design can have up to 14% (s.d. 0.7%) lower power when OLS is used compared to GLS. (Mumford, Hernandez-Garcia et al. 2006). An added benefit is that the GLS approach also enables us to estimate both BOLD and ASL effects simultaneously, which may be desirable in some analyses.

A final note is that the above quantification method using the GLM can be used whether the data are differenced and undifferenced, as long as the design matrix is constructed carefully to reflect the differencing scheme.

11. Challenges facing ASL

So far, we have illustrated the utility of ASL in functional MRI, but we must reiterate that ASL is not a panacea, though. Arterial spin labeling techniques pose a number of challenges. These challenges must be overcome so that ASL techniques can take a more central role in the study of brain function.

ASL is challenged by low SNR, since less than 10% of the water in a given voxel is contributed by blood (Pawlik, Rackl et al. 1981; Weiss, Buchweitz et al. 1982) and the label decays quickly. This problem can be partially alleviated by continuous arterial spin labeling, which increases the amount of label that is introduced into the tissue. Gains in the SNR have also been made with the development of pulse sequences that employ background suppression pulses, as these dramatically reduce the noise contribution from stationary tissue (Garcia, Duhamel et al. 2005). However, these background suppression pulses interfere with the ability to quantify perfusion, so they limit the measurement to relative CBF measures (Shin, Liu et al. 2011).

Another major challenge to ASL techniques is low temporal resolution because ASL measurements take from 3 to 6 seconds depending on the amount of time spent labeling the arterial blood. Collecting multi-slice data can be challenging because the imaging RF pulses can interfere with the inversion label of the arterial water (Silva, Zhang et al. 1995). Lastly, all slices must be acquired within a short period of time in order to sample the label before it clears from

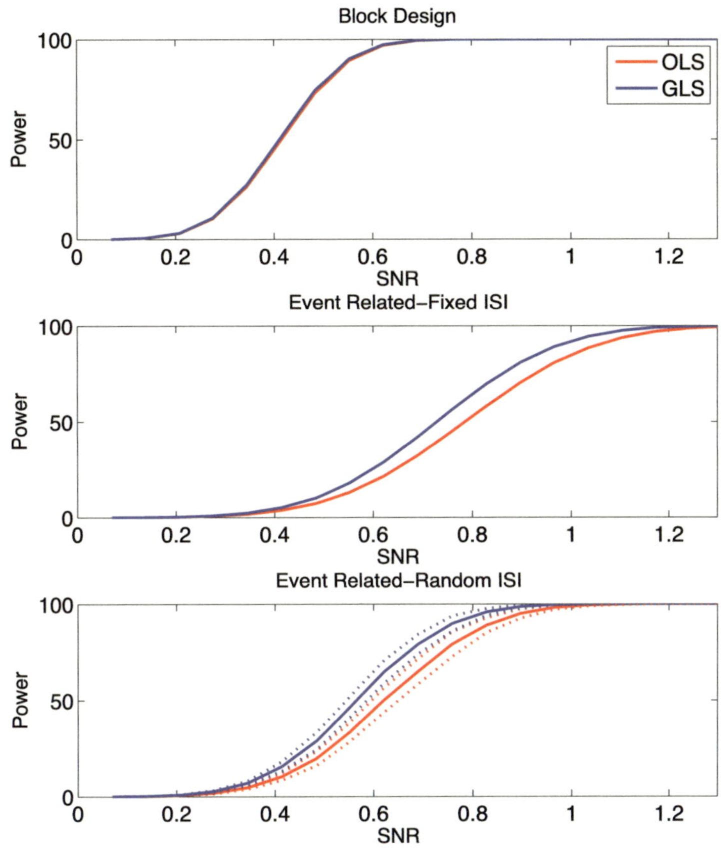

Figure 14. Whether we use an ordinary least squares (OLS) approach on differenced data or a generalized least squares approach (GLS) on raw data can make a significant difference in the statistical power of the analysis. It's not important for blocked designs, but in the case of event related designs, whose frequency content is relatively higher, it makes a significant difference. Reproduced from (Mumford, Hernandez-Garcia et al. 2006).

the tissue of interest. As we have discussed earlier, functional MRI experiments often require the ability to scan the whole brain at a rapid rate in order to localize and characterize brief, subtle changes in cerebral activity (i.e., event related experiments). Collecting multi-slice ASL images in a rapid manner is challenging, given the SNR and clearance rate of the label. While one can collect images at a rate roughly equivalent to the transit time of the subject (roughly 1.5 seconds) the number of slices at this rate is limited to less than five at the present time (Hernandez-Garcia, Lee et al. 2004; Hernandez-Garcia, Lee et al. 2005).

12. Conclusions

In this chapter we have examined arterial spin labeling as technique for functional MRI in depth. As we have seen, there are some clear advantages to collecting ASL data for functional studies in some circumstances, but other times it may not be beneficial at all. ASL is really well suited for longitudinal studies and studies with long blocks of activation, but it may not be so well suited for event related experiments. ASL offers quantitative measures of perfusion at rest and activation. On a per subject basis, the technique's limitations are low SNR and temporal resolution, but there are significant gains in terms of population variance and contrast between resting and active states.

Author details

Luis Hernandez-Garcia and Hesamoddin Jahanian

University of Michigan, FMRI Laboratory, USA

References

[1] Aguirre, G., J. Detre, et al. (2002). "Experimental design and the relative sensitivity of BOLD and perfusion fMRI." *Neuroimage* 15(3): 488-500.

[2] Ances, B. M., O. Leontiev, et al. (2008). "Regional differences in the coupling of cerebral blood flow and oxygen metabolism changes in response to activation: implications for BOLD-fMRI." *Neuroimage* 39(4): 1510-1521.

[3] Bagarinao, E., K. Matsuo, et al. (2003). "Estimation of general linear model coefficients for real-time application." *Neuroimage* 19(2): 422-429.

[4] Bandettini, P. A., E. C. Wong, et al. (1992). "Time course EPI of human brain function during task activation." *Magn Reson Med* 25(2): 390-397.

[5] Bernstein, M. A., K. F. King, et al. (2004). *Handbook of MRI pulse sequences*, Elsevier Academic Press.

[6] Bevington, P. R. and D. K. Robinson (2003). *Data reduction and error analysis for the physical sciences*. Boston, McGraw-Hill.

[7] Boynton, G., S. Engel, et al. (1996). "Linear systems analysis of functional magnetic resonance imaging in human V1." *J Neurosci* 16(13): 4207-4221.

[8] Buxton, R. and L. Frank (1997). "A model for the coupling between cerebral blood flow and oxygen metabolism during neural stimulation." *J Cereb Blood Flow Metab* 17(1): 64-72.

[9] Buxton, R., L. Frank, et al. (1998). "A general kinetic model for quantitative perfusion imaging with arterial spin labeling." *Magn Reson Med* 40(3): 383-396.

[10] Buxton, R., E. Wong, et al. (1998). "Dynamics of blood flow and oxygenation changes during brain activation: the balloon model." *Magn Reson Med* 39(6): 855-864.

[11] Caria, A., R. Veit, et al. (2007). "Regulation of anterior insular cortex activity using real-time fMRI." *Neuroimage* 35(3): 1238-1246.

[12] Changwei, W. W., C. Kai-Hsiang, et al. (2008). "Vascular space occupancy-dependent functional MRI by tissue suppression." *Journal of Magnetic Resonance Imaging* 28(1): 219-226.

[13] Chen, J. J., M. Wieckowska, et al. (2008). "Cerebral blood flow measurement using fMRI and PET: a cross-validation study." *International journal of biomedical imaging* 2008: 516359.

[14] Christopher, B. G., A. S. Ronald, et al. (2009). "Estimating cerebral blood volume with expanded vascular space occupancy slice coverage." *Magnetic Resonance in Medicine* 61(5): 1193-1200.

[15] Cohen, E. R., K. Ugurbil, et al. (2002). "Effect of basal conditions on the magnitude and dynamics of the blood oxygenation level-dependent fMRI response." *J Cereb Blood Flow Metab* 22(9): 1042-1053.

[16] Cox, R. W., A. Jesmanowicz, et al. (1995). "Real-time functional magnetic resonance imaging." *Magn Reson Med* 33(2): 230-236.

[17] Davis, T. L., K. K. Kwong, et al. (1998). "Calibrated functional MRI: mapping the dynamics of oxidative metabolism." *Proc Natl Acad Sci U S A* 95(4): 1834-1839.

[18] deCharms, R. C. (2008). "Applications of real-time fMRI." *Nat Rev Neurosci* 9(9): 720-729.

[19] deCharms, R. C., K. Christoff, et al. (2004). "Learned regulation of spatially localized brain activation using real-time fMRI." *Neuroimage* 21(1): 436-443.

[20] Detre, J., J. Leigh, et al. (1992). "Perfusion imaging." *Magn Reson Med* 23(1): 37-45.

[21] Duong, T., E. Yacoub, et al. (2002). "High-resolution, spin-echo BOLD, and CBF fMRI at 4 and 7 T." *Magn Reson Med* 48(4): 589-593.

[22] Edelman, R. and Q. Chen (1998). "EPISTAR MRI: multislice mapping of cerebral blood flow." *Magn Reson Med* 40(6): 800-805.

[23] Fernandez-Seara, M. A., J. Wang, et al. (2007). "Imaging mesial temporal lobe activation during scene encoding: comparison of fMRI using BOLD and arterial spin labeling." *Hum Brain Mapp* 28(12): 1391-1400.

[24] Fernandez-Seara, M. A., Z. Wang, et al. (2005). "Continuous arterial spin labeling perfusion measurements using single shot 3D GRASE at 3 T." *Magn Reson Med* 54(5): 1241-1247.

[25] Gai, N. D., S. L. Talagala, et al. (2011). "Whole-brain cerebral blood flow mapping using 3D echo planar imaging and pulsed arterial tagging." *J Magn Reson Imaging* 33(2): 287-295.

[26] Garcia, D., C. de Bazelaire, et al. (2005). *Pseudo-continuous Flow Driven Adiabatic Inversion for Arterial Spin Labeling*. Proceedings International Society for Magnetic Resonance in Medicine, Miami, FLA.

[27] Garcia, D. M., G. Duhamel, et al. (2005). "Efficiency of inversion pulses for background suppressed arterial spin labeling." *Magn Reson Med* 54(2): 366-372.

[28] Gembris, D., J. G. Taylor, et al. (2000). "Functional magnetic resonance imaging in real time (FIRE): sliding-window correlation analysis and reference-vector optimization." *Magn Reson Med* 43(2): 259-268.

[29] Gollub, R. L., H. C. Breiter, et al. (1998). "Cocaine decreases cortical cerebral blood flow but does not obscure regional activation in functional magnetic resonance imaging in human subjects." *Journal of cerebral blood flow and metabolism : official journal of the International Society of Cerebral Blood Flow and Metabolism* 18(7): 724-734.

[30] Gu, H., H. Lu, et al. (2006). "Noninvasive quantification of cerebral blood volume in humans during functional activation." *Neuroimage* 30(2): 377-387.

[31] Gunther, M., K. Oshio, et al. (2005). "Single-shot 3D imaging techniques improve arterial spin labeling perfusion measurements." *Magn Reson Med* 54(2): 491-498.

[32] Haacke, E. M. (1999). Magnetic resonance imaging : physical principles and sequence design. New York, Wiley.

[33] Hermes, M., D. Hagemann, et al. (2007). "Reproducibility of continuous arterial spin labeling perfusion MRI after 7 weeks." *Magma* 20(2): 103-115.

[34] Hernandez-Garcia, L., H. Jahanian, et al. (2011). "Real-time functional MRI using pseudo-continuous arterial spin labeling." Magnetic resonance in medicine : official journal of the Society of Magnetic Resonance in Medicine / Society of Magnetic Resonance in Medicine 65(6): 1570-1577.

[35] Hernandez-Garcia, L., H. Jahanian, et al. (2010). "Quantitative Analysis of Arterial Spin Labeling FMRI Data Using a General Linear Mode." *Mangetic Resonance Imaging* 28(7): 919-927.

[36] Hernandez-Garcia, L., G. R. Lee, et al. (2004). "Fast, pseudo-continuous arterial spin labeling for functional imaging using a two-coil system." *Magn Reson Med* 51(3): 577-585.

[37] Hernandez-Garcia, L., G. R. Lee, et al. (2005). "Quantification of perfusion fMRI using a numerical model of arterial spin labeling that accounts for dynamic transit time effects." *Magn Reson Med* 54(4): 955-964.

[38] Hernandez-Garcia, L., D. Lewis, et al. (2007). "Magnetization transfer effects on the efficiency of flow-driven adiabatic fast passage inversion of arterial blood." *NMR Biomed* 20: 733-742.

[39] Hoge, R. D., J. Atkinson, et al. (1999). "Linear coupling between cerebral blood flow and oxygen consumption in activated human cortex." *Proceedings of the National Academy of Sciences of the United States of America* 96(16): 9403-9408.

[40] Jaeggi, S. M., B. E. Studer, et al. (2009). *Improving Fluid Intelligence – Single N-back Is As Effective As Dual N-back*. Annual Meeting of the Psychonomic Society, Boston, MA.

[41] Jahanian, H., D. C. Noll, et al. (2011). "B0 field inhomogeneity considerations in pseudo-continuous arterial spin labeling (pCASL): effects on tagging efficiency and correction strategy." *NMR in Biomedicine*: n/a-n/a.

[42] Jin, T. and S.-G. Kim (2008). "Improved cortical-layer specificity of vascular space occupancy fMRI with slab inversion relative to spin-echo BOLD at 9.4†T." *Neuroimage* 40(1): 59-67.

[43] Kemeny, S., F. Q. Ye, et al. (2005). "Comparison of continuous overt speech fMRI using BOLD and arterial spin labeling." *Human Brain Mapping* 24(3): 173-183.

[44] Kety, S. S. and C. Schmidt (1945). "THE DETERMINATION OF CEREBRAL BLOOD FLOW IN MAN BY THE USE OF NITROUS OXIDE IN LOW CONCENTRATIONS." *American Journal of Physiology*(143): 53-66.

[45] Kim, J., J. Whyte, et al. (2006). "Continuous ASL perfusion fMRI investigation of higher cognition: quantification of tonic CBF changes during sustained attention and working memory tasks." *Neuroimage* 31(1): 376-385.

[46] Kim, S. (1995). "Quantification of relative cerebral blood flow change by flow-sensitive alternating inversion recovery (FAIR) technique: application to functional mapping." *Magn Reson Med* 34(3): 293-301.

[47] Kim, S. and N. Tsekos (1997). "Perfusion imaging by a flow-sensitive alternating inversion recovery (FAIR) technique: application to functional brain imaging." *Magn Reson Med* 37(3): 425-435.

[48] Kwong, K., J. Belliveau, et al. (1992). "Dynamic magnetic resonance imaging of human brain activity during primary sensory stimulation." *Proc Natl Acad Sci U S A* 89(12): 5675-5679.

[49] Lassen, N. A. and W. Perl (1979). *Tracer kinetic methods in medical physiology*. New York, Raven Press.

[50] Leontiev, O. and R. B. Buxton (2007). "Reproducibility of BOLD, perfusion, and CMRO2 measurements with calibrated-BOLD fMRI." *Neuroimage* 35(1): 175-184.

[51] Liu, T. and E. Wong (2005). "A signal processing model for arterial spin labeling functional MRI." *Neuroimage* 24(1): 207-215.

[52] Lu, H., M. Law, et al. (2005). "Novel approach to the measurement of absolute cerebral blood volume using vascular-space-occupancy magnetic resonance imaging." *Magn Reson Med* 54(6): 1403-1411.

[53] Lund, T., K. Madsen, et al. (2006). "Non-white noise in fMRI: does modelling have an impact?" *Neuroimage* 29(1): 54-66.

[54] Mulderink, T. A., D. R. Gitelman, et al. (2002). "On the use of caffeine as a contrast booster for BOLD fMRI studies." *Neuroimage* 15(1): 37-44.

[55] Mumford, J. A., L. Hernandez-Garcia, et al. (2006). "Estimation efficiency and statistical power in arterial spin labeling fMRI." *Neuroimage* 33(1): 103-114.

[56] Nakai, T., E. Bagarinao, et al. (2006). "Dynamic monitoring of brain activation under visual stimulation using fMRI--The advantage of real-time fMRI with sliding window GLM analysis." *Journal of Neuroscience Methods* 157(1): 158-167.

[57] Nencka, A. S. and D. B. Rowe (2007). "Reducing the unwanted draining vein BOLD contribution in fMRI with statistical post-processing methods." *Neuroimage* 37(1): 177-188.

[58] O'Gorman, R. L., V. Kumari, et al. (2006). "Personality factors correlate with regional cerebral perfusion." *Neuroimage* 31(2): 489-495.

[59] Olafsson, V. T., D. C. Noll, et al. (2008). "Fast Joint Reconstruction of Dynamic R2* and Field Maps in Functional MRI." *Medical Imaging, IEEE Transactions on* 27(9): 1177-1188.

[60] Olson, I. R., H. Rao, et al. (2006). "Using perfusion fMRI to measure continuous changes in neural activity with learning." *Brain and Cognition* 60(3): 262-271.

[61] Pawlik, G., A. Rackl, et al. (1981). "Quantitative capillary topography and blood flow in the cerebral cortex of cats: an in vivo microscopic study." *Brain Res* 208(1): 35-58.

[62] Pfeuffer, J., G. Adriany, et al. (2002). "Perfusion-based high-resolution functional imaging in the human brain at 7 Tesla." *Magn Reson Med* 47(5): 903-911.

[63] Phan, K. L., D. A. Fitzgerald, et al. (2004). "Real-time fMRI of cortico-limbic brain activity during emotional processing." *Neuroreport* 15(3): 527-532.

[64] Posse, S., D. Fitzgerald, et al. (2003). "Real-time fMRI of temporolimbic regions detects amygdala activation during single-trial self-induced sadness." *NeuroImage* 18: 760-768.

[65] Rao, H., S. J. Gillihan, et al. (2007). "Genetic variation in serotonin transporter alters resting brain function in healthy individuals." *Biological Psychiatry* 62(6): 600-606.

[66] Rao, H., J. Wang, et al. (2007). "Imaging brain activity during natural vision using CASL perfusion fMRI." *Human Brain Mapping* 28(7): 593-601.

[67] Raoult, H., J. Petr, et al. (2011). "Arterial spin labeling for motor activation mapping at 3T with a 32-channel coil: Reproducibility and spatial accuracy in comparison with BOLD fMRI." *Neuroimage* 58(1): 157-167.

[68] Shin, D., H. Liu, et al. (2011). *Effect of background suppression on CBF quantitation in pseudo continuous arterial spin labeling*. International Society for Magnetic Resonance in Medicine, Montreal, QC, Canada.

[69] Silva, A., W. Zhang, et al. (1995). "Multi-slice MRI of rat brain perfusion during amphetamine stimulation using arterial spin labeling." *Magn Reson Med* 33(2): 209-214.

[70] Smith, A. M., B. K. Lewis, et al. (1999). "Investigation of Low Frequency Drift in fMRI Signal." *Neuroimage* 9(5): 526-533.

[71] Talagala, S., F. Ye, et al. (2004). "Whole-brain 3D perfusion MRI at 3.0 T using CASL with a separate labeling coil." *Magn Reson Med* 52(1): 131-140.

[72] Tjandra, T., J. C. W. Brooks, et al. (2005). "Quantitative assessment of the reproducibility of functional activation measured with BOLD and MR perfusion imaging: Implications for clinical trial design." *Neuroimage* 27(2): 393-401.

[73] Vazquez, A. and D. Noll (1998). "Nonlinear aspects of the BOLD response in functional MRI." *Neuroimage* 7(2): 108-118.

[74] Vazquez, A. L., G. R. Lee, et al. (2006). "Application of selective saturation to image the dynamics of arterial blood flow during brain activation using magnetic resonance imaging." *Magn Reson Med* 55(4): 816-825.

[75] Voyvodic, J. T. (1999). "Real-time fMRI paradigm control, physiology, and behavior combined with near real-time statistical analysis." *Neuroimage* 10(2): 91-106.

[76] Wang, J., G. Aguirre, et al. (2003). "Empirical analyses of null-hypothesis perfusion FMRI data at 1.5 and 4 T." *Neuroimage* 19(4): 1449-1462.

[77] Wang, J., G. Aguirre, et al. (2003). "Arterial spin labeling perfusion fMRI with very low task frequency." *Magn Reson Med* 49(5): 796-802.

[78] Wang, J., G. K. Aguirre, et al. (2003). "Arterial spin labeling perfusion fMRI with very low task frequency." *Magn Reson Med* 49(5): 796-802.

[79] Wang, J., D. Alsop, et al. (2002). "Comparison of quantitative perfusion imaging using arterial spin labeling at 1.5 and 4.0 Tesla." *Magn Reson Med* 48(2): 242-254.

[80] Wang, J., H. Rao, et al. (2005). "Perfusion functional MRI reveals cerebral blood flow pattern under psychological stress." *Proceedings of the National Academy of Sciences of the United States of America* 102(49): 17804-17809.

[81] Wang, J., Y. Zhang, et al. (2005). "Amplitude-modulated Continuous Arterial Spin-labeling 3.0-T Perfusion MR Imaging with a Single Coil: Feasibility Study." *Radiology* 235(1): 218-228.

[82] Weiss, H., E. Buchweitz, et al. (1982). "Quantitative regional determination of morphometric indices of the total and perfused capillary network in the rat brain." *Circ Res* 51(4): 494-503.

[83] Williams, D., J. Detre, et al. (1992). "Magnetic resonance imaging of perfusion using spin inversion of arterial water." *Proc Natl Acad Sci U S A* 89(1): 212-216.

[84] Wise, R. G. and I. Tracey (2006). "The role of fMRI in drug discovery." *Journal of magnetic resonance imaging : JMRI* 23(6): 862-876.

[85] Wong, E., R. Buxton, et al. (1997). "Implementation of quantitative perfusion imaging techniques for functional brain mapping using pulsed arterial spin labeling." *NMR Biomed* 10(4-5): 237-249.

[86] Wong, E., R. Buxton, et al. (1998). "Quantitative imaging of perfusion using a single subtraction (QUIPSS and QUIPSS II)." *Magn Reson Med* 39(5): 702-708.

[87] Wong, E., R. Buxton, et al. (1998). "A theoretical and experimental comparison of continuous and pulsed arterial spin labeling techniques for quantitative perfusion imaging." *Magn Reson Med* 40(3): 348-355.

[88] Wu, W. C., M. Fernandez-Seara, et al. (2007). "A theoretical and experimental investigation of the tagging efficiency of pseudocontinuous arterial spin labeling." *Magn Reson Med* 58(5): 1020-1027.

[89] Yoo, S. S., T. Fairneny, et al. (2004). "Brain-computer interface using fMRI: spatial navigation by thoughts." *Neuroreport* 15(10): 1591-1595.

[90] Yoo, S. S. and F. A. Jolesz (2002). "Functional MRI for neurofeedback: feasibility study on a hand motor task." *Neuroreport* 13(11): 1377-1381.

[91] Zhang, W., A. Silva, et al. (1995). "NMR measurement of perfusion using arterial spin labeling without saturation of macromolecular spins." *Magn Reson Med* 33(3): 370-376.

Chapter 4

Phase Variations in fMRI Time Series Analysis: Friend or Foe?

Gisela E. Hagberg and Elisa Tuzzi

Additional information is available at the end of the chapter

http://dx.doi.org/10.5772/31782

1. Introduction

Functional MRI studies (fMRI) are based on the blood-oxygenation-level-dependent effect (BOLD) that arises in brain areas where neuronal activity takes place (Ogawa et al., 1990, 1992; 1993). BOLD induces changes in the local magnetic susceptibility and these can be measured by Gradient Echo (GE) Echo-Planar-Imaging (EPI). The fMRI signal thus observed consists of a complex value, which is subdivided into a magnitude and a phase term. In most fMRI studies, the phase signal is discarded and only the magnitude changes are used to detect activated brain areas. In recent years, an increased interest in using both signal types has emerged, with the prospective of increasing both statistical power and spatial specificity.

This chapter describes the fMRI signal from a mathematical and biophysical view point, with emphasis on the phase signal. We will show which kind of information that the phase conveys and exemplify how it can be incorporated in the fMRI analysis pipeline. In section 2, we first describe the basics of an fMRI study, from the fundamental MR signal equations to how a whole brain image is generated in the scanner. We also describe the blood-oxygenation-level-dependent (BOLD) effect that follows neuronal activation. It alters the local magnetic susceptibility and the homogeneity of the local magnetic field and hereby alters the GE EPI signal.

Next, we discuss the phase signal in detail and we characterize the phase in comparison with the more familiar magnitude signal. We illustrate how the probability distributions of the magnitude and the phase values change as the signal-to-noise-ratio increases and how physical mechanisms, like paramagnetic and diamagnetic changes in the local field homogeneity, influence the magnitude and the phase signal in different ways. Regarding neuronal activity, we will review different biophysical models of BOLD with emphasis on the phase signal. Early models describe how the phase effect tend to cancel out for microvascular networks, while a detectable effect remains in larger vessels only (Yablonskiy and Haacke, 1994; Kiselev and Posse 1999; Menon, 2002). On the contrary, a more recent model, based on the Sphere of Lorentz effect, shows that a spatially extended, activated brain area produces a detectable change in the phase signal (Zhao et al., 2007; Feng et al., 2009).

An issue with the phase signal, addressed in the fourth section, is that the potentially valuable phase information is buried in strong phase variations that extend across the head, reflecting variations in the magnetic field homogeneity that vary both in space and in time. These noise factors are unrelated to the neuronal activity and affect the phase to a greater extent than the magnitude signal (Petridou et al., 2009; Hagberg et al., 2008).Methods that suppress or remove these unwanted effects represent a key factor fMRI analysis of phase data. We describe the sources of unwanted phase effects, how these can be modelled, and quantified. We also review several available post-processing methods that can be used to suppress or even remove these unwanted phase effects.

In the last section, we discuss emerging methods that utilize the information conveyed by the phase in the analysis of fMRI data. We will highlight a method that can be used to improve the analysis of the conventional magnitude signal. This method is derived by identifying physiological noise sources from the phase data, and by noise regression of the magnitude signal. The BOLD effect in phase images can also be usedwith the prospective of increasing statistical power and spatial specificity. For illustrative purposes, we show some phase and magnitude results obtained at the group-level in an fMRI study of a motor task performed at 3T.

2. Basics of fMRI physiology

Magnetic Resonance Imaging (MRI) is a non-invasive tomographic technique based on the principles of NMR. In virtue of being a tomographic technique it provides the reconstruction of sections of a 3D object. The capability to explore the structure of the human body non-invasively makes MRI widely employed in clinical imaging. Since the early 1990's, MRI is used not only to study brain structure but also to study brain function. Theso-called functional Magnetic Resonance Imaging (fMRI), which is based on the BOLD contrast, discovered by Ogawa et al. (1990, 1992, 1993). The great advantage of fMRI resides in its non–invasiveness and high availability, since standard equipment already existing in hospitals can be used.

Neuronal activity causes a local increase in the available oxygen due to neurovascular coupling, either due to synaptic input from other brain areas, or due to local signal processing (Logothetis, 2008). The neurovascular coupling mechanism consists of several steps, leading from an increase in the local neuronal activity to changes in blood flow. Initially, the blood oxygen concentration decreases, causing the initial dip phenomenon of the BOLD effect. As a compensatory mechanism, vasoactive substances are released by glia cells into the extracellular space. When these substances reach nearby blood vessels, the blood flow increases, and so does the relative amount of oxygenated haemoglobin, as a result of changes in the diameter of the blood vessels, the density of the red blood cells, and the amount of O_2 and CO_2. The hemodynamic BOLD response thus apparently over-compensates the oxygen use in the activated neurons, giving rise to blood oxygen levels which are higher than necessary to re-establish local tissue oxygen availability. The onset of this BOLD response is typically delayed by 1-2 seconds and peaks 4-6 seconds after onset of the neural response. Spatially, the fMRI signal spreads out due to bigger veins that drain the capillaries close to the neurons at work.

2.1. The fundamental MRI signal equations

Any atom holding a net nuclear magnetic moment, *I*, can generate an MR signal detectable by a scanner. Most MRI methods employ hydrogen atoms ^1H, owing to a high nuclear gyromagnetic moment and the high water concentrations present in the human body (ca 55M).

When hydrogen atoms are dipped into an external field, B_0, the spins tend to either align with the external field (lower energy state) or orient themselves in the opposite direction (upper energy state). At equilibrium, spins are in light excess in the former state, resulting in a non-zero net magnetic moment *M*that is aligned with the external field (Fig. 1). According to the rules of quantum mechanics, the nuclear magnetic moment cannot be exactly aligned with the external magnetic field, and are therefore oriented at an angle with respect to the field. This difference in orientation between the external field and the field of the spins causes a torque effect leading to a precession of the spins around the external field (see zoomed box, Fig. 1). The precession occurs at the Larmor frequency, ω , according to the fundamental equation of NMR, the Larmor equation:

$$\omega = \gamma B_0 \tag{1}$$

where γ is the gyromagnetic ratio (in units of radian/T/s), and $B_0$the intensity of the magnetic field.

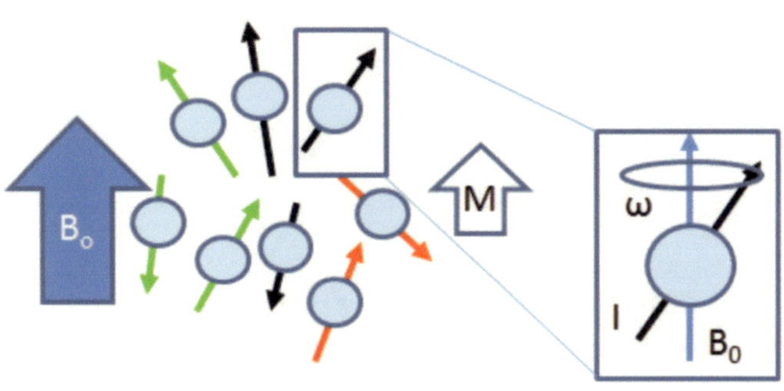

Figure 1. Net magnetic moment *M*of and ensemble of spins, each with an individual nuclear magnetic moment, *I*, placed in an external field B_0. In the zoomed inset, the precession of*I* around the B_0 axis at the Larmor frequency, ω is shown.

At equilibrium, no MR signal can be detected since the magnetization vector *M*is time independent. An MR signal can only be detected after the equilibrium condition has been perturbed, when the spin systems return to equilibrium and the spins flip from the higher to the lower energy level. The perturbation is made so that the spins can absorb the necessary energy to bridge the gap, ΔE, between the two spin states. A radiation field B_1 oscillating exactly at the Larmor frequency is used to excite the spins and induce transitions from the lower to the higher energy state. The perturbing B_1field is applied orthogonally with respect to B_0 (Fig. 2).

The induced transition to the upper energy level gives rise to a spin coherence, bringing the longitudinal equilibrium magnetization M_z into the transverse plane x-y, M_{xy}, in a fixed reference system (laboratory system) as shown in Fig. 2.

The transition from the longitudinal to the transverse direction is thus induced by a radio-frequency field B_1, or a so-called 90° pulse. When the $B_1$90° pulse is interrupted,*M* starts to precess around the B_0z-axis at the Larmor frequency, whilst re-establishing its initial equilibrium value.This causes the gradual recovery of the longitudinal component M_z and the complete extinction of M_{xy}. The temporal variation of the transverse component, M_{xy}, originates the MR signal, also known as Free Induction Decay (FID),oscillates at the Larmor frequency and is detectable as an induced voltage in the radio-frequency coil, according to Faraday's law of electromagnetic induction.

The evolution of the magnetization *M* of a spin system immersed in an external magnetic field, B_0, perturbed by an oscillating field, B_1, and the forces that rule the FID can be conjointly described by the phenomenological Bloch equations. The solutions to the Bloch equations govern the decay and recovery of the magnetization.

In the absence of local magnetic field inhomogeneities, the decay of the transverse component, M_{xy}, is described by the characteristic time constant, T_2, also known as the spin-spin relaxation time, while the characteristic time constant T_1, known as spin-lattice relaxation time, rules the recovery of the transverse component, M_z, according to:

$$M_{xy} = M_0 e^{-t/T_2} \tag{2}$$

$$M_z = M_0(1 - e^{-t/T_1}) \tag{3}$$

In the presence of local field homogeneities, that are not refocused by the imaging sequence the FID signaldecay is governed by the time constant $T_2{}^*$, or the effective transverse relaxation time:

$$M_{xy} = M_0 e^{-t/T^*{}_2} \tag{4}$$

This situation happens for gradient echo sequences used in fMRI studies. With this MR sequence, local field inhomogeneities are not refocused and in presence of a positive BOLD effect, a lengthening of T_2* arises due to a relative increase in the local blood concentration of oxyhemoglobin during activation. Oxyhemoglobin is less paramagnetic than deoxyhemoglobin and therefore generates a field shift which is diamagnetic relative to the baseline condition (see paragraph 4 and Fig 8). As will be discussed in greater detail further on, the presence of paramagnetic effects leads to an increase in the bulk magnetic field, with faster spin precession while the presence of diamagnetic effects causes a decreased local field, and a loss of angular precession velocity of the spins around B_o (see Eq. 1 and Fig. 2)

A diamagnetic or a paramagnetic shift in magnetic susceptibility causes variations in the local Larmor frequency and may lead to off-resonance and dephasing effects, $\varphi(t)$ depending on whether the field is uniform or varies across the voxel. Therefore a more general signal equation for a gradient echo sequences is:

$$\begin{aligned} M_{xy} &= M_o e^{-t/T^*_2} e^{-i\cdot\varphi(t)} \\ \varphi(t) &= \gamma \cdot \int_0^t B(\tau) d\tau \end{aligned} \qquad (5)$$

Implicitly, this equation takes into account the complex value of the MRI signal, which is described in more detail below. The decay term of [Eq. 5] is obtained from the magnitude component of the MR signal while the off-resonance dephasing effects, $\varphi(t)$are obtained from the phase.

2.2. Image generation and fMRI

The MRI signal is derived from the entire system of spins, which are excited by the RF pulse and precess at the Larmor frequency. In order to spatially distinguish between spins and to reconstruct an image, magnetic field gradients are added to the external magnetic field B_0. These gradients are applied in the three spatial directions (x,y,z), defining respectively the *slice-selection gradient*, G_z, the *phase-encode gradient*, G_y, and the*frequency-encode* (also *readout*) *gradient*, G_x. The first gradient is used to select the magnetization located inside a slice through the object, while the latter two gradients are used to encode the spatial position of this magnetization in 2D space. This information is sampled in a reciprocal space, termed k-space, and the image of the selected slice is reconstructed by a 2D Fourier Transform that maps the original magnetization distribution of the object from k-space back into object space.In an fMRI study, the signal changes caused by the BOLD response is followed by whole brain measurements that are obtained every 1-3 seconds, with typical voxel sizes between 3x3x3 and 1x1x1mm^3 in human brain studies. Due to the sluggishness of the BOLD functional brain studies generally employ gradient-echo echo-planar imaging (GE-EPI) sequences for data acquisition: a 2D image (or a slice) is acquired following a single excitation pulse, and 20-30 such slices are imaged yielding whole brain coverage within a 2-3 s long repetition time, with typical voxel sizes between 1x1x1 and 3x3x3 mm^3 (Schmitt et al., 1998). The GE-EPI sequence, in which the entire 2D k-space is filled using rapid gradient switching

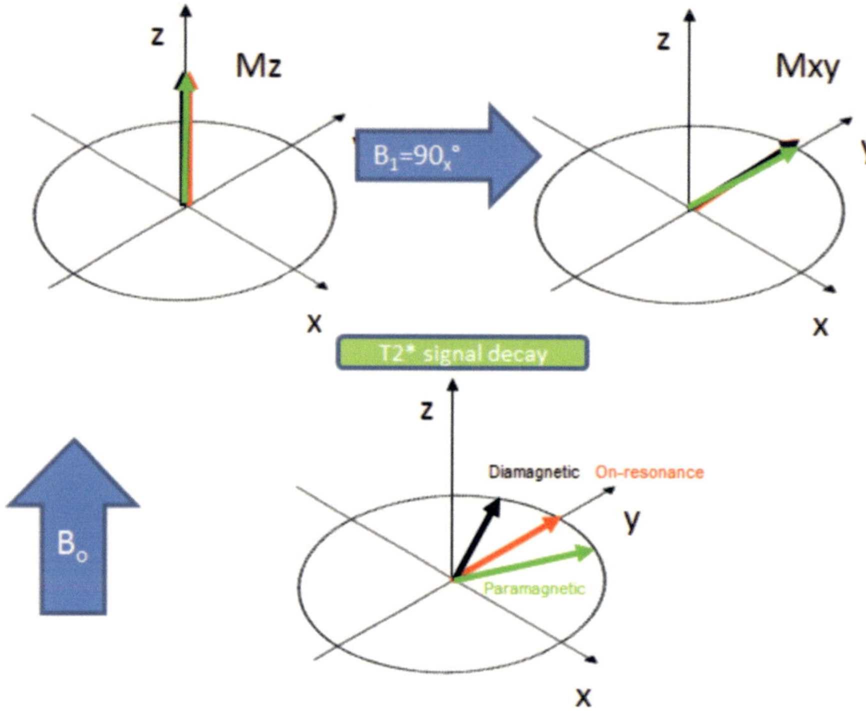

Figure 2. At equilibrium, the net magnetization is said to be longitudinal, M_z and is oriented along the direction of the external magnetic field, B_0. A radio frequency (RF) field B_1,oscillating at the Larmor frequency is applied orthogonally to B_0. This so-called 90° pulse, rotates the magnetization vector into the transverse orientation M_{xy}.During the echo time, spin coherence is lost leading to a T_2* signal decay, that depends on the local magnetic susceptibility. Spins in a diamagnetic environment precesses a little slower, while spins in a paramagnetic environment precesses a little faster than the on-resonance spins. The original equilibrium condition with a longitudinal net magnetization is recovered after a time that is about five times as long as the longitudinal T_1 relaxation time.

after a single excitation pulse, begins with a slice-selective excitation pulse with a flip angle α(i.e. an RF pulse that brings the Magnetization vector at an angle $\alpha$with respect to B_0), and is followed by an EPI readout of gradient echoes (i.e. a train of gradient echoes made up of readout gradientswith alternating polarity, see G_x in Fig.3). Before each readout gradient, a brief phase-encode blip (G_y) is applied in order to quickly sample the whole k-space. Alternatively, a full 3D volume can be acquired by so-called echo-volume imaging (EVI, Mansfield, 1977). In this case no slice-selection is necessary, instead an entire volume is excited by a single RF-pulse and 3D k-space is covered by phase encoding along two spatial directions.

The advantages of using gradient-echo EPI areitsrapid image generation with minimal energy deposition, sufficient spatial resolution and ease of acquisition, yielding whole brain coverage within a short imaging time. These advantages come however at the expense of a relative low signal-to-noise ratio, which is however combined with a great BOLD contrast-to-noise ratio.

http://dx.doi.org/10.5772/31782

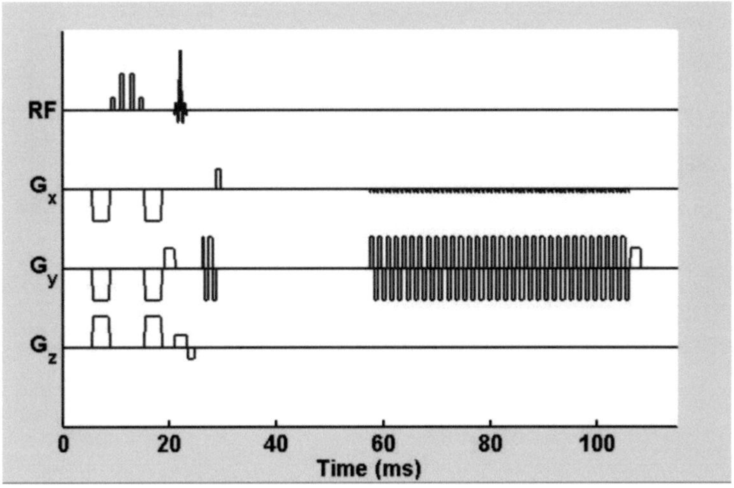

Figure 3. The gradient echo (GE) echo-planar-imaging (EPI) sequence. An RF pulse with flip angle α (first row) is applied together with a magnetic field gradient in the slice selection direction (G_z. last row). Hereby all the spins located inside the selected slice have theirmagnetization vector rotated by an angle $\alpha$with respect to B_0. k-space (the reciprocal of the image space) is sampled by a combination of brief phase-encoding blips (G_x second row) and readout gradientswith alternating polarity (G_y third row). Finally, a Fourier Transform is performed to recover the MR image of the selected slice.

To reach a sufficient BOLD contrast-to-noise ratio (CNR),several such GE-EPI images are acquired during presentation of selected stimuli, generating an fMRI time series.A typical stimulus paradigm for an fMRI experiment alternates "ON" states in which subjects are asked to perform a task, and "OFF" states (or "Control" state) in which they have to be at rest or perform a control task. Generally "ON" and "OFF" states are repeated every 20-40s, to maximize the BOLD signal, and several repetitions of such "ON"/"OFF" states are required. Typical durations of an fMRI time series is 5-10 min. An example of an fMRI experiment is shown in Fig. 4.

Besides the acquisition sequence and the stimulus paradigm, the success of an fMRI experiment depends uponthe statistical data analysis that is undertaken to individuate the cerebral regions where the variation of the magnitude MRI signal is stimulus correlated. The sensitivity of the statistical analysis depends on both the BOLD contrast and on the fMRI noise level. As we will show, the fMRI noise level can be reduced by using the information conveyed by the phase signal.

3. Characteristics of the MRI phase signal

As we will see, the MRI signal can be described by a complex number, consisting of a magnitude and a phase term. Most MRI methods only retain the magnitude term, but in some established applications, like phase-contrast angiography, Susceptibility Weighted Imaging

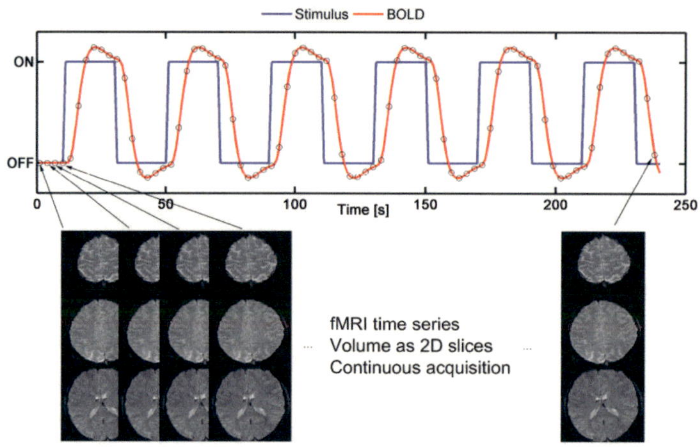

Figure 4. An fMRI experiment consisting of "OFF" and "ON" periods with a 20-40 seconds duration. Gradient-echo EPI images with whole brain coverage are acquired each 2-3 seconds, and 100 or more brain volumes make up the fMRI time series.

or anatomical phase imaging methods at high field strengths, the phase is used as a sole ingredient or in combination with the magnitude to generate image contrast.

The phase and the magnitude signals are like two sides of a coin for the MRI signal. While the magnitude signal is related to the dispersion of the underlying local magnetic field, the phase is sensitive to coherent magnetic field shifts. The reason behind this may become clear if we consider the MRI signal as a vector sum of single spin contributions located at different spatial points inside the voxel, precessing around the B_0 field according to the fundamental Larmor equation (e.g. Fig.1). In case of an ideal homogeneous magnetic field, without field dispersion or field shifts, all the vectors will precess 'on-resonance' at the same frequency generating a maximal magnitude signal at the echo time TE, and a zero phase. In the presence of a field inhomogeneity across the voxel, a field dispersion, each spin senses a slightly different field and precesses at a frequency which is slightly different from the 'on-resonance' condition. At TE, the vectors will therefore point into different directions and the vector sum generates a decreased magnitude signal, while the average phase is zero. In presence of a local deviation of the magnetic field which is uniform across the voxel, a field shift, all the vectors have the same frequency albeit different from the 'on-resonance' condition. At TE, there is a decreased magnitude signal, and an altered phase.

In the following we will discuss the different characteristics of the magnitude and phase of the MRI signal, with emphasis on some factors that are important for fMRI applications. Besides the signal and noise properties, we will describe three biophysical models for the effect of BOLD on the phase signal.

3.1. Definition of the magnitude and the phase of the fMRI signal

The MR detection systemis based on quadrature detection. The voltage induced by the oscillating net transversemagnetization$M_{xy}$is split into two channels that function with a phase difference with respect to each other. Two output signals are thus detected, the first which is 'in-phase' and the second which is 'in-quadrature' (ie phase shifted) with respect to the first. These signals can be described in mathematical terms by the real and the imaginary components of a complex vector, corresponding to the M_x and the M_y components of the detected net magnetization of the FID.

In the ideal case, the imaginary component of this complex vector is zero and all the necessary image information is contained in the real component. If this is not the case, it should in principle be possible to recover the real-valued component by'phase correcting' the data. Thereby the complex vector would be rotated into the M_x direction, and an absorption mode image of the magnetization vector would be available, as suggested bysome promising attempts (Bernstein et al., 1989, Bretthorst, 2008). However in general, due to phase imperfections that are difficult to correct, it is convenient to use the absolute (magnitude) value of the MRI signal, corresponding to the length of the complex vector. SuchMRI images are produced by taking the square root of the sum of squares of the real and imaginary components of the measured magnetization vector. The phase of the complex vector, or the argument of the complex vector, is obtained by evaluating the arc-tangent of the ratio of the imaginary and the real components (see Fig. 5):

$$S = abs(M_{xy}) = \sqrt{M_x^2 + M_y^2} \tag{6}$$

$$\varphi = \arg(M_{xy}) = \arctan(\frac{M_y}{M_x}) \tag{7}$$

In general, the phase evolves linearly with the local magnetic fieldB_{loc}in a particular voxel during the time between the excitation pulse and signal read out, i.e. the echo time, *TE*. At the time of acquisition the phase is:

$$\Delta\varphi = \gamma \cdot B_{loc} \cdot TE \tag{8}$$

where γ is the gyromagnetic ratio (in units of radian/T/s) for the observed nuclear spins (see Eq.2).

The phase evolution is defined as positive (clockwise) for increasing magnetic fields (paramagnetic shifts), and negative (counterclockwise) for decreasing magnetic fields (diamagnetic shifts). However this sign convention based on the physical torque effect on the magnetic vector is not compatible with the mathematical sign convention of complex numbers (see Fig. 2 and Hagberg, et al.,2010) and therefore often a negative sign is used in front of the expression in [Eq. 8].

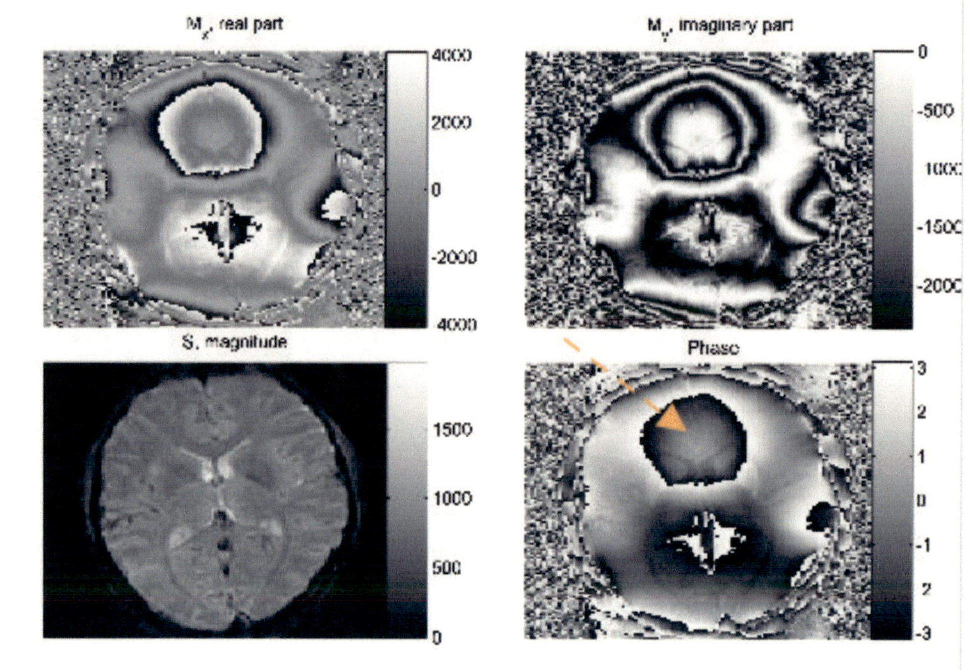

Figure 5. The MRI detection system subdivides the image information into a real and an imaginary component, corresponding to the M_x and the M_y components of the transverse magnetization. These components define a complex vector and can be rearranged to form the length [Eq. 6] and the argument [Eq.7] of the vector, corresponding to the MRI magnitude and phase signals, respectively. The red arrow indicates the area just above the frontal sinuses with strong magnetic field variations, due to the difference in susceptibility between the air in the frontal sinuses and the brain tissue.

3.2. Signal-to-noise ratio in magnitude and phase images

As we discussed above, the signal in the MRI detection system is acquired via two channels that correspond to the real and the imaginary components of a complex number. The electronic noise of each receiving channel has a Gaussian probability density function (PDF), with zero mean and variance determined by fluctuations around this mean. The PDF of the phase and magnitude noise are determined by the combined noise from both channels (Gudbjartsson and Patz 1996). The magnitude values are always greater than zero, and hence cannot have a zero mean. In presence of purely electronic noise, the noise in the magnitude image therefore follows the Rayleigh probability density function, while for SNR levels of 3 and above a Gaussian PDF is observed. The phase signal originating from pure electronic noise takes on any value in the $\pm\pi$ range with equal probability according to a uniform PDF (Eq. 9). As the SNR increases, the PDF of the phase signals evolve towards a Gaussian shape (Fig. 6). Since these the magnitude and the phase signals have a common origin in the real and imaginary components of the receiving channels, the SNR values are interrelated. For high SNR values, the PDF of the phase is obtained from the magnitude signal (Eq.9).

http://dx.doi.org/10.5772/31782

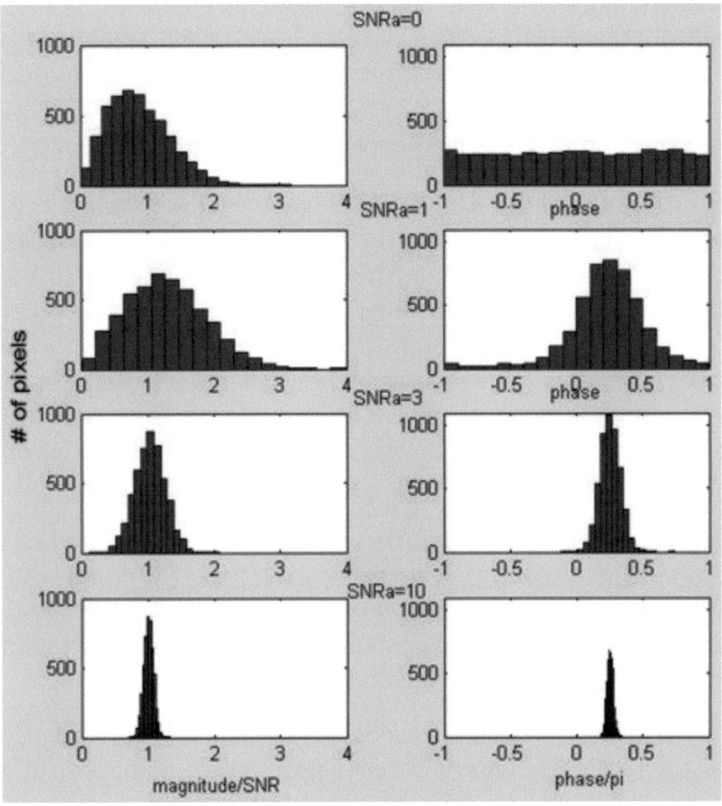

Figure 6. Histograms representing the signal-to-noise probability density functions (PDF) for the magnitude (left column) and phase (right column) signals. The following cases are shown. First row: with an SNR of zero (purely electronic noise), the magnitude values distribute according to a Rayleigh PDF, while the distribution of the phase values is uniform. In row 2-4, we show how the shape of the PDF changes as we go towards greater SNR. At low SNR levels of 1 (2nd row) a Rician PDF is observed for the magnitude data, while at an SNR of 3 (3rd row) and 10 (4th row) the shape is Gaussian. Above an SNR of 3, the phase signal also follows a Gaussian PDF, albeit with a more narrow shape, as the standard deviation of the phase signal is the inverse of the SNR observed for the magnitude signal.

$$SD_{\varphi} = \begin{cases} \dfrac{\pi}{\sqrt{3}} & \text{if SNR} = 0 \\ \\ \dfrac{1}{SNR} & \text{if SNR} >> 1 \end{cases} \tag{9}$$

These considerations evidence how the noise properties of magnitude and phase signals are interrelated at high SNR levels. Although the theory by Gudbjartsson and Patz was developed for an MRI image acquired at a single time point, their observations are also fundamental for describing the magnitude and phase fluctuations that occur during an fMRI time series, described in section 4 (see also Hagberg et al., 2008). Keeping these fluctuations as small as possible is important in fMRI, since the statistical T-values depend on them. Less fluctuations lead to higher BOLD contrast-to-noise ratios and greater statistical T-values, yielding an enhancement of the fMRI detection sensitivity.

3.3. The blood-oxygenation level dependent effect and the phase signal

The contrast in a GE-EPIsequence is generated by the signal evolution between a 90º RF excitation pulse and signal read-out (that is the echo time, TE). While the magnitude signal follows an exponential decay, the phase is linearly accrued during TE (see Eqs. 5, 8). The BOLD effect causesachange in the magnetic susceptibility difference between veins and the surrounding tissue. During the activated state, the oxygenation fraction Y is greater than during baseline states, and therefore the susceptibility difference is diminished. The effect of BOLD on the magnitude signal isdriven by a change in the dispersion (variation) of the magnetic field inside the voxel, while a change in phase is driven by a coherent shift (a net change) of the magnetic field.

The magnitude/phase evolution during TE is influenced not only by the BOLD effect per se but also by the vessel geometry and the spatial orientations of the vessels with respect to the orientation of the static magnetic field of the scanner (see Eq. 10, Fig. 7). Ideally, only the BOLD effect arising in the capillary bed close to the site of neuronal activation should be detected, however larger draining veins also contribute to the activation maps.

In order to calculate the phase variation during BOLD, let's consider the simple case of a single vessel crossinga voxel. The vessel can be modelled as an infinitely long cylinder with a different, blood-oxygenation and hematocrit dependent susceptibility than the surrounding brain tissue. In this case, the signal depends on the orientation of the vessel with respect to the static magnetic field of the scanner. Vessels oriented parallel to the field, do not 'oppose' the field as much as perpendicularly oriented vessels and hence produce a magnetic field shift that is localized inside the vessel, while in the perpendicular orientation, the vesselsproduce both extra- and intra-vascular effects. These are described by the following equations (Boxerman et al., 1995):

$$\begin{aligned} \Delta B_{vessel} &= iv + ev \\ iv &= 2\pi \cdot \Delta\chi_o \cdot (1-Y) \cdot Hct \cdot B_o \cdot \left(\cos^2\theta - \frac{1}{3} \right) \\ ev &= 2\pi \cdot \Delta\chi_o \cdot (1-Y) \cdot Hct \cdot B_o \cdot \left(\frac{R}{r} \right)^2 \cdot \sin^2\theta \cdot \cos(2\Psi) \end{aligned} \tag{10}$$

where $\Delta\chi_0$ is the susceptibility difference between the extra (*ev*) - and intravascular (*iv*) space (in cgs units), *Y* the oxygenation level of the blood, *Hct* is the haematocrit level, $B_0$is the static magnetic field strength, *R* the vessel radius and *r* the distance from the centre of the vessel cylinder, and θ, Ψ are the angles defined in Fig 7.

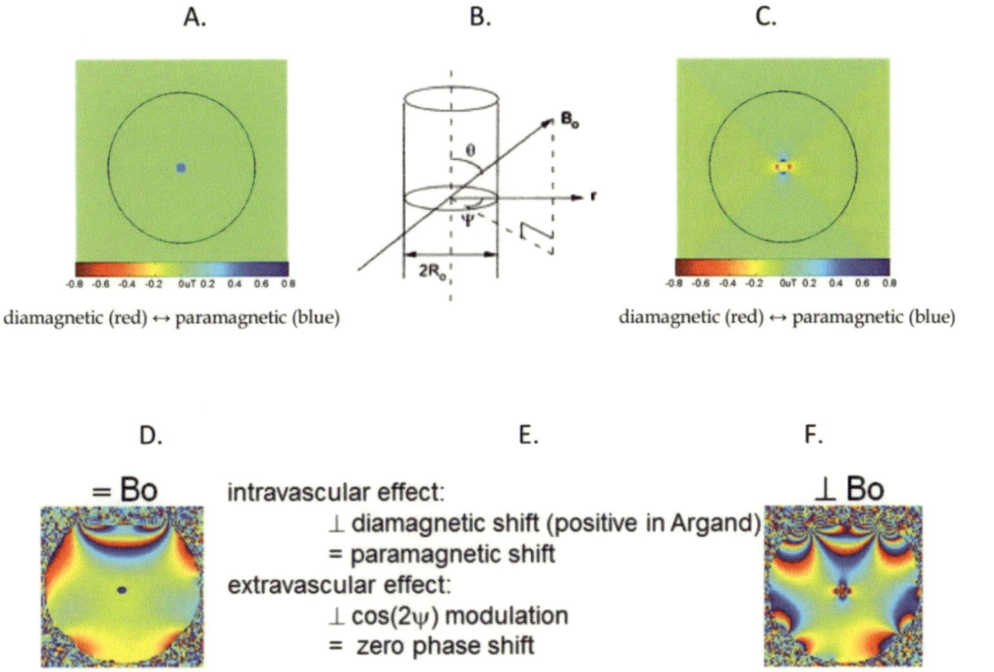

Figure 7. Biophysical cylinder model for a single vessel placed with its length axis oriented parallel (A,D, symbol: =) or perpendicular (C, F, symbol: ⊥) to the external magnetic field,B_0. B: The anglesθ, $\Psi$and the distance from the centre of the vessel, r,define the position for a point in space where a magnetic field is generated due to the blood-tissues susceptibility difference, as described in [Eq. 10]. A,C: Simulated magnetic field map for a cylinder placed in the centre of a sphere (first row: $\Delta\chi_0$=0.18ppm, hematocrit=0.4, oxygenation fraction Y=0.54). For a cylinder oriented parallel to the field, no phase modulation occur in the extravascular space (the field inside the cylinder is visible as a blue spot in the centre). For a cylinder oriented perpendicular to the external field, a cos2Ψ modulations occurs around the vessel. D,F: Measured MRI phase images (Magnetic field strength: 3T, GE-EPI, TE=20ms) for a cylinder with higher susceptibility than the surrounding water. The pattern corresponding to the dipole field is evident for the perpendicular orientation. Unwanted background variations of the field are present in the measured but not in the simulated data.E: Depending on the vessel orientation (= or ⊥) with respect to the external field, diamagnetic or paramagnetic field shifts can be observed.

In Fig. 7, phase images of a cylindrical vial containing a solution with higher magnetic susceptibility than the solution contained in the surrounding, outer area are shown for a parallel and a perpendicular orientation, with respect to the static magnetic field.The greatest extra-vascular effects are obtained for the perpendicular orientation, with a spatial cos2Ψ modulation around the vessel, and a diamagnetic phase shift inside the vessel. The size of the phase shifts will depend on the oxygenation fraction and on the geometrical factor, determined by the orientation of the vessels, as expressed by [Eq.10]. The expected phase values that occur during the resting and active conditions of an fMRI study can be obtained by straightforward calculations based on [Eq. 10]. For instance, during rest (when the oxygenation fraction is ca Y=0.54) the maximal field difference between the extra- and intravascular space amounts to ±0.62μT at 3T and appears in the tissue surrounding a perpendicular vessel (extravascular space). This value corresponds to a phase shift as high as ±4.98 radian if a

GE-EPI sequence with a TE of 30 ms is used. In presence of a BOLD response during the activated state (Y=0.68) the susceptibility difference between the blood and the tissue diminishes and therefore the maximal field difference becomes ±0.43μT (±3.45radian). In the intravascular space of a parallel vessel, the field undergoes a diamagnetic shift and is therefore reduced from 0.42 μT (3.37radian) at rest to 0.29 μT (2.33 radian) during activation. Consequently, the expected BOLD induced field change is maximal in the extravascular space for perpendicular vessels (±0.19μT, 1.52radian), while for the parallel case it is greatest in the intravascular space (-0.13μT, 1.04radian). These observations are also in line with previous work (Hoogenrad et al., 2001).

During an fMRI study the expected phase effect will be diminished due to the partial volume effects that arise when the vessel diameter is smaller than the measured voxel size (see Brainovich et al., 2010). In view of these phase shifts, Menon (2002) suggested that it should be possible to identify voxels dominated by draining vessels parallel to the field, due to the intravascular phase shift, while the contribution from perpendicular vessels should cancel out, due to signal modulation in the extravascular space that is averaged out as a consequence of partial volume effects.

The cortical region that encloses the activated neurons is characterized by a dense capillary network. Larger vessels that drain the activated region are more remotely located at the brain surface (e.g. Keller et al., 2011). The network consists of vessels with a range of different radii, and diffusion effects must also be considered to quantify the effect of the magnetic field distribution. This is not trivial to do and therefore calculations were initially done by Monte Carlo simulations (Yablonskiy and Haacke 1994) before an analytical expression was derived by Kiselev and Posse (1999, see Eqs. 5 and 25 in that paper). From these studies, it emerges that the strength of the BOLD- induced magnitude signal depends on the spatial vessel distribution, consistent with the fact that field dispersion dominates changes in the magnitude signal. On the other hand, if the single vessel model is used for the phase signal, positive and negative phase signals mutually cancel out for random vessel orientations, such as those found in the capillary network, while the phase signals add together for ordered orientations, as was recently explored in a simulation study (Chen and Calhoun, 2010).

These two models of the BOLD phase effect suggest that the interest of analyzing the phase signal resides in the possibility to exclude voxels dominated by draining veins, since predictions based on the summation of magnetic field changes in single vessels cancel out (Hoogenrad et al., 2001). A slightly different picture with regard to the BOLD effect and the variation of the phase signal emerges if the concept of the 'sphere ofLorentz effect' is taken into account (Durrant et al., 2003).Zhao et al., 2007 followed by Feng et al., 2009 showed by theory and by experiments that an extended activated brain area actually gives rise to a coherent phase shift. The sphere of Lorentz concept predicts the existence of two main effects of the BOLD response: local field changes caused by the red blood cells, and a bulk magnetic susceptibility shift which is further subdivided into a demagnetizing effect and a volume-averaged magnetization change (Durrant et al., 2003). The volume-averaged effect depends on the blood volume, the vessel geometry and the geometry of their ensuing magnetic field, as well as the size of the susceptibility effect in vessels located in the tissue surrounding the volume (Zhao et al., 2007). The spatial distribution of the phase effect that emerges from this model corresponds to a magnetic dipole and, similarly to the single vessel case, it depends on the geometry and orientation of the activated brain region with respect to the external field (Fig. 8).

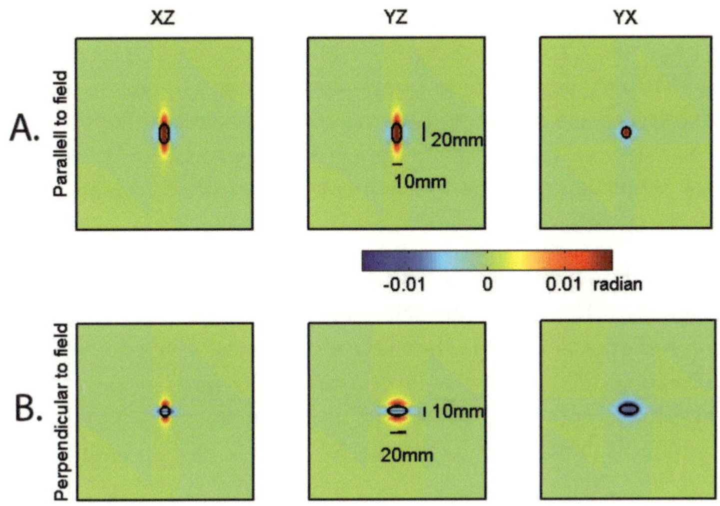

Figure 8. Simulation of the spatial pattern of the phase effect generated by the BOLD response in a brain area with an elliptic Gaussian shape (contours of the region are overlaid on the phase images). The external, static, magnetic field, B_0 is applied along the Z-direction, and the phase effect is shown in three planes: XZ, YZ, and YX. Similarly to the case of a single vessel, the phase effect depends on the orientation of the cortical regions with respect to B_0. A. When the longest axis of the region is oriented parallel to B_0, a paramagnetic shift that increases the local field occurs. B. When the longest axis of the region is oriented perpendicular to B_0, a diamagnetic effect that tends to decrease the field is generated.

In Fig. 7, phase images of a cylindrical vial containing a solution with higher magnetic susceptibility than the solution contained in the surrounding, outer area are shown for a parallel and a perpendicular orientation, with respect to the static magnetic field.The greatest extra-vascular effects are obtained for the perpendicular orientation, with a spatial $\cos 2\Psi$ modulation around the vessel, and a diamagnetic phase shift inside the vessel. The size of the phase shifts will depend on the oxygenation fraction and on the geometrical factor, determined by the orientation of the vessels, as expressed by [Eq.10]. The expected phase values that occur during the resting and active conditions of an fMRI study can be obtained by straightforward calculations based on [Eq. 10]. For instance, during rest (when the oxygenation fraction is ca Y=0.54) the maximal field difference between the extra- and intravascular space amounts to ±0.62μT at 3T and appears in the tissue surrounding a perpendicular vessel (extravascular space). This value corresponds to a phase shift as high as ±4.98 radian if a GE-EPI sequence with a TE of 30 ms is used. In presence of a BOLD response during the activated state (Y=0.68) the susceptibility difference between the blood and the tissue diminishes and therefore the maximal field difference becomes ±0.43μT (±3.45radian). In the intravascular space of a parallel vessel, the field undergoes a diamagnetic shift and is therefore reduced from 0.42 μT (3.37radian) at rest to 0.29 μT (2.33 radian) during activation. Consequently, the expected BOLD induced field change is maximal in the extravascular space for perpendicular vessels (±0.19μT, 1.52radian), while for the parallel case it is greatest in the intravascular space (-0.13μT, 1.04radian). These observations are also in line with previous work (Hoogenrad et al., 2001).

During an fMRI study the expected phase effect will be diminished due to the partial volume effects that arise when the vessel diameter is smaller than the measured voxel size (see Brainovich et al., 2010). In view of these phase shifts, Menon (2002) suggested that it should be possible to identify voxels dominated by draining vessels parallel to the field, due to the intravascular phase shift, while the contribution from perpendicular vessels should cancel out, due to signal modulation in the extravascular space that is averaged out as a consequence of partial volume effects.

The cortical region that encloses the activated neurons is characterized by a dense capillary network. Larger vessels that drain the activated region are more remotely located at the brain surface (e.g. Keller et al., 2011). The network consists of vessels with a range of different radii, and diffusion effects must also be considered to quantify the effect of the magnetic field distribution. This is not trivial to do and therefore calculations were initially done by Monte Carlo simulations (Yablonskiy and Haacke 1994) before an analytical expression was derived by Kiselev and Posse (1999, see Eqs. 5 and 25 in that paper). From these studies, it emerges that the strength of the BOLD- induced magnitude signal depends on the spatial vessel distribution, consistent with the fact that field dispersion dominates changes in the magnitude signal. On the other hand, if the single vessel model is used for the phase signal, positive and negative phase signals mutually cancel out for random vessel orientations, such as those found in the capillary network, while the phase signals add together for ordered orientations, as was recently explored in a simulation study (Chen and Calhoun, 2010).

These two models of the BOLD phase effect suggest that the interest of analyzing the phase signal resides in the possibility to exclude voxels dominated by draining veins, since predictions based on the summation of magnetic field changes in single vessels cancel out (Hoogenrad et al., 2001). A slightly different picture with regard to the BOLD effect and the variation of the phase signal emerges if the concept of the 'sphere ofLorentz effect' is taken into account (Durrant et al., 2003).Zhao et al., 2007 followed by Feng et al., 2009 showed by theory and by experiments that an extended activated brain area actually gives rise to a coherent phase shift. The sphere of Lorentz concept predicts the existence of two main effects of the BOLD response: local field changes caused by the red blood cells, and a bulk magnetic susceptibility shift which is further subdivided into a demagnetizing effect and a volume-averaged magnetization change (Durrant et al., 2003). The volume-averaged effect depends on the blood volume, the vessel geometry and the geometry of their ensuing magnetic field, as well as the size of the susceptibility effect in vessels located in the tissue surrounding the volume (Zhao et al., 2007). The spatial distribution of the phase effect that emerges from this model corresponds to a magnetic dipole and, similarly to the single vessel case, it depends on the geometry and orientation of the activated brain region with respect to the external field (Fig. 8).

Simulations of the BOLD effect thatincludes the sphere of Lorentz effect can be performed by a three-dimensional Fourier transform of the magnetic field distribution (Marques and Bowtell, 2005). Similar to the approach used by Feng et al. (2009), we performed such calculations for the BOLD response in a cortical region with an elliptic Gaussian field distribution described by:

$$\Delta B = -2\pi \cdot \Delta\chi_o \cdot \Delta Y_{cap} \cdot Hct \cdot B_o \cdot \left(\exp(-\frac{X^2}{2\cdot\sigma_x^2} - \frac{Y^2}{2\cdot\sigma_y^2} - \frac{Z^2}{2\cdot\sigma_z^2} \right)$$
$$\sigma_i = \frac{FWHM_i}{2\cdot\sqrt{2\cdot\ln(2)}} \quad (11)$$

where ΔY_{cap} is the change in oxygenation in the capillaries (0.08) due to BOLD, *X*, *Y* and *Z* are the Cartesian coordinates in 3D space, $FWHM_i$ expresses the spatial extent of the area in terms of the full width half maximum value for the coordinates in the *i*-direction, where *i=X,Y,Z;* and the other parameters are the same as in [Eq.10] and in Fig. 7. We varied the size of the activated region between 5-40mm (FWHM of the shortest length axis) and the strength of the BOLD response. Values for the BOLD induced magnetic field change were close to those observed above for the single vessel model (± 0.05-0.25 μT). The results from these simulations are shown in Figs. 8 and 9. We found that both the spatial extent of the activated area and the strength of the BOLD response influence the size of phase change. For instance, magnetic field shifts of ±0.1 μT lead to phase shifts in the range of 0.01-0.04 radian, depending on the spatial extent of the region. As shown in Fig. 9, the greatest phase effect was found in small areas with a strong BOLD response. The size of these expected phase changes are thus about two orders of magnitude lower than those found for fully sampled large draining vessels, despite similar size of the BOLD induced magnetic field change. This fact highlights the difference between biophysical models.

With regard to the time evolution of the phase effect, studies have shown that it closely follows that of the magnitude signal (Zhao et al., 2007). This observation is in agreement with a common BOLD origin for both the phase and magnitude signals.

As we have seen, three differentbiophysical models for describing the phase signal changes caused by the BOLD effect are available in the literature. The first modelis based onthe signal behaviour in large draining veins, the second in capillary networks, while the third model describes the BOLD phase effect caused by a spatially extendedcortical region. The spatial pattern of the BOLD effect is different for the phase than for the magnitude images, but from the literature we know that their temporal evolution is similar (Zhao et al., 2007; Feng et al., 2009).

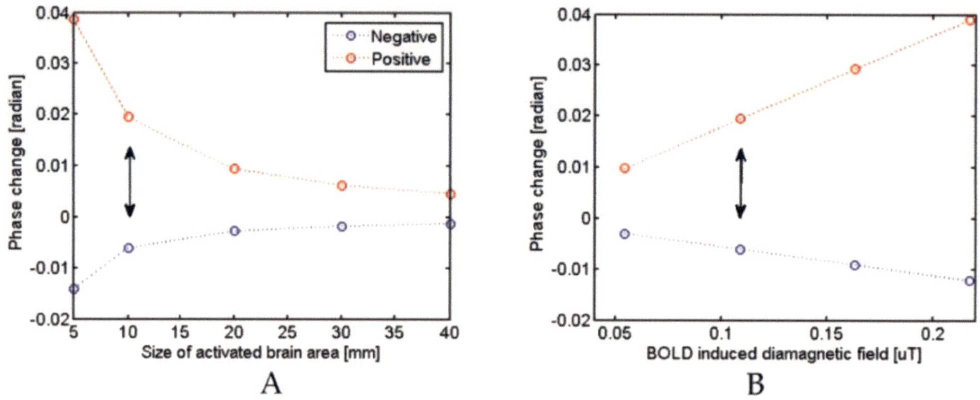

Figure 9. Simulations of the phase effect with positive (red) and negative (blue) sign generated by the BOLD response in a brain area, with an ellipsoid Gaussian shape. As in Fig. 9, the longest axis is oriented along B_0.A: Variation of the phase effect with the spatial extent of the activated area (indicated as the FWHM of the shortest length axes, see Eq. 11 and Fig.8). The BOLD response is fixed (±0.10 μT). The phase effect is greatest when the size of the region is small, but diminishes as the size of the activated region increases. B: Variation of the phase effect with the strength of the BOLD responses. The spatial extent of the activated brain area is fixed (FWHM=10mm). The positive (negative) phase increases (decreases) with increasing BOLD field changes. The arrow indicates equivalent points in Fig. 9 A and 9B where the FWHM is 10mm and the BOLD response has a field change of ±0.10 μT.

4. Unwanted phase effects and their removal

The phase value can be considered a fingerprint of the voxel averaged magnetic field. Therefore any effect that causes local or global dishomogeneities of the field leads to phase variations in space. Furthermore, during breathing and motion, the phase will vary from scan to scan in an EPI time series (Hagberg et al., 2008; Petridou et al., 2009). Since the expected BOLD-related phase changes are weak, correcting all unwanted phase fluctuations is of utmost importance.

MRI phase images are characterized by phase wrapping effects dueto an inherent limit of the MRI detection system: the range of phases that can be detected is limited to $\pm\pi$ and therefore phase wraps occur if this limit is exceeded. The true underlying phase value is thus equal to the observed phase ± an integer number of revolutions on the unit circle:

$$\varphi_{\text{true}} = \varphi_{\text{obs}} \pm 2\pi \cdot k \tag{12}$$

where k can take on any integer value. Such phase wraps occurin space and in time (Hagberg et al., 2008). For instance, close to air-tissue borders where the magnetic susceptibility difference is important, and therefore the phase value also varies greatly, especially if long TE times are used for the measurements, leading to phase jumps in the vicinity of such air-tissue borders (see [Eq.8], and Fig. 5). Temporal phase wraps may also occur in fMRI time series, as the magnetic field may vary with time due to physiological and scanner fluctuations.

4.1. Physiological and scanner noise in phase fMRI images

Unwanted field non-uniformities can be subdivided into static and dynamic magnetic susceptibility differences considering the duration of a 2D image acquisition and they are caused both by the subject and by the scanner (see Hagberg et al., 2008). Static effects are caused by factors that are considered to remain unchanged during the acquisition of a single image, while dynamic effects are caused by factors with a varying influence from slice to slice. Examples of static effects are imperfect shimming and slowly evolving field drifts, examples of dynamic effects are breathing, motion and system vibrations.

The magnetic susceptibility varies greatly across the brain, due to air-tissue bordersas well as due to more local susceptibility differences between the vessels,the sub-cortical structures, the grey and the white matter (Collins et al., 2002, see also the strong phase variations in brain regions just above the frontal sinuses in Fig. 5). When the brain is placed in an external magnetic field, these variations in magnetic susceptibility lead to field dishomogeneties. The scanner hardware compensates for some dishomogeneities by a procedure known as 'shimming', i.e. we try to make the field more uniform by activating the shim-coils. How good the shim can get depends on the hardware, and on some clinical scanners only linear correction terms are available, while other commercial systems are capable of performing corrections up to the 2nd order, therefore higher order field imperfections remain, despite the shimming procedure. Since the shim-values are generally set at the beginning of the fMRI experiment they do not compensate for field variations during scanning caused by respiration, that modifies the susceptibility difference between the lungs and the brain, see Ray et al., 2000, and by small micromovements of the brain during the respiratory and cardiac cycles (Le and Hu 1996; van Gelderen et al., 2007). These effects can be noticeable also in the fMRI magnitude signal (see Fig. 10), and must therefore be regressed out if more subtle effects are to be studied, like for instance in resting state studies. New hardware developments may in the future make such corrections unnecessary. One possibility is to change the shim-currents on-the-fly during the fMRI experiment, so that the effects caused by respiration are compensated (van Gelderen et al., 2007). Another possibility is to use magnetic field monitoring during image acquisition (Barmet et al., 2009). This information can either be used for image post-processing or for re-calibration of the shim-currents (Vannesjo et al., 2012).

Not only the subject, but also the scanner contributes to static effects, linked with the aforementioned imperfect shimming issue and slow magnetic field drifts. Time-dependent dynamic effects are caused by helium-pump related system vibrations, and concomitant field gradients arisingfrom Maxwell effects, eddy-current fields, and gradient non-linearities (Hagberg et al., 2008). In a typical fMRI time-series, the scan-to-scan variability of most of these factors can be neglected, except for system vibration, magnetic field drift and heating of the gradient system (Foerster et al., 2005).

4.2. Removal of unwanted phase effects

The unwanted phase effects that originate from different sources can be removed by different post-processing approaches that will be briefly described. We exemplify the effects of these methods on some data sets we acquired (Hagberg et al., 2008; 2012).

Phase wraps, defined in [Eq. 12], can be efficiently removed by performing phase unwrapping. Unwrapping is a straightforward process if it is applied across the time dimension, independently for each voxel in the fMRI time series (Hagberg et al., 2008). Alternatively, spatial phase unwrapping can be applied (e.g. Cusack and Papadakis, 2002; Jenkinson, 2003). Finally, the reference phase method has been proposed, that effectively removes phase wraps by referencing subsequent phase images to the first phase image acquired in the fMRI time series (Tomasi and Caparelli, 2007).

Logically, sources of unwanted phase effects must be unambiguously identified and characterized before removal. Recent studies have shown, that the dominating source of phase variations during fMRI time courses are induced by respiration and the cardiac beat (Petridou et al., 2009; Hagberg et al., 2008; 2012). The importance of respiratory and cardiac effects on the phase data becomes clear from Fourier analysis of the fMRI time series data (Fig.10). After spatio-temporal phase unwrapping, the physiological noise components are slightly less pronounced than after temporal unwrapping alone, while the reference phase method seems more efficient for removal of the respiratory component. The magnitude data is also affected by respiration and the heartbeat, and a very popular method for removal of physiological noise from magnitude data is RETROICOR (Glover et al, 2000). It uses the respiratory and cardiac traces to create regressors for noise removal. The benefit of this approach for phase fMRI time series acquired at 7T has been demonstrated (Petridou et al., 2009).

For long it has been known that information regarding respiration and heartbeat can be extracted from the phase value of the central k-space point (Hu et al., 1995; Le and Hu 1996). One possibility is thus toperform noise regression based on the phase of the central k-space point, instead or in combination with RETROICOR (Hagberg et al., 2012).

Other techniques specifically aim to compensate the scan-to-scan off-resonance effects in k-space. Available techniques are the DORK (acronym for dynamic off-resonance in k-space) and TOAST (temporal off-resonance alignment of single-echo time series technique) methods. In DORK, the global field-changes are estimated by the phase evolution of the central k-space point between subsequent TRs and then each read-out line is corrected in k-space assuming a linear phase accrual in time (Pfeuffer et al., 2002). In TOAST, phase rewinding is performed based on a voxel wise estimate of the B_0 changes from the smoothed difference between the phase at each single time point and the average phase in the entire fMRI time series (Hahn et al., 2009). In the next paragraph, we will show how these methods compare in terms of eliminating physiological noise.

Finally, an alternative method that we have explored is based on spatial high-pass filtering. In the human brain at 3T, we observed that the unwanted phase effects were characterized by a low spatial frequency (Hagberg et al., 2012). This characteristic is present also for phantom measurements as can be seen in Fig. 7. We found that the phase fluctuations became

http://dx.doi.org/10.5772/31782

similar to the ones observed in magnitude data in healthy human subjects,after application of Gaussian filters that removed all components extending 20mm or more (Hagberg et al., 2012). Fourier analysis also shows that the scanner Helium pump artefact at 2Hz and the respiratory artifact are completely removed by spatial filters, while some cardiac pulsation effects remained (Fig. 10, Hagberg et al., 2012).

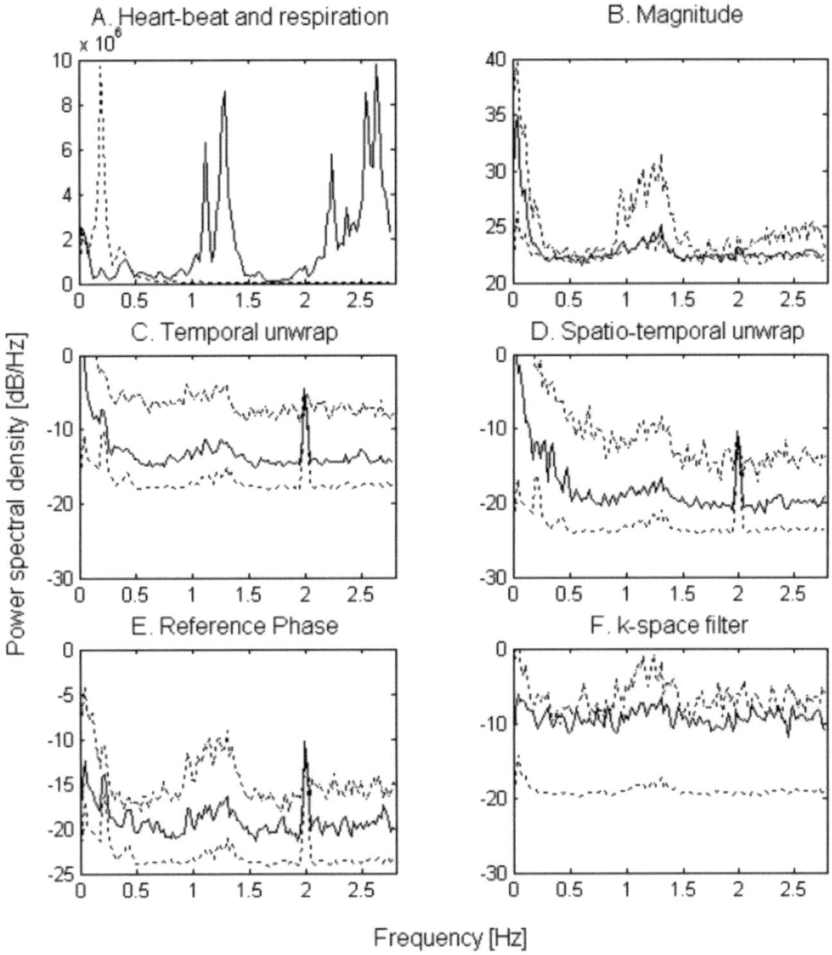

Figure 10. Fourier analysis of the heart beat and the respiratory cycle (A) and their relation to the MRI signal in magnitude (B) and phase (C-F) images. A: Respiration occurs between 0.12 and 0.25 Hz, and the first and second harmonics of the heart-beat around 1.2 and 2.4 Hz, respectively. B: In magnitude images, respiration induced fluctuationscause the typical 1/f decrement at low frequencies, but instabilities are also prominent around the first cardiac harmonic. Note the relative absence of disturbances around 2Hz, caused by the scanner Helium pump and visible in most of the phase time-series. Results are shown for gray matter (solid line), white matter (dotted line), and CSF (dashed line). C: temporal unwrap and detrending of the phase time-series (Hagberg et al., 2008); D: spatial unwrapping followed by temporal unwrapping and detrending, (Hagberg et al., 2008); E: Reference phase method (Tomasi and Caparelli, 2007); F: k-space filtering (Hagberg et al., 2012).

The efficiency of each technique for removal of phase instabilities can be evaluated and compared with results from the magnitude data. Krueger and Glover (2001) proposed a model that quantifies the available temporal stability of magnitude data, *tSNR* based on the factor λ_M that expresses the proportionality between the magnitude signal and the physiological noise. Basically this model implies that the available temporal signal-to-noise level reaches a maximum limiting value for high signal-to-noise-ratios at a single time point, SNR_0, calculated from the average signal in each voxel across the GE-EPI timecourse, *S*, divided by the Rayleigh corrected noise level, *N*:

$$SNR_o = \frac{S}{1.53 \cdot N}$$
$$tSNR = \frac{SNR_o}{\sqrt{1+\lambda_M{}^2 \cdot SNR_o{}^2}} \tag{13}$$

In accordance with the Gudbjartsson and Patz relationship between the noise in magnitude and phase images (Eq. 9), we can derive three models for the temporal fluctuations in phase images. In the first model, the noise components induce greater fluctuations in the phase than in the magnitude images, in the second model the contributions are similar in both image types and in the last model the noise is less pronounced in phase than in magnitude images, following the relationship (Hagberg et al., 2008):

$$tSD_\varphi = \frac{1}{tSNR} = \frac{\sqrt{1+\lambda_\varphi{}^2 \cdot SNR_0{}^2}}{SNR_0} \tag{14}$$

where $\lambda_M < \lambda_\varphi$; $\lambda_M = \lambda_\varphi$; or $\lambda_M > \lambda_\varphi$, respectively for the three noise models.

The *tSNR* obtained for the magnitude signal and $1/tSD_\varphi$ images obtained after different phase processing methods are shown in Fig. 11 and illustrates that the phase noise is generally greater than the magnitude noise, but can be efficiently reduced by adequate post-processing methods. (Fig. 11, see also Hagberg et al., 2012). Spatial high-pass filtering currently represents the best approach for removal of unwanted phase fluctuations. If adequately adapted, similar or even reduced λ_φ values are obtained for the phase compared to magnitude data. In general, the amount of filtering must be adapted to the field strength of the scanner, the echo time and voxel size of the GE-EPI sequence. For instance, at 3T, a TE of 30ms, and a voxel size of $3x3x3mm^3$ we found that a Gaussian filter with a 20mm FWHM suffices to reach the a similar stability in both phase and magnitude time series (Hagberg et al., 2012). Filters with a FWHM close to the voxel size, reduces the phase noise to less than half the magnitude noise. For most other post-processing methods, including spatio-temporal unwrapping, regression based on the phase evolution central k-space point (NVR), RETROICOR, and TOAST, the phase noise was always greater than the magnitude noise.

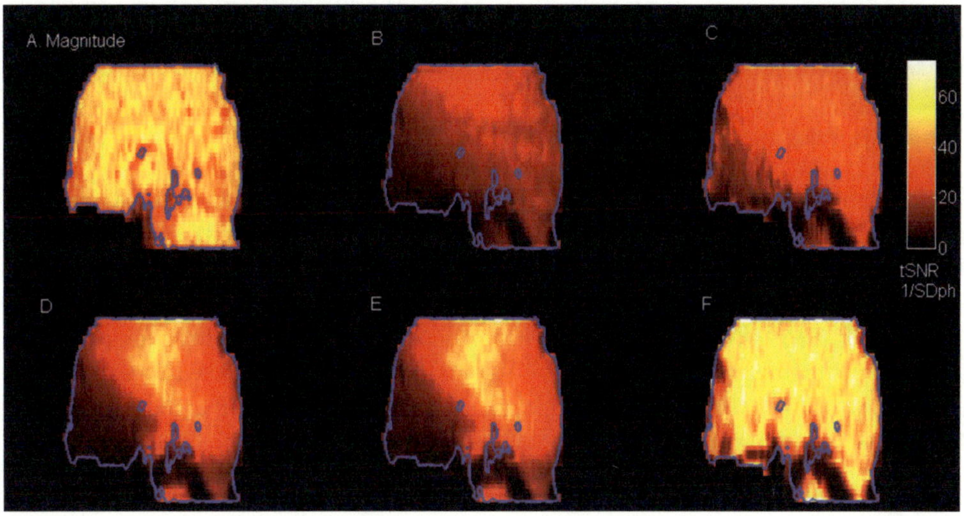

Figure 11. Temporal stability in magnitude (A) and phase (B-F) images. The phase images were post-processed by different methods prior to calculation of the inverse of the standard deviation of the phase across the time series (see Eq. 14). (B) RETROICOR; (C) TOAST; (D) noise regression based on the phase of the central k-space point; (E) combined approach based on both thecentral k-space point and RETROICOR; (F) : k-space High pass filtering of the phase data by a Homodyne filter based on a Gaussian kernel in k-space (Hagberg et al., 2012).

5. On the use of phase information in fMRI studies

The phase signal possesses many interesting features that may be used to extract valuable information without the need for additional scanning time. The question that remains is howthe phase information best is used in order to improve currently available standard fMRI methods. At the time being, no clear-cut answer emerges in the literature. Ongoing work will probably give us an answer in the near future. What follows is a brief overview of some possible approaches that are currently being explored. In the last paragraph, we will also show an example of the outcome of an fMRI analysis performed for both the magnitude and the phase signals within the framework of the standard General Linear Model, using different phase post-processing methods.

We have seen how unwanted phase fluctuations due to respiration or heartbeat are well visible in the phase data. The phase can thus be used to identify and characterize the temporal evolution of different noise sources. This information can be used to remove contributions from common noise sources from the magnitude data sets.

We have seen how the BOLD response influences the phase signal as described by different biophysical models. The dipole of the ensuing magnetic field effect is manifest at different length-scales, depending on whether draining veins or microvascular networks drive the response. Indeed, a phase effect arises both at the level ofsingledraining vessels with diameters of 100-500μm and inextended cortical regionswith BOLD activation. Therefore, the phase signal may be used to exclude single voxels dominated by draining veins (Menon, 2002) or to potentiate the statistical fMRI analysis based on magnitude data (Rowe and Logan, 2004).

5.1. Improving the magnitude signal: noise reduction

Physiological noise is a known limiting factor in fMRI studies (Krüger and Glover, 2001). Improving temporal stability is desirable, as it directly influences the statistical T-values. As pointed out above, studies show that the phase evolution at the centre of k-space in fMRI time series closely reflects field variations due to respiration and the cardiac beat (Hu et al, 1995, Le et al., 1996). Therefore the phase signal can be used as a proxy for physiological (and scanner) fluctuations. The magnitude signal could then be de-noised by simply regressing out unwanted fluctuations (Hagberg et al., 2012). This kind of use of the phase data may thus improve the outcome of fMRI analysis based on magnitude data.

The improvement in temporal stability of the magnitude signal was evaluated for different post-processing strategies (Hagberg et al., 2012). The standard RETROICOR method is based on signal regression (Glover, 2000). The onset of respiration and heart beat is obtained from measurements with a respiration belt and a pulse oxymeter synchronized to the acquisition of the fMRI time series. A set of 8 or more regressors are then created and used for signal regression and de-noising of the magnitude signal. In our approach, that we termed phase based nuisance variable regression (NVR, Hagberg et al., 2012), we used the phase evolution in the centre of the k-space as the only noise regressor. We also evaluated the possibility to combine the RETROICOR noise regressors with the regressor based on the phase evolution in central k-space (Table 1 and Fig. 12). We found that the temporal stability of the magnitude fMRI time course improved by 5% after NVR, which was slightly less than RETROICOR (Hagberg et al., 2012). It should be noted that RETROICOR is based on the cardiac and respiratory traces that must be acquired in synchrony with the fMRI time course, while NVR can be applied without this information. In addition, we found that a combination of RETROICOR and phase-based magnitude correction could improve the overall signal stability more than RETROICOR alone.

Tissue	Raw magnitude data	RETROICOR	NVR	RETROICOR and NVR combined
Gray matter voxels	0.0102±0.0042	0.0083±0.0041	0.0093±0.0041	0.0079±0.0039
White matter voxels	0.0056±0.0022	0.0040±0.0024	0.0050±0.0022	0.0035±0.0026
CSF voxels	0.0199±0.0076	0.0159±0.0066	0.0176±0.0065	0.0150±0.0058

Table 1 Stability in magnitude images in terms of the physiological noise parameter λ_M (see Eq. 13) for raw images and after noise-regression.

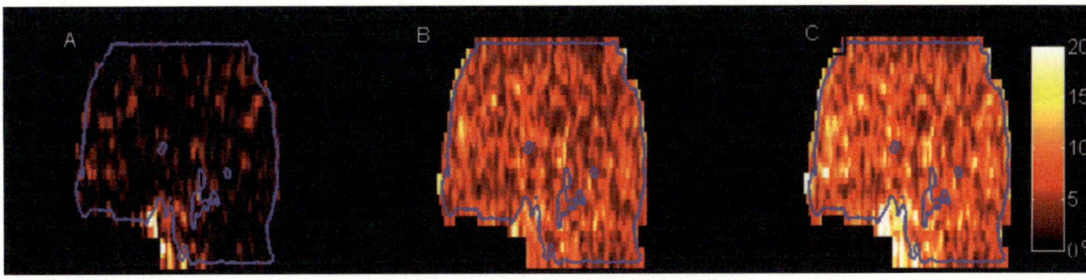

Figure 12. Increase in temporal stability of the magnitude signal for different magnitude post-processing methods. The data shown is the percent increase in temporal signal-to-noise ratio with respect to no post-processing for (A)after NVR; noise regression based on the evolution at the centre of k-space; (B) after RETROICOR; and (C) after a combined approach based on both NVR and RETROICOR methods. Data were obtained during a 5min acquisition of resting-state fMRI data. A sagittal plane showing the results for all acquired transverse slices is shown.

5.2. Statistical analysis methods incorporating the phase effect

Basically, two methods that incorporate phase signals in the fMRI analysis have been suggested. The first method aims at excluding those voxels from the analysis where the phase changes are great. These voxels are most probably related to draining veins (Menon 2002). The second method is the constant phase model of Rowe and Logan (2004) that extends the General Linear Model GLM to complex values. This model was further refined by Rowe (2005) who combined the real and the imaginary terms of the MRI signal into a single matrix expression. The General Linear Model which is used to evaluate magnitude images is:

$$abs(y_t) = X \cdot \beta + \eta \tag{15}$$

For complex numbers, GLM can be rewritten as (Rowe, 2005):

$$\begin{bmatrix} \mathrm{Re}(y_t) \\ \mathrm{Im}(y_t) \end{bmatrix} = \begin{bmatrix} \cos\varphi_t & 0 \\ 0 & \sin\varphi_t \end{bmatrix} \cdot \begin{bmatrix} x & 0 \\ 0 & x \end{bmatrix} \cdot \begin{bmatrix} \beta^{re} \\ \beta^{im} \end{bmatrix} + \begin{bmatrix} \eta^{re}{}_t \\ \eta^{im}{}_t \end{bmatrix} \tag{16}$$

where y_t is the complex valued fMRI time course, φ_t is the phase at each time point, x, $X$are the design matrices for the complex vector and the magnitude signals, respectively;$\beta_t{}^{re}$, $\beta_t{}^{im}$, β the parameter estimates obtained after model fitting; and$\eta_t{}^{re}$, $\eta_t{}^{im}$, η the noise of the real, imaginary and magnitude signals.

Examples of using this model indicate the difficulties that may arise when assessing fMRI results (Rowe 2005). Indeed at least three sets of maps showing activated voxels can be identified, dependent on which signals show significant activations: magnitude only (MO), phase only (PO), or activations in both magnitude and phase images (M&P). Considering the dipolar pattern that the BOLD response has (see Fig. 8), both positive and negative phase changes occur. Therefore we actually end up with as much as 5 sets of activation maps: MO; PO_positive; PO_negative, M&P_negative; M&P_positive (Rowe, 2005; Arja et al., 2010). At the time being, no method is available that brings the analysis of these maps one step further, so that a single activation map showing where in the brain activation takes place with a high sensitivity and specificity, is available. For instance, it is not clear what a PO activation which is visible in the phase but not in the magnitude images actually means. As we have seen in the earlier paragraphs, such an effect could be generated by a draining vessel, or by the sphere of Lorentz effect of the activated cortical patch, or perhaps by an unwanted phase effect that was not completely removed prior to analysis. It is desirable that methods that allow us to combine the activation maps are developedin the future. Such an approach could be based on the available biophysical models for the BOLD effect in phase images, perhaps even by performing quantitative susceptibility mapping (Balla et al., 2012), and may in the future bring us closer to the goal of correctly assigning the cerebral networks where the neuronal activity has its origin at the highest specificity and sensitivity.

5.3. On the use of phase images in an fMRI study

In the following, we will show an example of an fMRI study that incorporates both phase and magnitude images. The goal of the analysis was to investigate possible global phase effects according to the models that describe the phase signal changes caused by the BOLD effect. In order to make sure that we adequately handle unwanted phase effects, different approaches for removal of these were explored and their impact on the final activation maps were investigated.

Five healthy right handed subjects (three females and two males, 24-27 years) volunteered to participate in the fMRI study (approved by the local IRB). They performed index finger tapping in synchrony with a visual stimulus flashing at 1.7 Hz, in a block design paradigm: left hand tapping for 18s, 9s rest; right hand tapping for 18s, 9s rest; repeated 4 times. Complex valued BOLD fMRI time series were collected (Gradient echo EPI, 75 volumes, TR=3s, TE=39 ms, slice thickness: 3 mm and a 50% gap between 24 slices, bandwidth: 3396 Hz/px, voxel size: 3 x 3 x 4 mm3). Phase wraps and unwanted phase effects related to spatial static magnetic field gradients were removed by three different methods: a) spatio-temporal phase unwrapping, b) pixel wise noise regression based on the phase of the central k-space point and c) spatial high-pass filtering. For the spatial filtering method, a homodyne k-space approach was used based on the following three window functions: Hanning window, Gaussian window (with three different window widths: 5, 10 and 20mm) and a spline window, that was adapted to the magnitude k-space data of each image slice and each timepoint in the fMRI time series (Hagberg et al., 2012). Normalization to the MNI brain space was based on the magnitude image and subsequently applied to the phase images, and conventional spatial smoothing of the pre-processed images was applied to both phase and magnitude images, prior to statistical analysis using SPM5. Statistical analysis was first performed at the single-subject level, then the contrast images were included in a second-level group analysis to reveal the common activation pattern. We thus did not specifically use the Rowe and Logan model in [Eq. 16], but could rely on standard statistical analysis within the framework of the General Linear Model, using standard statistical methods.

Activation patterns for magnitude and phase images achieved after post-processing and statistical analysis are shown in Fig. 13. No significant activations were found in the phase data after a) spatio-temporal unwrapping or b) phase-regression. Only after post-processing by c) spatial homodyne high-pass filtering, significant activation patterns were obtained. Voxels with significant magnitude changes were located in the primary motor cortex. Some of these voxels also showed significant positive changes of the phase, however most voxels with a significant phase effect were located outside M1. In general, the parameter estimates for the phase signal were smaller than for the activations revealed in the magnitude data, leading to lower statistical Z and T-values (Table 2). The T-values for the magnitude images reached levels of 15, while the T-values for the phase data depended on which spatial homodyne window function that had been used. Satisfactory values of the statistical T-value (≥ 8) were obtained with the Homodyne approach using a Hamming window and a Gaussian function with FWHM=20mm, which most probably removes all unwanted phase effects at this field strength and for this echo time (Hagberg et al., 2012). The statistical parametric maps (SPMs) achieved with three of the five filters could be thresholded with $p<0.001$ (uncorrected for multiple comparison). The other two sets of data were thresholded with $p<0.1$ uncorrected, in order to reveal the significant activations (Fig. 13).

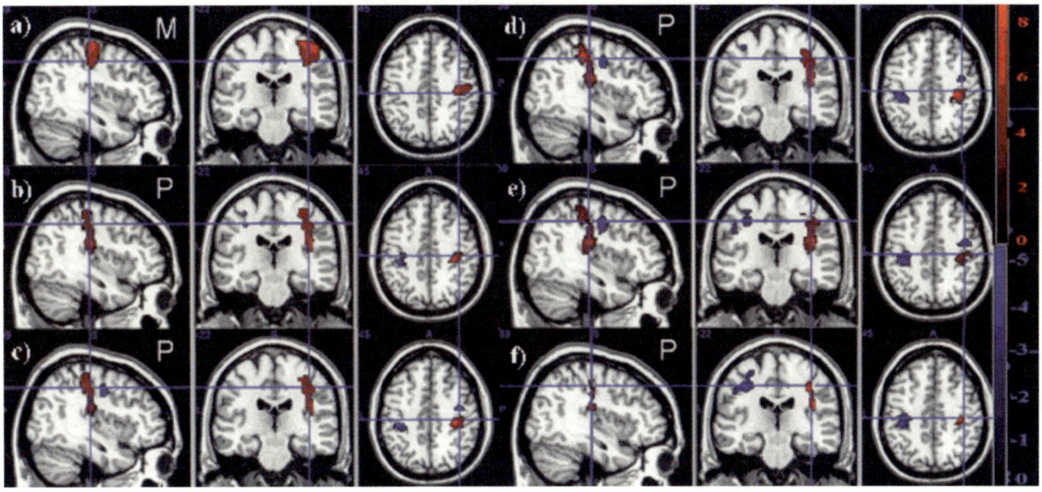

Figure 13. Activation patterns for a motor task obtained from a) magnitude (M) and b-f) phase (P) images. Unwanted phase effects were removed by five different homodyne (spatial high-pass) filters: b) Hamming window; c) Gaussian 20mm; d) Gaussian 10mm; e) Gaussian 5mm and f) k-space spline window. The statistical maps have been thresholded at a statistical significance level of $p<0.001$(uncorrected for multiple comparisons). Color bars indicate statistical T-value ranges.

	Cont	# voxels in cluster	Tvalue	Zvalue	p-value	MNI coordinates
			Phase			
Hamming n=32	Pos	Clmax 659	8.57	5.08	p_u<0.001	34, -26, 46
	Neg	85	5.84	4.16	p_u<0.001	38, -6, 42
Gauss FWHM 20mm	Pos	Clmax 792	8.04	4.93	p_u<0.001	32, -26, 48
	Neg	92	5.74	4.11	p_u<0.001	38, -6, 40
Gauss FWHM 10mm	Pos	Clmax 841	7.80	4.86	p_u<0.01	42, -20 22
	Neg	Clmax 238	5.38	3.96	p_u<0.01	38, -6, 44
Gauss FWHM 5mm	Pos	Clmax 1064	6.61	4.46	p_u<0.01	34, -26, 44
	Neg	Clmax 381	5.02	3.79	p_u<0.01	-40, -30, 52
k-spline filter	Pos	Clmax 606	8.62	5.10	p_u<0.001	42, -20, 18
	Neg	Clmax 213	5.28	3.90	p_u<0.001	-40, -30, 42
			Magnitude			
Raw data	Pos	Clmax 950	15.06	6.37	FWE=0.5	52, -14, 54

Table 2 Results obtained in a second level analysis. T and Z statistical values and size of the activated clusters (agglomerate of activated voxels) for magnitude and phase images are reported. Corresponding p-values and Montreal Neurological Institute (MNI) coordinates of their centre are shown for positive (Pos) and negative (Neg) phase changes, and for the positive magnitude activation.

After identification of voxels with significant signal changes in the second-level group analysis, the timeseries were extracted from these voxels for each subject. In Fig. 13, the extracted phase and magnitude timeseries, in brain regions with significant, stimulus correlated signal changes, are shown. In agreement with the work by Zhao et al. (2007), the time evolution of the phase and magnitude signals closely follow each other. The maximal change of the magnitude signal is 1-2% while the maximal phase change is 0.005-0.010 radians. The contrast-to-noise ratio was smaller for the phase than the magnitude data, both for the negative and for the positive phase changes. With respect to the size of the observed phase changes these were in agreement with those found for the simulated data (Figs. 8 and 9). The size of the activated clusters was 1000 voxels or less, corresponding to a region with an extension (FWHM) of 12mm. For this size, the observed phase changes of 0.005-0.01 radian would correspond to a diamagnetic field shift of ca 0.05-0.10 μT or less, which in turns corresponds to a change in the oxygenation fraction of ca 0.04-0.08 in a capillary network.

This preliminary study suggests how the phase information can be incorporated into the fMRI analysis. In a previous study, (Hagberg et al., 2008) we showed that it is possible to identify voxels with large phase changes in draining veins, by strong positive and negative phase changes close to the brain surface, where the largest vessels are found (Menon 2002). In the present study we used the same set of data but with a different post-processing approach, that included phase post-processing by Homodyne high pass filtering followed by normalization to a standard brain space and spatial smoothing of the final images. This approach allowed to remove contributions from strong positive and negative phase changes in draining vessels and enabled the detection of quite subtle phase changes. These results are consistent with a BOLD response in an extended brain area that generates a magnetic dipole effect (Durrant et al., 2003; Zhao et al., 2007).

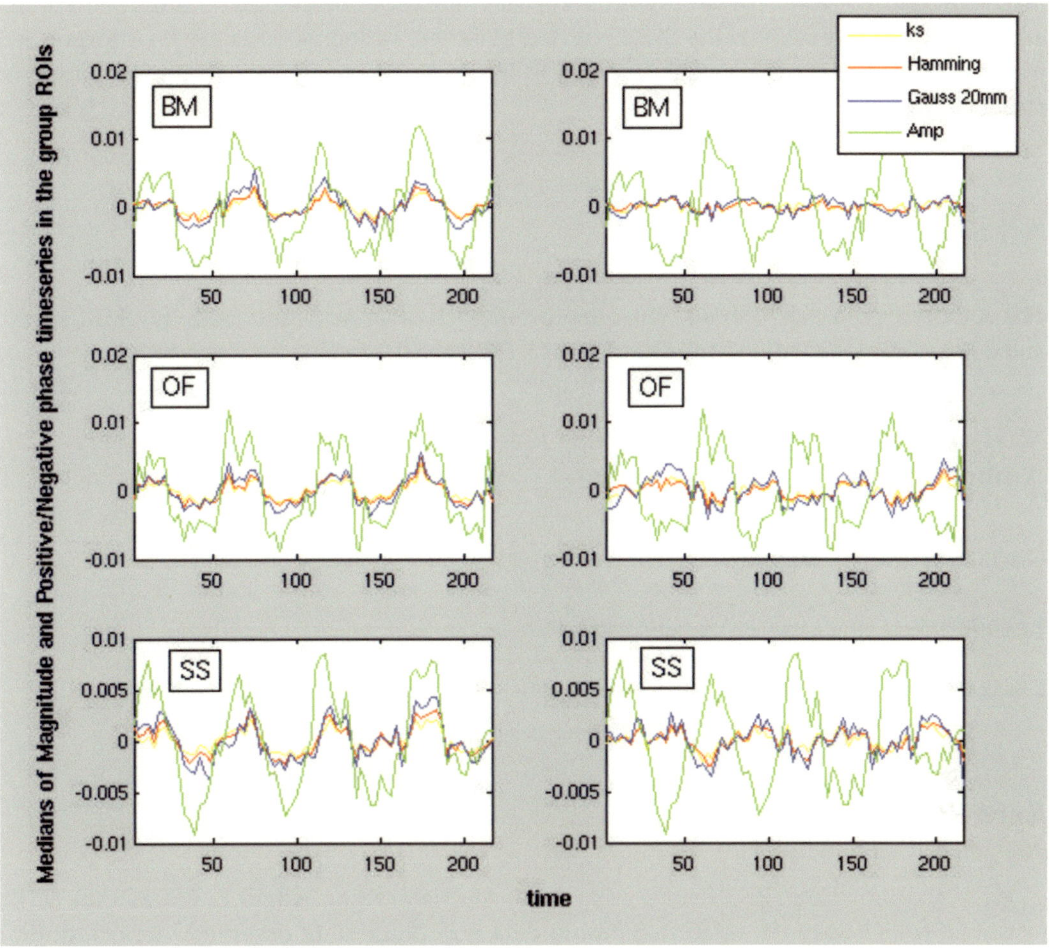

Figure 14. Timeseries extracted from brain regions with significant, stimulus correlated signal changes in three different subjects (SS, BM, and OF). Positive (left) and negative (right) phase signals are shown, overlaid on the magnitude signal (Amp, green). The phase signal clearly has a lower contrast-to-noise ratio than the magnitude signal. The temporal evolution of the phase and magnitude timeseries are similar.

6. Conclusion

Functional MRI methods based on the evaluation of magnitude image data may benefit from the inclusion of the information conveyed by the phase images. For instance, it is possible to directly improve the quality of the magnitude data by noise regression derived from the phase signal. In order to reveal subtle BOLD related activations in phase images, the many unwanted phase effects must first be accounted for. Spatial high-pass filtering that removes the spatially slowly evolving B_0 signal components currently represent the best approach for achieving this goal. We show that phase changes consistent with a dipolar activation pattern caused by an extended cortical region can be detected. Future work must address the question of how the BOLD related phase activation patterns can be incorporated in the statistical analysis pipeline, in order to improve the specificity and the sensitivity of fMRI.

Acknowledgements

The authors would like to thank Valentina Brainovich, who helped us with the fMRI study, and a grant from the Italian Ministry of Health (RF05.103).

Author details

Gisela E. Hagberg[1,2] and Elisa Tuzzi[1]

1 Fondazione Santa Lucia IRCCS, Rome, Italy

2 Max Planck Institute for Biological Cybernetics, Tübingen, Germany

References

[1] Arja, S. K., Feng, Z., Chen, Z., Caprihan, A., Kiehl, K. A., Adali, T., & Calhoun, V. D. (2010). Changes in fMRI magnitude data and phase data observed in block-design and event-related tasks. *Neuroimage*, 49(4), 3149-3160.

[2] Balla, D., Sanchez-panchuelo, R. M., Wharton, S., Hagberg, G. E., Scheffler, K., Francis, S. T., & Bowtell, R. (2012). Functional quantitative susceptibility mapping (fQSM). Proc Intl Soc Mag Reson Med. 20 :325.

[3] Barmet, C., De Zanche, N., & Pruessmann, K. P. (2008). Spatiotemporal magnetic field monitoring for MR. Magn Reson Med. , 60(1), 187-97.

[4] Bernstein, M. A., Thomasson, D. M., & Perman, W. H. (1989). Improved detectability in low signal-to-noise ratio magnetic resonance images by means of a phase-corrected real reconstruction. Med Phys. , 16(5), 813-817.

[5] Boxerman, J. L., Hamberg, L. M., Rosen, B. R., & Weisskoff, R. M. (1995). MR contrast due to intravascular magnetic susceptibility perturbations Magn Reson Med.; , 34(4), 555-566.

[6] Brainovich, V., Sabatini, U., & Hagberg, G. E. (2009). Advantages of using multiple-echo image combination and asymmetric triangular phase masking in magnetic resonance venography at 3 T. Magn Reson Imaging 27(1) , 23 EOF-37 EOF.

[7] Bretthorst, G. L. (2008). Automatic phasing of MR images. Part II: Voxel-wise phase estim. ationJournal of Magnetic Resonance , 191-193.

[8] Chen, Z., & Calhoun, V. D. Magnitude and Phase Behavior of Multiresolution BOLD Signal. (2010). Conc Magn Reson B 37B(3) , 129 EOF-145 EOF.

[9] Cusack, R., & Papadikis, N. (2002). New robust 3-D phase unwrapping algorithms: application to magnetic field mapping and undistorting echoplanar images. *Neuroimage*, 16, 754-764.

[10] Durrant, C. J., Hertzberg, M. P., & Kuchel, P. W. (2003). Magnetic susceptibility: further insights into macroscopic and microscopic fields and the sphere of Lorentz. ConcMagn. Reson. 18A(1): 72- 95.

[11] Feng, Z., Caprihan, A., Blagoev, K. B., & Calhoun, V. D. (2009). Biophysical modeling of phase changes in BOLD fMRI. *Neuroimage*, 47(2), 540-548.

[12] Foerster, B. U., Tomasi, D., & Caparelli, E. C. (2005). Magnetic field shift due to mechanical vibration in functional magnetic resonance imaging. Magn Reson Med 54(5) , 1261 EOF-7 EOF.

[13] Glover, G. H. (1999). Deconvolution of Impulse Response in Event-Related BOLD fMRI. *NeuroImage*, 9, 416-429.

[14] Glover, G. H., Li, T. Q., & Ress, D. (2000). Image-based method for retrospective correction of physiological motion effects in fMRI: RETROICOR. Magn Reson Med 44(1), , 162 EOF-7 EOF.

[15] Gudbjartsson, H., & Patz, S. (1995). The Rician distribution of noisy MRI data. Magn Reson Med 34(6) , 910 EOF-4 EOF.

[16] Hagberg, G. E., Bianciardi, M., Brainovich, V., Cassarà, A. M., & Maraviglia, B. (2008). The effect of physiological noise in phase functional *magnetic resonance imaging*from blood oxygen level-dependent effects to direct detection of neuronal currents. Magn Reson Imaging 26(7), , 1026 EOF-1040 EOF.

[17] Hagberg, G. E., Welch, E. B., & Greiser, A. (2010). The sign convention for phase values on different vendor systems: definition and implications for susceptibility-weighted imaging. Magn Reson Imaging. , 28(2), 297-300.

[18] Hagbcrg, G. E., Bianciardi, M., Brainovich, V., Cassarà, A. M., & Maraviglia, B. (2012). Phase stability in fMRI time series: effect of noise regression, off-resonance correction and spatial filtering techniques. *Neuroimage*, 59(4), 3748-3761.

[19] Hahn, A. D., Nencka, A. S., & Rowe, D. B. (2009). Improving robustness and reliability of phase-sensitive fMRI analysis using temporal off-resonance alignment of single-echo timeseries (TOAST). *Neuroimage*, 44(3), 742-52.

[20] Hoogenraad, F. G., Pouwels, P. J., Hofman, M. B., Reichenbach, J. R., Sprenger, M., & Haacke, E. M. (2001). Quantitative differentiation between BOLD models in fMRI. Magn Reson Med. , 45(2), 233-246.

[21] Hu, X., Le, T. H., Parrish, T., & Erhard, P. (1995). Retrospective estimation and correction of physiological fluctuation in functional MRI. Magn Reson Med 34(2) , 201 EOF-12 EOF.

[22] Jenkinson, M. (2003). Fast, automated N-dimensional phase-unwrapping algorithm. Magn Reson Med. 49(1), 193-197.

[23] Kiselev, V. G., & Posse, S. (1999). Analytical model of susceptibility-induced MR signal dephasing: effect of diffusion in a microvascular network. Magn Reson Med. , 41(3), 499-509.

[24] Krüger, G., & Glover, G. H. (2001). Physiological Noise in Oxygenation-Sensitive Magnetic Resonance Imaging. Magn Reson Med 46(4) , 631 EOF-7 EOF.

[25] Le, T. H., & Hu, X. (1996). Retrospective estimation and correction of physiological artifacts in fMRI by direct extraction of physiological activity from MR data. Magn Reson Med 35(3) , 290 EOF-8 EOF.

[26] Logothetis, N. K. (2008). What we can do and what we cannot do with fMRI. *Nature*, 453(7197), 869-78.

[27] Mansfield, P. (1977). Multi-planar image formation using NMR spinechoes. 1. Phys. C. 10:LL58., 55.

[28] Marques, J. P., & Bowtell, R. (2005). Application of a Fourier-based method for rapid calculation of field inhomogeneity due to spatial variation of magnetic susceptibility. Concepts Magn Reson Part B 25B , 65-78.

[29] Menon, R. (2002). Postacquisition Suppression of Large-Vessel BOLD Signals in High-Resolution fMRI Magn Reson Med 47(1), 1-9.

[30] Ogawa, S., Lee, T. M., Nayak, A. S., & Glynn, P. (1990). Oxygenation-sensitive contrast in magnetic resonance image of rodent brain at high magnetic fields. Magn Reson Med. , 14(1), 68-78.

[31] Ogawa, S., Tank, D. W., Menon, R., Ellermann, J. M., Kim, S. G., Merkle, H., & Ugurbil, K. (1992). Intrinsic signal changes accompanying sensory stimulation: functional

brain mapping with magnetic resonance imaging. Proc Natl Acad Sci U S A. , 89(13), 5951-5955.

[32] Ogawa, S., Menon, R. S., Tank, D. W., Kim, S. G., Merkle, H., Ellermann, J. M., & Ugurbil, K. (1993). Functional brain mapping by blood oxygenation level-dependent contrast magnetic resonance imaging. A comparison of signal characteristics with a biophysical model. Biophys J , 64(3), 803-812.

[33] Pfeuffer, J. Van de Moortele P.-F, Ugurbil K, Hu X, Glover GH, (2002). Correction of Physiologically Induced Global Off-Resonance Effects in Dynamic Echo-Planar and Spiral Functional Imaging. Magn Reson Med. , 10, 344-353.

[34] Petridou, N., Schäfer, A., Gowland, P., & Bowtell, R. (2009). Phase vs. magnitude information in functional *magnetic resonance imaging*time series: toward understanding the noise. Magn Reson Imaging 27(8) , 1046 EOF-57 EOF.

[35] Raj, D., Paley, D. P., Anderson, A. W., Kennan, R. P., & Gore, J. C. (2000). A model for susceptibility artefacts from respiration in functional echo-planar magnetic resonance imaging. Phys Med Biol 45(12) , 3809 EOF-20 EOF.

[36] Rowe, D. B., & Logan, B. R. (2004). A complex way to compute fMRI activation. *Neuroimage*, 1078 EOF-1092 EOF.

[37] Rowe, D. B. (2005). Modeling both the magnitude and phase of complex-valued fMRI data. *Neuroimage*, 25, 1310-1324.

[38] Schmitt, F., Stehling, M. K., & Turner, R. (1998). Echo-Planar Imaging: Theory, Technique and Application. Springer Verlag

[39] Van Gelderen, P., De Zwart, J. A., Starewicz, P., Hinks, R. S., & Duyn, J. H. (2007). Real-time shimmingto compensate for respiration induced Bo fluctuations. Magn Reson Med , 57, 362-368.

[40] Tomasi, D. G., & Caparelli, E. G. (2007). Macrovascular cobtribution in activation patterns of working memory. J Cereb Blood Flow Metab , 27, 33-42.

[41] Vannesjo, SJ, Haeberlin, M, Kasper, L, Pavan, M, Wilm, BJ, Barmet, C, & Pruessmann, KP. (2012). . Gradient system characterization by impulse response measurements with a dynamic field camera. Magn Reson Med. doi: 10.1002/mrm.24263.

[42] Yablonskiy, D. A., & Haacke, E. M. (1994). Theory of NMR signal behaviour in magnetically inhomogeneous tissues: the static dephasing regime. Magn Reson Med , 32(6), 749-763.

[43] Zhao, F., Jin, T., Wang, P., Hu, X., & Kim, S. G. (2007). Sources of phase changes in BOLD and CBV-weighted fMRI. Magn Reson Med , 57, 520-527.

Chapter 5

Simultaneous Measurement of fMRI and EEG – Principles and Applications

Yeji Han, Sung Suk Oh, Joong Koo Kang and
HyunWook Park

Additional information is available at the end of the chapter

http://dx.doi.org/10.5772/31194

1. Introduction

Functional MRI (fMRI) and electroencephalogram (EEG) are the most widely used neuroimaging techniques which assist neuroscientists in investigating human brain function. FMRI measures the induced magnetic field change, usually the change in the blood oxygenation level dependent (BOLD) contrast, which reflects physiological changes of neuronal activity (Ogawa et al., 2000). EEG measures the summed electrical activities of neurons by detecting the electric potential difference at the scalp. As the sources of the signal measured by fMRI and EEG are different, the spatial and temporal information of the two techniques are also different. For example, EEG has millisecond resolution which is much higher than fMRI's temporal resolution, characterized by the lag of the hemodynamic response of 6-7 seconds (Friston et al., 1998). Low temporal resolution of fMRI is inevitable not only because the BOLD contrast is developed over seconds by accumulated changes of the magnetic field, resulting from neuronal activities and vascular changes (Rosenkranz & Lemieux, 2010), but also because several seconds are required to perform each MRI measurement. Instead, fMRI has excellent spatial resolution which is much higher than EEG. Moreover, the spatial as well as the temporal resolution of fMRI is expected to further increase as the field strength of MRI system is still increasing (Ogawa et al., 2000; Regatte et al., 2007). Thus, EEG and fMRI are complementary for brain research in regard to the types of information they can provide, and it is important to utilize both EEG and fMRI signals to get a more comprehensive view of brain activities.

To combine the information provided by fMRI and EEG, one can consider performing data fusion after separately acquiring data using each modality, as it is done for many other multimodality approaches such as combining MRI with positron emission tomography (PET) or combining MRI with computed tomography (CT). Combining PET or CT images with MR images is rather straight-forward because only a correct image registration process is required for a successful fusion of multimodality data. Combining EEG and fMRI data is more complicated; the two types of data are different as fMRI data is presented in the image domain whereas EEG represents information using waveforms in the time domain. Thus, it is recommended to simultaneously acquire fMRI and EEG data so that the data can be more easily interpreted afterwards. In fact, when fMRI and EEG are simultaneously measured, a more accurate interpretation of the data is possible because the timing information is commonly available for both of the measurements, making it more concise to figure out which parts of the fMRI and EEG signals originated from the identical brain activity.

While simultaneous recording of the two modalities can provide high spatio-temporal information for brain research, there are several issues that have to be considered to successfully perform a simultaneous fMRI-EEG measurement, such as patient safety, fMRI image quality, and EEG quality. Patient safety and image quality can be usually coped by using a properly designed MR-compatible EEG instrument. However, degradation of EEG quality cannot be prevented during the acquisition process and several post-processing techniques have to be employed to remove artefacts in the acquired EEG data.

In this chapter, a practical approach for simultaneous fMRI and EEG measurement will be discussed in detail. In section 2, an actual measurement technique and required hardware/software will be presented. Then, common artefacts of EEG signal and removal techniques will be discussed in section 3. In section 4, a practical example of simultaneous fMRI-EEG experiment will be demonstrated.

2. Simultaneous measurement of EEG and fMRI

The first report on a combined use of fMRI and EEG was presented in 1993 (Ives et al., 1993), which showed that signals of brain activities could be obtained with both high spatial and high temporal resolution. In the following years, more experiments have been conducted to confirm the possibility of recording EEG inside an MR scanner (Huang-Hellinger et al., 1995; Lemieux et al., 1997). Nowadays, it has become more common to measure fMRI and EEG simultaneously and many practical applications have been introduced for epilepsy, sleep, and other brain functions.

When an experiment with a simultaneous measurement of fMRI and EEG is conducted, the first step is to prepare a proper EEG instrument. The EEG instrument is composed of electrodes which detect the EEG signal and a recording system which amplifies and digitizes the detected EEG signal. To record EEG signal within an MRI system, electrodes are attached to the patient head and placed inside an MR scanner while the recording system (the amplifier and the digitizer) is placed inside or outside the MR scanner according to different needs. Thus, it should be guaranteed that the electrodes are made with MRcompatible materials because anything placed inside an MR scanner should not contain ferrous material due to safety as well as potential susceptibility artifacts in the MR image. If an experiment requires the amplifier and digitizer placed inside the MR scanner, it should be ensured that they are

http://dx.doi.org/10.5772/31194

manufactured with MR-compatible materials and can operate properly in the high magnetic field.

For EEG quality, the number of electrodes located at the subject's brain scalp is a critical factor because it determines the overall spatial resolution (Reilly, 2005) and the numbers of electrodes in commercially available EEG instrument ranges from 32 to 256. Another factor which influences the EEG quality is the arrangement of electrodes and electrode wires connected to the amplifier/digitizer. To prevent possible signal artefact induced by the magnetic field change, they have to be set on the subject's head without any loops. To minimize the problems related to the electrodes and the electrode wires, an electrode cap is generally used (Bonmassar et al., 2002; Srivastava et al., 2005). In addition, immobilization of the subject's head, the electrodes, and the electrode wires are also important factors for the EEG quality, as well as for the fMRI quality, because the change of the loop area created by the electrode wires in the static field also induces artefacts. To reduce movements of the electrodes and the electrode wires, various techniques are being employed, such as weighing down the electrode wires using sandbags (Benar et al., 2003), padding (Hoffmann et al., 2000), and placing a tight bandage over the subject's head (Benar et al., 2003). Additionally, a vacuum cushion is widely used to reduce the subject's head motion (Benar et al, 2003).

For a simultaneous fMRI-EEG experiment, most of the published literature have conducted experiments using either BrainAmp (BrainProducts GmbH, Munich, Germany) or EMR32 (Schwarzer GmbH, Munich, Germany) because both of them allow the use of an amplifier, which is non-magnetic, shielded, battery-operated, and designed for use in the MRI scanning environment. The EEG signals are amplified, converted to digital signals, and transmitted from the MR scanning environment to a computer located in the control room via optical fibers. Figure 1 shows a typical example of the commercially available MRcompatible EEG instruments, which include electrode caps (32 and 64 channels), a vacuum cushion, and a sandbag.

Using a properly prepared EEG instrument, EEG, electrocardiogram (ECG), and electrooculogram (EOG) signals are recorded, usually with a sampling rate of 5 KHz to capture the rapid change of artefacts caused by switching the magnetic field gradients during fMRI acquisition (Allen et al., 2000). After the EEG signal is measured, artefacts have to be removed as explained in the following sections. Attenuation of the artefacts caused by the gradient fields is employed by initially filtering with a cut-off frequency, typically smaller than 70Hz. The resulting signals are then down-sampled to 200 Hz. Since 200Hz is the sampling rate used for the conventional EEG measurement, down-sampling can reduce the processing time without degrading the EEG quality (Allen et al., 2000).

When EEG and ECG signals are measured, the EEG signals can be represented with different montages by using different references (Fisch, 1999). One of the most widely used montages is the referential montage, which defines a reference electrode at an inactive position, such as the midline position between two hemispheres of the brain. The reference may also be defined at such positions, where the distances from the left channel and the corresponding right channel are equal, eg. earlobes. Thus, the referential montage represents the EEG signal compared to the background EEG signal. The bipolar montage refers to the potential difference between two adjacent electrodes, with longitudinally or transversely directed chains (Fisch,

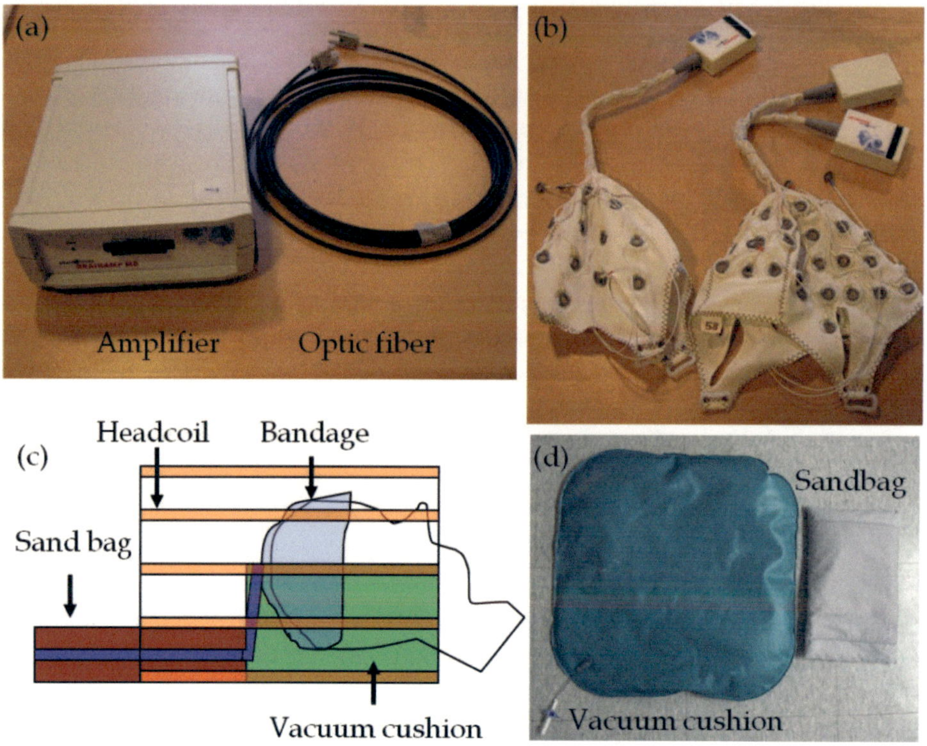

Figure 1. The EEG Instrument. (a) amplifier and optic fiber, (b) electrode caps, (c) illustration of EEG measurement inside an MRI system with an RF headcoil, and (d) aiding materials for movement reduction.

1999). This montage is known to be useful for localization. The third type of montage is the average reference montage, where the common reference is obtained by averaging all EEG signals. Lastly, the Laplacian montage averages several neighboring electrodes surrounding the measured EEG channel and uses the average signal as a reference (Fisch, 1999). For analysis of the EEG signal, any preferred montage can be constructed from the recorded signals in an effort to to clearly demonstrate waveform of EEG events as explained above.

For fMRI measurement, a conventional echo planar imaging (EPI) with the BOLD contrast is generally applied. However, some imaging parameters, such as repetition time (TR), should be carefully chosen to minimize the interference with EEG measurement (Mandelkow et al., 2006). In some cases where fMRI and EEG acquisition has to be recorded for an extended period of time, such as for sleep or epilepsy studies, the data size is too large that the data acquisition has to be performed in several separate sessions.

As briefly mentioned earlier, both the fMRI and the EEG signal can be contaminated when measured simultaneously. Thus, artefacts in the acquired data have to be eliminated before they can be analyzed. The types of artefacts encountered and the artefact reduction techniques that can be applied will be discussed in the following sections.

3. EEG artefacts and their reduction algorithms

When EEG and fMRI are simultaneously measured, both fMRI and EEG signals are influenced by each other. For example, MR image quality may be degraded because the main static field (B0) and the transverse rotational magnetic field (B1) can be altered due to EEG equipment inside the MR room (Mullinger et al., 2008). However, the quality of the MR image is not adversely degraded if the EEG equipment is properly designed and manufactured so that the field inhomogeneities caused by the EEG instrument are minimized. What should be more carefully investigated is the effect of fMRI measurement on the EEG signals, as the EEG quality can be significantly degraded with simultaneous fMRI acquisition, which can induce several types of artefacts, such as (i) image acquisition artefact (IAA); (ii) ballistocardiogram (BCG) artefact; and (iii) quantization error of the analogue-to-digital converter (ADC) (Garreffa et al., 2004). Therefore, when EEG is measured in combination with fMRI, a number of artefact reduction processes have to be employed before it can be used for clinical or research applications.

The two main EEG artefacts are IAA and BCG artefact. The IAA, caused by applied imaging gradient fields during fMRI measurements, is usually removed using periodicity and consistency of the applied gradient fields (Garreffa et al., 2004). The exact cause of the BCG artefacts is not figured out but it is generally known to be added to the EEG signal due to subtle movements of the subject's scalp or abrupt changes of the blood flow in the aortic arch during heartbeats (Ives et al., 1993). The IAA is easier to be removed than the BCG artefact because the former is relatively constant over time when the same imaging gradient is applied for every scan. However, it is rather difficult to remove the BCG artefact because the frequency, shape, and amplitude of the artefact vary from case to case. The third type of artefact is the quantization error. As the amplitude of the IAA can be more than two orders of magnitude higher than that of the EEG signal, the physiological signal in the EEG measurement may be lost if EEG is not measured with sufficient range (Allen et al., 2000).

In the following sections, a detailed explanation of the two main artefacts will be presented and the solutions for these two major artefacts will be discussed.

3.1. Image acquisition artefact (IAA)

The IAA is caused by the changes in the magnetic field during the scanning process, resulting from various factors such as the RF pulses, B^0 field inhomogeneities, body motion, and gradient switching. The switching gradient field, which is inevitable as manipulation of the applied gradient fields plays the main role in acquiring MR images, induces strong artefacts in EEG with an extremely large amplitude. Since the amplitude of IAA is 10~100 times larger than the amplitude of EEG signal, the actual EEG signal is almost completely obscured (Vanderperren et al., 2010). For that reason, earlier studies have performed aperiodic fMRI, where image acquisition is triggered after an EEG event of interest (Krakow et al., 2001). However, EEG-triggered fMRI or interleaved EEG-fMRI measurement is not an optimal solution because it reduces the flexibility of experimental design (Allen et al., 2000).

To remove the IAA from the EEG signal, researchers have focused on the properties of applied gradient fields. Since an MR imaging sequence, mostly the EPI sequence, is repeated during an fMRI experiment, identical changes take place in the gradient fields, inducing the IAA with a consistent shape for each EPI scan. Thus, most algorithms try to eliminate the IAA using the reproducibility and consistency of the IAA shape (Allen et al., 2000).

One of the first IAA reduction algorithms calculated the average IAA spectrum and subtracted the average IAA spectrum from the EEG spectrum (Sijbers et al., 1999; Hoffman et al., 2000). This artefact removal technique is integrated in FEMR software (Schwarzer GmbH, Munich, Germany), which eliminates unwanted signals that are outside the frequency range of the EEG (0.1-40Hz) using band-pass filtering. The artefacts within this frequency range have to be selectively removed. Hoffman's algorithm initially selects 10 different 10s-long sections of EEG acquired inside the magnet, which are unaffected by any artefacts, and uses them as baseline. Then, the frequency spectrum of the baseline is subtracted from the IAA-related spectrum embedded in the EEG signal (Hoffmann et al., 2000). Since IAA is periodic and its frequency range is limited, the removal can be performed in the frequency domain. However, there can be spectral overlap of the IAA and EEG signals in some cases, resulting in an excessive subtraction of the artefact spectrum from the EEG signal.

The other type of IAA reduction algorithm is based on Allen's algorithm (Allen et al., 2000), which is integrated in BrainVision Analyzer (Brain Products GmbH, Munich, Germany). Allen's IAA reduction algorithm consists of two stages: (i) the artefact waveform is averaged over a fixed number of gradient artefact epochs and subtracted from each epoch in the EEG signal; and (ii) the residual IAA is attenuated by a noise cancelling algorithm (Allen et al., 2000). Since template-based removal algorithms typically leave considerable residual noise, further reduction steps should be followed.

More recently, Allen's algorithm has been integrated in parts of many other IAA removal algorithms, including Niazy's algorithm (Niazy et al, 2005), which combines Allen's method with principal component analysis (PCA), and Mantini's algorithm (Mantini et al., 2007), which combines Allen's algorithm with independent component analysis (ICA). A new method to develop a more precise template using a time-continuous cubic spline model has been also presented (Koskinen et al. 2009). Figure 2 shows a typical example of EEG signal recorded inside a 3T MRI scanner (ISOL Technology, Korea). Fig.2 shows the EEG signal with IAA (top) and the signal after the IAA reduction using the BrainVision Analyzer (bottom).

In addition to the above-mentioned algorithms, triggering can be used in combination for detecting the starting point of every scan, which is then used as a marker for more precise averaging of the artefact waveform. Another approach is to perform an interleaved MRI acquisition, where MR acquisition is suspended at regular intervals, resulting in periods free of IAA on the EEG (Ritter et al, 2010).

In general, currently available hardware and software approaches to remove IAA from EEG signal provide satisfactory EEG signals. Moreover, it is less complicated than the removal of the cardiac pulse related artefacts which we discuss below because the timing and shape information of the IAA can be inferred from the applied gradients of MRI.

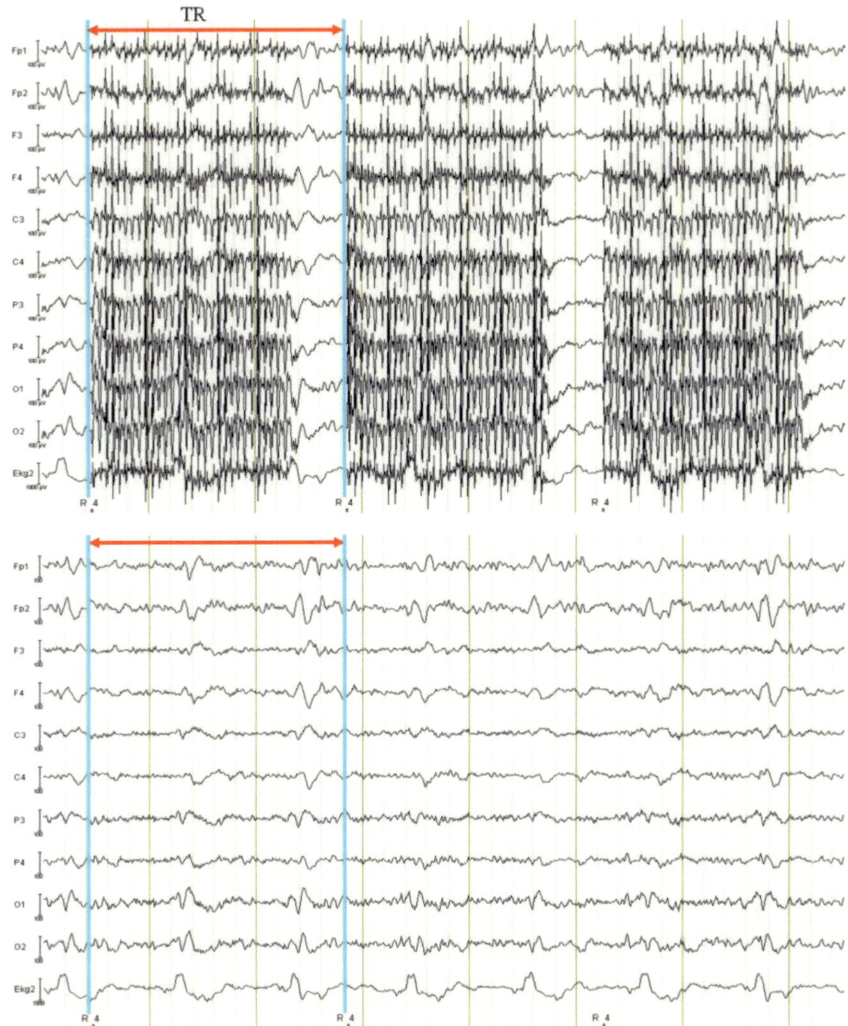

Figure 2. EEG signals recorded from 10 different channels (Fp1, Fp2, F3, F4, C3, C4, P3, P4, O1, O2) and ECG signal with the image acquisition artefact (top) and after the IAA reduction (bottom)

3.2. Cardiac-related artefact

Another EEG artefact that makes the interpretation of the EEG signal rather intricate is the ballistocardiogram artefact (BCG artefact), which is caused by alteration in physiological status. Although the exact cause of the BCG artefact is still being investigated, it is known to be induced by the movement of the patient's body as a result of acoustic vibrations due to the scanner as well as voluntary subject movements (Reilly, 1987). In this chapter, however, we refer to the small involuntary cardiac-related body movements (ballistocardiogram). The BCG artefact can be observed from the EEG signals measured outside an MR scanner but it is small and can be easily removed (Debener et al., 2008). However, the BCG artefact becomes much greater inside the scanner because of the interaction between the active cardiovascular system and the main static field (B0) (Allen et al., 1998). Consequently, it is observed in the EEG signals measured inside an MR scanner, even when EPI sequences are not applied.

A BCG artefact follows the shape of a cardiac cycle (Fig.3). In fact, the BCG artefact closely follows the cardiac cycle with a delay of approximately 210ms between the R peak of the ECG and the BCG artefact (Allen et al., 1998). Fig.3 also shows that the amplitude of the BCG artefact is comparable, if not slightly larger, to the EEG signal. However, it was predicted that the amplitude of the BCG artefact is proportional to the B0 field (Tenforde et al., 1983), and has been confirmed later (Debener et al., 2008). Debener et al. (2008) showed that the amplitude of the BCG artefact is a function of the static magnetic field strengthand the spatial variability of the BCG artefact substantially increased at higher magnetic field strengths. In reference to various BCG artefacts, Debener et al. has also demonstrated that individual subjects have different heart rates, peak latencies, and shapes (Debener et al., 2008). Moreover, the shapes of the BCG artefact of different EEG channels also vary. As a result, it is difficult to distinguish an 'event' from the EEG signal when the signal is distorted by the BCG artefact.

If the frequency of the BCG artefact is constrained in a certain range distinct from that of the EEG signal, one can think of filtering the artefact in the frequency domain. Although the frequency range of the BCG artefact is mainly in the theta (4-8 Hz) band, however, it can extend to the alpha (8-13 Hz) and delta (0.5-4 Hz) bands, thus overlapping with the EEG signals (Garreffa et al., 2004).

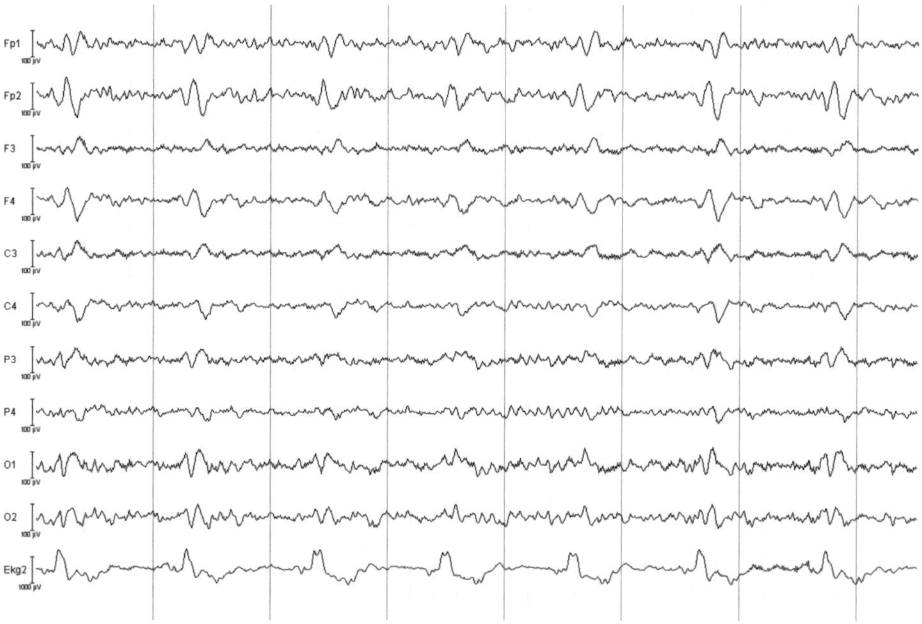

Figure 3. Shows ECG and EEG signals measured inside an MRI scanner without applying the imaging gradients. The BCG artefacts are embedded in each channel.

3.2.1. Average artefact subtraction (AAS)

There are also algorithms that remove the BCG artefacts from the EEG signal based on properties other than the frequency characteristic. One approach to remove the BCG artefact is based on the average artefact subtraction (AAS) algorithm (Allen et al., 1998), which creates an average BCG artefact template/waveform and subtracts it from the EEG signal. It is based on the observation that the shape and the occurrence rate of the BCG artefact is quite consistent across a number of successive heartbeats in each channel of the EEG. Thus, the AAS algorithm consists of two steps: generating an averaged BCG artefact template and subtracting it from the BCG artefact occurrences. First, reference points for each BCG artefact have to be defined. The average waveform is calculated for each referential EEG signal on a second-by-second basis and requires the previous 10 seconds of EEG and ECG signals. The R peak in the QRS complex of the ECG is detected for every heartbeat by thresholding the amplitude of the measured ECG signal using a predefined value and the onset of each BCG artefact is identified using a predefined delay (Allen et al., 1998). Once the onset of every cardiac cycle is detected, a BCG artefact template is generated by averaging the EEG signal over a predefined window size for each EEG channel as the BCG artefact exhibits different shapes in different EEG channels. Finally, the average BCG artefact template is subtracted from the measured EEG signal.

One of the most important steps for the AAS-based algorithms is to find the correct onset points of the BCG artefact because variations of the subject's heartbeat may alter the time delay between the R peak of the QRS complex in the ECG signal and the BCG artefact in the EEG signal. If the onset points are not correct, it is difficult to find the precise average artefact template and the exact reference point for subtraction and this leaves too much residual artefact after subtraction from the EEG signal. Although the AAS method assumes that the shape and occurrence of the BCG artefacts have little variance over time and the condition of the subject's heartbeat is stable, these assumptions may not be necessarily true. Thus, the AAS algorithm has a limitation that the corrected EEG signal can still have residual artefacts due to an inaccurate BCG artefact template.

To account for the shape variations of the BCG artefacts, more algorithms have been introduced to generate BCG artefact templates using other approaches, such as Gaussian weighted averaging (Goldman et al., 2000), median filtering (Sijbers et al., 2000; Ellingson et al., 2004), and wavelet-based adaptive filtering (Kim et al., 2004).

For the AAS-based methods, defining the template length is also an important issue. Since the actual R-R interval is not consistent, the BCG artefact can be subtracted twice if the R-R interval is shorter than the template length and insufficiently subtracted in the other case. To reduce the dependency of the AAS based algorithms on the template length, alternative algorithms have been introduced. One of these scales the template length to a certain percentage of the mean R-R period (Ellington et al. 2004). A plug-in for EEGLAB in FMRIB (http://www.fmrib.ox.ac.uk/eeglab/fmribplugin/), incorporates the BCG artefact data for all R-R period lengths in the current moving average window and the template is adaptively applied to each QRS period based on its R-R period (Brain Vision Analyzer software, http://www.brain-products.com/). Other algorithms account for the template duration problem using the Kalman filter (Bonmassar et al., 2002), the EOG signals (In et al., 2006), the recursive least squares motion recording (Masterton et al., 2007), and a moving general linear model (mGLM) (Vincent et al., 2007).

More recently, the optimal basis set (OBS) (Niazy et al., 2005) method was proposed to use a few principal components as representations of several distinct pulse artifact templates. Since these templates include most of the BCG artefact in any given EEG channel and can be used to regress the BCG artefact, the OBS method can account for a greater temporal variation in the BCG artefact shape (Debener et al., 2010). For the OBS method, however, the number of principal components that create the template have to be selected by the user, and thus, affect the performance of this method.

The major advantage of the AAS based algorithms, in comparison with the blind source separation (BSS) algorithm, is low complexity. However, obtaining satisfactory results can be quite difficult because finding the precise onset points of the cardiac cycle for an accurate template is rather complicated as demonstrated by the variants of the AAS method mentioned above.

3.2.2. Blind source separation (BSS)

The AAS based and the OBS based algorithms focus on the fact that the BCG artefact of each EEG-channel has different shapes. However, the BCG artefact can be also characterized by a number of typical topographies (Bénar et al., 2003). Thus, another algorithm to remove the BCG artefacts is based on the blind source separation (BSS) algorithm, which assumes that the characteristics of the BCG artefacts are independent from or orthogonal to the EEG signals (Bénar et al., 2003). The principal or independent components are obtained using the orthogonal or independent characteristics, respectively, between the BCG artefact and EEG activity of the brain. There are several ways to perform the separation of the BCG artefacts and the EEG signal. Early approaches of BSS algorithms are based on manual selection of the independent components, including the original algorithm using ICA (Bénar et al., 2003). In Bénar's algorithm, the components contributing to the BCG artefact are manually identified and taken out from the components. Then, reconstruction of the EEG signal with the remaining free of artefact components can produce EEG signal with reduced artefacts.

More ICA algorithms have been introduced, including Informax ICA, which uses the minimization of the mutual information (Bell & Sejnowski, 1995), FastICA, which uses the maximization of the non-Gaussianity (Hyvärinen & Oja, 2000), joint approximate diagonalization of eigenmatrices (JADE) (Cardoso & Souloumiac, 1993), and second order blind identification (Belouchrani et al., 1993). ICA methods are implemented in the EEGLAB software (http://sccn.ucsd.edu/eeglab/). Since the ICA algorithms reduce the BCG artefact by removing the subjectively defined components that are correlated to the BCG artefact after the ICA processing, the performance may depend on users.

To solve this problem, algorithms for automatic identification of components related to the BCG artefact have been also proposed. Some of the selection criteria for the automatic identification include correlation with the ECG signal (Srivastava et al., 2005; Mantini et al.,

2007), frequency content related to the heartbeat rate and its harmonics (Vanderperren et al., 2007), and autocorrelation of the EEG signal (Deburchgrave et al., 2008; Vandersperren et al., 2010). The BCG-related components can be distinguished by the peaks in their autocorrelation. In case of frequency related component selection, the fact that BCG artefacts can be characterized by peaks at the heart rate frequency and its harmonics is used. This property can be exploited as a tool to select the artifact-related components from the ICA decomposition, where the heart rate frequency is determined by computing the inverse of the average distance between consecutive QRS peaks. However, these algorithms require a delicate selection of parameters to achieve better performance (Vanderperren et al., 2010). Moreover, ICA method is inefficient to account for the spatial variation of the BCG artefact since independent components are obtained from EEG signals measured at different positions of the brain. As an alternative, OBS-ICA method can be used, which initially removes a major part of the BCG artifact using the OBS method and then removes the residual BCG artefact using the ICA method. (Debener et al., 2007)

The main advantage of the BSS algorithms is that, unlike the AAS based algorithms, the exact onset of the BCG artefact is not required to be identified. However, the BSS algorithms have higher computational complexity than the AAS algorithms. In addition, the BSS algorithms may not work properly for the EEG signals measured in higher field MRI scanners, i.e., 3T and 7T, as deviations in the spatial information of the BCG artefact become greater in higher field MRI systems (Debener et al., 2008).

3.2.3. Modified average artefact subtraction (MAAS)

As discussed in the previous sections, both ASS and BSS algorithms have advantages and disadvantages. However, for 3T or higher field MRI systems, the ASS algorithms can be more efficient (Vanderperren et al., 2010). To optimize the ASS algorithm for higher field MRI systems, a modified AAS (MAAS) algorithm has been proposed to improve the construction and the subtraction of the BCG template using a fully automatic process (Oh et al., 2012). The MAAS algorithm is composed of two steps; it initially finds an accurate reference point based on the ECG signal to use for the BCG artefact subtraction and the template generation. Then, an accurate BCG artefact subtraction algorithm is applied, which adaptively subtracts the BCG artefact template using different delay and window sizes.

To find the reference points for the BCG artefacts, R peaks of the ECG signal have to be accurately identified. According to Debener (Debener et al., 2007), a negative swing occurs between the S and T states of the P_QRS_T wave due to an axial blood flow momentum when the ECG signal is recorded in a high field MR system. This negative swing is used to automatically identify the R peak from ECG in the MAAS algorithm (Oh et al., 2012). Since an R peak and the following negative peak produce the maximum signal difference, the difference of positive and negative peaks is calculated to deduce the R peak. For each segment of the ECG signal with a predefined duration, a sliding window of 0.3s duration is defined to reflect the physiological characteristic (Debener et al., 2007). Within the sliding window, a positive peak and the following negative peak are identified and the difference between the peaks is calculated. After eliminating the positive and negative peak pairs whose difference is smaller than the average, the maximum peak among the remaining positive peaks is selected as the R peak candidate of the current window. Then, the false R peaks are eliminated from the candidate peaks using the resting heartbeat rate (Sharieff, 2006). Since the distances between adjacent R peaks should be within 0.5-1.3 seconds (for 46-120 bpm), the R peak candidate is eliminated if its distance from the previous R peak or the following R peak does not lie in this range. After elimination, the point which has the maximum ECG signal is assigned as a new R peak in sections where the distance of adjacent R peaks is larger than 1.3 seconds.

To detect the BCG artefact from the contaminated EEG signal, a BCG clip is defined for the EEG signal using the R peak as shown in Fig.4. After extracting the BCG clips from the current EEG segment, the BCG clips suitable for a correlation template (CT) generation are classified; if the absolute mean of a BCG clip is larger than the absolute mean of BCG from all previous EEG segments, it is not used for CT generation. Then, the remaining BCG clips are averaged to generate the CT, and then correlation coefficients between each BCG clip and the CT are calculated to figure out the true peak of the current BCG artefact. By considering the delay time between an R peak and the following BCG artefact, the BCG artefact is classified into a normal and a deformed BCG artefact.

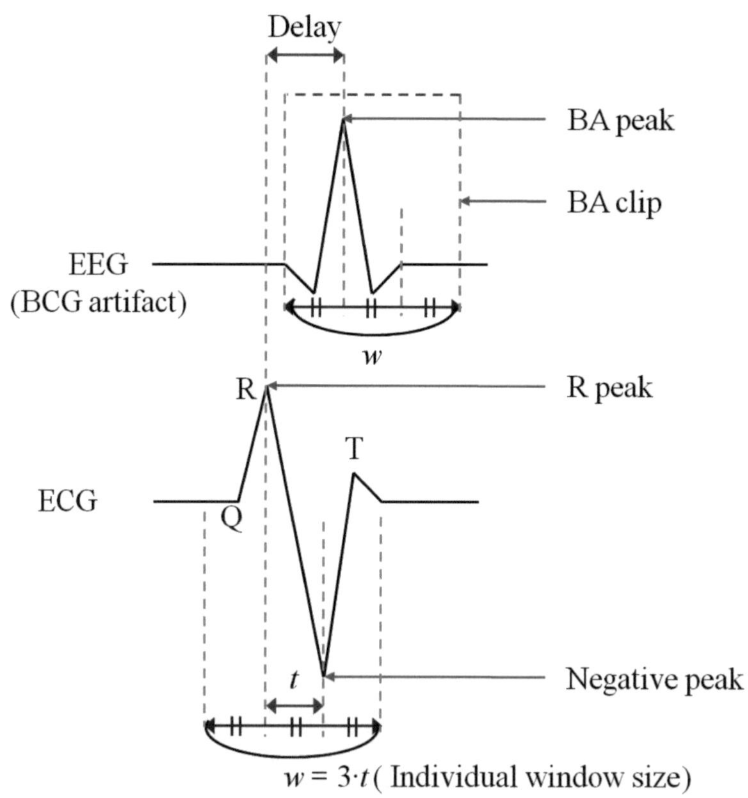

Figure 4. An illustration of the ballistocardiogram (BCG) artifact.

The BCG artefact template is simply obtained by averaging more than two normal BCG artefacts within the current segment, and its length is defined by the maximum window size of the normal BCG artefacts. For a normal BCG artefact, the contaminated EEG signal is corrected by subtracting the BCG artefact template using the individual delay and the individual window size. In case of a deformed BA artefact, the BCG artefact template is subtracted using the average window size and the average delay.

Figure 5 shows the contaminated EEG signal with the BCG artefact and the EEG signal after applying BCG reduction using the OBS, ICA, and MAAS algorithms. Since the MAAS algorithm detects R peaks accurately and uses variable window and delay properties, it can neatly remove the BCG artefacts. Since the OBS method uses the mean R-R interval as the window size of the BCG artefact template, it can result in distortion if the current R-R interval is shorter than the mean R-R interval. In this study, some of the epilepsy signals are overly removed since the mean R-R interval is used. The error is also partly due to the fact that the selected components of the OBS are related to epilepsy signals. For ICA, the estimated BCG artefact cannot handle the variation of the BCG artefacts for every EEG channel and the correction results can have residual artefacts.

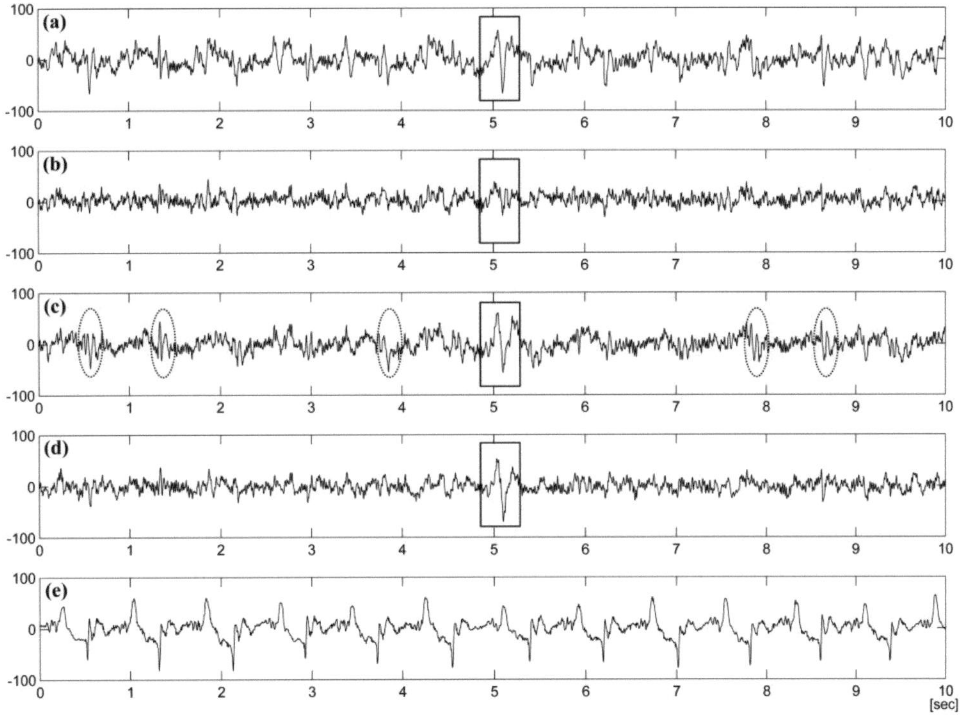

Figure 5. The BCG artefact and its corrected signals. (a) EEG signal on the F7 channel before correction, (b) corrected EEG signal by OBS, (c) corrected EEG signal by ICA, (d) corrected EEG signal by MAAS, and (e) scaled ECG signal. In this study, some of the epilepsy signals are overly removed when the OBS algorithm is used (marked with squares) and the ICA algorithm leaves too much residual BCG artefacts (marked with dotted ovals).

4. Application of fMRI+EEG measurements

The possibility of recording the EEG inside an MR scanner has opened a wide variety of new research areas. Generally, there are three application areas for simultaneous fMRI and EEG studies: neurovascular coupling, cognitive and systems neuroscience, and clinical studies. However, the most straightforward application would be to search for brain activation and deactivation areas related to specific events of the EEG signal. This type of research has been carried on via correlation between the BOLD time series and a postulated EEG-derived model of haemodynamic changes, generated by the general linear models (GLM). In other words, the temporal information acquired from the EEG signals can be used as timing information for the stimulus cue in fMRI. Then, the fMRI analysis can be performed using the MR images corresponding to the cue timing as images of stimulus conditions and the rest as images of reference conditions.

Typical application areas of simultaneous fMRI-EEG experiments include resting state studies such as sleep studies, brain rhythm studies, and activation studies for pain research, auditory research, visual research, and cognition research (Horovits et al., 2008; Christmann et al., 2007; Laufs et al., 2003). In this section, a practical example of a simultaneous fMRIEEG experiment for epilepsy patient is demonstrated. Moreover, three different BCG artefact reduction algorithms are compared using fMRI analysis results.

The development of simultaneous EEG and fMRI measurement was initially motivated out of interest in identifying the BOLD changes related to interictal epileptiform discharges (IED) (Rosenkranz et al., 2010). Although the timing and the shape of epilepsy waveform can be acquired using EEG, spatial information of EEG is not as accurate because the EEG signal is measured at certain locations on the surface of a brain only. Thus, fMRI analysis is used to provide more accurate spatial information of an epilepsy patient by providing not only the cortical but also the subcortical information.

In this experiment, three patients with intractable partial epilepsy were recruited and a written consent was signed by every patient prior to the experiments. A neurologist supervised the experiments to ensure the safety of the patients. EEG recording was performed inside a 3T MRI system (ISOL Technology, South Korea) using BrainAmp and BrainCap (BrainProducts GmbH, Germany). With BrainCap, 21 EEG channels (Fp1, Fp2, F3, F4, C3, C4, P3, P4, O1, O2, F7, F8, T7, T8, P7, P8, Fz, Cz, Pz, TP9, and TP10 in the 10-20 channel system), two ECG channels, and an EOG were recorded during 5 to 7 sessions, each lasting for 11 minutes. The international 10-20 system is an electrode placement scheme mostly applied for epilepsy studies, which is based on the relationship between the location of an electrode and the underlying area of the cerebral cortex. The 10 and 20 represent the distances between adjacent electrodes which are 10% and 20% of the total front-back and right-left distance of the skull (Fisch, 1999). The reference electrode of the cap was located at the mid-points (Cz and Fz), and EEG/ECG signals were recorded with a sampling rate of 5kHz to prevent aliasing.

At the same time, fMRI data was simultaneously acquired using the following imaging parameters: gradient-echo echo-planar-imaging (GE-EPI) sequence, TR/TE=3000/37 msec, matrix size=64×64, FOV=220×220 mm^2, slice thickness=4 or 5mm, number of slices=20 or 30, and flip angle=70°.

After the measurement, the EEG signals were filtered by a phase-shift-free Butterworth filter having a band-pass range of 0.5 to 55 Hz. IAA was then eliminated using the BrainVision Analyzer (BrainProducts) and the filtered signals were down-sampled to 500 Hz (Mandelkow et al., 2006). To remove the BCG artefacts, ICA, OBS and MAAS algorithms were separately applied to the EEG signals. For fMRI analysis, Statistical Parameter Mapping (SPM8, http://www.fil.ion.ucl.ac.uk/spm/) was used with the pre-processing steps of 'realign' and 'smoothing with a Gaussian filter having a full width at half maximum of 8mm'. For the t-test, p-value of 0.001 and a cluster analysis of 4mm was applied.

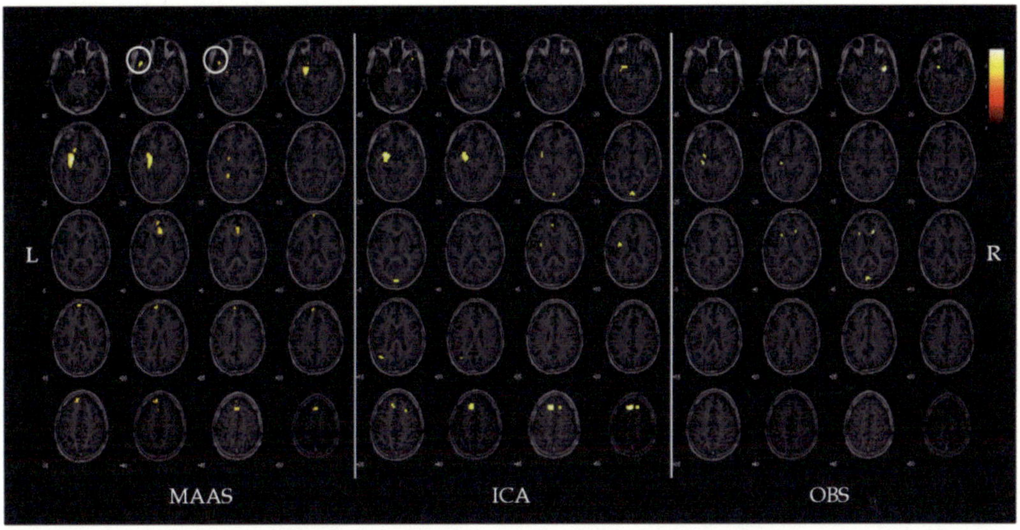

Figure 6. Activation regions from fMRI analysis after removing the BCG artefacts from the measured EEG signal using the MAAS, ICA, and OBS algorithms. The activation regions corresponding to the PEZ is marked with circles.

After post-processing, neurologists detected IEDs from the BCG-artefact removed EEG signals. To localize the IED regions in the brain, fMRI analysis was performed using SPM8 with the detected IED timing information. Then, the activated regions from the fMRI were compared to the presumed epileptogenic zone (PEZ), which was also determined by analyzing the EEG signals, PET images and anatomical MR images.

Since the activated regions from the fMRI are analyzed using the detected IEDs, the performance of the BCG artefact removal algorithms can be indirectly evaluated by comparing the fMRI activation regions. In this experiment, the IED regions in the subjects' brain were localized using the regions revealed by the fMRI analysis (Fig.6). IED regions of subject A is shown in Fig. 6. The localized activation regions of three subjects are shown in Table 1. While the detected fMRI activation regions from all three methods produced the same results for subjects B and C, coincident with PEZ, data of subject A produced different results when the

MAAS algorithm was applied. As shown in Fig.6, the left temporal pole of subject A could be only detected when the BCG artefact of the EEG signal was removed by the MAAS algorithm. When the fMRI activation results were compared to the PEZ of subject A, the activation region was coincident with the PEZ only when the MAAS algorithm was applied.

sub	PEZ	activated regions from fMRI		
		MAAS	ICA	OBS
A	l. ant -mid temporal area	l. limbic lobe, parahipocamppal gyrus	l. frontal lobe, superior frontal gyrus	r. temporal lobe, superior temporal lobe
		l. temporal pole	l. limbic lobe, parahipocamppal gyrus	r. limbic lobe, anterior cingulate
		r. limbic lobe, anterior cingulate	r. frontal lobe, superior frontal gyrus	l. limbic lobe, parahipocamppal gyrus
		l. frontal lobe, inferior frontal gyrus	r. occipital lobe, lingual gyrus	l. occipital lobe, cuneus
		left frontal lobe, superior frontal gyrus		
B	r. mid - temporal area	r. frontal lobe, middle frontal gyrus		
		r. frontal lobe, inferior frontal gyrus		
		r. temporal lobe, middle temporal gyrus		
C	bifrontal area	r. frontal lobe, middle frontal gyrus		
		r. frontal lobe, superior frontal gyrus		

Table 1. Localization of the IEDs based on the fMRI activation regions

As the results demonstrate, the MAAS algorithm outperforms the other algorithms in this study with epilepsy patients. In spite of its complex calculation, the ICA algorithm cannot clearly detect and remove the BCG artefacts because it cannot sufficiently account for the deviations of spatial information of the BCG artefacts (Debener et al., 2007). On the other hand, the timing information of every R peak is used to estimate the onset and duration of the corresponding BCG artefact more accurately in the MAAS algorithm. Since variation of a subject's heartbeat may alter the delay time of the BCG artifact in the EEG signal from the R peak in the ECG signal (Shin, et al., 2009; Assecondi et al., 2009), the BCG artefacts cannot be fully removed if a fixed delay is used for the BCG artefact detection, such as in the OBS method. Another advantage of the MAAS algorithm is that it uses individual window sizes when subtracting the BCG artefact template from the EEG signal. Since the OBS method uses the mean R-R interval as the window size for the BCG template, the correction result can be distorted if the current R-R interval is shorter than the mean R-R interval. In other words, some

http://dx.doi.org/10.5772/31194

regions may be subtracted twice as it experiences an overlapping of successive subtraction windows.

The simultaneous fMRI-EEG experiment for epilepsy patient enables various types of analysis for research and diagnosis. In clinical fields, simultaneous fMRI-EEG measurement can be used for localization of the epileptic source in presurgical evaluation (Zijlmans et al., 2007) or in epilepsy surgery (Thornton et al., 2010). For research purposes, it can be used to study absence seizures (Moeller et al., 2009), temporal lobe interictal spikes heterotopias (Kobayashi et al, 2006a), startle epilepsy (Fernández et al., 2011), and gray matter heterotopias (Kobayashi et al, 2006b). However, the BOLD changes are related to the detected IEDs with only about 67% accuracy (Al et al. 2003; Krakow, et al., 2001; Salek-Haddadi et al., 2006), possibly due to significant difference in hemodynamic response function (HRF) of epilepsy and the standard HRF model (Bénar et al., 2001). To account for this problem, various HRFs have been developed for epilepsy (Bagshaw et al., 2004; Grouiller et al., 2010).

5. Conclusion

In this chapter, principles and applications for simultaneous fMRI and EEG measurement were discussed. The simultaneous fMRI and EEG measurements can both provide high spatial and temporal resolution, and thus, can generate valuable data for experimental and clinical neuroscientists. However, there are certain problems, such as the IAA and the BCG artefacts, which have to be carefully handled to make a full use of the methodology. If a proper elimination method is employed to reduce possible artefacts in the fMRI and EEG signals, the simultaneous measurement can become a useful tool in various clinical and system neuroscience studies. Thus, a combined analysis of fMRI and EEG will provide a more promising future for elucidating the mechanisms of brain than the separate application of these two tools.

Acknowledgements

This work was partly supported by National Research Foundation of Korea (NRF) grant funded by the Korea government (MEST) (No. 2012-0000125).

Author details

Yeji Han[1], Sung Suk Oh[1], Joong Koo Kang[2] and HyunWook Park[1]

1 Korea Advanced Institute of Science and Technology, Republic of Korea

2 Asan Medical Center, Republic of Korea

References

[1] Al-Asmi, A., Benar, C.G., Gross, D.W., Khani, Y.A., Andermann, F., Pike, B., Dubeau, F., & Gotman, J. (2003). fMRI Activation in Continuous and Spike-Triggered EEG-fMRI Studies of Epileptic Spikes. *Epilepsia,* Vol.44, pp.1328–1339, ISSN 0013-9580

[2] Allen, P.J., Polizzi, G., Krakow, K., Fish, D.R., & Lemieux, L. (1998). Identification of EEG Events in the MR Scanner: the Problem of Pulse Artefact and a Method for its Subtraction. *NeuroImage,* Vol.8, No.3, pp.229-239, ISSN 1053-8119

[3] Allen, P.J., Josephs, O., & Turner, R. (2000). A Method for Removing Imaging Artifact from Continous EEG Recorded During Functional MRI. *Neuroimage,* Vol.12, pp.230–9, ISSN 1053-8119

[4] Assecondi, S., Hallez, H., Staelens, S., Bianchi, A.M., Huiskamp, G.M., & Lemahieu, I. (2009). Removal of the Ballistocardiographic Artifact from EEG-fMRI Data: a Canonical Correlation Approach. *Physics in Medicine and Biology,* Vol.54, pp.1673-1689, ISSN 0031-9155

[5] Bagshaw, A.P., Aghakhani, Y., Bénar C.G., Kobayashi, E., Hawco, C., Dubeau, F., Pike, G.B., & Gotman, J. (2004). EEG-fMRI of Focal Epileptic Spikes: Analysis with Mmultiple Haemodynamic Functions and Comparison with Gadolinium-Enhanced MR Angiograms, *Human Brain Mapping,* Vol.22, pp.179–192, ISSN 1065-9471

[6] Bell, A.J. & Sejnowski, T.J. (1995). An Information-Maximization Approach to Blind Separation and Blind Deconvolution. *Neural Computation,* Vol.7, No.6, pp.1129-1159, ISSN 0899-7667

[7] Belouchrani, A., Abed-Meraim, K., Cardoso, J.F., & Moulines, E. (1993). Second-Order Blind Separation of Temporally Correlated Sources. *Proceedings of International Conference on Digital Signal Processing,* Cyprus, pp.346-351.

[8] Bénar, C.G., Gross, D.W., Wang, Y., Petre, V., Pike, B., Dubeau, F., & Gotman, J. (2001). The BOLD Resoponse to Interictal Epileptiform Discharge, *NeuroImage,* Vol. 17, pp.1182-1192, ISSN 1053-8119

[9] Bénar, C., Aghakhani, Y., Wang, Y., Izenberg, A., Al-Asmi, A., Dubeau, F., & Gotman, J. (2003). Quality of EEG in Simultaneous EEG-fMRI for Epilepsy. *Clinical Neurophysiology,* Vol.114, No.3, pp.569-580, ISSN 1388-2457

[10] Bonmassar, G., Purdon, P.L., Jaaskelainen, I.P., Chiappa, K., Solo, V., Brown, E.N., & Belliveau, J.W. (2002). Motion and Ballistocardiogram Artifact Removal for Interleaved Recording of EEG and EPs during MRI. *NeuroImage,* Vol.16, No.4, pp. 1127-1141, ISSN 1053-8119

[11] Cardoso, J.F. & Souloumiac, A. (1993). Blind Beamforming for Non-Gaussian Signals. *IEE Proceedings F Radar and Signal Processing,* Vol.140, No.6, pp.362-370, ISSN 0956-375X

[12] Debener, S., Strobel, A., Sorger, B., Peters, J., Kranczioch, C., Engel, A.K., & Goebel, R. (2007). Improved Quality of Auditory Event-Related Potentials Recorded Simultaneously with 3T fMRI: Removal of the Ballistocardiogram Artefact. *NeuroImage,* Vol. 34, pp.587–597, ISSN 1053-8119

[13] Debener, S., Mullinger, K.J., Niazy, R.K., & Bowtell, R.W. (2008). Properties of the Ballistocardiogram Artefact as Revealed by EEG Recordings at 1.5, 3 and 7 Tesla Static Magnetic Field Strength. *International Journal of Psychophysiology,* Vol.67, No.3, pp. 189–199, ISSN 0167-8760

[14] Debener, S., Kranczioch, C., & Gutberlet, I. (2010). EEG Quality: Origin and Reduction of the EEG Cardiac-Related Artefact, In: *EEG-fMRI Physiological Basis, Technique, and Applications,* Christoph Mulert and Louis Lemieux, pp.135-151, ISBN 978-3-540-87918-3, Germany

[15] Deburchgraeve, W., Cherian, P.J., De Vos, M., Swarte, R.M., Blok, J.H., Visser, G.H., Govaert, P., & Van Huffel, S. (2008). Automated Neonatal Seizure Detection Mimicking a Human Observer Reading EEG. *Clinical Neurophysiology,* Vol.119, No.11, pp. 2447–2454, ISSN 1388-2457

[16] Ellingson, M.L., Liebenthal, E., Spanaki, M.V., Prieto, T.E., Binder, J.R., & Ropella, K.M. (2004). Ballistocardiogram Artifact Reduction in the Simultaneous Acquisition of Auditory ERPS and fMRI. *NeuroImage,* Vol.22, No.4, pp.1534-1542 ISSN 1053-8119

[17] Fernández, S., Donaire, A., Maestro, I., Seres, E., Setoain, X., Bargalló, N., Rumià, J., Boget, T., Falcón, C., & Carreño, M. (2011). Functional Neuroimaging in Startle Epilepsy: Involvement of a Mesial Frontoparietal Network. *Epilepsia,* Early View, ISSN 0013-9580

[18] Fisch, B. J. (1999). *Fisch and Spehlmann's EEG Primer: Basic Principles of Digital and Analog EEG* (3rd Edition), Elsevier, ISBN 978-0-444-82148-5, Amsterdam.

[19] Garreffa, G., Bianciardi, M.D., Hagberg, G.E., Macaluso, E., Marciani, M.G., Maraviglia, B., Abbafati, M., Carni, M., Bruni, I., & Bianchi, L. (2004). Simultaneous EEG–fMRI Acquisition: How Far Is It from Being a Standardized Technique?. *Magnetic Resonance Imaging,* Vol.22, No.10, pp.1445-1455, ISSN 0730-725X

[20] Goldman, R.I., Stern, J.M., Engel Jr., J., & Cohen, M.S. (2000). Acquiring Simultaneous EEG and Functional MRI. *Clinical Neurophysiology,* Vol.111, No.11, pp.1974-1980, ISSN 1388-2457

[21] Grouiller, F., Vercueil, L., Krainik, A., Segebarth, C., Kahane, P., & David, O. (2010). Characterization of the Hemodynamic Modes Associated with Interictal Epileptic Activity Using a Deformable Model-Based Analysis of Combined EEG and Functional MRI Recordings. *Human Brain Mapping,* Vol.31, pp.1157–1173 ISSN 1065-9471

[22] Hoffmann, A., Jaejer, L., Werhahn, K.F., Jaschke, M., Noachtar, S., & Reiser, M. (2000). Electroencephalography During Functional Echo-Planar Imaging: Detection

of Epileptic Spikes Using Post-processing Methods. *Magnetic Resonance in Medicine,* Vol.44, pp.791-798, ISSN 0740-3194

[23] Huang-Hellinger, F., Breiter, H. C., McCormack, G., Cohen, M. S., Kwong, K. K., Sutton, J. P., Savoy, R. L., Weisskoff, R. M., Davis, T. L., Baker, J. R., Belliveau, J. W., & Rosen, B. R. (1995). Simultaneous Functional Magnetic Resonance Imaging and Electrophysiological Recording. *Human Brain Mapping,* Vol.3, pp.13–23, ISSN 1065-9471

[24] Hyvärinen, A. & Oja, E. (2000). Independent Component Analysis: Algorithms and Applications. *Neural Networks,* Vol.13, No.4-5, pp.411-430 ISSN 0893-6080

[25] In, M.H., Lee, S.Y., Park, T.S., Kim, T.S., Cho, M.H., & Ahn, Y.B. (2006). Ballistocardiogram Artifact Removal from EEG Signals Using Adaptive Filtering of EOG Signals. *Physiological Measurement,* Vol.27, pp.1227-1240

[26] Ives, J.R., Warach, S., Schmitt, F., Edelman, R.R., & Schomer, D.L. (1993). Monitoring the Patient's EEG During Echo Planar MRI. *Electroencephalography and Clinical Neurophysiology,* Vol.87, No.6, pp.417–20, ISSN 1388-2457

[27] Kim, K.H., Yoon, H.W., & Park, H.W. (2004). Improved Ballistocardiac Artifact Removal from the Electroencephalogram Recorded in fMRI. *Journal of Neuroscience Methods,* Vol.135, No.1-2, pp.193-203

[28] Kobayashi, E., Bagshaw, A.P., Benar, C.G., Aghakhani, Y., Andermann, F., Dubeau, F., & Gotman, J. (2006a). Temporal and Extratemporal BOLD Responses to Temporal Lobe Interictal Spikes. *Epilepsia,* Vol.47, pp.343–354, ISSN 0013-9580

[29] Kobayashi, E., Bagshaw, A.P., Grova, C., Gotman, J., & Dubeau, F. (2006b). Grey Matter Heterotopia: What EEG-fMRI Can Tell Us About Epileptogenicity of Neuronal Migration Disorders, *Brain,* Vol.129, pp.366–374.

[30] Koskinen, M. & Varitiainen, N. (2009). Removal of Imaging Artifacts in EEG during Simultaneous EEG/fMRI Recording: Reconstruction of a High-Precision Artefact Template. *Neuroimage,* Vol.46, pp.160-167, ISSN 1053-8119

[31] Krakow, K., Lemieux, L., Messina, D., Scott, C.A., Symms, M.R., Duncan, J.S., & Fish, D.R. (2001). Spatio-Temporal Imaging of Focal Interictal Epileptiform Activity Using EEG-Triggered Functional MRI. Epileptic Disorders, Vol.3, pp.67–74, ISSN 1950-6945

[32] Lemieux, L., Allen, P.J., Franconi, F., Symms, M.R., & Fish, D.R. (1997). Recording of EEG During fMRI Experiments: Patient Safety. *Magnetic Resonance in Medicine,* Vol. 38, No.6, pp.943–52, ISSN 0740-3194

[33] Mantini, D., Perrucci, M.G., Cugini, S., Ferretti, A., Romani, G.L., & Del Gratta, C. (2007). Complete Artifact Removal for EEG Recorded During Continuous fMRI Using Independent Component Analysis. *NeuroImage,* Vol.34, pp.598–607, ISSN 1053-8119

[34] Masterton, R.A., Abbott, D.F., Fleming, S.W., & Jackson G.D. (2007). Measurement and Reduction of Motion and Ballistocardiogram Artefacts from Simultaneous EEG and fMRI Recordings. *Neuroimage*, Vol.37, pp.202-211, ISSN 1053-8119

[35] Moeller, F., LeVan, P., Muhle, H., Stephani, U., Dubeau, F., Siniatchkin, M., & Gotman, J. (2010). Absence Seizures: Individual Patterns Revealed by EEG-fMRI, *Epilepsia*, Vol.51, No.10, pp.2000-10, ISSN 0013-9580

[36] Mullinger, K., Debener, S., Coxon, R., Bowtell, R. (2008). Effects of Simultaneous EEG Recording on MRI Data Quality at 1.5, 3 and 7 tesla. *International Jounal of Psychophysiology*, Vol.67, No.3, pp.178-188, ISSN 0167-8760

[37] Niazy, R.K., Beckmann, C.F., Iannetti, G.D., Brady, J.M., & Smith, S.M. (2005). Removal of FMRI Environment Artifacts From EEG Data Using Optimal Basis Sets. *NeuroImage*, Vol.28, pp.720–737, ISSN 1053-8119

[38] Reilly, E.L. (2005). EEG Recording and Operation of the Apparatus. In: *Electroencephalography: Basic Principles, Clinical Applications and Related Fields*. Ernst Niedermeyer and Frenando Lopes da Silva, pp.139-160, Lippincott Williams & Wilkins, ISBN 0-7817-5126-8, Philadelphia.

[39] Ritter, P., Becter, R., Freyer, F., & Villringer, A. (2010). EEG Quality: The Image Acquisition Artefact, In: *EEG-fMRI Physiological Basis, Technique, and Applications*, Christoph Mulert and Louis Lemieux, pp.153-171, ISBN 978-3-540-87918-3, Germany

[40] Rosenkranz, K. & Lemieux, L. (2010). Present and Future of Simultaneous EEG-fMRI. *Magnetic Resonance Materials in Physics, Biology and Medicine*, Vol.23. pp.309-316, ISSN 0968-5243

[41] Salek-Haddadi, A., Diehl, B., Hamandi, K., Merschhemke, M., Liston, A., Friston, K., Duncan, J.S., Fish, D.R., & Lemieux, L. (2006). Hemodynamic Correlates of Epileptiform Discharges: An EEG-fMRI Study of 63 Patients with Local Epilepsy. *Brain Research*, Vol.1088, No.1, pp.148-166, ISSN 0006-8993

[42] Sharieff, G.Q. & Rao, S.O. (2006). The Pediatric ECG, *Emergency Medicine Clinics of North America*, Vol.24, pp.195-208, ISSN 0733-8627

[43] Shin, J.H., Lee, K.M., & Park, K.S. (2009). Non-Constrained Monitoring of Systolic Blood Pressure on a Weighing Scale, *Phyiological Mesurement*, Vol.30, pp.679-693, ISSN 0967-3334

[44] Sijbers, J., Michiels, I., Verhoye, M., Van Audekerke, J., Van der Linden, A., & Van Dyck, D. (1999). Restoration of MR-Induced Artifacts in Simultaneously Recorded MR/EEG Data. *Magnetic Resonance Imaging*, Vol.17, No.9, pp.1383–1391, ISSN 0730-725X

[45] Sijbers, J., Van Audekerke, J., Verhoye, M., Van der Linden, A., & Van Dyck, D. (2000). Reduction of ECG and Gradient Related Artifacts in Simultaneously Recorded

Human EEG/fMRI Data. *Magnetic Resonance Imaging*, Vol.18, No.7, pp.881-886, ISSN 0730-725X

[46] Srivastava, G., Crottaz-Herbette, S., Lau, K.M., Glover, G.H., & Menon, V. (2005). ICA-Based Procedures for Removing Ballistocardiogram Artifacts from EEG Data Acquired in the MRI Scanner. *NeuroImage*, Vol.24, No.1, pp.50-60, ISSN 1053-8119

[47] Tenforde, T.S., Gaffey, C.T., Moyer, B.R., & Budinger, T.F. (1983). Cardiovascular Alterations in Macaca Monkeys Exposed to Stationary Magnetic Fields: Experimental Observations and Theoretical Analysis. *Bioelectromagnetics*, Vol.4, No.1, pp.1-9, ISSN 1521-186X

[48] Thornton, R., Laufs, H., Rodionov, R., Cannadathu, S., Carmichael, D.W., Vulliemoz, S., Salek-Haddadi, A., McEvoy, A.W., Smith, S.M., Lhatoo, S., Elwes, R.D., Guye, M., Walker, M.C., Lemieux, L., & Duncan, J.S. (2010). EEG Correlated Functional MRI and Postoperative Outcome in Focal Epilepsy. *Journal of Neurology, Neurosurgery, and Psychiatry*, Vol.81, pp.922–927, ISSN 0022-3050

[49] Vanderperren, K., De Vos, M., Ramauter, J., Novitskiy, N., Mennes, M., Assecondi, S., Vanrumste, B., Stiers, P., Van den Bergh, B., Wagemans, J., Lagae, L., Sunaert, S. & Van Huffel, S. (2010). Removal of BCG Artifacts from EEG Recordings Inside the MR Scanner: A Comparison of Methodological and Validation-Related Aspects. *NeuroImage*, Vol.50, No.3, pp.920-934, ISSN 1053-8119

[50] Vincent, J.L., Larson-Prior, L.F., Zempel, J.M., Snyder, A.Z. (2007). Moving GLM Ballistocardiogram Artifact Reduction for EEG Acquired Simultaneously with fMRI. *Clinical Neurophysiology*, Vol.118, No.5, pp.981-998, ISSN 1388-2457

[51] Zijlmans, M., Huiskamp, G., Hersevoort, M., Seppenwoolde, J.H., van Huffelen, A.C., & Leijten, F.S. (2007). EEG-fMRI in the Preoperative Work-Up for Epilepsy Surgery. Brain, Vol.130, pp.2343–2353, ISSN 1460-2156

Chapter 6

Functional MRI of Awake Behaving Macaques Using Standard Equipment

Reza Farivar and Wim Vanduffel

Additional information is available at the end of the chapter

http://dx.doi.org/10.5772/31413

1. Introduction

Functional magnetic resonance imaging (fMRI) is perhaps the most important development in the measurement of human brain function in recent neuroscience history. The technique allows for noninvasive whole-volume measurements of brain function in humans and repeatable non-invasive measurement of activity over the whole brain, providing a correlate of neural activity without use of tracers or electrodes. Well-developed commercial MRI systems with built-in support for functional imaging are widely available, as are a plethora of tools, both free and commercial, for the analysis of human functional brain images. The combination of non-invasiveness and the broad availability of equipment and tools to perform fMRI have made this technique popular and broadly adopted.

While human fMRI can go a long way to answering fundamental questions about organization of function in the brain, much of the information we obtain from fMRI is correlative—for example, while activity in an area may correlate with a task, it is impossible to assume that the activity *causes* the observed behaviour. In order to draw causal links between brain function and behavioural performance, it is necessary to directly alter the brain either prior to or during fMRI. While some coarse techniques exist to manipulate human brain function, such as transcranial magnetic stimulation, non-invasive methods comparable to localized chemical deactivations (Liu, Yttri et al. 2010) and local electrical microstimulation (Tolias, Sultan et al. 2005; Ekstrom, Roelfsema et al. 2008; Moeller, Freiwald et al. 2008; Ekstrom, Roelfsema et al. 2009) in the monkey do not exist for human imaging, and direct microstimulation or temporary deactivation by cortical infusion in a normal human subject is ethically unthinkable.

Studying brain function in humans alone is also limiting in that very little is known about the structure of the human brain—at least when compared to what we know about the macaque brain, for example. Axonal tracing requires local injection of a tracer molecule followed by death and histological analysis, all of which is ethically infeasible in humans. In monkeys, this data has provided for a rich map of connectivity that can not only aid us in interpreting single-cell and fMRI results, but also formalize relationships between brain regions that can then be tested directly with temporary deactivation. For example, recently Schmid, Mrowka et al. (2010) used fMRI in awake-behaving macaques to demonstrate the neural basis for blind-sight —the residual visual capacity following loss of primary visual cortex (see (Leopold 2012 for a review). The anatomical basis for blind-sight has long been hypothesized to be the branching connections of the lateral geniculate nucleus of the thalamus (LGN) to other visual areas—in other words, while the V1 input to the cortical visual hierarchy is lost following ablation of V1, other visual areas receive indirect input via the thalamus. This hypothesis had not been directly tested, until Schmid et al. (2010)—that study was made possible by the fact that the projections of the LGN to the cortex are well known in the monkey.

Functional MRI in awake-behaving macaques thus affords us the possibility of studying primate brain function in a well-defined model (i.e., well-defined structure and connectivity) and to perturb the cortex using causal manipulations such as temporary deactivation, electrical microstimulation, or optogenetic manipulation (Boyden, Zhang et al. 2005; Gerits, Farivar et al. 2012). However, a number of challenges must be overcome to carry out a successful awakebehaving macaque fMRI study. Some aspects are general to animal neurophysiology studies— need for restraint, training, behavioural monitoring, etc.—while other aspects are unique to performing studies in the MRI scanner. These include the need to refrain from use of metallic material near the head, even if non-magnetic due to potential eddy current artifacts (Bernstein, King, et al. 2004), as well as issues pertaining to radio frequency (RF) noise, animal movement during scanning, motivation to perform specific behavioral tasks, environmental constraints for these tasks, etc. In this chapter, we begin with a comparison of human and monkey fMRI to highlight some of the differences and continue to discuss some of our solutions to the challenges of functional MRI with awake-behaving macaques. We close by discussing some unique possibilities afforded by fMRI in awake-behaving macaques.

1.1. Primate imaging and human comparison

A typical fMRI session with a human subject may follow this pattern: the participant is debriefed on both the experiment and aspects of the MR imaging, particularly noise issues and the importance of remaining still in the scanner. The participant's questions are answered and the experimenter, with the help of another individual typically, sets up the subject in a commercially available, clinical-grade head coil, places cushions around the subject's head and sometimes restrains the subject further by using a bite bar. The subject's behaviour may be monitored during the scanning as the subject tries to cooperate with the instructions given, and after each run, the experimenter talks with the subject to make sure all is well. During the scan, the subject does his utmost not to move and to pay attention to the task or stimuli, as requested by the experimenter.

A typical scan with an awake monkey may go a little differently: the experimenter restrains the monkey from his cage—good pole and collar training can go a long way to reduce this first step in a long line of frustrating steps. The animal is placed in an MR-compatible chair, and his head is subsequently restrained using an implanted headpost that is then attached to the chair. Custom-made RF coils are then placed around the animal's head and tested on a vector network analyzer to ensure adequate performance. The animal is water restricted to increase its motivation to participate, but the level of motivation will vary depending on various factors, such as time of day being scanned, frequency of being scanned, amount of prior training in a simulator box, etc. During the scan, the animal receives juice reward for performing the required task, whether central fixation or more complicated tasks requiring response by way of eye movements or hand manipulations. While a well-trained monkey will perform this task well above 90% of the time, the animal will make many fast movements during a given scan. During some runs, for unclear reasons, the animal may choose not to behave according to training, and often after a short break, good behaviour returns to standard.

In short, a great deal changes between scanning humans and monkeys, and they begin with the fact that one has a cooperative human subject, and an innately uncooperative monkey who may do what you ask (most of the time) in return for juice. The monkey makes many more movements during a scan despite the fact that his head is rigidly fixed to the chair, and this has a strong negative influence on image and data quality, regardless of whether he performs his trained task.

1.2. Issues of particular relevance to imaging awake non-human primates

In this section, we will briefly discuss some of the demands of functional imaging in awake animals, and then discuss some of our solutions in meeting those demands.

First, we need a stable preparation. By stability we mean two different things—that the animal moves as little as possible, and that its performance during a scan is consistent. The movement of the animal requires means of head restraint that balance animal comfort with robust restraint against movements. Importantly, the means of head restraint must be fully MRcompatible, precluding use of metals anywhere near the brain. The head restraint method depends on a stable platform, typically the primate chair, which also needs to be non-magnetic and mechanically stable. The aspect of stability pertaining to the performance of the animal requires extensive training outside of the magnet which can last months to a year, depending on the specifics of the task, with conditions as comparable to the MRI conditions as possible— i.e. with simulated MRI noise at the same level as that of the MRI, same distance to screen, etc. The training regimen should also aim to create a stable pattern of performance for a reasonable amount of time (typically 2-6 hours). In other words, the animal should be occupied with a specific task throughout an entire scan, otherwise in absence of a task at hand, it may resort to fast movements which are detrimental to the quality of the images.

Second, we need high performance receive coils to pick up the small signal changes acquired during functional imaging scans. The coil design and layout are limited by the first requirement outlined above, namely that the coils must work around the head restraint employed. Importantly, given the drop in the signal-to-noise ratio (SNR) as a function of the distance between the coil and the subject (Mispelter, Lupu et al. 2006) the coils must be placed as close to the animal's head (brain) as possible. There is the added challenge of the monkey's jaw muscles that add considerable distance between the coils placed at the sides and the brain.

Third, we need to image at high spatial and temporal resolutions. Both require fast imaging methods which can be realized with high-speed and high-powered gradient systems and parallel imaging using phased array receive RF coils (Roemer, Edelstein et al. 1990; Kimmlingen, Eberlein et al. 2004). While standard gradients (i.e., G ≈ 45mT/m and SR ≈ 200mT/m/s) on most 3-Tesla scanners can be sufficient to image volumes as small as 1.5x1.5x1.5mm, faster (>400mT/m/s) and stronger (>70mT/m) gradients allow us to encode voxels as small as 0.5x0.5x0.5mm (Janssens, Keil et al. 2012; Kimmlingen, Eberlein et al. 2004). Multi-channel Phased array RF coils allow us to speed up the acquisition of a single image even further by acquiring multiple image fragments simultaneously (Roemer, Edelstein et al. 1990; Kimmlingen, Eberlein et al. 2004). These increases in speed and performance are important both to minimize image acquisition artifacts and to mitigate the effects of subject movement.

Finally, given that functionally-relevant signal fluctuations are a small (<5%) of the overall signal and their fidelity may be further compromised by the increase in spatial and temporal resolution, we will need to find ways to enhance the strength of our functional signals. While some of this may be achieved through specific experimental designs, MR contrast agents—specifically MION (mono-iron oxide nanoparticles; Mandeville, Marota et al. 1998; Vanduffel, Fize et al. 2001, Leitte et al 2002)—can boost the functional signal drastically, by 3-4 fold at 3-Tesla.

1.3. Access to specialized equipment

A major appeal of our approach is that it does not require large-scale specialized equipment, such as a dedicated animal MRI. Our techniques can be easily incorporated in any standard clinical scanner, preferably 3-Tesla system with multiple receive channels. While dedicated animal-only MRI systems certainly have their own merit, the cost of such systems, above the costs of clinical-grade scanners that are available at many research facilities, prohibits their purchase by all but the wealthiest institutions.

2. Our approach

Below we outline the basic setup used for primate imaging in our group. We begin with a discussion of animal restraint because other aspects of the setup are restricted by this component.

2.1. Animal restraint

A now-common method of head restraint for neurophysiology studies involves attaching a metal post directly to the skull with titanium screws which tend to fuse to bone tissue (Betelak, Margiotti et al. 2001; Adams, Economides et al. 2007). Metallic objects close to the head would preclude any attempt to image the brain due to strong susceptibility artifacts (Matsuura, Inoue et al. 2005), and thus this method has been modified to make it appropriate for use in the MRI environment (Logothetis, Guggenberger et al. 1999; Vanduffel, Fize et al. 2001). In place of titanium screws, especially made ceramic screws (zirconia) are used that result in very little noticeable susceptibility artifacts (Logothetis, Guggenberger et al. 1999; Scherberger, Fineman et al. 2003). The screws form a contact with the skull and the headpost is not directly held in place with these screws but instead it is embedded in acrylic that also embeds the ceramic screws. We have found that when used as a base material around the screws, a highly pure poly(methyl methacrylate) formulation (C & B METABOND) can result in a much more rigid and stable headpost preparation. In this way, a rigid connection is established between the headpost and skull (Betelak, Margiotti et al. 2001).

The head is then held in place using the headpost which is attached to the chair. The chair and the headpost holder are all made of non-metalic material, usually acrylic sheets and ultem components where rigidity and high tolerance against cracking is necessary—acrylic, while very rigid, cracks under force. Ultem components retain rigidity while being more resistant to cracking and breaking under pressure. It is important to note that several alternative methods of invasive head restraint have been proposed and used (e.g. Pigarev, Nothdurft et al. 1997; Keliris, Shmuel et al. 2007).

The design of the chair must take into account the placement of the RF coils, eye illumination, juice reward, as well as any other experimental attachments necessary. These items must be rigidly attached such that they interact very little with one another and are not disturbed by animal movements. As we will discuss further below, the rigidity of the RF coils is key to stable measurements, and the chair design must take this into consideration (AppliedPrototype 2011).

In our experience, an approach that balances animal comfort with strong restraint is key. Many researchers may find that too much restraint, such as additional restraint on limbs, may make the animal uncomfortable and result in reduced cooperation and thus performance. A welcomed contribution to this area is the development of non-invasive head-restraint methods, such as that recently developed by Srihasam and colleagues (Srihasam, Sullivan et al. 2010). This method does not require implanting of headposts while promising an adequate restrain on the head for animals without additional head apparatuses, such as recording wells.

2.2. Coil designs

Magnetic resonance signals are weak—the ratio of nuclei in the low energy state to high energy state, which is directly proportional to the strength of the measured resonance signal, is on the order of 0.0031% even at very high field strengths (e.g., 9.4T; De Graaf 2007, p. 7). This means that accurate and robust measurement of resonance signals requires sensitive RF coils. Coil sensitivity is affected by a number of factors, such as choice of conductor, size of the coil, and most importantly, distance of the coil to the brain. With reduced coil size (i.e, ~4cm diameter instead of >10cm diameter for human head coils) and close proximity to the brain (a tight-fitting coil rather than a one-size-fits-all helmet as used for humans), one can optimize coil sensitivity, but this comes at the cost of reduced field of view (FOV; e.g., Logothetis, Merkle et al. 2002). This reduced FOV is compensated by using an array of coils such that each coil becomes sensitive to one part of the brain, and the combination of all the coils together image the entire brain (Roemer, Edelstein et al. 1990). An added benefit of so-called multi-channel phased arrays is that image acquisition time can also be accelerated by a method analogous to image-tiling (Pruessmann, Weiger et al. 1999; Sodickson, Griswold et al. 1999). While this does reduce the image SNR by up to 33% (Dietrich, Raya et al. 2007), it benefits us by allowing for faster imaging which (a) reduces image deformations, (b) allows us greater temporal resolution, and (c) grants us a greater degrees of freedom in our statistical analysis. In short, with parallel imaging we can make more measurements in less time with a small sacrifice in image SNR (de Zwart, van Gelderen et al. 2006).

We now almost exclusively use phased-array coils in our studies because of the advantages listed above. Phase-array coils for human applications typically make use of a helmet design for a complete sampling of the head (e.g., Wiggins, Triantafyllou et al. 2006), but in our experience helmet designs for monkeys pose difficulties due to obstruction by the headpost. In place of a helmet, a two-paddle design has many advantages, such as being easy to place over the head of many different monkeys and allowing them to be placed tightly around the head, thus maximizing the sensitivity of the coils. We now routinely use both 4-channel (two paddles with two channels each; e.g. Khachaturian 2010) and 8-channel coils (two paddles with 4 channels each), with the latter allowing for up to 3-fold acceleration as well. The receive channels are paired with a local transmit coil placed just above the head.

Because fMRI measurements can be heavily affected by temporal noise in addition to image SNR, it is imperative that the RF coils are rigidly placed so as to minimize movement or vibration (Roemer, Edelstein et al. 1990; Mispelter, Lupu et al. 2006). This applies to both the transmit coils and the receive channels. Vibration of the transmit coil can produce variations in the transmit field, adding temporal noise to our measurements (Mispelter, Lupu et al. 2006). Vibrations of the receive channels also adds temporal noise because movement of the receive coils away from the brain has the effect of changing the intensity of the signal detected by the coil, thus bringing about large changes in image intensity (Mispelter, Lupu et al. 2006). The receive channels are designed for use at a specific distance to the brain (Mispelter, Lupu et al. 2006), and changes in this distance will negatively affect the sensitivity of the coil which will result in both a decrease in signal and an increase in noise.

2.3. Sequence setup

The ideal functional imaging sequence would grant us high resolution, geometrically veridical images with high SNR at very short intervals. There is unfortunately a delicate balancing act as parameters that increase our resolution also decrease our SNR and prolong our imaging time while decreasing image veridicality. It is impossible to prescribe imaging parameters as this will vary depending on the setup at hand, the types of coils used (single loop or phased arrays), the performance of the gradients, the desired resolution, and finally, the type of signal measured. Discussion of these parameters is beyond the scope of this chapter, but a few general guidelines are helpful in animal imaging.

Given the size of the monkey brain as compared to the human brain (80 compared to 1250cm^3; Gonen, Liu et al. 2008), voxel sizes used in humans (typically 2-3mm isotropic) would tend to be far too large for use in primate imaging as they would obstruct many small details of importance in the functional maps. While one may strive towards voxel sizes of 1mm or less, slightly above that is acceptable for imaging macaque brains. The size of the voxel is dictated by the size of the imaging matrix, which in turn dictates how many voxels are to be measured in a given slice. Naturally, the larger the matrix size, the longer it will take to complete an image. The time to acquire one 2-D image depends on a number of factors, including matrix size, echo time, bandwidth, partial Fourier encoding, parallel imaging, amongst others (Bernstein, King et al. 2004). Thus it is not feasible to provide a formula to calculate acquisition time of one slice as it depends on a number of factors, many of which are also dependent on available hardware specifications. In general, however, increasing the matrix size increases echo time (TE) and the slice acquisition time—the increasing TE causes increased image blurring and image deformation and increasing slice acquisition time decreases temporal resolution (Bernstein, King et al. 2004). While for BOLD imaging at 3T the optimal TE is around 30 msecs (Fera, Yongbi et al. 2004), with MION contrast agents, one can decrease the TE to below 20 msecs (Leite, Tsao et al. 2002; Leite and Mandeville 2006), allowing for faster slice acquisition, which decreases image deformations and permits higher temporal resolutions.

Because even a well-restrained animal will move during acquisitions, the effect of movements on image acquisition must be carefully considered (Goense, Whittingstall et al. 2010). For example, as is typical of many fMRI studies, we routinely use interleaved acquisition whereby odd slices are acquired first prior to even slices. However, if an animal moves during single frame, then the odd and even slices are misaligned and standard 3d registration methods will be unable to correct for this. Worse, the misalignment can then propagate through the time series if slice-time correction is performed on the dataset (Sladky, Friston et al. 2011). Fortunately, a simple solution to this problem is to image in plane of head motion. For example, head restrained animals generally make very few lateral movements or rotations, but can use their bodies to push and pull their heads—thus the plane of motion is mainly sagittal. If one were to acquire horizontal or coronal slices, then such motion would pass through multiple slices and it would be impossible to visualize the motion and correct for them, since each slice is also acquired at a different time point. However, by acquiring sagittal slices, one can take optimal images of the movements and correct for them by applying 2-D slice-by-slice registration such as that offered by 2dImgReg of the AFNI toolset (Cox 1996). This correction can be done prior to slice-time correction so as to minimize the effects of head shifts in the sagittal plane on the rest of the dataset.

3. Preprocessing steps

A typical preprocessing stream for human functional data involves slice-time correction, 3D motion correction, and 3 Gaussian smoothing of the data prior to statistical analysis (Ashby 2011; Strother 2006). While the same steps may be used for processing of primate data, a few optimizations can be made.

As described above, by taking sagittal slices one can image the movements of the monkey head more adequately and correct for the 2-D movement using freely available tools from the AFNI toolkit (namely 2dImgReg). This software tool allows the specification of one image to serve as the "base" image—an image where no motion was present between the two interleaved passes. Individual slices from the series are then registered to their target slice in this base image using only 2-D image registration. The result generates x,y and pitch rotation for each slice. Slice-based correction can improve 3-D registration because the latter cannot deal with situations where motion occurs during the acquisition of a volume and this can introduce errors in the 3-D correction (Kim, Boes et al. 1999). Thus while we have only begun to use 2-D in-plane realignment to correct for rapid head motion of the monkey, its utility has been previously demonstrated in human fMRI (Kim, Boes et al. 1999), although it is not widely used in that field.

Because it is customary to include motion-realignment parameters in the general linear model as nuisance variables, the output of 2-D in-plane registration per slice poses a problem because we typically have 50+slices, and thus would have over a 150 parameters to include in our model. This impractical setup can be avoided by reducing the movement parameters using principal components analysis—since the motion in all the slices will likely be largely correlated, reducing them with principle components analysis will likely yield a good estimate of 2-D head motion that can then be used in the general linear model.

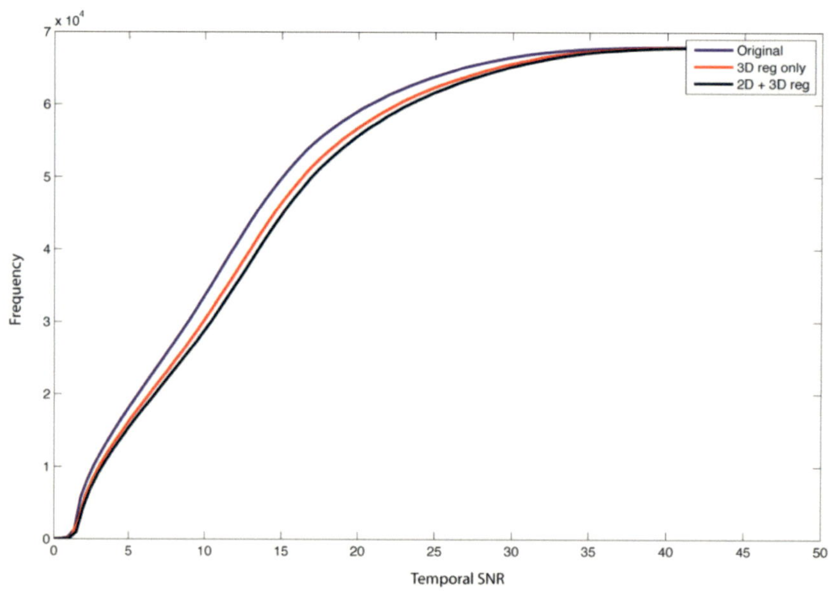

Figure 1. Cumulative histogram of voxel temporal SNR across different realignment strategies. The blue line depicts cumulative SNR without any motion correction. The red line depicts standard 3-D rigid motion correction, while the black line depicts the additional gain made by in-plane 2-D followed by 3-D rigid motion correction. 2-D in-plane registration can therefore improve temporal SNR, while the largest gain is made by 3-D rigid registration.

We can describe the potential sensitivity gain of such preprocessing steps by estimating temporal SNR, defined as time-course mean divided by time-course standard deviation

(Constantinides, Atalar et al. 1997). The cumulative histogram below (Figure 1) describes such a gain when using 2-D image realignment with sagittal slices followed by 3-D realignment, as opposed to the classic 3-D realignment typically performed. Over the entire volume of the brain (a crude measure), there is a gain of approximately 9% in temporal SNR by performing 3-D realignment, but this increases to an approximately 14% gain when it follows 2-D registration. Thus 2-D in-plane registration of sagittal slices followed by 3-D registration is a more appropriate strategy for image registration than classic 3-D realignment for motion correction.

Slice-time correction follows this 2-D motion correction because now the slices in the time series should be correctly aligned in space, permitting a meaningful correction to be performed in time. It is common to use sinc interpolation to correct for slice-time differences, but this method can propagate errors such as sudden changes in voxel values through the entire time series. It is thus advisable to use slower, more robust interpolators, such as Fourier or spline interpolators. Again, the AFNI tools offer a range of interpolators that can be used in place of sinc, as realized by the 3dTshift tool.

Three-dimensional movements, including lateral movements (if any) can then be corrected using any of a number of motion-correction tools—almost every package incorporates a motion correction algorithm, while a recent comparison (Oakes, Johnstone et al. 2005) suggests that those used by two of the more popular packages, namely SPM and AFNI, are quite comparable in performance, while the latter is substantially faster. Indeed, the AFNI tool 3dvolreg substantially outperforms SPM in 3-D motion realignment in speed and yields comparable results while not requiring the Matlab software package.

4. Functional-anatomical registration: Image deformation correction

Echo planar imaging (EPI), the method of choice for functional imaging, is prone to image deformations that can make the registration of the images to anatomical targets challenging (Farzaneh, Riederer et al. 1990; Andersson, Hutton et al. 2001). This is an issue that affects all EPI methods and animal imaging is no different. A number of solutions exist that either minimize deformations by acquiring segments of images more rapidly, or correct those deformations using informed or uninformed methods (i.e., using additional information about the imaging conditions obtained during the acquisition, or not).

First, a common method used to reduce deformation is to segment a single image into two or more pieces, and acquire each piece separately (Butts, Riederer et al. 1994). This method of segmented EPI is implemented prior to image reconstruction and involves acquiring interleaved lines of *k*-space in separate passes. The effect is that in each pass, very little time is spent acquiring the given lines, thus deformations are minimized. By acquiring multiple separate segments, the entire image is constructed out of minimally deformed acquisitions and the resulting image has high anatomical veridicality (Logothetis, Merkle et al. 2002; Goense, Whittingstall et al. 2010). The method, however, prolongs overall acquisition time substantially and is strongly vulnerable to even slight movements that would set the k-space lines out of alignment—thus it is very challenging to perform this with awake behaving animals. Using a more conventional single-shot method, one could also acquire an entire image with a smaller matrix—thus a lower resolution—which would reduce deformations but the decreased resolution has many other disadvantages including reduced anatomical veridicality (Bernstein, King et al. 2004).

Because of the challenges of effectively reducing deformations at the acquisition side, a number of efforts have been made to correct for deformations after acquisition. Here, two categories of approaches can be considered—the informed and uninformed methods. An informed method makes use of independent additional information about the deformations, such as a field map (Jezzard and Balaban 1995; Andersson, Hutton et al. 2001) to correct for them. The main idea behind the field map method is to acquire another un-deformed image at the same resolution and position as the deformation-afflicted images, and then calculate the degree of deformation between the ideal and acquired images. This deformation field map can then be used to correct deformations in all the functional images. The field-map method has proven generally satisfactory in human imaging, but for animal imaging it poses a serious challenge because it of the difficulty of making a stable field-map in an awake animal (see Jezzard 2012 for a review).

An interesting new development involves estimating deformations by measuring them in opposite directions (Holland, Kuperman et al. 2010). In echo-planar imaging, deformations are mostly in the phase-encoding direction (Jezzard and Balaban 1995). The method of Holland et al. (2010) involves acquiring two images in opposite phase-encoding directions. This means that each image will have deformation in the direction opposite to the other image (compression in one image will correspond to expansion in the other image). The balance between these two images would then be the un-deformed image, and the deformation map estimated from these two images can then be applied to the entire dataset. Furthermore, entire datasets or sessions can also be acquired in opposite phase-encoded directions to minimize imaging bias and combined so as to maximize overall image SNR— by combining minimally deformed voxels from forward and reverse phase-encoding directions.

4.1. 3-D Diffeomorphic demons correction

Image deformations can also be corrected by using the information available in the image itself in absence of secondary sources. In general, linear registration methods, such as affine registration methods that allow for translation, rotation, scale and shear, do not adequately correct for image deformations and one needs to make use of non-linear registration methods to achieve a good match between an EPI stack and an anatomical image. There are many non-linear methods that can be used to improve registration (for a review, see Zitova and Flusser 2003), but a danger of over-fitting is always present, as is the danger of creating new features —for example, if a gyrus in the functional image happens to fall closer to a sulcus after affine registration, some non-linear registration methods may force a dent in the gyrus to match the target image. We have found that non-linear registration using diffeomorphic demons (Vercauteren, Pennec et al. 2009) produces satisfactory and efficient results. The original demons algorithm (Thirion 1998) treats the registration problem as one of heat diffusion (an analogy to Maxwell's demons). In this formulation, the boundaries of the target image are considered permeable membranes that allow for the diffusion of the source (moving) image through the permeable interfaces by effectors (agents) in the membrane or image contour. The source (moving) image is warped on a deformable grid, which maintains spatial relations between image elements. When combined with diffeomorphism (Chef d'Hotel, Hermosillo et al. 2002; Vercauteren, Pennec et al. 2009), which treats two images as manifolds with a differentiable one-to-one displacement field between them, the diffeomorphic demons algorithm (Vercauteren, Pennec et al. 2009) provides for an efficient non-linear multimodal (i.e., EPI to T1-weighted) registration method. This registration method ensures that the spatial order of voxels is maintained during registration, so as to prevent excessive compression or expansion of features in the image. Figure 2 demonstrates the advantage of 3-D diffeomorphic registration of an EPI volume to an anatomical target, as opposed to a 12-parameter affine registration.

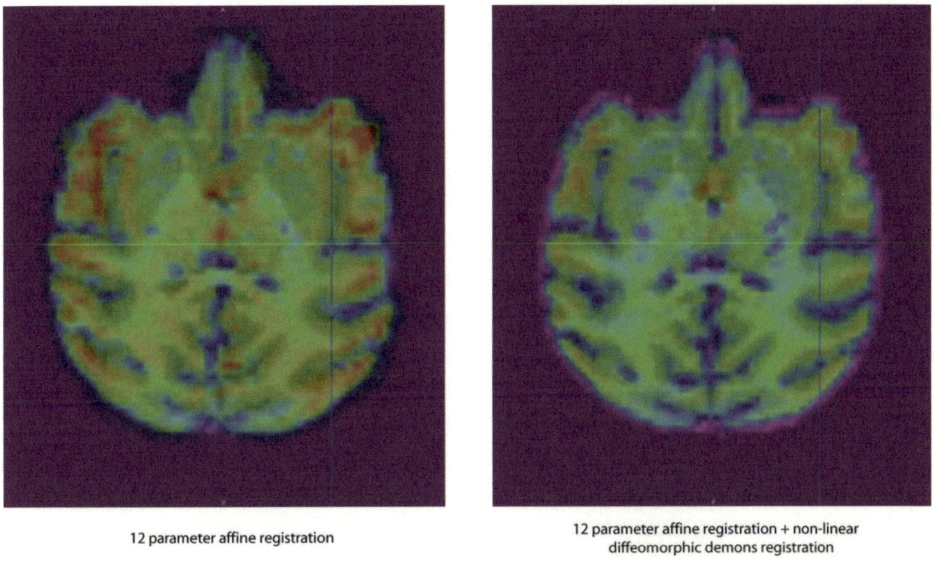

Figure 2. Comparison of 12-parameter affine registration (translation (x,y,z), rotation (pitch, roll, yaw), scale (ax,ay,az) and shear across x,y & z) with or without non-linear diffeomorphic registration for an EPI acquired with left-right phase encoding. The areas in red are regions that are poorly registered.

While EPI images in general have a very different contrast compared to MPRAGE anatomical images, images acquired with MION are even more different. This mismatch of the intensity distributions in the image can negatively affect the image registration. Hence it is useful to either use a mutual-information minimization strategy that avoids intensitybased matching, or modify the contrast profile of the T1-weighted anatomical such that it has a contrast comparable to the MION EPI images. In other words, we substitute a different image intensity look-up table such that the T1-weighted image looks more similar to a MION EPI image. The contrast profile or look-up table of an image may be modified through any of a number of image processing tools, including the Image Processing toolbox in Matlab or with the freely available Slicer package and its Histogram Matching module.

5. Statistical analysis

For the most part, statistical analysis of macaque functional images follows the same methods as those used in human imaging (Friston, Holmes et al. 1995). Specifically, the general linear model is used to make estimates of differential brain activity by convolving a known impulse response function, in our case the MION response function (Leite, Tsao et al. 2002; Leite and Mandeville 2006), with the onset of the stimulation events and adjusted for the duration of those events. The model response profile for all conditions are then fitted to the high-pass filtered data and the beta values estimated for two conditions are compared and tested for statistical significance with various corrections in place to reduce the possibility of Type I errors, such as Bonferroni correction, Gaussian Random Field (Worsley and Friston 1995), or false discovery rate (Benjamini and Hochberg 1995). However, a few additional steps can go a long way to decrease Type II errors and increase our statistical power, and these are discussed below.

5.1. Artifact rejection

The most overlooked but important aspect of functional image analysis is to simply look at the raw data and verify data quality. This may come as a daunting task, given the seemingly large amount of data. Yet a number of popular packages offer very simple methods of quickly assessing data quality with visual inspect and/or by quantitative methods. For example, the volume viewer in the FSL package (fslview) allows one to quickly load and animate through an entire functional volume time series while checking through in multiple orthogonal views for artefacts. The AFNI viewer allows one to rapidly click through voxels to visually check for unexpected signal changes. It is almost a fact that image fluctuations will occur when working with an awake and behaving animal, but the severity of the changes can be quickly assessed with these tools. Indeed, it is of great value to familiarize oneself with these tools to also evaluate the efficacy of the entire preprocessing stream.

A number of formal methods can be used to select and remove outliers from the dataset on the basis of additional data, such as behaviour of the animal, or based on movement estimates obtained from motion-correction tools (Goense, Whittingstall et al. 2010). In general it is preferable to exclude "bad" volumes from the analysis rather than correct for them, because the degree and effectiveness of any correction is hard to evaluate and indeed, some "corrections" may actually add more noise to the data than they remove. At least both AFNI's and SPM's GLM analysis tools (the latter via the ART repair toolkit; Mazaika, Hoeft et al. 2009) allows for exclusion of outliers. The ART repair tool readily allows for exclusion (or de-weighting) of the outlier volumes from the GLM analysis, with outliers selected on the basis of movement parameters and average signal intensity. In contrast, AFNI's GLM tool, 3dDeconvolve, allows for exclusion of volumes on any basis, allowing the experimenter to select volumes based on external measurements such as animal behaviour as well as on image realignment parameters, etc.

5.2. Behavioural and motion regressors

The exclusion of outliers can have drastic effects on reducing the noise in a dataset. Another approach is to reduce non-white noise by providing estimates of noise sources and incorporating this in the GLM analysis as regressors of no interest (Friston, Williams et al. 1996; Lund, Madsen et al. 2006). It is common and quite effective to include movement estimates obtained during motion correction in the GLM model. Additional regressors can be included, such as behaviours of no interest (e.g. saccades made during a fixation task) or external estimates of motion or jaw movement. A challenge with including regressors of no interest is that one does not know a priori how the regressors correspond to changes in voxel intensity values. For example, head movements will likely affect voxel values, but whether the direction of change or its magnitude or even its pattern would correspond to the movement estimates is not clear. In general, inclusion of additional estimates of noise sources does reduce variability in a dataset and is therefore recommended.

An approach that we are currently investigating is the inclusion of variability estimates obtained from non-brain tissue. In this approach, a brain mask is inverted to only select nonbrain voxels. Using principle components analysis, we then estimate a limited number of components that explain the majority of non-brain time-course variance. Finally, we include these estimates as regressors of no interest in the GLM analysis. In effect, we treat non-brain tissue as samples of noise, and estimate the main sources of this noise by reducing the data with principle components analysis. We assume that there are fewer noise sources than nonbrain voxels, and that the various noise sources result in correlated extraneous signals across multiple voxels. Based on these assumptions, we can then seek to estimate the noise sources by estimating principle components of the time-course of non-brain voxels, which will highlight noise components that may be distributed across a number of voxels. Noise signals may arise from eye, head, or jaw movement, or instabilities in the RF coil. The figure below (Figure 3) describes the amount of variance explained in the brain by non-brain components. Note that the components well capture the large fluctuations in the jaw muscles and the eyes, but also capture a significant amount of variability in the brain. This method is still under investigation, but it seems a quite promising method to reduce variability in time course data.

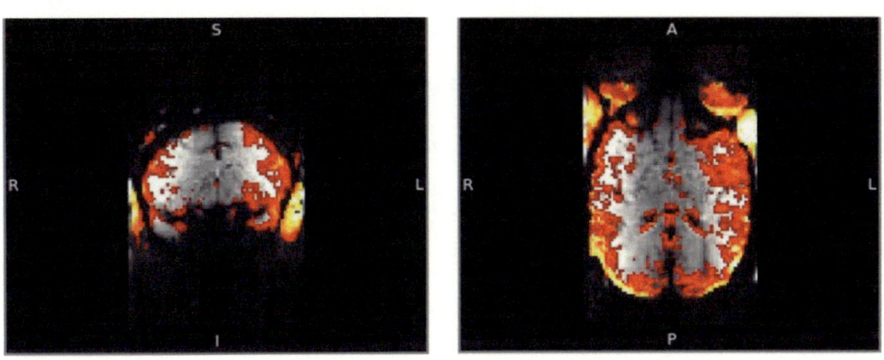

Figure 3. Variance explained by 20 non-brain principle components (corrected FWE < 0.05)

5.3. Experimental design strategies to minimize movement of an uncooperative animal subject

A major challenge of working with animals is that one has to motivate them to perform the task at hand. Typically this is done by restricting their fluid intake so that they are motivated to work for a liquid reward during the scans. However, even then it can be difficult or impossible to completely ensure the animal complies with the experimenter's wishes. The optimal solution involves motivating the animal so that it works for fluids, habituating it to the scanner environment by training it frequently in a simulated scanner environment, and minimizing the effect that its movements can have on the experiment itself. We routinely habituate our animals for months before the actual scans, and even then they are habituated in the scanner for a number of scans before their performance compares to that obtained outside of the scanner. The animal restraint, as discussed above, must be balanced against animal comfort given that the animal may be restrained for several hours at a time.

6. Combining methods with fMRI

While functional imaging in macaques bridges a crucial gap in comparing the function of human and non-human primate brains (Vanduffel, Fize et al. 2002; Orban, Fize et al. 2003; Orban, Claeys et al. 2005; Nelissen, Vanduffel et al. 2006; Kolster, Mandeville et al. 2009), a major advantage of working with an animal model is the ability to directly manipulate the brain to assess the causal interactions of brain networks. Our lab and others have now successfully implemented electrical microstimulation of the brain simultaneous with functional imaging to assess the role of FEF microstimulation on the cortical response to simple and complex visual stimuli (Ekstrom, Roelfsema et al. 2008; Ekstrom, Roelfsema et al. 2009), stimulation of face-selective patches in the macaque cortex (Moeller, Freiwald et al. 2008) and role of microstimulation in the propagation of information in the cortex more generally (Tolias, Sultan et al. 2005; Logothetis, Augath et al. 2010). The information obtained by this combination could not have been garnered using other methods, such as single-unit recordings, as this would require simultaneously measuring over large swaths of cortex—a task that is virtually impossible.

Another important development is the temporary deactivation of a brain area to assess its role in a more complex network. The application of muscimol, a GABA agonist, to temporarily shut down an area is now common and many labs have successfully applied this to measure the effect of deactivation on behaviour and single-unit responses. Our lab has now extended this to use with fMRI to allow us to measure both the brain response and its correlated behavioural change (Gerits, Wardak et al. 2009).

In addition to temporary deactivation, a number of groups have successfully combined local cortical lesions with fMRI. Pinsk, Moore, et al. (2005) were first to demonstrate feasibility of performing fMRI in awake-behaving macaques with straiate cortex lesions. Lesions to the striate cortex in humans lead to a phenomenon termed "blindsight" (see Leopold, 2012, for a recent review). Blindsight patients are essentially unaware of visual stimulation, yet they are able to respond to such stimuli without conscious awareness. Pinsk, Moore et al. (2005) were first to demonstrate that following striate cortex lesions, the extrastriate visual cortex continues to be responsive to visual stimuli as measured by fMRI. This finding was later corroborated by Schmid, Panagiotaropoulos et al. (2009) who showed that V2 and V3 in awake-behving macaques with focal striate lesions continue to be responsive to visual stimulation, albeit at 70% the pre-lesion rate, when measured with fMRI. They further demonstrated that retinotopy in V2 and V3 corresponding to the lesion in V1 did not change, suggesting that post-lesion activity in these areas was not driven by peri-lesion V1. This further highlighted the possibility that blindsight is mediated by sub-cortical inputs. Shmid et al (2010) provided the conclusive evidence that blindsight is mediated by sub-cortical inputs to extrastriate cortex—specifically the LGN. Using fMRI in awake-behving macaques with striate lesions, they first corroborated earlier findings of Pinsk et al (2005) and Schmid et al. (2009) and demonstrated that extrastriate cortex remains active following striate cortex, but at a lower level. They then temporarily deactivated the LGN with a GABA agonist while measuring the responses of the visual cortex of the macaques using fMRI. They found that deactivation of the LGN in striate-lesioned macaques completely abolished the residual response of the extrastriate cortex to visual inputs as well as residual behavioural function. This was the first direct evidence that projections from LGN to extrastriate cortex allow for blindsight to occur, and demonstrates the power of combining fMRI with causal manipulations to answering fundamental questions in systems neuroscience.

An exciting new development by our group is the integration of optogenetic methods (Boyden, Zhang et al. 2005) with fMRI in awake-behaving macaques (Gerits, Farivar et al. 2012). We have demonstrated that optogenetic stimulation of the frontal eye fields results in activation in a number of cortically-connected nodes, and for the first time, that optogenetic stimulation in primates can affect behavioural performance—in our case, saccade latency in a visually-guided task (Gerits, Farivar et al. 2012). These results are important for the further development and refinement of optogenetic methods to select sub-population of neurons in the cortex and evaluate their contribution to global brain activity as well as behaviour.

While fMRI provides images of population neural activity based on indirect measures, it is possible to combine it with simultaneous electrophysiological recordings (Logothetis, Pauls et al. 2001; Logothetis, Kayser et al. 2007; Oeltermann, Augath et al. 2007). This combination helps overcome an important challenge in interpreting the fMRI response and provide a confirmation and an additional metric of the underlying neural processes. However, a more practical approach is to use fMRI to guide subsequent electrophysiology outside of the MRI (e.g. Tsao, Freiwald et al. 2006; Freiwald and Tsao 2010). Of course, this does not allow for direct links to be made between BOLD signal fluctuations and underlying spiking activity, but it does allow for more refined examination of a patch of tissue identified by fMRI using single-cell recordings.

7. Impact and future directions

Functional MRI is useful for measuring activity over the whole brain, for identifying regions of importance for subsequent analysis, for assessing brain-wide effects of lesions, deactivations, or selected stimulations. Performing fMRI in awake-behaving macaques is not trivial and requires reconsideration of a number of factors, including restraint, training, coil design, image acquisition, data pre-processing, and additional refinements of the statistical analysis steps. Implementation of the solutions described above contributes to the feasibility of performing fMRI in awake-behving macaques, which opens the door to answering a large number of questions in systems neuroscience. The power of this utility is best realized when combined with temporary causal manipulations of cortical function, such as combining with electrical or optogenetic stimulation or with temporary and/or permanent deactivation, or in conjunction with single-cell recordings to understand in detail the contribution of single cell responses to behavioural outcomes.

Author details

Reza Farivar[1,2] and Wim Vanduffel[1,3]

1 Massachusetts General Hospital and Harvard Medical School, USA

2 McGill Vision Research Unit, McGill University, Canada

3 Katholieke Universiteit Leuven, Belgium

References

[1] Adams, D. L., J. R. Economides, et al. (2007). "A biocompatible titanium headpost for stabilizing behaving monkeys." J Neurophysiol 98(2): 993-1001.

[2] Andersson, J. L., C. Hutton, et al. (2001). "Modeling geometric deformations in EPI time series." Neuroimage 13(5): 903-919.

[3] Applied Prototype. (2011). "Vanduffel/HMS fMRI chair." from http://appliedprototype.com/AppliedPrototype/Vanduffel.html.

[4] Ashby, F. G. (2011). Statistical analysis of fMRI data. Cambridge, Mass., MIT Press.

[5] Benjamini, Y. and Y. Hochberg (1995). "Controlling the false discovery rate: A practical and powerful approach to multiple testing." Journal of the Royal Statistical Society, Series B (Methodological) 57: 289-300.

[6] Bernstein, M. A., K. F. King, et al. (2004). Handbook of MRI pulse sequences. Amsterdam ; Boston, Academic Press.

[7] Betelak, K. F., E. A. Margiotti, et al. (2001). "The use of titanium implants and prosthodontics techniques in the preparation of non-human primates for long-term neuronal recording studies." J Neurosci Methods 112(1): 9-20.

[8] Boyden, E. S., F. Zhang, et al. (2005). "Millisecond-timescale, genetically targeted optical control of neural activity." Nat Neurosci 8(9): 1263-1268.

[9] Butts, K., S. J. Riederer, et al. (1994). "Interleaved echo planar imaging on a standard MRI system." Magn Reson Med 31(1): 67-72.

[10] Chef d'Hotel, C., G. Hermosillo, et al. (2002). Flows of diffeomorphisms for multimodal image registration. IEEE International Symposium on Biomedical Imaging. Constantinides, C. D., E. Atalar, et al. (1997). "Signal-to-noise measurements in magnitude images from NMR phased arrays." Magn Reson Med 38(5): 852-857.

[11] Cox, R. W. (1996). "AFNI: software for analysis and visualization of functional magnetic resonance neuroimages." Comput Biomed Res 29(3): 162-173.

[12] De Graaf, R. A. (2007). In vivo NMR spectroscopy : principles and techniques. Chichester, West Sussex, England ; Hoboken, NJ, John Wiley & Sons.

[13] de Zwart, J. A., P. van Gelderen, et al. (2006). "Accelerated parallel imaging for functional imaging of the human brain." NMR Biomed 19(3): 342-351.

[14] Dietrich, O., J. G. Raya, et al. (2007). "Measurement of signal-to-noise ratios in MR images: influence of multichannel coils, parallel imaging, and reconstruction filters." J Magn Reson Imaging 26(2): 375-385.

[15] Ekstrom, L. B., P. R. Roelfsema, et al. (2008). "Bottom-up dependent gating of frontal signals in early visual cortex." Science 321(5887): 414-417.

[16] Ekstrom, L. B., P. R. Roelfsema, et al. (2009). "Modulation of the contrast response function by electrical microstimulation of the macaque frontal eye field." J Neurosci 29(34):10683-10694.

[17] Farzaneh, F., S. J. Riederer, et al. (1990). "Analysis of T2 limitations and off-resonance effects on spatial resolution and artifacts in echo-planar imaging." Magn Reson Med 14(1): 123-139.

[18] Fera, F., M. N. Yongbi, et al. (2004). "EPI-BOLD fMRI of human motor cortex at 1.5 T and 3.0 T: sensitivity dependence on echo time and acquisition bandwidth." J Magn Reson Imaging 19(1): 19-26.

[19] Freiwald, W. A. and D. Y. Tsao (2010). "Functional compartmentalization and viewpoint generalization within the macaque face-processing system." Science 330(6005): 845-851.

[20] Friston, K. J., A. P. Holmes, et al. (1995). "Statistical parametric maps in functional imaging: A general linear approach." Human Brain Mapping 2(4): 189-210.

[21] Friston, K. J., S. Williams, et al. (1996). "Movement-related effects in fMRI time-series." Magn Reson Med 35(3): 346-355.

[22] Gerits, A., C. Wardak, et al. (2009). Behavioral and brain-wide functional consequences of reversible LIP inactivation during visual search. Society for Neuroscience Annual Meeting, Chicago, IL, Online.

[23] Gerits, A., R. Farivar, et al. (2012). "Optogenetically Induced Behavioral and Functional Network Changes in Primates." Curr Biol.

[24] Goense, J. B., K. Whittingstall, et al. (2010). "Functional magnetic resonance imaging of awake behaving macaques." Methods 50(3): 178-188.

[25] Gonen, O., S. Liu, et al. (2008). "Proton MR spectroscopic imaging of rhesus macaque brain in vivo at 7T." Magn Reson Med 59(4): 692-699.

[26] Hackensack, NJ, Imperial College Press ; Distributed by World Scientific.

[27] Holland, D., J. M. Kuperman, et al. (2010). "Efficient correction of inhomogeneous static magnetic field-induced distortion in Echo Planar Imaging." Neuroimage 50(1): 175-183.

[28] Janssens, T., B. Keil, et al. (2012). "An implanted 8-channel array coil for high-resolution macaque MRI at 3T." Neuroimage 62(3): 1529-1536.

[29] Jezzard, P. (2012). "Correction of geometric distortion in fMRI data." Neuroimage 62(2): 648-651.

[30] Jezzard, P. and R. S. Balaban (1995). "Correction for geometric distortion in echo planar images from B0 field variations." Magn Reson Med 34(1): 65-73.

[31] Keliris, G. A., A. Shmuel, et al. (2007). "Robust controlled functional MRI in alert monkeys at high magnetic field: effects of jaw and body movements." Neuroimage 36(3): 550-570.

[32] Khachaturian, M. H. (2010). "A 4-channel 3 Tesla phased array receive coil for awake rhesus monkey fMRI and diffusion MRI experiments." J Biomed Sci Eng 3(11): 1085-1092.

[33] Kim, B., J. L. Boes, et al. (1999). "Motion correction in fMRI via registration of individual slices into an anatomical volume." Magn Reson Med 41(5): 964-972.

[34] Kimmlingen, R., E. Eberlein, et al. (2004). An easy to exchange high performance head gradient insert for a 3T whole body MRI system: First results. Procedings of International Society for Magnetic Resononacne Medicine. 11.

[35] Kolster, H., J. B. Mandeville, et al. (2009). "Visual field map clusters in macaque extrastriate visual cortex." J Neurosci 29(21): 7031-7039.

[36] Leite, F. P. and J. B. Mandeville (2006). "Characterization of event-related designs using BOLD and IRON fMRI." Neuroimage 29(3): 901-909.

[37] Leite, F. P., D. Tsao, et al. (2002). "Repeated fMRI using iron oxide contrast agent in awake, behaving macaques at 3 Tesla." Neuroimage 16(2): 283-294.

[38] Leopold, D. A. (2012). "Primary visual cortex: awareness and blindsight." Annu Rev Neurosci 35: 91-109.

[39] Liu, Y., E. A. Yttri, et al. (2010). "Intention and attention: different functional roles for LIPd and LIPv." Nat Neurosci 13(4): 495-500.

[40] Logothetis, N. K., C. Kayser, et al. (2007). "In vivo measurement of cortical impedance spectrum in monkeys: implications for signal propagation." Neuron 55(5): 809-823.

[41] Logothetis, N. K., H. Guggenberger, et al. (1999). "Functional imaging of the monkey brain." Nat Neurosci 2(6): 555-562.

[42] Logothetis, N. K., J. Pauls, et al. (2001). "Neurophysiological investigation of the basis of the fMRI signal." Nature 412(6843): 150-157.

[43] Logothetis, N. K., M. Augath, et al. (2010). "The effects of electrical microstimulation on cortical signal propagation." Nat Neurosci 13(10): 1283-1291.

[44] Logothetis, N., H. Merkle, et al. (2002). "Ultra high-resolution fMRI in monkeys with implanted RF coils." Neuron 35(2): 227-242.

[45] Lund, T. E., K. H. Madsen, et al. (2006). "Non-white noise in fMRI: does modelling have an impact?" Neuroimage 29(1): 54-66.

[46] Mandeville, J. B., J. J. Marota, et al. (1998). "Dynamic functional imaging of relative cerebral blood volume during rat forepaw stimulation." Magn Reson Med 39(4): 615-624.

[47] Matsuura, H., T. Inoue, et al. (2005). "Quantitative analysis of magnetic resonance imaging susceptibility artifacts caused by neurosurgical biomaterials: comparison of 0.5, 1.5, and 3.0 Tesla magnetic fields." Neurol Med Chir (Tokyo) 45(8): 395-398; discussion 398-399.

[48] Mazaika, P., F. Hoeft, et al. (2009). Methods and Software for fMRI Analysis for Clinical Subjects. Human Brain Mapping.

[49] Mispelter, J., M. Lupu, et al. (2006). NMR probeheads for biophysical and biomedical experiments : theoretical principles & practical guidelines. London

[50] Moeller, S., W. A. Freiwald, et al. (2008). "Patches with links: a unified system for processing faces in the macaque temporal lobe." Science 320(5881): 1355-1359.

[51] Nelissen, K., W. Vanduffel, et al. (2006). "Charting the lower superior temporal region, a new motion-sensitive region in monkey superior temporal sulcus." J Neurosci 26(22): 5929-5947.

[52] Oakes, T. R., T. Johnstone, et al. (2005). "Comparison of fMRI motion correction software tools." Neuroimage 28(3): 529-543.

[53] Oeltermann, A., M. A. Augath, et al. (2007). "Simultaneous recording of neuronal signals and functional NMR imaging." Magn Reson Imaging 25(6): 760-774

[54] Orban, G. A., D. Fize, et al. (2003). "Similarities and differences in motion processing between the human and macaque brain: evidence from fMRI." Neuropsychologia 41(13): 1757-1768.

[55] Orban, G. A., K. Claeys, et al. (2005). "Mapping the parietal cortex of human and non-human primates." Neuropsychologia.

[56] Pigarev, I. N., H. C. Nothdurft, et al. (1997). "A reversible system for chronic recordings in macaque monkeys." J Neurosci Methods 77(2): 157-162.

[57] Pinsk, M. A., T. Moore, et al. (2005). "Methods for functional magnetic resonance imaging in normal and lesioned behaving monkeys." J Neurosci Methods 143(2): 179-195.

[58] Pruessmann, K. P., M. Weiger, et al. (1999). "SENSE: sensitivity encoding for fast MRI." Magn Reson Med 42(5): 952-962.

[59] Roemer, P. B., W. A. Edelstein, et al. (1990). "The NMR phased array." Magn Reson Med 16(2): 192-225.

[60] Scherberger, H., I. Fineman, et al. (2003). "Magnetic resonance image-guided implantation of chronic recording electrodes in the macaque intraparietal sulcus." J Neurosci Methods 130(1): 1-8.

[61] Schmid, M. C., S. W. Mrowka, et al. (2010). "Blindsight depends on the lateral geniculate nucleus." Nature 466(7304): 373-377.

[62] Schmid, M. C., T. Panagiotaropoulos, et al. (2009). "Visually driven activation in macaque areas V2 and V3 without input from the primary visual cortex." PLoS One 4(5): e5527.

[63] Sladky, R., K. J. Friston, et al. (2011). "Slice-timing effects and their correction in functional MRI." Neuroimage.

[64] Sodickson, D. K., M. A. Griswold, et al. (1999). "Signal-to-noise ratio and signal-to-noise efficiency in SMASH imaging." Magn Reson Med 41(5): 1009-1022.

[65] Srihasam, K., K. Sullivan, et al. (2010). "Noninvasive functional MRI in alert monkeys."Neuroimage 51(1): 267-273.

[66] Strother, S. C. (2006). "Evaluating fMRI preprocessing pipelines." IEEE Eng Med Biol Mag 25(2): 27-41.

[67] Thirion, J. P. (1998). "Image matching as a diffusion process: an analogy with Maxwell's demons." Med Image Anal 2(3): 243-260.

[68] Tolias, A. S., F. Sultan, et al. (2005). "Mapping cortical activity elicited with electrical microstimulation using FMRI in the macaque." Neuron 48(6): 901-911.

[69] Tsao, D. Y., W. A. Freiwald, et al. (2006). "A cortical region consisting entirely of face-selective cells." Science 311(5761): 670-674.

[70] Vanduffel, W., D. Fize, et al. (2001). "Visual motion processing investigated using contrast agent-enhanced fMRI in awake behaving monkeys." Neuron 32(4): 565-577.

[71] Vanduffel, W., D. Fize, et al. (2002). "Extracting 3D from motion: differences in human and monkey intraparietal cortex." Science 298(5592): 413-415.

[72] Vercauteren, T., X. Pennec, et al. (2009). "Diffeomorphic demons: efficient non-parametric image registration." Neuroimage 45(1 Suppl): S61-72.

[73] Wiggins, G. C., C. Triantafyllou, et al. (2006). "32-channel 3 Tesla receive-only phased-array head coil with soccer-ball element geometry." Magn Reson Med 56(1): 216-223.

[74] Worsley, K. J. and K. J. Friston (1995). "Analysis of fMRI time-series revisited--again."Neuroimage 2(3): 173-181.

[75] Zitova, B. and J. Flusser (2003). "Image registration methods: a survey." Image and Vision Computing 21(11): 977-1000.

Chapter 7

Brain Functional Networks in the Developing Brain Using Resting BOLD

Wei Gao, Hongtu Zhu, Kelly S. Giovanello,
J. Keith Smith, Dinggang Shen, John H. Gilmore and
Weili Lin

Additional information is available at the end of the chapter

http://dx.doi.org/10.5772/30640

1. Introduction

One of the most astounding transformations that takes place in the living world is the transformation of human brain from an earthworm like neural tube in the embryo to a well organized system consisting of billions of neurons, trillions of synapses, as well as supporting cells/structures. Equally astounding is the emergence of various complex functions that accompany this remarkable structural transformation. In the long query to shed light on brain's functional development process, two questions – (1) how and when specific functional modules emerge; and (2) how different functional modules integrate/coordinate to form an efficient brain system, lie in the core. Traditional behavioral studies (Amsterdam, 1972; Cepeda et al., 2001; Diesendruck et al., 2004) have provided valuable information about the developmental process, but such studies do not offer insights into the anatomical and functional development of the brain in relation to the observed behavioral maturation. Neuroimaging studies, in contrast, (including positron emission tomography (PET), magnetic resonance imaging (MRI), etc) enable either direct or indirect elucidation of the underlying neuronal activity, potentially, unveiling the neurophysiological mechanisms guiding the developmental process. Insights from such studies are also critical for better understanding of the neuropathology underlying various developmental disorders, whose early diagnosis/intervention/treatment may benefit greatly from the quantitative and objective nature of these investigations.

Traditional task-based functional magnetic resonance imaging (fMRI) techniques have been a powerful and non-invasive tool for adult brain imaging for decades (Bandettini et al., 1992;

Belliveau et al., 1991; Ogawa et al., 1990) and its application in developmental investigation has also been long practiced. The feasibility of fMRI in newborns was first demonstrated by Born et al in 1996 (Born et al., 1996), who reported a differential visual activation pattern between infants and older children. Next, Hykin et al (Hykin et al., 1999) successfully conducted fMRI experiments in fetuses in 1999. Other groups have subsequently reported different functional activation patterns during both pre-and postnatal periods (Anderson et al., 2001; Born et al., 1998; Dehaene-Lambertz et al., 2002; Fulford et al., 2003). However, constrained by the inability to "perform tasks," only visual, auditory or tactile stimuli could be applied in such investigations, greatly limiting the scope of brain functions that were explored using this technique.

The recent advance of resting state functional connectivity MRI (fcMRI) technique (Biswal et al., 1995; Lowe et al., 1998), however, has greatly expanded the scope of fMRI beyond taskbased approaches and provided a unique opportunity to answer fundamental questions about the brain's functional network organization. Initially described in Biswal et al's seminal paper (Biswal et al., 1995), fcMRI measures temporal synchronization of spontaneous blood oxygen level dependent (BOLD) fluctuations at rest, eliminating the need for external stimuli/tasks. Thus, fcMRI greatly simplifies the experimental design and provides unique information regarding the brain's functional organization that is not available through traditional task-based fMRI analysis. In adult subjects, various resting state functional networks (RSNs) have been consistently reported including a motor-sensory network (Biswal et al., 1995), a visual network (Lowe et al., 1998), attention networks (Fox et al., 2006), a memory network (Vincent et al., 2006), and a network mediating self-related thinking (i.e., the brain's default-mode network) (Greicius et al., 2003; Raichle et al., 2001; Shulman GL, 1997). Results from these RSNs are highly consistent with the known anatomical connection and/or functional coactivation patterns (Biswal et al., 1995). More importantly, fcMRI is particularly suited for studies where the target population is not able to perform specific cognitive tasks as required by traditional fMRI, such as studies of early brain development and neurological diseases (Greicius et al., 2004; Liu et al., 2008; Seeley et al., 2009).

Although fcMRI has been successfully applied in numerous studies aimed at unveiling the brain's functional organization (e.g., development, aging, disease), several issues could affect the robustness/interpretation of the obtained results. Among them, the physiological underpinnings of fcMRI signal remain elusive. Since the BOLD signal provides only an indirect measure of the underlying neuronal activity, a pressing need for understanding of the neuronal correlates of the fMRI signal has been widely recognized. The consensus is that task-driven BOLD signal is highly correlated with local field potential (LFP), which is a complex signal arising from integrated electrical activity in the pre-and postsynaptic terminals recorded in the vicinity of embedded microelectrodes in the cortex (Logothetis, 2003; Logothetis et al., 2001; Raichle and Mintun, 2006). However, neuronal dynamics resulting from external tasks are likely fundamentally different from that observed in fcMRI signal. As a result, a direct translation from the previous findings of task-driven fMRI to fcMRI interpretation may be untenable. To directly delineate the underlying neural correlates of spontaneous BOLD signal fluctuations, several groups have made important discoveries. First, using an anesthetized

http://dx.doi.org/10.5772/30640

monkey model, Shmuel and Leopold (Shmuel and Leopold, 2008) demonstrated a significant, time-lagged correlation between spontaneous BOLD fluctuations and the power of neuronal activity in the local filed potential gamma band, multi-unit activity, and spiking activity, directly supporting the notion that fcMRI correlation reflects the underlying neuronal interplay between remote cortical regions. Second, by extracting signal from the depth electrodes placed in the auditory cortex of epilepsy patients, Nir et al (Nir et al., 2008) found significant inter-hemispheric correlations in the slow modulations of neuronal firing rates and gamma (40~100Hz) LPF power with a clear spatial selectivity, further revealing the underlying neuronal correlates of fcMRI signal. Additionally, He et al (He et al., 2008) also examined the correlation structure of slow cortical potential (SCP) signal recorded from the implanted electrodes in a similar epilepsy patient study and showed a consistent correlation structure with that from fcMRI signals, providing evidence of the possible contribution of SCP in the observed fcMRI correlation structure. Overall, although some discrepancies exist among the studies examining the neuronal correlates of the observed fcMRI signal, the common finding of significant neuronal activity correlation strongly supports the notion that functional organization structures observed in fcMRI experiments reflect underlying neuronal interplay. Future studies are still needed to further discern the physiological origins of fcMRI signal.

While fcMRI has been utilized to examine a wide variety of neurological diseases (Fair et al., 2010; Greicius et al., 2004), aging (Achard and Bullmore, 2007), as well as brain development (Fair et al., 2008; Fransson et al., 2007; Gao et al., 2009), emphasis in this chapter is given to how fcMRI can be used to explore temporal and spatial evolution of early brain functional development. In particular, we will focus on the first two years of life – the most critical time period for postnatal brain development (Tau and Peterson, 2010). As early as the fifth week after conception, the first synapses begin forming in the fetus's spinal cord which permits the first fetal movements such as spontaneous arches and curls of the whole body (Joseph, 1999). More complicated and coordinated movements such as stretching, yawning, sucking, swallowing, and grasping soon follow in the first and second trimesters. Toward the end of the prenatal period, fetuses begin to respond to familiar odors and sounds (Joseph, 1999). It is clear that nature ensures such vital and basic bodily functions are in place by the time a baby emerges from the supporting womb.

After prenatal development, the first two years of life are commonly viewed as the most critical time period for postnatal brain development (Tau and Peterson, 2010). At birth, the brain is approximately one-quarter of its adult size and quickly grows to an adult-like volume (around 80%) by the end of the second year. The dramatic increase in the number of dendrites and synapses is partly responsible for such a fast volume increase. In fact, the first two years of life are an exuberant period for synapse growth and cerebral cortex produces most of its synaptic connections. By the end of the second year, a toddler's cerebral cortex contains more than one thousand trillions of synapses, which are far more than that in an adult's brain (around 50 percent more) and will experience a prolonged, experiencedependent pruning process to achieve its matured configuration (Johnson, 2000, 2001; Tau and Peterson, 2010). Axonal elongation, thickening, and myelination processes are also important growth factors during this period (Gao et al., 2008; Haynes et al., 2005). It has been documented that long axonal connections are nearly complete around 9 months of age (Conel, 1939-1963; Tau and Peterson, 2010). Moreover, continuous myelination of white matter during the first two years of life greatly improves the brain's electrical signal transfer efficiency (Haynes et al., 2005). Concurrent with the above mentioned dramatic structural development is the parallel improvement in brain function. Decasper and Fifer (DeCasper and Fifer, 1980) demonstrated that the sensorimotor experiences of fetus and voice of mother (Kolata, 1984) can be recognized. Additionally, Fivush and Hamond (Fivush, 1990) showed that 2yr olds can already retrieve much detail about a trip to the zoo. Studies on toddlers also revealed that 18-to 24-month-olds are able to use speaker's gaze direction (Baldwin, 1993) and affective expression (Tomasello et al., 1996) as cues leading to speaker's communicative purposes. Akhtar and Tomasello (Akhtar, 2000) further proposed that children are able to infer the meaning of words through an understanding of people's minds (Diesendruck et al., 2004). More importantly, one of the unique properties of human self-awareness has also been documented to undergo major development during the first two years of life (Amsterdam, 1972). The gradual emergence of such higher order cognitive functions leads to the hypothesis of parallel establishment of corresponding functional networks during this critical period. Moreover, as more complex functions emerge, the brain's information transfer efficiency is also expected to improve. Leveraging on graph theory analysis (Bullmore and Sporns, 2009), brain's information transfer efficiency can be effectively measured based on fcMRI time series data.

Although this chapter primarily focuses on the applications of fcMRI in normal brain development, fcMRI studies in developmental disorders are also briefly reviewed. This chapter is organized as follows: Section 2 provides a review of data acquisition, preprocessing and various analysis approaches for fcMRI study; Section 3 illustrates recent findings on the emergence of various functional networks; Section 4 shifts attention from individual networks to the whole brain network; Section 5 reviews recent applications of fcMRI on various developmental disorders; Section 6 provides a discussion of the limitations and future directions of the use of fcMRI in developmental studies; and finally, Section 7 provides a conclusion.

2. Data acquisition and analysis

The acquisition of resting state fcMRI data is identical to that of traditional task-based fMRI (except no task/stimuli are required during the acquisition) for which echo planer imaging (EPI) is the most frequently used imaging sequence due to its fast scan time. Using an EPI sequence, a 2D slice can be acquired in approximately 40 to 100ms and 2 to 3 seconds for whole brain coverage, which alleviates motion-related artifacts during imaging. However, this sequence is prone to other artifacts related to either hardware (e.g., eddy current effect) or imaging sample (e.g., T2* related susceptibility effect). Moreover, although EPI is already a fast imaging technique, the time resolution (~2 seconds) is still far from sufficient to characterize the dynamic of underlying neuronal activity and unable to faithfully sample several physiological factors such as respiratory/cardiac cycles resulting in potential noise contamination. Therefore, further improvement in imaging acquisition methods for faster scanning is greatly desired (Feinberg et al., 2010). Another important factor in fcMRI experiments is the required duration of acquisition. Van Dijk et al. (Van Dijk et al., 2010) compared the correlation strengths resulting from different lengths of data acquisitions (from ~4min to ~12min) and reported that they stabilized at 5-6 min of data acquisition. More recently, Whitlow et al. (Whitlow et al., 2011) reported similar findings and further proposed that graph theory metrics of network connectivity might be accurately calculated from as little as 1.5~2 min of fcMRI data. Besides these technical considerations, fcMRI study requires that subjects are at a "resting" state. In adult studies, different ways of keeping subjects at resting are utilized such as fixation, eyes closed resting (ECR), and eyes open resting (EOR). Van Dijk et al. (Van Dijk et al., 2010) compared the functional connectivity strength within default and attention networks during these different resting states and reported significant differences between ECR and fixation, as well as between ECR and EOR, but not between EOR and fixation, indicating a significant effect of eyes opening versus eyes closing. However, similar comparisons on other functional networks are lacking and the conclusion might not be directly generalized. In developmental studies, keeping the subjects "resting" and still within the scanner at wake condition proves to be extremely challenging. Therefore, most developmental studies, especially those focusing on the first two years of life, scan their subjects during sleeping state (either natural sleeping or sedation-induced sleeping). For older children, as keeping them sleeping within the scanner becomes harder, watching movie is an alternative for keeping them still in the scanner, but the brain state difference between sleeping and movie watching should be considered (Gao and Lin, 2012).

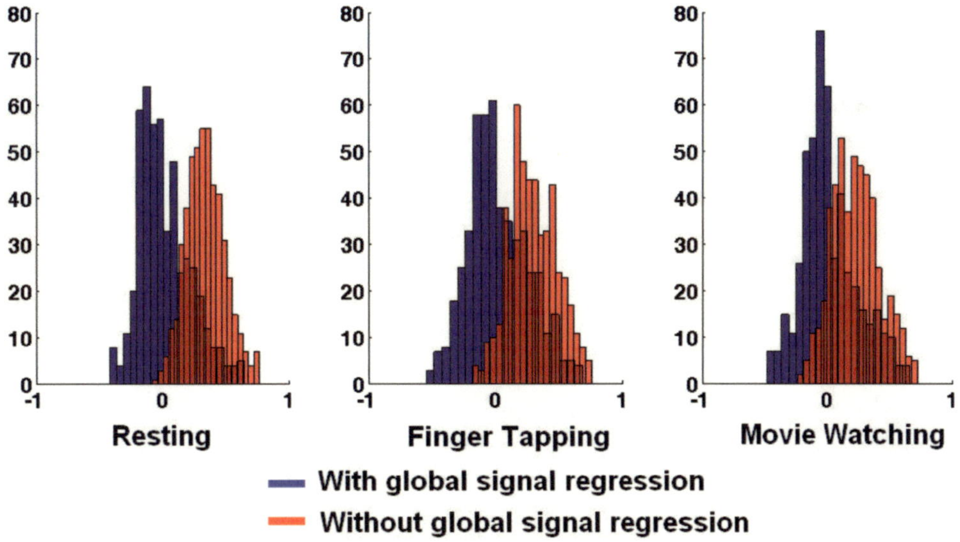

Figure 1. The histograms of the group mean correlation matrices (16 subjects, 32 ROIs covering five predefined functional networks, 496 correlations) with (blue) and without (red) global signal regression during three experimental conditions-resting, finger tapping, and movie watching. This figure was reproduced with the permission from Gao and Lin. (2012).

While acquiring fcMRI images is rather straightforward and EPI is the most widely employed imaging sequence, the means through which fcMRI data is processed to discern brain functional connectivity is less consistent across different research labs. Typically, fcMRI is preprocessed, including time shifting, motion correction, spatial smoothing, linear trend removal, low pass filtering (typically <0.1Hz) and nuisance signal regression. Although minor discrepancies exist regarding the pre-processing procedures among different researchers, most of the steps are fairly consistent and widely accepted with the exception of nuisance signal regression, especially global signal removal, which is designed to minimize the potential sources of physiological noise such as respiration and cardiac motion (Chang and Glover, 2010; Fox et al., 2009; Murphy et al., 2009). Such a regression forces the resulting correlation strength distribution (from a seed-based analysis) to be bellshaped centered at zero. Fig. 1 shows the distribution of correlation coefficients with and without global signal removal during three different experimental conditions-resting, finger tapping, and movie watching (Gao and Lin, 2012). It is immediately apparent that the removal of global signal leads to the presence of negative correlations. One immediate question is whether the observed negative correlation is "introduced" by global signal removal or is "brought" into light after removing the nuisance factors (global signal) obscuring the true relationships. Great controversy exists concerning the interpretation of negative correlations (Fox et al., 2009; Murphy et al., 2009). Fox et al (Fox et al., 2009) have shown that after removing the networks in question from the global signal mask (essentially eliminating the mathematical constraints on the two systems), certain anti-correlation can still be observed. Moreover, in an independent study, Chang and Glover (Chang and Glover, 2010) showed evidence of the presence of anti-correlation between the default network and the task-positive dorsal attention network, both with and without controlling for physiological noise variation induced by respiration and heart beating (although physiological noise correction did increase the spatial extent and magnitude of negative correlations). Such findings partially support the "authenticity" of the observed negative correlation, but future studies are needed to further delineate the nature of the global signal and the interpretation of the negative correlations after their removal.

After the pre-processing steps, three categories of methods generally exist for further processing fcMRI data to examine brain functional connectivity: (1) seed-based method; (2) independent component analysis based method; and (3) graph-theory analysis based method. The seed-based approach is the most widely adopted likely due to its simplicity in both implementation and interpretation. By designating a seed region for obtaining the reference signal, a voxel-wise whole brain correlation analysis can be conducted to acquire individual-based functional connectivity maps, which can then be pooled together to obtain a group map after statisitcal testing of each voxel's connectivity strength. Such a map represents the architecture of the functional network dictated by the seed region, enabling easy interpretation of the results. However, this approach could potentially be biased by the choice of the seed region and can only determine the brain functional connectivity specifically associated with the chosen seed. Most of the reported studies detemine the seed region either based on anatomical features or published coordinates of previous fMRI activation studies. Then a spherical ROI is defined centered at the seed with a certain radius, where both the location and size can vary from study to study. This variability poses a potential problem for direct comparisons between different studies. This problem is worsened in developmental studies since both the structural and functional brain organization might differ from those of adults, making the choice of seed regions based on the readily available adult literatures impractical. Mathematically, seed-based correlation analysis measures only the pair-wise correlation between a seed region and the voxel in question while ignoring the multivariate relationship among all brain voxels, which may be suboptimal especially when one intends to define the architectures of several networks simultaneously (Joel et al., 2011).

The independent component analysis (ICA) approach, on the other hand, largely alleviates the above mentioned problems. First, since it does not need pre-specified seed regions but extracts spatially/temporally independent components in a data-driven fashion, the biases associated with choosing seed regions are largely avoided. Second, depending on whether spatial or temporal ICA is applied, this approach maximizes the spatial/temporal independence among all identified components simultaneously by considering all voxel's time courses together, thus alleviating the "pair-wise" correlation problem embedded in the seed-based approach. Given these advantages, numerous studies (Damoiseaux et al., 2006; Fransson et al., 2010; Gao et al., 2009) have applied ICA and the results are fruitful. However, these advantages come at a cost. Since there is no inherent "ranking" available for ICA-extracted components, selecting components reflecting brain functions is a matter of subjectivity through either visual inspection or template matching. Although quantitative methods (De Martino et al., 2007; Perlbarg et al., 2007; Tohka et al., 2008) aiming to solve this issue are emerging, their effectiveness and robustness need to be validated by independent studies.

Another widely used strategy is to parcellate the entire brain into a set of ROIs and to perform graph-theory based analyses to delineate the whole brain's functional topology. More detailed reviews of graph theory based analysis can be found in Bullmore and Sporns (Bullmore and Sporns, 2009). The small-world related measures (Latora and Marchiori, 2001; Watts and Strogatz, 1998) are especially popular and have been used to quantify the brain's development (Supekar et al., 2009), aging (Achard and Bullmore, 2007), as well as various disease processes (Liu et al., 2008). However, similar to a seed-based approach, there is no standard approach as to how the whole brain should be parcellated. In practice, different anatomical feature based parcellation schemes including AAL (Tzourio-Mazoyer et al., 2002), ANIMAL (Collins et al., 1995) have been used. However, as described by Fornito et al (Fornito et al., 2010) and Wang et al. (Wang et al., 2009), different graph analytic measures may arise from different parcellation resolutions, making direct comparisons between studies untenable. Moreover, anatomical parcellation approaches do not explicitly consider functional similarity within the defined regions. As a result, spatial intra-region inhomogeneity of the BOLD signals may lead to spuriously averaged signals, and hence, inaccurate graph analytic measures. Data driven functional similarity based brain parcellation schemes are particularly desired in studies where the architecture of brain functional regions may deviate from that in normal adults. Such methods (Nelson et al., 2010; Neumann et al., 2006; Shen et al., 2010) directly utilize Pearson's correlation or canonical correlation between BOLD time series as similarity measure aiming at defining functionally sensible parcellations. As such strategies could effectively address the functional inhomogeneity concern inherent in anatomical-defined parcellation schemes, potential issues also exist. Among them, the most challenging one might be the number of regions to define. As most current methods relying on predefining a fixed number based on prior information, more data-driven approaches are highly desired and deserve further effort.

In addition, another limit for graph-theory based methods is that there is currently no optimal way to compare networks with different number of nodes and/or edges (van Dijk et al., 2010). A summary of different fcMRI analysis methods can be found in Table. 1.

Methods	Brief Description	Advantage	Disadvantage
Seed-based method	By designating a seed region for obtaining the reference signal, a voxel-wise whole brain correlation analysis is conducted to define a particular functional network.	Easy implementation ; Straightforward Interpretation.	Seed choice bias; Focusing on pair-wise correlation while ignoring multivariate relationship among multiple networks.
Independent component analysis (ICA)	Functional networks are extracted through blind source separation technique-ICA.	Data-driven way to define functional networks; Simultaneous extraction of multiple functional networks based on their multivariate relation.	Subjective selection of independent components for subsequent analysis.
Graphtheory based analysis	The brain is first pacellated into a set of ROIs after which graphtheory based analysis is performed to evaluate brain's functional topology using various network measures.	Whole brain analysis enabling assessment of brain's overall information transferring property through various graph-theory based measures.	Anatomically defined parcellation; Currently no standard ways to compare networks with different numbers of nodes and/or edges.

Table 1. Typical methods applied in fcMRI study.

3. The emergence of individual functional networks during the first two years of life

A mature brain works as a unified system involving various functionally specialized networks each dealing with an array of specialized functions. Generally, depending on the nature of the functions, brain networks are classified into primary functional networks such as sensory, motor, and visual networks or higher order cognitive networks such as attention, memory, executive control, emotion, etc. The developmental trajectory of each individual functional network is of great interest. Successful interaction with the outside world rests on the proper functioning of the primary networks and it has been suggested that they are essential for the tuning and establishment of the higher order cognitive networks (Johnson, 2000, 2001). Therefore, it is anticipated that primary functional networks will develop faster and earlier than those higher order cognitive functional networks.

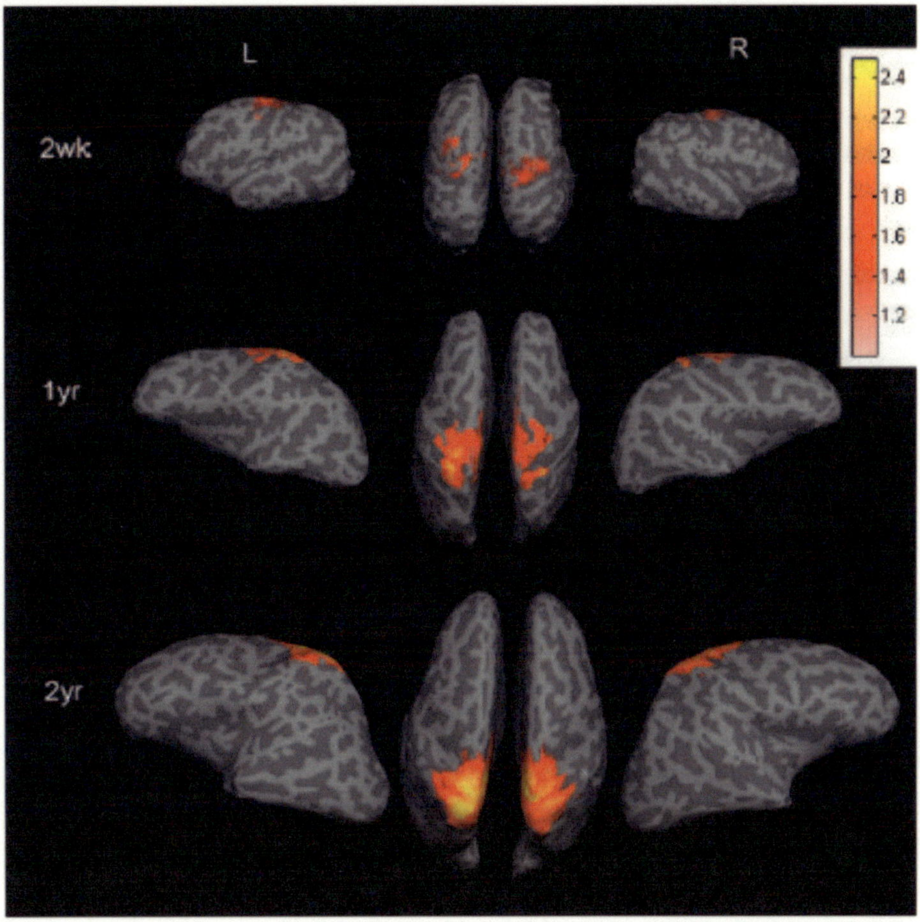

Figure 2. The functional connectivity maps of motor-sensory network at 2wks, 1yr and 2yrs of age, respectively. The color bar represents z-score values. This figure was reproduced with the permission from Lin et al. (2008).

Using a seed-based approach, we (Lin et al., 2008) have reported the developmental patterns of both the motor-sensory and visual networks during the first two years of life (Fig. 2 and Fig. 3). The seeds were manually drawn around bilateral motor-sensory cortex and visual cortex for the two networks, respectively. The most noticeable feature is that starting at neonates, the two networks already demonstrate a bilateral-symmetric functional connectivity pattern similar to that of 2yr olds, which is also reasonably consistent with their adult configuration (Biswal et al., 1995). Consistent with this finding, Fransson et al (Fransson et al., 2007) also reported the existence of adult-like motor-sensory, visual, as well as auditory networks in preterm infants scanned at term equivalent age. A longitudinal study by Smyser et al (Smyser et al., 2010) further delineated the dramatic development of such primary sensory networks during the prenatal period from 26 weeks postmenstrual age to the term equivalent age supporting the network structures observed in neonatal subjects. The early synchronization of these primary sensory functional networks most likely indicates an evolutional optimization to ensure early survival as well as effective interaction with the outside world for proper experience-based tuning of other higher order networks. However, despite of the common feature of early synchronization, Fig. 2 and 3 show that the motor-sensory network demonstrates more development during the first year of life, while visual network shows more increase during the second year of in terms of normalized volume of cortical connectivity, which indicates certain level of differential growth trajectories for different primary networks.

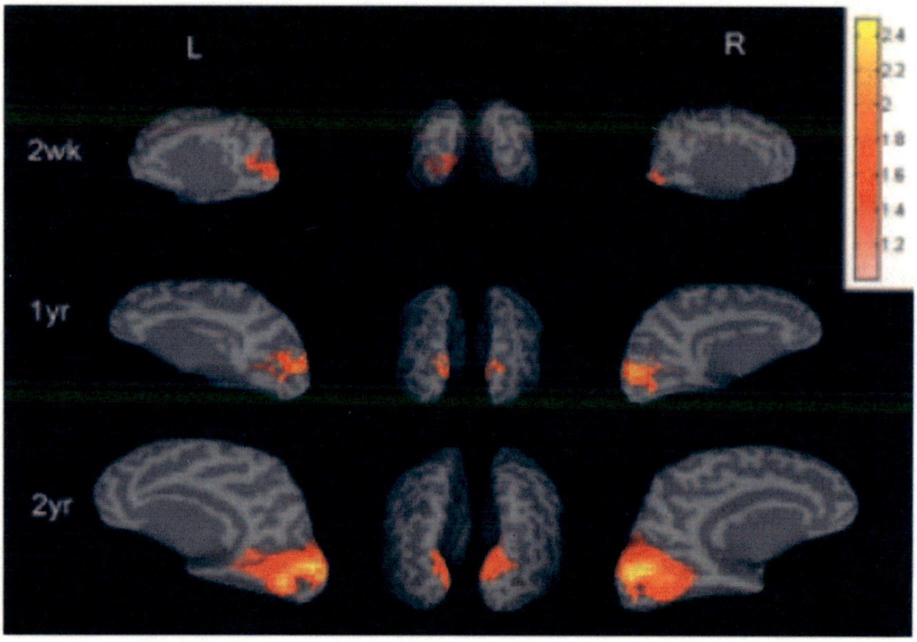

Figure 3. The functional connectivity maps of visual network at 2wks, 1yr and 2yrs of age, respectively. The color bar represents z-score values. This figure was reproduced with the permission from Lin et al. (2008).

While the primary brain functional networks appear to be reasonably developed immediately after birth, the growth trajectories of other higher order cognitive functional networks likely differ. Recently, a novel brain functional network was proposed by Raichle et al (Raichle et al., 2001) where the involved brain regions exhibit down-regulated activity during various goal-directed external tasks but up-regulated activity during uncontrolled resting state (Shulman GL, 1997). Various studies have been conducted in an attempt to probe the exact function of this network (Buckner et al., 2008; Buckner and Carroll, 2007; Greicius and Menon, 2004; Gusnard et al., 2001; Gusnard and Raichle, 2001). Among them, the consensus is that this

http://dx.doi.org/10.5772/30640

network mainly supports self-related internal cognition (Andrews-Hanna et al., 2010; Buckner et al., 2008; Buckner and Carroll, 2007; Gusnard et al., 2001). From a developmental perspective, the establishment of self-consciousness has to be the preface of such self-referential thinking process and hence is likely related to the integrity of this network. Interestingly, Amsterdam (Amsterdam, 1972) vividly documented the emerging process of self-consciousness during the first two years of life. He showed that neonates could not recognize themselves from their mirror images but infants from 6 through 12 months of age demonstrate prolonged and repeated reaction to their mirror images as a sociable playmate. Wariness, withdrawal, self-admiring, and embarrassed behaviors start at 14 months and have been observed in 75% of the children after 20 months of age. From 20 to 24 months of age, the majority of subjects demonstrate recognition of their mirror images. If the default network is indeed responsible for the self-consciousness based internal thinking process, we would expect a gradual orchestration of this network during the first two years of life. Fig. 4 shows the default network of 3wks, 1yr and 2yr olds, respectively. As shown, a rather primitive/incomplete default network consisting of 6 brain regions is observed in neonates, among which only medial prefrontal cortex (MPFC) and posterior cingulate cortex (PCC) regions are typically observed in its adult configuration. The finding of an "incomplete" default network is consistent with results from several other studies (Fransson et al., 2007; Smyser et al., 2010). However, at 1yr old, a total of 13 regions are observed and 10 of them are consistent with that observed in adults, including ventral/dorsal MPFC, PCC/retrosplenial (Rsp), bilateral lateral temporal cortex (LTC), bilateral inferior parietal lobule (IPL), and the hippocampal formation (HF) (Buckner et al., 2008). Such dramatic synchronization reflects fast orchestration of this network during the first year of life. In 2yr olds, the default network continues to grow by covering more regions (13 regions) consistent with adult's configuration. However, 6 non-typical regions including the orbital frontal, anterior cingulate cortex (ACC), right and medial parietal, and bilateral superior temporal regions are also observed.

The spring embedding plots (Ebbels et al., 2006) further demonstrate the topology of the default network throughout the first two years of life (Fig. 5). These graphs are arranged such that the most strongly connected regions were drawn close to each other and placed in the center of the graph and vice versa. One notable feature is that both PCC/Rsp and MPFC are consistently located in the center of each graph with the exception of the neonate group (only PCC/Rsp), implying that these two regions are most strongly connected with other regions. Moreover, the PCC/Rsp and MPFC reliably exhibit the highest mean connection strengths across all ages (Fig. 6). Such finding is consistent with adult studies (Buckner et al., 2008) reporting the "hub" role of these two regions in this distributed network. The betweenness centrality (BC) (Brandes, 2001) – a measure of node importance in graph theory further confirm that PCC/RSp and MPFC indeed serve as the hubs in the pediatric default network (Fig. 7). These results suggest that the PCC/Rsp may be the major hub of the default network whereas the MPFC subsequently emerges, potentially, as the secondary hub starting at 1yr of age.

Overall, a gradual emergence of the default network is observed during the first two years of life and the PCC/Rsp and MPFC emerge as two potential hubs. This finding is consistent with previous reports of the emergence of self-consciousness (Amsterdam, 1972), as well as other

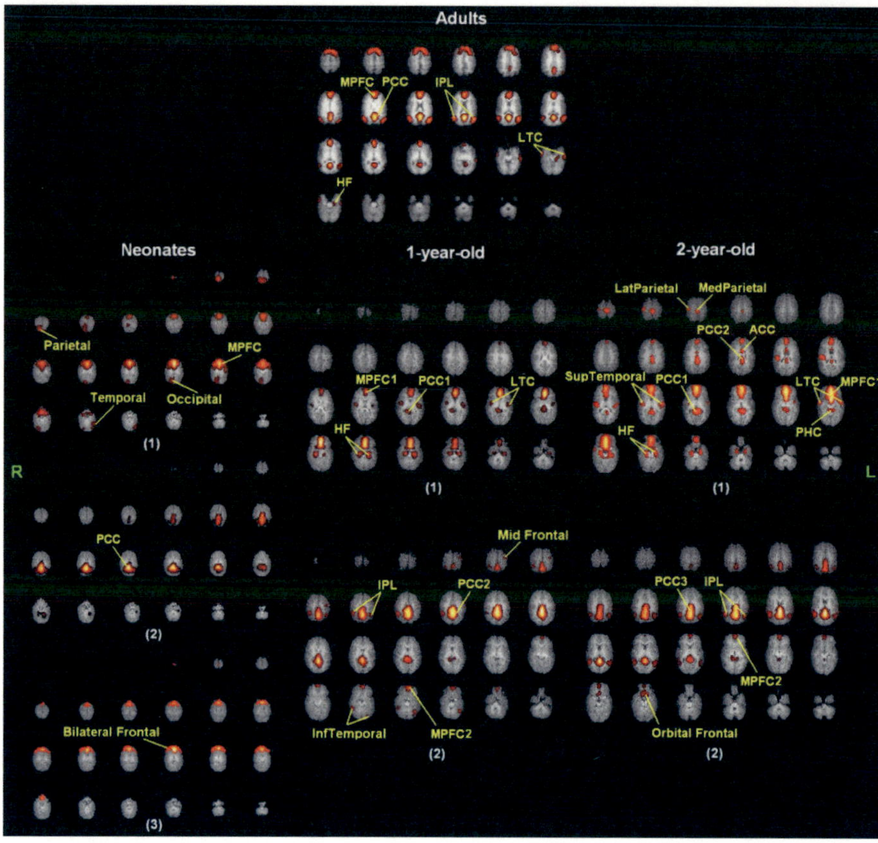

Figure 4. The independent components associated with brain's default network for neonates, 1yr olds, 2yr olds and adults. Individual regions are labeled for each map. ACC: anterior cingulate cortex; HF: hippocampal formation; InfTemporal: inferior temporal cortex; IPL: inferior parietal lobule; LTC: lateral temporal cortex; LatParietal: lateral parietal cortex; MedParietal: medial parietal cortex; MidFrontal: middle frontal cortex; MPFC: medial prefrontal cortex; PCC: posterior cingulate cortex; PHC: para-hippocampal cortex; SupTemporal: superior temporal cortex. This figure was reproduced with the permission from Gao et al. (2009).

default network related functions such as episodic memory (Fivush, 1990; Kolata, 1984). Moreover, the appearance of non-typical regions in all three pediatric groups' default network indicates further pruning/specialization process needed for final maturation. Therefore, in contrast to the primary networks (e.g., motor-sensory, visual) for which the major development is done prenatally, the establishment of the higher order default network takes prolonged postnatal development to achieve its functioning configuration.

Up to now, the discussions are largely focused on individual network level. As our brain is an integrated system consisting of many functionally specialized networks, it is of critical importance to further understand the interaction among different networks. Five networks including the previously described motor-sensory, visual, default network and two other higher order networks-dorsal attention and fronto-parietal control network (Vincent et al., 2008) were simultaneously studied based on an extended dataset including 51 neonates (27

http://dx.doi.org/10.5772/30640

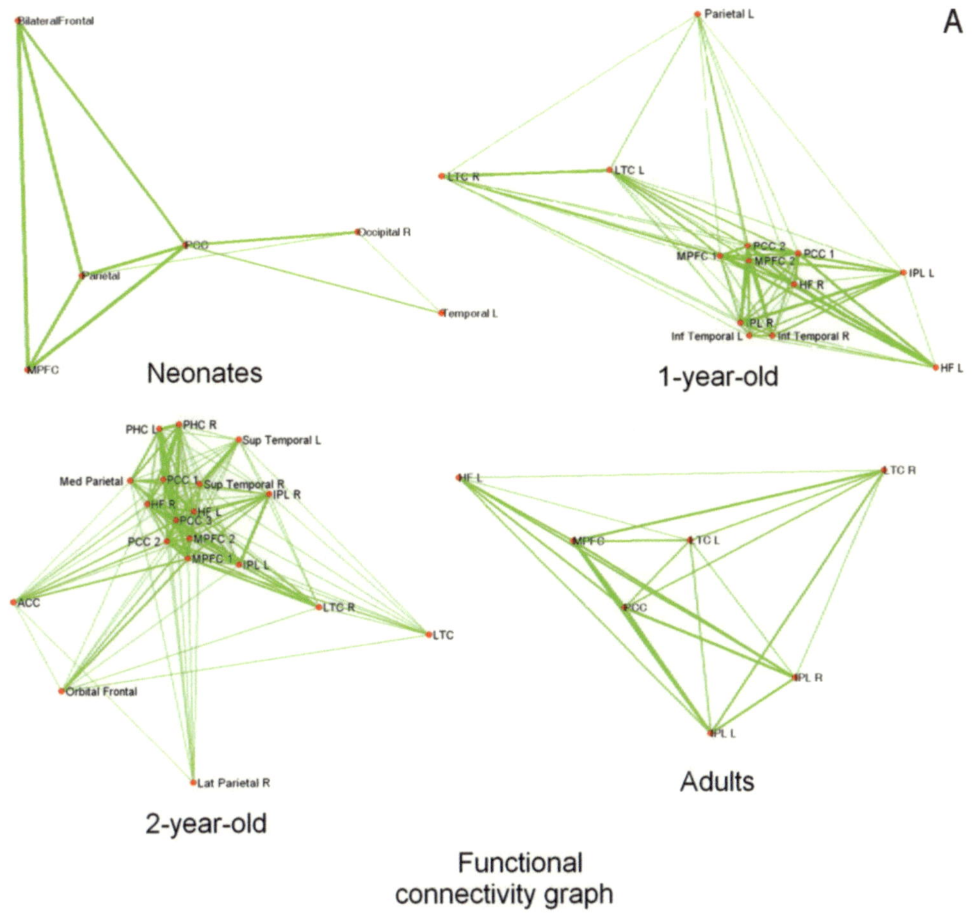

Figure 5. The spring embedding plots of the group-mean functional correlation patterns within the default network for neonates, 1yr olds, 2yr olds and adults. This figure was reproduced with the permission from Gao et al. (2009).

male, 23± 12days (SD)), 50 1yr olds (27 male, 13± 1 months) and 46 2yr olds (28 male, 24± 1 months). Specifically, 5, 6, 6, 6, and 9 ROIs (8mm-radius sphere, altogether 32 ROIs) were defined based on the MNI coordinates published in adult studies (Vincent et al., 2008) to construct the five networks, respectively, from which the group mean correlation matrices were constructed (32×32) and visualized using spring-embedding plot (Fig. 8). In this figure, the segregation/synchronization of the five networks across the first two years of life can be simultaneously characterized. In neonates, except for the motor-sensory network (red) and the visual network (blue), which are already well organized into cohesive clusters, the remaining three higher-order cognitive networks are rather scattered (i.e., three separated clusters for default network (yellow); two for the dorsal attention network (magenta) and two for fronto-parietal control network (brown).

This finding is highly consistent with results shown in Fig. 2-4, which again indicates early maturation of primary functional networks. In fact, the motor-sensory and visual networks

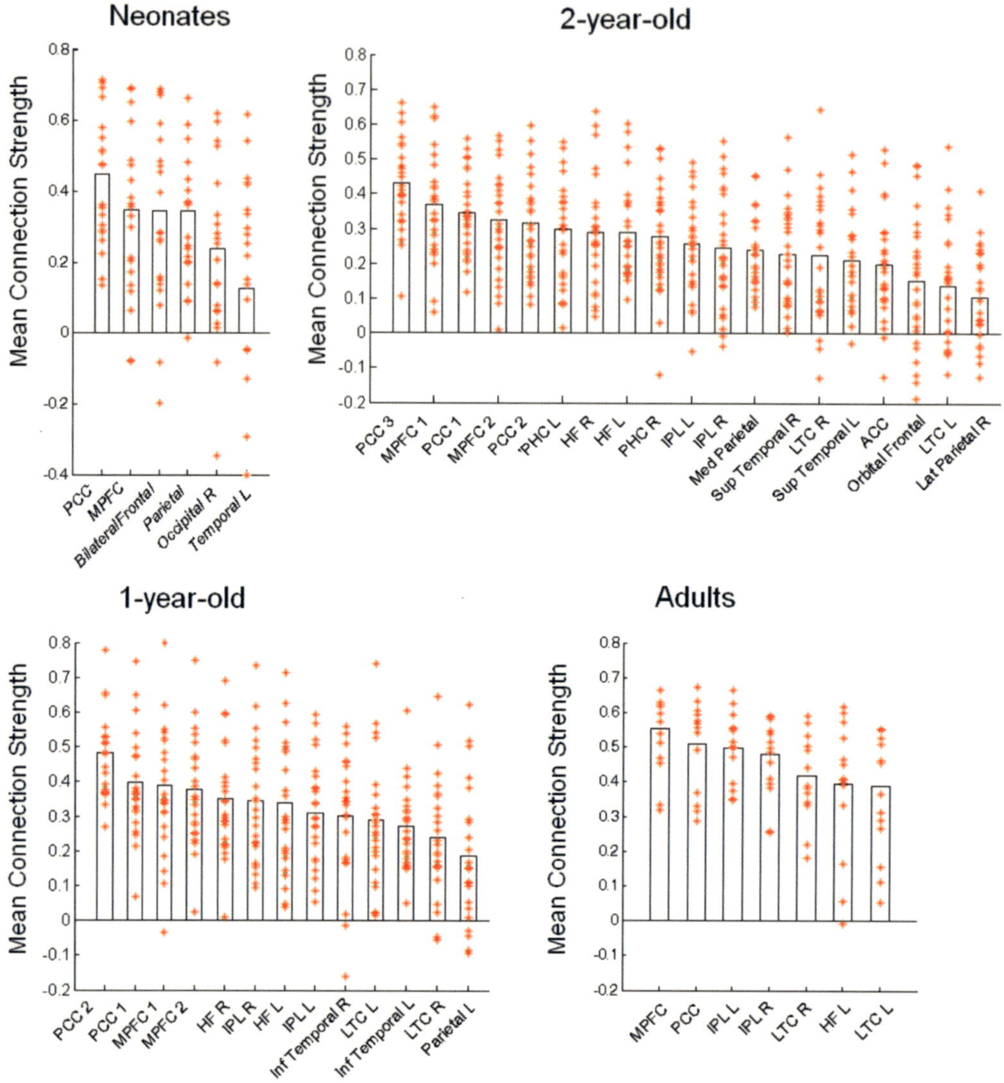

Figure 6. The mean connectivity strength of each node within the default mode network for neonates, 1yr olds, 2yr olds and adults. Each bar represents the group mean of each node's mean connectivity strength within the default network. Each asterisk represents an individual subjects' value. This figure was reproduced with the permission from Gao et al. (2009).

remain as cohesive clusters throughout the different age groups examined (i.e., neonates, 1yr olds, 2yr olds, and adults, Fig. 8). In great contrast, the synchronization of higher-order cognitive networks takes a more prolonged time. Specifically, the default network starts to be synchronized into one cluster in 1yr olds and remains so thereafter. However, although the other two networks (i.e., the dorsal attention network and the fronto-parietal control network) demonstrate certain long-distance synchronization during the first two years of life (Fig. 8), it

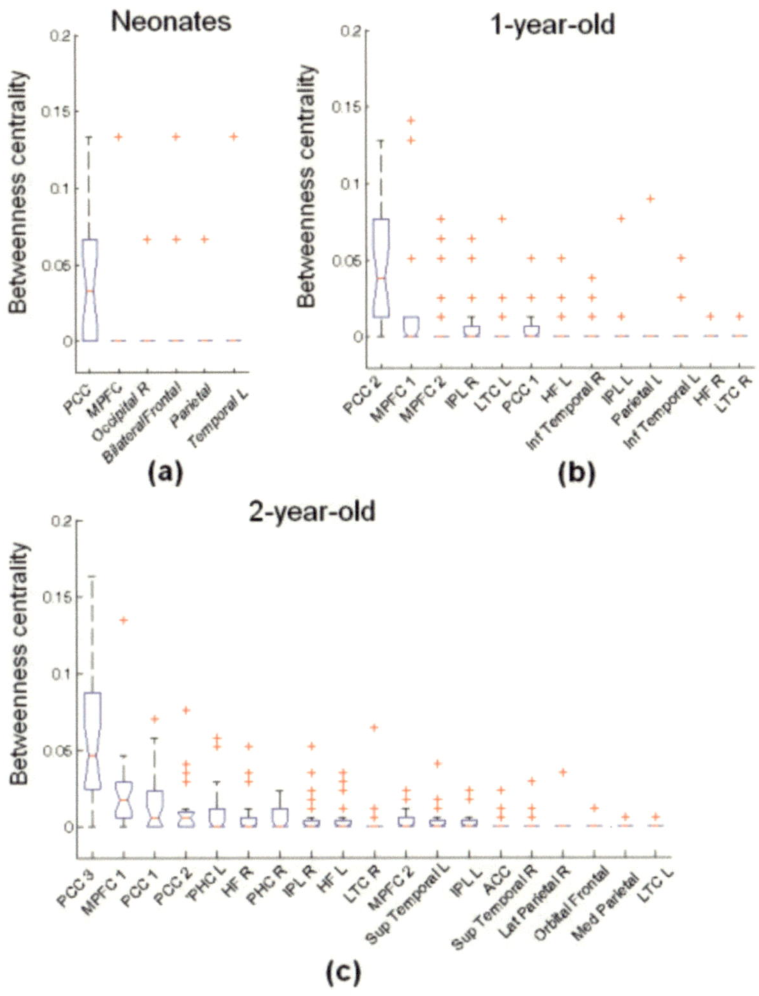

Figure 7. The betweenness centrality value for each node in the default network for neonates, 1yr olds, and 2yr olds. This figure was reproduced with the permission from Gao et al. (2009).

is not until adults are they organized into selfcontained clusters. Clearly, more quantitative characterization, as well as additional sampling between 2yr olds and adults, is highly desirable for more specific delineation of the complete growth trajectories of these five networks. Nevertheless, one trend is again clear in Fig. 8; primary networks such as motor-sensory and visual are wired prenatally while the synchronization of higher-order cognitive networks is achieved during postnatal life (Fransson et al., 2007; Lin et al., 2008; Smyser et al., 2010). Interestingly, from a spatial perspective, higher order cognitive networks are usually spatially distributed but primary sensory networks are mostly locally wired. This local organization might partially contribute to their early synchronization. Specifically, it is likely that long distance connections are not established immediately after birth (discussed below) and that the anatomical proximity facilitates formation of short-distance connections so close

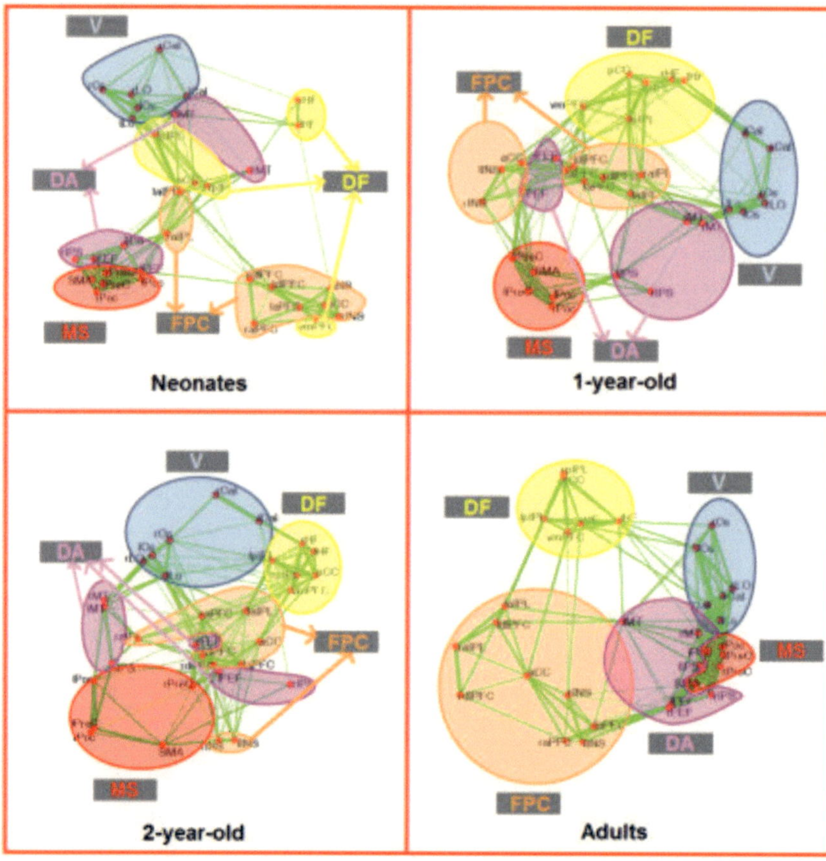

Figure 8. Spring embedding plots of the group-mean correlation pattern among five predefined functional networks: motor-sensory, visual, default, dorsal attention, and fronto-parietal control network. The spatial evolution of each network is highlighted by different colors. DA: dorsal attention network; DF: default network; FPC: fronto-parietal control network; MS: motor-sensory network; V: visual network.

interaction immediately after birth can be achieved. This property is likely attributed by evolutional optimization so that early survival can be ensured. Moreover, this evolution process promotes the development of experience based fine tuning and maturation of higher order networks (Johnson, 2000, 2001). Overall, existing results support the notion that brain's matured organization is optimized according to the developmental process which not only ensures early functioning of primary functions but also allows sufficient environmental tuning of higher order functions for better adaptation.

4. The evolution of whole brain's efficient topology

The examination of individual functional networks provides vital information about the early development of brain's sub-systems. Yet, how the whole brain topology evolves during this critical time remains elusive. By designating different brain regions as nodes and connections between them as edges, recent graph theory advancement, especially the smallworld topology (Watts and Strogatz, 1998) appears well suited to quantitatively characterize the complex network structure of the human brain. Small world topology is characterized by a high level of local cliquishness as well as sufficient long-range direct connections bridging the local clusters, leading to excellent local and global wiring efficiency for information transfer. The regions that bridge disparate and local clusters are important information transfer portals, usually termed network hubs (Barabasi and Albert, 1999). While many studies have confirmed the existence of small world characters in matured brain networks (Achard et al., 2006; Hilgetag et al., 2000; Salvador and E., 2005; Watts and Strogatz, 1998), Fransson et al (Fransson et al., 2010) showed that the infant (1 week old) brain also exhibits a small world topology. However, how the whole brain network topology continues to evolve during the first years of life remains elusive. In particular, given the dramatic axonal elongation/thickening and myelination during this period (Conel, 1939-1963; Fair et al., 2008; Tau and Peterson, 2010), the resulting enhancement in long-distance connections must have profound implications in achieving a highly efficient brain network. To test these hypotheses, 51 neonates (27 male, 23±12days (SD)), 50 1yr olds (27 male, 13± 1 months) and 46 2yr olds (28 male, 24± 1 months) with successful fcMRI scan were retrospectively identified and graph theoretical analyses were conducted (same as the data used in Fig. 7) (Gao et al., 2011).

A whole brain parcellation approach (as described above) was employed by parcellating the brain into 90 ROIs and correlation matrices from individual subjects were constructed after standard preprocessing (Section 2). Local efficiency (LE), global efficiency (GE) (Latora and Marchiori, 2001), and betweenness centrality measures (Brandes, 2001) were calculated. The significant correlation matrices ($p<0.05$, FDR corrected) are presented in Fig. 9 for neonates, 1yr and 2yr olds. Dramatically increased connection density (Fig. 9b) can be observed from neonates to 1yr olds. Moreover, the anatomical distance histogram plot (Fig. 9c) of these significant connections clearly demonstrates a rightward shift, suggesting increased longrange connections. More importantly, those connections exhibiting increased strength with age are always associated with significantly longer connection distances than those decreasing strength with age (Fig. 9d).

To further examine the topological changes as resulting from the emerging long-distance connections, the top 15% of most significant connections from the group mean correlation matrices were visualized on a brain surface in Fig. 10. Imposing a fixed cost ensures that the topological changes arise not from different number of connections but instead from the rearrangement of connections. Different nodes are classified into different functional modules (represented by different colors) based on modularity analysis of their functional relationship (Newman, 2006). Interestingly, the neonates group's modules are largely local and agree well with the boundaries of different lobes. However, 1yr and 2yr olds' modular structures become dramatically more distributed. For example, in 1yr olds, the cyan-colored module spans across temporal and parietal lobes; the red-colored module covers both the parietal and subcortical regions; while the magenta-colored module includes regions from the parietal, frontal, and temporal lobes. Similarly, the red-colored module covers parietal and subcortical regions; the

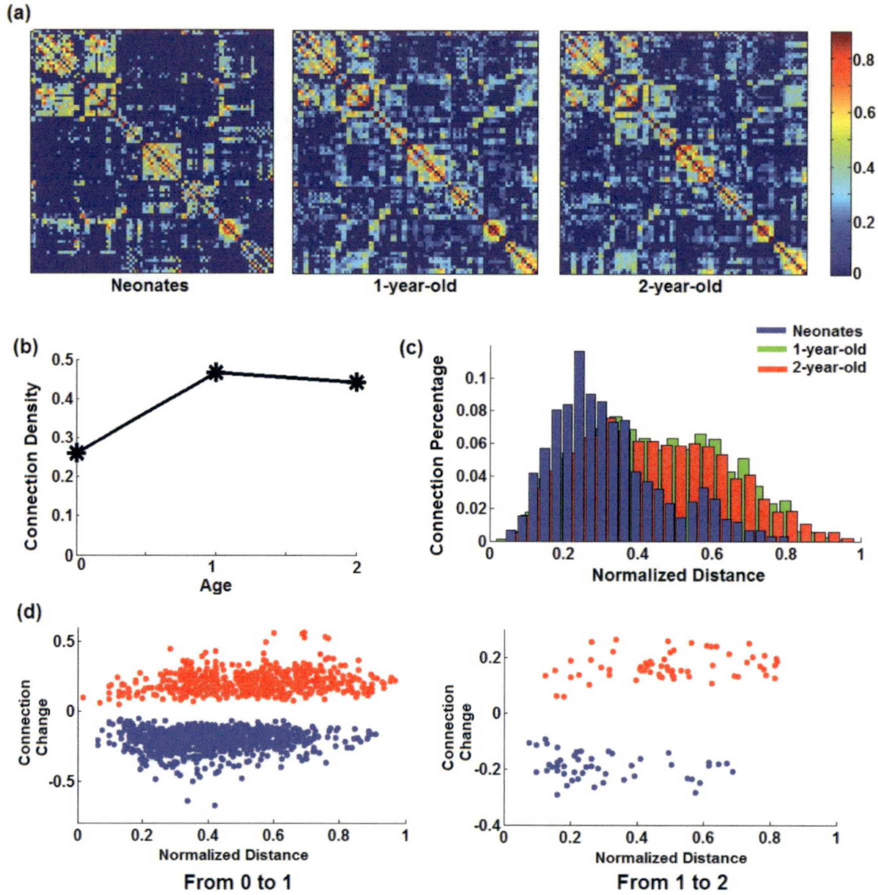

Figure 9. Assessment of age-dependent change of brain network topology. (a) significant connectivity matrices ($p<0.05$, FDR corrected) for neonates, 1yr and 2yr olds; (b) connection density change in the three age groups; (c) the histogram of the normalized anatomical distance associated with the significant connections in each age group's network; (d) comparison of the anatomical distances associated with connections showing age-dependent increase/decrease in strength. Increasing connections with age (red) consistently possess longer anatomical distances than decreasing connections (blue, $p<1e-10$ from 0 to 1and $p=1e-6$ from 1 to 2).

cyan-colored module spans across the frontal and temporal lobes; while the blue-colored module is distributed between the frontal and parietal lobes in 2yr olds, demonstrating a largely distributed functional modular structure. The apparent increase of long-range connections (shown in Fig. 10) highlights the modular structure changes from local to distributed network during the first two years of life.

One of the main characteristics of graph theory analysis is the determination of brain functional hubs for information transfer. The brain's functional hubs (top 10 regions showing highest betweenness centrality values at cost 15%) demonstrate an interesting evolution pattern during the first two years of life (see Fig. 11). There are two apparent clusters of hubs, anterior and posterior clusters, in neonates, suggesting that major functional developments may have

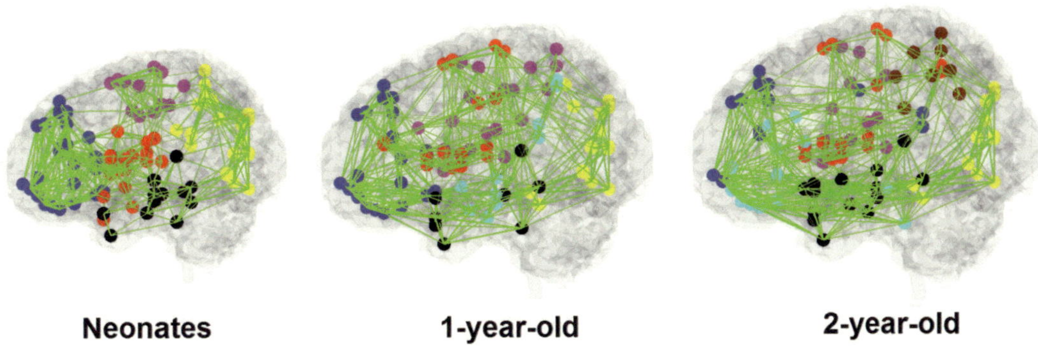

Figure 10. Changes of modular structure during the first two years of life. Different colors represent different modules in each age group.

already occurred in both the frontal and occipital lobes. In addition, most of the connections between the hubs are interhemispherical with no clear connections between the two clusters. In 1 yr olds, the hubs are well connected and more anatomically centered with few hubs located in the anterior and posterior areas of the brain. Finally, the distribution of hubs becomes more spatially uniform in 2yr olds and the connections associated with the hubs provide a fairly uniform covering of the whole brain (Fig. 11).

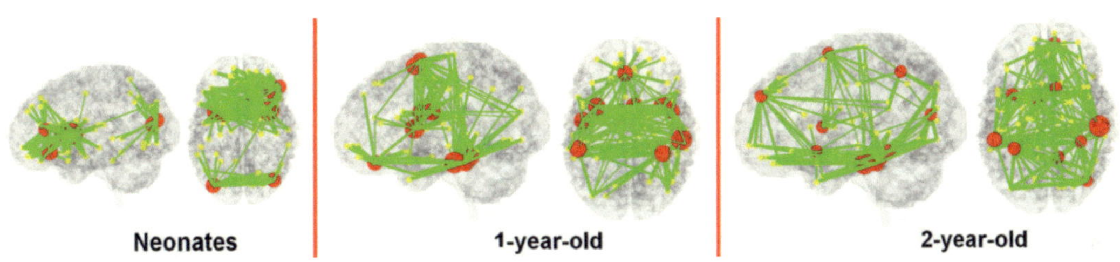

Figure 11. The distribution of functional hubs in the three age groups based on connection matrices at cost 15%. Hub regions together with connections (at cost 15%) are visualized on the corresponding brain surface.

Together, with the aforementioned brain functional network development during the first two years of life, it is highly plausible that brain efficacy for information transfer also undergoes a similar evolution process. To quantitatively examine the efficiency improvement resulting from such dramatic topological changes, local (LE) and global efficiency (GE) for the whole network were calculated for each individual subject at the same cost of 15%. The results are presented in Fig. 12. Significant increases of LE and GE are observed from 3wks to 1yr of age (red asterisks, p=4e-7 and 6e-8 for LE and GE, respectively) but neither measure is statistically different between 1 and 2 yrs of age. Note when comparing neonates with 2yr olds, 2yr olds also demonstrate significantly higher LE (p=1e-8) and GE (p=7e-9).

In summary, the results reported in the literature have consistently indicated that the neonatal brain already exhibits a small-world network with reasonable information transfer efficiency

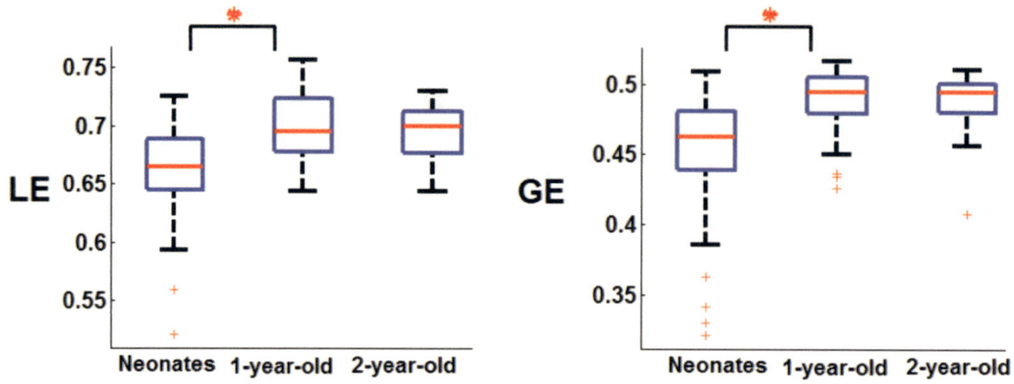

Figure 12. Quantitative measures of whole brain network's local efficiency (LE) and global efficiency (GE) at cost 15% based on individual subject's correlation matrices. Red asterisks represent significant difference at p<0.05 (FDR corrected).

(Fransson et al., 2010; Gao et al., 2011). The well documented prenatal axonal development (Haynes et al., 2005) might set the foundation for an efficient connectivity network hence explains this finding. However, significant improvements of both LE and GE are observed from neonates to 1yr olds, whereas no statistical difference between 1 and 2yr olds is observed (Fig. 12). This non-linear developing pattern of brain wiring efficiency, particularly GE, is intriguing. Neonates demonstrate an anatomically ordered architecture (Fig. 10), in which neighboring clusters are largely sequentially connected, leading to a longer path length, while 1 and 2yr olds show considerably better wired brain topologies with abundant bridging connections even between distant clusters, suggesting that anatomical long-range connections may play a critical role in the observed improvement. Indeed, quantitative compassions of connectivity strength show remarkable growth of longdistance connections (Fig. 9) strongly supporting the anatomical basis of network efficiency and suggesting that the establishment of a higher GE rests critically on the development of long-distance connections. An increase in long-distance connections has been reported when comparing school-aged children with adults (Fair et al., 2008; Fair et al., 2009; Kelly et al., 2009). Nevertheless, given the age differences between our study cohort and that reported in the literature, the biological underpinnings associated with the observed increased longrange connections most likely differ. Specifically, since the axonal properties (e.g., length, diameter, density) are most likely well established in school-aged children, the observed increase in long-range connectivity strength might primarily arise from the continuing myelination process, improving information transfer efficiency between distant brain regions (Tau and Peterson, 2010). In contrast, although some of the major long-range axonal connections (especially callosal fibers) are present in neonates (Fransson et al., 2007; Tau and Peterson, 2010), continuing axonal elongation/thickening is expected at this age to facilitate forming connections with other distant brain functional areas (e.g., in the anterior-posterior direction), especially during the first year of life (Tau and Peterson, 2010). Moreover, the myelination process is also documented to experience the most dramatic growth during the first two years of life (Gao et al., 2008; Haynes et al.,

2005; Tau and Peterson, 2010). Therefore, a combination of axonal growth and myelination likely accounts for the changes observed in our study.

The observation of no significant changes between 1yr and 2yr olds in whole brain smallworld properties echoes the findings by Supekar et al (Supekar et al., 2009) and Fair et al (Fair et al., 2009), who also failed to detect significant whole brain efficiency changes when comparing school-aged children with adults. However, significant regional wiring changes (LE/GE) have been documented in previous studies (Fair et al., 2009; Supekar et al., 2009; Gao et al., 2011). In considering these findings, it is plausible that the whole brain reaches an adultlike "small-world" topology from 1-year-old on while after that although significant local changes continue to reshape the brain for the development of specific functions, the whole brain efficiency property remains un-disturbed.

5. Explorations into developmental disorders

Studies utilizing fcMRI to discern potential neuropathological underpinnings of various developmental disorders are emerging (Church et al., 2009; Di Martino et al., 2010; Dickstein et al., 2010; Fair et al., 2010; Liston et al., 2011; Qiu et al., 2010; Uddin et al., 2008). For example, Church et al (Church et al., 2009) examined youth subjects (aged between 9 to 15) with Tourette syndrome (TS), a developmental disorder characterized by irresistible stereotyped movements and vocalizations called "tics". They specifically evaluated functional connectivity associated with two putative control networks – a "fronto-parietal network" likely involved in online control and a "cingulo-opercular network" important for set-maintenance. Specifically, they found that most of the connections with abnormal connection strengths (less-matured) were associated with the fronto-parietal network indicating more pronounced disruption of adaptive online control function in TS subjects. Di Martino et al (Di Martino et al., 2010) investigated autism spectrum disorders (ASD), a developmental disorder characterized by impaired social interaction and communication. They compared the striatal functional connectivity between children with ASD (aged between 7-and 13-year-old) and normal controls and found widespread hyper-connectivity among all striatal regions as well as between striatal regions and other associative regions such as insular and right superior temporal gyrus. Based on these findings, they postulated that developmental derangement rather than immaturity of functional circuits might underlie ASD. Dickstein et al (Dickstein et al., 2010) studied pediatric bipolar disorder (BD) in youth subjects between the age of seven and seventeen and found greater negative functional connectivity between left dorsal lateral prefrontal cortex and right superior temporal gyrus in BD subjects when compared with normal controls, indicating abnormal development of this important circuit for working memory and learning.

Attention-deficit/hyperactivity disorder (ADHD) characterized by attentional problems and hyperactivity has also attracted great interest in resting-state fMRI studies (Liston et al., 2011). Among these, Fair et al (Fair et al., 2010) examined the functional connectivity within the default network and found generally reduced connectivity in children with ADHD comparing with normal controls, reflecting atypical circuit maturation of this network. Interestingly, using the independent component analysis approach, Qiu et al (Qiu et al., 2010) reported both decreased and increased functional connectivity within this network (i.e., increased functional connectivity in anterior cingulate cortex, posterior cingulate cortex, lateral prefrontal cortex, precuneus and thalamus but increased connectivity in posterior medial frontal cortex). In a separate study, Uddin et al (Uddin et al., 2008) also found decreased network homogeneity in the precuneus node of the default network. Besides default network, other studies targeting at ADHD have also investigated into abnormal functional connectivity in other parts of the brain. For a recent review of fcMRI studies on ADHD, please refer to Liston et al (Liston et al., 2011).

As described above, most of existing studies focused on the population of children no younger than 7-year-old. However, the first two years of life is perhaps a critical time period where certain developmental disorders start to develop such as autism syndrome disorder ASD (Levy et al., 2009). Other disorders such as ADHD may have a late onset as school-aged children but the underlying neurophysiologic changes may occur well before the onset of behavioral symptoms (Ilott et al., 2010; Kieling et al., 2008). Hence the understanding of the underlying neuropathology during the first years of life is vital to early diagnosis and potential treatment of such disorders. As a result, resting-state fMRI studies focusing on a much younger age period are highly warranted for successful delineation of the underlying neurophysiological mechanisms of different developmental disorders. In addition, most of the existing studies focus on specific circuits/connections, potentially limiting the findings of novel abnormal circuits/connections that may underlie such disorders. Therefore, more studies with an exploratory nature are greatly needed in future research.

6. Technical issues in the application of fcMRI for early brain development

There are potential issues that are of particular importance for the utilization of fcMRI in the developmental brain. The first and foremost is the definition of ROIs for pediatric subjects, especially for those younger than 1 year of age when both brain's structural morphology and functional organization is largely different from its matured configuration. The current practice of applying either a parcellation template or functional activation coordinates from adult studies to the pediatric populations facilitates direct comparison between the two populations. However, such a practice may lead to potential bias against pediatric subjects given the known organizational difference. To a certain extent, this may be a circular problem since it is this development related reorganization that we are studying. Fortunately, the utilization of a data-driven method such as ICA could potentially alleviate this problem given that no prior ROIs are needed. However, as mentioned previously, the choice of components reflecting functional connectivity is subjective. Thus, quantitative comparisons of connection strength are less straightforward for ICA-based approaches. Clearly, methodological improvements are greatly needed to solve such issues.

But before that, developmental studies should choose different strategies according to the specific question(s) involved and the results should be interpreted with these issues in mind.

Viewing the brain's functional organization at the network level has been shown to be important in brain disease studies (Greicius et al., 2004; Seeley et al., 2009). For example, Seely et al (Seeley et al., 2009) proposed the "network degeneration hypothesis" where they reported that neurodegeneration may be related to network-level dysfunction and suggested the need for developing new network-based diagnostic assays. Logically, the interaction between different functional networks might also have significant influence on the developmental process. Different methods aimed at directly quantifying network-level interactions are emerging. Jafri et al (Jafri et al., 2008) described a method to calculate the functional connectivity between networks defined by spatial ICA and examined the differences between Schizophrenia and normal subjects. Gao et al (Gao et al., 2010) have proposed a different method based on canonical correlation to directly quantify the mutual dependence between two functional networks and applied it to delineate the dynamics of network level interactions under different brain states. In the future, we expect an increasing interest in both the development and application of network based analysis due to the intrinsic organization of the brain as individual functional modules.

Finally, in most cases the infants are sleeping inside the scanner with or without sedation in developmental investigations. The effect of sedation on resulting connectivity measurement should be better addressed. Even without sedation, different stages of sleep could also have an effect on the resulting functional connectivity measures. Therefore, studies incorporating sleep stage monitoring (e.g., through simultaneous EEG recording) would be beneficial to minimize experimental variability.

7. Conclusion

In this chapter, we have reviewed several results delineating the normal brain functional development process during the first two years of life using resting state functional connectivity. The application of fcMRI in various developmental disorders was also discussed. The uncertainty of the physiological underpinnings of fcMRI signal, as well as other technical issues related to its practical application are also briefly discussed pointing to the pressing need for improvement in future studies. The results support the notion that the brain's primary functional networks are in place and partially functioning immediately after birth, but the maturation of higher order cognitive networks takes prolonged experience based tuning during postnatal life. Moreover, dramatic evolution of the brain's functional topology is observed during this critical period, which greatly changes the brain's modular structure and distribution of hubs, as well as improves whole brain information transfer efficiency at both local and global levels. Finally, the evidence shows that less maturation and/or derangement of certain functional circuits might underlie different developmental disorders.

Author details

Wei Gao[1], Hongtu Zhu[2], Kelly S. Giovanello[3], J. Keith Smith[4], Dinggang Shen[1], John H. Gilmore[5] and Weili Lin[1]

1 Department of Radiology and Biomedical Research Imaging Center, UNC-Chapel Hill, USA

2 Department of Biostatistics and Biomedical Research Imaging Center, UNC-Chapel Hill, USA

3 Department of Psychology and Biomedical Research Imaging Center, UNC-Chapel Hill, USA

4 Department of Radiology, UNC-Chapel Hill, USA

5 Department of Psychiatry, UNC-Chapel Hill, USA

References

[1] Achard, S., Bullmore, E., 2007. Efficiency and cost of economical brain functional networks. PLoS Comput Biol 3, e17.

[2] Achard, S., Salvador, R., Whitcher, B., Suckling, J., Bullmore, E., 2006. A resilient, low-frequency, small-world human brain functional network with highly connected association cortical hubs. J Neurosci 26, 63-72.

[3] Akhtar, N., & Tomasello, M., 2000. The social nature of words and word learning. Oxford University Press., Oxford.

[4] Amsterdam, 1972. Mirror self-image reactions before age two. Developmental Psychology. Psychology 5, 297-305.

[5] Anderson, A.W., Marois, R., Colson, E.R., Peterson, B.S., Duncan, C.C., Ehrenkranz, R.A., Schneider, K.C., Gore, J.C., Ment, L.R., 2001. Neonatal auditory activation detected by functional magnetic resonance imaging. Magn Reson Imaging 19, 1-5.

[6] Andrews-Hanna, J.R., Reidler, J.S., Sepulcre, J., Poulin, R., Buckner, R.L., 2010. Functional-Anatomic Fractionation of the Brain's Default Network. Neuron 65, 550-562.

[7] Baldwin, D.A., 1993. Early referential understanding: young children' s ability to recognize referential acts for what they are. Developmental Psychology 29, 832-843.

[8] Bandettini, P.A., Wong, E.C., Hinks, R.S., Tikofsky, R.S., Hyde, J.S., 1992. Time course EPI of human brain function during task activation. Magn Reson Med 25, 390-397.

[9] Barabasi, A.L., Albert, R., 1999. Emergence of scaling in random networks. Science 286, 509-512.

[10] Belliveau, J.W., Kennedy, D.N., Jr., McKinstry, R.C., Buchbinder, B.R., Weisskoff, R.M., Cohen, M.S., Vevea, J.M., Brady, T.J., Rosen, B.R., 1991. Functional mapping of the human visual cortex by magnetic resonance imaging. Science 254, 716-719.

[11] Biswal, B., Yetkin, F.Z., Haughton, V.M., Hyde, J.S., 1995. Functional connectivity in the motor cortex of resting human brain using echo-planar MRI. Magn Reson Med 34, 537-541.

[12] Biswal, B.B., Mennes, M., Zuo, X.N., Gohel, S., Kelly, C., Smith, S.M., Beckmann, C.F., Adelstein, J.S., Buckner, R.L., Colcombe, S., Dogonowski, A.M., Ernst, M., Fair, D., Hampson, M., Hoptman, M.J., Hyde, J.S., Kiviniemi, V.J., Kotter, R., Li, S.J., Lin, C.P., Lowe, M.J., Mackay, C., Madden, D.J., Madsen, K.H., Margulies, D.S., Mayberg, H.S., McMahon, K., Monk, C.S., Mostofsky, S.H., Nagel, B.J., Pekar, J.J., Peltier, S.J., Petersen, S.E., Riedl, V., Rombouts, S.A., Rypma, B., Schlaggar, B.L., Schmidt, S., Seidler, R.D., Siegle, G.J., Sorg, C., Teng, G.J., Veijola, J., Villringer, A., Walter, M., Wang, L., Weng, X.C., Whitfield-Gabrieli, S., Williamson, P., Windischberger, C., Zang, Y.F., Zhang, H.Y., Castellanos, F.X., Milham, M.P., 2010. Toward discovery science of human brain function. Proc Natl Acad Sci U S A 107, 4734-4739.

[13] Born, P., Leth, H., Miranda, M.J., Rostrup, E., Stensgaard, A., Peitersen, B., Larsson, H.B., Lou, H.C., 1998. Visual activation in infants and young children studied by functional magnetic resonance imaging. Pediatr Res 44, 578-583.

[14] Born, P., Rostrup, E., Leth, H., Peitersen, B., Lou, H.C., 1996. Change of visually induced cortical activation patterns during development. Lancet 347, 543.

[15] Brandes, U., 2001. A faster algorithm for betweenness centrality. J Math Sociol 25, 163-177.

[16] Buckner, R.L., Andrews-Hanna, J.R., Schacter, D.L., 2008. The brain's default network: anatomy, function, and relevance to disease. Ann N Y Acad Sci 1124, 1-38.

[17] Buckner, R.L., Carroll, D.C., 2007. Self-projection and the brain. Trends Cogn Sci 11, 49-57.

[18] Bullmore, E., Sporns, O., 2009. Complex brain networks: graph theoretical analysis of structural and functional systems. Nat Rev Neurosci 10, 186-198.

[19] Cepeda, N.J., Kramer, A.F., Gonzalez de Sather, J.C., 2001. Changes in executive control across the life span: examination of task-switching performance. Dev Psychol 37, 715-730.

[20] Chang, C., Glover, G.H., Time-frequency dynamics of resting-state brain connectivity measured with fMRI. Neuroimage 50, 81-98.

[21] Church, J.A., Fair, D.A., Dosenbach, N.U., Cohen, A.L., Miezin, F.M., Petersen, S.E., Schlaggar, B.L., 2009. Control networks in paediatric Tourette syndrome show immature and anomalous patterns of functional connectivity. Brain 132, 225-238.

[22] Collins, D.l., Holmes, C.J., Peters, T.M., Evans, A.C., 1995. Automatic 3-D model-based neuroanatomical segmentation. Hum Brain Mapp 3, 190-208.

[23] Conel, J.L., 1939. The Postnatal Development of Human Cerebral Cortex. Harvard University Press, Cambridge.

[24] Damoiseaux, J.S., Rombouts, S.A., Barkhof, F., Scheltens, P., Stam, C.J., Smith, S.M., Beckmann, C.F., 2006. Consistent resting-state networks across healthy subjects. Proc Natl Acad Sci U S A 103, 13848-13853.

[25] De Martino, F., Gentile, F., Esposito, F., Balsi, M., Di Salle, F., Goebel, R., Formisano, E., 2007. Classification of fMRI independent components using IC-fingerprints and support vector machine classifiers. Neuroimage 34, 177-194.

[26] DeCasper, A.J., Fifer, W.P., 1980. Of human bonding: newborns prefer their mothers' voices. Science 208, 1174-1176.

[27] Dehaene-Lambertz, G., Dehaene, S., Hertz-Pannier, L., 2002. Functional neuroimaging of speech perception in infants. Science 298, 2013-2015.

[28] Di Martino, A., Kelly, C., Grzadzinski, R., Zuo, X.N., Mennes, M., Mairena, M.A., Lord, C., Castellanos, F.X., Milham, M.P., 2010. Aberrant striatal functional connectivity in children with autism. Biol Psychiatry 69, 847-856.

[29] Dickstein, D.P., Gorrostieta, C., Ombao, H., Goldberg, L.D., Brazel, A.C., Gable, C.J., Kelly, C., Gee, D.G., Zuo, X.N., Castellanos, F.X., Milham, M.P., 2010. Fronto-temporal spontaneous resting state functional connectivity in pediatric bipolar disorder. Biol Psychiatry 68, 839-846.

[30] Diesendruck, G., Markson, L., Akhtar, N., Reudor, A., 2004. Two-year-olds' sensitivity to speakers' intent: an alternative account of Samuelson and Smith. Dev Sci 7, 33-41.

[31] Ebbels, T.M., Buxton, B.F., Jones, D.T., 2006. springScape: visualisation of microarray and contextual bioinformatic data using spring embedding and an 'information landscape'. Bioinformatics 22, e99-107.

[32] Fair, D.A., Cohen, A.L., Dosenbach, N.U., Church, J.A., Miezin, F.M., Barch, D.M., Raichle, M.E., Petersen, S.E., Schlaggar, B.L., 2008. The maturing architecture of the brain's default network. Proc Natl Acad Sci U S A 105, 4028-4032.

[33] Fair, D.A., Cohen, A.L., Power, J.D., Dosenbach, N.U., Church, J.A., Miezin, F.M., Schlaggar, B.L., Petersen, S.E., 2009. Functional brain networks develop from a "local to distributed" organization. PLoS Comput Biol 5, e1000381.

[34] Fair, D.A., Posner, J., Nagel, B.J., Bathula, D., Dias, T.G., Mills, K.L., Blythe, M.S., Giwa, A., Schmitt, C.F., Nigg, J.T., 2010. Atypical default network connectivity in youth with attention-deficit/hyperactivity disorder. Biol Psychiatry 68, 1084-1091.

[35] Feinberg, D.A., Moeller, S., Smith, S.M., Auerbach, E., Ramanna, S., Glasser, M.F., Miller, K.L., Ugurbil, K., Yacoub, E., 2010. Multiplexed echo planar imaging for sub-second whole brain FMRI and fast diffusion imaging. PLoS One 5, e15710.

[36] Fivush, R., Hamond, N.R, 1990. Autobiographical memory across the preschool years: Toward reconcepualizing childhood amnesia. In: J.A.Hudson, R.F.a. (Ed.), Knowing and remembering in young children. Cambridge Univ. Press, New York:.

[37] Fornito, A., Zalesky, A., Bullmore, E.T., 2010. Network scaling effects in graph analytic studies of human resting-state FMRI data. Front Syst Neurosci 4, 22.

[38] Fox, M.D., Corbetta, M., Snyder, A.Z., Vincent, J.L., Raichle, M.E., 2006. Spontaneous neuronal activity distinguishes human dorsal and ventral attention systems. Proc Natl Acad Sci U S A 103, 10046-10051.

[39] Fox, M.D., Zhang, D., Snyder, A.Z., Raichle, M.E., 2009. The global signal and observed anticorrelated resting state brain networks. J Neurophysiol 101, 3270-3283.

[40] Fransson, P., Aden, U., Blennow, M., Lagercrantz, H., 2010. The Functional Architecture of the Infant Brain as Revealed by Resting-State fMRI. Cereb Cortex.

[41] Fransson, P., Skiold, B., Horsch, S., Nordell, A., Blennow, M., Lagercrantz, H., Aden, U., 2007. Resting-state networks in the infant brain. Proc Natl Acad Sci U S A 104, 15531-15536.

[42] Fulford, J., Vadeyar, S.H., Dodampahala, S.H., Moore, R.J., Young, P., Baker, P.N., James, D.K., Gowland, P.A., 2003. Fetal brain activity in response to a visual stimulus. Hum Brain Mapp 20, 239-245.

[43] Gao, W., Lin, W., 2011. Frontal parietal control network regulates the anti-correlated default and dorsal attention networks. Hum Brain Mapp (In Press).

[44] Gao, W., Lin, W., Chen, Y., Gerig, G., Smith, J.K., Jewells, V., Gilmore, J.H., 2008. Temporal and Spatial Development of Axonal Maturation and Myelination of White Matter in the Developing Brain. AJNR Am J Neuroradiol.

[45] Gao, W., Zhu, H., Giovanello, K., Lin, W., 2010. Multivariate network-level approach to detect interactions between large-scale functional systems. Med Image Comput Comput Assist Interv 13, 298-305.

[46] Gao, W., Zhu, H., Giovanello, K.S., Smith, J.K., Shen, D., Gilmore, J.H., Lin, W., 2009. Evidence on the emergence of the brain's default network from 2-week-old to 2-year-old healthy pediatric subjects. Proc Natl Acad Sci U S A 106, 6790-6795.

[47] Greicius, M.D., Krasnow, B., Reiss, A.L., Menon, V., 2003. Functional connectivity in the resting brain: a network analysis of the default mode hypothesis. Proc Natl Acad Sci U S A 100, 253-258.

[48] Greicius, M.D., Menon, V., 2004. Default-mode activity during a passive sensory task: uncoupled from deactivation but impacting activation. J Cogn Neurosci 16, 1484-1492.

[49] Greicius, M.D., Srivastava, G., Reiss, A.L., Menon, V., 2004. Default-mode network activity distinguishes Alzheimer's disease from healthy aging: evidence from functional MRI. Proc Natl Acad Sci U S A 101, 4637-4642.

[50] Gusnard, D.A., Akbudak, E., Shulman, G.L., Raichle, M.E., 2001. Medial prefrontal cortex and self-referential mental activity: relation to a default mode of brain function. Proc Natl Acad Sci U S A 98, 4259-4264.

[51] Gusnard, D.A., Raichle, M.E., 2001. Searching for a baseline: functional imaging and the resting human brain. Nat Rev Neurosci 2, 685-694.

[52] Haynes, R.L., Borenstein, N.S., Desilva, T.M., Folkerth, R.D., Liu, L.G., Volpe, J.J., Kinney, H.C., 2005. Axonal development in the cerebral white matter of the human fetus and infant. J Comp Neurol 484, 156-167.

[53] He, B.J., Snyder, A.Z., Zempel, J.M., Smyth, M.D., Raichle, M.E., 2008. Electrophysiological correlates of the brain's intrinsic large-scale functional architecture. Proc Natl Acad Sci U S A 105, 16039-16044.

[54] Hilgetag, C.C., Burns, G.A., O'Neill, M.A., Scannell, J.W., Young, M.P., 2000. Anatomical connectivity defines the organization of clusters of cortical areas in the macaque monkey and the cat. Philos Trans R Soc Lond B Biol Sci 355, 91-110.

[55] Hykin, J., Moore, R., Duncan, K., Clare, S., Baker, P., Johnson, I., Bowtell, R., Mansfield, P., Gowland, P., 1999. Fetal brain activity demonstrated by functional magnetic resonance imaging. Lancet 354, 645-646.

[56] Ilott, N.E., Saudino, K.J., Asherson, P., 2010. Genetic influences on attention deficit hyperactivity disorder symptoms from age 2 to 3: a quantitative and molecular genetic investigation. BMC Psychiatry 10, 102.

[57] Jafri, M.J., Pearlson, G.D., Stevens, M., Calhoun, V.D., 2008. A method for functional network connectivity among spatially independent resting-state components in schizophrenia. Neuroimage 39, 1666-1681.

[58] Joel, S.E., Caffo, B.S., van Zijl, P.C., Pekar, J.J., 2011. On the relationship between seed-based and ICA-based measures of functional connectivity. Magn Reson Med.

[59] Johnson, M.H., 2000. Functional brain development in infants: elements of an interactive specialization framework. Child Dev 71, 75-81.

[60] Johnson, M.H., 2001. Functional brain development in humans. Nat Rev Neurosci 2, 475-483.

[61] Joseph, R., 1999. Fetal Brain & Cognitive Development. Developmental Review 20, 81-98.

[62] Kelly, A.M., Di Martino, A., Uddin, L.Q., Shehzad, Z., Gee, D.G., Reiss, P.T., Margulies, D.S., Castellanos, F.X., Milham, M.P., 2009. Development of anterior cingulate functional connectivity from late childhood to early adulthood. Cereb Cortex 19, 640-657.

[63] Kieling, C., Goncalves, R.R., Tannock, R., Castellanos, F.X., 2008. Neurobiology of attention deficit hyperactivity disorder. Child Adolesc Psychiatr Clin N Am 17, 285-307, viii.

[64] Kolata, G., 1984. Studying learning in the womb. Science 225, 302-303.

[65] Latora, V., Marchiori, M., 2001. Efficient behavior of small-world networks. Phys Rev Lett 87, 198701.

[66] Levy, S.E., Mandell, D.S., Schultz, R.T., 2009. Autism. Lancet 374, 1627-1638.

[67] Lin, W., Zhu, Q., Gao, W., Chen, Y., Toh, C.H., Styner, M., Gerig, G., Smith, J.K., Biswal, B., Gilmore, J.H., 2008. Functional connectivity MR imaging reveals cortical functional connectivity in the developing brain. AJNR Am J Neuroradiol 29, 1883-1889.

[68] Liston, C., Cohen, M.M., Teslovich, T., Levenson, D., Casey, B.J., 2011. Atypical prefrontal connectivity in attention-deficit/hyperactivity disorder: pathway to disease or pathological end point? Biol Psychiatry 69, 1168-1177.

[69] Liu, Y., Liang, M., Zhou, Y., He, Y., Hao, Y., Song, M., Yu, C., Liu, H., Liu, Z., Jiang, T., 2008. Disrupted small-world networks in schizophrenia. Brain 131, 945-961.

[70] Logothetis, N.K., 2003. The underpinnings of the BOLD functional magnetic resonance imaging signal. J Neurosci 23, 3963-3971.

[71] Logothetis, N.K., Pauls, J., Augath, M., Trinath, T., Oeltermann, A., 2001. Neurophysiological investigation of the basis of the fMRI signal. Nature 412, 150-157.

[72] Lowe, M.J., Mock, B.J., Sorenson, J.A., 1998. Functional connectivity in single and multislice echoplanar imaging using resting-state fluctuations. Neuroimage 7, 119-132.

[73] Murphy, K., Birn, R.M., Handwerker, D.A., Jones, T.B., Bandettini, P.A., 2009. The impact of global signal regression on resting state correlations: are anti-correlated networks introduced? Neuroimage 44, 893-905.

[74] Neumann, J., von Cramon, D.Y., Forstmann, B.U., Zysset, S., Lohmann, G., 2006. The parcellation of cortical areas using replicator dynamics in fMRI. Neuroimage 32, 208-219.

[75] Newman, M.E., 2006. Modularity and community structure in networks. Proc Natl Acad Sci U S A 103, 8577-8582.

[76] Nir, Y., Mukamel, R., Dinstein, I., Privman, E., Harel, M., Fisch, L., Gelbard-Sagiv, H., Kipervasser, S., Andelman, F., Neufeld, M.Y., Kramer, U., Arieli, A., Fried, I., Malach, R., 2008. Interhemispheric correlations of slow spontaneous neuronal fluctuations revealed in human sensory cortex. Nat Neurosci 11, 1100-1108.

[77] Ogawa, S., Lee, T.M., Nayak, A.S., Glynn, P., 1990. Oxygenation-sensitive contrast in magnetic resonance image of rodent brain at high magnetic fields. Magn Reson Med 14, 68-78.

[78] Perlbarg, V., Bellec, P., Anton, J.L., Pelegrini-Issac, M., Doyon, J., Benali, H., 2007. CORSICA: correction of structured noise in fMRI by automatic identification of ICA components. Magn Reson Imaging 25, 35-46.

[79] Qiu, M.G., Ye, Z., Li, Q.Y., Liu, G.J., Xie, B., Wang, J., 2010. Changes of Brain Structure and Function in ADHD Children. Brain Topogr.

[80] Raichle, M.E., MacLeod, A.M., Snyder, A.Z., Powers, W.J., Gusnard, D.A., Shulman, G.L., 2001. A default mode of brain function. Proc Natl Acad Sci U S A 98, 676-682.

[81] Raichle, M.E., Mintun, M.A., 2006. Brain work and brain imaging. Annu Rev Neurosci 29, 449-476.

[82] Salvador, R., Suckling, J., Coleman, M., Pickard, J. D., Menon, D., Bullmore,, E., 2005. Neurophysiological architecture of functional magnetic resonance images of human brain. Cerebral Cortex 34, 387-413.

[83] Seeley, W.W., Crawford, R.K., Zhou, J., Miller, B.L., Greicius, M.D., 2009. Neurodegenerative diseases target large-scale human brain networks. Neuron 62, 42-52.

[84] Shmuel, A., Leopold, D.A., 2008. Neuronal correlates of spontaneous fluctuations in fMRI signals in monkey visual cortex: Implications for functional connectivity at rest. Hum Brain Mapp 29, 751-761.

[85] Shulman GL, F.J., Corbetta M, Buckner RL, Miezin FM, 1997. Common blood flow changes across visual tasks: II.: decreases in cerebral cortex. J. Cogn. Neurosci 9, 648-663.

[86] Smyser, C.D., Inder, T.E., Shimony, J.S., Hill, J.E., Degnan, A.J., Snyder, A.Z., Neil, J.J., 2010. Longitudinal Analysis of Neural Network Development in Preterm Infants. Cereb Cortex.

[87] Supekar, K., Musen, M., Menon, V., 2009. Development of large-scale functional brain networks in children. PLoS Biol 7, e1000157.

[88] Tau, G.Z., Peterson, B.S., 2010. Normal development of brain circuits. Neuropsychopharmacology 35, 147-168.

[89] Tohka, J., Foerde, K., Aron, A.R., Tom, S.M., Toga, A.W., Poldrack, R.A., 2008. Automatic independent component labeling for artifact removal in fMRI. Neuroimage 39, 1227-1245.

[90] Tomasello, M., Strosberg, R., Akhtar, N., 1996. Eighteen-month-old children learn words in non-ostensive contexts. J Child Lang 23, 157-176.

[91] Tzourio-Mazoyer, N., Landeau, B., Papathanassiou, D., Crivello, F., Etard, O., Delcroix, N., Mazoyer, B., Joliot, M., 2002. Automated anatomical labeling of activations in SPM using a macroscopic anatomical parcellation of the MNI MRI single-subject brain. Neuroimage 15, 273-289.

[92] Uddin, L.Q., Kelly, A.M., Biswal, B.B., Margulies, D.S., Shehzad, Z., Shaw, D., Ghaffari, M., Rotrosen, J., Adler, L.A., Castellanos, F.X., Milham, M.P., 2008. Network homogeneity reveals decreased integrity of default-mode network in ADHD. J Neurosci Methods 169, 249-254.

[93] Van Dijk, K.R., Hedden, T., Venkataraman, A., Evans, K.C., Lazar, S.W., Buckner, R.L., 2010. Intrinsic functional connectivity as a tool for human connectomics: theory, properties, and optimization. J Neurophysiol 103, 297-321.

[94] Vincent, J.L., Kahn, I., Snyder, A.Z., Raichle, M.E., Buckner, R.L., 2008. Evidence for a frontoparietal control system revealed by intrinsic functional connectivity. J Neurophysiol 100, 3328-3342.

[95] Vincent, J.L., Snyder, A.Z., Fox, M.D., Shannon, B.J., Andrews, J.R., Raichle, M.E., Buckner, R.L., 2006. Coherent spontaneous activity identifies a hippocampal-parietal memory network. J Neurophysiol 96, 3517-3531.

[96] Wang, J., Wang, L., Zang, Y., Yang, H., Tang, H., Gong, Q., Chen, Z., Zhu, C., He, Y., 2009. Parcellation-dependent small-world brain functional networks: a resting-state fMRI study. Hum Brain Mapp 30, 1511-1523.

[97] Watts, D.J., Strogatz, S.H., 1998. Collective dynamics of 'small-world' networks. Nature 393, 440-442.

[98] Whitlow, C.T., Casanova, R., Maldjian, J.A., 2011. Effect of resting-state functional MR imaging duration on stability of graph theory metrics of brain network connectivity. Radiology 259, 516-524.

Chapter 8

Big Challenges from the "Little Brain" — Imaging the Cerebellum

John Schlerf, Tobias Wiestler,
Timothy Verstynen and Jörn Diedrichsen

Additional information is available at the end of the chapter

http://dx.doi.org/10.5772/50746

1. Introduction

The cerebellum is a compact structure that has about one-ninth of the volume of the neocortex, but comprises more than half of the neurons in the central nervous system. While the cerebellum has historically been understood to assist with movement and balance (e.g., Holmes, 1917, 1939), today we know that the cerebellum is also involved in non-motor functions (for review, see Stoodley & Schmahmann, 2009). Indeed, the cerebellum appears to play a role in nearly as many separate functions as the neocortex. This functional diversity within a restricted area poses unique challenges for the study of the cerebellum using functional magnetic resonance imaging (fMRI).

While no additional equipment or specialized pulse sequences are required to obtain fMRI data from the cerebellum, most so-called "whole brain" fMRI experiments focus on the study of the neocortex, often excluding inferior aspects of the cerebellum from data acquisition altogether. Furthermore, most common analysis methods are designed with the neocortex in mind and may be suboptimal for the analysis of cerebellar data, thus necessitating novel approaches to obtain insights about cerebellar function.

In this chapter we will highlight the current "best practices" to cerebellar fMRI experiments and address methodological challenges such as anatomical realignment, cerebellar atlases, cardiac artifacts on functional data, and the physiological source of the cerebellar blood-oxygen-level-dependent (BOLD) signal. We also review important research findings about the cerebellum, highlighting the role of fMRI studies in shaping our understanding of cerebellar function.

2. Methodological concerns for cerebellar imaging

2.1. Anatomical methods

Hypothesis testing using fMRI often requires data from multiple individuals; therefore we need a common anatomical frame of reference. Most widely-used normalization methods result in poor alignment of regions within the cerebellar cortex, which weakens statistical power (Diedrichsen, 2006). The deep cerebellar nuclei (the output of the cerebellum) are small and lie adjacent to the fourth ventricle and vascular supply, and therefore pose a special challenge to spatial normalization.

2.2. Localization within the cerebellar cortex

Like the neocortex, the cerebellar cortex is intricately folded. There are clearly defined lobules and sub-lobules within this cortex, separated by a number of fissures that can be reliably identified across individuals (Figure 1). Each lobule has its own pattern of connectivity with the neocortex (Kelly & Strick, 2003), and likely its own functional role (Stoodley & Schmahmann, 2009). Since each lobule is relatively small (5-10mm), inaccuracies in normalization will lead to functional data being averaged across neighboring structures. Indeed, if one uses standard spatial normalization procedures (which result in satisfactory overlap of structures within the cerebrum), the location of the primary fissure in the cerebellum (which forms the boundary between the anterior and posterior lobes) can be mis-localized by 1-2 cm (Diedrichsen, 2006). The top row of Figure 2 illustrates this overlap following the standard nonlinear SPM normalization to the MNI group template. When averaging anatomical scans across participants after normalization, most spatial details of the underlying anatomy are lost.

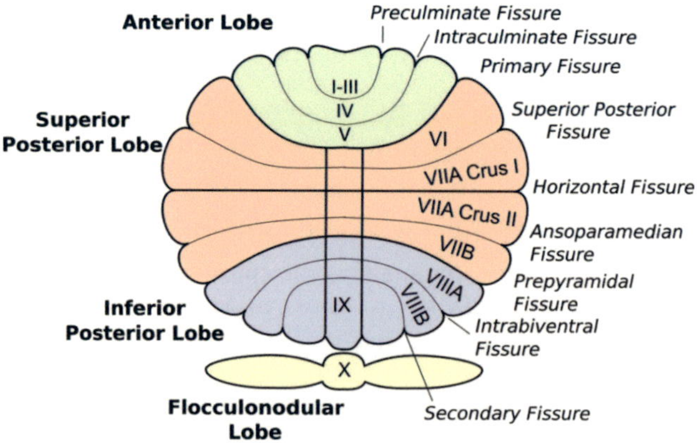

Figure 1. Anatomy of the cerebellar cortex. Traditionally, the cerebellum is split into the Anterior (green), Posterior (red and blue), and Flocculonodular (yellow) Lobes. Since the posterior lobe in the human cerebellum is large, researchers often subdivide it into Superior (red) and Inferior (blue) aspects. The cortex is subdivided longitudinally into lobules (Roman Numerals) by deep fissures. The current consensus on anatomical nomenclature is based on work by Larsell (1952, 1953), as laid out (along with excellent translations of prior nomenclature systems) by Schmahmann and colleagues (1999).

http://dx.doi.org/10.5772/50746

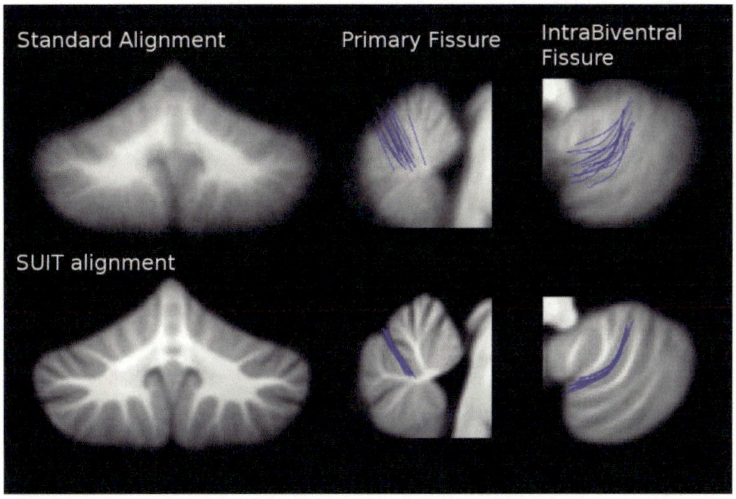

Figure 2. Comparison of normalization procedures. The top row demonstrates the normalization and coregistration of 21 individuals using the SPM standard normalization procedure to the MNI 152 template, which is not specialized for the cerebellum. Note the loss of detail, and the poor alignment of the Primary and Intrabiventral fissures. The second row shows the same 20 brains normalized using the SUIT template.

An equally serious problem is anatomical bias. Even within the same neuroimaging coordinate system (e.g., MNI152), different normalization methods may lead to substantially different locations of the same cerebellar structure in atlas space (Diedrichsen et al., 2009). For example, it has become common to identify the anatomical location of cerebellar coordinates using the MRI atlas of the human cerebellum (Schmahmann et al., 1999). While this atlas provides a framework for subdividing the structure into individual lobules, the atlas is based on a single individual with a unique anatomy. Examination of 20 participants after normalization in SPM suggests that coordinate-based assignment using this atlas is correct for only 38% of the voxels in the cerebellar cortex (Diedrichsen et al., 2009).

Better normalization should be achievable, as cerebellar fissures are stereotyped and consistent across individuals (Makris et al., 2003, 2005). This makes it possible to use higher resolution anatomical normalization methods combined with a template that preserves anatomical detail within the cerebellum, implemented in the SUIT toolbox (Diedrichsen, 2006). This template is spatially unbiased relative to the space defined by the MNI 152 template, so each cerebellar structure is located where it would end up – on average-after affine alignment to the MNI 152 template. Thus, the resultant coordinates can be directly compared to results obtained in other studies. The use of the toolbox dramatically improves overlap of individual fissures (Figure 2, bottom row).

Since the development of the SUIT method, other methods have been published that use similar high-resolution warps and detailed nonlinear templates. While these methods improve the alignment of cerebellar structures, SUIT still improves the overlap of individual lobules by about 8% (Diedrichsen et al., 2009). The advantage of the SUIT approach likely arises from the initial segmentation into cerebellum and neocortex, as only subtentorial structures drive the normalization. Segmentation also reduces the likelihood that activity from nearby neocortex will be misattributed to the cerebellum.

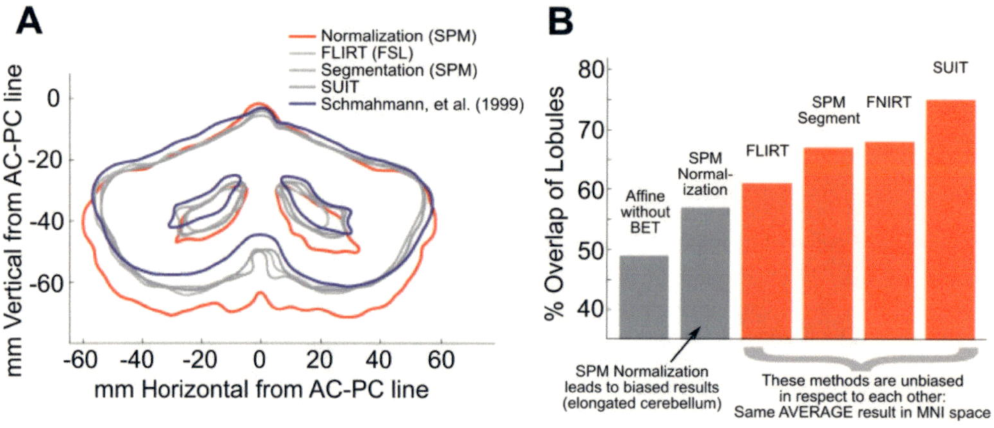

Figure 3. Overlap and bias of different normalization methods. A) When the outline of the cerebella are overlaid, the standard SPM normalization method (red outline) is clearly biased, particularly with respect to the Schmahmann atlas (blue outline). The other methods (including SUIT) are unbiased with respect to each other. B) Comparing the percent overlap of individual lobules, we can see some differences even among methods which result in the same overall spatial location.

To overcome bias in labeling the anatomical locations of activated regions, a probabilistic atlas for the cerebellum has been developed (Diedrichsen et al., 2009) to allow for accurate anatomical description of cerebellar activations observed in group analyses. This atlas can also facilitate analyses that eschew the group-overlap approach in favor of analysis based on anatomical regions of interest (ROIs). While it is possible to identify cerebellar lobules manually to constrain the anatomical extent of analysis (Makris et al., 2003, 2005), doing this by hand is tedious, and requires specialized knowledge to be reliable. The use of independently validated probabilistic atlases is much more efficient.

2.2.1. *Imaging the nuclei*

While SUIT-normalization and the associated probabilistic atlas provide sufficient accuracy for fMRI studies of the cerebellar cortex, more specialized techniques are required for the alignment of the deep cerebellar nuclei (DCN). These structures can be difficult to identify using conventional anatomical scans, as they can appear contiguous with the cerebellar white matter. Due to the high concentration of iron in these nuclei, they can be visualized more easily in T1-weighted images in older individuals (Dimitrova et al., 2002b) or by using T2-weighted images (Dimitrova et al., 2006).

Because the deep cerebellar nuclei are small, the inter-individual overlap using typical normalization methods is quite poor. Good superposition of the nuclei in group analyses is essential, especially because the T2* signal from the DCN is lower than the surrounding white matter and the nearby gray matter (Dimitrova et al., 2006). Since variability in fMRI is proportional to the average raw signal, this means that the functionally induced signal variations in the dentate nucleus are very low compared with functional activation in nearby gray matter. If these signals are averaged together with the signal from areas where both the raw signal and noise is twice as large (a consequence of spatial smoothing), it is extremely unlikely that they will be detected. This problem is exacerbated by high field strength, where the differences in T2* are even more pronounced.

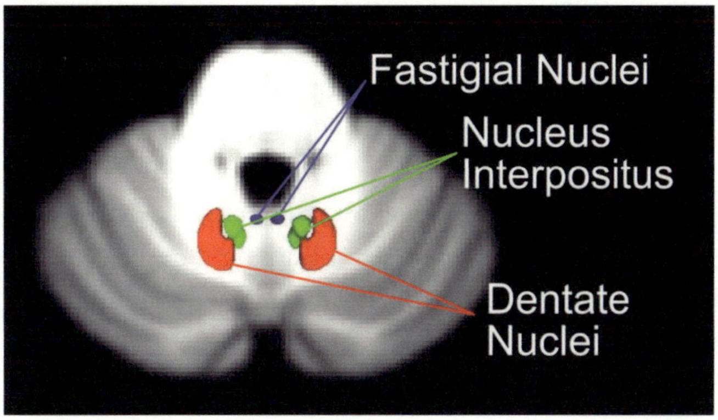

Figure 4. Overview and approximate location of the deep cerebellar nuclei. Adapted from (Diedrichsen et al., 2011).

A refined method for the normalization of the cerebellum and the dentate nucleus has recently been developed in conjunction with a probabilistic atlas for the DCN (Diedrichsen et al., 2011). As with the cerebellar cortical atlas (Diedrichsen et al., 2009), inter-individual anatomical variability depends on the normalization method chosen. Traditional MNI-alignment methods perform the worst, while segmentation-based methods perform slightly better (Diedrichsen et al., 2011). The best overlap for the deep cerebellar nuclei was obtained by using a variation of the SUIT-normalization method, combining white matter segmentation with an anatomical ROI that covers the deep cerebellar nuclei. Currently, this ROI must be traced by hand, and is most reliable when traced onto the phase image of susceptibility-weighted scans collected at 7T (Diedrichsen et al., 2011).

Using these techniques, Timmann and colleagues have shown that the dentate nucleus can be subdivided into cognitive and sensorimotor components (Küper et al., 2011a). They were able to demonstrate a somatotopic gradient between hand and foot movement in the motor portion of the dentate nucleus at 7T (Küper et al., 2011b). These studies provide important and exciting advances into our understanding of the human dentate nucleus. They also, however, highlight the limits of current fMRI techniques and the need for further development. Both magnetic inhomogeneity artifacts as well as physiological noise near arterial vessels make the detection of these weak signals challenging. As high-field scanners become more accessible and functional imaging at this field strength matures, we hope to see these techniques used for testing specific hypotheses about the deep cerebellar nuclei.

2.3. Interpretation of the cerebellar bold signal

FMRI involves repeatedly imaging the brain and analyzing small changes caused by variations in blood oxygenation. It is important to know which neurophysiological processes drive the cerebellar BOLD signal. Neural activity increases the cerebral metabolic rate of oxygen consumption ($CMRO_2$), which causes an increase in the local concentration of deoxy-hemoglobin, whose magnetic properties decrease BOLD. Neuronal activity also leads to a compensatory increase in cerebral blood flow (CBF), which raises the local concentration of oxygenated hemoglobin and thus increases the BOLD signal. In the neocortex, these two quantities have been measured and the relationship between them appears to be relatively fixed: the increased oxygenated hemoglobin from increased CBF outstrips the increased deoxy-hemoglobin from increased $CMRO_2$ by ratio of 2:1 (Hoge et al., 1999). As a result, neural activity in the cerebral cortex increases the BOLD signal. However, the architecture of the cerebellar cortex is dramatically different from that of the neocortex, and the same ratios may not apply. Careful thought and experimental work are needed to illuminate the relationship between neural activity and BOLD in the cerebellum.

2.3.1. Neural sources

There are two primary input systems to the cerebellar cortex (Figure 5). The first is the mossy fiber inputs that project to granule cells, which give rise to parallel fibers. Each parallel fiber synapses onto the dendritic trees of thousands of Punkinje cells-the only output cell of the cerebellar cortex. A single Purkinje cell can receive weak synapses from tens of thousands of parallel fibers. The second input system is the climbing fibers that arise from the inferior olive. Each Purkinje cell receives a very strong synaptic input from a single climbing fiber. Stimulation of each of these two pathways leads to time-locked increases in blood flow at the cerebellar cortex (Mathiesen et al., 2000), which by inference suggests that activity in both pathways may be contributing to the BOLD signal.

To understand the source of the cerebellar BOLD signal under physiologically plausible conditions, it is informative to understand how much energy (and hence oxygen) neural processes within the cerebellum demand. Calculations based on cell numbers, spontaneous firing rates and membrane potentials (Howarth et al., 2009) lead to the estimation that approximately 18% of the energy in the rat cerebellar cortex is consumed by Purkinje cells. In contrast, granule cells use 67% of the energy, most of which is consumed by the process of relaying mossy fiber input to cells in the cerebellar cortex. During sensory stimulation, mossy fibers have been shown to fire bursts of action potentials exceeding 700 Hz, compared to baseline firing rates at rest on the order of 3 Hz (Rancz et al., 2007). If the proportional increases of blood flow and energy use observed in the cerebral cortex hold for the cerebellum, we would be tempted to conclude that activity-dependent BOLD signal changes are mostly caused by increased signaling in the mossy fiber input pathway.

http://dx.doi.org/10.5772/50746

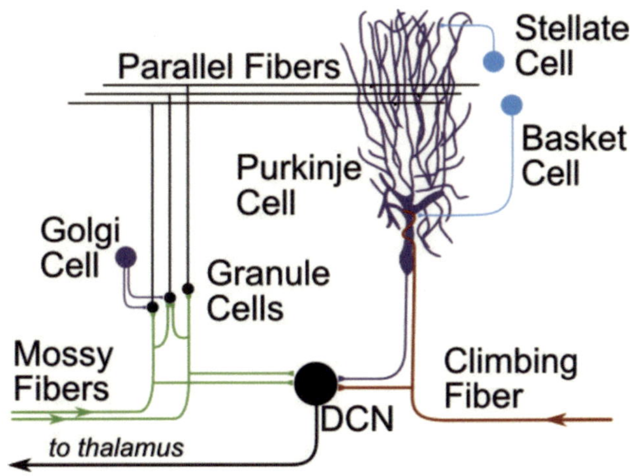

Figure 5. Overview of the cerebellar circuit. The input pathways are the mossy and climbing fibers, which are excitatory. Mossy fibers (green) synapse onto granule cells, which give rise to excitatory parallel fibers which synapse onto Purkinje cells, the output of the cerebellar cortex. Climbing fibers (red) synapse directly onto the Purkinje cells. Stellate, Basket, Golgi, and Purkinje cells are all inhibitory. Both input pathways, as well as Purkinje cells, synapse on the Deep Cerebellar Nuclei (DCN), the output of the cerebellum. (Adapted from Ito, 2001)

When the parallel fibers of anesthetized rats are electrically stimulated, blood flow to the region increases (Iadecola et al., 1996). This increased blood flow outstrips the increased $CMRO_2$, thereby leading to increased tissue oxygen (Thomsen et al., 2009). By inference, this suggests that increased activity in the mossy fiber/parallel fiber system will be tightly coupled with an increase in the BOLD signal. In line with energy use predictions outlined in the previous paragraph, increased neural activity within Purkinje cells alone does not seem to be strong enough to increase blood flow (Thomsen et al., 2004).

The possible response of the BOLD signal to activity within the climbing fiber system is less clear. After lesions of the inferior olive, stimulation of the whiskers of a rat produced only 42% of the blood flow increase observed in control animals (Zhang et al., 2003). This suggests that climbing fibers may be responsible for approximately half of the activity-dependent modulation of the BOLD signal. However, when complex (and simple) spike firing rates in Purkinje cells are increased by pharmacologically removing inhibitory input from interneurons, increases are observed in $CMRO_2$, but not in CBF (Thomsen et al., 2009). According to these results, increasing the complex spike rate by increasing the climbing fiber input could actually decrease the BOLD signal.

Clearly, the question of which neural processes are reflected in the cerebellar BOLD signal remains unanswered. While it appears likely that increased mossy/parallel fiber input would lead to robust increases in the BOLD signal at the cerebellar cortex, it remains unclear how climbing fiber input would modulate BOLD, if at all. Parallel electrophysiological recording and BOLD measurements would constitute an important step forward.

2.3.2. Physiological artifacts

Since the BOLD signal depends on the rate of blood flow and oxygenation, it should come as no surprise that it is sensitive to changes in respiration and cardiac function. This is especially true in the cerebellum and brainstem, which are surrounded by a dense vascular bed. Problematically, these physiological processes have been shown to change in response to behavior. Systematic heart rate changes are known to occur in many tasks involving the control of action (Damen & Brunia, 1987; Jennings et al., 1991, 1992, 1993; Jennings & van der Molen, 2002) as well as feedback processing in cognitive (Crone et al., 2004, 2005, 2006) and motor tasks (Schlerf et al., 2012). Thus, we may misinterpret BOLD signal changes caused by physiological processes as changes caused by neural activity.

Fortunately, physiological signals (respiration, heart rate) are easily recorded. The main challenge, then, becomes one of data analysis: how should one best remove physiological artifacts from the fMRI signals using these recordings? There are two principal ways that physiological processes can influence the BOLD signal. First, the time of slice acquisition relative to the physiological event is important (Figure 6A and B). The rate of blood flow varies for different phases of the cardiac cycle, and the brainstem shows movement with every heartbeat. The rise and fall of the chest during breathing induces subtle changes in the magnetic field. An efficient way to deal with changes related to the phase of these processes without any *a priori* knowledge of how the BOLD signal will change with these events is to use the RETROICOR method (Glover et al., 2000). This method deals with this by including both the sine and the cosine-and higher harmonics-of the phase of the physiological process as nuisance covariates within a standard GLM analysis. Thus, variability within a voxel's time series due to phasic changes of the signals measured will be modeled, instead of merely adding residual noise to the estimate of the BOLD response. This method was developed for the cerebral cortex, but provides efficient noise reduction for sub-tentorial structures as well. Within the brainstem, it has been demonstrated that the BOLD signal contains physiological noise which can be best modeled by three harmonics of respiration, two harmonics of heart rate, as well as a multiplicative term (Harvey et al., 2008). For areas around the cerebellum and brain stem, our own observations suggest that more than 30% of the variance in the raw fMRI signal can be explained by the first two harmonics of heart and breathing rate (Figure 6C). More recent methods have also been proposed for modeling other physiological artifacts in the BOLD signal, like spontaneous variations in breathing rate (Birn et al., 2006, 2008) and cardiac output (Chang & Glover, 2009). This approach can be quite important for correctly identifying neurally relevant portions of the cerebellar BOLD signal (Schlerf et al. 2012). While the proper balance between noise reduction and over-fitting has to be determined in each individual case, recording heart rate and breathing should become commonplace for cerebellar imaging studies to ensure that the effects of interest do not correlate systematically with physiological processes.

http://dx.doi.org/10.5772/50746

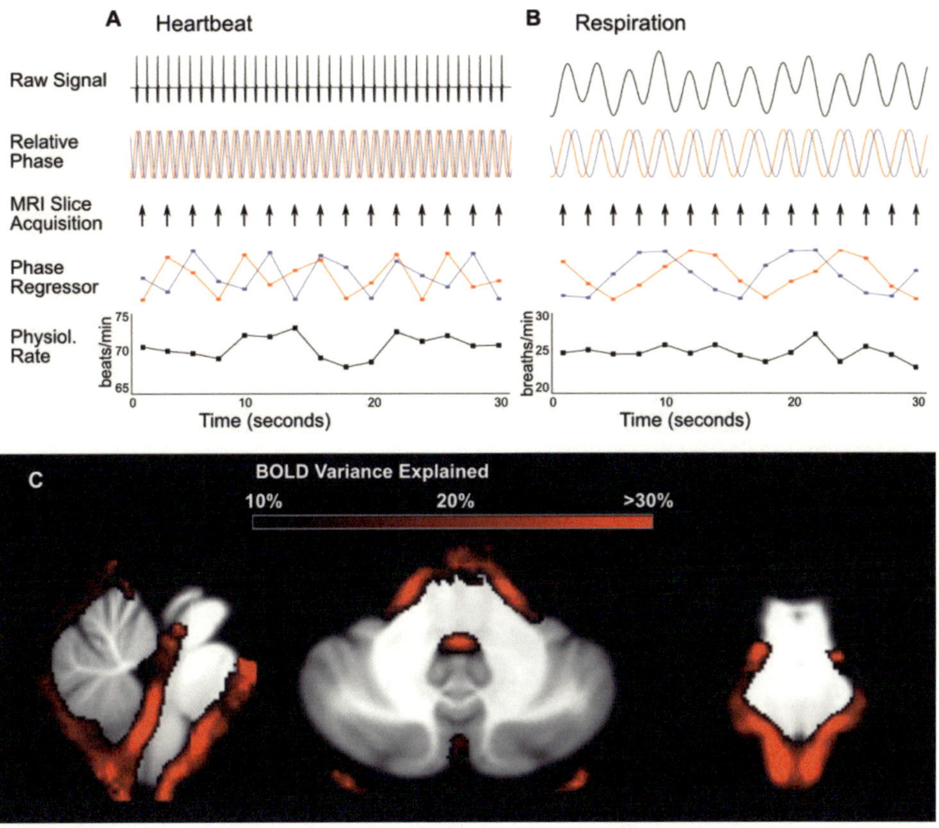

Figure 6. Influence of heart rate onto the BOLD signal. A) As a demonstration, a representative individual's heartbeat is shown. The top row shows the raw EKG trace of the hearbeat, and the next shows the first Fourier expansion (both sine and cosine waves) of the relative phase of the heartbets. Since fMRI data is acquired at regular intervals, BOLD artifacts may arise related to the phase of the heartbeat the data was acquired. The bottom row demonstrates heart rate effects. B) A representative individuals breathing is shown. The top row is the raw trace from a pneumatic compression belt, and the next rows show the instantaneous phase, as well as the phase and rate regressors. Note that breathing rate is substantially slower than heart rate. C) Data from a study conducted of the cerebellum (Schlerf et al., 2012) showing a high level of variability (greater than 30% of BOLD variance) accounted for by cardiac and respiratory phase regressors in CSF and arteries surrounding the cerebellum.

3. Functional studies of the cerebellum

3.1. Sensorimotor function

Because damage or removal of the cerebellum results first and foremost in movement impairments, the cerebellum has historically been categorized as a sensorimotor structure (Holmes, 1917, 1939). While it is now understood that the functional domain of the cerebellum extends beyond movement, sensorimotor processing remains one of the most active and important areas of cerebellar research.

Movement is one of the most efficient tasks for modulating BOLD-signal within the cerebellum. Below we review a few important fMRI results that have shaped our thinking of the function of this structure.

3.1.1. Maps of the body

Physiological work has suggested the presence of multiple body maps within the cerebellum. Two somatotopic maps were initially observed in the non-human primate cerebellum: one in the anterior lobe and the other in the inferior posterior lobe (Snider & Eldred, 1952). Both of these maps are somatotopically organized on a gross level (Figure 7). This organization has also been shown in humans using functional imaging. Indeed, the first published imaging study of the cerebellum (using PET) revealed spatially specific activations in response to finger and eye movements in the cerebellum (Fox et al., 1985).

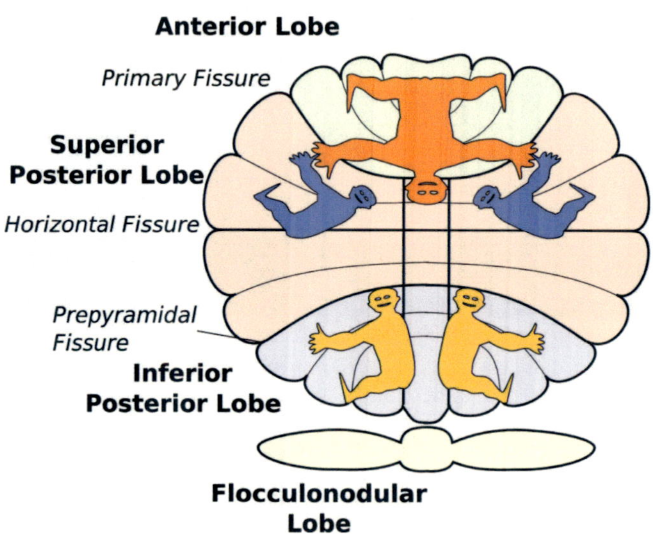

Figure 7. Multiple body representations within the cerebellum. The lobes are shaded to match the regions in Figure 1. Shown are the anterior (top, orange) and inferior posterior (bottom, yellow) body representations, as well as the newly identified superior posterior representation (blue). (Adapted from Snider & Eldred, 1952; Grodd et al., 2001; Schlerf et al., 2010)

Using fMRI, researchers have confirmed and extended these results. Rijntjes and colleagues (1999) confirmed that both sensory-motor representations show a somatopic gradient using finger and toe movements. Grodd et al. (2001) built upon this finding, demonstrating anterior and posterior representations for the hand, lips, tongue, and feet. These distinct maps were observed in the anterior and posterior lobes, in good agreement with animal physiology. The representation within the anterior lobe (lobules I-V; see Schmahmann et al., 1999) is generally unilateral, with each cerebellar hemisphere representing the movement of the ipsilateral limb. Within the somatomotor representation in the posterior lobe (chiefly lobule VIII), the observed activations are usually also stronger on the ipsilateral side, but bilateral activation is commonly observed (Grodd et al., 2001; Schlerf et al., 2010).

A number of studies have tried to dissociate the function of the anterior and posterior body maps. Thickbroom et al. (2003) explored the functional response to both active and passive movements of single fingers. Activation was observed to both types of movement in both the anterior lobe and the inferior aspect of the posterior lobe, suggesting that these regions have both sensory and motor functions. Habas et al. (2004a, 2004b) explored the response of the anterior map and the inferior posterior map during bimanual movements. Contrasting with the Thickbroom study, they found that passive bimanual movements did not activate the representation in lobule VIII, while active bimanual movements did (Habas et al., 2004a). A second study found that the inferior representation is more strongly activated by out-of-phase bimanual movements than in-phase movements, suggesting that this region may be specifically engaged by movements requiring the coordination of multiple effectors (Habas et al., 2004b).

A recent study, designed to explore the cerebellar response to coordination between individual fingers, tested simple rhythmic movements of all digits (fingers or toes) as well as complex sequential movements of individual digits (Schlerf et al., 2010). The body map within the anterior lobe was not strongly affected by movement complexity. Interestingly, there was evidence for a new map in lobules VI and VIIA Crus I, which was much more strongly activated for sequential movements of both the fingers (medially) and toes (laterally, see Figure 7). This suggests that these regions (thought to have cognitive functions) may be more involved in motor behavior than previously thought (Schlerf et al., 2010).

The aforementioned fMRI studies reveal somatotopic organization on a gross anatomical level, with neighboring voxels activating in response to different body parts. Physiological experiments, however, suggest that neighboring patches of cerebellar cortex respond to stimulation of quite different areas of the skin within the same bodypart. The term "fractured somatotopy" has been coined to capture this disjointed organization (Shambes et al., 1978; Bower & Woolston, 1983; Kassel et al., 1984). This organization poses a challenge to cerebellar imaging, as each voxel averages the activity over several of these patches of cerebellar cortex. Thus, conventional fMRI analyses will reach a resolution limit. A recent study was able to explore finer resolution maps of individual fingers using advanced analysis techniques (Wiestler et al., 2011). The analysis approach rests on the assumption that even if the average level of BOLD signal change across a region is not sensitive to individual finger movement, the spatial pattern of activity across multiple voxels may be specific for fingers and can be detected by multivariate analyses. The authors used local voxel patterns to train a linear classifier, and then predicted which finger was moved or stimulated in an independent test dataset. Better-than-chance prediction indicates that the region contains a finger representation. Two finger maps were identified: one in lobule V, and a second in lobule VIII. The activation patterns in these two representations were equally informative of finger identity for movement and tactile stimulation. This suggests that motor as well as sensory events are encoded with equal fidelity in the cerebellum. In contrast to the finger representations in the neocortex, which are ~3-4mm in size, the authors showed that the cerebellar representations are ~1-2mm in size, smaller than an individual voxel. Such "hyper-acuity" of multivariate analysis has also been demonstrated for the case of orientation columns in visual cortex (Swisher et al., 2010). Thus, multivariate analysis promises to be an excellent tool to study fractured somatotopy within the cerebellum.

3.1.2. Motor learning

In addition to studying topographic organization, fMRI has been used to explore the cerebellar involvement in motor learning. A large body of evidence shows that the cerebellum is involved in the process of error-based learning (Martin et al., 1996; Straube et al., 2001; Maschke et al., 2004; Pisella et al., 2005; Smith & Shadmehr, 2005; Morton & Bastian, 2006; Tseng et al., 2007; Alahyane et al., 2008). One example of such a task is saccade adaptation, in which the length of an eye-movement is adjusted on a trial-by-trial basis, based on whether the eye movement under-or over-shot the target during the previous saccade. This rapid learning process is impaired in cerebellar patients (Golla et al., 2008).

A seminal model proposed by Marr (1969) and Albus (1971) suggests that the architecture of the cerebellar cortex is ideally suited for such error-based learning. The model suggests that context information about the planned or ongoing movement is signaled over the mossy-fiber-parallel-fiber system. The climbing fiber input to the cerebellar cortex then provides an error signal that leads to plasticity at the parallel-fiber-Purkinje cell synapses. Through this mechanism, the cerebellar output associated with the same movement (parallel fiber input) is rapidly adjusted.

Researchers have used error-based learning tasks to test the involvement of the cerebellum in motor adaptation, albeit with mixed results. Generally, two types of learning responses can be predicted for the blood oxygenation level dependent (BOLD) signal. The first, based directly on the Marr/Albus model (Marr, 1969; Albus, 1971), is that the BOLD response should reflect an error-signal, and therefore should be high early in learning when errors are large and then taper off as performance improves. The second type of activity change is related to the storage of the learned representation. This idea predicts that as participants learn to produce more accurate movements (often referred to by motor control researchers as "acquiring an internal model") the BOLD response in regions that support those accurate movements should increase, because they perform more, or more accurate computations. Early-learning related activity was observed during a study of skill learning, with increased activity in the ipsilateral anterior lobe early during a pinch-force training task (Floyer-Lea & Matthews, 2005). An adaptation study conducted by Imamizu and colleagues (2000) reported activation patterns during a visuomotor adaptation task which were consistent with both error-processing and storage. With learning, activity decayed in broad regions throughout the anterior and lateral cerebellum, while the most lateral regions (bilaterally) increased in activation as the new rule was stored. A more recent study using a similar adaptation task observed chiefly storage related activity in the ipsilateral anterior lobe (Luauté et al., 2009). Storage related activity in PET has also been observed for saccade adaptation, with bilateral cerebellar lobules VI and VII showing increased activity when performing the adapted movement (Desmurget et al., 2000).

Learning is a slow process when compared with the time-course of the BOLD signal measured in fMRI, and it is not easy to repeat, which reduces the reliability of the measured brain response. Finally, many behavioral changes can be conflated with learning. For example, Seidler and colleagues (2002) argue that storage-related changes in cerebellar activation reflect performance parameters rather than learning per se. Similarly, the error-related component of the learning response may result from increased corrective movements when errors are prevalent. A study looking for signatures of error processing using a design containing random perturbations found no cerebellar regions which were significantly active in a contrast of trials containing errors and corrective movements against trials containing corrective movements without errors, while cerebral voxels in the postcentral gyrus were active in this contrast (Diedrichsen et al., 2005). A very recent study used a task in which corrective movements were impossible, and observed proprioceptive error-related activity in lobules V and VI (Schlerf et al., 2012).

The cerebellum may also play a role in other forms of learning. The serial reaction time task (SRTT) is a classic paradigm used to study motor sequence learning (Nissen & Bullemer, 1987). Typically, a set of stimuli appear on a computer screen, telling participants which keys to press. If these cues appear in a sequential order, responses will get faster as participants learn the sequence. Patients with cerebellar damage have consistently shown impairments in learning during the SRTT (Doyon et al., 1997; Molinari et al., 1997; Gómez-Beldarrain et al., 1998; Shin & Ivry, 2003). Yet, imaging studies using fMRI and PET have consistently shown either no change in cerebellar responses during learning (Seidler et al., 2002; van der Graaf et al., 2006) or slight decreases in the anterior lobe ipsilateral to the moving hand (Hazeltine et al., 1997; Doyon et al., 2002). Thus, while patient studies suggest some role of the cerebellum in sequence learning, evicence from brain imaging is much more limited.

3.1.3. Timing and coordination

Patients with cerebellar lesions show consistent impairment in tasks that require timing and coordination (in addition to deficits in error-based learning). The relationship between these two functions, and the cerebellar involvement in both, has been a very active area of research that has sparked substantial controversies in the last years.

There is a longstanding debate about the cerebellar contribution to movement timing, and fMRI has provided evidence for both sides. Work on eyeblink conditioning, where a tone is predictably delivered prior to an aversive stimulus delivered to the eyelid, has historically provided strong evidence for the role of the cerebellum in timing (Gellman & Miles, 1985; Steinmetz et al., 1986; Ivry & Keele, 1989). This task has been successfully adapted to the MR environment, revealing activation in lobules VI and VIIA Crus I during eye-blink conditioning (Ramnani et al., 2000; Dimitrova et al., 2002a; Miller et al., 2003; Gerwig et al., 2007; Cheng et al., 2008). Timed movement tasks, such as the production of rhythmic finger movements, reveal strong cerebellar activation (Lutz et al., 2000; Dhamala et al., 2003; Garraux et al., 2005; Jantzen et al., 2005; Spencer et al., 2007). Studies have also examined the response to timing without movement, and found cerebellar activation in response to sensory estimates of temporal delay in left lobules VI and VIIA Crus I (Harrington et al., 2004) and broad activation throughout the anterior lobe in response to deviations from predictable rhythmic sequences (Liu et al., 2008).

While the link between timing and the cerebellum may be somewhat controversial (Rao et al., 2001; Hinton et al., 2004), the link between the cerebellum and coordination is not. A cardinal symptom of cerebellar disease is ataxia, or the lack of movement coordination. This deficit is especially pronounced when the movement task involves the coordination of movement across multiple joints (Thach et al., 1992; Bastian et al., 1996; Thach, 1998). In healthy participants, increased cerebellar activation in fMRI studies is commonly observed when the coordination requirements of the task increase. For example, during eye-hand coordination the cerebellum increases its activation parametrically with increased temporal offset between the required eye and hand movements (Miall et al., 2001).

Many researchers assume that the precise sequential activation of separate muscles in the context of a coordinated movement relies on the same neural mechanisms that underlie explicit timing tasks (Ivry et al., 2002; Ivry & Spencer, 2004). In this view, deficits in timing are the cause of coordination deficits. Coordinating across effectors (e.g., the shoulder and hand muscles when pitching a baseball) would thus require accurately timing the serial order of events. Incorrect timing of the release of a baseball during the shoulder swing of a pitch would result in highly uncoordinated and inaccurate pitching. However, recent behavioral studies have shown that the motor system controls the moment of ball release based on a predictive estimate of the arm position, rather than "timing" the ball release based on the state of an internal clock triggered by a particular event (Hore & Watts, 2005). An imaging study comparing the role of the cerebellum in timing and coordination using a single task showed that these are dissociable processes with lateral lobule V being more strongly engaged during coordination than timing (Diedrichsen et al., 2007).

3.2. Nonmotor function

While the cerebral neocortex is divided into cytoarchitectonic regions that delineate functional specialization, the cerebellar cortex is nearly uniform. Functional specificity is likely determined by the connections between the cerebellum and other parts of the brain (Schmahmann, 2004). Virus tracer studies in primates have demonstrated connections between the cerebellum and cortical association areas, such as the prefrontal and parietal cortices (Strick et al., 2009). FMRI connectivity studies confirm the existence of these networks in humans (Krienen & Buckner, 2009; Buckner et al., 2011), with connections largely involving the lateral cerebellar hemispheres, especially lobules VI and VIIA (Crus I and II). Since the prefrontal and parietal association areas within the neocortex are believed to play important roles in executive function and language, connected regions within the cerebellum should contribute to these functions as well.

3.2.1. Executive control

FMRI has provided important evidence for the role of the cerebellum in executive functions such as attention and working memory. Attention is the cognitive process of focusing on a particular feature in the environment. Working memory is the short-term storage (on the order of seconds) of information for later processing. Humans with damaged cerebella have been observed to have deficits in both attention (e.g., Akshoomoff & Courchesne, 1992) and working memory (e.g., Ravizza et al., 2006).

An early fMRI study of the cerebellar role in attention asked participants to attend to visual stimuli and count certain targets, or make repetitive finger movements without attention (Allen et al., 1997). Motor responses were mostly contained within lobule V, while attentional responses were more posterior, chiefly lobules VI and VIIA, which are connected with the prefrontal cortex. A third task combined these processes (asking participants to press a button every time the target stimulus appeared), and revealed activation in both of these regions (Allen et al., 1997). This finding corroborated an early patient study of the cerebellar role in attention (Akshoomoff & Courchesne, 1992). At any given point, participants were asked to respond to either colored flashes or tones, detecting a target item in that series. During a "switch" condition, participants were asked to wait for a target in the other sensory modality following the successful identification of a target. Patients with damage to the cerebellum responded particularly slowly during this condition, suggesting that attention switching may be an important function of the cerebellum (Akshoomoff & Courchesne, 1992). An fMRI study was conducted using a similar task, asking participants to attend to either the shape or color of a visual stimulus. Increased and reproducible cerebellar activation near right lobule VI was observed during the attention switching condition compared to the sustained attention condition (Le et al., 1998). Bischoff-Grethe and colleagues (2002) pointed out that these switching tasks did not control for motor demands, as separate stimulus-response mappings must be maintained and switched in addition to attention. This study replicated the previous task of Le et al (1998), and included a control condition where participants performed the same attention switch but only responded to one of the two features. Since the control condition involved identical attentional processes, any differences would support a motor explanation of this effect. Analysis of this data found increased activity in right lobule VI when two response mappings were required, suggesting that response reassignment was more important than attention switching (Bischoff-Grethe et al., 2002).

Individuals with cerebellar damage have impaired verbal working memory, but intact spatial working memory (Ravizza et al., 2006). Transcranial magnetic stimulation over right lobule VI selectively impairs response time on verbal working memory tasks (Desmond et al., 2005). FMRI has also supported a cerebellar role in working memory. Studies have compared activation during the Sternberg working memory task (Sternberg, 1966) to a control task of silent rehearsal, where participants continuously read a list of items for later matching. Cerebellar activity in bilateral lobules VI and VIIa crus I is observed during both tasks, while Lobule VIII is specifically activated in response to working memory (Desmond et al., 1997; Chen & Desmond, 2005a). Activity in all of these regions, bilateral lobules VI and VIIa Crus I and right lobule VIII, increases parametrically with working memory load (Kirschen et al., 2005).

The Sternberg task (Sternberg, 1966) can be used in event-related fMRI studies, where distinct components of working memory can be explored. Lobules VI and VIIA Crus I are chiefly involved during the encoding phase, while lobules VIIB and VIIIA were involved in encoding as well as maintenance across delay (Chen & Desmond, 2005b).

3.2.2. Language

While overt language disturbances such as aphasia are not commonly observed following cerebellar damage, speech disturbances have been documented (e.g., Holmes, 1917), which may mask subtle linguistic effects. FMRI has become an important tool for exploring the cerebellar contribution to language.

The first study to observe cerebellar activation in a cognitive task detected activity in the right cerebellum during verb generation (Petersen et al., 1989). Verb generation has since been successfully used to demonstrate functional activation of bilateral lobules VI and right VIIA Crus I by a number of researchers (e.g., Seger et al., 2000; Frings et al., 2006). A similar task is word-stem completion, where words are generated to complete a command prompt of a few letters. An early study observed that the right cerebellum (chiefily lobule VIIA Crus I, extending into midline lobules VI through VIIB) is more active when there are a limited number of correct completions of the word compared to when there are many (Desmond et al., 1998). This suggests that the search for correct words requires more cerebellar computation than selection among a number of easily identified options.

Despite the ease of observing activation during linguistic tasks, the role of the cerebellum in language is not without controversy. Participants with cerebellar degeneration do not have impairments in verb-generation tasks (Richter et al., 2004). Furthermore, while most fMRI studies control for speech articulation using noun repetition tasks, some investigators argue that this is not a strong enough control. Ackermann and colleagues (1998) observed cerebellar activity in right lobule VI by asking participants to perform a self-guided repetition task by rehearsing the months of the year repeatedly. This suggests that the cerebellum's role may still be fundamentally articulatory, as well-learned sequences of words are not demanding to recite. While the cerebellar contribution to language is not the only example of the difficulty in reconciling brain imaging studies with clinical investigation, it does underscore the need for elegant experimental manipulations when exploring cerebellar activity.

4. Conclusion

This chapter has provided a brief overview of some of the specialized techniques for imaging this particular part of the brain. We discussed some specialized anatomical methods for normalization and inter-individual alignment, as well as region-of-interest identification. We also discussed the importance of removing physiological nuisance covariates, and the promise of future innovation by multivariate analysis methods. Brain scientists who are interested in the cerebellum should become familiar with these methods.

We also discussed literature covering the functional specificity of this structure. Cerebellar regions likely play a role in many different functions, ranging from sensorimotor functions to higher cognitive functions. We provided an overview of the major areas of research, attempting to highlight some of the current controversies with regard to the role of the cerebellum. As MRI scanners and pulse sequences improve and researchers make better use of specialized

http://dx.doi.org/10.5772/50746

cerebellar analysis techniques, some of these controversies may be resolved. Fortunately, new controversies will undoubtedly emerge to take their place.

The cerebellum is a fascinating structure that poses unique challenges to MRI. As new solutions to some of these challenges come of age, we expect that imaging the "little brain" will continue to provide fresh and important insights to cerebellar function for years to come.

Author details

John Schlerf[1], Tobias Wiestler[3], Timothy Verstynen[2] and Jörn Diedrichsen[3]

1 Johns Hopkins University, USA

2 Carnegie Mellon University, USA

3 University College London, UK

References

[1] Ackermann H, Wildgruber D, Daum I, Grodd W. Does the cerebellum contribute to cognitive aspects of speech production? A functional magnetic resonance imaging (fMRI) study in humans. *Neurosci Let* 247: 187–190, 1998.

[2] Akshoomoff N, Courchesne E. A new role for the cerebellum in cognitive operations. *Behav Neurosci* 106: 731–738, 1992.

[3] Alahyane N, Fonteille V, Urquizar C, Salemme R, Nighoghossian N, Pelisson D, Tilikete C. Separate Neural Substrates in the Human Cerebellum for Sensory-motor Adaptation of Reactive and of Scanning Voluntary Saccades. *Cerebellum* 7: 595–601, 2008.

[4] Albus J. A theory of cerebellar function. *Math Biosci* 10: 25–61, 1971.

[5] Allen G, Buxton RB, Wong EC, Courchesne E. Attentional Activation of the Cerebellum Independent of Motor Involvement. *Science* 275: 1940 –1943, 1997.

[6] Bastian AJ, Martin TA, Keating JG, Thach WT. Cerebellar ataxia: abnormal control of interaction torques across multiple joints. *J Neurophysiol* 76: 492 –509, 1996.

[7] Bischoff-Grethe A, Ivry R, Grafton S. Cerebellar involvement in response reassignment rather than attention. *J Neurosci* 22: 546–553, 2002.

[8] Bower JM, Woolston DC. Congruence of spatial organization of tactile projections to granule cell and Purkinje cell layers of cerebellar hemispheres of the albino rat: vertical organization of cerebellar cortex. *J Neurophysiol* 49: 745 –766, 1983.

[9] Buckner RL, Krienen FM, Castellanos A, Diaz JC, Yeo BTT. The Organization of the Human Cerebellum Estimated by Intrinsic Functional Connectivity. *J Neurophysiol* 106: 2322–2345, 2011.

[10] Chen SHA, Desmond JE. Cerebrocerebellar networks during articulatory rehearsal and verbal working memory tasks. *NeuroImage* 24: 332–338, 2005a.

[11] Chen SHA, Desmond JE. Temporal dynamics of cerebro-cerebellar network recruitment during a cognitive task. *Neuropsychologia* 43: 1227–1237, 2005b.

[12] Cheng DT, Disterhoft JF, Power JM, Ellis DA, Desmond JE. Neural substrates underlying human delay and trace eyeblink conditioning. *PNAS* 105: 8108–8113, 2008.

[13] Crone E, Bunge S, de Klerk P, van der Molen M. Cardiac concomitants of performance monitoring: context dependence and individual differences. *Cog Brain Res* 23: 93–106, 2005.

[14] Crone E, Jennings R, Van der Molen M. Developmental change in feedback processing as reflected by phasic heart rate changes. *Dev Psych* 40: 1228–1238, 2004.

[15] Crone E, Somsen R, Zanolie K, Van der Molen M. A heart rate analysis of developmental change in feedback processing and rule shifting from childhood to early adulthood. *J Exp Child Psych* 95: 99–116, 2006.

[16] Damen E, Brunia C. Changes in Heart Rate and Slow Brain Potentials Related to Motor Preparation and Stimulus Anticipation in a Time Estimation Task. *Psychophysiol* 24: 700–713, 1987.

[17] Desmond J, Chen S, Shieh P. Cerebellar transcranial magnetic stimulation impairs verbal working memory. *Ann Neurol* 58: 553–560, 2005.

[18] Desmond J, Gabrieli J, Glover G. Dissociation of Frontal and Cerebellar Activity in a Cognitive Task: Evidence for a Distinction between Selection and Search. *NeuroImage* 7: 368–376, 1998.

[19] Desmond J, Gabrieli J, Wagner A, Ginier B, Glover G. Lobular patterns of cerebellar activation in verbal working-memory and finger-tapping tasks as revealed by functional MRI. *J Neurosci* 17: 9675–9685, 1997.

[20] Desmurget M, Pélisson D, Grethe JS, Alexander GE, Urquizar C, Prablanc C, Grafton ST. Functional adaptation of reactive saccades in humans: a PET study. *Exp Brain Res* 132: 243–259, 2000.

[21] Dhamala M, Pagnoni G, Wiesenfeld K, Zink C, Martin M, Berns G. Neural correlates of the complexity of rhythmic finger tapping. *NeuroImage* 20: 918–926, 2003.

[22] Diedrichsen J, Balsters JH, Flavell J, Cussans E, Ramnani N. A probabilistic MR atlas of the human cerebellum. *NeuroImage* 46: 39–46, 2009.

[23] Diedrichsen J, Criscimagna-Hemminger S, Shadmehr R. Dissociating Timing and Coordination as Functions of the Cerebellum. *J Neurosci* 27: 6291–6301, 2007.

[24] Diedrichsen J, Hashambhoy Y, Rane T, Shadmehr R. Neural correlates of reach errors. *J Neurosci* 25: 9919–9931, 2005.

[25] Diedrichsen J, Maderwald S, Küper M, Thürling M, Rabe K, Gizewski ER, Ladd ME, Timmann D. Imaging the deep cerebellar nuclei: A probabilistic atlas and normalization procedure. *NeuroImage* 54: 1786–1794, 2011.

[26] Diedrichsen J. A spatially unbiased atlas template of the human cerebellum. *NeuroImage* 33: 127–138, 2006.

[27] Dimitrova A, Weber J, Maschke M, Elles H, Kolb FP, Forsting M, Diener H, Timmann D. Eyeblink-related areas in human cerebellum as shown by fMRI. *Hum Brain Map* 17: 100–115, 2002a.

[28] Dimitrova A, Weber J, Redies C, Kindsvater K, Maschke M, Kolb F, Forsting M, Diener H, Timmann D. MRI atlas of the human cerebellar nuclei. *NeuroImage* 17: 240–255, 2002b.

[29] Dimitrova A, Zeljko D, Schwarze F, Maschke M, Gerwig M, Frings M, Beck A, Aurich V, Forsting M, Timmann D. Probabilistic 3D MRI atlas of the human cerebellar dentate/interposed nuclei. *NeuroImage* 30: 12–25, 2006.

[30] Doyon J, Gaudreau D, Laforce R Jr, Castonguay M, Bédard PJ, Bédard F, Bouchard JP. Role of the striatum, cerebellum, and frontal lobes in the learning of a visuomotor sequence. *Brain Cogn* 34: 218–245, 1997.

[31] Doyon J, Song AW, Karni A, Lalonde F, Adams MM, Ungerleider LG. Experience-dependent changes in cerebellar contributions to motor sequence learning. *PNAS* 99: 1017–1022, 2002.

[32] Floyer-Lea A, Matthews P. Distinguishable brain activation networks for short- and long-term motor skill learning. *J Neurophysiol* 94: 512–518, 2005.

[33] Fox P, Raichle M, Thach W. Functional mapping of the human cerebellum with positron emission tomography. *PNAS* 82: 7462–7466, 1985.

[34] Frings M, Dimitrova A, Schorn CF, Elles H-G, Hein-Kropp C, Gizewski ER, Diener HC, Timmann D. Cerebellar involvement in verb generation: An fMRI study. *Neurosci Let* 409: 19–23, 2006.

[35] Garraux G, McKinney C, Wu T, Kansaku K, Nolte G, Hallett M. Shared brain areas but not functional connections controlling movement timing and order. *J Neurosci* 25: 5290–5297, 2005.

[36] Gellman R, Miles F. A new role for the cerebellum in conditioning? *Trends Neurosci* 8: 181–182, 1985.

[37] Gerwig M, Kolb FP, Timmann D. The involvement of the human cerebellum in eyeblink conditioning. *Cerebellum* 6: 38–57, 2007.

[38] Glover G, Li T-Q, Ress D. Image-based method for retrospective correction of physiological motion effects in fMRI: RETROICOR. *Magn Reson Med* 44: 162–167, 2000.

[39] Golla H, Tziridis K, Haarmeier T, Catz N, Barash S, Thier P. Reduced saccadic resilience and impaired saccadic adaptation due to cerebellar disease. *Eur J Neurosci* 27: 132–144, 2008.

[40] Gómez-Beldarrain M, García-Moncó JC, Rubio B, Pascual-Leone A. Effect of focal cerebellar lesions on procedural learning in the serial reaction time task. *Exp Brain Res* 120: 25–30, 1998.

[41] van der Graaf FHCE, Maguire RP, Leenders KL, de Jong BM. Cerebral activation related to implicit sequence learning in a Double Serial Reaction Time task. *Brain Res* 1081: 179–190, 2006.

[42] Grodd W, Hülsmann E, Lotze M, Wildgruber D, Erb M. Sensorimotor mapping of the human cerebellum: fMRI evidence of somatotopic organization. *Hum Brain Map* 13: 55–73, 2001.

[43] Habas C, Axelrad H, Cabanis E. The cerebellar second homunculus remains silent during passive bimanual movements. *Neuroreport* 15: 1571–1574, 2004a.

[44] Habas C, Axelrad H, Nguyen T, Cabanis E. Specific neocerebellar activation during out-of-phase bimanual movements. *Neuroreport* 15: 595–599, 2004b.

[45] Harrington D, Boyd L, Mayer A, Sheltraw D, Lee R, Huang M, Rao S. Neural representation of interval encoding and decision making. *Cog Brain Res* 21: 193–205, 2004.

[46] Harvey A, Pattinson K, Brooks J, Mayhew S, Jenkinson M, Wise R. Brainstem functional magnetic resonance imaging: Disentangling signal from physiological noise. *J Magn Reson Imaging* 28: 1337–1344, 2008.

[47] Hazeltine E, Grafton ST, Ivry R. Attention and stimulus characteristics determine the locus of motor-sequence encoding. A PET study. *Brain* 120 (Pt 1): 123–140, 1997.

[48] Hinton S, Harrington D, Binder J, Durgerian S, Rao S. Neural systems supporting timing and chronometric counting: an FMRI study. *Cog Brain Res* 21: 183–192, 2004.

[49] Hoge R, Atkinson J, Gill B, Crelier G, Marrett S, Pike G. Linear coupling between cerebral blood flow and oxygen consumption in activated human cortex. *PNAS* 96: 9403–9408, 1999.

[50] Holmes G. The symptoms of acute cerebellar injuries due to gunshot injuries. *Brain* 40: 461–535, 1917.

[51] Holmes G. The cerebellum of man. *Brain* 62: 1–30, 1939.

[52] Hore J, Watts S. Timing Finger Opening in Overarm Throwing Based on a Spatial Representation of Hand Path. *J Neurophysiol* 93: 3189 –3199, 2005.

[53] Howarth C, Peppiatt-Wildman C, Attwell D. The energy use associated with neural computation in the cerebellum. *J Cereb Blood Flow Metab* 30: 403–414, 2009.

[54] Iadecola C, Li J, Xu S, Yang G. Neural mechanisms of blood flow regulation during synaptic activity in cerebellar cortex. *J Neurophysiol* 75: 940–950, 1996.

[55] Imamizu H, Miyauchi S, Tamada T, Sasaki Y, Takino R, Putz B, Yoshioka T, Kawato M. Human cerebellar activity reflecting an acquired internal model of a new tool. *Nature* 403: 192–195, 2000.

[56] Ito M. Cerebellar Long-Term Depression: Characterization, Signal Transduction, and Functional Roles. *Physiol Rev* 81: 1143–1195, 2001.

[57] Ivry R, Keele S. Timing Functions of The Cerebellum. *J Cogn Neurosci* 1: 136–152, 1989.

[58] Ivry R, Spencer R, Zelaznik H, Diedrichsen J. The cerebellum and event timing. *Ann N Y Acad Sci* 978: 302–317, 2002.

[59] Ivry R, Spencer R. The neural representation of time. *Curr Opin Neurbiol* 14: 225–232, 2004.

[60] Jantzen K, Steinberg F, Kelso J. Functional MRI reveals the existence of modality and coordination-dependent timing networks. *NeuroImage* 25: 1031–1042, 2005.

[61] Jennings J, van der Molen M, Brock K, Somsen R. Response inhibition initiates cardiac deceleration: evidence from a sensory-motor compatibility paradigm. *Psychophysiol* 28: 72–85, 1991.

[62] Jennings J, van der Molen M, Brock K, Somsen R. On the synchrony of stopping motor responses and delaying heartbeats. *J Exp Psychol [Hum Percept]* 18: 422–436, 1992.

[63] Jennings J, van der Molen M, Brock K, Somsen R. How are tonic and phasic cardiovascular changes related to central motor command? *Biol Psych* 35: 237–254, 1993.

[64] Jennings R, van der Molen M. Cardiac timing and the central regulation of action. *Psychol Res* 66: 337–349, 2002.

[65] Kassel J, Shambes GM, Welker W. Fractured cutaneous projections to the granule cell layer of the posterior cerebellar hemisphere of the domestic cat. *J Comp Neurol* 225: 458–468, 1984.

[66] Kelly R, Strick P. Cerebellar loops with motor cortex and prefrontal cortex of a nonhuman primate. *J Neurosci* 23: 8432–8444, 2003.

[67] Kirschen MP, Chen SHA, Schraedley-Desmond P, Desmond JE. Load- and practice-dependent increases in cerebro-cerebellar activation in verbal working memory: an fMRI study. *NeuroImage* 24: 462–472, 2005.

[68] Krienen F, Buckner R. Segregated fronto-cerebellar circuits revealed by intrinsic functional connectivity. *Cereb Cortex* 19: 2497, 2485, 2009.

[69] Küper M, Dimitrova A, Thürling M, Maderwald S, Roths J, Elles HG, Gizewski ER, Ladd ME, Diedrichsen J, Timmann D. Evidence for a motor and a non-motor domain in the human dentate nucleus -- An fMRI study. *NeuroImage* 54: 2612–2622, 2011a.

[70] Küper M, Thürling M, Stefanescu R, Maderwald S, Roths J, Elles HG, Ladd ME, Diedrichsen J, Timmann D. Evidence for a motor somatotopy in the cerebellar dentate nucleus—an fMRI study in humans. *Hum Brain Map*. .

[71] Larsell O. The morphogenesis and adult pattern of the lobules and fissures of the cerebellum of the white rat. *J Comp Neurol* 97: 281–356, 1952.

[72] Larsell O. The cerebellum of the cat and the monkey. *J Comp Neurol* 99: 135–199, 1953.

[73] Le TH, Pardo JV, Hu X. 4 T-fMRI Study of Nonspatial Shifting of Selective Attention: Cerebellar and Parietal Contributions. *J Neurophysiol* 79: 1535 –1548, 1998.

[74] Liu T, Xu D, Ashe J, Bushara K. Specificity of Inferior Olive Response to Stimulus Timing. *J Neurophysiol* 100: 1557–1561, 2008.

[75] Luauté J, Schwartz S, Rossetti Y, Spiridon M, Rode G, Boisson D, Vuilleumier P. Dynamic changes in brain activity during prism adaptation. *J Neurosci* 29: 169–178, 2009.

[76] Lutz K, Specht K, Shah NJ, Jäncke L. Tapping movements according to regular and irregular visual timing signals investigated with fMRI. *Neuroreport* 11: 1301–1306, 2000.

[77] Makris N, Hodge SM, Haselgrove C, Kennedy DN, Dale A, Fischl B, Rosen BR, Harris G, Caviness VS, Schmahmann JD. Human Cerebellum: Surface-Assisted Cortical Parcellation and Volumetry with Magnetic Resonance Imaging. *J Cogn Neurosci* 15: 584–599, 2003.

[78] Makris N, Schlerf J, Hodge S, Haselgrove C, Albaugh M, Seidman L, Rauch S, Harris G, Biederman J, Caviness V, Kennedy D, Schmahmann J. MRI-based surface-assisted parcellation of human cerebellar cortex: an anatomically specified method with estimate of reliability. *NeuroImage* 25: 1146–1160, 2005.

[79] Marr D. A theory of cerebellar cortex. *J Physiol* 202: 437–470, 1969.

[80] Martin T, Keating J, Goodkin H, Bastian A, Thach W. Throwing while looking through prisms. I. Focal olivocerebellar lesions impair adaptation. *Brain* 119: 1183–1198, 1996.

[81] Maschke M, Gomez C, Ebner T, Konczak J. Hereditary Cerebellar Ataxia Progressively Impairs Force Adaptation During Goal-Directed Arm Movements. *J Neurophysiol* 91: 230–238, 2004.

[82] Mathiesen C, Caesar K, Lauritzen M. Temporal coupling between neuronal activity and blood flow in rat cerebellar cortex as indicated by field potential analysis. *J Physiol* 523: 235–246, 2000.

[83] Miall RC, Reckess GZ, Imamizu H. The cerebellum coordinates eye and hand tracking movements. *Nat Neurosci* 4: 638–644, 2001.

[84] Miller MJ, Chen N, Li L, Tom B, Weiss C, Disterhoft JF, Wyrwicz AM. fMRI of the Conscious Rabbit during Unilateral Classical Eyeblink Conditioning Reveals Bilateral Cerebellar Activation. *J Neurosci* 23: 11753 –11758, 2003.

[85] Molinari M, Leggio M, Solida A, Ciorra R, Misciagna S, Silveri M, Petrosini L. Cerebellum and procedural learning: evidence from focal cerebellar lesions. *Brain* 120: 1753–1762, 1997.

[86] Morton S, Bastian A. Cerebellar Contributions to Locomotor Adaptations during Splitbelt Treadmill Walking. *J Neurosci* 26: 9107–9116, 2006.

[87] Nissen MJ, Bullemer P. Attentional requirements of learning: Evidence from performance measures. *Cogn Psych* 19: 1–32, 1987.

[88] Petersen SE, Fox PT, Posner MI, Mintun M, Raichle ME. Positron Emission Tomographic Studies of the Processing of Singe Words. *J Cogn Neurosci* 1: 153–170, 1989.

[89] Pisella L, Rossetti Y, Michel C, Rode G, Boisson D, Pélisson D, Tilikete C. Ipsidirectional impairment of prism adaptation after unilateral lesion of anterior cerebellum. *Neurology* 65: 150–152, 2005.

[90] Ramnani N, Toni I, Josephs O, Ashburner J, Passingham RE. Learning- and expectation-related changes in the human brain during motor learning. *J Neurophysiol* 84: 3026–3035, 2000.

[91] Rancz EA, Ishikawa T, Duguid I, Chadderton P, Mahon S, Hausser M. High-fidelity transmission of sensory information by single cerebellar mossy fibre boutons. *Nature* 450: 1245–1248, 2007.

[92] Rao S, Mayer A, Harrington D. The evolution of brain activation during temporal processing. *Nat Neurosci* 4: 317–323, 2001.

[93] Ravizza S, Mccormick C, Schlerf J, Justus T, Ivry R, Fiez J. Cerebellar damage produces selective deficits in verbal working memory. *Brain* 129: 306–320, 2006.

[94] Richter S, Kaiser O, Hein-Kropp C, Dimitrova A, Gizewski E, Beck A, Aurich V, Ziegler W, Timmann D. Preserved verb generation in patients with cerebellar atrophy. *Neuropsychologia* 42: 1235–1246, 2004.

[95] Rijntjes M, Buechel C, Kiebel S, Weiller C. Multiple somatotopic representations in the human cerebellum. *Neuroreport* 10: 3653–3658, 1999.

[96] Schlerf J, Ivry RB, Diedrichsen J. Encoding of Sensory Prediction Errors in the Human Cerebellum. *J Neurosci* 32: 4913–4922, 2012.

[97] Schlerf JE, Verstynen TD, Ivry RB, Spencer RMC. Evidence of a Novel Somatopic Map in the Human Neocerebellum During Complex Actions. *J Neurophysiol* 103: 3330–3336, 2010.

[98] Schmahmann JD, Doyon J, McDonald D, Holmes C, Lavoie K, Hurwitz AS, Kabani N, Toga A, Evans A, Petrides M. Three-Dimensional MRI Atlas of the Human Cerebellum in Proportional Stereotaxic Space. *NeuroImage* 10: 233–260, 1999.

[99] Schmahmann JD. Disorders of the cerebellum: ataxia, dysmetria of thought, and the cerebellar cognitive affective syndrome. *J Neuropsychiatry Clin Neurosci* 16: 367–378, 2004.

[100] Seger CA, Desmond JE, Glover GH, Gabrieli JD. Functional magnetic resonance imaging evidence for right-hemisphere involvement in processing unusual semantic relationships. *Neuropsychology* 14: 361–369, 2000.

[101] Seidler R, Purushotham A, Kim S, Uğurbil K, Willingham D, Ashe J. Cerebellum activation associated with performance change but not motor learning. *Science* 296: 2043–2046, 2002.

[102] Shambes GM, Gibson JM, Welker W. Fractured Somatotopy in Granule Cell Tactile Areas of Rat Cerebellar Hemispheres Revealed by Micromapping; pp. 116– 140. *Brain Behav Evol* 15: 116–140, 1978.

[103] Shin JC, Ivry RB. Spatial and Temporal Sequence Learning in Patients with Parkinson's Disease or Cerebellar Lesions. *J Cogn Neurosci* 15: 1232–1243, 2003.

[104] Smith M, Shadmehr R. Intact Ability to Learn Internal Models of Arm Dynamics in Huntington's Disease But Not Cerebellar Degeneration. *J Neurophysiol* 93: 2809–2821, 2005.

[105] Snider R, Eldred E. Cerebrocerebellar relationships in the monkey. *J Neurophysiol* 15: 27–40, 1952.

[106] Spencer RMC, Verstynen T, Brett M, Ivry R. Cerebellar activation during discrete and not continuous timed movements: an fMRI study. *Neuroimage* 36: 378–387, 2007.

[107] Steinmetz JE, Rosen DJ, Chapman PF, Lavond DG, Thompson RF. Classical conditioning of the rabbit eyelid response with a mossy-fiber stimulation CS: I. Pontine nuclei and middle cerebellar peduncle stimulation. *Behav Neurosci* 100: 878–887, 1986.

[108] Sternberg S. High-speed scanning in human memory. *Science* 153: 652–654, 1966.

[109] Stoodley CJ, Schmahmann JD. Functional topography in the human cerebellum: A meta-analysis of neuroimaging studies. *NeuroImage* 44: 489–501, 2009.

[110] Straube A, Deubel H, Ditterich J, Eggert T. Cerebellar lesions impair rapid saccade amplitude adaptation. *Neurology* 57: 2105 –2108, 2001.

[111] Strick P, Dum R, Fiez J. Cerebellum and Nonmotor Function. *Annu Rev Neurosci* 32: 413–434, 2009.

[112] Swisher JD, Gatenby JC, Gore JC, Wolfe BA, Moon C-H, Kim S-G, Tong F. Multiscale Pattern Analysis of Orientation-Selective Activity in the Primary Visual Cortex. *J Neurosci* 30: 325 –330, 2010.

[113] Thach W. A role for the cerebellum in learning movement coordination. *Neurobiol Learn Mem* 70: 177–188, 1998.

[114] Thach WT, Goodkin HP, Keating JG. The Cerebellum and the Adaptive Coordination of Movement. *Annu Rev Neurosci* 15: 403–442, 1992.

[115] Thickbroom G, Byrnes M, Mastaglia F. Dual representation of the hand in the cerebellum: activation with voluntary and passive finger movement. *NeuroImage* 18: 670–674, 2003.

[116] Thomsen K, Offenhauser N, Lauritzen M. Principal neuron spiking: neither necessary nor sufficient for cerebral blood flow in rat cerebellum. *J Physiol* 560: 181–189, 2004.

[117] Thomsen K, Piilgaard H, Gjedde A, Bonvento G, Lauritzen M. Principal Cell Spiking, Postsynaptic Excitation, and Oxygen Consumption in the Rat Cerebellar Cortex. *J Neurophysiol* 102: 1503–1512, 2009.

[118] Tseng Y-W, Diedrichsen J, Krakauer JW, Shadmehr R, Bastian AJ. Sensory prediction errors drive cerebellum-dependent adaptation of reaching. *J Neurophysiol* 98: 54–62, 2007.

[119] Wiestler T, McGonigle DJ, Diedrichsen J. Integration of sensory and motor representations of single fingers in the human cerebellum. *J Neu*rophysiol 105: 3042–3053, 2011.

[120] Zhang Y, Forster C, Milner T, Iadecola C. Attenuation of activity-induced increases in cerebellar blood flow by lesion of the inferior olive. Am J Physiol Heart Circ Physiol 285: H1177–H1182, 2003.

Section 2

Advances in Visual and Auditory Cortical Network Applications

Chapter 9

A Probabilistic Atlas of Human Visual Areas and Information-Theoretic Analysis of Individual Variability in Their Loci

Hiroki Yamamoto

Additional information is available at the end of the chapter

http://dx.doi.org/10.5772/30760

1. Introduction

The human visual cortex consists of multiple functionally distinct visual areas that have been individually localized on the cortical surface by brain imaging of their associated retinotopic activity. Thus, their locations can be quantitatively compared across individuals once they are expressed in a common coordinate space. In several studies (Dougherty et al. 2003, Fischl et al. 1999b, Hasnain et al. 1998), such comparisons have been performed using the Talairach stereotactic coordinate system (Talairach et al. 1967, Talairach & Tournoux 1988) as the common anatomical space. Because the Talairach system has been commonly used as a standard to describe the loci of brain activation or lesions, and to normalize the functional data between individual brains, evaluation of the interindividual variability within the Talairach space is essential for reliably interpreting a broad range of brain function data.

The Talairach system considers interindividual variations in overall brain size and shape but not the variations of smaller scale structures, such as the cortical sulci and gyri, which differ substantially and heterogeneously between individuals (Ono et al. 1990). Consequently, interindividual variability within the Talairach space is not only extensive, but also complex in morphology. Analysis of this variability is therefore not straightforward and requires conceptual and practical simplification.

Recently, we have proposed that inconsistency and uncertainty are fundamental concepts characterizing the interindividual variability within the visual areas (Yamamoto et al. 2011). This inconsistency refers to the disparity between the individuals regarding the location of a particular visual area. The inconsistency in the Talairach space has been assessed in several studies by calculating the variance of the position of a representative point within a visual area

(Dougherty et al. 2003, Hasnain et al. 1998) or by constructing a probabilistic map in which each point is associated with a probability that the visual area resides there (Amunts et al. 2000, Fischl et al. 1999b). Although the inconsistency provides a good basis for analyzing a single, isolated area, it is inherently insufficient for regulating multiple visual areas with pairwise adjacency. When transforming multiple areas together into Talairach space, their adjacency causes overlaps between the neighboring areas of different individuals. Therefore, in evaluating the interindividual variability in multiple visual areas, we must address another variable, uncertainty.

This uncertainty relates to the difficulty of knowing which visual areas reside at a given position where the greater the number of overlaps between the different areas, the greater the degree of uncertainty. We should emphasize that uncertainty is conceptually dissociated from inconsistency. Even if there is no area that has a high probability of being present at an anatomical point, the uncertainty at the point is zero if only a single area has a non-zero probability of being present. Conversely, even if an area has a high probability of being present, the uncertainty is large if other areas also have a high probability of being present. When brain activation is observed at a region described as having a high probability but also has a high uncertainty area, we cannot confidently attribute this region to a specific anatomical and functional area. In addition, when the individual brain activations are pooled in Talairach space, caution is warranted regarding the source of activation at such points because the activations of different areas are probably highly confounded. These considerations suggest that quantification of the uncertainty is crucial for reliable interpretation of the functional data in Talairach space. However, this issue has not been investigated.

In the present study, we applied a probabilistic and information theoretical framework to quantify the uncertainty and the inconsistency of multiple human visual areas that were localized based on functional MRI (fMRI) retinotopy measurements. Inconsistency was estimated by generating a probability map based on 10 hemispheres of five subjects, where each point was associated with a probability that it belonged to each retinotopic area. We used the Shannon entropy (Shannon 1948) as an uncertainty measure and generated a novel map we called an entropy map, where each point is associated with a Shannon entropy computed from the probability that the point belongs to each of the visual areas (Eq. (1) in the section 2.3). Generally, the Shannon entropy serves as a useful measure to quantify the uncertainty of a situation (Cencov 1982). In our study, the relevance is when one wants to predict or judge a visual area at a given Talairach point in clinical practice or scientific research, where higher entropy equates to a higher probability of incorrect classification. Using the probability and entropy maps, we analyzed the inconsistency and uncertainty of the locations of retinotopic areas registered in Talairach space.

Although Talairach registration was originally described as piecewise linear transformations, there are a number of methods used, where these include two main types; namely volume-based methods using linear, piecewise-linear, or nonlinear transformations and surface-based methods, allowing for a more precise alignment of the sulcal and gyral patterns (see Gholipour et al. 2007 for a recent comprehensive review). The choice of the transformation method is an important consideration that influences the results of our analysis. In our recent study

http://dx.doi.org/10.5772/30760

(Yamamoto et al. 2011), we employed a sophisticated surface-based method (Fischl et al. 1999b). In the present study, we used a single linear transformation (nine parameters, three each for translation, rotation, and orthogonal scaling) for the entire brain (Desmond & Lim 1997). Many more sophisticated, methods such as nonlinear volume methods (Ashburner 2007, Woods et al. 1998) and surface-based methods (Drury et al. 1996, Fischl et al. 1999b, Thompson et al. 1997) subsume and therefore outperform this method. Thus, the probability and entropy maps created using a single linear transformation can provide a baseline for the evaluation of the inconsistency and uncertainty in Talairach space. Understanding the limits of the transformation is important for performing reliable group analysis and interpreting the published Talairach coordinates of the activation foci because the linear transformation is still commonly used in both basic and clinical neuroscience.

2. Methods

2.1. Imaging

We studied ten hemispheres from five normal subjects (one female, four males; mean 33 years). The subjects were in good health with no past history of psychiatric or neurological disease. The study protocols were approved by the ethical committees of our institutions. All subjects provided written, informed consent prior to study enrollment. The structural and functional MRI measurements were performed using a standard clinical 1.5 Tesla scanner (GE, Signa NV/i). Prior to the experimental scans, high-contrast T1-weighted three-dimensional (3D) SPGR structural images (echo time (TE) = 3.0 ms, repetition time (TR) = 56 ms, flip angle (FA) = 55°; excitations (NEX) = 1, voxel size (VS) = 0.781 × 0.781 × 1.4 or 0.938 × 0.938 × 1.4 mm) of the whole brain were recorded for each subject, which was used for reconstructing the individual brain surface. For each subject, three types of images were obtained on each scan day, with a standard flexible surface coil placed at the occipital pole. First, T1-weighted Inversion Recovery 3D Fast SPGR structural images (TE = 2.7 ms, TR = 6.0 ms, inversion time (TI) = 600 ms, FA = 15°; NEX = 1, VS = 0.781 × 0.781 × 1.4 mm) were acquired for anatomical registration. Second, a set of 16 (three subjects) or 17 (two subjects) adjacent high-resolution T1-weighted spin echo anatomical slices (TE = 9 ms, TR = 420 ms, NEX = 2, VS = 0.781 × 0.781 × 4 mm) was obtained. Finally, multiple functional scans (6 – 8) were obtained in the same slice positions as the anatomical slices while the subject viewed visual stimuli (e.g., a flickering checkerboard ring), using a T2*-weighted two-dimensional gradient echo, echo planar imaging (TE = 50 ms, TR = 2000 ms, FA = 90°, VS = 1.563 × 1.563 × 4 mm).

2.2. Localization of retinotopic areas

After reconstructing each individual's cortical surface, locations of the retinotopic visual areas were identified with fMRI and standard retinotopic mapping procedures (Table 1). The surface was reconstructed to lie approximately in the middle of the gray matter, using a method that was a hybrid of volume segmentation (Drury et al. 1999) and surface deformation (Dale & Sereno 1993). We used mrGray software (Teo et al. 1997) for volume segmentation and a custom software for surface generation. The software generated an initial surface at the border of the white and gray matter using a marching cube algorithm (Lorensen & Cline 1987) and then deformed it to lie in the middle of the gray matter using the deformable template algorithm (Dale & Sereno 1993).

Technique	Method	Software
Cortical surface reconstruction	Step 1. Anatomical volume segmentation (cerebrospinal fluid, gray matter, white matter)	mrGray (Teo et al., 1997)
	Step 2. Triangular tessellation with the marching cube algorithm (Lorensen & Clein, 1987)	in-house software called BrainFactory (Yamamoto et al., 2002, 2008, 2011) written in C++ & Tcl/Tk using Visualization Toolkit (Kitware inc.)
	Step 3. Deformation of the tessellated surface with the deformable template algorithm (Dale & Sereno, 1993)	BrainFactory
Localization of retinotopic areas	Phase-encoding mapping with fMRI (Engel et al., 1994) Visual stimuli: a rotating wedge for visual field angular mapping and a expanding (fovea-16 deg eccentricity) annulus for eccentricity mapping, 60s /rotation(expansion) × 6 cycles Parcellation of visual areas: visual inspection of the retinotopy and the visual filed sign maps (Sereno et al., 1995)	BrainFactory
Brain registration	A linear rigid volume-based method (Desmond & Lim, 1997) Parameters: 9 parameters (3 scalings, 3 rotations, 3 translations) determined by visual inspection of anatomical landmarks Target brain:10 hemispheres from 5 subjects	BrainFactory

Table 1. Summary of key techniques

The surface regions delimiting areas V1d, V2d, V3d, V3B, V3A, V7, LOc, MT+, V1v, V2v, V3v, V4v, and V8 were determined by phase-encoded retinotopic mapping methods (DeYoe et al. 1994, Engel et al. 1997, Sereno et al. 1995), allowing visualization of the polar angle (Fig. 1A, upper panel) and eccentricity (Fig. 1A, lower panel) components of the retinotopic map. The retinotopic map was constructed using a phase-encoding technique where the receptive field centers were temporally coded using polar coordinates (Engel et al. 1994). The polar angle component of the map was measured by performing the fMRI while the subject viewed a wedge-shaped checkered pattern (24° center angle) that rotated counter-clockwise around the fixation point, making one rotation in 60 s. The eccentricity component was measured while the subject viewed a checkered annulus (2° width) that expanded from the fovea to 16° peripherally over 50 s and then disappeared for 10 s. Each stimulus underwent color (black/white, red/green, or blue/yellow) pattern reversal (1 Hz) and was presented in six cycles, evoking a periodic response at a given point on the retinotopy map, where the corresponding position in the visual field was encoded in the phase of the response. The response phase for each of the polar angles or eccentricity components was computed using Fourier analysis and mapped onto the cortical surface (Fig. 1A). More details of the surface reconstruction, the retinotopy mapping, and the parcellation schema have been described elsewhere (H. Yamamoto et al. 2008, Yamamoto et al. 2011).

http://dx.doi.org/10.5772/30760

2.3. Generation of the probability and entropy maps

Quantification of the inconsistency and uncertainty of the locations of the human retinotopic areas in the Talairach space comprised three basic steps: 1) construction of volumetric models of the retinotopic areas, 2) generation of the probability map, and 3) generation of the entropy map, as schematized in Fig. 1. In the first step, volumetric models of the retinotopic areas were constructed for each of the 10 hemispheres by first localizing them on the cortical surface (Fig. 1B) and then converting each area's surface to a volumetric model of 3 mm thickness (Fig. 1C). This conversion began by computing the distance from the surface of each area to the points of an output volume. The distance data was then thresholded at 1.5 mm (half the distance of the assumed thickness) to produce a draft version of the volumetric model for each area. Finally, the draft models of the different areas in one hemisphere were compared to each other to detect where the dilation caused the overlap among them, and the overlapped voxels detected were removed from the models, producing a final volumetric model for each area. The volumetric model is a 3D binary array of voxels, with each voxel having a label indicating the presence or absence of the area. We confirmed that each model did not overlap with the others and had no topological defects, such as discontinuities and holes, by visual inspection of the 3D volume and surface rendered models.

In the second step, the inconsistency associated with each area was assessed as its occurrence probability; that is, in the form of the probability map. For each area, we transformed 10 volumetric models constructed from the 10 hemispheres into Talairach space (Fig. 1D, E) and then counted the number of overlapping models for each anatomic point. We used a single homogeneous transformation (Table 1), which consists of nine parameters, three translations, three rotations, and three scalings with respect to the axes of a Cartesian frame. The homogeneous transformation was determined using a standard method (Desmond & Lim 1997). First, the translation and rotation components of the matrix were computed from the locations of the anatomical landmarks (AC, PC, and mid-sagittal plane) identified by visual inspection of the structural volume. The alignment was checked and corrected by graphically comparing the X-, Y-, and Z-axes with the three orthogonal slices of the standard volume and the reconstructed surface. Next, the scale components were determined by measuring the size of the brain along each of the three axes as the bounding box dimensions of the surface and then computing the scaling factors to match the size to that of the 1988 Talairach atlas brain (X dimension: 136 mm; Y: 172 mm; Z: 118 mm). The scaling factor was determined separately for the left and right hemispheres. Because the Talairach atlas only contains a right hemisphere, the volumetric models for the left hemispheres were mirrored around the Y-axis and treated as though they were in right hemispheres. The occurrence probability of each area of Talairach space was computed by dividing the number of overlaps by the total number of hemispheres ($N = 10$). This computation was repeated every 1 mm in the Talairach space covering the visual cortex, yielding a 3D probability map for each retinotopic area (Fig. 1F). Furthermore, the probability maps for all the retinotopic areas were integrated into a maximum probability map (Fig. 1G) where each voxel was assigned a label indicating which area had the greatest probability of being present there and was given the maximum value.

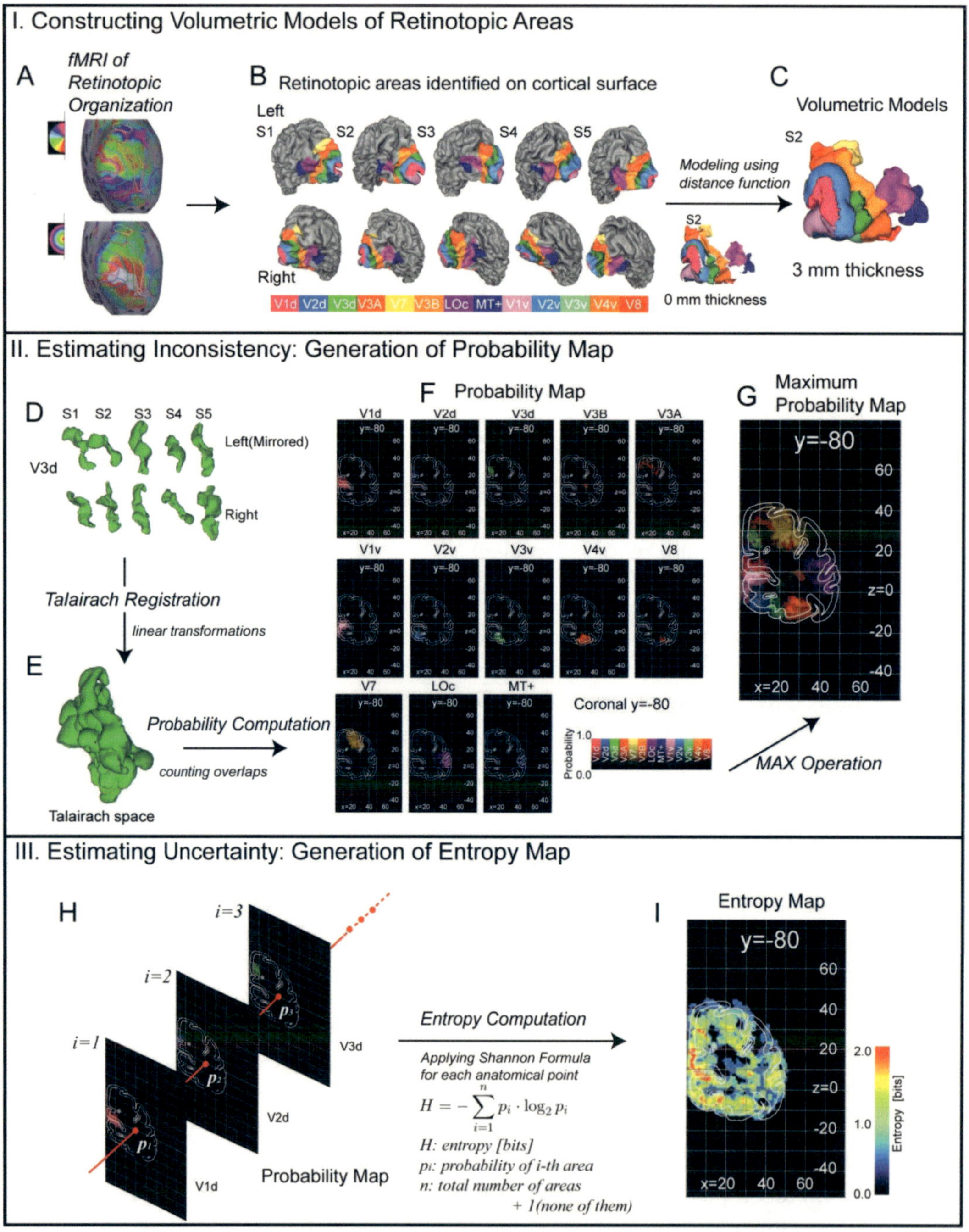

Figure 1. Overview of probabilistic atlas generation. See text for further explanation.

In the final step, the uncertainty associated with the retinotopic areas was assessed in the form of an entropy map (Fig. 1I) by calculating the Shannon entropy (Shannon 1948) from the occurrence probabilities of the different retinotopic areas (Fig. 1H), where this yields a quantitative measure of the uncertainty represented by a probability distribution. If no uncertainty exists in the distribution, the entropy equals zero, but if the entropy is greater than zero, it is maximal when all events are equally probable. The Shannon entropy (H) of the retinotopic areas can be expressed in terms of the occurrence probabilities (pi) of the different retinotopic areas and is given by

$$H = -\sum_{i=1}^{n} p_i \cdot \log_2 p_i \left[\text{bits}\right], \tag{1}$$

where n is the total number of candidate areas. By applying this equation to the probability maps of the retinotopic areas, we calculated the entropy for each point in the Talairach space covering the visual cortex (Fig. 1H, I). In the present analysis, $n = 14$; each of the 13 retinotopic areas = p_1, p_2,...p_{13} ; plus none of them, p_{14}. Thus, the maximum theoretical value for entropy is

$$-\sum_{i=1}^{14} \frac{1}{14} \log_2 \frac{1}{14} = \log_2 14 \cong 3.8 \left[\text{bits}\right], \tag{2}$$

2.4. Estimating two sources of uncertainty

We estimated two other kinds of entropies underlying the entropy about 14 possible states (retinotopic areas, and non-retinotopic area,) as computed above. The first was the entropy of two possible states, whether a voxel is located anywhere in the retinotopic areas (pr) or located outside of the areas (pnr), which was computed as follows:

$$H_r = -p_r \cdot \log_2 p_r - p_{nr} \cdot \log_2 p_{nr} \left[\text{bits}\right], \tag{3}$$

where

$$p_r = \sum_{i=1}^{13} p_i = 1 - p_{14} \text{ and } p_{nr} = p_{14}\,. \tag{4}$$

Because H_r is binomial, its maximum theoretical value is 1 bits. This binomial entropy would account for the uncertainty about whether a voxel is located within the visual cortex or is in the white matter, outside the brain or in the non-retinotopic cortex.

The second entropy type was the entropy of retinotopic areas computed on the probability distribution conditioned on a voxel being anywhere in the retinotopic areas, and was computed as follows:

$$H_c = -\sum_{i=1}^{13} p_{i/r} \cdot \log_2 p_{i/r} \left[\text{bits}\right], \tag{5}$$

where the conditional probability $p_{i/r} = p_i\, p_r$ and H_c was computed for voxels with $p_r \geq 0.2$, that is, for voxels that fell within any retinotopic area at least in two of ten hemispheres because it is impossible for areal overlap to occur for a voxel with $p_r = 0.1$. As the conditional entropy H_c has 13 possible states, its maximum theoretical value is

$$-\sum_{i=1}^{13} \frac{1}{13} \log_2 \frac{1}{13} = \log_2 13 \cong 3.7 \left[\text{bits}\right], \tag{6}$$

Please note, unlike the unconditional entropy H, H_c was computed on the probability distribution that did not comprise the event of "a voxel being in non-retinotopic area" and was rather conditional on the event of "a voxel being in retinotopic area". This conditional entropy would thus not be affected by the misalignment of cortical surfaces across hemispheres, and thereby reflect the uncertainty due to variability in the cortical locations of the areas in Talairach space.

2.5. Analysis and visualization software

The analyses were performed and visualized using BrainFactory software (Yamamoto 2011, H. Yamamoto et al. 2008, Yamamoto et al. 2002) written in VTK (Kitware, Clifton Park, NY) and MATLAB (Mathworks, Natick, MA), which has been successfully applied to cortical surface-based analysis of fMRI data (Ban et al. 2006, Ejima et al. 2007, Maeda et al. 2010, Yamamoto et al. 2011, T. Yamamoto et al. 2008) and cortico-cortical evoked potentials (Matsumoto et al. in press).

3. Results

To analyze the interindividual variability in the locations of the visual areas, we created a 3D probability map of the visual areas (e.g., for a coronal slice, see Fig. 1F) where each voxel was associated with the occurrence probabilities (N = 10) for each of the areas. It is difficult to display all such multivariate volume data in only two dimensions, so only the essence of the data is graphically presented in Fig. 2A in the form of a maximum probability map, where each point is color-coded according to the visual area with the greatest frequency (the maximum probability area), and the brightness represents the probability that the area resides at that voxel. The regions of maximal consistency are shown with maximal brightness, and the regions of minimal consistency are shown with minimal brightness. Figure 2A displays this information using the same coronal serial slices as the

http://dx.doi.org/10.5772/30760

1988 atlas of Talairach and Tournoux (1988), overlaid on the Talairach brain, the gray matter of which is outlined using white lines. The probability data in the horizontal and sagittal slices are also shown in Figs. 3A and 4A, respectively.

3.1. Inconsistency in the locations of visual areas: Probability map

The maximum probability area changes within the slices in the same hierarchical order as in the individual hemispheres. This topographic pattern is clearly illustrated in Fig. 5A, which shows the maximum probability map overlaid on the surface representation of the Talairach brain. The topographic preservation can be confirmed, except for the island-like V3 regions within V3B. In contrast to the near complete preservation of the topographic relations of the maximum probability areas, the probability maps of the corresponding areas (e.g., for a coronal slice, see Fig. 1F) revealed substantial inconsistency. As can be partially seen in Figs. 1F, 2A, 3A, and 4A, the probabilistic volume for a corresponding area (voxels having non-zero probabilities) is blurred so strongly that its extent is much wider than the thickness of the cortical gray matter. The strength of this tendency seems to vary among the visual areas, being strongest for area V7/IPS-0, which is located anterior to V3A (Tootell et al. 1998).

Positional inconsistency was quantitatively analyzed, and the results are shown in Fig. 6. Figure 6A illustrates the overall positional inconsistency, showing the distribution of maximum probability values within the voxels containing a non-zero probability. Most of the voxels were associated with small probabilities, with approximately half (47.3%) of them being associated with a value of 0.1 (only a single case representing a visual area) and 96.5% being associated with values less than 0.5 (0.2: 27.0%; 0.3: 15.1%; 0.4: 7.1%). The remaining 3.5% of the voxels were associated with probabilities equal to or larger than 0.5, with the maximum value being not 1.0 but 0.8 (0.5: 2.7%; 0.6: 0.67 %; 0.7: 0.14%; 0.8: 0.007%).

Separate analyses for the individual visual areas were also performed by spatially averaging all of the probability values within each area's probabilistic volume. The average probability is compared among the visual areas in Fig. 6B. The differences were not large, but the average probabilities were slightly lower for the dorsal areas (V3d, V7, and LOc), indicating a higher alignment inconsistency. Area V7 had the lowest values, as expected from the maximum probability map. The overall average of the average probabilities of the areas was 0.166 (SD: 0.013). The coefficient of variance (the SD divided by the mean) was 7.6%.

The analogous alignment inconsistency was evident when using another measure, percent blurring (Fischl et al. 1999b), which is defined as the percentage increase in the probabilistic volume of an area as compared with the mean volume of the individual areas. The percent blurring, which is compared among the visual areas in Fig. 6C, shows a pattern that is nearly reversed compared with the average probability (Fig. 6B), indicating a similar relative inconsistency among the visual areas. The average percent blurring for the group of areas was 460%.

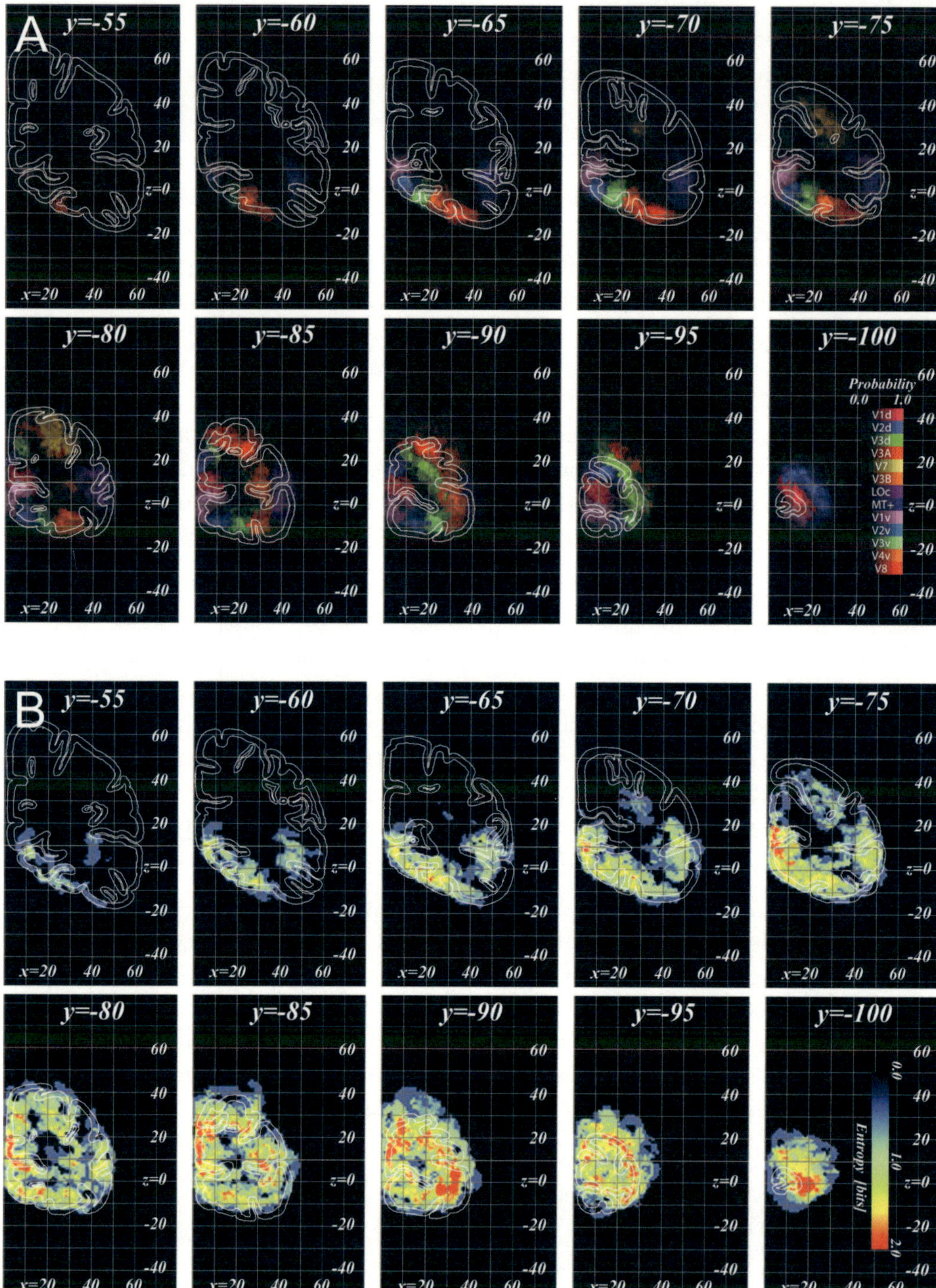

Figure 2. Maximum probability map (A) and entropy map (B) on serial coronal sections.

http://dx.doi.org/10.5772/30760

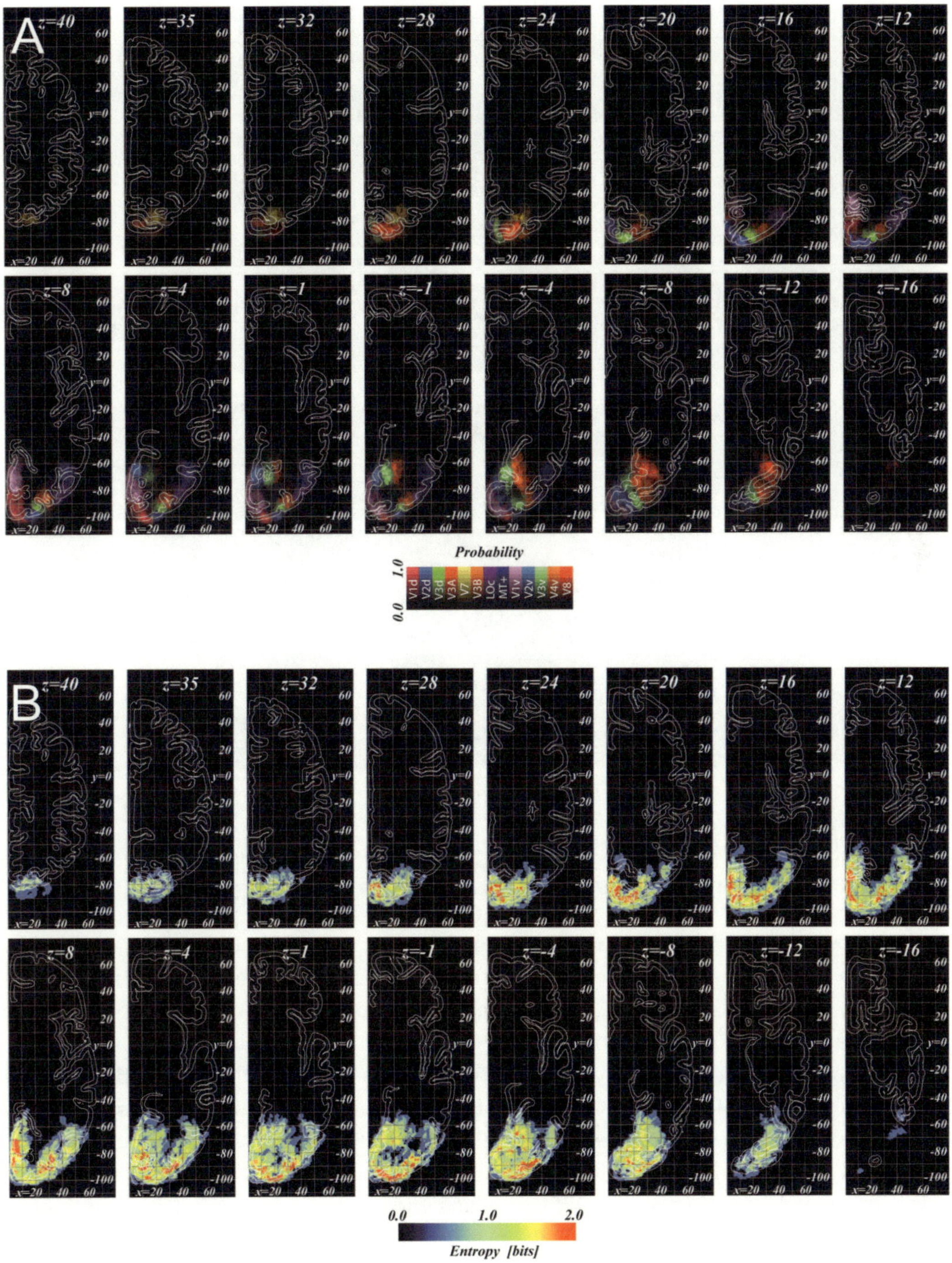

Figure 3. Maximum probability map (A) and entropy map (B) on serial horizontal sections.

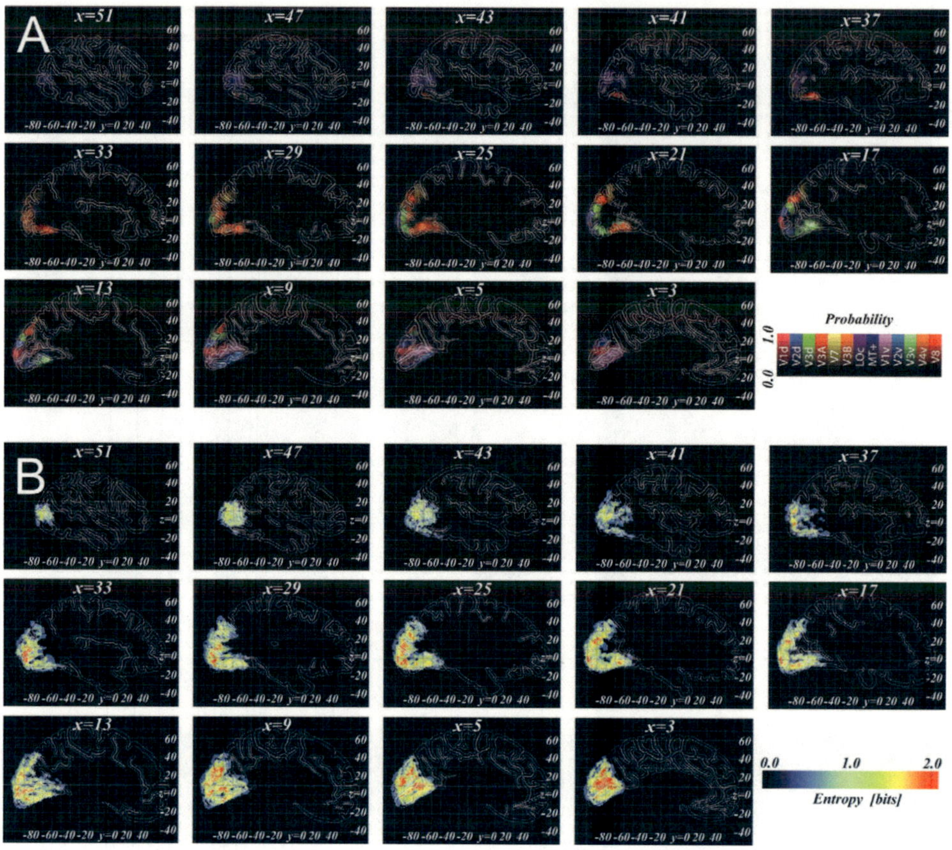

Figure 4. Maximum probability map (A) and entropy map (B) on serial sagittal sections.

3.2. Uncertainty in the locations of visual areas: Entropy map

Next, we performed a series of quantitative analyses regarding the uncertainty about which visual area actually resides at each voxel in Talairach space. As expected, registration of the individual visual areas into the Talairach space led to substantial overlaps among different visual areas. Figure 7A demonstrates this by showing the distribution of the number of overlaps among the areas. When considering the overlaps for all voxels associated with a non-zero probability (71804 voxels), these occurred in nearly half (46.7%) of the voxels and ranged from two to five overlapping areas per voxel (2: 35.4%; 3: 9.8%; 4: 1.5%; 5: 0.04%). Because it is impossible for overlap to occur in a voxel with a 0.1 probability (being in a visual area only for one hemisphere), we also computed the overlaps within the set of voxels with a probability greater than 0.1 (48116 voxels), finding the proportion of overlaps increased to occupy 69.6% of the voxels (2: 52.8%; 3: 14.6%; 4: 2.2%, 5: 0.05%).

To analyze the uncertainty resulting from the overlaps in an information theory framework (Shannon 1948), we introduced the Shannon entropy, a measure of the amount of uncertainty

http://dx.doi.org/10.5772/30760

represented by a probability distribution. In our study, this distribution is the probability of occurrence of the 13 delineated areas plus the probability of occurrence of none of the areas at each voxel in Talairach space. If any of the probabilities is 1.0, such that there is no uncertainty, the Shannon entropy of the visual area is zero, but if not, the entropy is greater than zero. The entropy takes a maximum value of 3.8 bits when the probabilities of all events occurring are the same, such as when we are very uncertain of which of the 14 events has occurred. By applying Shannon's formula to the occurrence probabilities (Fig. 1H, I), we created a 3D map of the entropy of the visual areas within the same occipital region of Talairach space as the probability map.

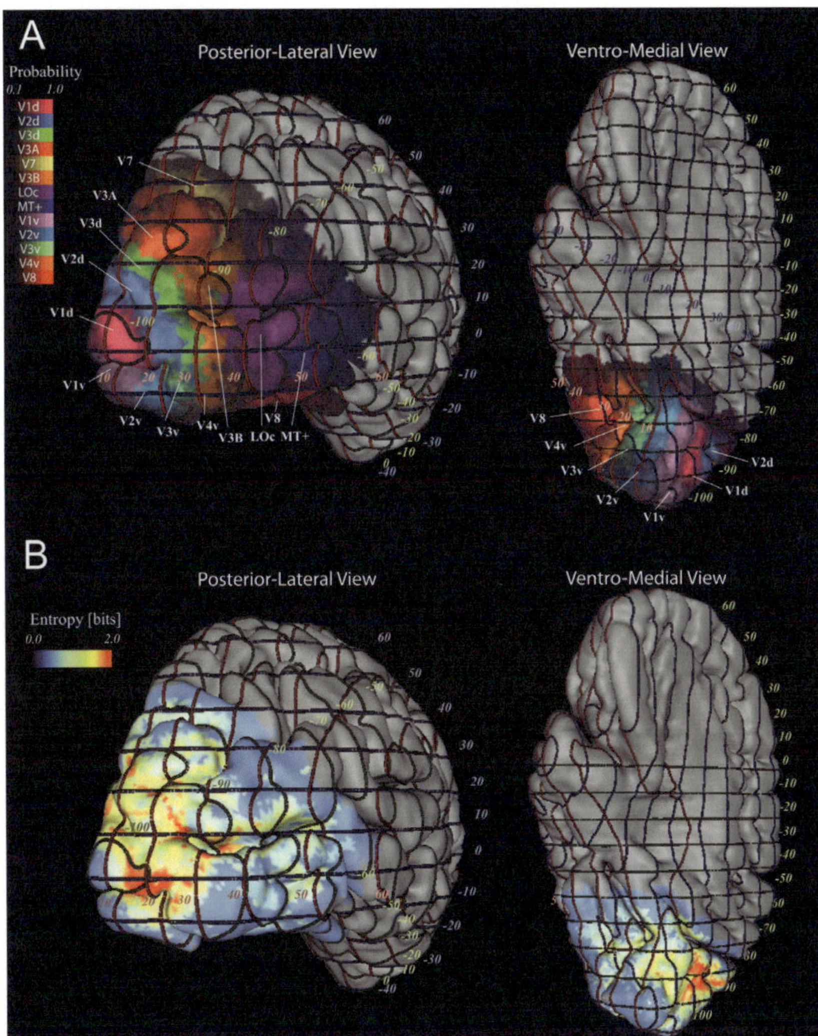

Figure 5. Surface representation of the maximum probability (A) and entropy map (B) of the areas.

Figure 2B/3B/4B shows the entropy map overlaid on the Talairach brain with the same coronal/horizontal/sagittal slice positions as the maximum probability map, where the entropy of the visual area at each voxel is color-coded from blue to red. Figure 5B also shows the entropy map overlaid on the surface representation of the Talairach brain in the same way as the maximum probability map (Fig. 5A). The entropy was less than 2 bits for most voxels, as evidenced by the predominance of blue and yellow regions. As shown in Fig. 7B, the larger the entropy, the smaller the number of voxels. The entropy values ranged from 0.47 to 2.45 bits, with a median value of 0.92 bits.

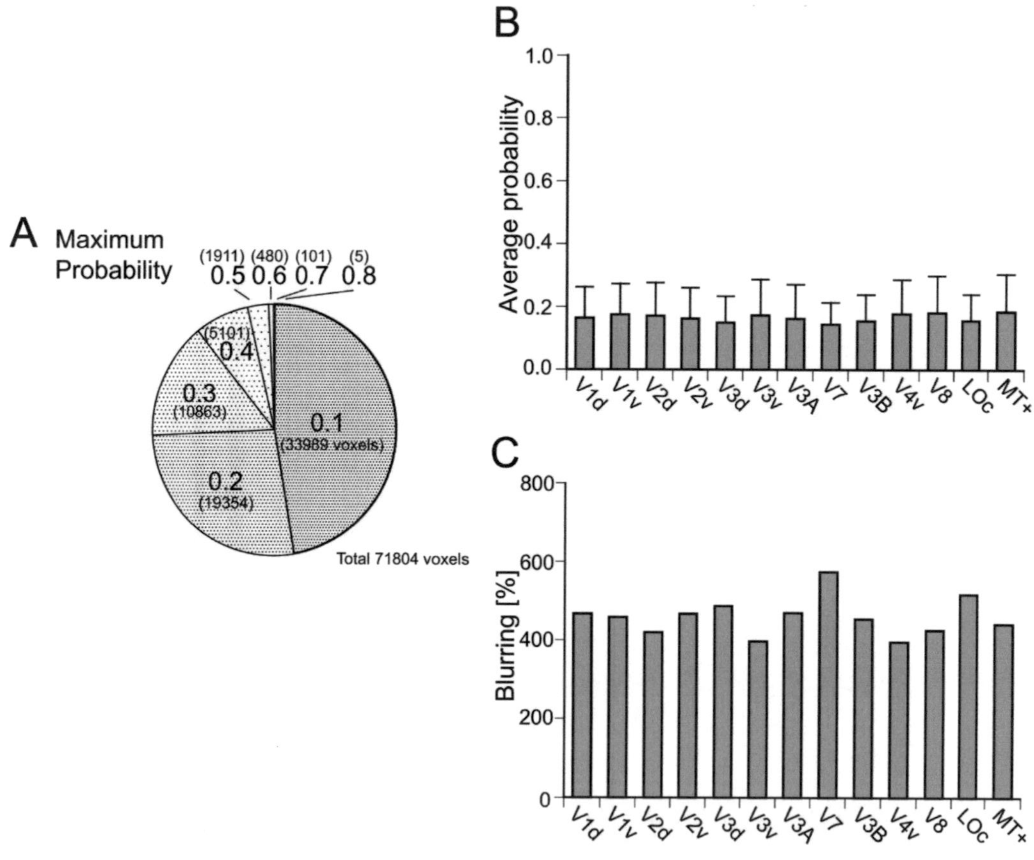

Figure 6. (A) Proportion of the maximum probabilities. The numbers of voxels are indicated in parentheses. (B) Average probability for each visual area across all the area voxels (probabilistic volume) in Talairach space, which is defined as the voxels being classified as an area for at least one hemisphere. Error bars denote SD. (C) Percent blurring of each visual area. This metric measures the amount of increase between the total volume of the area's voxels in its probability map and the mean volume of the area across hemispheres. If registration is perfect, the total volume in Talairach space will be equal to the mean volume, and percent blurring will be zero. For example, a value of 400% means that the total volume of the area's voxels is 5 times as big as the mean volume.

The comparison of the entropy map with the maximum probability map shows a topographical relationship between these. Regions with high entropy values (red) are generally located near

http://dx.doi.org/10.5772/30760

places where multiple visual areas would be located in close proximity, such as the foveal confluence separating the dorsal and ventral region (e.g., [X,Y,Z]=[20,-100,0]) and the borders of areas (e.g., for V3d/V3A/V3B, [20,-90,20]). Conversely, the regions with low entropy values (blue) tend to lie outside of the crowded fields.

Entropy is conceptually and computationally distinct from maximum probability, such that its map has a unique importance. However, if these measures were empirically related to each other, the entropy map would become less important. For a potential relationship, one would expect that high entropy would generally be associated with a low probability of belonging to one area. Therefore, we examined the relationship between entropy and maximum probability on a voxel-by-voxel basis, but found no correlation consistent throughout their range (Fig. 8). Although the expected negative correlation was observed for the higher maximum probability levels (0.4—0.8), the number of voxels in the range occupied only 11% of all the area's voxels. The entropies of the voxels with lower probability levels, which occupied the remaining 89% of the voxels, dispersed widely and showed rather positive correlations with maximum probabilities. Consequently, the entropy provides information distinct from that conveyed by the maximum probability, which emphasizes the importance of the entropy map.

We also quantified the uncertainty at the level of the visual areas by averaging all of the entropy values within the probabilistic volume of each area. The average entropy is compared among the areas in Fig. 7C. The average entropy was smaller, less than 1 bit, for areas V7, V8, LOc and MT+, and was at a minimum for MT+. For other areas, the average entropy was greater than 1 bit and was at a maximum for V2d. The difference between the maximum and minimum was 0.47 bits. The average entropy for the group of areas was 1.07 bits (SD: 0.175). The coefficient of variance was 16.4%, which is about two-fold larger than the probability.

Comparison of the average entropy (Fig. 7C) with the average occurrence probability (Fig. 6B) and blurring (Fig. 6C) shows the dissociation between uncertainty and inconsistency at the cortical area level. For example, areas V7 and V8 display opposite patterns: V7 has the smallest average probability, the largest blurring, and the second smallest entropy value, whereas V8 has a relatively large average probability, small blurring, and comparable entropy with V7; thus, V7 exhibits relatively strong inconsistency with relatively small uncertainty, whereas V8 exhibits relatively weak inconsistency with relatively small uncertainty. As complementary examples, V3d exhibits relatively strong inconsistency with relatively large uncertainty, whereas V2d exhibits relatively small inconsistency with relatively large uncertainty.

3.3. Two sources of uncertainty in the locations of visual areas

Finally, we analyzed the potential sources of the areal uncertainty. There are at least two sources of entropy when computed over the probability distribution of 14 possible states (13 areas plus none of them), as stated above. The first is the uncertainty as to whether a voxel is located within the visual cortex or lies within the white matter, outside the brain, or in non-retinotopic areas. We estimated this cortical registration factor by computing the entropy over the probability distribution of two events of the "voxel being in any of retinotopic areas" or not. Figure 9A compares the estimated entropy (theoretical maximum, 1 bit) among the retinotopic areas. Although the binomial entropy was a little smaller for area V7, it was nearly

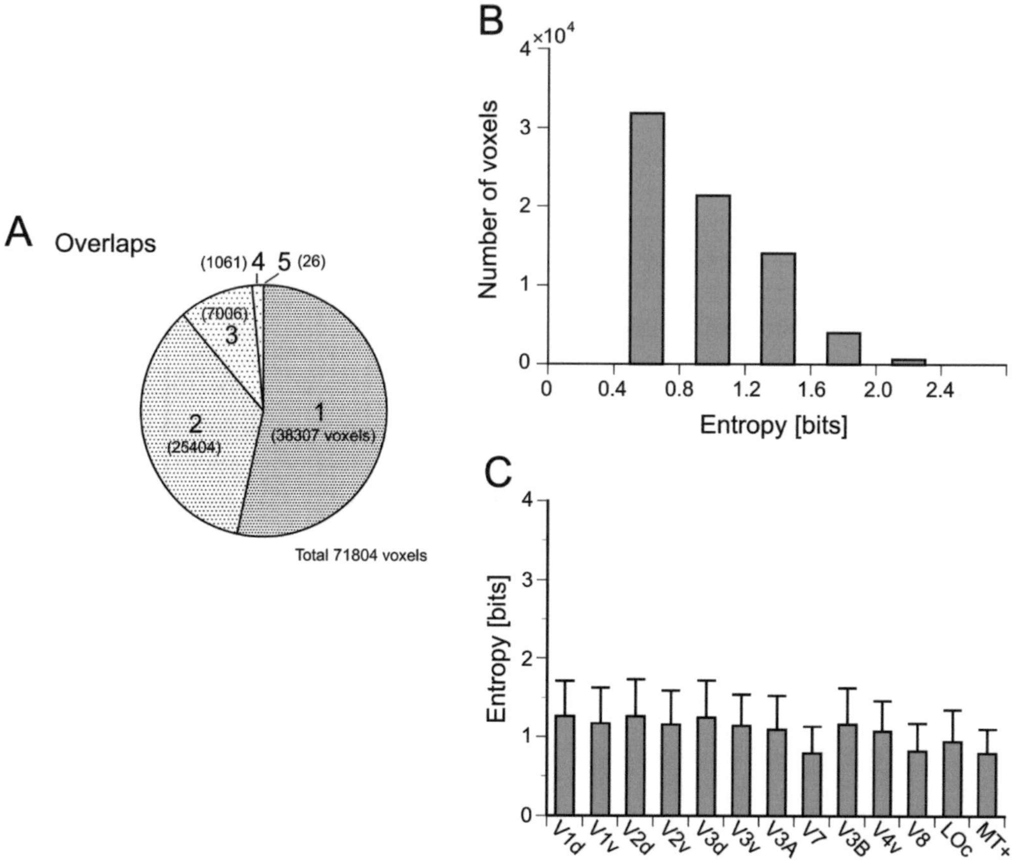

Figure 7. (A) Proportion of the number of overlaps among different visual areas in Talairach space. For example, an "overlap" of 1 means that these voxel correspond uniquely to a single area, whereas overlaps of 2 means that these voxel correspond two areas (one area for some hemispheres but another area for the other hemispheres). (B) Histogram of the entropy values. (C) Average entropy of each visual area across all the area's voxels (probabilistic volume). Error bars denote SD.

constant across the areas, as indicated by the fact that the standard deviation (0.022) was only 2.5% of the mean (0.87). The coefficient of variance was much smaller, by a factor of 6.4, compared with the result of the entropy over the full distribution of the 14 possible states (Fig. 5C). Therefore, this factor of uncertainty alone cannot explain the entropy difference among retinotopic areas.

The second factor was the uncertainty that remains even after the removal of the first uncertainty due to the registration, such as after knowing a voxel was being registered to some part of the retinotopic visual cortex. We estimated this by computing the entropy over the probability distribution of the retinotopic areas conditioned on a voxel belonging to any of the areas. This conditional entropy would reflect the underlying intersubject variability in the location on the cortical surface. Figure 9B compares the estimated conditional entropy (theoretical

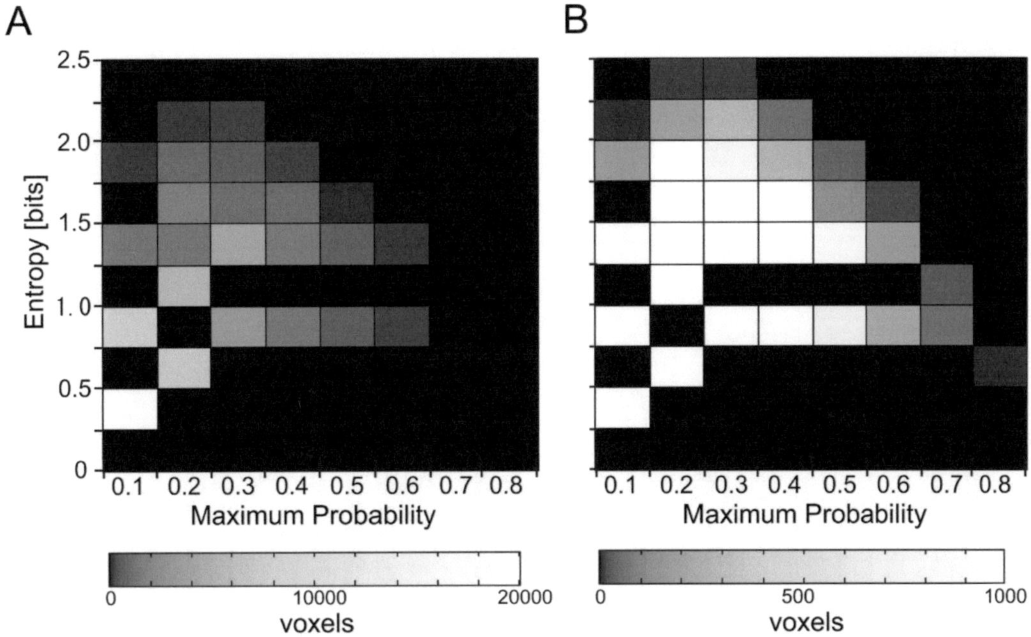

Figure 8. Two-dimensional histogram of entropy vs. maximum probability for all the area voxels. The histogram assigns a brightness value to each bin based on the number of voxels. The brighter the bin, the greater the number of voxels in that range. The brightness scale ranges from 0 to 20000 voxels. (B) Same histogram image as in A with a different, narrower scale to emphasize small differences in the number of voxels. The scale ranges from 0 to 1000 voxels.

maximum, 3.7 bits) among the retinotopic areas. Unlike the first binomial entropy due to cortical registration (Fig. 9A), the conditional entropy showed a substantial difference among the retinotopic areas, as indicated by a coefficient of variance of 27.6% (mean, 0.89; SD, 0.246), which was much larger, by a factor of 10.9, than the first entropy. The approximate ten-fold difference in the two entropy measures remained even when the comparison was limited to lower retinotopic areas, excluding the higher areas, V7, V8, LOc, and MT+, which are enclosed by more non-adjacent borders (coefficient of variance, 10.9% vs. 1.1%). Notably, the pattern of the difference among the areas (Fig. 9B) was similar to that found for the non-conditional entropy (Fig. 7C) but with more pronounced variability. The coefficient of variance and the difference between the maximum and minimum were larger, by factors of 1.7 and 1.6, respectively, than those obtained for the non-conditional entropy.

4. Discussion

The Talairach coordinate system has inherent limits in reducing the interindividual variability in the complex topography of the brain structure. The purpose of the present study was to apply a probabilistic and information theoretical framework for analyzing the residual interindividual variability of human visual area loci in Talairach space. The key feature of the framework is its ability to quantify the complementary aspects of the interindividual variability, inconsistency and uncertainty by constructing probability and entropy maps. The inconsistency and uncertainty of the locations of retinotopic areas found by analysis using this framework will be discussed below. In addition, we will discuss the applications of the maps and the framework.

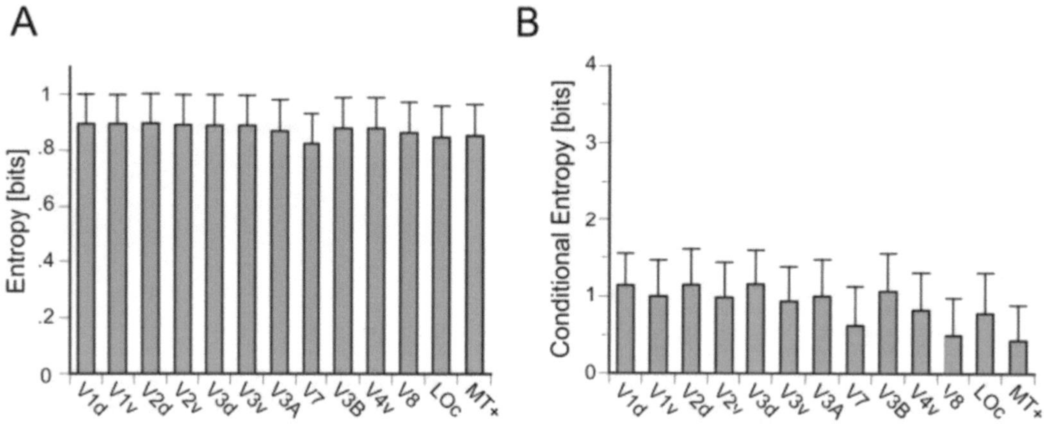

Figure 9. (A) Entropy computed on probability distribution whether a voxel is located in any retinotopic area or not. (B) Entropy computed on areal probability distribution *conditioned* on a voxel being in the visual cortex (retinotopic area). Error bars denote SD.

4.1. Inconsistency in the locations of visual areas: Probability map

In the present study, the alignment inconsistency of retinotopic areas (V1d/v, V2d/v, V3d/v, V4v, V8, V3A, V3B, V7, LOc, and MT+) was analyzed using a probabilistic approach (Mazziotta et al. 1995, Roland & Zilles 1994). The comprehensive application of this approach to visual areas was pioneered by Van Essen et al. (2001), who constructed probability maps of multiple areas (V1d/v, V2d/v, V3d/v, V3A, V4v, V8, MT+) identified using fMRI data from four hemispheres, using nonlinear surface warping in terms of the landmark coregistration rather than the standard linear Talairach transformation. Recently, we have also created probability maps using a similar surface based-method (Yamamoto et al. 2011). In this study, we generated probability maps of multiple retinotopic areas using standard linear Talairach transformation. Such maps have been made for the cytoarchitectonic definitions of areas V1 and V2 in five (Roland et al. 1997) and 10 brains (Amunts et al. 2000) and areas V3v and V4v in ten brains (Rottschy et al. 2007). Cytoarchitectonic studies of V1 and V2 have demonstrated that the probability volume of V2 surrounds that of V1 with substantial overlap, implying that the inconsistency was large, but not so large as to grossly violate the positional relationship. Our probability maps extend the observation to multiple functionally-defined visual areas.

Regarding the comparison across areas, it has been noted in cytoarchitectonic studies (Amunts et al. 2000) that the degree of inconsistency is smaller for V1 than for V2. However, such a pattern was not detected using the percent blurring measure in another study using fMRI

(Fischl et al. 1999b), where the comparison was made among V1d/v, V2d/v, and MT+. Consistent with this, the percent blurring measured in this study did not differ greatly among these areas. The discrepancy between anatomical and functional results may be resolved by examining them together using a common measure. One such integrated approach has been recently applied to areas V1 and V2 using probability maps, showing consistent results between the anatomical and functional probability maps (Wohlschlager et al. 2005). However, there appears to be a tendency for interindividual variability of the areas to increase in the anatomical map (see Fig. 5 in the study). Further studies are needed to elucidate the reason for this small discrepancy.

We also made a comparison across areas, introducing an intuitive measure of the average probability. Although the difference was not marked, we found slightly smaller average probabilities for areas V3d, V7, and LOc than for the other areas. A possible explanation is that as these areas are often buried in deep sulci, namely the intraparietal sulcus and lateral occipital sulcus, and that small differences in the courses of the sulci may prevent these areas from overlapping between the different subjects. The grand average across all areas was found to be only a little under 0.2, indicating substantial inconsistency.

4.2. Uncertainty in the locations of visual areas: Entropy map

Whereas the probability map allowed us to determine how probable it is that a particular visual area resides at a set of Talairach coordinates, the entropy map allowed us to quantify the uncertainty about the set of probability events. It should be stressed that even if events are highly probable, the entropy is large when the different events occur simultaneously, such as when the different areas overlap. The significance of the entropy map is in this dissociation. Indeed, our voxel-wise analysis of the relationship between the entropy and the maximum probability maps (Fig. 8) revealed that these statistics are largely unrelated, empirically supporting the dissociation of human visual areas in Talairach space. Although this dissociation has not been explicitly examined, the overlap between different structures has been noted previously in the context of the brain registration. Roland et al. (1997) identified areas V1 and V2 cytoarchitectonically and reported that the overlap between V1 and V2 is as large as the overlap between the different V1 areas. Van Essen (2005) analyzed the degree of overlap between the neighboring brain sulci using a probability map. Importantly, these studies all disregarded the occurrence probabilities of the overlapping structures, despite the uncertainty being maximal when each event is equiprobable. The present study extends these earlier analyses by relating the probabilities to the concept of uncertainty using entropy.

We found that the overall average of the entropy of retinotopic areas is approximately 1 bit in Talairach space, indicating that there are, on average, 1 bits of uncertainty about which retinotopic area is present at a point when its Talairach coordinates are specified. This means that there is an equal chance of each area being present for a point where two areas overlap. As for the spatial pattern of entropy distribution, the entropy map is not isotropic; it contains larger values around the borders of retinotopic areas and smaller values around their non-adjacent borders, which is expected given the possibility of overlaps between the different areas. In addition, we found that there is also a variation in entropy among retinotopic areas.

4.3. Two sources of uncertainty in the locations of visual areas

What causes the areal entropy and its variation in Talairach space? One possibility might be the poor alignment of the cortex under the simple Talairach volumetric registration because the probability distribution on which the entropy is computed contains information about whether a voxel belongs to the cortex. Indeed, the weakness of Talairach normalization is commonly agreed upon, and a prior probabilistic atlas of the human visual areas has been created using surface-based registration to partly overcome this problem (Van Essen et al. 2001, Yamamoto et al. 2011). The present study estimated the uncertainty caused by the inadequacy of Talairach registration by using the binomial entropy as either 'a voxel being in retinotopic area' or not and found that the measure is nearly constant among retinotopic areas. This suggests that the uncertainty due to registration is not a major factor underlying the uncertainty about retinotopic areas in Talairach space, thereby suggesting that the uncertainty due to intersubject variability in the relative size or cortical location of distinct areas (Dougherty et al. 2003, Ejima et al. 2003, Van Essen et al. 2001, Yamamoto et al. 2011) would be a significant factor.

This source of variability is, however, is not well understood. Therefore, we estimated this by computing the entropy for the probability distribution conditioned on a voxel in the cortex. Another approach could apply our non-conditional entropy measure to a surface-based group averaged brain, but this is beyond the scope of the present work. In addition, this method would lose some generality because the surface registration generally requires a set of anatomical or functional landmarks, and their selection would affect entropy. In contrast, our conditional entropy only assumes the association of a voxel to the cortex, such that it would constitute a baseline measure of the uncertainty remaining after surface-based normalization. The conditional entropy averaged across retinotopic areas was found to decrease by approximately 10%, supporting the general advantage of the surface-based normalization over Talairach normalization. This improvement would increase for the existing surface-based methods (Drury et al. 1996, Fischl et al. 1999b, Thompson et al. 1997) because they apply to the individual variability of gyral patterning. We also found that the difference in entropy across areas was amplified upon conditioning. This amplification underlines the fact that entropy provides information distinct from probability and confirms that the uncertainty due to intersubject variability in the layout of the cortical areas is a primary cause of the areal uncertainty in Talairach space. The much smaller observed conditional entropy of several higher areas confirms this idea because they are enclosed by many non-adjacent borders. However, with the exception of the apparent adjacency, we cannot specify the exact reason for the modest variations observed in lower retinotopic areas.

4.4. Preliminary comparison among registration methods

Three types of registration methods are commonly used, namely the linear volume-based, nonlinear volume-based, and surface-based methods. Among these, we evaluated a linear volume-based method for 5 subjects in the present study and a surface-based method for 16 subjects in a recent study (Yamamoto et al. 2011). Extending these studies, we initiated a comparison study among the three types of methods. Table 2 shows a preliminary summary

http://dx.doi.org/10.5772/30760

of the results obtained for 16 subjects. As used in the present study, the linear volume-method is a single linear transformation; the nonlinear volume-based method used is an automated image registration (AIR) method (Woods et al. 1998) where a 12 parameter affine linear transformation was followed by a second-order 30 parameter, nonlinear polynomial model; the surface-based method used is an Automated Spherical Warping (Fischl et al. 1999b) of the FreeSurfer software package (Dale et al. 1999, Fischl et al. 1999a), which displaced the vertices of each labeled surface to match its folding pattern with that of the FreeSurfer average template surface. As expected, the surface-based method gave the best results for probability. However, surprisingly, the surface-based method lost its advantage for entropy (average entropy: ~ 1 bits). This suggests that the current technology is not perfect and should therefore be used with caution regarding the limits.

	Linear volume-based method	Nonlinear volume-based method	Surface-based method
Average probability	0.089 (0.016)	0.10 (0.013)	0.27 (0.074)
Average entropy [bits]	1.01 (0.18)	1.07 (0.20)	1.15 (0.23)

The average probability(entropy) over 12 areas (V1, V2d, V2v, V3d, V3v, V3A, V7, V3B, V4v, V8, LOc, MT+) was listed. Parenthesis indicates SD across the areas. 16 subjects' brain were coregistered with the three method (see text for description of the methods).

Table 2. Preliminary comparison among registration methods

4.5. Applications

The present probabilistic and information theoretical framework has a variety of applications related to interindividual variability in the human brain. Probability and entropy maps of retinotopic areas can readily benefit any application where it is necessary to predict which areas are present at an anatomical point with a high probability and to establish the uncertainty associated with the prediction given its Talairach coordinates. Using these maps, the researchers and clinicians can infer, with a known degree of uncertainty, the visual area that is most likely present at an activation site or a lesion site within a target brain for which any form of the direct identification is impossible. The non-conditional entropy is a useful measure of the uncertainty observed in a practical situation, such as to average functional activations for different subjects or regimens for stereotaxic surgical treatment, whereas the conditional entropy allows a more fundamental appreciation of uncertainty regarding the cortical location of visual areas when analyzing the layout and network of cortical areas. The other main application is in the field of brain registration and brain warping (Toga 1999). As demonstrated in this study, Talairach registration reduced the entropy of visual areas from approximately 4 bits to 1 bit. In this way, the performance of various registration methods can be compared quantitatively using the degree of reduction in the entropy of the structures to be coregistered. Finally, the present work on visual areas can be generalized to not only other functional structures, but also other anatomical structures. Thus, an important direction for future research is to extend the present framework to address brain structure-function relationships, which are the focus of an ongoing large-scale international project that aims to develop a probabilistic brain atlas representing many types of information on brain function and structure (Mazziotta et al. 2001). One such approach is quantifying relationships using the Shannon mutual information based on simultaneous probability maps of structural and functional areas (Yamamoto et al. 2003).

5. Conclusion

We made two kinds of multisubject maps (10 hemispheres from five subjects) to characterize interindividual variability in the positions of human visual areas (V1/2/3, V3A, V3B, V7, V4v, V8, LOc, and MT+), which were localized using fMRI and coregistered into standard Talairach space by a single linear transformation. The first is a probability map representing the amount of alignment inconsistency for each area, where each voxel in space is associated with a probability affiliated with a given area. The second, a novel map termed an entropy map where each voxel is associated with a Shannon entropy computed from the probabilities, represents the amount of uncertainty that exists regarding which area resides there and is maximal when all areas are equally probable. We demonstrate that the entropy map provides distinctly different information from the probability map, highlighting the significance of the entropy map. The overall average probability and entropy values were ~0.2 and ~1.0 bits, respectively, with some area differences. The difference increased when computing the entropy for the probability distribution conditioned on a voxel located anywhere in the visual cortex, suggesting that a major source of the observed uncertainty is not due to the poor alignment of the cortex, but to individual variability in the cortical locations of the visual areas. The probability and entropy maps generated in this study will readily benefit any application where it is necessary to predict the areas that are most likely present at an anatomical voxel and to establish the uncertainty associated with this prediction.

Acknowledgements

This work was supported by Grant-in-Aids for Global COE (Centers of Excellence) Program (D07), Scientific Research on Innovative Areas (No. 22135007), and Scientific Research (C) (No. 22530793) of Japan's Ministry of Education, Culture, Sports, Science and Technology.

Author details

Hiroki Yamamoto*

Graduate School of Human and Environmental Studies, Kyoto University, Japan

http://dx.doi.org/10.5772/30760

References

[1] Amunts K, Malikovic A, Mohlberg H, Schormann T & Zilles K. (2000). Brodmann's areas 17 and 18 brought into stereotaxic space-where and how variable? *NeuroImage,* Vol.11, No.1, pp. 66-84, ISSN 1053-8119

[2] Ashburner J. (2007). A fast diffeomorphic image registration algorithm. *NeuroImage,* Vol.38, No.1, pp. 95-113, ISSN 1053-8119

[3] Ban H, Yamamoto H, Fukunaga M, Nakagoshi A, Umeda M, Tanaka C & Ejima Y. (2006). Toward a common circle: interhemispheric contextual modulation in human early visual areas. *The Journal of neuroscience,* Vol.26, No.34, pp. 8804-9, ISSN 0270-6474

[4] Cencov NN. (1982). *Statistical decision rules and optimal inference,* American Mathematical Society, ISBN 0821845020, Providence, R.I.

[5] Dale AM, Fischl B & Sereno MI. (1999). Cortical surface-based analysis. I. Segmentation and surface reconstruction. *NeuroImage,* Vol.9, No.2, pp. 179, ISSN 1053-8119

[6] Dale AM & Sereno MI. (1993). Improved localization of cortical activity by combining EEG and MEG with MRI cortical surface reconstruction: A linear approach. *J Cogn Neurosci,* Vol.5, No.2, pp. 162-76, ISSN 1530-8898

[7] Desmond JE & Lim KO. (1997). On- and offline Talairach registration for structural and functional MRI studies. *Human Brain Mapp,* Vol.5, No.1, pp. 58-73, ISSN 1097-0193

[8] DeYoe EA, Bandettini P, Neitz J, Miller D & Winans P. (1994). Functional magnetic resonance imaging (FMRI) of the human brain. *J Neurosci Methods,* Vol.54, No.2, pp. 171-87, ISSN 0165-0270

[9] Dougherty RF, Koch VM, Brewer AA, Fischer B, Modersitzki J & Wandell BA. (2003). Visual field representations and locations of visual areas V1/2/3 in human visual cortex. *J Vis,* Vol.3, No.10, pp. 586-98, ISSN 1534-7362

[10] Drury HA, Van Essen DC, Anderson CH, Lee CW, Coogan TA & Lewis JW. (1996). Computerized mappings of the cerebral cortex: a multiresolution flattening method and a surface-based coordinate system. *Journal of Cognitive Neuroscience,* Vol.8, No.1, pp. 1-28, ISSN 1530-8898

[11] Drury HA, Van Essen DC, Corbetta M & Snyder AZ. (1999). Surface-based analyses of the human cerebral cortex., In: *Brain Warping* AW Toga, pp. 337-63, Academic Press, ISBN 0126925356, San Diego

[12] Ejima Y, Takahashi S, Yamamoto H, Fukunaga M, Tanaka C, Ebisu T & Umeda M. (2003). Interindividual and interspecies variations of the extrastriate visual cortex. *Neuroreport,* Vol.14, No.12, pp. 1579-83, ISSN 0959-4965

[13] Ejima Y, Takahashi S, Yamamoto H & Goda N. (2007). Visual Perception of Contextual Effect and Its Neural Correlates In: *Representation and Brain*, S Funahashi, pp. 3-20, Springer Verlag, ISBN 978-4-431-73020-0, Tokyo

[14] Engel SA, Glover GH & Wandell BA. (1997). Retinotopic organization in human visual cortex and the spatial precision of functional MRI. *Cereb Cortex*, Vol.7, No.2, pp. 181-92, ISSN 1047-3211

[15] Engel SA, Rumelhart DE, Wandell BA, Lee AT, Glover GH, Chichilnisky EJ & Shadlen MN. (1994). fMRI of human visual cortex. *Nature*, Vol.369, No.6481, pp. 525, ISSN 0028-0836

[16] Fischl B, Sereno MI & Dale AM. (1999a). Cortical surface-based analysis. II: Inflation, flattening, and a surface-based coordinate system. *NeuroImage*, Vol.9, No.2, pp. 195-207, ISSN 1053-8119

[17] Fischl B, Sereno MI, Tootell RB & Dale AM. (1999b). High-resolution intersubject averaging and a coordinate system for the cortical surface. *Hum Brain Mapp*, Vol.8, No.4, pp. 272-84, ISSN 1097-0193

[18] Gholipour A, Kehtarnavaz N, Briggs R, Devous M & Gopinath K. (2007). Brain Functional Localization: A Survey of Image Registration Techniques. *Medical Imaging, IEEE Transactions on*, Vol.26, No.4, pp. 427-51, ISSN 0278-0062

[19] Hasnain MK, Fox PT & Woldorff MG. (1998). Intersubject variability of functional areas in the human visual cortex. *Hum Brain Mapp*, Vol.6, No.4, pp. 301-15, ISSN 1097-0193

[20] Lorensen W, E. & Cline H, E. . (1987). *Marching cubes: A high resolution 3D surface construction algorithm.* Proceedings of the 14th annual conference on Computer graphics and interactive techniques, ISBN 0897912276, Anaheim, California, July 27-31, 1987

[21] Maeda K, Yamamoto H, Fukunaga M, Umeda M, Tanaka C & Ejima Y. (2010). Neural correlates of color-selective metacontrast in human early retinotopic areas. *Journal of Neurophysiology*, Vol.104, No.4, pp. 2291-301, ISSN 0022-3077

[22] Matsumoto R, Ikeda A, Fumuro T, Mikuni N, Miyamoto S, Fukuyama H, Takahashi R, Najm I, Shibasaki H & Luders H. (in press). Parieto-frontal network in humans studied by cortico-cortical evoked potential. *Human Brain Mapping*, ISSN 1097-0193

[23] Mazziotta J, Toga A, Evans A, Fox P, Lancaster J, Zilles K, Woods R, Paus T, Simpson G, Pike B, Holmes C, Collins L, Thompson P, MacDonald D, Iacoboni M, Schormann T, Amunts K, Palomero-Gallagher N, Geyer S, Parsons L, Narr K, Kabani N, Le Goualher G, Boomsma D, Cannon T, Kawashima R & Mazoyer B. (2001). A probabilistic atlas and reference system for the human brain: International Consortium for Brain Mapping (ICBM). *Philos Trans R Soc Lond B Biol Sci*, Vol.356, No.1412, pp. 1293-322, ISSN 0962-8436

[24] Mazziotta JC, Toga AW, Evans A, Fox P & Lancaster J. (1995). A probabilistic atlas of the human brain: theory and rationale for its development. The International Consortium for Brain Mapping (ICBM). *NeuroImage,* Vol.2, No.2, pp. 89-101, ISSN 1053-8119

[25] Ono M, Kubik S & Abernathey C. (1990). *Atlas of the Cerebral Sulci,* Thieme, ISBN 0865773629, Stuttgart, Germany

[26] Roland PE, Geyer S, Amunts K, Schormann T, Schleicher A, Malikovic A & Zilles K. (1997). Cytoarchitectural maps of the human brain in standard anatomical space. *Human Brain Mapping,* Vol.5, No.4, pp. 222-27, ISSN 1097-0193

[27] Roland PE & Zilles K. (1994). Brain atlases--a new research tool. *Trends Neurosci,* Vol. 17, No.11, pp. 458-67, ISSN 0166-2236

[28] Rottschy C, Eickhoff SB, Schleicher A, Mohlberg H, Kujovic M, Zilles K & Amunts K. (2007). Ventral visual cortex in humans: Cytoarchitectonic mapping of two extrastriate areas. *Hum Brain Mapp,* Vol.28, No.10, pp. 1045-1059, ISSN 1065-9471

[29] Sereno MI, Dale AM, Reppas JB, Kwong KK, Belliveau JW, Brady TJ, Rosen BR & Tootell RB. (1995). Borders of multiple visual areas in humans revealed by functional magnetic resonance imaging. *Science,* Vol.268, No.5212, pp. 889-93, ISSN 0036-8075

[30] Shannon CE. (1948). The mathematical theory of communication, I and II. *Bell System technical journal,* Vol.27, pp. 379-443

[31] Talairach J, Szikla G, Tournoux P, Prosalentis A, Bordas-Ferrier M, Covello L, Iacob M & Mempel E. (1967). *Atlas d'anatomie stereotaxique du telencephale.,* Masson, Paris

[32] Talairach J & Tournoux P. (1988). *Co-Planar Stereotactic Atlas of the Human Brain,* Thieme, ISBN 978-0865772939, Stuttgart/New York

[33] Teo PC, Sapiro G & Wandell BA. (1997). Creating connected representations of cortical gray matter for functional MRI visualization. *IEEE Transactions on Medical Imaging,* Vol.16, No.6, pp. 852-63, ISSN 0278-0062

[34] Thompson PM, MacDonald D, Mega MS, Holmes CJ, Evans AC & Toga AW. (1997). Detection and mapping of abnormal brain structure with a probabilistic atlas of cortical surfaces. *J Comput Assist Tomogr,* Vol.21, No.4, pp. 567-81, ISSN 0363-8715

[35] Toga AW. (1999). *Brain warping,* Academic Press, ISBN 0126925356, San Diego

[36] Tootell RB, Hadjikhani N, Hall EK, Marrett S, Vanduffel W, Vaughan JT & Dale AM. (1998). The retinotopy of visual spatial attention. *Neuron,* Vol.21, No.6, pp. 1409-22, ISSN 0896-6273

[37] Van Essen DC. (2005). A Population-Average, Landmark- and Surface-based (PALS) atlas of human cerebral cortex. *NeuroImage,* Vol.28, No.3, pp. 635-62, ISSN 1053-8119

[38] Van Essen DC, Lewis JW, Drury HA, Hadjikhani N, Tootell RB, Bakircioglu M & Miller MI. (2001). Mapping visual cortex in monkeys and humans using surface-based atlases. *Vision Res,* Vol.41, No.10-11, pp. 1359-78, ISSN 0042-6989

[39] Wohlschlager AM, Specht K, Lie C, Mohlberg H, Wohlschlager A, Bente K, Pietrzyk U, Stocker T, Zilles K, Amunts K & Fink GR. (2005). Linking retinotopic fMRI mapping and anatomical probability maps of human occipital areas V1 and V2. *NeuroImage,* Vol.26, No.1, pp. 73-82, ISSN 1053-8119

[40] Woods RP, Grafton ST, Watson JD, Sicotte NL & Mazziotta JC. (1998). Automated image registration: II. Intersubject validation of linear and nonlinear models. *J Comput Assist Tomogr,* Vol.22, No.1, pp. 153-165, ISSN 0363-8715

[41] Yamamoto H. (2011) BAVIEW: software for visualization and analysis of probabilistic atlases of human visual areas. *ESMRMB 2011,* Leipzig, DE, Oct. 6-8, 2011

[42] Yamamoto H, Ban H, Fukunaga M, Umeda M, Tanaka C & Ejima Y. (2008). Large- and Small-Scale Functional Organization of Visual Field Representation in the Human Visual Cortex., In: *Visual Cortex: New Research.,* TA Portocello, RB Velloti, pp. 195-226, Nova Science Publisher, ISBN 1604565306, New York

[43] Yamamoto H, Fukunaga M, Takahashi S, Mano H, Tanaka C, Umeda M & Ejima Y. (in press). Inconsistency and uncertainty of the human visual area loci following surface-based registration: Probability and Entropy Maps. *Human Brain Mapping,* doi: 10.1002/hbm.21200, ISSN 1065-9471

[44] Yamamoto H, Fukunaga M, Takahashi S, Tanaka C, Ebisu T, Umeda M & Ejima Y. (2002) BrainFactory: an integrated software system for surface-based analysis of fMRI data. *Human Brain Mapping 2002,* Sendai, Japan, June 2-6, 2002

[45] Yamamoto H, Fukunaga M, Tanaka C, Ebisu T, Umeda M & Ejima Y. (2003) A New Method for Quantifying Brain Structure-Function Relationships Based on Simultaneous Probability Map and Information Theory. *Society for Neuroscience 33rd Annual Meeting,* New Orleans, USA, Nov. 8-12, 2003

[46] Yamamoto T, Takahashi S, Hanakawa T, Urayama S, Aso T, Fukuyama H & Ejima Y. (2008). Neural correlates of the stereokinetic effect revealed by functional magnetic resonance imaging. *J Vis,* Vol.8, No.10, pp. 141-17, ISSN 1534-7362

Chapter 10

The Contribution of fMRI in the Study of Visual Categorization and Expertise

Natasha Sigala

Additional information is available at the end of the chapter

http://dx.doi.org/10.5772/36031

1. Introduction

We recognize and categorize objects around us within a fraction of a second and in a number of different ways, depending on context, our experience with them, and the purpose of the categorization. For example the same animal can be *a dog, a bow wow* or *a bulldog, a mammal or a Canis lupus familiaris.* We are also able to recognize it in a variety of lighting conditions, orientations and positions, despite the large number of two dimensional images that every three dimensional object generates. It is therefore not surprising that our extraordinary ability to recognize objects has fascinated philosophers for a very long time. In *Categories* Aristotle made an attempt to categorize everything, mainly by analysing patterns of language and speech, answering questions like "τί ἐστί", "what is it?", "what like is it?" (Cross, 1959), and by describing the defining qualities that all instances of a particular category share, e.g. all soft things share the quality of softness.

Real progress in describing and understanding such qualities (features) was made by Fred Attneave in experiments of visual perception during his Ph.D. research at Stanford University (Attneave, 1957). Specifically his results showed that subjective ratings of the perceived dissimilarities between stimuli (letter stings and shapes) and the frequency of errors while learning those stimuli, could both be explained in terms of distances between stimuli represented as points in a "psychological space". The idea of describing subjective psychological experiences with geometry influenced Roger Shepard, who applied and extended multidimensional scaling (MDS), the representation of objects (e.g. shapes, words, faces) as points in space, so that the distances between the points represent the perceived similarities between the objects (Richardson, 1938), (Torgerson, 1952), (Shepard, 1958), (Shepard, 1962a), (Shepard, 1962b), (Shepard, 1987).

The exact ways in which similarity of a perceived object and its mental representation is measured have been the focus of research for over half a century. According to the theory of prototypes, we categorize new objects by comparing them to abstracted representations that are the central tendency of all the examples of the categories that we have experienced (Posner and Keele, 1968), e.g. a prototypical triangle or square. According to exemplar theories, the perceived objects are compared with stored representations of exemplars, grouped by category (Medin and Schaffer, 1978), e.g. memories of actual triangles and squares we have experienced. Although the early exemplar models postulated equal weights for all stimulus features, later contributions included an attention-optimisation hypothesis, allowing the perceived distance (similarity) between objects to vary with context and task demands (Generalised Context model, (Nosofsky, 1986)). For example this implies that for a layperson the representations of a trout and bass may be close together (similar), but for an ichthyologist or fisherman those representations are further apart (less similar) because they can identify a number of features that make the two types of fish different.

Eleanor Rosch analyzed language patterns and visual perception across cultures and did the experiments that I suspect Aristotle would have liked to perform himself (Heider, 1972, Rosch et al., 1976a). She showed that the categorizations we make rely on features of high validity (e.g. feathers and wings for birds) which form prototypes that aid cognitive economy and reflect perceived world structure (Boden, 2006), pp. 520-521. In other words "robin" is a more typical bird example, than "ostrich", and this helps us recognise and make decisions about objects faster, and also reflects the likelihood to encounter such objects in our environment. Rosch named the hierarchical level where category prototypes are found as the basic level of categorization (e.g. bird, fish, tree), which is the one that provides the most useful information, the first to be named by children, and the most necessary in language (Rosch, 1976, Rosch et al., 1976b). The level below (subordinate) contains more specific information (e.g. robin, sparrow), while the level above (superordinate) (e.g. animal, plant) less specific information. These results explained how we structure information about the world around us, to a large extent in a universal, rather than arbitrary or culture-dependent way. The opposite conclusion would render the following "ancient Chinese" animal taxonomy, presented by Jorge Luis Borges (1966) (in (Boden, 2006), p. 519) entirely possible:

"[Animals] are divided into (a) those that belong to the Emperor, (b) embalmed ones, (c) those that are trained, (d) suckling pigs, (e) mermaids, (f) fabulous ones, (g) stray dogs, (h) those that are included in this classification, (i) those that tremble as if they were mad, (j) innumerable ones, (k) those drawn with a very fine camel's hair brush, (l) others, (m) those that have just broken a flower vase, (n) those that resemble flies from a distance".

Crucially for the question of expertise, Rosch's basic level of categorization, can be and is modified by experience. Based on their observation that an airplane mechanic answered questions about airplanes differently to other (non-expert) participants, Rosch and colleagues suggested that one way to study what part a person's previous knowledge plays in categorization, would be a systematic variation of the participants' level of expertise and the object categories (Rosch et al., 1976a). They also speculated that an ichthyologist would use as a basic level, what a novice would consider a more specific, subordinate one (e.g. trout or salmon, instead of fish). Tanaka and Taylor performed such an experiment, and compared the categorization performance of dog and bird experts in listing distinctive features, as well as speed and level of categorizing animals in and out of their area of expertise. They found that the most informative category for the experts was the more specific, subordinate level, rather than the basic one (Tanaka and Taylor, 1991), e.g. crow or robin, rather than bird for the bird experts, and beagle or collie, rather than dog for the dog experts.

Despite long-running debates about the usefulness of the concept of similarity in explaining categorization (e.g. Goodman, 1972), and the merits of prototype, exemplar or other theories, most researchers agree that categorization is based on an expanded notion of similarity, an overall similarity, which encompasses physical, functional and overall features, reflecting a person's theoretical understanding of the world, e.g. (Murphy and Medin, 1985) in (Ahn and Dennis, 2001). The concept of similarity has furthermore inspired computational theories of vision, where "the representation of similarity is taken to be the goal of the visual system" (Edelman, 1998), modern extensions of which are particularly appropriate for relating neural data, behaviour and models across species and neuroscientific methods (Kriegeskorte, 2009).

In this chapter I will first review some background work that sets the context for the subsequent discussion of a subset of fMRI studies that have shed light on the questions of perceptual and cognitive categorization and expertise over the last 15 years.

2. Computational and experimental work

During the end of the 20th and the beginning of the 21st century, the idea that neuronal populations code representations which reflect both physical stimulus similarity and perceived stimulus similarity, as shaped by task demands, received significant experimental support. In order to bridge the species gap between human and non-human primates, and study meaningfully the neuronal representations in the macaque brain, it was first necessary to test if the categorizations that monkeys make can also be explained with MDS, and prototype and/or exemplar theories. This involved the creation of novel, parametrically designed stimuli with a known number of varying dimensions (features). The premise here was inspired by computational theories of vision that called for representations that were two-dimensional viewpoint-dependent snapshots of three-dimensional objects (Poggio and Edelman, 1990) that preserved the geometry of similarity amongst the objects (Edelman, 1998) (as opposed to stored three-dimensional representations of the objects themselves (Marr and Nishihara, 1978) (Biederman, 1987)). The first study to show that MDS was a useful tool for understanding object categorization in the monkey was described in (Sands et al., 1982). Crucially, they did not provide categorization training to the animals, and they explored how macaques perceived and represented pictures of various natural categories (faces, fruit, colours). They reached the conclusion that macaques treat pictorial stimuli categorically cf. (Sigala, 2009). Several years later (Sugihara et al., 1998) employed novel stimuli (computer-generated 3D animals) to test the usefulness of MDS in the study of object recognition, following a previous demonstration with the human visual system (Cutzu and Edelman, 1996). After systematic training of the monkeys to report their perception of the stimuli, Sugihara et al. showed that the psychophysical representation of the novel stimuli, as revealed by MDS, captured the similarities built in the stimulus space. This means that the monkey visual system could successfully recover a two-dimensional configuration of the stimuli that were originally built in a high-dimensional space (set of 56 variables), and reliably capture their relative similarities. This was the first piece of evidence for the non-human primate visual system that representations of the stimuli might be representations of the similarities of the stimuli.

Following this psychophysical experiment, Sigala and colleagues also created two parametrically designed stimulus sets (schematic faces and fish) with four varying dimensions (Sigala et al., 2002). The participants (both humans and non-human primates) first had to report how similar the stimuli were to them, then were trained to categorize them at the subordinate level based on the combination of two features, and then reported how similar the stimuli were to them a second time. By collecting similarity ratings before and after the categorization training, it was possible to show that the perceptual similarity of the stimuli changed, particularly for the macaques, after they had learnt to categorize the stimuli based on two features (ignoring the other two equally varying features) (Fig. 1). The initial MDS solution of the similarity ratings showed that the stimulus features were not consistently used when comparing the stimuli (Fig. 1a).

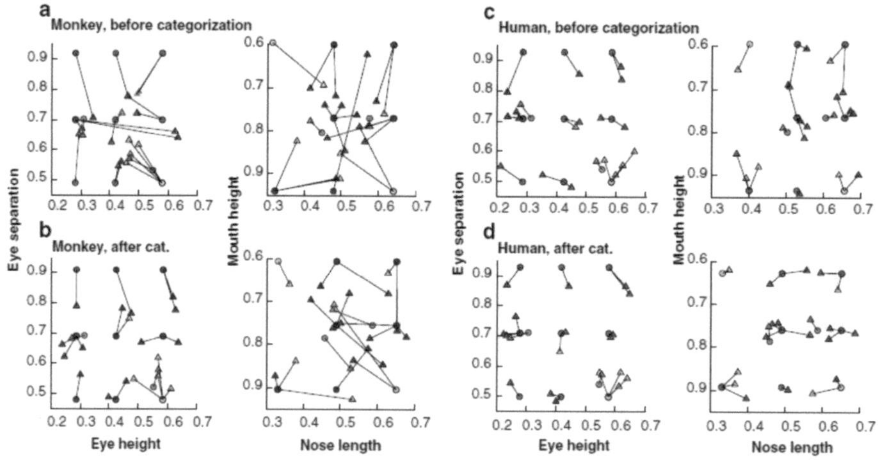

Figure 1. Psychophysical representation of schematic faces in a non-human (a, b) and a human (c, d) participant, as revealed by Multi Dimensional Scaling (MDS) on similarity ratings of the stimuli before and after categorization training. (Triangles: psychophysical representation based on similarity ratings; circles: physical stimulus values). 20 schematic faces are represented by: red and yellow circles, for category 1 and 2 exemplars used in training respectively; blue circles for the test exemplars used in the transfer phase of categorization; purple circles: prototypes for categories 1 and 2. The two categories were linearly separable along the Eye Height, Eye Separation dimensions, but not along the Nose Length, Mouth Height. Lines connect matching physical stimulus representations. When several patterns share the same combination of physical dimension values, multiple triangles are connected to the same circle. Longer lines correspond to less faithful psychophysical representation of corresponding physical stimulus values. Reproduced from (Sigala et al., 2002) with permission.

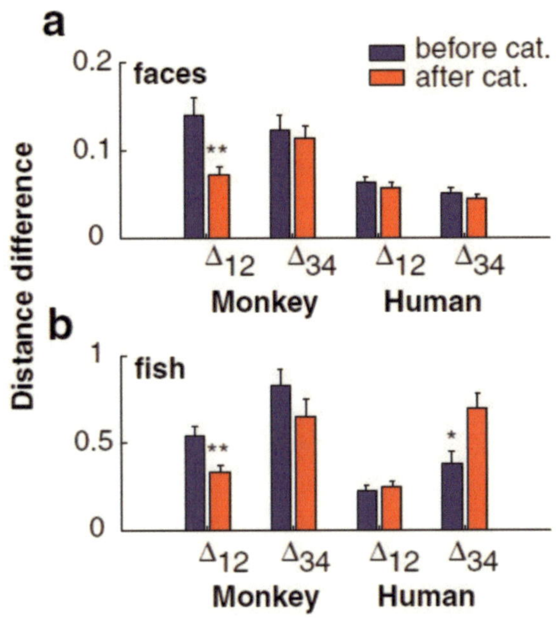

Figure 2. a) Mean distance difference between psychophysical and physical stimuli along the dimensions that were diagnostic for the categorization task (Δ12) and the dimensions that also varied, but were not diagnostic for the categorization task (Δ34), before and after training, averaged over 20 schematic faces (error bars are standard error of mean). b) Average distance difference along the non-diagnostic and diagnostic dimensions before and after categorization for schematic fish stimuli. Data before and after categorization taken from two different monkeys and a single human subject. Significance levels (t-test): (∗∗) corresponds to P<0.005 and (∗) to P<0.01. Reproduced from (Sigala et al., 2002) with permission.

However training to categorize the stimuli based on a subset of their features, changed the way they perceived and represented them, even in the context of a different task (similarity ratings vs. categorization) (Fig. 1b, Fig. 2, Fig. 3). Looking for the neuronal underpinnings that supported this perceptual change (Sigala and Logothetis, 2002) found that cells in the anterior inferior temporal cortex selectively represented the values of the features that were important (diagnostic) for the categorization task, over the values of the features that were not (Fig. 4). This was clear evidence that perceptual expertise correlated with selective tuning of cells in the temporal cortex, which presumably developed over the course of training, since the stimuli were unlike anything the animals encountered in their normal environment. It was also evidence in favour of the Generalised Context Model (Nosofsky, 1986), according to which selective attention processes make the perceptual multidimensional stimulus space to shrink or expand reflecting the importance of the most relevant stimulus dimensions (Fig. 5) (Gauthier and Palmeri, 2002). It is clear that if one input the firing rates of the recorded cells in an MDS analysis, the result would resemble the solution recovered for the behavioural data after training (Fig. 1b), where the diagnostic feature values (that separate the categories) are perceived as dissimilar and end up further away from each other, while the non-diagnostic feature values (that don't separate the categories) are perceived as similar and end up close together in space. A similar finding was reported in a single-unit study of inferior temporal neurons, where the stimulus space of parametrically designed shapes was recovered both by psychophysical and neurophysiological measurements (Op de Beeck et al., 2001).

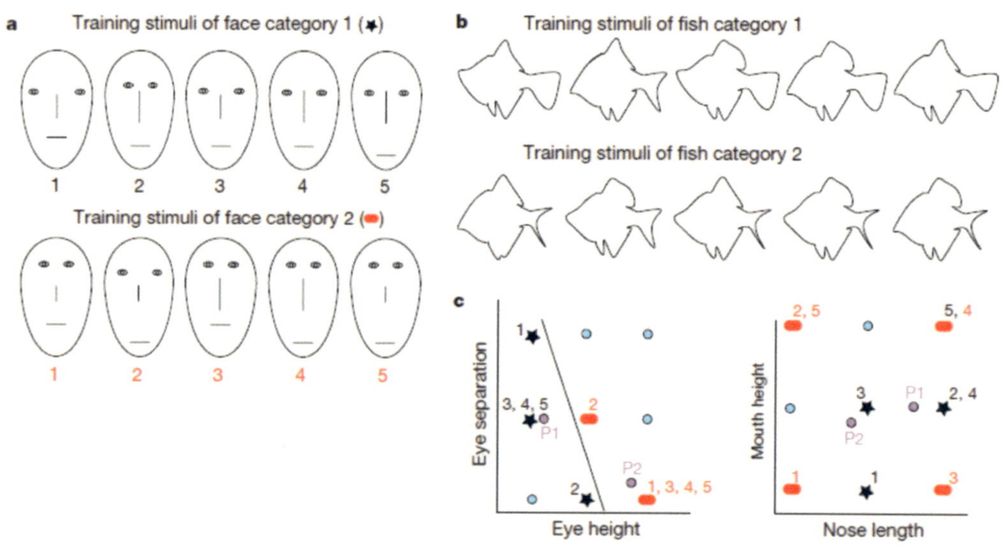

Figure 3. Stimuli and category structure used by (Sigala et al., 2002) and (Sigala and Logothetis, 2002). a. The first stimulus set consisted of line drawings of faces with four varying features: eye height, eye separation, nose length and mouth height. b, The second stimulus set consisted of fish outlines with four varying features: the shape of the dorsal fin, tail, ventral fins and mouth. In both stimulus sets, each feature could take one of three discrete values. The categories were separable along two of the four stimulus features, but information about only one of the diagnostic features was insufficient for optimal performance. The monkeys were presented with one stimulus at a time. c, Two-dimensional representation of the stimulus space. Black stars represent the stimuli of the first category and red ovals represent the stimuli of the second category. Each number indicates the position of one corresponding stimulus from a. As the stimuli differ along four dimensions, the two-dimensional representations in this figure result in overlap of stimuli that are distinct. The purple circles (P1 and P2) represent the prototypes. Cyan circles represent test exemplars that did not belong to a fixed category. The two categories were linearly separable along the eye height, eye separation (EH, ES) dimensions (solid line) but not along the nose length, mouth height (NL, MH) dimensions. Reproduced from (Sigala and Logothetis, 2002) with permission.

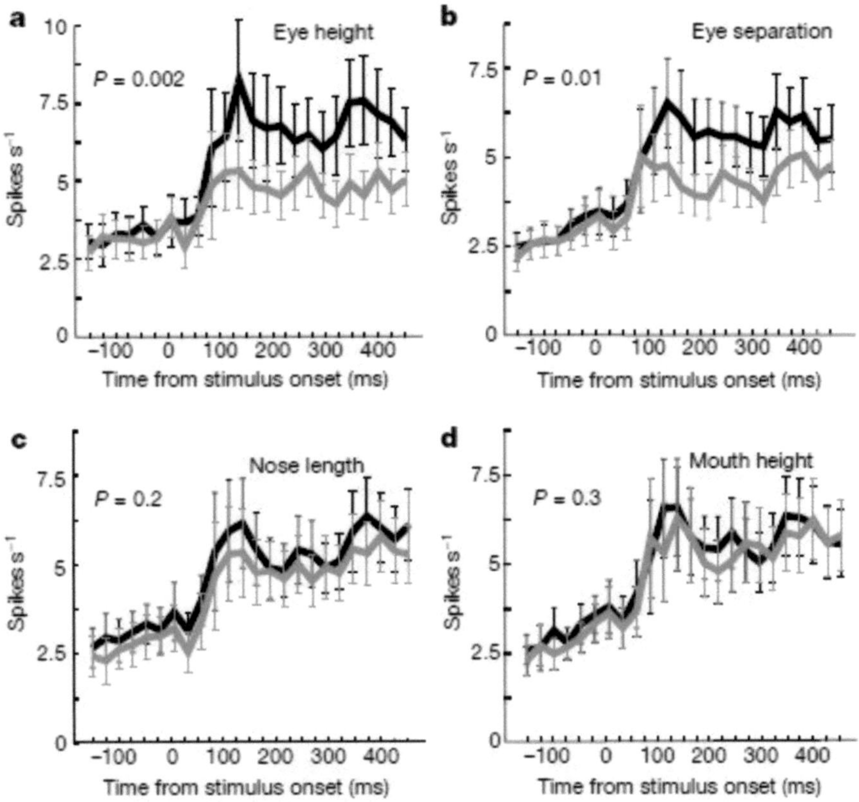

Figure 4. Population average of neuronal tuning to the features that were diagnostic for the categorization task in (Sigala and Logothetis, 2002) (reproduced with permission). The average activity is sorted by stimulus feature (eye height (a), eye separation (b), nose length (c) and mouth height (d)). Black traces indicate average responses to the best feature value; grey traces indicate average responses to the worst feature value. The bars indicate standard error of the mean. For each feature (a–d) a t-test was performed for the time window 100–475 ms after stimulus presentation, testing the hypothesis that the mean firing rate to the best feature was higher than the mean firing rate to the worst feature. The hypothesis could not be rejected in the case of the diagnostic features (eye height (a) and eye separation (b)).

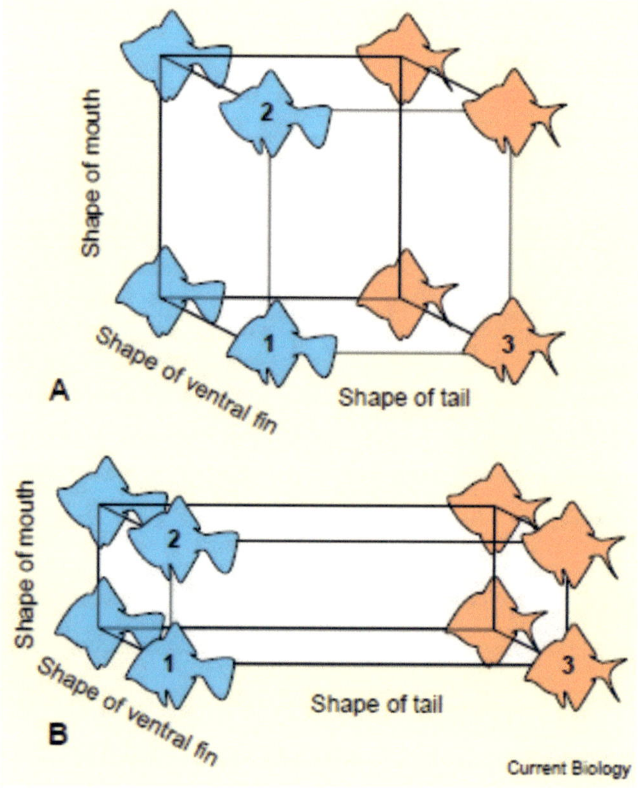

Figure 5. Stretching and compression of the stimulus space by selective attention according to the task demands. As the subjects learn to categorize fish into two groups the multidimensional space becomes relatively stretched along a diagnostic dimension (here shape of the tail) relative to the non-diagnostic dimensions. Before categorization (A), object 1 is equally similar to objects 2 and 3. During categorization (B), object 1 becomes more similar to object 2 than to object 3 through selective attention to the shape of the tail. Reproduced from (Gauthier and Palmeri, 2002) with permission.

3. The brave new era of fMRI

Until the nineties the study of the neural substrates of object recognition in the human brain relied mainly on patients with brain damage, e.g. (Farah, 1992). However behavioural and event-related potential studies of normal subjects, as well as behavioural, lesion and single-

http://dx.doi.org/10.5772/36031

unit recording studies in macaque monkeys, had provided a wealth of evidence for the functional organisation of the primate visual system, and of object recognition in particular. With the advent of functional imaging (initially PET, but mainly of fMRI) it became possible to see the normal human brain in action for the first time. Two of the very first questions people wanted to ask concerned the localisation and functional organisation of object recognition: where does recognition happen in the brain, and are there specialized cortical modules for different object categories. These were really important question to ask because a) single-unit recordings (Desimone et al., 1984) and cortical field recordings in human patients (Allison et al., 1994) did not have the global coverage necessary to show the extent of clusters of inferior temporal neurons tuned to faces or other trained object classes e.g. paperclips (Logothetis and Pauls, 1995), and to assess hemispheric laterality; b) structural models of vision and object recognition (Marr and Nishihara, 1978), (Biederman, 1987) did not suggest different representations for different types of objects, or multiple levels of recognition; but c) neuropsychological studies had provided evidence for at least three different brain modules, specific for faces, common objects and words, see (Farah, 1992) for a review.

In this chapter, I present some experimental and theoretical progress that has followed from the literature of perceptual similarity, and will review contributions of the fMRI literature.

(Sergent et al., 1992) provided the first neuroimaging evidence for category-selective responses for face stimuli, with PET imaging of normal subjects. They showed a dissociation between face processing in the right ventromedial hemisphere, and object processing in the left occipito-temporal cortex. The first study to employ fMRI to address the localisation question was by (Puce et al., 1995). They compared faces vs. scrambled faces and reported a number of bilateral activations, with the strongest in the right fusiform gyrus. But it was the study by (Kanwisher et al., 1997) (see (Kanwisher et al., 1996) for preliminary results) that thoroughly tested the response specificity of a proposed face module for the first time, and coined the acronym, the Fusiform Face Area (FFA) that has been with us for almost 15 years now (Fig. 6).

Nancy Kanwisher and her colleagues first localised the area that was significantly more active for faces than objects, then tested if the activity of the area could be explained by a number of variables: low level features; the fact that all faces belonged in the same level category, while the objects were from different categories, by comparing with houses; the fact that faces were compared with inanimate objects, by comparing with hands; attentional factors, by comparing passive fixation with a one-back task. That study was important for several reasons: a) because it showed a clear selectivity for faces over other object categories; b) it pointed to the need for different theories of computation for object and face recognition; c) brought the discussion about whether faces were special, or this specificity could be expected to develop for every class of objects that one was expert in (Diamond and Carey, 1986), in the neuroimaging field.

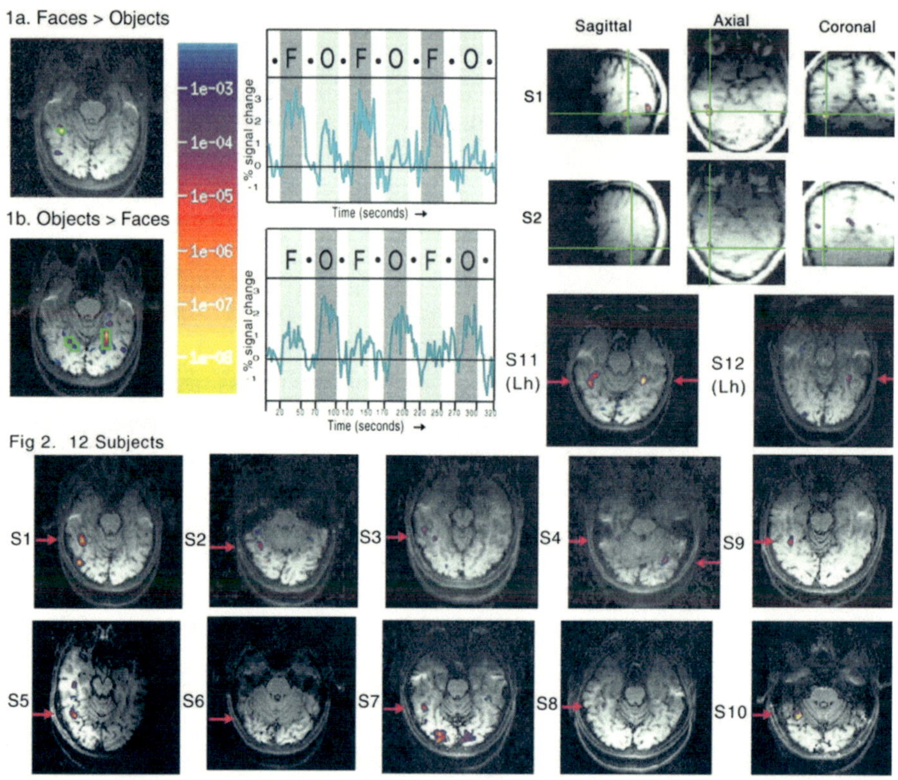

Figure 6. The Fusiform Face Area first reported in (Kanwisher et al., 1997) (reproduced with permission). Results from one subject. The right hemisphere appears on the left. The brain images at the left show in colour the voxels that produced a significantly higher MR signal intensity during the epochs containing faces than during those containing objects (1a) and vice versa (1b) for 1 of the 12 slices scanned. These significance images are overlaid on a T1-weighted anatomical image of the same slice. In each image, an ROI (Region Of Interest) is shown outlined in green, and the time course of raw percentage signal change over the 5 min 20 sec scan (based on unsmoothed data and averaged across the voxels in this ROI) is shown at the right. Epochs in which faces were presented are indicated by the vertical gray bars marked with an "F"; gray bars with an "O" indicate epochs during which assorted objects were presented; white bars indicate fixation epochs.

4. Car, bird and greeble experts

Gauthier and her colleagues had also been working on the question of functional specialization using fMRI (Gauthier et al., 1996a) (Gauthier et al., 1996b), and had preliminary evidence that FFA may also be recruited in subordinate categorizations (e.g. for the basic level "bird", the subordinate level –requiring additional perceptual processing-could be "skylark" or "blackbird") (Gauthier et al., 1997). Their fMRI work with greebles, novel objects created to provide a control set equivalent for faces (Fig. 7), indicated that the FFA may be an area involved in visual expertise in general and not just for faces (Gauthier et al., 1999). This proposal has become known as the expertise hypothesis.

http://dx.doi.org/10.5772/36031

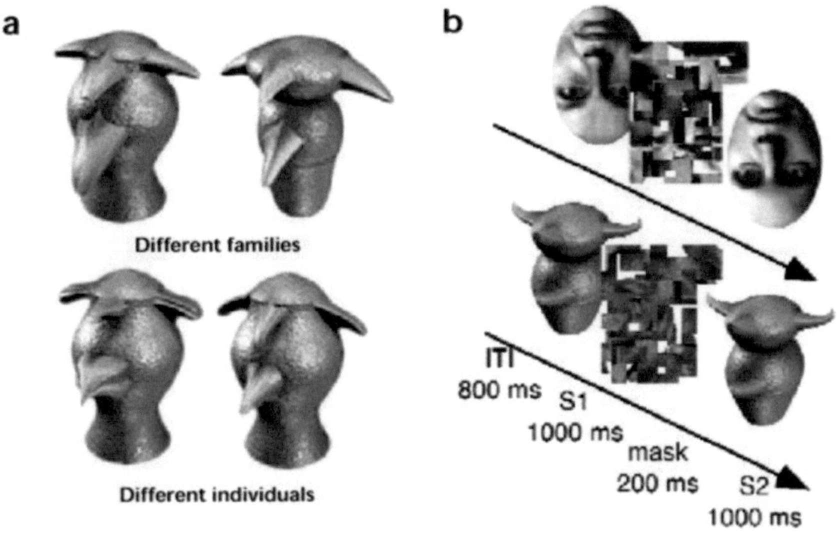

Figure 7. Meet the Greebles. Two greebles from different 'families', as defined by the shape of the large central part, as well as two individual greebles from the same family, differing only in the shape of the smaller parts. Adapted from (Gauthier et al., 1999) with permission

The greebles however have been criticised as being too similar to faces (in terms of symmetry, features, configural processing), which would make them less than ideal controls for the face specificity hypothesis of FFA (see below for neuro-imaging evidence). In a different experiment, Gauthier and her colleagues, recruited car and bird experts in order to test whether FFA was recruited during categorization of objects in their domain of expertise (Gauthier et al., 2000). The results showed that while the FFA response was still greater for faces, rather than other objects, it was nonetheless involved in processing the objects in the participants' area of expertise as well (e.g. birds for bird experts). This finding has been subsequently replicated by other laboratories (Xu, 2005), as well as in different domains of expertise (e.g. with moths and butterflies) (Rhodes et al., 2004). The BOLD signal intensity of the FFA in these studies is greater for faces than other stimuli, so there is specificity, but not exclusivity, for face stimuli, a domain of expertise that most humans share.

Another way to test the expertise hypothesis, instead of recruiting real world experts, is to effectuate expertise in the lab following intense training until a certain discrimination criterion has been reached, and/or a certain number of training hours (e.g. 10) have been completed. These paradigms, not only allow to test the responses of FFA or any other brain area, but also to study other changes as a result of training. For example (Op de Beeck et al., 2006) tested changes in cortical activations before and after 10 hours of training to discriminate novel 3D shapes at a subordinate level in a match-to-sample task. The authors used three novel shape categories, which they named "smoothies", "spikies" and "cubies" based on the type of protrusions they had (Fig. 8).

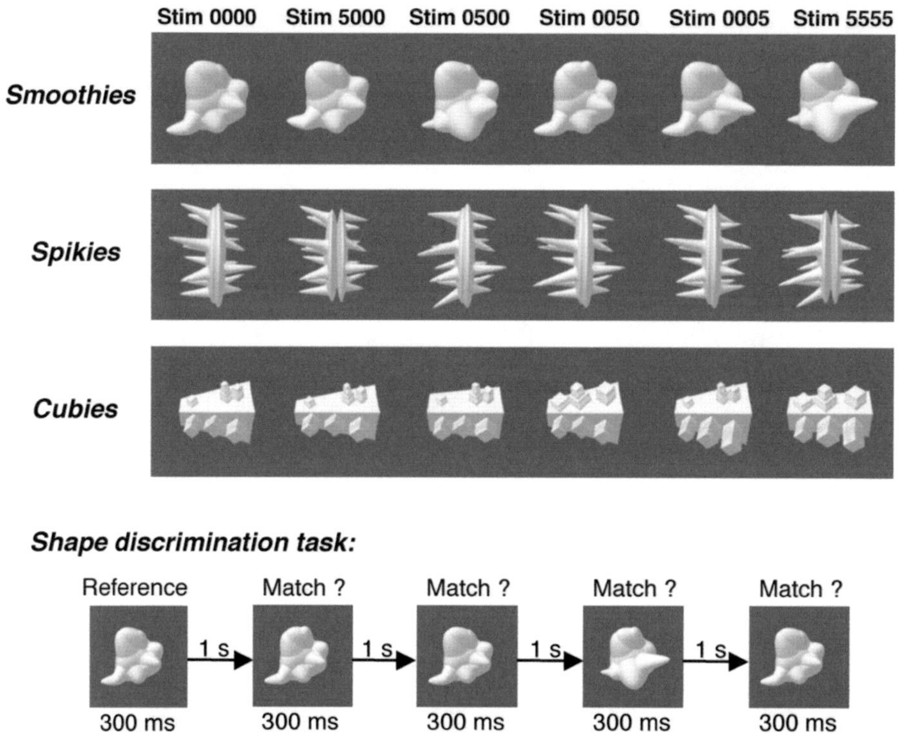

Figure 8. Stimuli and task used in (Op de Beeck et al., 2006), reproduced with permission. For each of the three classes (smoothies, spikies, and cubies), exemplars were constructed from a four-dimensional object space. Each exemplar had a value from 0 to 5 on each of four shape dimensions. The top three rows show exemplars from each class: value 0 on each dimension (far left), value 5 on one dimension and value 0 on the other dimensions (middle four exemplars), and value 5 on each dimension (far right). The bottom half of the figure shows the task used to train subjects in shape discrimination.

The approach of using artificial stimuli avoids confounds of prior, uncontrolled experience of the participants with the stimuli, and should lead to new cortical representations that are formed during the course of training. The participants were scanned before and after the training sessions, and while performing an orthogonal task to shape discrimination (colour change detection task), so any observed changes in the cortical activations could not be attributed to changes in task performance or differences of other cognitive factors e.g. attention, task difficulty etc. This study showed that the behavioural improvement in shape discrimination (one trained class for each participant) was accompanied by the following changes in cortical activations: 1) more voxels, therefore more groups of neurons, responded to the trained rather untrained stimuli in the post-training scan; 2) these voxels did not form a large cluster, but multiple small clusters in the extrastriate visual cortex (Fig. 9); 3) the Lateral Occipital Cortex (LOC) showed more activity to the trained stimuli compared to the FFA and early retinotopic visual areas, but also showed more activity to the stimuli before the training had taken place, especially in the right hemisphere; 4) the right FFA did not show increased responsiveness to the trained stimuli, except in one participant trained with "smoothies", who

http://dx.doi.org/10.5772/36031

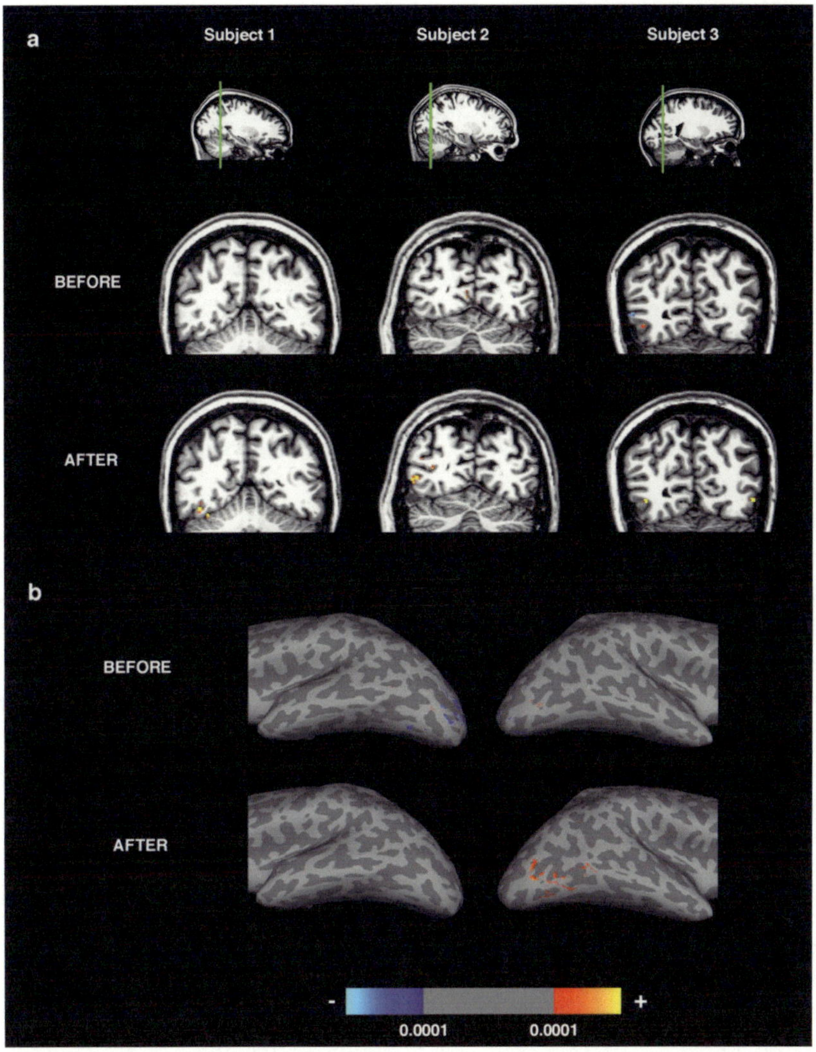

Figure 9. fMRI evidence that categorization training changes object representations in human extrastriate cortex, reproduced from (Op de Beeck et al., 2006) with permission. Activations (significance maps thresholded at $p < 0.0001$, uncorrected) are shown for the contrast [trained > untrained], with red/yellow indicating positive contrast and blue indicating negative contrast. a, Functional activation overlaid on a coronal anatomical slice for three subjects. The left, middle, and right subjects were trained with the smoothies, spikies, and cubies, respectively. These subjects were representative in the size of training effects seen across the population. Slices are shown with right hemisphere at the left. b, Functional activation overlaid on a ventrolateral view of the inflated brain of a fourth subject (trained with the smoothies).

during debriefing reported interpreting the stimuli as "women wearing hats", in other words face-like (Fig. 9).

This last observation shed light as to what could be driving the FFA responses in the studies with greebles, which are also face-like stimuli. This was further tested in a recent study by

(Brants et al., 2011) which replicated the design of the original greeble fMRI study (Gauthier et al., 1999) and extended it by scanning the participants both before and after training with the greebles. The main finding of this experiment is that even before training the participants showed an inversion effect[1] with the Greebles (that did not increase after training), indicating that these objects are interpreted as living entities with faces, and the cortical activations elicited by them may not be due to expertise.

Two more studies that contrasted the effect of mere exposure vs. categorization training (van der Linden et al., 2008) and a categorization vs. individuation task (Wong et al., 2009) with computer-generated, parametrically designed stimuli also showed increased cortical activations after training in areas near and somewhat overlapping with the right FFA.

5. The extent of the brain areas involved in categorization revealed with fMRI

Lack of increased right FFA response to trained stimuli (morphed cars) within a subordinate category was also reported by (Jiang et al., 2007) in a study of categorization training that revealed a large number of areas, including prefrontal, insular, parietal and inferior temporal cortices, as well as the thalamus that responded to the trained stimuli vs. a fixation baseline (Table 1). A similar network of cortical and subcortical areas, including the striatum, was also revealed by a study of categorization of dynamic moving patterns (Li et al., 2007), contrasted with a rotation detection task of the same stimuli. This study indicated a different role for temporal and parietal areas, and for prefrontal and striatal areas. For the former areas, the selectivity to diagnostic visual features was tuned according to task demands. Prefrontal and striatal areas are thought to provide top-down control by shaping the neural representations in the cortical areas involved in the relevant stimulus feature processing (e.g. inferior temporal cortex for shape). This study replicated and extended a number of key finding from separate single unit studies in the macaque, partly because of the global brain coverage that fMRI affords, but also thanks to careful parametric design of the stimulus space, and the sensitivity of Multi Voxel Pattern Analysis (MVPA) methods that capitalize on differences across distributed patterns of activity in the brain, e.g. (Cox and Savoy, 2003), (Norman et al., 2006) (Fig. 10).

MVPA, an approach that uses a variety of powerful pattern classification algorithms to decode the information in multi-voxel patters of activity, is particularly well suited for the study of distributed cortical representations. Although single unit studies had already pointed to the distributed nature of information coding e.g. (Young and Yamane, 1992), it is the excellent spatial coverage and non-invasive nature of fMRI that can provide information about where categories are represented in the human brain. The advantages of MVPA include: a) the

1 n fMRI experiments the inversion effect is associated with higher activity in FFA for upright than for inverted stimuli. The inversion effect is a hallmark of holistic processing (the integration of the stimulus parts into a whole representation), which may be specific to faces, as opposed to part-based processing for other types of stimuli, like animals or artifacts.

http://dx.doi.org/10.5772/36031

Region	mm^3	T	MNI Corrrdinates X	Y	Z
Stimuli > baseline					
Left Visual Cortex	26912	10.53	-44	-88	-8
		9.26	-22	-102	-2
		7.13	-26	-92	-14
Right Visual Cortex	33024	9.06	26	-105	4
		8.70	32	-100	8
		8.55	20	-100	-6
Left Inf/Sup Parietal	12280	8.25	-30	-60	50
		7.80	-38	-42	44
		7.00	-24	-68	48
R/L Sup Mot	7504	7.45	4	6	52
		7.12	-6	2	56
		6.94	-2	12	48
R Cerebelum	544	5.12	18	-64	-46
R Sup Parietal	1232	5.12	28	-66	54
		4.35	26	-52	48
L Brainstem	1584	4.93	-16	-18	12
		4.48	-14	-16	2
L Thalamus	216	4.85	-10	-26	-12
R Thalamus	728	4.78	16	-18	8
L Cerebelum	528	4.76	-16	-54	-20
L Precentral	656	4.65	-48	6	32
		3.87	-50	0	42
R Mid/Sup Frontal	408	4.65	32	-2	58
L Mid Frontal	240	4.60	-50	32	32
R Inf Parietal/Supramarginal	328	4.59	44	-36	46
L Sup/Mid Frontal	1008	4.57	-22	-8	44
		4.07	-26	-4	54
L Insula	568	4.49	-32	20	2
L Cerebum	280	4.46	-6	-28	26
		3.94	-8	-20	28
L Cerebelum	368	4.45	-16	-64	-50
R Cerebelum	480	4.43	36	-18	54
L Cerebelum	472	4.43	-8	-74	-20
L Rolandic	264	4.40	-46	-6	10
R Postcentral	208	4.34	62	-18	50
R Sup/Mid Occipital	760	4.33	28	-74	32

Table 1. Table of brain areas that are activated by the trained stimuli (cars) in (Jiang et al.,2007) (Supplementary Table 1, reproduced with permission).

sensitivity to signals that can be ignored in univariate approaches, where each voxel is treated independently and its activity modulation must reach a statistical significance threshold; b), brain activity and behaviour can be related on a trial-by-trial basis (as in single unit studies) rather than hundreds of averaged trials; and c) most importantly, it allows to characterize the structure of neural representations in individual subjects with or without the use of pattern classifiers (Norman et al., 2006). The earliest attempt to examine and relate the entire distributed pattern of voxel activations in object-related cortical areas to object categories was pioneered by Edelman and colleagues (Edelman et al., 1998), and showed promising results in recovering stimulus spaces with MDS. Pictures of natural stimuli, human and animal faces (Figure 11), and computer-generated stimuli resembling fighter planes, human bodies, sharks, four-legged animals, cars and vans (Figure 12), separated in meaningful categories based on the fMRI activity patterns. This study pioneered the multivariate analysis approach with fMRI

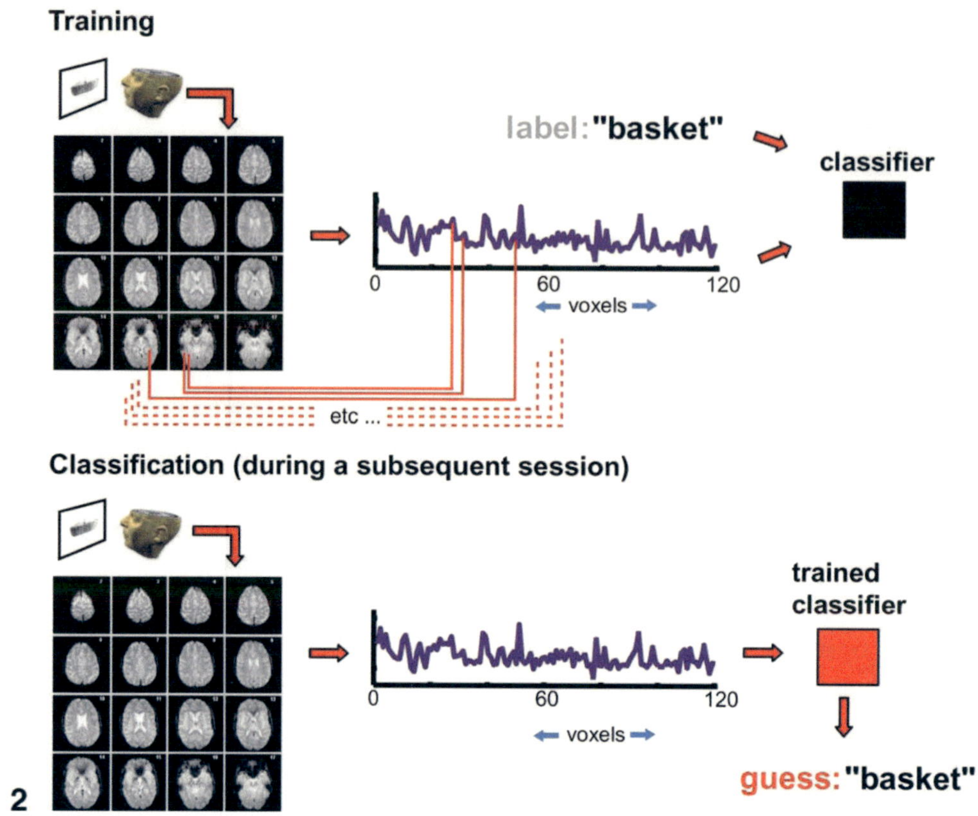

Figure 10. Basic classification scheme in MVPA (adapted from (Cox and Savoy, 2003) with permission). Each participant viewed block of images belonging to 1 of 10 categories. The pattern of activity over a previously selected subset of voxels, based on a feature selection procedure, is treated as a high-dimensional vector, shown here as a profile plot (the mapping between the voxels in the brain and points in the profile plot is symbolized by red lines). These vectors, along with labels corresponding to the category, are given to a classifier, which learns statistical regularities in the patterns, and maps between brain patterns and experimental conditions. In a subsequent session (separated in time by as much as a month), the same subject is shown the same category of objects. fMRI data are collected with the same spatial sampling, and the pattern over the same voxel subset is extracted. This pattern is given to the trained classifier that infers the category the subject was viewing, based on the decision boundary it extracted. The same steps can be followed with data collected in a single session, where half the data is used to train the classifier, and half is used in the testing phase.

data, and further suggested that both perceptual and cortical representations may be representations of similarity, which can vary according to the importance of the stimulus features and level of experience with the stimuli, e.g. (Schyns and Rodet, 1997), (Op de Beeck H, 2000), (Sigala, 2004).

The finding of distributed, rather than category-specific clusters of activation in the cortex, was later confirmed with a now seminal study by (Haxby et al., 2001) who showed that despite a small number of areas specialized for specific categories (see section: Other specific modules: Bodies, Places, Words), the representations of objects in the ventral temporal cortex are very distributed and overlapping. This means that the representation of a particular object, or object

http://dx.doi.org/10.5772/36031

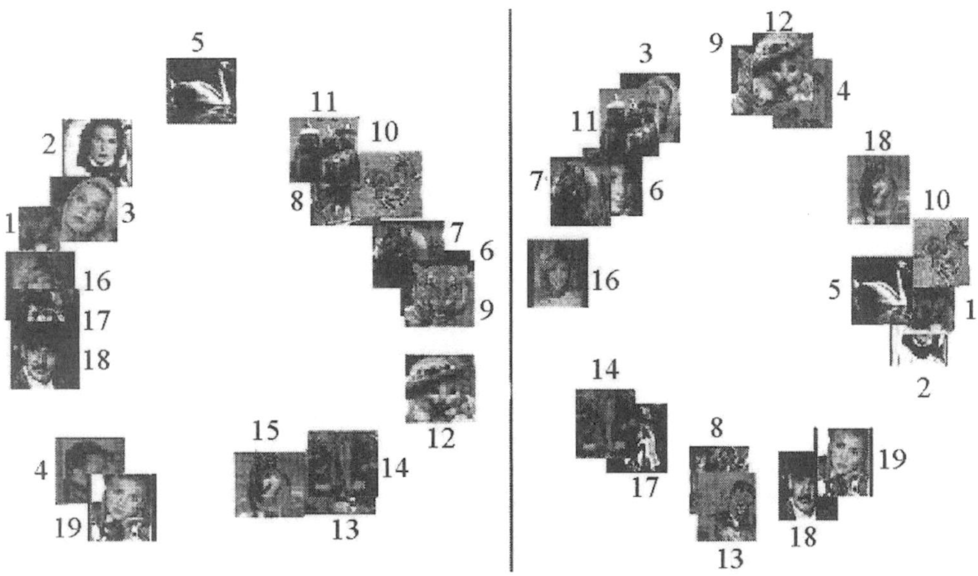

Figure 11. The left panel illustrates the stimulus space configuration, as it was derived with MDS from seven participants' most significant voxels that responded preferentially to whole objects, but not to their scrambled versions. Those voxels were bilaterally in the Lateral Occipital Complex (area LO).The number labels indicate the serial number of the picture in the epoch. The main result is the separation of faces and animals in two linearly separable clusters. The panel on the right shows the result of a bootstrap procedure, where MDS was applied to randomly permuted time courses. This result shows no clustering that corresponds to specific stimulus categories. (Figure from Edelman et al. 1998, reproduced with the permission of the Psychonomic Society.)

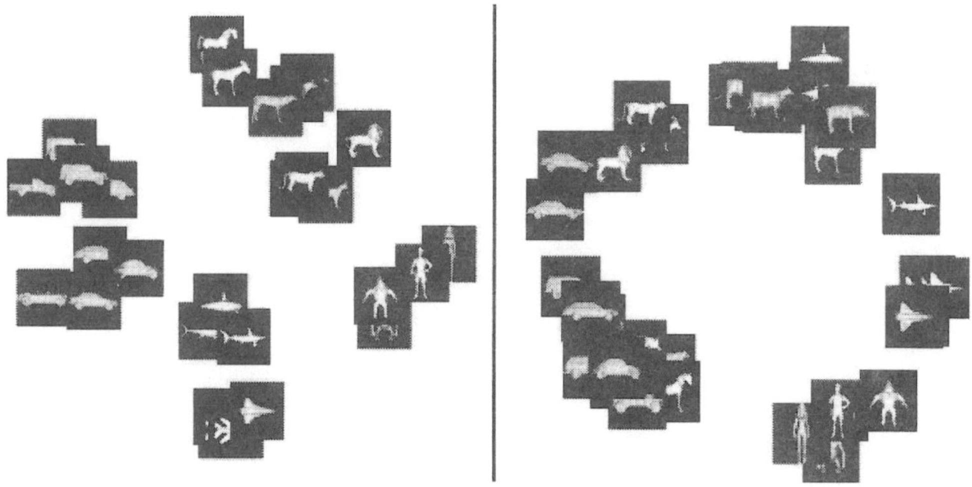

Figure 12. Comparison of the stimulus space derived with MDS applied on perceptual judgments of similarities (left panel) and fMRI activation patterns (right panel). (Figure from Edelman et al. 1998, reproduced with the permission of the Psychonomic Society.)

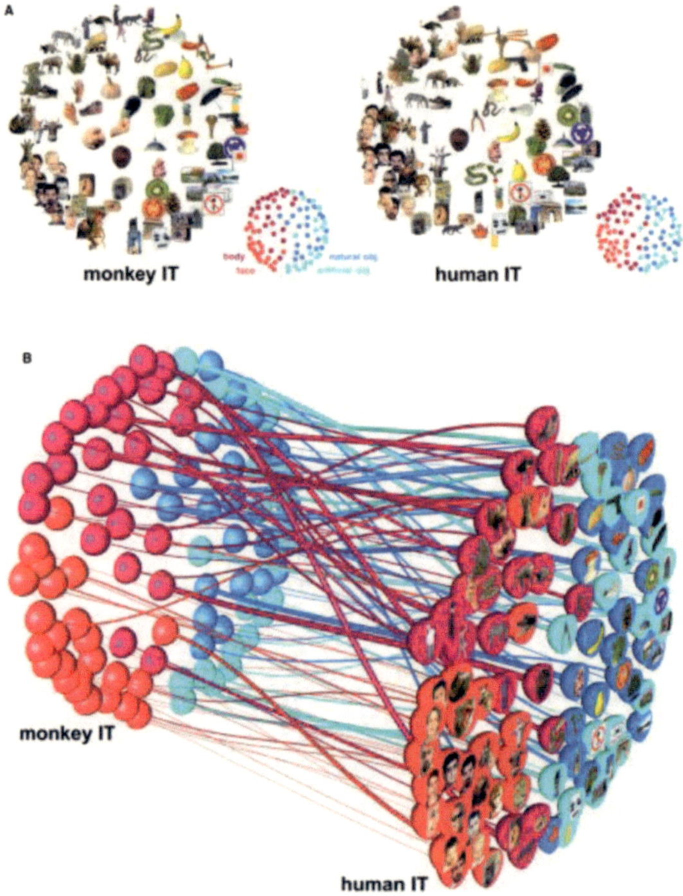

Figure 13. Similarity of representations in human and monkey Inferior Temporal Cortex (IT). Reproduced from (Kriegeskorte et al., 2008b) with permission. (A) Stimulus space reflecting the similarity of responses in IT. The stimuli have been arranged such that their pairwise distances approximately reflect response-pattern similarity (MDS, dissimilarity: 1 – Pearson r, criterion: metric stress). In each arrangement, images placed close together elicited similar response patterns. Images placed far apart elicited dissimilar response patterns. The arrangement is unsupervised: it does not presuppose any categorical structure. (B) Fiber-flow visualization emphasizing the interspecies differences. This visualization combines all the information from (A) and links each pair of dots representing a stimulus in monkey and human IT by a "fiber." The thickness of each fiber reflects to what extent the corresponding stimulus is inconsistently represented in monkey and human IT. The interspecies consistency ri of stimulus i is defined as the Pearson correlation between vectors of its 91 dissimilarities to the other stimuli in monkey and human IT. The thickness of the fiber for stimulus i is proportional to (1 – ri)2, emphasizing the most inconsistently represented stimuli.

category, is encoded in the activity pattern created by a large number of neurons, and each of these neurons takes part in instantiating multiple object representations.

In the most recent and visually compelling demonstrations of the power of combining the notion of similarity with multivariate analyses, (Kriegeskorte et al., 2008b) demonstrated the close correspondence of the activity patterns in human and non-human IT cortex, and revealed a mostly orderly representation of object super-ordinate categories (e.g. animate, inanimate),

http://dx.doi.org/10.5772/36031

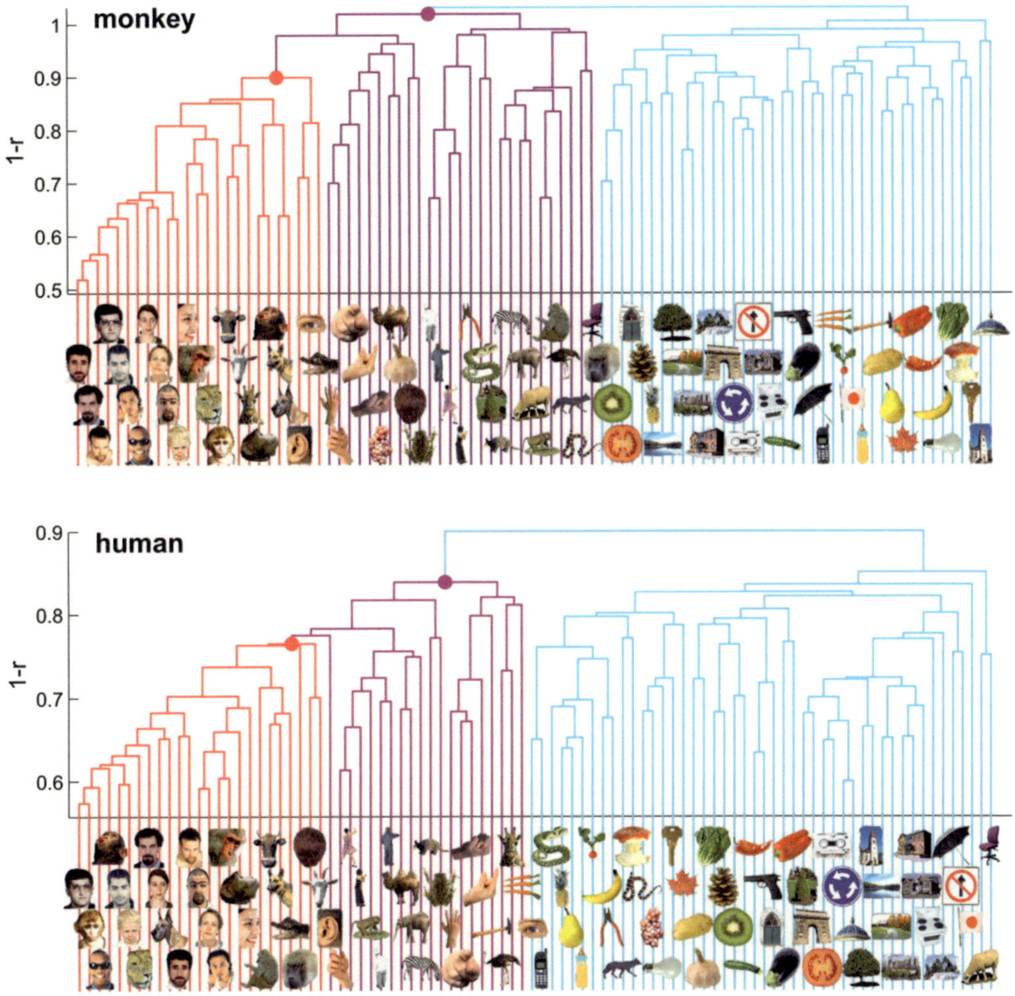

Figure 14. Hierarchical clustering of Inferior Temporal Cortex (IT) response patterns. Reproduced from (Kriegeskorte et al., 2008b) with permission. This analysis proceeds from single-image clusters (bottom of each panel) and successively combines the two clusters closest to each other in terms of the average response-pattern dissimilarity, so as to form a hierarchy of clusters (tree structure in each panel). The vertical height of each horizontal link indicates the average response-pattern dissimilarity (the clustering criterion) between the stimuli of the two linked subclusters (dissimilarity: 1 – r). The cluster trees for monkey and human are the result of completely independent experiments and analysis pipelines. This data-driven technique reveals natural-category clusters that are consistent between monkey and human. For easier comparison, subcluster trees were colored (faces, red; bodies, magenta; inanimate objects, light blue).

as well as a continuous representations of categories and exemplars (Fig.13, 14). This approach promises to help bridge the gap of relating the different approaches (computational, imaging, electrophysiological and psychophysical) and species (mainly human and non-human primates) that have figured in the field of object recognition and categorization over the decades (Kriegeskorte et al., 2008a).

6. Other selective modules: Bodies, places, words

The answer to the question "are faces special" is probably yes – the question refers to visual processing mechanisms that may be unique to faces. The case study of a boy who sustained brain damage on the first day he was born, resulting in an inability to recognise individual faces (while his object recognition is largely intact) even after acquired experience and plasticity, as he was tested at the age of 16 (Farah et al., 2000), indicates that the brain is equipped to cope with face recognition at birth and this core ability may not be restored with experience. The FFA therefore may be the result of natural selection for copying with a common and important stimulus for survival, which occupies a dense stimulus/similarity space. This is because all faces look largely similar in terms of number and configuration of features, and multiple categorizations need to be applied on them in a very short time (e.g. identity, sex, emotion). This does not mean however that experience is not important, since the fine tuning of face recognition ability also depends on experience, as seen e.g. in the other-race effect[2] (Lindsay et al., 1991), (Feng et al.).

Other modules selective for processing certain categories of stimuli that have been revealed with fMRI are the Parahippocampal Place Area (PPA), which is selective for the geometry of the local environment including pictures of houses (Epstein R, 1998), the Extrastriate Body Area (EBA) selective to human body parts (Downing et al., 2001), as well as the letter string area, which has been elegantly shown to form as a result of experience (Baker et al., 2007).

Imaging work in the non-human primate has also revealed the full extent of the representation of faces and other categories in the brain. The equivalent of the human FFA, seems to be still elusive (Ku et al., 2011), but a network of areas that respond more to faces than to other objects has been repeatedly shown with fMRI in the macaque brain including the temporal and prefrontal cortex, and the amygdala (Logothetis et al., 1999), (Tsao et al., 2003), (Hoffman et al., 2007), (Tsao et al., 2008b), (Tsao et al., 2008a), (Ku et al., 2011). The exact role of each face patch is not clear, but what we do know is that it only takes a small amount of micro-stimulation of a small group of face selective neurons to affect the perception of a macaque categorizing face vs. non-face stimuli and bias the response towards the face category (Afraz et al., 2006). A direct comparison of the reported locations of these face patches from different groups, reveals such a dense network of selective cell groups, which clearly points to a very distributed representation of faces in the non-human primate brain (Fig. 15). However, it should be noted that the majority of these face patches are also active in the anaesthetised preparation (Logothetis et al., 1999), (Ku et al., 2011), complicating the interpretation of the findings and the role of these activations in conscious perception.

---

2 he other race effect (or own-race bias): the phenomenon describes the fact that, despite our impressive ability to effortlessly recognize and remember faces, we make more errors when we remember and recognize faces from another, less familiar race (O'Toole et al., 1994).

http://dx.doi.org/10.5772/36031

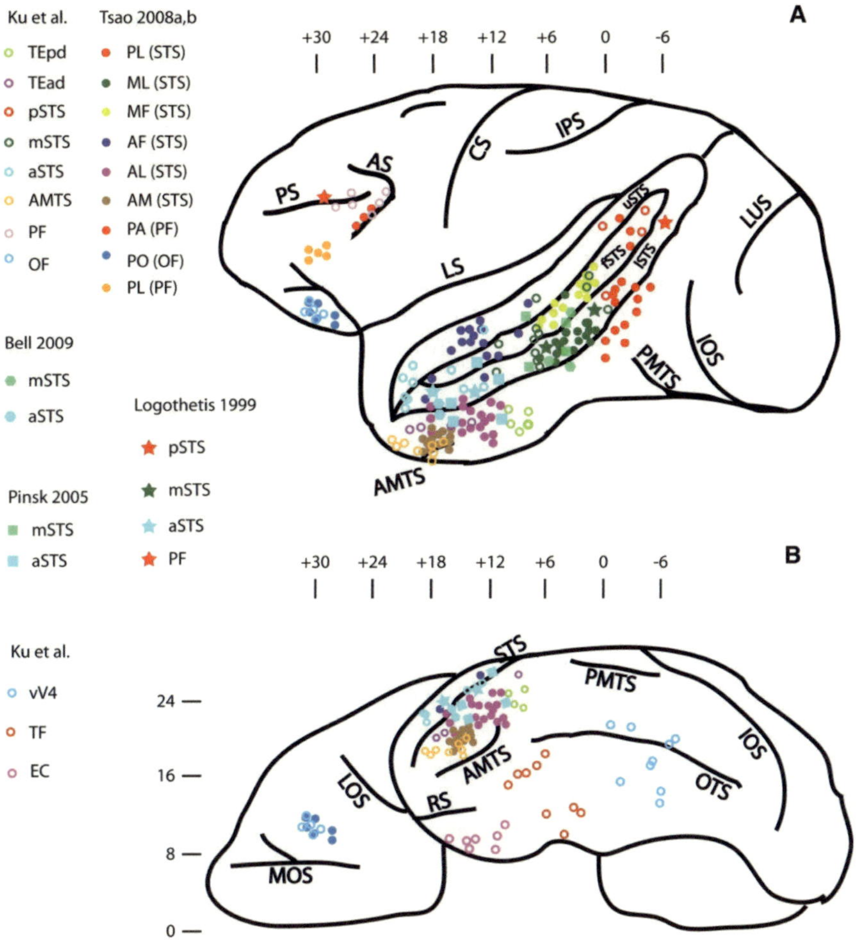

Figure 15. Comparison of Face-Selective Activation Found in the Current Study with Face-Selective Activation Described in the Literature Superimposed on a Side and Ventral View of the Brain. (A) Side view of the brain. The locations of face-selective patches found in the literature ((Bell et al., 2009), (Logothetis et al., 1999), (Pinsk et al., 2005), (Tsao et al., 2008a) and (Tsao et al., 2008b) are marked by closed symbols and locations found in the current study are indicated by the open circles. For the activated areas described by (Tsao et al., 2008a) and (Tsao et al., 2008b), naming conventions used by the authors were retained with the locations in parentheses. Locations were estimated based on AP positions when given; otherwise positions were estimated by comparing the coronal slices to the atlas by (Saleem and Logothetis, 2006). In cases where activation extended over multiple slices the average position was taken. The locations shown are after normalization to the macaque template (McLaren et al., 2009). Note that not all studies (including the present one) make a distinction between STS patches located on the lip and the fundus. (B) Ventral view of face-selective activation in this study (open circles) and the literature (closed symbols). Abbreviations: LOS: lateral orbital sulcus; MOS: medial orbital sulcus; OTS: occipitotemporal sulcus; PMTS: posterior middle temporal sulcus; RS: rhinal sulcus. Figure reproduced from (Ku et al., 2011) with permission.

7. Conclusion

It would be unfortunate if the reader was left with the impression that fMRI has only revealed locations and representations of stimuli that are compatible with certain computational and theoretical accounts of visual perception and long term memory. Other important contributions of fMRI in the literature of expertise relate to evidence for the improvement of working memory abilities in experts (Moore et al., 2006), as well as the demands of encoding strategies (e.g. chunking) that can make certain tasks easier (Bor et al., 2003), and the possibility that a frontoparietal cortical network may be a general purpose expertise-based network, e.g. (Bor and Owen, 2007).

Despite its relatively young age, the fMRI community has engaged with and made important contributions to most of the questions that had been keeping single-unit electrophysiologists busy for decades, regarding for example functional specialisation, local vs. distributed processing, hierarchical representations and the malleability of those representations. The emphasis in the field has slowly but surely shifted from where functions are taking place to how the underlying computations are achieved.

The unprecedented ability to see the whole human brain in action has reassured us about its similarities with non-human primate brains, e.g. (Orban et al., 2004), but also pinpointed differences, e.g. (Petit et al., 1997), and has revealed the extent to which the brain operates as a functional network, e.g. (Vogel et al., 2010, Van Dijk KR et al., 2010).

In the field of visual categorization and expertise, fMRI has revealed a number of specialised areas for processing biologically and culturally important categories of stimuli, like faces and letter strings. At the same time fMRI has revealed how distributed the representations of most object categories are, and how these may be organised in a hierarchical way that breaks up the complexity of the world in manageable chunks, governed by perceptually and cognitively defined similarity rules, that take into account task demands.

Author details

Natasha Sigala*

Clinical Imaging Sciences Centre, Brighton and Sussex Medical School, University of Sussex, United Kingdom

References

[1] Afraz, S. R., Kiani, R. & Esteky, H. (2006). Microstimulation of inferotemporal cortex influences face categorization. *Nature* 442, 692-695.

[2] Ahn, W. & Dennis, M. J. (2001) Dissociation between categorization and similiarity judgement: differential effect of causal status on feature weights. In: *Similarity and Categorization*, (eds.) U. Hahn & M. Ramscar, pp. 87-107. Oxford: Oxford University Press.

[3] Allison, T., Ginter, H., McCarthy, G., Nobre, A. C., Puce, A., Luby, M. & Spencer, D. D. (1994). Face recognition in human extrastriate cortex. *J Neurophysiol* 71, 821-825.

[4] Attneave, F. (1957). Transfer of experience with a class-schema to identification-learning of patterns and shapes. *Journal of experimental psychology* 54, 81-88.

[5] Baker, C. I., Liu, J., Wald, L. L., Kwong, K. K., Benner, T. & Kanwisher, N. (2007). Visual word processing and experiential origins of functional selectivity in human extrastriate cortex. *Proc Natl Acad Sci U S A* 104, 9087-9092. doi:0703300104 [pii] 10.1073/pnas.0703300104.

[6] Bell, A. H., Hadj-Bouziane, F., Frihauf, J. B., Tootell, R. B. & Ungerleider, L. G. (2009). Object representations in the temporal cortex of monkeys and humans as revealed by functional magnetic resonance imaging. *J Neurophysiol* 101, 688-700. doi:90657.2008 [pii] 10.1152/jn.90657.2008.

[7] Biederman, I. (1987). Recognition-by-Components: a theory of human image understanding. *Psychological review* 94, 115-147.

[8] Boden, M. A. (2006) *Mind as machine: a history of cognitive science*. Oxford: Oxford University Press.

[9] Bor, D., Duncan, J., Wiseman, R. & Owen, A. M. (2003). Encoding strategies dissociate prefrontal activity from working memory demand. *Neuron* 37, 361-367.

[10] Bor, D. & Owen, A. M. (2007). Cognitive training: neural correlates of expert skills. *Curr Biol* 17, R95-97. doi:S0960-9822(07)00821-4 [pii] 10.1016/j.cub.2007.01.019.

[11] Brants, M., Wagemans, J. & Op de Beeck, H. P. (2011). Activation of Fusiform Face Area by Greebles is Related to Face Similarity but Not Expertise. *Journal of Cognitive Neuroscience*. doi:doi:10.1162/jocn_a_00072.

[12] Cox, D. D. & Savoy, R. L. (2003). Functional magnetic resonance imaging (fMRI) "brain reading": detecting and classifying distributed patterns of fMRI activity in human visual cortex. *Neuroimage* 19, 261-270.

[13] Cross, R. C. (1959). Category DIfferences. *Proceedings of the Aristotelian Society* 59, 255-270.

[14] Cutzu, F. & Edelman, S. (1996). Faithful representation of similarities among three-dimensional shapes in human vision. *Proc. Natl. Acad. Sci. USA* 93, 12046-12050.

[15] Desimone, R., Albright, T. D., Gross, C. G. & Bruce, C. (1984). Stimulus-selective properties of IT neurons in the macaque. *The Journal of Neuroscience* 4, 2051-2062.

[16] Diamond, R. & Carey, S. (1986). Why faces are and are not special: an effect of expertise. *Journal of Experimental Psychology: General* 115, 107-117.

[17] Downing, P. E., Jiang, Y., Shuman, M. & Kanwisher, N. (2001). A cortical area selective for visual processing of the human body. *Science* 293, 2470-2473. doi:10.1126/science.1063414, 293/5539/2470 [pii].

[18] Edelman, S. (1998). Representation is representation of similarities. *Behavioral & Brain Sciences* 21, 449-467.

[19] Edelman, S., Grill-Spector, K., Kushnir, T. & Malach, R. (1998). Towards direct visualization of the internal shape representation space by fMRI. *Psychobiology* 26, 309-321.

[20] Epstein R, K. N. (1998). A cortical representation of the local visual environment. *Nature* 392, 598-601.

[21] Farah, M. J. (1992). Is an object an object an object? Cognitive and neuropsychological investigations of domain-specificity in visual object recognition. *Current Directions in Psychological Science* 1, 164-169.

[22] Farah, M. J., Rabinowitz, C., Quinn, G. E. & Liu, G. T. (2000). Early commitment of neural substrates for face recognition. *Cogn Neuropsychol* 17, 117-123. doi:713751844 [pii], 10.1080/026432900380526.

[23] Feng, L., Liu, J., Wang, Z., Li, J., Li, L., Ge, L., Tian, J. & Lee, K. The other face of the other-race effect: An fMRI investigation of the other-race face categorization advantage. *Neuropsychologia*. doi:S0028-3932(11)00438-6 [pii], 10.1016/j.neuropsychologia.2011.09.031.

[24] Gauthier, I., Anderson, A. W., Tarr, M. J. & Gore, J. C. (1996a) Levels of categorization in visual object recognition studied with fMRI. In: *3rd Annual Meeting of the Cognitive Neuroscience Society*. San Fransisco.

[25] Gauthier, I., Anderson, A. W., Tarr, M. J., Skudlarski, P. & Gore, J. C. (1997). Levels of categorization in visual recognition studied using functional magnetic resonance imaging. *Current Biology* 7, 645-651.

[26] Gauthier, I., Behrmann, M., Tarr, M. J., Anderson, A. W., Gore, J. C. & McClelland, J. L. (1996b) Subordinate-level categorization in human inferior temporal cortex: Converging evidence from neuropsychology and brain imaging. In: *Society for Neuroscience Abstracts*.

[27] Gauthier, I. & Palmeri, T. J. (2002). Visual Neurons: Categorization-Based Selectivity. *Current Biology* 12, 282-284.

[28] Gauthier, I., Skudlarski, P., Gore, J. C. & Anderson, A. W. (2000). Expertise for cars and birds recruits brain areas involved in face recognition. *Nat Neurosci* 3, 191-197.

[29] Gauthier, I., Tarr, M. J., Anderson, A. W., Skudlarski, P. & Gore, J. C. (1999). Activation of the middle fusiform 'face area' increases with expertise in recognizing novel objects. *Nat Neurosci* 2, 568-573. doi:10.1038/9224.

[30] Haxby, J., Gobbini, M., Furey, M., Ishai, A., Schouten, J. & Pietrini, P. (2001). Distributed and overlapping representations of faces and objects in ventral temporal cortex. *Science* 293, 2425-2430.

[31] Heider, E. R. (1972). Universals in color naming and memory. *J Exp Psychol* 93, 10-20.

[32] Hoffman, K. L., Gothard, K. M., Schmid, M. C. & Logothetis, N. K. (2007). Facial-expression and gaze-selective responses in the monkey amygdala. *Curr Biol* 17, 766-772. doi:S0960-9822(07)01144-X [pii], 10.1016/j.cub.2007.03.040.

[33] Jiang, X., Bradley, E., Rini, R. A., Zeffiro, T., Vanmeter, J. & Riesenhuber, M. (2007). Categorization training results in shape- and category-selective human neural plasticity. *Neuron* 53, 891-903. doi:S0896-6273(07)00114-6 [pii], 10.1016/j.neuron.2007.02.015.

[34] Kanwisher, N., Chun, M. M., McDermott, J. & Ledden, P. J. (1996). Functional imaging of human visual recognition. *Brain Res Cogn Brain Res* 5, 55-67.

[35] Kanwisher, N., McDermott, J. & Chun, M. M. (1997). The fusiform face area: a module in human extrastriate cortex specialized for face perception. *Journal of Neuroscience* 17, 4302-4311.

[36] Kriegeskorte, N. (2009). Relating Population-Code Representations between Man, Monkey, and Computational Models. *Front Neurosci* 3, 363-373.

[37] Kriegeskorte, N., Mur, M. & Bandettini, P. (2008a). Representational similarity analysis - connecting the branches of systems neuroscience. *Front Syst Neurosci* 2, 4. doi: 10.3389/neuro.06.004.2008.

[38] Kriegeskorte, N., Mur, M., Ruff, D. A., Kiani, R., Bodurka, J., Esteky, H., Tanaka, K. & Bandettini, P. A. (2008b). Matching categorical object representations in inferior temporal cortex of man and monkey. *Neuron* 60, 1126-1141. doi:S0896-6273(08)00943-4 [pii], 10.1016/j.neuron.2008.10.043.

[39] Ku, S. P., Tolias, A. S., Logothetis, N. K. & Goense, J. (2011). fMRI of the face-processing network in the ventral temporal lobe of awake and anesthetized macaques. *Neuron* 70, 352-362. doi:S0896-6273(11)00205-4 [pii], 10.1016/j.neuron.2011.02.048.

[40] Li, S., Ostwald, D., Giese, M. & Kourtzi, Z. (2007). Flexible coding for categorical decisions in the human brain. *J Neurosci* 27, 12321-12330.

[41] Lindsay, D. S., Jack, P. C., Jr. & Christian, M. A. (1991). Other-race face perception. *J Appl Psychol* 76, 587-589.

[42] Logothetis, N. K., Guggenberger, H., Peled, S. & Pauls, J. (1999). Functional imaging of the monkey brain. *Nat Neurosci* 2, 555-562.

[43] Logothetis, N. K. & Pauls, J. (1995). Psychophysical and physiological evidence for viewer-centered object representations in the primate. *Cereb Cortex* 5, 270-288.

[44] Marr, D. & Nishihara, H. K. (1978). Representation and recognition of the spatial organization of three-dimensional shapes. *Proc R Soc Lond B Biol Sci* 200, 269-294.

[45] McLaren, D. G., Kosmatka, K. J., Oakes, T. R., Kroenke, C. D., Kohama, S. G., Matochik, J. A., Ingram, D. K. & Johnson, S. C. (2009). A population-average MRI-based atlas collection of the rhesus macaque. *Neuroimage* 45, 52-59. doi:S1053-8119(08)01176-2 [pii], 10.1016/j.neuroimage.2008.10.058.

[46] Medin, D. L. & Schaffer, M. M. (1978). Context theory of classification learning. *Psychological Review* 85, 207-238.

[47] Moore, C. D., Cohen, M. X. & Ranganath, C. (2006). Neural mechanisms of expert skills in visual working memory. *J Neurosci* 26, 11187-11196.

[48] Murphy, G. L. & Medin, D. L. (1985). The role of theories in conceptual coherence. *Psychol Rev* 92, 289-316.

[49] Norman, K. A., Polyn, S. M., Detre, G. J. & Haxby, J. V. (2006). Beyond mind-reading: multi-voxel pattern analysis of fMRI data. *Trends Cogn Sci* 10, 424-430.

[50] Nosofsky, R. (1986). Attention, similarity, and the identification-categorization relationship. *J of experimental psychology: general* 115, 39-57.

[51] Op de Beeck H, B. E., Wagemans J, Sunaert S, Van Hecke P (2000). The representation of shape in the context of visual object categorization tasks. *Neuroimage* 12, 28-40.

[52] Op de Beeck, H., Wagemans, J. & Vogels, R. (2001). Inferotemporal neurons represent low-dimensional configurations of parameterized shapes. *Nat Neurosci* 4, 1244-1252. doi:10.1038/nn767, nn767 [pii].

[53] Op de Beeck, H. P., Baker, C. I., DiCarlo, J. J. & Kanwisher, N. G. (2006). Discrimination training alters object representations in human extrastriate cortex. *J Neurosci* 26, 13025-13036. doi:26/50/13025 [pii], 10.1523/JNEUROSCI.2481-06.2006.

[54] Orban, G. A., Van Essen, D. & Vanduffel, W. (2004). Comparative mapping of higher visual areas in monkeys and humans. *Trends Cogn Sci* 8, 315-324.

[55] O'Toole, A. J., Deffenbacher, K. A., Valentin, D. & Abdi, H. (1994). Structural aspects of face recognition and the other-race effect. *Mem Cognit* 22, 208-224.

[56] Petit, L., Clark, V. P., Ingeholm, J. & Haxby, J. V. (1997). Dissociation of saccade-related and pursuit-related activation in human frontal eye fields as revealed by fMRI. *J Neurophysiol* 77, 3386-3390.

[57] Pinsk, M. A., DeSimone, K., Moore, T., Gross, C. G. & Kastner, S. (2005). Representations of faces and body parts in macaque temporal cortex: a functional MRI study.

Proc Natl Acad Sci U S A 102, 6996-7001. doi:0502605102 [pii], 10.1073/pnas. 0502605102.

[58] Poggio, T. & Edelman, S. (1990). A network that learns to recognize three-dimensional objects. *Nature* 343, 263-266.

[59] Posner, M. I. & Keele, S. W. (1968). On the genesis of abstract ideas. *Journal of Experimental Psychology* 77, 353-363.

[60] Puce, A., Allison, T., Gore, J. C. & McCarthy, G. (1995). Face-sensitive regions in human extrastriate cortex studied by functional MRI. *J Neurophysiol* 74, 1192-1199.

[61] Rhodes, G., Byatt, G., Michie, P. T. & Puce, A. (2004). Is the fusiform face area specialized for faces, individuation, or expert individuation? *J Cogn Neurosci* 16, 189-203. doi:10.1162/089892904322984508.

[62] Richardson, M. W. (1938) Multidimensional Psychophysics. In: *46th annual meeting of the American Psychological Association*, pp. 659. Colombus, Ohio: Psychological Bulletin.

[63] Rosch, E. (1976) Classification of real-world objects: origins and representations in cognition. In: *Thinking: readings in cognitive science*, (eds.) P. N. Johnson-Laird & P. C. Wason, pp. 212-277. Cambridge University Press.

[64] Rosch, E., Mervis, C. B., Gray, W. D., Johnson, D. M. & Boyes-Braem, P. (1976a). Basic objects in natural categories. *Cognitive Psychology* 8, 382-439.

[65] Rosch, E., Simpson, C. & Miller, R. S. (1976b). Structural bases of typicality effects. *J Exp Psychol: Hum Perception Performance* 2, 491-502.

[66] Saleem, K. S. & Logothetis, N. K. (2006) A Combined MRI and Histology Atlas of the Rhesus Monkey Brain. Academic Press.

[67] Sands, S. F., Lincoln, C. E. & Wright, A. A. (1982). Pictorial similarity judgments and the organization of visual memory in the rhesus monkey. *Journal of Experimental Psychology: General* 111, 369-389.

[68] Schyns, P. G. & Rodet, L. (1997). Categorization creates functional features. *Journal of Experimental Psychology: Learning, Memory & Cognition* 23, 681-696.

[69] Sergent, J., Ohta, S. & MacDonald, B. (1992). Functional neuroanatomy of face and object processing. A positron emission tomography study. *Brain* 115 Pt 1, 15-36.

[70] Shepard, R. (1958). Stimulus and response generalisation: tests of a model relating ganeralization to distance in psychological space. *Journal of Experimental Psychology* 55, 509.

[71] Shepard, R. (1962a). The analysis of proximities: Multidimensional scaling with an unknown distance function. Part I. *Psychometrika* 27, 125-140.

[72] Shepard, R. (1962b). The analysis of proximities: Multidimensional scaling with an unknown distance function. Part II. *Psychometrika* 27, 219-246.

[73] Shepard, R. (1987). Toward a universal law of generalization for psychological science. *Science* 237, 1317-1323.

[74] Sigala, N. (2004). Visual categorization and the inferior temporal cortex. *Behav Brain Res* 149, 1-7. doi:S0166432803002249 [pii].

[75] Sigala, N. (2009). Natural images: a lingua franca for primates? *The Open Neuroscience Journal* 3, 48-51. doi:10.2174/1874082000903010048.

[76] Sigala, N., Gabbiani, F. & Logothetis, N. K. (2002). Visual categorization and object representation in monkeys and humans. *J Cogn Neurosci* 14, 187-198. doi: 10.1162/089892902317236830.

[77] Sigala, N. & Logothetis, N. K. (2002). Visual categorization shapes feature selectivity in the primate temporal cortex. *Nature* 415, 318-320. doi:10.1038/415318a, 415318a [pii].

[78] Sugihara, T., Edelman, S. & Tanaka, K. (1998). Representation of objective similarity among 3D shapes in the monkey. *Biological Cybernetics* 78, 1-7.

[79] Tanaka, J. W. & Taylor, M. (1991). Object categories and expertise: is the basic level in the eye of the beholder? *Cognitive Psychology* 23, 457-482.

[80] Torgerson, W. (1952). Multidimensional scaling: I. Theory and method. *Psychometrica* 17, 401-419.

[81] Tsao, D. Y., Freiwald, W. A., Knutsen, T. A., Mandeville, J. B. & Tootell, R. B. (2003). Faces and objects in macaque cerebral cortex. *Nat Neurosci* 6, 989-995.

[82] Tsao, D. Y., Moeller, S. & Freiwald, W. A. (2008a). Comparing face patch systems in macaques and humans. *Proc Natl Acad Sci U S A* 105, 19514-19519. doi:0809662105 [pii], 10.1073/pnas.0809662105.

[83] Tsao, D. Y., Schweers, N., Moeller, S. & Freiwald, W. A. (2008b). Patches of face-selective cortex in the macaque frontal lobe. *Nat Neurosci* 11, 877-879. doi:nn.2158 [pii], 10.1038/nn.2158.

[84] van der Linden, M., Murre, J. M. & van Turennout, M. (2008). Birds of a feather flock together: experience-driven formation of visual object categories in human ventral temporal cortex. *PLoS One* 3, e3995. doi:10.1371/journal.pone.0003995.

[85] Van Dijk KR, Hedden T, Venkataraman A, Evans KC, Lazar SW & RL., B. (2010). Intrinsic functional connectivity as a tool for human connectomics: theory, properties, and optimization. *J Neurophysiol.* 103, 297-321.

[86] Vogel, A. C., Power, J. D., Petersen, S. E. & Schlaggar, B. L. (2010). Development of the brain's functional network architecture. *Neuropsychol Rev* 20, 362-375. doi:10.1007/s11065-010-9145-7.

[87] Wong, A. C., Palmeri, T. J., Rogers, B. P., Gore, J. C. & Gauthier, I. (2009). Beyond shape: how you learn about objects affects how they are represented in visual cortex. *PLoS One* 4, e8405. doi:10.1371/journal.pone.0008405.

[88] Xu, Y. (2005). Revisiting the role of the fusiform face area in visual expertise. *Cereb Cortex* 15, 1234-1242. doi:bhi006 [pii], 10.1093/cercor/bhi006.

[89] Young, M. P. & Yamane, S. (1992). Sparse Population Coding of Faces in the Inferotemporal Cortex. Science 256, 1327.

Chapter 11

Color Specificity in the Human V4 Complex – An fMRI Repetition Suppression Study

Tessa M. van Leeuwen, Karl Magnus Petersson,
Oliver Langner, Mark Rijpkema and Peter Hagoort

Additional information is available at the end of the chapter

http://dx.doi.org/10.5772/51143

1. Introduction

The hierarchy of color processing areas in the human brain starts from cone-opponent signals in the retina. Relayed by the lateral geniculate nucleus (LGN), color information arrives at the primary visual cortex (V1) and area V2, where color contrast is computed (Conway, 2009). Later visual area V4 and 'globs' of neurons in ventral occipital cortex (immediately anterior to V4) are deemed important for color constancy and for the luminance-invariant coding of individual hues. Not only electrophysiological data from non-human primate studies (Conway, Moeller, & Tsao, 2007; Kotake, Morimoto, Okazaki, Fujita, & Tamura, 2009; Kusunoki, Moutoussis, & Zeki, 2006; Matsumora, Koida, & Komatsu, 2008; Stoughton & Conway, 2008; Zeki, 1980), but also neuroimaging and lesion studies in humans implicate the 'V4 complex' in higher order color processing (Barbur & Spang, 2008; Bartels & Zeki, 2000; Bouvier & Engel, 2006; McKeefry & Zeki, 1997). It is known however, that V4 is not a color area per se, as it is also involved in shape and form processing (e.g. Schiller, 1995).

The posterior section of the human V4 complex (posterior fusiform gyrus) is referred to as V4; the anterior section (anterior fusiform gyrus) as V4α (Bartels & Zeki, 2000; McKeefry & Zeki, 1997). Other definitions have also been proposed for V4 (e.g. V8 (Hadjikhani, Liu, Dale, Cavanagh, & Tootell, 1998), VO1 (Brewer, Liu, Wade, & Wandell, 2005)). V4 is active in passive color tasks (Howard et al., 1998; McKeefry & Zeki, 1997; Sakai et al., 1995). Both subregions contribute to color constancy operations (Barbur & Spang, 2008; Bartels & Zeki, 2000). In tasks demanding active color manipulation, it is mainly V4α that is involved (Beauchamp, Haxby, Jennings, & DeYoe, 1999; Martin, Haxby, Lalonde, Wiggs, & Ungerleider, 1995; Morita et al., 2004; Simmons et al., 2007; Zeki & Marini, 1998). Murphey, Yoshor and Beauchamp (2008) have identified color-specific neurons in V4α in one patient, using electrophysiological methods. Altogether these findings suggest a functional distinction between V4 and V4α. Here, we investigate whether V4 and V4α exhibit different color selectivity properties.

We performed a functional magnetic resonance imaging (fMRI) experiment in which we applied repetition suppression. In repetition suppression, a stimulus is presented more than once (as in behavioral priming studies (e.g. Maccotta & Buckner, 2004). The process has been referred to as fMRI-adaptation (K. Grill-Spector & Malach, 2001), mnemonic filtering (Miller, Li, & Desimone, 1993); repetition suppression (Desimone, 1996); decremental responses (Brown & Xiang, 1998) and neural priming (Maccotta & Buckner, 2004). The blood-oxygen-level dependent (BOLD) response evoked by the second and later presentations of the stimulus is (in general) reduced compared to the response evoked by the first presentation (K Grill-Spector, Henson, & Martin, 2005; Henson, 2003). These priming-related repetition suppression effects can be elicited by different types of priming-paradigms, such as masked priming (Dehaene et al., 2001; Eddy, Schnyer, Schmid, & Holcomb, 2007), visual object priming (Cavina-Pratesi, Kentridge, Heywood, & Milner, 2010), and semantic priming (Luo et al., 2004). Depending on the type of stimulus that is processed, and on the stage of stimulus processing, different parts of the brain may show repetition suppression. Because repetition suppression would only occur in brain regions that are sensitive to the specific aspect of the stimulus that is being repeated, the technique allows for better spatial location determination than conventional fMRI studies. This can be illustrated by thinking about a paradigm in which familiar colored objects are visually presented to the participants (i.e. a red ball). Individually presented, these objects will activate all relevant brain areas that are involved in shape and color processing, as well as memory and semantic brain areas. However, when using the repetition suppression technique and showing a second stimulus in which for instance only the shape of the object is repeated, only shape areas will show a reduction in the BOLD response. In this way, repetition suppression may lead to enhanced precision in determining the location of relevant brain areas involved in specific processes of interest (K Grill-Spector, et al., 2005; K. Grill-Spector et al., 1999).

Here, we investigated the repetition of color. We hypothesized that in a collection of neurons with specific color selectivity, the repetition of the *same* color would reduce the BOLD response more than the presentation of two *different* colors: if the second color differs from the first color, this would recruit different neurons in case of very specific color selectivity of the neurons. Neurons with less specific color selectivity would respond to a wider range of colors and would not strictly require one particular color as input in order to become active (Conway, et al., 2007); hence, these neurons would show (some) repetition suppression not only when the same color, but also when two different colors are presented subsequently, because they would be (weakly) active for both presentations of colors.

http://dx.doi.org/10.5772/51143

We hypothesized that the color selectivity of neurons would increase when progressing from posterior (V4) to anterior (V4α) areas along the processing hierarchy.

Several fMRI studies have previously used repetition suppression to investigate color processing. Engel et al. (2005) deployed fMRI-adaptation to show repetition suppression effects for color in populations of both oriented and unoriented color-selective cells in visual cortex. Two other fMRI studies have aimed to identify regions in the ventral-occipital cortex that are selectively involved in color processing (and not texture or form processing) (Cant, Arnott, & Goodale, 2009; Cavina-Pratesi, et al., 2010). In both studies, 3D-objects were used of which the form, texture, and color could either stay identical from one stimulus presentation to the next (repetition suppression is expected), or be changed (no repetition suppression). Both studies found that the left anterior collateral sulcus, of which the stereotaxic coordinates correspond to left V4α, was more responsive when the color of the objects changed between subsequent stimulus presentations; in case of Cavina-Pratesi et al (2010), this effect was bilatereal, including also right anterior collateral sulcus. Cant et al. (2009) found additional, similar effects in right fusiform gyrus while Cavina-Pratesi et al. (2010) also found effects in left lingual gyrus. It should be noticed that in the study by Cant et al. (2009), the above mentioned regions were not solely selective to color as they also responded to changes in texture; it were primarily medial regions of the ventral stream along the fusiform gyrus that were showing selectivity to color changes. Taken together, these studies strongly suggest that (left) V4α contains neurons that are rather specifically tuned for color such that they reduce their activity when that same color is repeated. However, these studies yielded no clear evidence that the posterior fusiform gyrus (V4) exhibits similar color specificity, independent of texture information. In our study, we explicitly test whether area V4 shows similar repetition suppression effects for color as area V4α. We propose that neurons in V4α, situated later in the processing hierarchy, show greater specificity for color than neurons in V4 and that is why area V4α has been readily identified by repetition suppression paradigms and V4 has not been.

In our experiment we used simple patches of color (squares) that were always of the same form and texture, minimizing intrusion of these dimensions in the experiment. We applied a priming paradigm to obtain BOLD repetition suppression effects for color. The color of the prime square was either the *same* color (SC) as the target square, a *different* color (DC), or an *achromatic* color (AC) containing only grey values (r=g=b). In neurons with high color selectivity, the SC condition was hypothesized to reduce the BOLD response more than the DC condition. The achromatic color condition was included as a control condition for which we expected no or very little repetition suppression. We use an active task to engage the participants; the color of the target square had to be indicated with a button press and reaction times were analyzed. Because we wanted to investigate repetition suppression effects in both hemispheres and both sections of the V4-complex, we determined *a priori* regions of interest (ROIs) in anterior and posterior parts of the V4 complex to assess the effects in all of these regions (Figure 1). The ROIs were defined on the basis of findings from Bartels and Zeki (2000) and the coordinates of V4α corresponded to the reported effects by Cavina-Pratesi et al. (2010). We predicted that only in V4α, and not more posterior V4, the BOLD suppression effects would differ between repetitions of the same versus two different colors (SC and DC conditions).

2. Material and methods

2.1. Participants

Forty subjects aged 18-38 (mean age 26.2 years, *SD*=4.7 years, 4 males, 4 left-handed) participated. In both groups, 2 left handed individuals were present, making sure that any reaction time differences found between the groups were not influenced by handedness. All participants completed a pre-screening questionnaire to assess their medical history, handedness and MRI-compatibility. All had normal or corrected to normal vision and reported no color blindness, and were able to discriminate the stimulus colors. None reported a neurological or psychiatric disease. Informed consent was obtained (prior to scanning) after explanation of the experimental procedures. The study was approved by the local ethics committee, in accordance with the declaration of Helsinki. One participant was excluded prior to analysis after reassessment of her medical history, leaving 39 participants.

As the experiment was part of a larger study on synesthesia (van Leeuwen, Petersson, & Hagoort, 2010), 21 of the subjects were grapheme-color synesthetes (Hubbard & Ramachandran, 2005). The remaining 19 subjects were matched to the synesthetes on sex, age, handedness, and educational level. Differences between the two subgroups were not found and hence both groups were collapsed for this study.

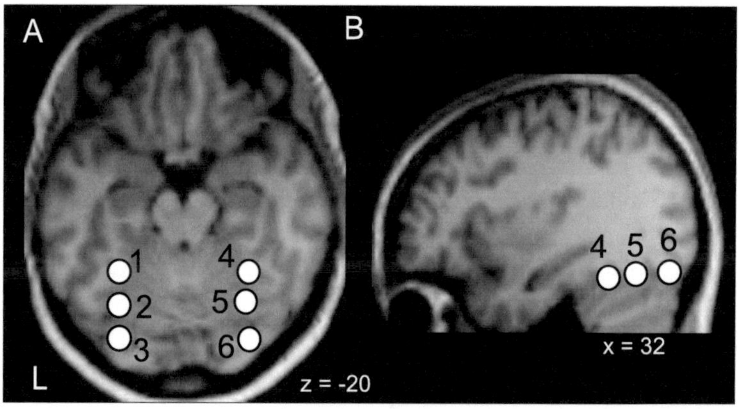

A) and B) Axial and sagittal brain slices, respectively, of template brain illustrating the position of the six ROIs in the V4 complex in the fusiform gyrus. See below for MNI coordinates (x,y,z) denoting the center of the 5 mm radius ROIs within different sections of the V4 complex. Coordinates are derived from Bartels and Zeki (2000, Table 1). 1. L V4α (-32,-50,-22); 2. L V4α/V4 (-30,-66,-22); 3. L V4 (-32,-82,-20); 4. R V4α (32,-50,-24); 5. V4α/V4 (30,-64,-22); 6. R V4 (32,-82,-20). L=left, R=right.

Figure 1. Location of regions of interest in the V4 complex.

http://dx.doi.org/10.5772/51143

2.2. Experimental design

We applied a priming paradigm to obtain BOLD repetition suppression effects for color. Both prime and target consisted of colored squares. The color of the prime was either the *same* color (SC) as the target, a *different* color (DC), or an *achromatic* color (AC). Different repetition suppression effects were predicted for each condition. In the SC trials (for instance a red square followed by another red square) we hypothesized that the repetition of the prime color in the target would lead to BOLD repetition suppression effects for the target square in neurons that are more selectively tuned for that color. The overlap between neuronal processing of the color of the prime and the color of the target would lead to fMRI adaptation effects, reducing the BOLD response (Henson, 2003). In the DC trials the color of the target differed from that of the prime (for instance a red square followed by a green square). We predicted that a repetition suppression effect in the DC condition would only occur if the different colors were (partly) processed by the same neuronal population; hence by neurons with a less specific color selectivity function (Conway, et al., 2007). In areas containing many neurons that exhibit specific color selectivity, DC trials would not lead to large repetition suppression effects; different neurons would code for the prime and target colors, inducing a relatively large BOLD response. We expected that only in V4α, and not in V4, the SC condition would lead to more BOLD suppression (and hence less activation) than the DC condition; this follows from our hypothesis that V4α contains neurons that are more specifically selective for color than neurons in V4.

We included the achromatic color condition as a control condition for which we expected no or very little repetition suppression. Here, the primes consisted of achromatic squares (grey values) while the targets were chromatic squares (for instance a grey square followed by a red square). In color sensitive regions (Beauchamp, et al., 1999; McKeefry & Zeki, 1997) and neurons (Conway, et al., 2007; Murphey, et al., 2008), the responses to achromatic stimuli are typically much weaker than responses to chromatic stimuli, although they may be higher than baseline. We therefore predicted that the responses to the achromatic primes would affect the responses to the chromatic targets relatively little; little or no repetition suppression would occur for the AC trials, resulting in a relatively strong BOLD response for the colored targets. For the SC and DC trials, we expected (some) repetition suppression to occur in all of the color-sensitive V4-complex. To assess the general responsiveness to color in the V4 complex we therefore compared the effects of the AC condition to the effects of the collapsed SC and DC conditions.

In addition to the color priming conditions, the design contained synesthetic priming conditions of which the trials were intermixed with the color priming trials. SC, DC, and AC manipulations were also present in the synesthetic priming, but the primes consisted of black graphemes (single letters, digits, or symbols) instead of squares. The synesthetic priming results will not be discussed in detail here and can be reviewed elsewhere (van Leeuwen, et al., 2010); synesthetic trials will be referred to as fillers.

During the experiment the participants indicated the color of the target square with a button press (one button for each color, hence a color decision was required): we used this active task because attention enhances the cortical responses to color (Chawla, Rees, & Friston, 1999;

Corbetta, Miezin, Dobmeyer, Shulman, & Petersen, 1991). On the basis of previous findings we expected faster reaction times for the same color trials, due to facilitating effects of the prime on target processing (DiPace, Marangolo, & Pizzamiglio, 1997; Marangolo, Dipace, & Pizzamiglio, 1993; Simon, 1988). Additionally, we expected the different color trials to cause reaction time interference (increased reaction times) due to the switch in color from prime to target; in case any facilitating processes already occurred for the prime, these would have to be overcome at the time of the response to the different target color. In the achromatic trials the achromatic prime was never informative of the target button to be pressed, because all the target colors were chromatic. We therefore expected that the reaction times for the different color trials would be slower than those for the achromatic trials due to more interference from chromatic color information from the prime.

2.3. Materials and apparatus

Color stimuli were derived from unique synesthetic hues of the synesthetic participants. The synesthetic colors were selected by the synesthetes prior to the practice session. After color selection, for each synesthete four letter and color pairs were chosen for which the letters elicited four clearly distinct and strong synesthetic colors. These synesthetically induced colors were used thoughout the experiment as primes and targets. Mainly the colors red, green, blue, and yellow were used. All colored squares had a size of 2.1º x 2.1º of visual angle and were presented using Presentation (version 10.2, *Neurobehavioral Systems Inc.*, www.neurobs.com). The background was light grey (full screen, 9.1 cd/m2). Stimuli were presented on a 44.5 x 33.5 cm display screen in the scanner tunnel, placed at a viewing distance of 60 cm (controlled by a Dell Pentium IV Windows XP computer, display mode 800 x 600 pixels at 60 Hz, projected by a EIKI X986 beamer).

All colors appeared equally often in both the SC and DC conditions, and equally often as prime and target stimuli. The mean luminance of the colored stimuli was 8.4 cd/m2 (*SD*=10.3 cd/m2). Black, dark grey, light grey, and white squares were used as achromatic primes in the AC trials and their mean luminance was 7.5 cd/m2 (*SD*=13.5 cd/m2), which did not differ significantly from that of the chromatic stimuli. Grey values (rgb) for the black, dark grey, light grey and white AC primes were (0,0,0), (80,80,80), (160,160,160), and (240,240,240), respectively. We used 4 different shades of achromatic squares because there were also 4 different colors present in the experiment. Hence also in the achromatic condition, we wanted to make use of 4 different prime stimuli to keep the same variation in the achromatic primes as in the colored primes. In other studies, for instance achromatic (=greyscale) Mondrian stimuli have been used as contrasts to colored Mondrian stimuli to map color areas (McKeefry & Zeki, 1997). Although the colored stimuli were used equally often in the SC and DC conditions (which constitutes our most important comparison), we wanted to avoid the possible confound of color luminance in our analyses. We therefore, for each trial, calculated the absolute difference in luminance between the stimuli and the background luminance. We modeled this luminance difference explicitly during analysis of the fMRI data by including it as a parametric modulation value with each trial, which could capture any effects that could solely be ascribed to variance in luminance.

http://dx.doi.org/10.5772/51143

2.4. Procedure

SC, DC, and AC trials appeared in a ratio of 1:2:1 (48:96:48), for a total of 192 trials (and 192 synaesthesia filler trials). The 1:2:1 ratio was chosen such that the expectancy of a same color trial matched the expectancy of any particular target color (25%) as closely as possible, to minimize behavioral strategy effects. The stimuli were divided into four identical runs, each containing 12 SC, 24 DC, and 12 AC trials for both the color priming and the filler trials, resulting in a total of 96 trials per run. Twenty-four null-events (20%, fixation only) were included in each run to avoid BOLD saturation. All stimuli were pseudo-randomized per run, with maximally 2 repetitions of prime type (SC, DC, or AC) and prime color, maximally 3 repetitions of the same target color, and maximally 5 repetitions of overall condition (color or filler).

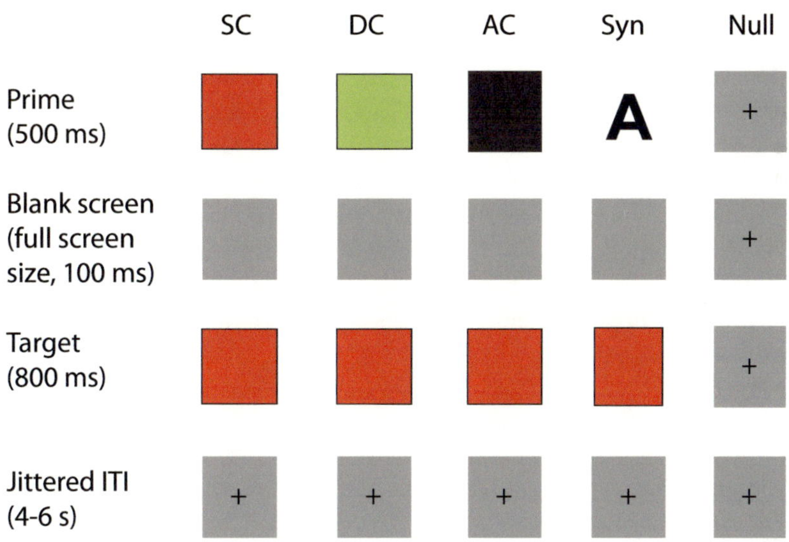

Figure 2. Experimental design. Each trial consisted of a prime stimulus followed by a blank screen, and a target stimulus (depicted from top to bottom). Prime and target stimuli were shown in the center of the screen and measured 2.1x2.1 degrees of visual angle only, surrounded by a grey background color. SC=same color, same color is repeated in prime and target (e.g. red (prime and target)); DC=different color, two different colors are used for prime and target (e.g. red for prime and green for target or green for prime and blue for target); AC=achromatic color, prime is always achromatic grey value (r=g=b) and target is in color. Syn, example of synaesthesia (filler) condition in which a letter was used as a prime; Null, depiction of null-event during which the fixation cross remained on the screen. In the synaesthesia (filler) condition, the letter could evoke synaesthesia in the target color (SCSyn), synaesthesia in a different color than the target (DCSyn) or evoke no synaesthesia (ACSyn); hence the same manipulations applied as in the color condition. The synaesthesia results are discussed elsewhere (van Leeuwen, et al., 2010). For number of trials and ratios see Section 2.4 Procedures.

Each trial consisted of a prime that was displayed for 500 ms, followed by a blank screen for 100 ms (light grey background color), and finally a target for 800 ms. During the jittered inter-trial-interval of 4-6 seconds a fixation cross was displayed. Participants were instructed to indicate the color of the target squares fast but accurately by pressing one of four response buttons with the associated finger of their right hand. Each experimental color was assigned to one of the response buttons.

First, each participant completed 16 practice trials outside of the scanner containing exemplars of all conditions. Response devices were a normal keyboard for the practice session and an MR-scanner compatible Lumitouch response box for the fMRI session. After the practice trials the fMRI session began, starting with a 5 minute structural scan to familiarize participants with the scanner noise. Next, the participants completed the first two runs of the priming experiment (12 minutes each, 380 images), and were then allowed to take a break for 10 minutes outside of the scanner. The last two priming runs followed after the break. During scanning participants wore sound-attenuating headphones to protect their hearing from the scanner noise.

2.5. Image acquisition parameters

MR data were acquired with a 3.0 Tesla Siemens TrioTim MR scanner and an 8-channel head array (Invivo). A single shot gradient echo-planar imaging (EPI) sequence was used to acquire functional MR images (33 slices, TE=30 msec, TR=2090 msec, flip angle=80º, 224 mm FOV, 64 x 64 matrix, 3.5 x 3.5 mm voxel size, 3.0 mm slice thickness,.5 mm slice gap). Atlas-based registration (AutoAlign, Siemens (Van der Kouwe et al., 2005)) was applied for all EPI runs to ensure the same slice positions across all functional runs of one subject (also across the break). A high-resolution T1-weighted structural image was acquired for each subject (MPRAGE, TE=2.96 msec, TR=2300 msec, 256 mm FOV, 256 x 256 matrix, 1 mm3 resolution, acquisition time 5 minutes, accelerated with factor 2 by GRAPPA parallel imaging (Griswold et al., 2002)).

2.6. Data analysis

2.6.1. Behavioral data

The reaction time data were analyzed with a repeated measures ANOVA. Only correct trials were included. Reaction times that were more than two standard deviations away from the subject and condition mean were considered to be outliers and removed from the analysis. Mauchly's test of sphericity was applied (as standardly implemented in SPSS) and in cases of non-sphericity, a Greenhouse-Geisser correction was used to adjust the degrees of freedom (uncorrected degrees of freedom are reported).

2.6.2. Imaging data

MR data were preprocessed and analyzed with SPM5 (Wellcome Department of Imaging Neuroscience, www.fil.ion.ucl.ac.uk/spm/software/spm5). Prior to analysis, the first five volumes of each subject were discarded to avoid transient T1 effects. To correct for head motion, the functional images of each subject were spatially realigned to the first image using a six parameter [translation in x, y, and z directions; rotations pitch, roll, and jaw] rigid body transformation for each image. Slice timing correction was applied and the images were normalized to the standard EPI template of SPM5 to allow for group inference. During normalization the images were resampled to a 2x2x2 mm resolution. Finally, all images were spatially filtered using a 10 mm FWHM isotropic Gaussian filter, as has been previously applied by others (van Leeuwen, et al., 2010; Van Wingen et al., 2008; Weiss, Zilles, & Fink, 2005; Wildgruber et al., 2005).

Statistical analyses were based on the General Linear Model (GLM) framework. For each subject, the design matrix was constructed and the BOLD signal was modeled by the canonical hemodynamic response function (HRF). A high-pass filter (128 sec cut-off) was used to remove low-frequency effects, and global scaling was applied to remove various global effects of no interest. The design matrix consisted of regressors modeling each of the six experimental conditions (SC, DC, and AC conditions for color and fillers) and one parametric modulation regressor for each of the six experimental regressors to model the luminance difference with the background. All events were modeled by the onset of the target squares. The six realignment parameters that were obtained during preprocessing were included in the model as covariates of no interest. Parameter estimates were obtained for each condition and each participant to generate relevant contrast images and allow for second-level random effects analysis. Coordinates are reported in MNI space (x, y, z).

2.7. Region of Interest analyses

To assess repetition suppression effects in anterior and posterior parts of the V4 complex, we determined regions of interest (ROIs) on the basis of a review by Bartels and Zeki (2000). In Table 1 of the review, the minimum and maximum extent (in x, y, and z directions) of V4 and V4α activations are listed. We determined two (left and right hemisphere) ROIs in the anterior part of V4α (y=-50), and two ROIs in the posterior part of V4 (y=-80), at intermediate x and z coordinate positions. We added two ROIs at the V4α/V4 border (y=-65) to complete our survey of the V4 complex. The center coordinates of each ROI were converted to MNI space (with tal2mni, derived from http://imaging.mrc-cbu.cam.ac.uk/imaging/MniTalairach) and ROIs (spheres with a 5 mm radius (10 mm diameter)) were subsequently created using MarsBaR (Brett, Anton, Valabregue, & Poline, 2002). Figure 1 shows the ROI positions on a template brain and the ROI center coordinates are listed in the figure legend.

To look at repetition suppression effects for color in the different ROIs, we extracted the mean values of the GLM parameter estimates for each priming condition (SC, DC, and AC) for each subject and for each selected ROI using MarsBaR (Brett, et al., 2002). First, we confirmed overall color sensitivity in the V4 complex (in the left and right hemisphere) by comparing the BOLD effects for the SC and DC conditions with the AC condition. We then tested the specific hypothesis that the SC trials would lead to less activation (due to more repetition suppression) than DC trials in V4α, but not in V4; leading to an interaction effect. The mean parameter estimates of the ROIs were therefore subjected to a repeated measures ANOVA with the within-subjects factors region (V4α and V4) and prime type (SC and DC), separately for each hemisphere.

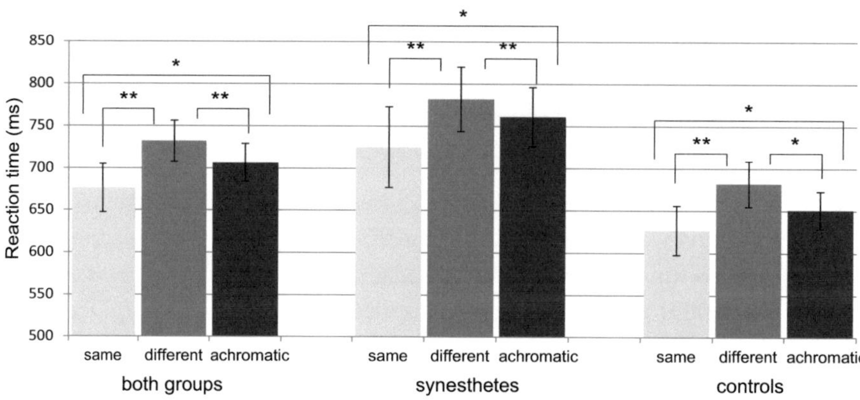

Figure 3. Behavioral effects of color priming. Mean reaction times in milliseconds for the same color (light grey bars), different color (dark grey bars) and achromatic color (near black bars) conditions. Error bars denote ± standard error of the mean. * (p<.05) and ** (p<.001) denote significant differences in reaction times of the repeated measures ANOVA with factor prime type (SC, DC, and AC), for both groups collapsed (N=38) depicted on the left. For illustrative purposes, the results per group are also depicted (synesthetes (N=19) in the middle and for controls (N=19) on the right. Here, * (p<.05) and ** (p<.001) denote the significance of the results of separate ANOVAs for each group.

3. Results

We excluded one participant from the analyses, due to an overall high error rate of 27.1% on the behavioral task (the average error rate of the remaining participants (N=38) was 3.9% (*SD*=4.3%)).

3.1. Behavioral data

We expected to see an effect of prime type (SC, DC, and AC) on the reaction times (RTs). Incorrect responses (for percentages see Figure 3) and outliers (5.8% total) were discarded. The average RTs are summarized in Figure 3, separated for the two subgroups of participants. Before looking at the within-subjects effects on the reaction times and error rates of the group of 38 participants, we verified there were no effects of subgroup (synesthetic and non-synesthetic). A main effect of subgroup was present in the reaction times ($F(1,36)=4.825$, $p<.05$); synesthetes showed longer reaction times overall. This was presumably caused by an increase in overall task difficulty for synesthetes due to the inclusion of the synesthetic priming condition. There was no subgroup x prime type interaction, indicating that the priming effects did not differ across subgroups. Therefore the RT data of all 38 participants were combined for further analyses and an ANOVA with only the factor prime type was run, including all 38 participants. For the error rates, no main effect of subgroup and no subgroup x prime type interaction were found. The average error rates across all participants were 4.3%, 4.6%, and 2.6% for the same, different, and achromatic color condition, respectively. A repeated measures ANOVA of the RTs with the within-subjects factor prime type revealed a significant effect of prime type ($F(2,74)=12.688$, $p<.001$). Planned comparisons showed that the RTs of the same color condition were much faster than the RTs of the different color condition ($F(1,37)=20.228$, $p<.001$, 56 msec faster) and also faster than the achromatic condition RTs ($F(1,37)=5.436$, $p<.05$, 30 msec faster). The RTs of the different color condition were also significantly slower than the RTs of the achromatic trials ($F(1,37)=13.959$, $p<.001$, 25 msec slower). The results are in line with previous findings on color priming (Marangolo, et al., 1993; Simon, 1988) and show that a repetition of the prime color in the target yields an RT advantage.

To check for speed-accuracy trade-offs, we calculated a repeated measures ANOVA with the within-subjects factor prime type for the error rates. We found a significant effect of prime type ($F(2,74)=6.829$, $p<.05$). Planned comparisons revealed an effect only for the AC condition in comparison to the SC ($F(1,37)=8.090$, $p<.01$) and DC conditions ($F(1,37)=13.228$, $p<.001$); there were less errors in the AC condition. SC and DC conditions did not differ in error rate ($F(1,37)=.306$, n.s.). The results indicate there was no speed-accuracy trade-off between the SC and the DC conditions, as there were no more errors in the SC condition.

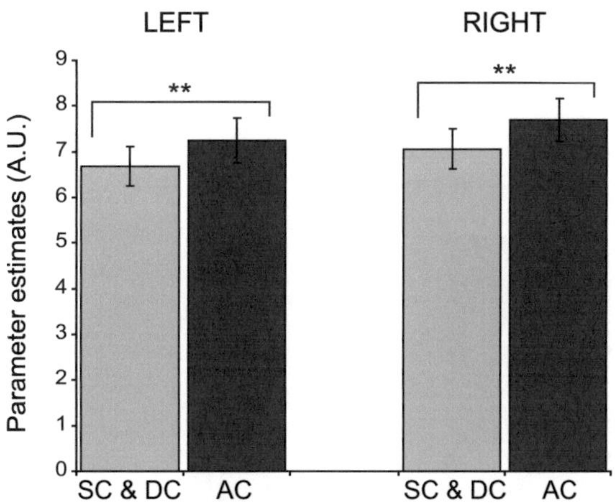

The difference between the AC trials (achromatic primes, no repetition suppression expected) and the collapsed SC and DC trials (colored primes, repetition suppression expected) in left and right hemisphere. A significant difference (**=p<.001) implies color sensitivity of the area. Mean parameter estimates (N=38) collapsed across the 3 ROIs are plotted. Error bars depict ± standard error of the mean.

Figure 4. Color sensitivity in the left and right V4 complex.

3.2. fMRI data

3.2.1. Region of Interest analyses

To look at repetition suppression effects for color in the different ROIs, we extracted the mean values of the parameter estimates for each priming condition (SC, DC, and AC) for each subject and for each selected ROI. First, we confirmed that color-induced repetition suppression was indeed occurring for the SC and DC conditions in the left and right V4 complex. We averaged the BOLD effects of these conditions for each participant and compared them to the effects of the AC condition (collapsed across the 3 ROIs within each hemisphere). In both the left and right hemisphere, there was a significantly reduced BOLD response for the averaged SC-DC conditions compared to the AC condition (left: $F(1,113)=13.634$, $p<.001$; right: $F(1,113)=22.563$, $p<.001$), which is illustrated in Figure 5.

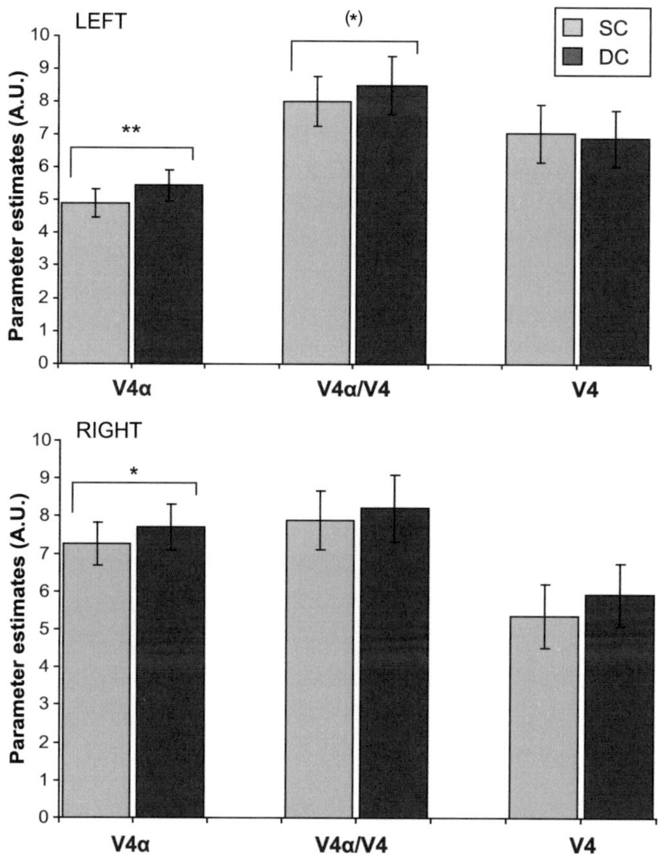

Mean parameter estimates (N=38) for each ROI are plotted for the SC and DC conditions. Error bars depict ± standard error of the mean. * (p<.05) and ** (p<.001) denote significant effects, (*) denotes a marginal effect.

Figure 5. Specificity of color selectivity in the V4 complex.

Next, in a repeated measures ANOVA with the factors region (V4α and V4) and prime type (SC and DC), we investigated a region by prime type interaction effect. In the left hemisphere, we found a significant interaction ($F(1,74)=4.123$, $p<.05$). There was no significant interaction in the right hemisphere ($F(1,74)=.089$, n.s.). Because we were interested in potential differences between the color selectivity properties of V4α and V4, we analyzed the ROIs in V4α and V4 separately. In left V4α, the SC condition showed a significantly lower BOLD effect than the DC condition ($F(1,37)=19.952$, $p<.001$), while in left V4, there was no effect of prime type ($F(1,37)=.256$, n.s.). The same pattern was found in the right hemisphere: right V4α

($F(1,37)=6.778$, $p<.05$), and right V4 ($F(1,37)=2.260$, n.s.). The mean BOLD effects for the SC and DC conditions of each ROI are summarized in Figure 5. A marginal difference between the SC and DC conditions is present at the V4α/V4 border in the left hemisphere ($F(1,37)=3.540$, $p<.068$). On the right, there are no effects at the V4α/V4 border ($F(1,37)=1.872$, n.s.).

3.2.2. *Whole brain analyses*

To illustrate the distribution of brain regions that showed an adaptation effect for color, the whole brain analysis of the contrast of AC trials compared to the averaged SC and DC conditions (overall effect of color-induced repetition suppression) is summarized in Figure 6A and Table 1. It can be seen that color-induced repetition suppression occurs in the bilateral V4 complex (fusiform gyrus, Figure 6A), as well as in the left cuneus and frontal regions. The effects in fusiform gyrus comprise V4α as well as more posterior V4. The comparison of the DC versus SC condition is also included (Figure 6B). This contrast was masked with brain regions that showed an overall adaptation effect for color, to exclude effects in areas that did not show an overall color adaptation effect. A significant effect in left fusiform gyrus was found, corresponding to the coordinates of area V4α, in line with previous repetition suppression studies (Cant, et al., 2009; Cavina-Pratesi, et al., 2010). In Figure 6C, the comparison of DC versus SC is shown for two representative individual subjects highlighting the anatomical position of the effects in fusiform gyrus. There were no significant whole brain effects of luminance.

Brain region	k	p	MNI coordinates (x, y, z)	T-value
AC > (SC & DC)				
L fusiform gyrus (BA 37/19)	1687	.000	-32, -50, -20	7.66
R fusiform gyrus (BA 37/19)	1805	.000	30, -52, -20	6.02
L cuneus/middle occ. gyrus (BA 18/19)	825	.001	-24, -82, 6	4.97
R cingulate gyrus (BA 31)	1170	.000	20, -40, 32	4.83
L medial frontal gyrus (BA 10)	312	.048	-14, 52, 8	4.53
L rolandic operculum	904	.000	-38, -16, 24	4.50
R medial frontal gyrus (BA 11)	495	.009	12, 56, -12	4.45
DC > SC*				
L fusiform gyrus (BA 37)	188	.017	-32, -48, -22	4.19
			-28, -56, -18	3.88
			-24, -66, -14	3.67

fMRI whole brain results of color priming (N=38, random effects analyses, whole brain threshold p<.001uncorrected, extent threshold 200 voxels). Cluster size (k), corrected P-values at cluster-level (p), MNI coordinates of local maxima and T-values are listed. Brodmann areas (BA) are in parentheses. R=right, L=left. *corrected for search volume (at whole brain threshold p<.001unc, extent threshold 100 voxels) with overall adaptation effect of AC > (SC & DC), to exclude effects in areas that did not show an overall color adaptation effect.

Table 1. Whole brain results of color priming

4. Discussion

Literature on color processing in human ventral occipital cortex suggests that a functional division exists between more anterior (V4α) and posterior (V4) sections of the human V4 complex. In this study, we show a functional segregation of V4α and V4 on the basis of the color adaptation properties of the neurons within the two sections. Our data suggest that in V4α, neurons are tuned for a smaller range of colors compared to neurons in V4. Although previous studies already reported repetition suppression for color in areas corresponding to left V4α (Cant, et al., 2009; Cavina-Pratesi, et al., 2010) which is similar to own our whole brain results, our region of interest approach made it possible to additionally assess repetition suppression in the right hemisphere and in the posterior V4 complex. We were able to show that repetition suppression also occurs in right hemisphere V4α but not in more posterior sections of the V4 complex. A stronger specificity for color in V4α is in line with the findings of Murphey et al. (2008), who identified color-selective neurons in V4α by means of electrophysiological recordings in a patient suffering from epilepsy. Electrical stimulation of the neurons that surrounded an inserted electrode led to the conscious perception of the same color for which these neurons were selective, suggesting rather color specific color selectivity in this area. Our results additionally demonstrate that this color specificity does not occur to the same degree in V4, an area where Murphey et al. (2008) did not record. Although our paradigm is not suitable to determine whether neurons in V4α are actually responsive to only *one* particular color, the localized effect that we see in V4α provides evidence for a functional subdivision between V4 and V4α.

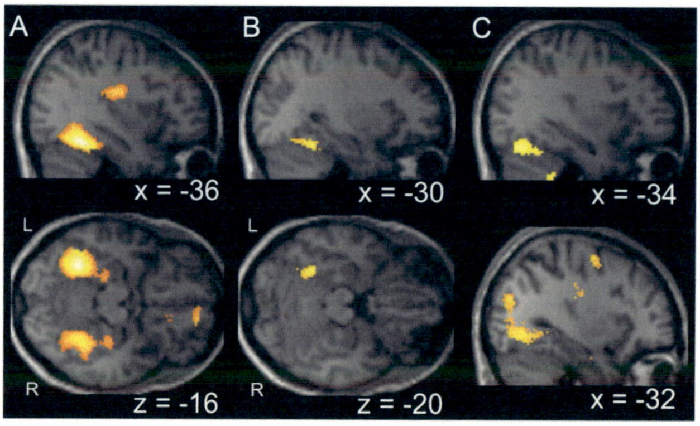

A) Sagittal (top) and axial (bottom) anatomical brain slices (template brain) with the group effects (N=38) of the contrast of the achromatic condition versus the averaged different color and same color conditions (AC > (DC & SC)). The position of the effects in left and right fusiform gyrus is shown (see Table 1). Random effects analyses, whole brain threshold p<.001unc, extent threshold 200 voxels. L=left, R=right. B) Sagittal (top) and axial (bottom) brain slices (template brain) with the group effect (N=38) of stronger adaptation for the same color condition than for the different color condition (DC > SC, also see Table 1). Random effects analyses, whole brain threshold p<.001unc, extent threshold 100 voxels, corrected for search volume with the contrast in A. C) Sagittal brain slices of two single participants with the same contrast as in B, showing individual activation patterns in left fusiform gyrus. Whole brain threshold p<.05unc, extent threshold 50 voxels.

Figure 6. Whole brain results of color priming.

Many studies report activity in area V4α when active color manipulations and tasks are involved, like color sequencing (Beauchamp, et al., 1999), object color (Martin, et al., 1995; Zeki & Marini, 1998) and color imagery (Howard, et al., 1998). A direct link between color awareness, perceptual knowledge about color, and specific color-selective neurons in the brain has been proposed (Barsalou, Simmons, Barbey, & Wilson, 2003; Goldberg, Perfetti, & Schneider, 2006; Simmons, et al., 2007). The findings of Murphey et al. (2008) in V4α strongly support this idea. Our data and set-up do not allow us to conclude that V4α is involved in color awareness or knowledge about color, but our results do suggest that the more narrow selectivity of color neurons in V4α may lie at the basis for the importance of this area in active color manipulations. Selective signaling for color is a requirement for the specific assignment of a color to an object or for making a decision about color identity; this process may take place in V4α.

It is unlikely that a lack of detection power caused the absence of color-specific BOLD reduction effects in area V4. First of all, the anatomical locations of all our ROIs were well within the extent of locations of visual areas as listed in Bartels and Zeki (2000). Also, data from a separate color viewing experiment in the same subjects indicate strong activation of overlapping fusiform areas (van Leeuwen, et al., 2010). It is therefore unlikely that a suboptimal location of the ROIs caused the absence of any effects. Furthermore, it was clear that V4 was engaged during our task. In the left hemisphere, there was no prime type effect in V4, but the overall BOLD response to the SC and DC conditions in V4 was actually higher than in V4α ($F(1,75)=12.775$, $p<.001$).

Importantly, the areas in which our ROIs were located were responsive to color overall (contrast of AC condition compared to average DC and SC conditions), as shown in Figure 4 and in the whole brain results (Figure 5 and Table 1). Our whole brain results show effects in bilateral fusiform gyrus which correspond very well to previous findings in the literature for the preference of chromatic over achromatic stimuli in ventral-occipital regions (e.g. Bartels & Zeki, 2000; Beauchamp, et al., 1999; Howard, et al., 1998; McKeefry & Zeki, 1997; Mullen, Dumoulin, McMahon, de Zubicaray, & Hess, 2007; Sakai, et al., 1995). In principle, we could have expected overall adaptation effects for color (AC > (DC & SC) in more visual regions than only V4 and middle occipital gyrus. However, our stimuli were specifically designed to induce effects in brain regions involved in higher-order color processing. The stimuli were rather small, and contained neither specific orientation, nor motion, nor texture, and the adaptation period was very brief; primary visual areas may not have been stimulated enough by our stimuli to show adaptation effects. Even though many visual regions are generally responsive to color, the above mentioned brain regions actually *prefer* chromatic stimuli over achromatic stimuli (e.g. Bartels & Zeki, 2000; Mullen, et al., 2007), which may explain why we find adaptation effects exactly in those regions.

It is also important to note that the pattern of brain activity in V4α was different from the pattern of behavioral priming in the reaction times. It is not the case that the repetition suppression effects in V4α are merely reflecting the behavioral task demands; the reaction times showed a clear interference effect for the different color trials (Figure 3), while the brain activity in bilateral V4α was strongest for the achromatic condition, as expected for the repetition suppression effect (see Figure 4 for the collapsed data). Also, when we included reaction times into our model as an additional regressor, the pattern of brain activity in visual areas did not change (data not shown). This evidence indicates that the effects of the reaction times were independent from the perceptual effects of stimulus repetition. Horner and Henson (2008) researched the effects of response learning and stimulus repetition in the brain, and found that perceptual repetition suppression in posterior brain areas is independent of the task that the subjects were performing. Our data support this finding.

In future studies, the challenge will be to learn more about the spatial organization of the human anterior color center. In macaque, Conway et al. (2007) have shown that color-biased cells in the inferior temporal cortex, anterior to area V4, are organized in 'globs' of luminance-invariant cells, and are alternated with 'interglob' regions that contain non-luminance-invariant neurons. Perhaps neurons in V4α are also organized in such a pattern. Kotake et al. (2009) and Conway and Tsao (2009) have demonstrated that in macaque area V4, neurons are spatially organized by color preference; a similar arrangement may be present in humans. A suggestion for future research would be to apply a repetition suppression paradigm in which the difference between the prime and target color was varied gradually, to investigate the spatial arrangement of perceptually similar colors. Another promising method are fMRI studies in which multi-voxel pattern analysis is applied (Haynes & Rees, 2006). With this method, Parkes, Marsman, Oxley, Goulermas and Wuerger (2009) have identified patches of cells with the same color preference in primary visual cortex of humans, and Seymour, Clifford, Logothetis and Bartels (2009) have shown that classification of voxels based solely on color preference is possible in all early visual areas except V5/MT+. Likewise, Brouwer and Heeger (2009) have successfully classified stimulus color in V1, V2, V3, V4, and VO1 (V4α is sometimes referred to as VO1 (Brewer, et al., 2005)). Voxels in fMRI contain a very large amount of neurons, which is why the classification results do not exclude a patch-like clustering of color responsive neurons in V4α. Importantly, Brouwer and Heeger (2009) also found that principal component scores (reflecting variation in the responses to different colors across voxels) in areas V4 and VO1 showed a progression through perceptual color space, which was not found for other color-responsive areas like V1. The results suggest a transformation in color processing from V1 to a more perceptual color representation in V4 and VO1, implying a spatial distribution of colors according to a perceptual similarity matrix.

5. Conclusion

To summarize, our results suggest that a functional division can be made between visual areas V4 and V4α within the human V4 complex, with more specific color selectivity taking place in V4α. This functional division may underlie the role of V4α in tasks that require active manipulation of color. Future studies will be able to reveal more details about the structural and functional organization of the human V4 complex; multi-voxel pattern analysis, high resolution (f)MRI, refined repetition suppression techniques, and studies combining color processing and behavior will be useful tools to advance our understanding of human color perception.

Acknowledgements

This work was supported by the Volkswagen-Foundation [grant number I/80 743].

Author details

Tessa M. van Leeuwen[1], Karl Magnus Petersson[1,2], Oliver Langner[1,3], Mark Rijpkema[1] and Peter Hagoort[1,2]

1 Donders Institute for Brain, Cognition and Behaviour, Radboud University Nijmegen, The Netherlands

2 Max Planck Institute for Psycholinguistics, Nijmegen, The Netherlands

3 Behavioural Science Institute, Radboud University Nijmegen, The Netherlands

References

[1] Barbur, J. L., & Spang, K. (2008). Colour constancy and conscious perception of changes of illuminant. *Neuropsychologia,* 46(3), 853-863.

[2] Barsalou, L. W., Simmons, W. K., Barbey, A. K., & Wilson, C. D. (2003). Grounding conceptual knowledge in modality-specific systems. *Trends in Cognitive Sciences,* 7(2), 84-91.

[3] Bartels, A., & Zeki, S. (2000). The architecture of the colour centre in the human visual brain: new results and a review. *European Journal of Neuroscience,* 12(1), 172-190.

[4] Beauchamp, M. S., Haxby, J. V., Jennings, J. E., & DeYoe, E. A. (1999). An fMRI version of the Farnsworth-Munsell 100-Hue test reveals multiple color-selective areas in human ventral occipitotemporal cortex. *Cerebral Cortex,* 9(3), 257-263.

[5] Bouvier, S. E., & Engel, S. A. (2006). Behavioral deficits and cortical damage loci in cerebral achromatopsia. *Cerebral Cortex,* 16(2), 183-191.

[6] Brett, M., Anton, J.-L., Valabregue, R., & Poline, J.-B. (2002). Region of interest analysis using an SPM toolbox [Abstract]. *8th International Conference on Functional Mapping of the Human Brain, Sendai, Japan.*

[7] Brewer, A. A., Liu, J., Wade, A. R., & Wandell, B. A. (2005). Visual field maps and stimulus selectivity in human ventral occipital cortex. [10.1038/nn1507]. *Nature Neuroscience,* 8(8), 1102-1109.

[8] Brouwer, G. J., & Heeger, D. J. (2009). Decoding and Reconstructing Color from Responses in Human Visual Cortex. *Journal of Neuroscience,* 29(44), 13992-14003.

[9] Brown, M. W., & Xiang, J. Z. (1998). Recognition memory: Neuronal substrates of the judgement of prior occurrence. [Review]. *Progress in Neurobiology,* 55(2), 149-189.

[10] Cant, J., Arnott, S., & Goodale, M. (2009). fMR-adaptation reveals separate processing regions for the perception of form and texture in the human ventral stream. *Experimental Brain Research,* 192(3), 391-405-405.

[11] Cavina-Pratesi, C., Kentridge, R. W., Heywood, C. A., & Milner, A. D. (2010). Separate Processing of Texture and Form in the Ventral Stream: Evidence from fMRI and Visual Agnosia. *Cerebral Cortex,* 20(2), 433-446.

[12] Chawla, D., Rees, G., & Friston, K. J. (1999). The physiological basis of attentional modulation in extrastriate visual areas. *Nature Neuroscience,* 2(7), 671-676.

[13] Conway, B. R. (2009). Color vision, cones, and color-coding in the cortex. *Neuroscientist,* 15(3), 274-290.

[14] Conway, B. R., Moeller, S., & Tsao, D. Y. (2007). Specialized color modules in macaque extrastriate cortex. *Neuron,* 56(3), 560-573.

[15] Conway, B. R., & Tsao, D. Y. (2009). Color-tuned neurons are spatially clustered according to color preference within alert macaque posterior inferior temporal cortex. *Proceedings of the National Academy of Sciences of the United States of America,* 106(42), 18034-18039.

[16] Corbetta, M., Miezin, F., Dobmeyer, S., Shulman, G., & Petersen, S. (1991). Selective and divided attention during visual discriminations of shape, color, and speed: functional anatomy by positron emission tomography. *Journal of Neuroscience,* 11(8), 2383-2402.

[17] Dehaene, S., Naccache, L., Cohen, L., Le Bihan, D., Mangin, J. F., Poline, J. B., et al. (2001). Cerebral mechanisms of word masking and unconscious repetition priming. [Article]. *Nature Neuroscience,* 4(7), 752-758.

[18] Desimone, R. (1996). Neural mechanisms for visual memory and their role in attention. [Article; Proceedings Paper]. *Proceedings of the National Academy of Sciences of the United States of America,* 93(24), 13494-13499.

[19] DiPace, E., Marangolo, P., & Pizzamiglio, L. (1997). Response bias in color priming. *Acta Psychologica,* 95(1), 3-14.

[20] Eddy, M. D., Schnyer, D., Schmid, A., & Holcomb, P. J. (2007). Spatial dynamics of masked picture repetition effects. [Article]. *NeuroImage,* 34(4), 1723-1732.

[21] Engel, S. A. (2005). Adaptation of oriented and unoriented color-selective neurons in human visual areas. *Neuron,* 45(4), 613-623.

[22] Goldberg, R. F., Perfetti, C. A., & Schneider, W. (2006). Perceptual knowledge retrieval activates sensory brain regions. *Journal of Neuroscience,* 26(18), 4917-4921.

[23] Grill-Spector, K., Henson, R. N. A., & Martin, A. (2005). Repetition and the brain: neural models of stimulus-specific effects. *Trends in Cognitive Sciences,* 10(1), 15-23.

[24] Grill-Spector, K., Kushnir, T., Edelman, S., Avidan, G., Itzchak, Y., & Malach, R. (1999). Differential processing of objects under various viewing conditions in the human lateral occipital complex. [Article]. *Neuron,* 24(1), 187-203.

[25] Grill-Spector, K., & Malach, R. (2001). fMR-adaptation: a tool for studying the functional properties of human cortical neurons. [Article]. *Acta Psychologica,* 107(1-3), 293-321.

[26] Griswold, M. A., Jakob, P. M., Heidemann, R. M., Nittka, M., Jellus, V., Wang, J., et al. (2002). Generalized autocalibrating partially parallel acquisitions (GRAPPA). *Magnetic Resonance in Medicine,* 47(6), 1202-1210.

[27] Hadjikhani, N., Liu, A. K., Dale, A. M., Cavanagh, P., & Tootell, R. B. H. (1998). Retinotopy and color sensitivity in human visual cortical area V8. [10.1038/681]. *Nature Neuroscience,* 1(3), 235-241.

[28] Haynes, J.-D., & Rees, G. (2006). Decoding mental states from brain activity in humans. *Nature Reviews Neuroscience,* 7, 523-534.

[29] Henson, R. N. A. (2003). Neuroimaging studies of priming. *Progress in Neurobiology,* 70, 53-81.

[30] Horner, A. J., & Henson, R. N. (2008). Priming, response learning and repetition suppression. *Neuropsychologia,* 46(7), 1979-1991.

[31] Howard, R. J., ffytche, D. H., Barnes, J., McKeefry, D., Ha, Y., Woodruff, P. W., et al. (1998). The functional anatomy of imagining and perceiving colour. *Neuroreport,* 9(6), 1019-1023.

[32] Hubbard, E. M., & Ramachandran, V. S. (2005). Neurocognitive mechanisms of synesthesia. *Neuron,* 48, 509–520.

[33] Kotake, Y., Morimoto, H., Okazaki, Y., Fujita, I., & Tamura, H. (2009). Organization of color-selective neurons in macaque visual area V4. *Journal of Neurophysiology,* 102(1), 15-27.

[34] Kusunoki, M., Moutoussis, K., & Zeki, S. (2006). Effect of background colors on the tuning of color-selective cells in monkey area V4. *Journal of Neurophysiology,* 95(5), 3047-3059.

[35] Luo, Q., Peng, D. L., Jin, Z., Xu, D., Xiao, L. H., & Ding, G. S. (2004). Emotional valence of words modulates the subliminal repetition priming effect in the left fusiform gyrus: an event-related fMRI study. [Article]. *NeuroImage,* 21(1), 414-421.

[36] Maccotta, L., & Buckner, R. L. (2004). Evidence for neural effects of repetition that directly correlate with behavioral priming. [Article]. *Journal of Cognitive Neuroscience,* 16(9), 1625-1632.

[37] Marangolo, P., Dipace, E., & Pizzamiglio, L. (1993). Priming effect in a color discrimination task. *Perceptual and Motor Skills,* 77(1), 259-269.

[38] Martin, A., Haxby, J. V., Lalonde, F. M., Wiggs, C. L., & Ungerleider, L. G. (1995). Discrete cortical regions associated with knowledge of color and knowledge of action. *Science,* 270(5233), 102-105.

[39] Matsumora, T., Koida, K., & Komatsu, H. (2008). Relationship between color discrimination and neural responses in the inferior temporal cortex of the monkey. *Journal of Neurophysiology,* 100(6), 3361-3374.

[40] McKeefry, D., & Zeki, S. (1997). The position and topography of the human colour centre as revealed by functional magnetic resonance imaging. *Brain,* 120(12), 2229-2242.

[41] Miller, E., Li, L., & Desimone, R. (1993). Activity of neurons in anterior inferior temporal cortex during a short- term memory task. *The Journal of Neuroscience,* 13(4), 1460-1478.

[42] Morita, T., Kochiyama, T., Okada, T., Yonekura, Y., Matsumura, M., & Sadato, N. (2004). The neural substrates of conscious color perception demonstrated using fMRI. *Neuroimage,* 21(4), 1665-1673.

[43] Mullen, K. T., Dumoulin, S. O., McMahon, K. L., de Zubicaray, G. I., & Hess, R. F. (2007). Selectivity of human retinotopic visual cortex to S-cone-opponent, L/M-cone-opponent and achromatic stimulation. [Article]. *European Journal of Neuroscience,* 25(2), 491-502.

[44] Murphey, D. K., Yoshor, D., & Beauchamp, M. S. (2008). Perception matches selectivity in the human anterior color center. *Current Biology,* 18(3), 216-220.

[45] Parkes, L. M., Marsman, J. B. C., Oxley, D. C., Goulermas, J. Y., & Wuerger, S. M. (2009). Multivoxel fMRI analysis of color tuning in human primary visual cortex. *Journal of Vision,* 9(1), 13.

[46] Sakai, K., Watanabe, E., Onodera, Y., Uchida, I., Kato, H., Yamamoto, E., et al. (1995). Functional mapping of the human color-center with echo-planar magnetic-resonance-imaging. *Proceedings of the Royal Society B-Biological Sciences,* 261(1360), 89-98.

[47] Schiller, P. H. (1995). Effect of lesions in visual cortical area V4 on the recognition of transformed objects. [10.1038/376342a0]. *Nature,* 376(6538), 342-344.

[48] Seymour, K., Clifford, C. W. G., Logothetis, N. K., & Bartels, A. (2009). The coding of color, motion, and their conjunction in the human visual cortex. *Current Biology,* 19(3), 177-183.

[49] Simmons, W. K., Ramjee, V., Beauchamp, M. S., McRae, K., Martin, A., & Barsalou, L. W. (2007). A common neural substrate for perceiving and knowing about color. *Neuropsychologia,* 45(12), 2802-2810.

[50] Simon, J. R. (1988). A priming effect in a choice reaction-time task. *Acta Psychologica,* 69(1), 45-60.

[51] Stoughton, C. M., & Conway, B. R. (2008). Neural basis for unique hues. *Current Biology,* 18(16), R698-R699.

[52] Van der Kouwe, A. J., Benner, T., Fischl, B., Schmitt, F., Salat, D. H., Harder, M., et al. (2005). On-line automatic slice positioning for brain MR imaging. *NeuroImage,* 27(1), 222-230.

[53] van Leeuwen, T. M., Petersson, K. M., & Hagoort, P. (2010). Synaesthetic Colour in the Brain: Beyond Colour Areas. A Functional Magnetic Resonance Imaging Study of Synaesthetes and Matched Controls. *PLoS ONE,* 5(8), e12074.

[54] Van Wingen, G. A., Van Broekhoven, F., Verkes, R. J., Petersson, K. M., Backstrom, T., Buitelaar, J. K., et al. (2008). Progesterone selectively increases amygdala reactivity in women. [Article]. *Molecular Psychiatry,* 13(3), 325-333.

[55] Weiss, P. H., Zilles, K., & Fink, G. R. (2005). When visual perception causes feeling: enhanced cross-modal processing in grapheme-color synesthesia. *NeuroImage,* 28(4), 859-868.

[56] Wildgruber, D., Riecker, A., Hertrich, I., Erb, M., Grodd, W., Ethofer, T., et al. (2005). Identification of emotional intonation evaluated by fMRI. *NeuroImage,* 24, 1233-1241.

[57] Zeki, S. (1980). The representation of colors in the cerebral cortex. *Nature,* 284(5755), 412-418.

[58] Zeki, S., & Marini, L. (1998). Three cortical stages of colour processing in the human brain. Brain, 121, 1669-1685.

Chapter 12

Developmental Plasticity: FMRI Investigations into Human Visual Cortex

Alyssa A. Brewer and Brian Barton

Additional information is available at the end of the chapter

http://dx.doi.org/10.5772/56340

1. Introduction

The ultimate goal of many fields of neuroscience research is to harness the ability for the human brain to reorganize, as an understanding of how to induce plasticity in cortex could foster the development of treatments of such devastating conditions as paralysis, neurodegenerative disease and stroke. The specifics of the timing and types of reorganization possible in cortical sensorimotor maps have generated tremendous interest, both in the adult and juvenile brain. While it is clear that cortical representations in human are able to undergo significant reorganization during development (an early critical period of life) (e.g., [1-4]), the extent of reorganization possible in the developing human brain is still not fully known. In addition, there are likely differences in the timing and degree of reorganization possible across the different sensory modalities [5-11]. This chapter will focus on examples of plasticity in the visual cortex of the developing human arising from congenital disorders, with specific attention to the use of functional magnetic resonance imaging (fMRI) in in vivo measurements of these cortical changes.

1.1. The definition of plasticity

The phenomenon we refer to as 'structural plasticity' will be defined as the long-term reorganization of human visual cortex due to a structural change such as the growth of new axons [5] (Table 1). This type of reorganization is generally expected to be permanent and to take weeks or months to accomplish [12]. 'Structural plasticity' is to be differentiated from 'adaptation', which we use here to refer to rapid changes in visual processing arising from the normal organization of the visual system and not a result of long-term reorganization (see [5] for a detailed discussion of terms).

Phenomena of adaptation are expected to be immediate and completely reversible. Furthermore, these two terms are to be separated from 'functional plasticity', which we will use to describe longer-term reorganization of human visual cortex, primarily due to re-weighting or unmasking of extant neural connections rather than to more extensive structural changes like the growth of new axons. Functional plasticity is expected to take days to weeks to accomplish, and it is still unclear how permanent it can be. At one point or another, each of these terms has been described using the term 'plasticity', and we would like to put an end to this non-specificity, as it leads to misunderstandings regarding each phenomenon.

1.2. Overview of developmental plasticity in the visual cortex of animal models

Many aspects of the development of the visual system have been extensively studied in animal models such as monkey and cat. The mammalian visual system develops in a balance between intrinsic, genetically-guided factors and activity-based experience [13-17]. The ability of the brain to wire and rewire itself based on visual input is known as experience-dependent plasticity (a form of structural plasticity) and plays a key role in determining the final organization of adult visual cortex [1, 12, 15, 18]. The following descriptions of visual cortex development in animal models can serve here as examples of what processes may underlie the visual cortex reorganization described in this chapter for the developing human.

In human and monkey, visual information travels from the retina through the lateral geniculate nucleus (LGN) of the thalamus to primary visual cortex (area V1) in the posterior occipital lobe [19]. In the adult primate, the inputs from each eye are organized into separate groups within V1 called ocular dominance (OD) columns [20]. Emerging from a literature which viewed the immature brain as a haphazardly wired structure that undergoes large-scale remodeling at birth to form OD columns, Hubel and Wiesel were surprised to find that the OD columns within primary visual cortex are actually organized at birth in an adult-like pattern [21-23]. Several subsequent studies investigated whether visual experience (e.g., light stimulation) is necessary for development of the mature pattern of OD columns seen in the infant brain [22-25]. Horton and Hocking [20] finally resolved the issue with heroic measurements of completely dark-reared infant macaque monkeys; the OD columns in these monkeys with no visual experience are organized like those measured in the adult macaque. Other features of cortical functional architecture, such as orientation tuning and orientation columns, are also present before any visual experience [15].

Further experiments investigating how this initial functional architecture is formed have unveiled activity-dependent processes that operate even before the retinal photoreceptors are functional and play a key role in determining the OD organization [16]. In utero, spontaneous coordinated waves of retinal ganglion cell action potentials are required for the development of normal OD and orientation columns [13, 15, 26, 27]. Thus, patterned retinal inputs to the cortex modify initially imprecise connections to produce an organized visual topography prior to the visual experience.

In turn, visual experience during an early critical period of life can profoundly alter the cortical visual topography formed by the in utero retinal activity [14, 18]. For example, if one eye is

Measurement	Structural Plasticity	Functional Plasticity	Adaptation
Temporal scale of cause	Longer than inciting factor	Longer than inciting factor	Short (~tracking input statistics)
Temporal scale of effect	Weeks to months	Days to weeks	Short (~tracking input statistics)
Anatomical connectivity	New connections	Likely re-weighted existing connections	Unlikely to change
Receptive field: Location	May change	May change	May change
Receptive field: Size	May change	May change	May change
Receptive field: Gain	May change	May change	May change
Reversibility	Not typical	Yes, in days to weeks	Yes, within a short time

Table 1. Characteristics of Plasticity and Adaptation. This table delineates core characteristics of three types of 'plasticity': structural plasticity, functional plasticity, and adaptation. Differentiating between each of the three can be difficult, because many measurements do not distinguish one type from another. In addition, there are not always sharp distinctions between the three phenomena. Confusion and controversy have often arisen from the non-specific use of the term 'plasticity' to describe these different cortical changes. This table is adapted and expanded from Box 1 in [5], a prominent review of plasticity and stability of the visual system.

deprived of vision for several weeks in infancy, then most of the mature visual cortical neurons are responsive only to stimuli presented to the eye that remained open; the cortical territory normally devoted to the closed eye is greatly diminished [18, 21]. This reorganization causes a loss of perception in the eye that was closed, as seen in cases of congenital cataracts leading to amblyopia and life-long blindness [28-31].

1.3. Using the topography of visual cortex to track changes in human development

These examples of visual cortical development were studied in animal models using such invasive techniques as electrophysiology and histology. How can we study cortical changes in the living, developing human? One technique that is proving very effective is to measure visual field maps (VFMs) in human visual cortex with functional magnetic resonance imaging (fMRI) to study cortical reorganization in response to abnormal visual input [5].

Human visual cortex is organized into distinct VFMs whose locations and properties provide important information about visual computations [32-35]. In each retinotopically-organized VFM, neurons whose visual receptive fields lie next to one another in visual space are located next to one another in cortex, forming one complete representation of contralateral visual space. By taking advantage of the knowledge of the retinotopic organization of visual input, multiple cortical VFMs can be measured using fMRI with respect to the two orthogonal dimensions needed to identify a unique location in visual space: eccentricity and polar angle (Figs. 3A, 4A, 5A). Each VFM subserves a specific computation or set of computations; locating these VFMs allows for the systematic exploration of these computations across visual cortex [33, 36].

Understanding the organization of these maps provides a baseline for studying reorganization following abnormal development or cortical damage. This chapter will review a series of studies of human developmental visual plasticity that capitalize on our ability to measure these VFMs with a high level of detail in individual subjects. We examine genetic, retinal, and cortical alterations that lead to reorganization of the cortical VFM organization. These examples of cortical developmental plasticity give us insight into the flexibility and constraints present in the human visual system.

2. Methods

2.1. General fMRI methods

Measurements of VFMs in human require very precise fMRI measurements and highly detailed, individual subject analysis [32, 34]. Before delving into the details of the two main types of fMRI experimental techniques used to measure VFMs, it is vital to remember that all fMRI techniques are empowered and limited by four common factors: the properties of fMRI in general, subject biology, experimental stimuli, and statistical analysis. Regardless of the cleverness of the experimental design, basic errors in any of these categories run the risk of rendering an entire study irrelevant due to inaccurate or misinterpreted results [37]. We will briefly review these factors with respect to their importance to visual field mapping.

2.1.1. Properties of fMRI

Let's begin with the basics of fMRI. When neurons are active consequent to task-induced or spontaneous neural metabolic modulations, the ratio of diamagnetic, oxygenated hemoglobin to paramagnetic, deoxygenated hemoglobin in nearby blood increases after a few seconds with a characteristic profile, called the hemodynamic response function (HRF; for reviews and more details, see [38-41]). fMRI is a technique that takes advantage of such blood oxygen-level dependent (BOLD) activity in the brain by applying a strong magnetic field and encoding positions in space through slight differences in magnetic field strength and phase [40, 42]. This allows for a very specific spatial readout of brain activity (on the order of $1mm^3$, often in the 1-4 mm^3 range), but the hemodynamic response is on the order of seconds, which limits the ability of fMRI to distinguish fine temporal differences [43].

Scanners suitable for human research commonly have a magnetic field strength of 1, 1.5, 3, or 4 Tesla (T), though access to 7T scanners is becoming more common. The field strength affects many aspects of the scanning parameters that can be chosen, but there is a basic trade-off between spatial resolution and signal-to-noise ratio (SNR), although different field strengths may also suffer from differing distortions or artifacts [44-46]. Lower field strength gives researchers a lower SNR, which forces researchers to use larger voxels. A larger voxel size increases SNR, because signal can be averaged across a greater number of active neurons within a voxel. However, large voxels also reduce spatial detail, which is important for advanced fMRI techniques such as those used for visual field mapping [34, 38, 47]. A key element for making careful, accurate measurements of plasticity in human cortex is to be able to acquire both high spatial resolution images and high SNR. Naturally, there are many other details to consider in the scanning parameters, but these are outside the scope of the chapter.

2.1.2. Subject cortical biology

The basic properties of fMRI scanners quickly become more complicated when biological factors are taken into account [39, 48, 49]. Air cavities, like the frontal sinuses or the ear canals, can create 'blow-out' artifacts in the images of the nearby portions of brain by locally distorting the magnetic fields [50]. Although less relevant for most measurements of visual cortex, it is a factor that should not be ignored. For example, if one were to compare visual activity in primary visual cortex to visual activity near these distorted regions, one may make incorrect inferences about activity level differences that have little to do with the experiment but instead are driven by these distortions [34, 38, 39, 48, 49, 51].

Another consideration is the arterial and venial physiology of individual subjects. Large draining veins may create distortions in the fMRI signal arising from neighboring cortex [52]. These distortions are unpredictable due both to the sensitivity of these distortions to the relative magnetic gradients and to the high variability of the venous system across individuals. For example, Winawer et al. [53] demonstrated the role for a 'venous eclipse' in the difficulties in measuring of the contiguous hV4 hemifield representation on the human ventral occipital cortex. These difficulties led to several years of confusion and controversy regarding the VFM organization in this region [33, 54-59].

It is also important to take the cortical anatomy of the individual subject into account. Although it is becoming more common practice in some fields to segment out white from gray matter and consider only activity occurring in gray matter, this is still not the standard approach [39]. As far as is currently known, the primary neural computations take place in gray matter; therefore, one should only consider activity in biologically relevant tissue, particularly when comparing regions of interest that could, without considering only gray matter, include different amounts of white matter as frequently occurs in typical slice-based analysis [60]. It should also be noted that not all white-gray matter segmentation is created equal—there are several algorithms used to parcel them out, but all require hand-editing for appropriate accuracy, which can be time intensive [32, 61-63].

2.1.3. Experimental stimuli

For any field of visual neuroimaging research, one must be incredibly careful in the choice of experimental stimuli, especially when contrasting activity between two or more types of stimuli. First, one must take into account the retinotopic nature of large portions of visual cortex [32-34, 64] and present stimuli in retinotopically consistent locations (unless activity is being measured for different locations of the visual field). Next, stimuli must be chosen that are equivalent on every aspect except those of interest. Unaccounted-for features can explain the results of interest when low-level features of stimuli are ignored in favor of interpreting data in terms of perceptual categories (e.g., [65, 66]).

2.1.4. Statistical analysis

In the unending search for statistical significance, researchers often use some variety of spatial averaging across subjects in order to increase their statistical power. There are two primary approaches to spatial brain averaging: morph individual subject brains onto a template brain or average locally based on local anatomical landmarks. From a behavioral experimental perspective, this may seem as reasonable as averaging together d-prime scores for individual subjects, but that is far from the case. One must take into account the inter-subject biological variability, not just in the location of veins, but of every aspect of the brain. Individual, neurologically normal subjects may have extra, missing, differently sized or differently folded gyri and sulci [67, 68]. The cortical surface of primary visual cortex, V1, can vary by a factor of ~2.5 among normal subjects, which not only affects the measurement of V1, but of all neighboring regions as a result of the displacement with respect to the underlying anatomy [69]. Although likely less problematic for vision, handedness has also been implicated in differences between hemispheric organization [70, 71]. Also, overall brain size can vary not only by gender, but by skull and body size. Additionally, skull shape can vary, shifting locations of individual brain regions due to the brain displacement relative to other subjects [72]. Obviously this is not an exhaustive list of normal variation, and it doesn't even take into account the wide variety of possibilities for damage in people who self-report being neurologically normal—or the differences between patients with similar neurological conditions within a particular study of visual plasticity.

Despite these issues, the vast majority of researchers use spatial averaging. The only truly viable solution is functional averaging, which is where regions of interest (ROIs) are defined based on functional data, then data from corresponding ROIs across subjects are averaged together. For example, if one wanted to study color activity in hV4, one would first use a visual field mapping technique to map hV4 in each subject (see Travelling Wave Retinotopy and Population Receptive Field Modeling sections below), then average together color activity for each subject from their individually defined hV4 maps [32, 54]. In this way, each subject's unique brain is taken into account rather than being distorted to approximate a template, and the relevant information is still averaged across subjects.

Finally, let us consider another common practice: spatial smoothing. This is when researchers average together activity in nearby voxels, often using a several-millimeter Gaussian kernel. What this allows researchers to do is to smooth the data for each voxel, reducing overall signal variability. However, such smoothing eliminates detailed information in the process. While this may be acceptable when one's technique cannot utilize detailed spatial information, it is disastrous when applied to detailed information such as that contained in visual field maps [33, 58].

2.2. Travelling wave retinotopy

Travelling Wave Retinotopy (TWR) was developed to measure retinotopically organized visual cortex (also known as visual field maps, VFMs). To do so, TWR takes advantage of knowledge gleaned from electrophysiological measurements of visual cortex in animal models about the structure and stimulus preferences of VFMs. In short, a VFM is a visual area with a complete representation of visual space, where neurons that represent adjacent locations on the retina (and thus visual space) are also adjacent in cortex. Because many computations are required to create our visual experience, our brains have many specialized VFMs which perform one or more of those computations across the entire visual scene (e.g., motion perception happens throughout our visual field, not just in the upper left quadrant).

Taking advantage of this knowledge, TWR uses two types of stimuli to measure the two orthogonal dimensions of visual space: eccentricity (i.e., center to periphery) and polar angle (i.e., around the clock) [32, 34, 73-76] (Figs. 3A, 4A). These stimuli tile across as much of the visual field as possible given field of view limitations with the MR scanner. The first stimulus is designed to elicit each voxel's preferred eccentricity by presenting a high-contrast flickering stimulus shaped like a ring, expanding in discrete even steps from the center of vision (fixation) to the periphery. The second is designed to elicit each voxel's preferred polar angle by presenting a high-contrast flickering stimulus shaped like a wedge that extends from the fovea to periphery, which rotates in discrete even steps around the fixation point. A checkerboard pattern commonly comprises the high contrast stimulus, as this pattern maximally stimulates V1; however, any pattern of interest can theoretically be used [77]. In each scan, only one type of stimulus is presented (rings or wedges), and all of visual space is cycled through several times with each stimulus (Fig. 1). Typically, several scans are then averaged together for each stimulus type. The type of activity driven by these stimuli physically appears to be a wave that travels from one end of the VFM to the other along iso-angle or iso–eccentricity lines, giving TWR its name.

The design of TWR presents all eccentricities or polar angles at a given frequency per scan (typically 6-8), which allows the use of Fourier analysis. First, TWR only considers activity that is at the signal frequency (typically 6-8 cycles/scan), excluding low-frequency physiological noise, among other things. A coherence analysis is then performed, where the strength at the signal frequency is divided by the strength at the other frequencies, and only voxels with coherence above a threshold are further evaluated (typically 0.15 to 0.30 coherence) [32, 34]. Finally, a phase is assigned to each voxel, which corresponds to a particular eccentricity or polar angle preference (Fig. 1).

Between these measurements of eccentricity and polar angle, TWR provides an estimate of the preferred RF center of the population of neurons with visual RFs in each voxel (Fig. 2). This technique is excellent at driving activity in early retinotopic areas (e.g., V1, V2, V3, hV4), in which neural RFs are relatively small [73, 74, 78, 79]. However, VFMs with larger neural RFs (e.g., regions in lateral-occipital cortex) are more difficult to measure, because the voxels respond to large swaths of the visual field [33, 78-80]. Even with large receptive fields, regions can still be retinotopically organized with retinotopically dispersed, preferred receptive field centers [32]. This large response field means that several TWR stimulus positions can similarly activate such a region. Because TWR analysis relies on measuring cortical responses in sync with the stimulus movement, regions that activate to a specific set of multiple stimulus positions may appear to have lower or no retinotopic responses, even with retinotopically dispersed RF centers.

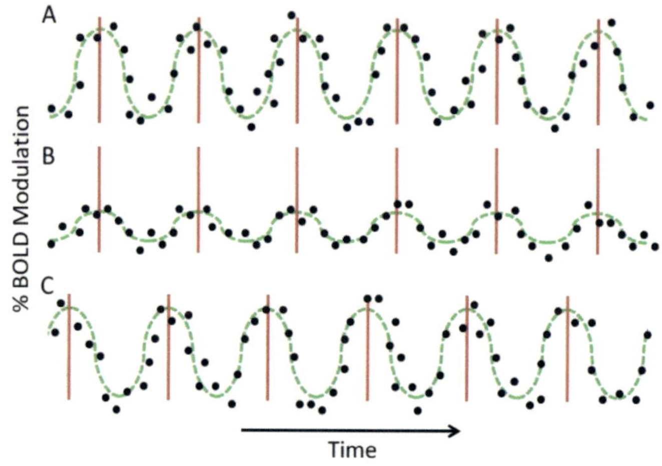

Figure 1. Visual Field Mapping Time Series Analysis. Each row represents the activity and analysis of a time series of a single 6-cycle scan of one type of experimental stimuli (expanding rings or rotating wedges) for a single voxel. Black dots indicate simulated raw data points of % BOLD modulation. The red lines indicate the peak activations per cycle for an imaginary set of voxels, which are the measurements used by the TWR analysis. The green dotted line represents a sinusoidal fit of the simulated data points, which are the measurements used by pRF modeling. Rows (A) and (B) represent time series of voxels with identical peak responses, indicating identical stimulus selectivity. However, (B) has much lower %BOLD modulation than (A), which could be due to two common factors: differences in local vasculature or broader receptive field tuning for (B) than (A). Rows (A) and (C) represent time series of voxels with identical %BOLD modulation, but different peak responses, which indicate different stimulus selectivity (i.e., different 'phases' of response). For example, (A) might represent a voxel with a preferred eccentricity tuning of 5° eccentric to fixation, whereas (C) might have a preferred tuning of only 2° eccentric to fixation.

Additionally, VFMs that have a larger degree of RF scatter within each voxel are more difficult to measure for the same reasons as large neural RFs. As a result of the power of TWR, it has become the standard workhorse for localizing and measuring the spatial details of VFMs. Relative to other neuroimaging methods like general condition contrast (GCC) approaches, TWR provides an unprecedented amount of detail [32, 33, 77]. For example, while the GCC may reveal the location of a large cluster of voxels that has been termed the human motion complex, hMT+, TWR reveals several distinct, highly organized VFMs within that large voxel cluster [64, 81].

2.3. Population receptive field modeling

The most recently developed technique of those covered in this chapter is population receptive field (pRF) modeling [78]. Once it had become clear that there were serious limits to the ability of TWR to deal with VFMs with large RFs, researchers at Stanford University decided that the best solution was to use a model of the pRFs of each voxel in VFMs [78]. This approach arose in part from previous work in several studies [30, 57, 79, 82, 83], which measured receptive fields in human and macaque visual cortex. The logic of pRF modeling is that, because VFMs are retinotopically organized, the population of RFs in each voxel of a VFM is expected to have similar preferred centers and sizes, allowing their combined pRFs to be estimated as a single, two-dimensional Gaussian RF (Fig. 2). Despite the fact that there is some variability in the neural RFs of each voxel in terms of their preferred center and size, termed RF scatter, the pRF provides a good, if somewhat slightly larger, estimate of the individual neural RFs in the voxel [78, 81, 84].

The advantages of the method are generally stated in comparison to the field standard for measuring VFMs, TWR. Not only does pRF modeling allow for the measurement of the preferred centers of the pRF of a voxel like TWR, but it also measures the spatial spread of the pRF, allows for the use of any stimulus that systematically tiles visual space, and takes less scan time to do, as it can use a single stimulus like a moving bar to measure both dimensions of visual space.

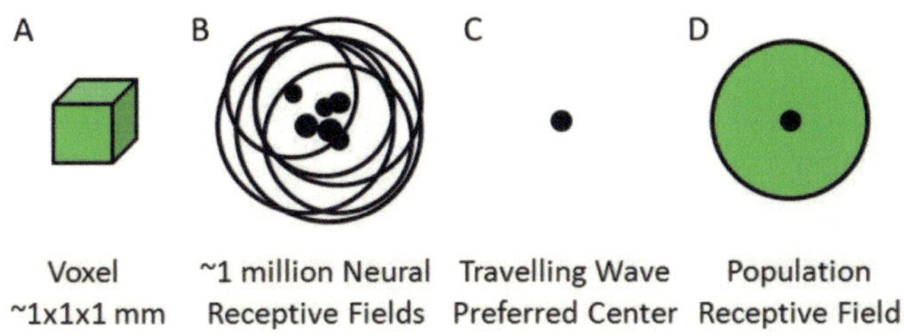

Figure 2. Measurements of an Individual Voxel. (A) A typical voxel recorded from a 3T MRI scanner is on the order of 1 mm3, though often slightly larger (2-3 mm3). (B) Within each typical voxel, there are on the order of ~1 million neurons, depending on the size of the voxel. For voxels in retinotopic visual cortex, the neurons each have similarly located spatial receptive fields (black outlines) with preferred centers (black dots). (C) TWR takes advantage of the fact that nearby neurons in retinotopic cortex have similar preferred centers in order to estimate a population preferred center for the population of neurons in a given voxel. (D) pRF modeling takes advantage of the fact that nearby neurons in retinotopic cortex have similar receptive fields in order to estimate not only a preferred center, but also a pRF for the population of neurons in a given voxel.

To accomplish this, the pRF model first creates a database of all possible pRF sizes and centers given the field of view of the stimulus used. Then, the model creates an expected HRF profile for each of the pRF possibilities based on an assumed or individually measured HRF. Finally, the model uses a least-squares fitting method to iteratively test each of the pRF possibilities for each voxel independently against the actual data collected [78]. Whichever pRF best fits the data is then assigned as the pRF for that voxel. Voxels are discarded if they do not have above a threshold of variance explained by the model. Although it is technically possible to use any stimulus that traverses the entire field of view, typically the stimuli take one of two forms. First is a slightly modified version of the TWR stimuli, in which neutral gray blank periods are inserted at an off-frequency from the stimulus frequency (e.g., 4 instead of 6-8 cycles/scan, so they are separable in the Fourier analysis). The second and increasingly common stimulus is a high-contrast flickering checkerboard moving bar stimulus that steps across the field of view in the 8 cardinal directions, again with several interspersed neutral gray blank periods. In theory, one could also tile visual space using any stimulus of interest, if the aforementioned stimuli do not drive the cortical region well. Since they were designed to drive activity in early visual cortex, it is possible other stimuli may perform better in higher-order VFMs [85, 86].

Because pRF modeling has proven so successful, we expect that it will eventually replace TWR as the standard method for measuring VFMs. Moreover, pRF modeling has an excellent future in the measurement of the details of pRFs, which is particularly important for the measurement of visual plasticity in humans. So far, the technique has primarily used a two-dimensional Gaussian profile for the pRF estimates, but researchers are working on the use of center-surround Gaussian pRFs, multiple location pRFs, and non-classical pRF shapes, which may allow for better pRF estimation as time continues [87]. For its purpose, the limitations of the pRF modeling method are essentially the measurement limitations of the scanner and stimuli used. In the future, it is likely that pRF modeling will be very successful when used in isolation, but also excellent to use in conjunction with other techniques.

3. Developmental plasticity following congenital lesions

Cortical reorganization is generally accepted by neuroscientists as a fact of life during development. Along the way to adulthood, neural structures grow, sprout, connect, and are pruned in a series of complex cascades and critical periods. In short, cortical reorganization not only happens during development, it's rampant. Perhaps that is why there is less controversy surrounding claims of cortical plasticity as a result of congenital abnormalities. As a rule, the developing brain is more plastic than the adult brain, but in both cases, the brain is striking a balance between plasticity and stability [5].

For the visual system, the balance is decidedly skewed towards stability, except in the earlier stages of development [88]. Stability is very useful for the visual system in the adult, because there are very few circumstances under which changes in properties of neurons in visual cortex such as their spatial representations is beneficial after development [6]. In fact, in most circumstances, it would be costly and destructive to change visual circuitry in the brain. In effect, each portion of visual space has its own processing pipeline through the visual hierarchy, which is possible only if independent processing channels for each portion of space are maintained. If the visual system were highly plastic, portions of the visual field would necessarily be cross-wiring these processing pipelines, disrupting the processing hierarchy in chaotic fashion wherever a crossing occurred. Contrast this with the memory systems of the brain, which could not possibly function without a high degree of plasticity to encode memories and associate them with one another (e.g., [89]). Of course, memory systems are not entirely plastic, as memories must also be maintained, but the balance point between plasticity and stability is quite different from that of the visual system.

In the congenital disorders described, developmental alterations of the organization of visual input to cortex are expected to drive the reorganization of visual representations in visual cortex. This reorganization of cortical visual representations might occur in one or more of the following ways (e.g., [5, 90]). First, major changes in the visual input to cortex may lead to a total disruption of retinotopic organization in visual cortex. Second, major retinal changes may

http://dx.doi.org/10.5772/56340

lead to a reorganization of a section of cortical visual representation in response to a loss of retinal activity for a portion of visual space. Third, altered visual pathways could lead to a restriction of the field of view as cortical maps reach a limit on how much of the contra- and ipsilateral hemifields could be represented in each hemisphere. Fourth, changes in the crossing of the visual pathways could lead to larger-than-normal, contiguous representations of visual space representing more than the normal contralateral hemifield. Finally, such changes in visual pathways could produce overlapped visual field representations, in which an abnormal representation of the ipsilateral visual field is superimposed on the representation of the normal contralateral visual field. In a single congenital disorder affecting the visual system, one or even several such changes in cortical visual representations could theoretically be present. Measurements of what types of reorganization actually appear in these disorders highlight the capabilities for adaptation to altered visual inputs in human visual cortex.

The abnormal drivers of developmental plasticity in the visual system can be grouped into three major categories. The first are retinal lesions, in which visual information about a location is never present in the visual system because of a functional failure in retinal circuitry. The second are altered visual pathways, in which visual information is misrouted from the retina to its intended thalamic or cortical targets. Finally, we have cortical lesions, in which visual information enters the visual system, but undergoes abnormal processing due to a disruption in the cortical processing hierarchy. The following sections will describe examples from each category that were studied in human subjects using fMRI.

3.1. Congenital retinal lesions can produce changes in cortical visuospatial representations.

3.1.1. Rod monochromatism

Two primary classes of photoreceptors tile the human retina: cones, which are maximally responsive under normal lighting (photopic) conditions, and rods, which are maximally responsive under very low lighting (scotopic) conditions [91]. The cones are highly concentrated in the fovea of each retina, whereas the concentration of rods drops dramatically and becomes zero within a radius of ~0.6-0.8° of visual angle about the eyes' fixation point [91-93]. While the rod-free fovea occupies <1% of the retina, the cortical representation of this retinal region occupies roughly 20% of primary visual cortex [94].

Under scotopic conditions, normal subjects perceive a scotoma ~2-4° of visual angle in diameter at the center of their vision [93, 95-97]. This rod scotoma, like the blind spot, is functionally equivalent to a retinal lesion—light falls on a location in the retina but is not processed due to a lack of active photoreceptors (Fig. 3AB). Recent research indicates that the rod scotoma in normal adults does not fill-in lines or patterns or color, as rods alone do not give rise to color [97], but does fill-in diffusely lit surfaces [98]. The differences in filling-in properties of the rod scotoma from those of the blind spot likely are due to the differences in their cortical receptive field sizes (small vs. large), their ocularity (binocular vs. monocular), their retinal locations (central vs. peripheral), and their physical exposures (infrequent vs. constant). Regardless, the cortical representation of the rod scotoma provides a natural testing ground for hypotheses regarding visual plasticity in response to retinal lesions.

Because we have invented many clever ways of maintaining photopic conditions outside of natural sunlight, modern humans are exposed to scotopic vision, and the rod scotoma, rather infrequently. Under normal conditions, occasional exposure to the rod scotoma, with cones inactive and rods active, throughout development and adulthood is not enough to induce cortical plasticity. For example, in adulthood, cortical representations of the fovea do not shift after prolonged exposure of one week to purely scotopic conditions. This was tested by Baseler et al. [95], who used TWR under photopic and scotopic conditions in normal subjects to map out the portions of V1 that represent the rod scotoma. They found no evidence for reorganization in these normal subjects either in the short term or after one week of adaptation. In fact, the region representing the rod-free fovea under photopic conditions was simply inactive under scotopic conditions, as expected from a lack of signal from the retina (Fig. 3AB).

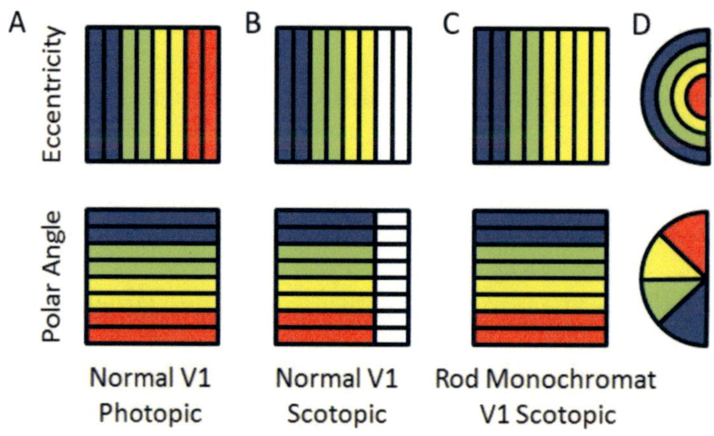

Figure 3. Visual Field Maps in Normal and Rod Monochromat Subjects. (A) Top: Eccentricity gradient for a VFM in a subject with normal vision viewing the stimulus under photopic (daylight) light conditions. Note the gradient running from the center to more peripheral eccentricities runs from right to left. This gradient would be orthogonal to the polar angle gradient in the bottom panel, such that each iso-eccentricity line has a representation of the full range of polar angles. The combination of the orthogonal gradients in top and bottom panels forms one complete representation of a hemifield of visual space. This forms one half of a complete VFM, with the corresponding half being located in the opposite hemisphere of the brain. Because the hemifield represented in this cartoon is the left (see legend in (D)), this map would be located in the right hemisphere. The colors denote positions in visual space represented in each cortical location. Note that the red colors depicting the center of space are represented in this cartoon VFM for a normal subject under photopic conditions. (B) Top: Eccentricity gradient for a VFM in a subject with normal vision viewing the stimulus under scotopic (starlight) light conditions. Bottom: Polar angle gradient of the same VFM, in the same subject under scotopic conditions. Note that the red colors depicting the center of space are not represented in the eccentricity gradient, nor are the corresponding locations of the polar angle gradient. The reason is that, under scotopic conditions, a normal subject has a "rod scotoma" at the center of their gaze where there are few-to-no rod photoreceptors; thus the portions of the VFM representing the center of gaze are not activated. (C) Top: Eccentricity gradient for a VFM in a rod monochromat under scotopic conditions. Bottom: Polar angle gradient for a VFM in a rod monochromat under scotopic conditions. Note that the red colors in the top panel of (A) have been replaced with yellow colors of (C), and that the corresponding locations in the lower panel of (C) are active, unlike the same location in the bottom panel of (B). The reason for this is that the rod monochromat's VFM has reorganized, such that regions of the VFM that would have represented the rod-free fovea now represents regions just outside. (D) Top: Visual space legend for eccentricity. Each color represents an iso-eccentricity line in the left visual hemifield. Bottom: Visual space legend for polar angle. Each color represents an iso-polar angle line in the left visual hemifield.

What if, instead, you were exposed to the rod scotoma for every moment your eyes were open since birth? Would the cortical region that would normally represent the rod-free fovea remain inactive, or would this region of cortex reorganize to represent a still-active, rod-driven region of the retina? These questions can be investigated in a particular group of subjects called rod monochromats, who lack retinal cone function. This loss of cone function is due to a specific genetic mutation in the cone cells that disrupts the phototransduction cascade in all three types of cone photoreceptors [99, 100]; the cones are present in the retina in normal locations and density, but do not respond to stimulation by light [101]. As a result, rod monochromats depend upon only their rod photoreceptors to see and, as such, have been constantly exposed to the rod scotoma throughout development and into adulthood.

Baseler et al. [95] also used TWR under photopic and scotopic conditions to map out the region of V1 devoted to processing the rod scotoma in rod monochromats. In contrast to the results in normal subjects, their measurements showed that the rod monochromats had significant activity under both photopic and scotopic conditions in the region of V1 that would be expected to represent the rod-free scotoma (Fig. 3C). In addition, this cortical activity now represented parafoveal input; this cortical region had reorganized in rod monochromats to receive input from the region of retina surrounding the rod-free fovea. This result indicates that constant exposure to the rod scotoma throughout development leads to cortical reorganization, while infrequent exposure does not.

3.2. Genetic mutations that alter the optic nerves lead to significant alterations in cortical visual field map organization.

3.2.1. Albinism

The most obvious abnormality associated with albinism is the lack of normal pigmentation of the skin, hair, and eyes due to a defect in melanin production [102]. A lesser-known abnormality is albinism's disruptive effect on the development of the visual pathways, in which abnormal projections from each temporal retina carry additional ipsilateral information to each V1 [103-105]. Typically, each cortical hemisphere primarily represents the one contralateral hemifield of visual space (Fig. 4A), but with the abnormal projections in albinism each hemisphere represents a greater span of the visual field from the ipsilateral eye [106] (Fig. 4B). The question, then, is what happens with that information? Is the abnormal ipsilateral information simply suppressed, such that a normal contralateral hemifield representation is maintained in visual cortex, but with, perhaps, abnormal visual perception from the missing ipsilateral coverage? Or, do the projections lead to integrated representations of the ipsilateral information from the temporal retina? If the latter, how are these ipsilateral visual representations integrated into the normal contralateral ones?

Previous electrophysiological research in non-primate albino mammals has suggested that the genetic mutations that underlie albinism can drive different degrees of topographical reorganization in cortex (for review, see [105]). In one organizational pattern measured primarily in Siamese cats and albino ferrets that is called the 'Boston' pattern, the abnormal pathways are reorganized within the projections from the thalamus to cortex, which leads to a contiguous

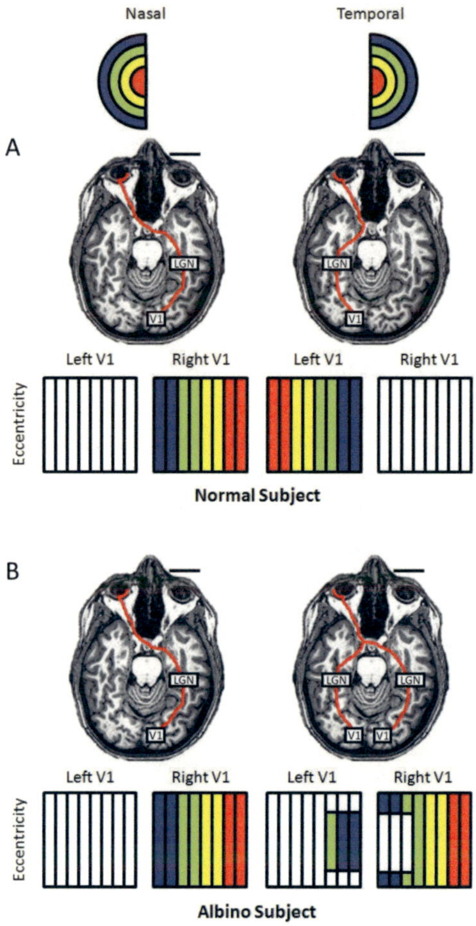

Figure 4. Visual Field Projections in Normal and Albino Subjects. In all panels, the right eye is covered with an eye patch, and only information from the left eye reaches visual cortex. The left column concerns cases where the left side of visual space, which corresponds to the nasal retina of the left eye, is stimulated. The right column concerns cases where the right side of visual space, which corresponds to the temporal retina of the left eye, is stimulated. Top row: Eccentricity visual space legend. Each color represents an iso-eccentricity line in the visual hemifield. (A) Normal subject. Top row: Information flow from the retina to visual cortex. Note that the left side of visual space (nasal retina, left column) is passed to the right hemisphere, whereas the right side of visual space (temporal retina, right column) is passed to the left hemisphere. Bottom row: Eccentricity gradient response for left and right V1. Colors represent locations in visual space denoted by the legend at the top of the figure. Note that the left side of visual space (nasal retina, left column) is represented in right V1, whereas the right side of visual space (temporal retina, right column) is represented in left V1. (B) Albino subject. Top row: Information flow from the retina to visual cortex. Note that the left side of visual space (nasal retina, left column) is passed to the right hemisphere, as in the normal subject in (A). However, unlike the normal subject in (A), the majority of the right side of visual space (temporal retina, right column) is passed to the right hemisphere. Also note that a weakened, residual normal pathway still projects a portion of the left side of visual space (temporal retina, right column) to the left hemisphere. Bottom row: Eccentricity gradient response for left and right V1. Colors represent locations in visual space denoted by the legend at the top of the figure. Note that the left side of visual space (nasal retina, left column) is represented in right V1, as in the normal subject in (A). However, unlike the normal subject in (A), the majority of the right side of visual space (temporal retina, right column) is represented in right V1. Some of the right visual field is still represented in left V1, but not at all at the scale as the normal subject in (A).

http://dx.doi.org/10.5772/56340

representation of the abnormal and normal representations of visual space within visual cortex [105, 107-109]. In a second pattern called 'Midwestern' that is also seen in these mammals, the abnormal visual projections continue from the thalamus into visual cortex without reordering into a contiguous map with concurrent cortical suppression of the abnormally projecting pathways [105, 110]. The visual perceptual dysfunction in these patterns appears in some measurements to include a lack of behavioral sensitivity to the temporal retina [107, 111]. Both patterns have actually been measured within the primary visual cortex of the same animal [112].

The third and rarest pattern, called the 'true albino' pattern, has been measured in one non-human primate as well as a few albino mammals. These electrophysiological measurements suggested that this pattern lacks the reordering of inputs into cortex, forming a superimposed set of inputs from nasal and temporal retinae, but unlike the cortical suppression suggested in the 'Midwestern' pattern, these superimposed inputs are functional [113].

The use of electrophysiology in these studies makes the analysis of the broad organization of visual space across a cortical region difficult. The differing organizational patterns within discrete sections of primary visual cortex in individual albino animals highlights the importance of being able to examine larger regions of the cortical topography than provided by electrophysiological measurements. Functional MRI, however, provides excellent spatial resolution for visual field measurements across a broad field of view, the type of measurement necessary to fully examine the cortical pattern of the representations of visual space across larger swaths of cortex than possible with electrophysiology. In addition, fMRI more easily allows for analysis across a greater number of subjects, important to fully understand the extent of variations possible across individual subjects. Finally, the non-invasive nature of fMRI permits investigation into the effects of albinism on human cortical development.

A recent fMRI study by Hoffman et al. [114] used TWR to map out the cortical representations of visual space in VFMs V1, V2, and V3 in human subjects with ocular or oculocutaneous albinism. They found that the ipsilateral information from the temporal retina is abnormally represented in cortex and forms a mirror-image map in V1 that is superimposed on top of the normal contralateral representation from the nasal retinal projections, following the 'true albino pattern' [113] (Fig. 4; also see [104, 106]). This causes each voxel in V1 to represent two visual field locations that are mirrored around the vertical meridian. A similar pattern of integration was measured in VFMs V2 and V3. Together with the previous study in non-human primate, these results suggest that the developmental mechanisms that subserve the reorganization of thalamocortical projections in non-primate albino mammals are not active in albino primate, leading to a more exclusive 'true albino' pattern of topographical integration in primates.

A few further studies have explored how this abnormal integration of the retinal projections into early visual cortex representations relates to the deficits in visual perception described in albinism. In contrast to the results seen in several animal studies [107, 111], Hoffman et al. [115, 116] and Klemen et al. [117] found no visual field deficits related to the abnormal visual projections and cortical representations. Thus, human albinism presents an excellent example of how intra-cortical self-organization during development allows abnormal representations to still subserve normal visual perception.

3.2.2. Achiasma

Albinism provided an example of how partial misrouting of the temporal retinal fibers can be successfully integrated into cortical topography as a mirror-symmetric map superimposed on the normal contralateral map. But what happens with a complete misrouting of the temporal retinal fibers? Is there a limit to the extent of such integration? Another relatively recently discovered congenital disorder called achiasma allows us to investigate these questions [118, 119].

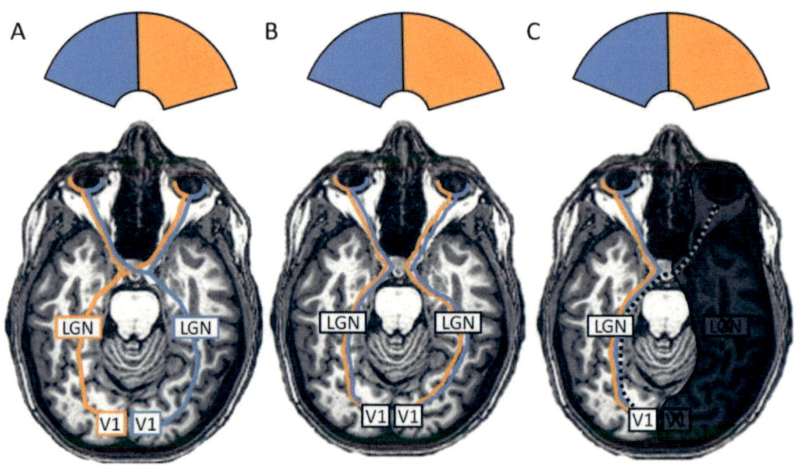

Figure 5. Visual Field Projections in Normal, Achiasmatic, and Missing Hemisphere Subjects. In all panels, blue represents the left half of visual space and tan represents the right side (top panel). (A) Normal subject. An image falling on the retinae is separated into nasal and temporal components in each eye. The part of the image within the left hemifield (blue) falls on the nasal retina in the left eye and the temporal retina in the right eye. The information from these two sections of each retina representing the left hemifield is transmitted through the optic nerve to the optic chiasm. To keep the correct hemifield information together for collective cortical processing, the axons from the nasal retina of the left eye cross over to the right optic tract at the optic chiasm, while the axons from the temporal retina of the right eye remain on the right side. Together, these sets of fibers carrying left hemifield image information then project first to the right lateral geniculate nucleus (LGN) of the thalamus and then through the optic radiations to V1 at the occipital pole of the right cortical hemisphere. Retinotopic organization is preserved along these pathways and then integrated into specific maps within visual cortex under the guidance of molecular cues and patterns of coordinated activity. The same pathways operate in reverse for the right hemifield (tan). Note that this results in one complete representation of visual space distributed across both hemispheres. (B) Achiasmatic subject. Because the fibers from the temporal retinae failed to properly cross at the expected region of the optic chiasm, information from each eye is passed only to the ipsilateral hemisphere. Thus, information from the left eye remains in the left hemisphere and information from the right eye remains in the right hemisphere. Note that this forms two complete representations of the full field of visual space, one for each eye, in each hemisphere. (C) Missing hemisphere subject. The development of the right hemisphere terminated in utero; therefore information from the left eye cannot possibly be passed to the right hemisphere. Instead, like achiasmatic subjects, the information remains in the ipsilateral hemisphere. Note that this means the entire visual field must be represented in the left hemisphere, resulting in one complete representation of visual space. The dotted black and white line represents possible remnants of connections to an undeveloped right eye.

Normally, molecular cues guide the axons of the retinal ganglion cells to either cross (temporal retinal fibers) to form the optic chiasm or to remain ipsilateral (nasal retinal fibers) (Figs. 4A, 5A). In individuals with achiasma, this crossing fails, with axons across each entire retina projecting through the ipsilateral thalamus to the ipsilateral cortical hemisphere (Fig. 5B) [118, 119]. The behavioral effects of this disorder can include deficits related to ocular instability (e.g., nystagmus, problems with stereopsis) [120 122].

Functional MRI was first used to investigate the topographical changes in achiasma by Victor et al. [120]. This study used a full field visual stimulus to show that stimulation of each eye produced functional activity within only the ipsilateral hemisphere. In addition, they found that the stimulation of each hemisphere produced overlapping activity within primary visual cortex. Their subjects had few visual deficits, other than problems related to ocular instability. These findings suggest successful integration of the misrouted visual pathways into the cortical topography. However, the pattern of cortical integration could not be determined from these methods.

With the development of higher resolution fMRI methods like TWR [34, 74] and pRF modeling [78], researchers have since been able to investigate more details about the topographical integration as well as changes in receptive field properties of regions within V1. A very recent study by Hoffman et al. [122] used both TWR and pRF to reveal that the congenital changes in achiasma drive developmental reorganization much like that seen in albinism. The complete misrouting of temporal retinal fibers was able to be successfully integrated as an overlapping, functional map of the ipsilateral visual field on top of that of the contralateral field (Fig. 5B). Further, the pRF measurements gave the first glimpse into human receptive field changes in these congenital disorders of visual pathway misrouting. Their results demonstrate pRFs that represent bilateral regions of the visual field in both striate and extrastriate VFMs along the medial occipital lobe.

Hoffman et al. [122] further used diffusion tensor imaging (DTI) to measure the general topography of the geniculostriate and occipital callosal connections in their achiasmatic subjects. Previous studies in achiasmatic dogs showed that the misrouted fibers formed retinotopic maps of opposing hemifields in adjacent layers of the LGN of the thalamus, suggesting no regorganization of geniculostriate pathways. The DTI measurements in human confirmed both that there were no significant changes in the input tracts to visual cortex as well as grossly normal topography in the inter-hemispheric callosal connections between the VFMs along the calcarine sulci. Together with the fMRI topographical and receptive field measurements, these findings by Hoffman et al. [122] suggest that the majority of the developmental reorganization in achiasma occurs in the intra-cortical local wiring within visual cortex.

3.3. How does the visual cortex reorganize following a massive cortical lesion in utero?

3.3.1. Missing hemisphere

In our last section, a dramatic recent example of a cortical lesion in utero allows us to examine the effects of a massive structural change to cortex on the development of cortical visual field representations. Muckli et al. [123] used TWR to study VFMs in a girl whose right cerebral hemisphere was completely absent at birth (Fig. 5C). She has only a rudimentary right eye and right optic nerve, but no right optic tract. Despite this extensive cortical loss and absent right eye, this subject has mostly normal visual function in both hemifields, with a mild left hemiplegia. Her spared bilateral visual functions suggest an enormous degree of developmental cortical reorganization [124].

In this very unusual case, the right hemisphere failed to develop around the seventh week of gestation due to an in utero lesion. Therefore, no crossing of temporal retinal fibers to the contralateral LGN and cerebral hemisphere was possible. Instead, the axons from the left retina carrying information from the entire visual scene (both hemifields) could only project to the occipital lobe of the left hemisphere (Fig. 5C).

In this situation, what happens with the integration of the retinal fibers from the single eye? As with the development of the visual pathways in congenital disorders with grossly intact cortical structures, four primary possibilities exist. The first could be termed 'normal' development, in which the contralateral visual field develops as usual and no ipsilateral representation is created, leading to a full left hemianopsia. The second is the opposite of the first, such that only the ipsilateral field is represented, leading to total right hemianopsia following suppression of the normal contralateral representation. The third is that contiguous VFMs, representing both hemifields, develop either as complete maps or in patches in place of contralateral hemifields. Finally, superimposed or interdigitated representations of the contralateral and ipsilateral hemifields may develop, as seen in albinism and achiasma. In addition, it is possible that more than one of these organizational structures could exist concurrently among VFMs, as seen in some non-primate albino mammals.

Interestingly, the TWR measurements by Muckli et al. [123] demonstrate such a concurrent existence of more than one organizational pattern in her intact left hemisphere. In V1, contiguous representations of both the contra- and ipsilateral visual fields were superimposed within the calcarine sulcus. However, discrete islands of ipsilateral quarterfield representations were also measured along the vertical meridian of V2 and V3, with a distortion of the contralateral visual field representations.

Why in this case do we see full field maps in addition to the superposition found in albinism and achiasma? Perhaps reorganization at an earlier stage of the visual pathways has occurred here. Muckli et al. [123] examined the organization of the visual field representations in the remaining left LGN. Unlike the reorganization seen in albinism [114] and achiasma [122] in which the contra- and ipsilateral representations remained segregated in the LGN, the measurements in this subject appeared to be fully overlapping, suggesting a greater degree of reorganization at this stage of visual processing. Similar LGN reorganization was measured in some albino cats and ferrets, which also had at least sections of contiguous full field visual field representations in their primary visual cortex [107-110].

The organization of the visual field representations is reflected in the extent of her preserved visual function. First, the fact that she has intact vision in the ipsilateral, right visual field demonstrates that the novel ipsilateral visual field representations are functionally relevant.

http://dx.doi.org/10.5772/56340

Second, visual perimetry mapping showed that both the right and left visual fields were centrally intact, but that both were somewhat restricted. Interestingly, the insertion of the ipsilateral islands into the contralateral representations resulted in both distortions of the cortical contralateral representations as well as concurrent distortions of her functional contralateral visual field. This restriction was more prominent in the upper visual field of both fields, in accord with the greater distortion of the contralateral visual field representation on the ventral cortical surface. The different patterns of self-organization in these VFMs in this unique case demonstrate a surprisingly flexible developmental restructuring in response to massive cortical loss.

4. Conclusion

This chapter has examined fMRI measurements of naturally-occurring cortical reorganization due to congenital abnormalities in the developing human visual system. These examples of retinal and genetic changes and cortical loss highlight the roles of plasticity and stability in the developing visual system. In these congenital conditions, the abnormally represented visual information has become available for grossly normal visual perception, despite limited compensatory alterations in the relatively stable geniculostriate projections. Together, these studies suggest that instead of large-scale reorganizations of the afferent visual pathways, relatively less drastic intra-cortical mechanisms underlie such developmental plasticity.

The next essential expansion of these findings is to apply our understanding of developmental plasticity to the controversial field of plasticity in adult sensory systems. Several aspects of cortical organization are thought to remain plastic into adulthood, allowing cortical sensori-motor maps to be modified continuously by experience. This dynamic nature of cortical circuitry is important for learning, as well as for repair following nervous system injury. Studies of the extent of cortical reorganization in adult primate visual cortex in response to changes in sensory input have produced quite disparate results, however (e.g., [1, 2, 125-138]). The ability to control cortical reorganization in the adult either by harnessing existing adult mechanisms for plasticity or by reactivating mechanisms of developmental plasticity would be a tremendous advancement in the treatment of cortical damage and neurodegenerative disease. The understanding of the balance of plasticity and stability in the human visual system arising from the examination of such congenital visual disorders as described here will be a key element of our progress towards this goal.

Acknowledgements

This work was supported by University of California, Irvine, startup funds to A.A.B.

Author details

Alyssa A. Brewer* and Brian Barton

*Address all correspondence to: aabrewer@uci.edu

Center for Cognitive Neuroscience and Department of Cognitive Sciences, University of California, Irvine, CA, USA

References

[1] J. C. Horton and D. R. Hocking, "Timing of the critical period for plasticity of ocular dominance columns in macaque striate cortex," J Neurosci, vol. 17, pp. 3684-709, May 15 1997.

[2] J. C. Horton and D. R. Hocking, "Effect of early monocular enucleation upon ocular dominance columns and cytochrome oxidase activity in monkey and human visual cortex," Vis Neurosci, vol. 15, pp. 289-303, Mar-Apr 1998.

[3] D. D. O'Leary, N. L. Ruff, and R. H. Dyck, "Development, critical period plasticity, and adult reorganizations of mammalian somatosensory systems," Curr Opin Neurobiol, vol. 4, pp. 535-44, Aug 1994.

[4] T. K. Hensch, "Critical period plasticity in local cortical circuits," Nat Rev Neurosci, vol. 6, pp. 877-88, Nov 2005.

[5] B. A. Wandell and S. M. Smirnakis, "Plasticity and stability of visual field maps in adult primary visual cortex," Nat Rev Neurosci, vol. 10, pp. 873-84, Dec 2009.

[6] S. M. Smirnakis, A. A. Brewer, M. C. Schmid, A. S. Tolias, A. Schuz, M. Augath, W. Inhoffen, B. A. Wandell, and N. K. Logothetis, "Lack of long-term cortical reorganization after macaque retinal lesions," Nature, vol. 435, pp. 300-7, May 19 2005.

[7] D. V. Buonomano and M. M. Merzenich, "Cortical plasticity: from synapses to maps," Annu Rev Neurosci, vol. 21, pp. 149-86, 1998.

[8] M. M. Merzenich, R. J. Nelson, M. P. Stryker, M. S. Cynader, A. Schoppmann, and J. M. Zook, "Somatosensory cortical map changes following digit amputation in adult monkeys," J Comp Neurol, vol. 224, pp. 591-605, Apr 20 1984.

[9] J. H. Kaas, M. M. Merzenich, and H. P. Killackey, "The reorganization of somatosensory cortex following peripheral nerve damage in adult and developing mammals," Annu Rev Neurosci, vol. 6, pp. 325-56, 1983.

[10] L. G. Cohen, P. Celnik, A. Pascual-Leone, B. Corwell, L. Falz, J. Dambrosia, M. Honda, N. Sadato, C. Gerloff, M. D. Catala, and M. Hallett, "Functional relevance of cross-modal plasticity in blind humans," Nature, vol. 389, pp. 180-3, Sep 11 1997.

[11] K. Fox and R. O. Wong, "A comparison of experience-dependent plasticity in the visual and somatosensory systems," Neuron, vol. 48, pp. 465-77, Nov 3 2005.

[12] C. D. Gilbert, W. Li, and V. Piech, "Perceptual learning and adult cortical plasticity," J Physiol, vol. 587, pp. 2743-51, Jun 15 2009.

[13] C. J. Shatz, "Emergence of order in visual system development," Proc Natl Acad Sci U S A, vol. 93, pp. 602-8, Jan 23 1996.

[14] D. H. Hubel, T. N. Wiesel, and S. LeVay, "Plasticity of ocular dominance columns in monkey striate cortex," Philos Trans R Soc Lond B Biol Sci, vol. 278, pp. 377-409, Apr 26 1977.

[15] B. Chapman, I. Godecke, and T. Bonhoeffer, "Development of orientation preference in the mammalian visual cortex," J Neurobiol, vol. 41, pp. 18-24, Oct 1999.

[16] M. Weliky and L. C. Katz, "Correlational structure of spontaneous neuronal activity in the developing lateral geniculate nucleus in vivo [see comments]," Science, vol. 285, pp. 599-604, 1999.

[17] L. C. Katz and C. J. Shatz, "Synaptic activity and the construction of cortical circuits," Science, vol. 274, pp. 1133-8, Nov 15 1996.

[18] C. J. Shatz and M. P. Stryker, "Ocular dominance in layer IV of the cat's visual cortex and the effects of monocular deprivation," J Physiol, vol. 281, pp. 267-83, Aug 1978.

[19] D. C. Van Essen, "Behind the optic nerve: an inside view of the primate visual system," Trans Am Ophthalmol Soc, vol. 93, pp. 123-33, 1995.

[20] J. C. Horton and D. R. Hocking, "An adult-like pattern of ocular dominance columns in striate cortex of newborn monkeys prior to visual experience.," Journal of Neuroscience, vol. 16, pp. 1791-1807, 1996.

[21] D. H. Hubel and T. N. Wiesel, "Receptive Fields of Cells in Striate Cortex of Very Young, Visually Inexperienced Kittens," J Neurophysiol, vol. 26, pp. 994-1002, Nov 1963.

[22] D. H. Hubel, T. N. Wiesel, and S. LeVay, "Functional architecture of area 17 in normal and monocularly deprived macaque monkeys," Cold Spring Harb Symp Quant Biol, vol. 40, pp. 581-9, 1976.

[23] T. N. Wiesel and D. H. Hubel, "Ordered arrangement of orientation columns in monkeys lacking visual experience," J Comp Neurol, vol. 158, pp. 307-18, Dec 1 1974.

[24] S. LeVay, M. P. Stryker, and C. J. Shatz, "Ocular dominance columns and their development in layer IV of the cat's visual cortex: a quantitative study.," Journal of Comparative Neurology, vol. 179, pp. 223-244, 1978.

[25] P. Rakic and M. S. Lidow, "Distribution and density of mnoamine receptor s in the primate visual cortex devoid of retinal input from early embryonic stagesemut f," J. Neuroscience, vol. 15, pp. 256-2574, 1995.

[26] D. Stellwagen, C. J. Shatz, and M. B. Feller, "Dynamics of retinal waves are controlled by cyclic AMP," Neuron, vol. 24, pp. 673-85, Nov 1999.

[27] M. P. Stryker and W. A. Harris, "Binocular impulse blockade prevents the formation of ocular dominance columns in cat visual cortex," J Neurosci, vol. 6, pp. 2117-33, Aug 1986.

[28] C. Zetterstrom, A. Lundvall, and M. Kugelberg, "Cataracts in children," J Cataract Refract Surg, vol. 31, pp. 824-40, Apr 2005.

[29] S. Awaya and S. Miyake, "Form vision deprivation amblyopia: further observations," Graefes Arch Clin Exp Ophthalmol, vol. 226, pp. 132-6, 1988.

[30] X. Li, S. O. Dumoulin, B. Mansouri, and R. F. Hess, "The fidelity of the cortical retinotopic map in human amblyopia," Eur J Neurosci, vol. 25, pp. 1265-77, Mar 2007.

[31] X. Li, S. O. Dumoulin, B. Mansouri, and R. F. Hess, "Cortical deficits in human amblyopia: their regional distribution and their relationship to the contrast detection deficit," Invest Ophthalmol Vis Sci, vol. 48, pp. 1575-91, Apr 2007.

[32] A. A. Brewer and B. Barton, "Visual field map organization in human visual cortex," in Visual Cortex - Current Status and Perspectives, S. Molotchnikoff and J. Rouat, Eds., ed: InTech, 2012, pp. 30-60.

[33] B. A. Wandell, A. A. Brewer, and R. F. Dougherty, "Visual field map clusters in human cortex," Philos Trans R Soc Lond B Biol Sci, vol. 360, pp. 693-707, Apr 29 2005.

[34] B. A. Wandell, S. O. Dumoulin, and A. A. Brewer, "Visual field maps in human cortex," Neuron, vol. 56, pp. 366-83, Oct 25 2007.

[35] D. C. Van Essen, "Organization of Visual Areas in Macaque and Human Cerebral Cortex," in The Visual Neurosciences, L. M. Chalupa and J. S. Werner, Eds., ed Boston: Bradford Books, 2003.

[36] J. H. Kaas, "Topographic Maps are Fundamental to Sensory Processing," Brain Research Bulletin, vol. 44, pp. 107–112, 1997.

[37] S. Zeki, "Improbable areas in the visual brain," Trends Neurosci, vol. 26, pp. 23-6, Jan 2003.

[38] B. A. Wandell and J. Winawer, "Imaging retinotopic maps in the human brain," Vision Res, Aug 6 2010.

[39] N. K. Logothetis and B. A. Wandell, "Interpreting the BOLD signal," Annu Rev Physiol, vol. 66, pp. 735-69, 2004.

[40] S. Ogawa, T. M. Lee, A. R. Kay, and D. W. Tank, "Brain magnetic resonance imaging with contrast dependent on blood oxygenation," Proc Natl Acad Sci U S A, vol. 87, pp. 9868-72, Dec 1990.

[41] D. J. Heeger and D. Ress, "What does fMRI tell us about neuronal activity?," Nat Rev Neurosci, vol. 3, pp. 142-51, Feb 2002.

[42] P. A. Bandettini, E. C. Wong, R. S. Hinks, R. S. Tikofsky, and J. S. Hyde, "Time course EPI of human brain function during task activation," Magn Reson Med, vol. 25, pp. 390-7, Jun 1992.

[43] A. K. Liu, J. W. Belliveau, and A. M. Dale, "Spatiotemporal imaging of human brain activity using functional MRI constrained magnetoencephalography data: Monte Carlo simulations," Proc Natl Acad Sci U S A, vol. 95, pp. 8945-50, Jul 21 1998.

[44] M. B. Hoffmann, J. Stadler, M. Kanowski, and O. Speck, "Retinotopic mapping of the human visual cortex at a magnetic field strength of 7T," Clin Neurophysiol, vol. 120, pp. 108-16, Jan 2009.

[45] J. D. Swisher, J. C. Gatenby, J. C. Gore, B. A. Wolfe, C. H. Moon, S. G. Kim, and F. Tong, "Multiscale pattern analysis of orientation-selective activity in the primary visual cortex," J Neurosci, vol. 30, pp. 325-30, Jan 6 2010.

[46] C. Triantafyllou, R. D. Hoge, G. Krueger, C. J. Wiggins, A. Potthast, G. C. Wiggins, and L. L. Wald, "Comparison of physiological noise at 1.5 T, 3 T and 7 T and optimization of fMRI acquisition parameters," Neuroimage, vol. 26, pp. 243-50, May 15 2005.

[47] E. Disbrow, T. P. Roberts, D. Slutsky, and L. Krubitzer, "The use of fMRI for determining the topographic organization of cortical fields in human and nonhuman primates," Brain Res, vol. 829, pp. 167-73, May 22 1999.

[48] N. K. Logothetis, "What we can do and what we cannot do with fMRI," Nature, vol. 453, pp. 869-78, Jun 12 2008.

[49] Y. Ejima, S. Takahashi, H. Yamamoto, M. Fukunaga, C. Tanaka, T. Ebisu, and M. Umeda, "Interindividual and interspecies variations of the extrastriate visual cortex," Neuroreport, vol. 14, pp. 1579-83, Aug 26 2003.

[50] N. K. Chen and A. M. Wyrwicz, "Correction for EPI distortions using multi-echo gradient-echo imaging," Magn Reson Med, vol. 41, pp. 1206-13, Jun 1999.

[51] M. Lauritzen, "Relationship of spikes, synaptic activity, and local changes of cerebral blood flow," J Cereb Blood Flow Metab, vol. 21, pp. 1367-83, Dec 2001.

[52] A. Seiyama, J. Seki, H. C. Tanabe, I. Sase, A. Takatsuki, S. Miyauchi, H. Eda, S. Hayashi, T. Imaruoka, T. Iwakura, and T. Yanagida, "Circulatory basis of fMRI signals: re-

lationship between changes in the hemodynamic parameters and BOLD signal intensity," Neuroimage, vol. 21, pp. 1204-14, Apr 2004.

[53] J. Winawer, H. Horiguchi, R. A. Sayres, K. Amano, and B. A. Wandell, "Mapping hV4 and ventral occipital cortex: the venous eclipse," J Vis, vol. 10, p. 1, 2010.

[54] A. A. Brewer, J. Liu, A. R. Wade, and B. A. Wandell, "Visual field maps and stimulus selectivity in human ventral occipital cortex," Nat Neurosci, vol. 8, pp. 1102-9, Aug 2005.

[55] K. A. Hansen, K. N. Kay, and J. L. Gallant, "Topographic organization in and near human visual area V4," J Neurosci, vol. 27, pp. 11896-911, Oct 31 2007.

[56] A. R. Wade, A. A. Brewer, J. W. Rieger, and B. A. Wandell, "Functional measurements of human ventral occipital cortex: retinotopy and colour," Philos Trans R Soc Lond B Biol Sci, vol. 357, pp. 963-73, Aug 29 2002.

[57] J. Larsson and D. J. Heeger, "Two retinotopic visual areas in human lateral occipital cortex," J Neurosci, vol. 26, pp. 13128-42, Dec 20 2006.

[58] R. B. Tootell and N. Hadjikhani, "Where is 'dorsal V4' in human visual cortex? Retinotopic, topographic and functional evidence," Cereb Cortex, vol. 11, pp. 298-311, Apr 2001.

[59] N. Hadjikhani, A. K. Liu, A. M. Dale, P. Cavanagh, and R. B. Tootell, "Retinotopy and color sensitivity in human visual cortical area V8," Nat Neurosci, vol. 1, pp. 235-41, Jul 1998.

[60] B. A. Wandell, S. Chial, and B. Backus, "Visualization and Measurement of the Cortical Surface," J. of Cognitive Neuroscience, vol. 12, pp. 739-52, 2000.

[61] P. C. Teo, G. Sapiro, and B. A. Wandell, "Creating connected representations of cortical gray matter for functional MRI visualization," IEEE Trans Med Imaging, vol. 16, pp. 852-63, Dec 1997.

[62] D. C. Van Essen, J. W. Lewis, H. A. Drury, N. Hadjikhani, R. B. Tootell, M. Bakircioglu, and M. I. Miller, "Mapping visual cortex in monkeys and humans using surface-based atlases," Vision Res, vol. 41, pp. 1359-78, 2001.

[63] A. M. Dale, B. Fischl, and M. I. Sereno, "Cortical surface-based analysis. I. Segmentation and surface reconstruction," Neuroimage, vol. 9, pp. 179-94, Feb 1999.

[64] H. Kolster, R. Peeters, and G. A. Orban, "The retinotopic organization of the human middle temporal area MT/V5 and its cortical neighbors," J Neurosci, vol. 30, pp. 9801-20, Jul 21 2010.

[65] X. Yue, B. S. Cassidy, K. J. Devaney, D. J. Holt, and R. B. Tootell, "Lower-level stimulus features strongly influence responses in the fusiform face area," Cereb Cortex, vol. 21, pp. 35-47, Jan 2011.

[66] T. Liu, F. Pestilli, and M. Carrasco, "Transient attention enhances perceptual performance and FMRI response in human visual cortex," Neuron, vol. 45, pp. 469-77, Feb 3 2005.

[67] J. S. Peper, R. M. Brouwer, D. I. Boomsma, R. S. Kahn, and H. E. Hulshoff Pol, "Genetic influences on human brain structure: a review of brain imaging studies in twins," Hum Brain Mapp, vol. 28, pp. 464-73, Jun 2007.

[68] W. F. Baare, H. E. Hulshoff Pol, D. I. Boomsma, D. Posthuma, E. J. de Geus, H. G. Schnack, N. E. van Haren, C. J. van Oel, and R. S. Kahn, "Quantitative genetic modeling of variation in human brain morphology," Cereb Cortex, vol. 11, pp. 816-24, Sep 2001.

[69] R. F. Dougherty, V. M. Koch, A. A. Brewer, B. Fischer, J. Modersitzki, and B. A. Wandell, "Visual field representations and locations of visual areas V1/2/3 in human visual cortex," J Vis, vol. 3, pp. 586-98, 2003.

[70] M. P. Bryden, H. Hecaen, and M. DeAgostini, "Patterns of cerebral organization," Brain Lang, vol. 20, pp. 249-62, Nov 1983.

[71] R. A. Harshman, E. Hampson, and S. A. Berenbaum, "Individual differences in cognitive abilities and brain organization, Part I: Sex and handedness differences in ability," Can J Psychol, vol. 37, pp. 144-92, Mar 1983.

[72] J. H. Kaas, "The evolution of the complex sensory and motor systems of the human brain," Brain Res Bull, vol. 75, pp. 384-90, Mar 18 2008.

[73] S. A. Engel, G. H. Glover, and B. A. Wandell, "Retinotopic organization in human visual cortex and the spatial precision of functional MRI," Cereb Cortex, vol. 7, pp. 181-92, Mar 1997.

[74] S. A. Engel, D. E. Rumelhart, B. A. Wandell, A. T. Lee, G. H. Glover, E. J. Chichilnisky, and M. N. Shadlen, "fMRI of human visual cortex," Nature, vol. 369, p. 525, Jun 16 1994.

[75] M. I. Sereno, A. M. Dale, J. B. Reppas, K. K. Kwong, J. W. Belliveau, T. J. Brady, B. R. Rosen, and R. B. Tootell, "Borders of multiple visual areas in humans revealed by functional magnetic resonance imaging," Science, vol. 268, pp. 889-93, May 12 1995.

[76] E. A. DeYoe, G. J. Carman, P. Bandettini, S. Glickman, J. Wieser, R. Cox, D. Miller, and J. Neitz, "Mapping striate and extrastriate visual areas in human cerebral cortex," Proc. Natl. Acad. Sci. (USA), vol. 93, pp. 2382-2386, 1996.

[77] S. A. Engel, "The development and use of phase-encoded functional MRI designs," Neuroimage, vol. 62, pp. 1195 - 200, Aug 15 2012.

[78] S. O. Dumoulin and B. A. Wandell, "Population receptive field estimates in human visual cortex," Neuroimage, vol. 39, pp. 647-60, Jan 15 2008.

[79] A. T. Smith, K. D. Singh, A. L. Williams, and M. W. Greenlee, "Estimating receptive field size from fMRI data in human striate and extrastriate visual cortex," Cereb Cortex, vol. 11, pp. 1182-90, Dec 2001.

[80] B. A. Wandell and J. Winawer, "Imaging retinotopic maps in the human brain," Vision Res, vol. 51, pp. 718-37, Apr 13 2011.

[81] K. Amano, B. A. Wandell, and S. O. Dumoulin, "Visual field maps, population receptive field sizes, and visual field coverage in the human MT+ complex," Journal of Neurophysiology, vol. 102, pp. 2704-18, Nov 2009.

[82] J. D. Victor, K. Purpura, E. Katz, and B. Mao, "Population encoding of spatial frequency, orientation, and color in macaque V1," J Neurophysiol, vol. 72, pp. 2151-66, Nov 1994.

[83] R. B. Tootell, J. D. Mendola, N. K. Hadjikhani, P. J. Ledden, A. K. Liu, J. B. Reppas, M. I. Sereno, and A. M. Dale, "Functional analysis of V3A and related areas in human visual cortex," J Neurosci, vol. 17, pp. 7060-78, Sep 15 1997.

[84] B. M. Harvey and S. O. Dumoulin, "The relationship between cortical magnification factor and population receptive field size in human visual cortex: constancies in cortical architecture," J Neurosci, vol. 31, pp. 13604-12, Sep 21 2011.

[85] M. I. Sereno, H. H. Yu, and E. K. Vogel, "Mapping Retinotopy in Higher Level Extrastriate Visual Areas in Humans," in Society for Neuroscience, San Diego, 2001.

[86] M. I. Sereno, S. Pitzalis, and A. Martinez, "Mapping of contralateral space in retinotopic coordinates by a parietal cortical area in humans," Science, vol. 294, pp. 1350-4, Nov 9 2001.

[87] W. Zuiderbaan, B. M. Harvey, and S. O. Dumoulin, "Modeling center-surround configurations in population receptive fields using fMRI," J Vis, vol. 12, p. 10, 2012.

[88] H. J. Neville and D. Bavelier, "Variability of developmental plasticity," in Mechanisms of Cognitive Develoment: Behavioral and Neural Perspectives, J. L. McClelland and R. S. Siegler, Eds., ed New Jersey: Lawrence Erlbaum Associates, Inc., 2001.

[89] M. M. Merzenich and K. Sameshima, "Cortical plasticity and memory," Curr Opin Neurobiol, vol. 3, pp. 187-96, Apr 1993.

[90] P. Sinha and M. Meng, "Superimposed hemifields in primary visual cortex of achiasmic individuals," Neuron, vol. 75, pp. 353-5, Aug 9 2012.

[91] C. A. Curcio, K. R. Sloan, R. E. Kalina, and A. E. Hendrickson, "Human photoreceptor topography," J Comp Neurol, vol. 292, pp. 497-523, Feb 22 1990.

[92] C. A. Curcio, K. R. S. Jr., O. Pakcer, A. E. Hendrickson, and R. E. Kalina, "Distribution of cones in human and monkey retina: individual variaibility and radial asymmetry," in Science, 1987, pp. 579-581.

[93] K. R. Duffy and D. H. Hubel, "Receptive field properties of neurons in the primary visual cortex under photopic and scotopic lighting conditions," Vision Res, vol. 47, pp. 2569-74, Sep 2007.

[94] J. C. Horton and W. F. Hoyt, "The representation of the visual field in human striate cortex. A revision of the classic Holmes map," Arch Ophthalmol, vol. 109, pp. 816-24, Jun 1991.

[95] H. A. Baseler, A. A. Brewer, L. T. Sharpe, A. B. Morland, H. Jagle, and B. A. Wandell, "Reorganization of human cortical maps caused by inherited photoreceptor abnormalities," Nat Neurosci, vol. 5, pp. 364-70, Apr 2002.

[96] N. Hadjikhani and R. B. Tootell, "Projection of rods and cones within human visual cortex," Hum Brain Mapp, vol. 9, pp. 55-63, 2000.

[97] D. H. Hubel, "Vision in dim light," Nature, vol. 388, pp. 32-3, Jul 3 1997.

[98] D. H. Hubel, P. D. Howe, A. M. Duffy, and A. Hernandez, "Scotopic foveal afterimages," Perception, vol. 38, pp. 313-6, 2009.

[99] S. Kohl, B. Baumann, M. Broghammer, H. Jagle, P. Sieving, U. Kellner, R. Spegal, M. Anastasi, E. Zrenner, L. T. Sharpe, and B. Wissinger, "Mutations in the CNGB3 gene encoding the beta-subunit of the cone photoreceptor cGMP-gated channel are responsible for achromatopsia (ACHM3) linked to chromosome 8q21," Hum Mol Genet, vol. 9, pp. 2107-16, Sep 1 2000.

[100] S. Kohl, T. Marx, I. Giddings, H. Jagle, S. G. Jacobson, E. Apfelstedt-Sylla, E. Zrenner, L. T. Sharpe, and B. Wissinger, "Total colourblindness is caused by mutations in the gene encoding the alpha-subunit of the cone photoreceptor cGMP-gated cation channel," Nat Genet, vol. 19, pp. 257-9, Jul 1998.

[101] M. Glickstein and G. G. Heath, "Receptors in the monochromat eye," Vision Res, vol. 15, pp. 633-6, Jun 1975.

[102] J. L. Rees, "Genetics of hair and skin color," Annu Rev Genet, vol. 37, pp. 67-90, 2003.

[103] P. Hedera, S. Lai, E. M. Haacke, A. J. Lerner, A. L. Hopkins, J. S. Lewin, and R. P. Friedland, "Abnormal connectivity of the visual pathways in human albinos demonstrated by susceptibility-sensitized MRI," Neurology, vol. 44, pp. 1921-6, Oct 1994.

[104] A. B. Morland, M. B. Hoffmann, M. Neveu, and G. E. Holder, "Abnormal visual projection in a human albino studied with functional magnetic resonance imaging and visual evoked potentials," J Neurol Neurosurg Psychiatry, vol. 72, pp. 523-6, Apr 2002.

[105] R. W. Guillery, "Neural abnormalities of albinos," Trends in Neurosciences, vol. 9, pp. 364-367, 1986.

[106] A. B. Morland, H. A. Baseler, M. B. Hoffmann, L. T. Sharpe, and B. A. Wandell, "Abnormal retinotopic representations in human visual cortex revealed by fMRI," Acta Psychol (Amst), vol. 107, pp. 229-47, Apr 2001.

[107] R. W. Guillery, V. A. Casagrande, and M. D. Oberdorfer, "Congenitally abnormal vision in Siamese cats," Nature, vol. 252, pp. 195-9, Nov 15 1974.

[108] D. H. Hubel and T. N. Wiesel, "Aberrant visual projections in the Siamese cat," J Physiol, vol. 218, pp. 33-62, Oct 1971.

[109] K. Huang and R. W. Guillery, "A demonstration of two distinct geniculocortical projection patterns in albino ferrets," Brain Res, vol. 352, pp. 213-20, Jun 1985.

[110] J. H. Kaas and R. W. Guillery, "The transfer of abnormal visual field representations from the dorsal lateral geniculate nucleus to the visual cortex in Siamese cats," Brain Res, vol. 59, pp. 61-95, Sep 14 1973.

[111] E. I. Elekessy, J. E. Campion, and G. H. Henry, "Differences between the visual fields of Siamese and common cats," Vision Res, vol. 13, pp. 2533-43, Dec 1973.

[112] M. L. Cooper and G. G. Blasdel, "Regional variation in the representation of the visual field in the visual cortex of the Siamese cat," J Comp Neurol, vol. 193, pp. 237-53, Sep 1 1980.

[113] R. W. Guillery, T. L. Hickey, J. H. Kaas, D. J. Felleman, E. J. Debruyn, and D. L. Sparks, "Abnormal central visual pathways in the brain of an albino green monkey (Cercopithecus aethiops)," J Comp Neurol, vol. 226, pp. 165-83, Jun 20 1984.

[114] M. B. Hoffmann, D. J. Tolhurst, A. T. Moore, and A. B. Morland, "Organization of the visual cortex in human albinism," J Neurosci, vol. 23, pp. 8921-30, Oct 1 2003.

[115] M. B. Hoffmann, P. S. Seufert, and L. C. Schmidtborn, "Perceptual relevance of abnormal visual field representations: static visual field perimetry in human albinism," Br J Ophthalmol, vol. 91, pp. 509-13, Apr 2007.

[116] M. B. Hoffmann, B. Lorenz, A. B. Morland, and L. C. Schmidtborn, "Misrouting of the optic nerves in albinism: estimation of the extent with visual evoked potentials," Invest Ophthalmol Vis Sci, vol. 46, pp. 3892-8, Oct 2005.

[117] J. Klemen, M. B. Hoffmann, and C. D. Chambers, "Cortical plasticity in the face of congenitally altered input into V1," Cortex, Mar 23 2012.

[118] P. Apkarian, L. Bour, and P. G. Barth, "A unique achiasmatic anomaly detected in non-albinos with misrouted retinal fugal projections," Eur J Neurosci, vol. 6, pp. 501-7, Mar 1 1994.

[119] P. Apkarian, L. J. Bour, P. G. Barth, L. Wenniger-Prick, and B. Verbeeten, Jr., "Non-decussating retinal-fugal fibre syndrome. An inborn achiasmatic malformation associated with visuotopic misrouting, visual evoked potential ipsilateral asymmetry and nystagmus," Brain, vol. 118 (Pt 5), pp. 1195-216, Oct 1995.

[120] J. D. Victor, P. Apkarian, J. Hirsch, M. M. Conte, M. Packard, N. R. Relkin, K. H. Kim, and R. M. Shapley, "Visual function and brain organization in non-decussating retinal-fugal fibre syndrome," Cereb Cortex, vol. 10, pp. 2-22, Jan 2000.

[121] S. Prakash, S. O. Dumoulin, N. Fischbein, B. A. Wandell, and Y. J. Liao, "Congenital achiasma and see-saw nystagmus in VACTERL syndrome," J Neuroophthalmol, vol. 30, pp. 45-8, Mar 2010.

[122] M. B. Hoffmann, F. R. Kaule, N. Levin, Y. Masuda, A. Kumar, I. Gottlob, H. Horiguchi, R. F. Dougherty, J. Stadler, B. Wolynski, O. Speck, M. Kanowski, Y. J. Liao, B. A. Wandell, and S. O. Dumoulin, "Plasticity and stability of the visual system in human achiasma," Neuron, vol. 75, pp. 393-401, Aug 9 2012.

[123] L. Muckli, M. J. Naumer, and W. Singer, "Bilateral visual field maps in a patient with only one hemisphere," Proc Natl Acad Sci U S A, vol. 106, pp. 13034-9, Aug 4 2009.

[124] A. A. Brewer, "Visual maps: To merge or not to merge," Curr Biol, vol. 19, pp. R945-7, Nov 3 2009.

[125] C. Darian-Smith and C. D. Gilbert, "Axonal sprouting accompanies functional reorganization in adult cat striate cortex.," Nature, vol. 368, pp. 737-740, 1994.

[126] G. C. DeAngelis, A. Anzai, I. Ohzawa, and R. D. Freeman, "Receptive field structure in the visual cortex: does selective stimulation induce plasticity?," Proc Natl Acad Sci U S A, vol. 92, pp. 9682-6, Oct 10 1995.

[127] J. C. Horton and D. R. Hocking, "Monocular core zones and binocular border strips in primate striate cortex revealed by the contrasting effects of enucleation, eyelid suture, and retinal laser lesions on cytochrome oxidase activity," J Neurosci, vol. 18, pp. 5433-55, Jul 15 1998.

[128] J. H. Kaas, "Plasticity of sensory and motor maps in adult mammals.," Annual Review of Neuroscience, vol. 14, pp. 137-167, 1991.

[129] C. I. Baker, E. Peli, N. Knouf, and N. G. Kanwisher, "Reorganization of visual processing in macular degeneration," J Neurosci, vol. 25, pp. 614-8, Jan 19 2005.

[130] C. D. Gilbert and T. N. Wiesel, "Receptive field dynamics in adult primary visual cortex," Nature, vol. 356, pp. 150-2, Mar 12 1992.

[131] Y. M. Chino, J. H. Kaas, E. L. Smith, 3rd, A. L. Langston, and H. Cheng, "Rapid reorganization of cortical maps in adult cats following restricted deafferentation in retina," Vision Res, vol. 32, pp. 789-96, May 1992.

[132] C. Darian-Smith and C. D. Gilbert, "Topographic reorganization in the striate cortex of the adult cat and monkey is cortically mediated," J Neurosci, vol. 15, pp. 1631-47, Mar 1995.

[133] A. Das and C. D. Gilbert, "Topography of contextual modulations mediated by short-range interactions in primary visual cortex," Nature, vol. 399, pp. 655-61, Jun 17 1999.

[134] A. Das and C. D. Gilbert, "Long-range horizontal connections and their role in cortical reorganization revealed by optical recording of cat primary visual cortex," Nature, vol. 375, pp. 780-4, Jun 29 1995.

[135] J. S. Sunness, T. Liu, and S. Yantis, "Retinotopic mapping of the visual cortex using functional magnetic resonance imaging in a patient with central scotomas from atrophic macular degeneration," Ophthalmology, vol. 111, pp. 1595-8, Aug 2004.

[136] J. H. Kaas, L. A. Krubitzer, Y. M. Chino, A. L. Langston, E. H. Polley, and N. Blair, "Reorganization of retinotopic cortical maps in adult mammals after lesions of the retina," Science, vol. 248, pp. 229-31, Apr 13 1990.

[137] H. A. Baseler, A. B. Morland, and B. A. Wandell, "Topographic organization of human visual areas in the absence of input from primary cortex," J Neurosci, vol. 19, pp. 2619-27, Apr 1 1999.

[138] C. D. Gilbert, "Adult cortical dynamics," Physiol Rev, vol. 78, pp. 467-85, Apr 1998.

Chapter 13

Learning-Based Cross-Modal Plasticity in the Human Brain: Insights from Visual Deprivation fMRI

Lora Likova

Additional information is available at the end of the chapter

http://dx.doi.org/10.5772/54835

1. Introduction

Neuroimaging, and functional Magnetic Resonance Imaging (fMRI) in particular, has brought dramatic changes in all fields of brain studies, one of the notable fields under intensive development being that of brain plasticity. To overcome the artificiality of typical experimental paradigms, this chapter presents a set of interrelated investigations focused on the reorganization of working memory and sensorimotor mechanisms as a result of training in a complex 'real-life' paradigm. The novel paradigm that I have introduced is based on the spatiomotor task of drawing, because drawing – from artistic to technical – is a 'real-life' task that uniquely incorporates diverse aspects of perceptual, cognitive and motor skills, thus activating the full *perception-cognition-action 'loop'*. It is important to recognize that drawing engages a wide range of spatial manipulation abilities (e.g., spatio-constructional decisions, coordinate transformations, geometric understanding and visualization), together with diverse mental representations of space, conceptual knowledge, motor planning and control mechanisms, working and long-term memory, attentional mechanisms, as well as, empathy, emotions and forms of embodied cognition (Grossi and Trojano, 1999, for review; Trojano et al., 2009; Likova, 2012a; 2013). Furthermore, the development of a novel training method in non-visual forms of drawing, the Cognitive-Kinesthetic Drawing Method (Likova, 2010a), makes it possible to use drawing for studying training-based cross-modal reorganization in visual areas activated by non-visual tasks.

A further advanced aspect of the studies is that they were run across subject populations with diverse levels of visual experience, ranging from sighted (under blindfolded conditions), through late-onset blind with past visual experience, to congenitally blind who have never had any visual input to stimulate the visual system.

A problem, specific to the studies on brain plasticity in the blind, is that usually such studies do not include training, but simply compare activation in the blind to that in the sighted. The studies in this chapter have taken a different , *training-based* approach, which allows us to go beyond a mere task/activation correlation to *causal* inferences and also to investigate the *evolution* of plastic changes *within the same* brain. Again, this was possible thanks to our novel Cognitive-Kinesthetic Method for teaching people to draw without vision, coupled with a novel *memory paradigm* based on drawing. The pre/post-training brain changes were assessed by fMRI; the respective behavioral changes were evaluated by pre/post-training comparisons of the drawing performance. Cross-group analyses between the diverse subject populations allowed us to evaluate novel hypotheses on cross-modal brain plasticity, and moreover, suggested a need for re-conceptualization of basic principles of the functional architecture of the brain.

1.1. The novel approach

Conceptual Framework

1. *Space transcends any specific sensory modality.* As emphasized by the phenomenon of drawing by the blind (e.g., Kennedy, 1993, 2000, 2003, 2006; Ponchilla, 2008; Likova, 2010 a,b; 2012a,b; 2013), space and spatial structure are not represented solely by the visual modality. The visual system may be the modality best suited to process spatial information, but it is not the only one. Thus, when deprived of visual input, the brain is capable of employing the 'free' visual resources in the most relevant way. (As there is an ambiguity in the use of the term 'spatial', particularly in the working memory and imagery literature, we note that when used in this paper, 'spatial' refers to the perception of any spatial structure – 2D or 3D, static or dynamic – *independently* of the sensory modality exploring it. For example, a face can be recognized by exploring its spatial structure with the hands, or a geometric function can be 'seen' by audio-graphics, etc.) Our view is that *drawing* deals with spatial structures in this general sense, and consequently it has the advantage that can readily be 'translated' from a visual into a tactile form.

2. *Closing the perception-cognition-action loop is a powerful amplifier for learning* , so the *task* selection is critical. *Drawing* is a complex task precisely orchestrating multiple brain mechanisms, thus our hypothesis is that, consequently, it is suitable for an integrative, perception-cognition-action paradigm.

3. *Training in highly engaging unfamiliar tasks* that provide fun and inspiring outcomes is a fruitful paradigm for driving brain reorganization and assessing its earlier stages. *Drawing* is such a task, particularly when studied under the unusual circumstances of visual deprivation.

4. *Tasks demanding detailed re-expression of memory representations force the development of precise and robust memory. Drawing* from memory demands such explicit *re-expression* through the motor loop, and consequently it demands 'high-resolution' internal representations to be communicated back through the drawing act.

http://dx.doi.org/10.5772/54835

5. *A direct memory 'readout' would be a big advantage for memory-studying paradigms. Drawing* from memory provides such direct memory 'readout' in the form of the drawn image, as it ensures an explicit expression of the remembered information by externalization of the mental representation that guides the drawing hand.

Drawing as a Novel Memory Paradigm Incorporating the Framework Principles

Since drawing uniquely incorporates the principles in our conceptual framework outlined above, we used drawing to develop a paradigm for both *studying* and *training* memory (e.g., Likova, 2010b; 2012a,b). We hypothesized that *memory-guided* drawing, in particular, has a powerful potential for training because it forces the learners not to just passively explore the stimuli, but to develop detailed and stable *memory* representations (*Drawing/Memory Hypothesis*). Furthermore, we conceptualized that, as our novel paradigm can be adapted from the visual to the tactile modality, it can become an effective tool for studying *learning-based cross-modal brain reorganization.*

The Need for the Cognitive-Kinesthetic Training Method

The visual appreciation of line drawings (e.g., Fig. 1) seems such an effortless process that we may not be aware of the invisible work of powerful high-level cognitive mechanisms that provide the artist with the ability to transform a 3D object into its 2D projection by abstracting just the right contours into a line drawing, and we may not appreciate how our brains make the 'inverse transformation' to the immediate understanding of what such 2D drawings represent.

Figure 1. Matisse: Lithographes No. 54: ' *Marie'* (left), No. 35: ' *Ma Maîtresse'* (center), No. 45: ' *Les Colombes Amoureuses'* (right), from the illustrations for Ronsard's *'Florilège des Amours'* (1948).

It came as a surprise, therefore, to find that, when exposed for the first time to 2D raised-line drawings, many blind participants had tremendous difficulty in the tactile recognition and comprehension of the depicted 3D objects. This 'negative' finding, however, provided us with the opportunity to employ a learning paradigm *in adults* , thus enabling for the first time the investigations of the developmental process of key cognitive components important in drawing, such as *spatial memory.*

The inherently visual nature of drawing, however, is an obvious barrier for the implementation of the drawing-based memory paradigm to research and rehabilitation of people with blindness or profound visual impairment. This issue was successfully overcome by implementing the C-K method that I have conceptualized and developed for training freehand drawing without vision, guided solely by *non-visually acquired* memory. This unique method has already shown that it can enhance detailed working memory, and in particular, the generation and maintenance of robust spatial-memory representations (Likova, 2012a,b; 2013), which are crucial for replacing the *'eye-hand'* coordination (lost under blindness) by *'non-visual memory-hand'* coordination (Fig. 2). Moreover, I have proposed that an effective approach to blindness requires wide-spectrum blindness rehabilitation through a unified, single task, and have shown that the C-K method is highly promising to fulfill that demand.

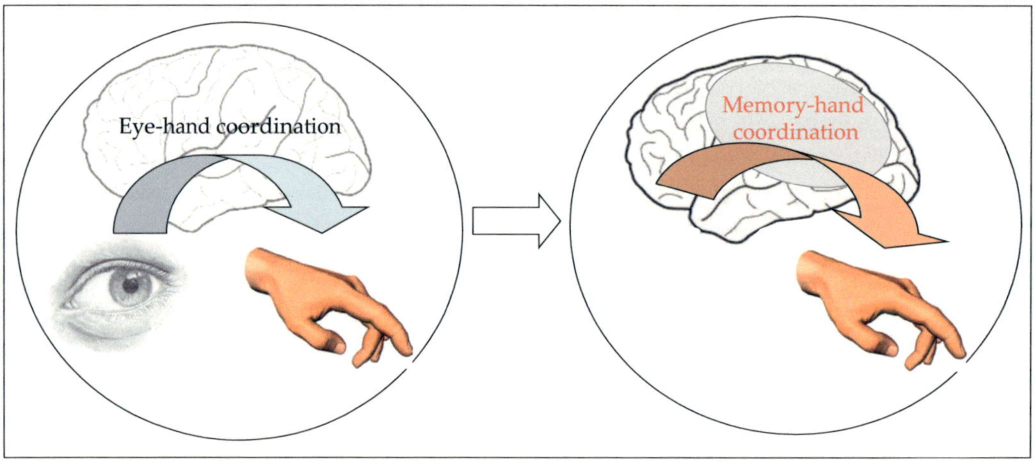

Figure 2. Illustration of our conceptualization that targeted training for development of robust nonvisual memory representations for spatiomotor coordination is needed when the 'eye-hand' coordination is lost or impaired. This conceptualization underlies the Cognitive-Kinesthetic Drawing Method and its rehabilitation potential.

The Cognitive-Kinesthetic training proved to be very effective in the successful training even of people who are totally blind. In only a week of 1 - 1.5 hour/day training sessions, the participants learn to draw complex structures, such as faces and objects (Fig. 3A).

Moreover, to provide for pre/post training comparisons of brain activation, we have developed an innovative fMRI platform, which includes an MRI-compatible multisensory drawing tablet, with a stylus incorporating a fiber-optic motion-capture system to record the drawing movements to 1 mm accuracy. The C-K Method, with the technological advances of the innovative experimental platform, allowed us to run the first neuroimaging studies on *learning* to draw and on *cross-modal memory reorganization*.

http://dx.doi.org/10.5772/54835

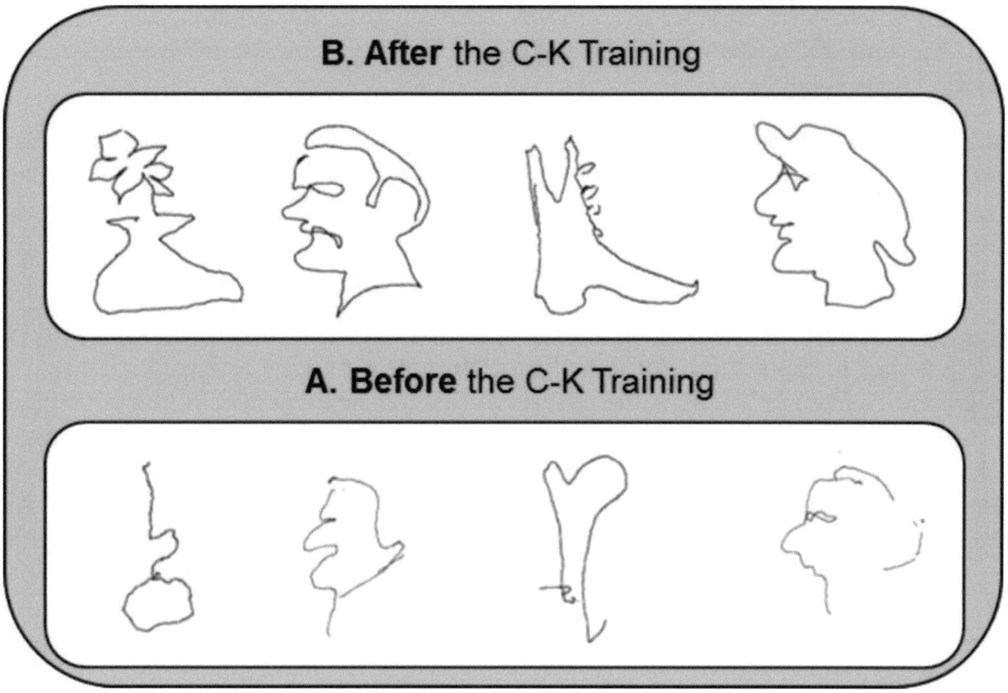

Figure 3. Examples of drawings from different totally blind participants recorded by our motion-capture system in the fMRI scanner. A. Bottom row: The same faces and objects drawn by the same subjects before the training.B.Top row:Blind subjects drawings after a week of the Cognitive-Kinesthetic training.

1.2. The working memory 'Sketchpad' puzzle

More than a century ago, Helmholtz (1871) had pointed out that artists possess not only advanced observational capabilities, but also an enhanced *memory* of the observed images. While the first part of this statement is often cited in vision science, the memory-related aspect has been widely neglected. Our recent results, however, come fully in support of Helmholtz' observation, and are consistent with the *Drawing/Memory Hypothesis* that we propose.

Freehand memory-guided drawing, in particular, challenges the neural system to provide encoding, retrieval and effective 'projection' of memory representations back onto some form of an internal high-resolution ' *projection screen*' , so as to guide the spatial trajectory of the drawing-hand with the requisite precision. It thus closes the full processing 'loop' from perception through memory to precisely controlled motor action (Likova, 2012a).

We suggest that the classic model of working memory provides a framework for understanding the memory processes involved in the drawing task. In this model, working memory (WM) is the "ability to retain information for several seconds to guide goal-directed behaviors" (Baddeley and Hitch, 1974; Baddeley, 1986, 2000, 2003). The WM model has provided major insights into functional neuroimaging on memory, and conversely, has been in turn updated based on neuroimaging data. Recently, for example, it has been suggested that WM processes are not restricted to short-term memory but also reflect activation of long-term memory (LTM) representations (e.g., Ranganath, 2009; Buchsbaum and D'Esposito, 2009; Ishai, 2009). Our treatment of WM is consistent with this recently expanded interpretation.

Baddeley's model consists of four major components, the one we find of interest here being the visuo-spatial 'sketchpad', which conveys the idea that memory is 'drawing' or generating a mental sketch of the memorized object, an internal representation that can be further spatially manipulated. Thus, this 'sketchpad' concept provides much of the functionality of the internal 'projection screen' introspectively needed for drawing.

However, where in the brain the WM 'sketchpad' might be implemented? The neural substrate required to serve such 'sketchpad' function has to provide a large, high-resolution topographic 'screen' or 'map'. Although the full neural implementation of the sketchpad concept is still a puzzle, these requirements point out to the primary visual cortex, area V1, as the best candidate. A few theoretical and neurophysiological studies in non-human primates (e.g., Mumford, 1992; Super et al., 2001a,b; Lee and Mumford, 2003; Super, 2003) had already suggested that V1 may provide the high-resolution visuo-spatial 'sketchpad' function, but there has been little supporting evidence in human. Traditionally, all regions of the early visual cortex have been considered predominantly bottom-up, purely sensory, regions, specifically devoted to the visual modality. However, it is now known that this is not the case, as many of these regions are also subject to *top-down visual* processes such as visual imagery (Ishai and Sagi, 1995; Kastner et al. 1998; Ishai et al., 2000; Kosslyn et al., 1996, 1997, 1999; Kosslyn and Thompson 2003; Kreiman et al., 2000; Lambert et al., 2004; Mechelli et al., 2004; O'Craven and Kanwisher, 2000; Ganis et al., 2004; 2005; Amedi et al., 2005; Slotnick et al. 2005), and - most recently – to short-term visual memory (e.g., William at al., 2008; Harrison and Tong, 2009). Beyond visual top-down effects, different regions of the visual cortex have been also activated cross-modally in a number of auditory and tactile *sensation* tasks (e.g., Sadato et al., 1996; Cohen et al., 1997; Zangaladze et al., 1999; Pascual-Leone and Hamilton, 2001; Block, N., 2003). Moreover, two recent studies have found V1 activation during verbal and episodic memory tasks (Amedi et al., 2003; Raz et all, 2005). No investigation has been done, however, on cross-modal *tactile memory* involvement in these regions.

1.3. Can blind drawing help in solving the puzzle? If there is working memory sketchpad, may it be amodal?

Recently, I hypothesized that the early visual areas, and V1, in particular, can support *modality-independent (i.e., amodal)* 'sketchpad' in human (Likova, 2010b; 2012a; 2013). This hypothesis, which is of importance for models of memory and functional brain architecture, is investigated under our novel drawing-based paradigm. To critically test the amodal hypothesis, the studies in this chapter focus on groups of participants with different forms of total ocular visual deprivation – congenitally blind, late-onset blind and blindfolded – of brain reorganization is studied by fMRI before and after C-K training. To our knowledge, there are no previous studies on the *learning* to draw in the blind and on the *corresponding dynamic phases* of adult brain reorganization. The data showed strong, well-structured V1 activation in a *tactilely-acquired memory* task. These results are consistent with V1 implementing the

http://dx.doi.org/10.5772/54835

memory sketchpad concept, however, as no vision is involved, they imply a reconceptualization of the sketchpad from being 'visuo-spatial' to being ' *'amodal-spatial'*.

2. General methods

2.1. Innovative experimental platform

As there are no preceding neuroimaging studies of such training, we had to develop a unique experimental platform integrating a number of innovations, such as: (i) the *Cognitive-Kinesthetic Method* , (ii) a *multisensory MRI-Compatible Drawing Tablet (for both tactile and visual drawing)* , incorporating a motion-capture stylus, (iii) a *Method for Estimating Topographic Maps in the Blind,* (iv) as well as the first implementation of *Visual Probabilistic Maps* in blind individuals.

2.2. Experimental design

A battery of raised-line models of faces and objects was developed as the drawing targets, and a three-task block fMRI paradigm, with task duration being 20 s, interleaved with a 20 s null intervals (Fig. 4 A). The tasks were as follows: *Explore/Memorize, E/M—* perceptual exploration and memorization of the model to be drawn; *MemoryDraw, MD—* a memory-guided non-visual drawing task; and *Scribble, S—* a motor-control and 'negative' memory-control task. Importantly, as opposed to the traditional null intervals, the subjects not only rested motionless during these intervals *(NullInterval, NI)* but were instructed (and practiced) to clear any memory or image structures from awareness ('mind blank'). The start of each task or null interval was prompted by an auditory cue. The whole three-task sequence with interleaved null intervals (*NI, E/M, NI, MD, NI, S, NI)* was repeated 12 times in each 1-hour fMRI session, using a new drawing model for each repeat.

In *Explore/Memorize,* using the left hand only, the subjects had to explore tactually a raised-line drawing model on the left slot of the drawing tablet continuously for 20 s, and to develop a full memory representation of the image in preparation for the *MemoryDraw* task. Then the model was removed and the subjects rested motionless for 20 s with no image in mind (*NullInterval*). In the following *MemoryDraw* phase, the fiber-optic stylus was used to draw the image (from tactile memory) on the right slot of the tablet using the right hand only. In the control task *Scribble* , the subjects moved the stylus with the right hand in a random trajectory matching the extent and rate of the drawing movements but with 'mind-blank'.

This design ensures that the drawing was performed exclusively by the right hand while keeping the left hand motionless during *MemoryDraw* ; moreover, because the virtual stylus was used to draw on the plastic surface of the tablet slot, there was no tactually perceivable trace and no possibility for tactile tracing with the left hand during drawing in the scanner. The lack of left hand movement during drawing was also confirmed by the lack of BOLD activation in the left hand motor area. These restrictions thus enforce the encoding of robust *memory* representations in order to guide the drawing trajectory.

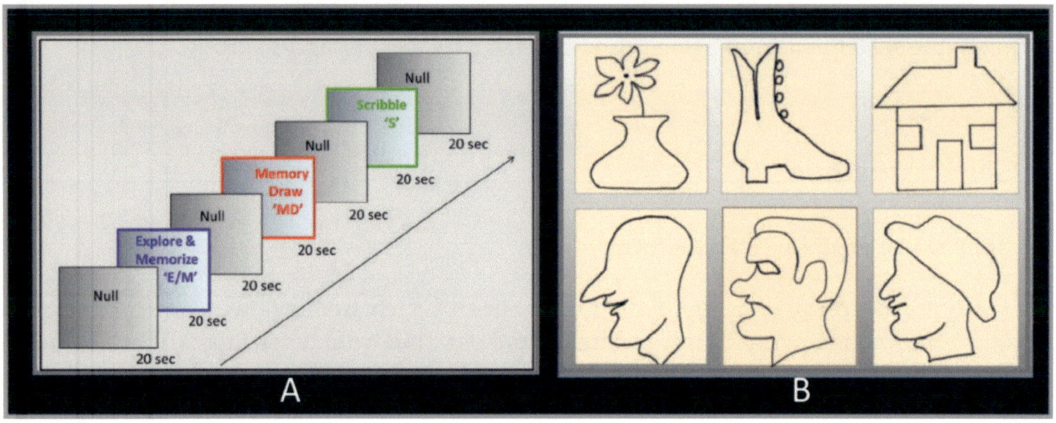

Figure 4. Experimental design. (A) Drawing was investigated in a three-phase paradigm consisting of a memory-guided drawing task, abbreviated as "MemoryDraw" *(MD),* plus two control tasks: a motor and "negative" memory control task "Scribble" *(S),* and a task of perceptual exploration and memorization of the model to be drawn "Explore/ Memorize" *(E/M).* Each task's duration was 20 s, with 20 s null intervals interposed between the tasks, the whole 140 s trial sequence being repeated 12 times in each of the 1-hour pre- and post-training fMRI sessions using a new image for each repeat. (B) Raised-line drawings of realistic faces and objects were presented as templates to be explored by the subject using her left hand in the *E/M* phase. Two repetitions of each stimulus were run. The quality of the reproductions was assessed by a masked rating procedure, based on recognition and similarity to the templates. Eexamples of reproduction are shown in in Figs. 3 & 6B.

2.3. Tactile stimulus presentation and hand-movement control

A custom-built, multisensory MRI-compatible drawing system. To run drawing studies within the scanner is not a conventional protocol and faced a lot of challenging technological problems. The special-purpose drawing system (Fig. 5) that we have developed (1) is MRI-compatible, (2) is ergonomically adaptable to the small space available inside the scanner bore, (3) allows multiple tactile models to be presented sequentially in the scanner, (4) captures and records the drawing trajectory with high precision, and (5) can provide real-time visual feedback of the drawn trajectory (when appropriate in *visual* drawing). It incorporates a dual-slot drawing tablet that is height/distance adjustable for the subject's arm length and a specially-adapted version of a fiber-optic motion-capture system for on-line recording of the drawing movements. This multisensory system supports fMRI investigations of *tactually-* guided drawing by providing for the presentation of multiple raised-line models in the scanner without the need of any operator assistance. It also allows us to record relevant behavioral events and to correlate them to the brain activation for full off-line analysis.

Auditory cue presentation. The auditory stimuli were presented through Resonance Technologies Serene Sound earphones (Resonance Technologies, Salem, MA). To reduce scanner noise, this equipment employs external ear protectors with perforated earplugs that conduct the auditory cues directly into the auditory passage while blocking much of the scanner noise.

http://dx.doi.org/10.5772/54835

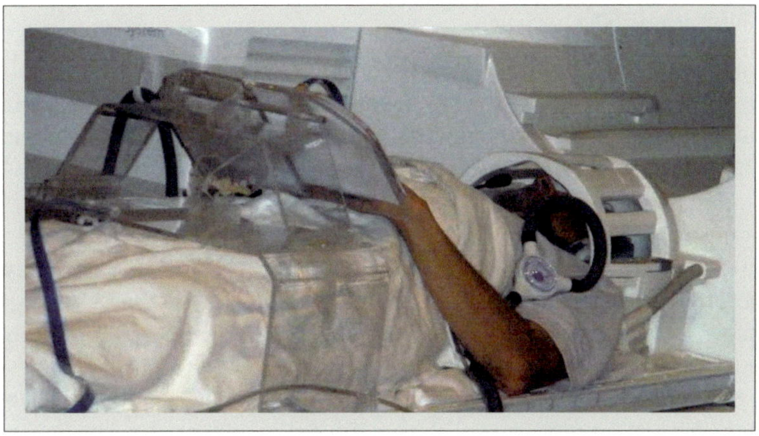

Figure 5. A subject on the scanner bed operating our novel multimodal MRI-compatible drawing device. The plexiglass gantry supports a drawing tablet while a fiber-optic drawing stylus captures and records the drawing movements with high precision. The motion capture information synchronized with the fMRI allows the effect of behavioral events to be analyzed to high precision.

2.4. Computerized Recognizability Index (CR Index)

The C-K training results in dramatic drawing improvements (e.g., Fig. 3B, top row; Fig. 7B) in all forms of visual deprivation, but to be able to quantify that improvement we developed the CR Index, which is based on an optimized spatial-correlation fit across the full spectrum of affine transformations (translation, rotation, scaling and shear). It is applied to each recorded drawing with respect to the set of template images to be drawn, cumulating the proportion of the image that contains matching contours (with sparse-matrix correction for incomplete drawings). The total CR Index of the drawings can then be compared before and after training, to quantify the improvement in the quality of the drawing skill that has been achieved. The CR Index was validated by a masked rating procedure for the quality of the drawing reproductions based on perceptual similarity to the templates.

2.5. MRI data collection, analysis, and visualization

FMRI acquisition

MR data were collected on a Siemens Trio 3T scanner equipped with 8-channel EXCITE capability, a visual stimulus presentation system and response buttons. A high-resolution anatomical (T1-weighted) volume scan of the entire brain was obtained for each observer (0.8 × 0.8 × 0.8 mm voxels). The fMRI blood-oxygenation-level-dependent (BOLD) responses were collected with EPI acquisition from the whole head coil. There were 34 axial slices at 2 s TR, with TE of 28 ms and flip angle of 80°, providing 3.0 × 3.0 × 3.5 mm voxels throughout the brain. The functional activations were processed for slice-time correction and motion correction. An anatomical segmentation algorithm (mrGray, part of the VISTA software package specified below) was applied to the T1 scan, ensuring localization of the signal within the cortical gray matter close to the activated neurons.

FMRI time course analyses

The analysis software was Stanford Vision and Imaging Science and Technology (VISTA, Stanford University) software. The data were analyzed to estimate the effective neural activation amplitudes (e.g., Friston et al., 1994) for each task across the 12 repeats of the 3-task sequence in the one-hour scan by the following procedure. A General Linear Model (GLM) consisting of a (3+1)-parameter boxcar neural activation model convolved with an estimated hemodynamic response function (HRF) was fitted to the BOLD responses for each 3-task sequence, combined with a 1-parameter boxcar model of the 8 auditory cue presentations and an additive 4th-order polynomial to capture low-frequency drift in the BOLD signal. (The HRF was determined once per session by optimizing it to a subset of gray matter voxels identified as those whose average modulation (as a result of the task alternation sequence across the E/M, MD and S tasks) exceeded the statistical level of $z = 3$ in each voxel ($p < 0.001$).) Thus, the parameters of the activation model consisted of the boxcar activation amplitudes for the three task periods, relative to the remainder of the 140 s scan duration.

Voxel-wise parametric maps

For each task, statistical parametric maps were generated, based on the estimated activation amplitudes from the above GLM in each voxel. As is standard for GLM, the boxcar neural activation model for each 20 s task period was contrasted with the entire remainder of the 140 s scan duration. All three task-models were optimized jointly to the detrended BOLD waveform. These maps could be viewed in the 3D volume or projected onto 3D views of the inflated cortex or on flatmaps of cortical regions of particular interest.

ROI activation analysis

The effective neural activation amplitudes (bar graphs) for each condition in each region of interest (ROI) were estimated by the same GLM procedure but now applied to the average signal across all voxels within the ROI. This procedure also provided high-quality time courses for evaluation of the response dynamics and its comparison across tasks and stages of training.

The confidence intervals and error bars were defined by the amplitude variability of the 12 repeats of the 3-task sequence in each one-hour scan. The dashed lines represent the 99% "zero" confidence interval within which the activation amplitudes are not significantly different from zero (at $p < 0.05$, corrected for multiple applications within each figure). The error bars are "difference" confidence intervals designed to illustrate the t-test for the significance of differences between any two activation levels.

In the text, the significance of all ROI-comparisons are specified by the t-test using a statistical criterion threshold of $p < 0.05$ corrected for multiple comparisons.

Retinotopic maps and motion localizer in the sighted

There is a topographic 'projection' or 'mapping' of the visual input information to the retinal surface, and from the retinal surface to corresponding points on the surface of the visual cortex of the brain, thus forming what are known as retinotopic cortical maps. The boundaries of the retinotopic projection areas in the occipital cortex V1, V2d, V2v, V3d and V3v were established as described in Sereno et al. (1995); Tootell et al. (1996) and Engel, Glover and Wandell (1997).

Retinotopic projection areas V3A, V3B, hV4 and V7 were specified in accordance with Tyler et al. (2005) and Schira et al (2010). The hMT+ motion complex was identified using the standard stimuli of an expanding and contracting motion vector field of low-contrast random dots, alternating with a field of static dots as previously implemented in Likova and Tyler, 2007.

Topographic maps in the blind

On the one hand, no informed analysis of the visual cortex could be done without knowledge of its retinotopic and functional organization; on the other hand, no retinotopic mapping is possible in the blind (other than cerebral blindness, which does not apply to any of our participants), so it is a challenge to localize any specific visual area. To resolve this issue and determine the borders of different 'visual' areas in the blind participants, we combined three different approaches. First, we used an innovative 14-step procedure (Likova, 2010a) to warp the brains of sighted and blind subjects to the same MNI brain coordinates. Second, for the primary visual cortex in particular, the location of the V1 ROI was confirmed by its intersection with its anatomical markers (the calcarine sulcus). Additionally, we also compared the result with the Freesurfer probability map of V1, which aligned sufficiently accurately (within 87% overlap of the voxels) with those from the other methods.

3. Where may the amodal spatial memory be implemented? Blindfolded studies and reconceptualization of the WM sketchpad

3.1. Beyond the traditional view of modality-specific cortices specialized for sensory processing

According to the traditional view of brain architecture, the so-called 'sensory' cortical areas are modality-specific and specialized for sensory processing only. Although this view is still prevalent in textbooks, increasing evidence suggests that the function of these areas is not restricted to sensory processing, but (as discussed in the Introduction) is also subject to a number of top-down processes such as modality-specific imagery; furthermore, many of these 'sensory' areas are activated in cross-modal tasks. A body of experimental data shows that in blind individuals, brain areas commonly associated with the processing of visual information can undergo reorganization and be recruited in tasks such as Braille reading, naming, tactile object perception or tactile motion (e.g., Sedato et al., 1996; Uhl et al., 1991, 1993; Cohen et al., 1997; DeVolder et al., 1997; Buechel et al., 1998; Hamilton et al., 2000; Burton et al., 2002, 2003, 2004; Gizewski et al. 2003; Theoret et al., 2004; Merabet et al., 2005; Merabet, 2008; Pascual-Leone et al., 2005; Amedi et al., 2003, 2004, 2008; Goyal et al., 2006; Borowsky et al., 2007 ; Ptito et al., 2008; Deibert et al. 1999; Pietrini et al.; Cohen et al., 2004; Hagen et al., 2002; Matteau et al., 2010) and during drawing in blind individuals (Amedi et al., 2008; Likova, 2010a,b; 2012a,b). Transcranial magnetic stimulation (TMS) of the visual cortex has been found to disrupt Braille reading or verb-generation in the blind (Cohen et al., 1997; Hamilton & Pascual-Leone, 1998; Amedi et al., 2004).

This section probes deeper levels of cross-modal interactions that are of importance for advanced models of functional architecture of the brain.

3.2. Subjects and methods

A group of six subjects with normal vision were blindfolded throughout the experimental procedures. The subjects ranged in age from 25 to 59 and were 4 females and 2 males, right-handed with no formal training in visual art and with no cognitive disorders. All subjects gave informed consent for the experimental protocol approved by the local research ethics committee. Each subject underwent training in blindfolded drawing (1 to 1.5 hour/day for a week with our Cognitive-Kinesthetic Method.)

3.3. Results

Functional MRI, run before and after the blindfolded training, revealed strong activation in the primary visual cortex (area V1) for the tactile memory-guided drawing task (Memory-Draw, Fig. 6A), while, by contrast, no significant V1 activation was recorded for the non-memory drawing task (Scribble, Fig. 6B).

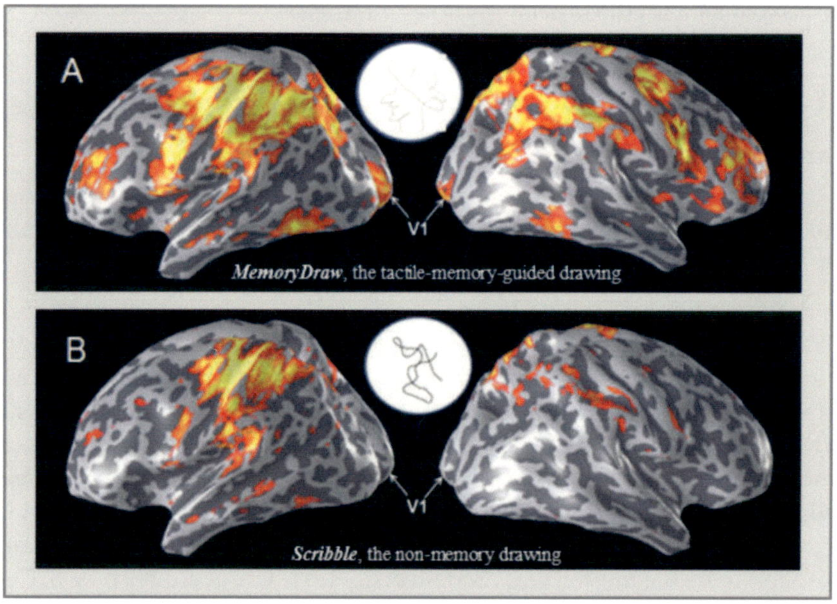

Figure 6. Comparison of the activation networks in the two types of drawing:(A) MemoryDraw (the tactile memory-guided drawing): Successful recall from memory of the specific tactile template explored 20 s earlier during E/M is critical for guiding the detailed movements of the drawing hand. (B) Scribble (the tactile non-memory, 'mind-less'drawing) was used as motor control and 'negative memory' control condition. Note the dramatic difference between the two types of drawing activity (Avs B): The memory-guided drawing activates an expanded bilateral network; remarkably, there is a well-localized activation in the primary visual cortex, V1. In contrast, the non-memory drawing predominantly activates the left hemisphere within the right hand region; most importantly, it does not activate the visual cortex at all.

Remarkably, without any visual involvement at any stage of the learning or drawing process, the C-K training enable the blindfolded participants to memorize and draw well complex images in only 20 s, guided solely by memory acquired during haptic exploration! Note how well coordinated and recognizable these drawings are (see Fig. 7 B).

3.3.1. Retinotopic maps for memory-guided drawing in the blindfolded

The best way to look in detail at the puzzling activation in the visual cortex is to present the data in a flatmap format with the retinotopic maps delineated (Fig. 7 A).

MemoryDraw in the blindfolded produced an unexpected, unique occipital response. Despite the lack of any visual (or even tactile) stimulation, there was extensive activation of the primary retinotopic area, V1, while the rest of the visual hierarchy showed a pronounced suppression of the BOLD response relative to the baseline level.

The other two activated regions seen on the flat map are the caudal intraparietal sulcus (cIPS) and an additional locus at the temporo-occipital border in LOtv. The latter site, although in close proximity and often partially overlapping with the ventro-anterior visual LOC, is shifted anteriorly towards the temporal lobe. These two non-occipital loci have been previously reported to be involved in tactile perception (Prather & Sathian, 2002; Amedi et al., 2001, 2002; James et al., 2002; Stoesz et al., 2003; Pietrini et al., 2004; Reed et al., 2004; Merabet et al., 2005; Van Boven et al., 2005; Weisser et al., 2005; Zhang et al., 2005).

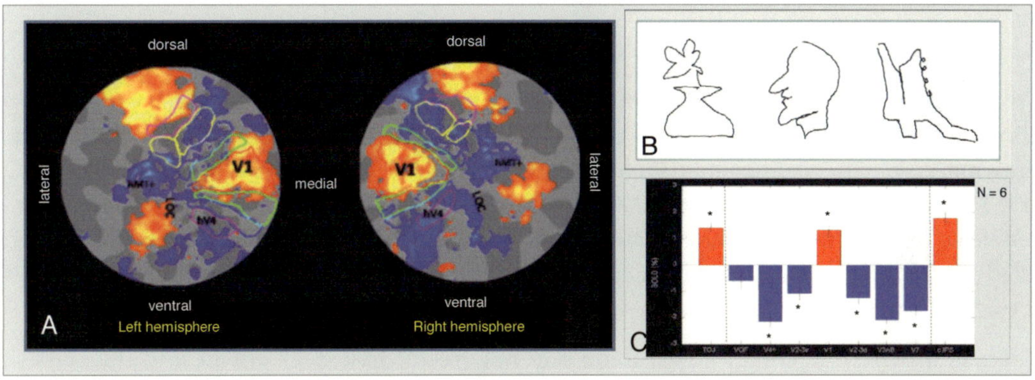

Figure 7. A. MD flat maps centered on the occipital pole. ROIs for the retinotopic hierarchy are indicated by colored outlines. The post-training MD map shows a unique 'triad' of three activation regions (orange-yellow coloration). Note in particular that the (non-visually-stimulated) primary visual cortex, V1, forms an isolated 'island' of activation surrounded by a 'sea' of suppression in the adjacent retinotopic areas. The other two activated regions seen on the flat map are the caudal intraparietal sulcus (cIPS) dorsally and an additional locus at the temporo-occipital border (LOtv). B.Post-training examples of blindfolded drawings demonstrating the level of detail the participants became able to complete in 20 s in the MRI scanner. Note that no vision was involved at any stage of the learning or drawing process! Recorded by the virtual stylus in the scanner are drawings of a vase with a flower, a face profile and a boot, each of which is easy to recognize (the corresponding templates are the 1st and the 2nd on the top row, and the 1st on the bottom row in Fig. 4B). C. Average response amplitude with standard errors for blindfolded memory-guided drawing in a group of 6 subjects, showing positive signal in the triad of areas – primary visual area V1, cIPS and LOtv; these three 'islands' of positive activation are separated by strong deactivation throughout boththe ventral and the dorsal extrastriate areas. Error bars represent 1 standard error of the means. Asterisks indicate activations significant at $p < 0.01$. [After Likova, 2013].

3.4. Discussion

3.4.1. Reconceptualization of the memory sketchpad

Tactile memory processing in the primary visual cortex?

Employing a pioneering memory paradigm based on non-visual drawing demonstrated that V1 can be strongly activated in a non-visual working memory task in the blindfolded. The V1 activation we have found is remarkable in many respects: i) the retinotopic analysis shows that the activation is rather well localized within V1, ceasing rapidly at a specific eccentricity; ii) surprisingly, its extension of about 10 deg (in terms of the retinotopic mapping) corresponds approximately to the angular subtense at the face that the drawings would have if projected optically; and iii) most remarkably – there is a massive deactivation surrounding V1, thus effectively 'cutting-off' any signal propagation from or through the extrastriate cortex. These results have major implications as discussed below.

To our knowledge, this study is the first evidence of involvement of the primary visual cortex in a tactile-memory task. We also know no other neuroimaging study either on the learning to draw, or on the effect of training on tactile-memory under visual deprivation. In MemoryDraw, the tactile template under the fingers of the left hand is memorized and then removed, so the drawing movements of the right hand are guided solely by that tactilely-acquired memory with no concurent tactile input from the drawn template (Penfield & Boldrey, 1937; Penfield & Rasmussen, 1950).

Amodal spatial memory vs visual imagery interpretation: Distinct patterns

To put the present results in a neurobiological perspective, we need a relevant concept about memory encoding and retrieval for the kinds of detailed spatial images used in the drawing task. The visuo-spatial sketchpad component of the classic WM model (Baddeley and Hitch, 1974; Baddeley, 1986, 2000, 2003) conveys the idea that working memory incorporates a detailed mental sketch of the retrieved object that can be further spatially manipulated (Fig. 8). This model provided us with a basis to generating specific hypothesis about the activation components in the drawing task.

There is a current tendency to attribute any new result that does not fit into the classical feedforward model of brain architecture, to a narrow number of top-down processes and - most recently - to 'visual imagery'. However, the pattern that we found of strong V1 activation surrounded by deactivation in the higher extrastriate areas (Fig. 7 A,B) is distinct, essentially the inverse of the known visual imagery pattern. In direct contrast to our results, the known neural substrate for visual imagery has been found to exhibit a gradient of activation decreasing from the higher visual areas toward V1, where the activation is much weaker or even lacking entirely. Although a weak level of V1 activation has sometimes been reported during imagery (Kosslyn et al., 1993; Le Bihan et al., 1993; Sabbah et al., 1995; Chen et al., 1998; Thompson et al., 2001; Ishai, Haxby, and Ungerleider, 2002; Lambert, Sampaio, Scheiber, and Mauss, 2002 ; Ganis et al., 2004), a larger number of other studies did not find any V1 activation at all (Goldenberg et al., 1991; Charlot et al., 1992; D'Esposito et al., 1997; Mazard et al., 2004; Mellet et al., 1995, 1998a, 1998b; Ishai, et al., 2000; Knauff et al., 2000;

Trojano et al., 2000; Wheeler et al., 2000; Formisano et al., 2002; Sack et al., 2002; Kaas et al., 2010). This decreasing gradient of activation implies that the imagery signal propagates from higher cortical regions in a top-down manner through the visual hierarchy towards V1.

In summary, while substantial imagery activation in the higher occipital areas has been consistently found across studies, this has not been the case with V1; furthermore, imagery does not produce any significant negative signal in the occipital cortex (Ganis et al., 2004; also see that paper for an overview of the imagery areas throughout the brain).

Our approach allows for testing between the following hypotheses: Hypothesis I: If the memory-based activation is due to the visual imagery mechanism, the predicted response-profile in the visual cortex would have the top-down 'imagery signature' of activation significantly decreasing from the higher extrastriate areas toward V1 (as reviewed above). Hypothesis II: If V1 was being employed as a WM sketchpad independently from the visual imagery process, it would be activated by a separate pathway external to the visual hierarchy, together with activation of WM-related sites beyond the occipital lobe.

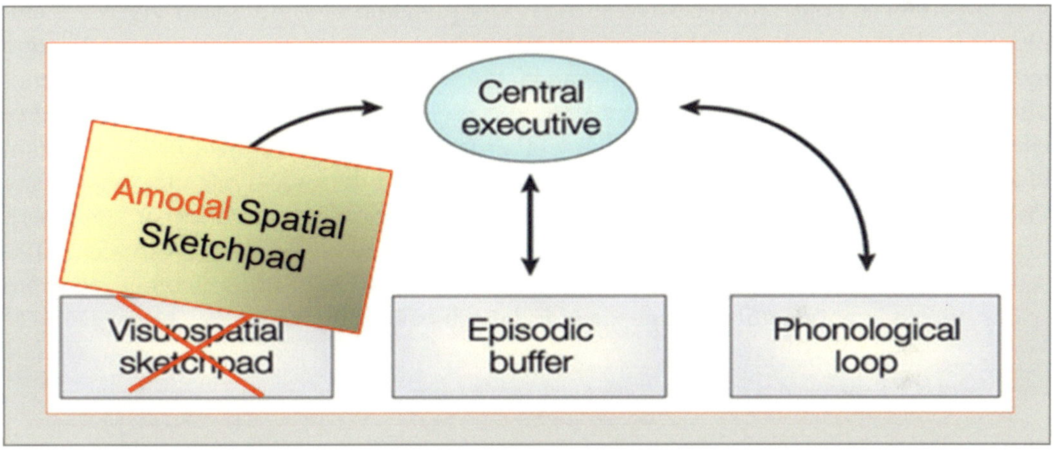

Figure 8. Our proposal for a re-conceptualization of the visuo-spatial sketchpad as an amodal-spatial sketchpad. Modified schematic of the main modules of Baddeley's classic model of working memory including the visuo-spatial sketchpad [after Baddeley, 2003], where the added "Amodal Spatial Sketchpad" block depicts our re-conceptualization of the visuo-spatial sketchpad as being accessible to any sensory modality. [From Likova, 2012a]

The imagery hypothesis seems not to be supported by the data because we find the inverse gradient of strong activation in the iconic visual area V1, and not only decreased but negative response in the higher occipital areas. The deactivation means that it is implausible the signal to be 'delivered' through the visual hierarchy, because this hierarchy is suppressed, i.e., V1 is 'cut-off' from the higher-level cortical regions that could generate and propagate imagery signals. Thus, this activation pattern is incompatible with the main principle of visual imagery as a top-down process propagating through the visual pathway towards V1. Instead, this pattern seems to be more consistent with Hypothesis II that V1 is operating as a working memory component, such as the hypothetical spatial memory buffer/sketchpad of the composite working memory model.

Re-conceptualization of the visuo-spatial sketchpad as an amodal-spatial sketchpad

The overall pattern of our results suggests that, even under complete visual deprivation, the high-resolution, topographically organized early visual cortex can operate as the conscious 'projection screen' for working memory, even when evoked by non-visual tasks. The spatiotopic organization and inherently parallel processing of the largest cortical map, V1, and most importantly, the small receptive fields providing the high-spatial resolution of the V1 map (e.g., Hubel & Wiesel, 1968; van Essen & Maunsell, 1980; van Essen et al., 1984; Mumford, 1992), makes it a very attractive candidate for precise spatial and spatial-memory processing, in particular, once it is freed up from visual processing, as in blindfolding or blindness (see sections 4 & 5). The data described in this chapter provide a direct evidence supporting such an implementation of a spatial sketchpad in human V1; most notably, our results make another step forward implying for a first time a generalization of the 'visuo-spatial' sketchpad concept beyond the visual modality, into a modality-independent or amodal spatial sketchpad.

It is important to note, however, that such an interpretation does not mean that V1 is employed for long-term storage of the memory trace, but merely for its operational activation in working memory to meet the task demands. Moreover, in contrast to the usual format of a baseline condition, we instructed and practiced the subjects to eliminate any rehearsal of either the just-explored templates or of any other memory images for the full 20 s duration of each null interval, which was also too long to account for the known retention time of any visual or iconic image of the memory trace, which is of the order of a second or less (Sperling, 1963; Di Lollo, 1980; Loftus et al., 1992). Since the drawings were not experienced as spatial images during the null interval, they were evidently held in some other, non-conscious storage location until it was needed for the subsequent drawing task.

3.4.2. Why is V1 surrounded by extrastriate suppression?

The current analysis was not designed to directly answer the question of the role of the extrastriate suppression (Fig. 7), but it allows several general considerations to be drawn. For instance, our results demonstrate a unique activation/suppression pattern produced without either visual or tactile stimulation in the occipital cortex. This fact eliminates potential sensory mechanisms, such as a direct drive of the primary visual cortex through direct connections from the primary somatosensory cortex.

Why is this suppression needed? One possible explanation is the existence of a mechanism homologous to what we have found previously in a figure/ground paradigm: a topographically precise, suppressive top-down feedback to the ground projection in V1 (Likova and Tyler, 2008). We suggested that such suppression of the less relevant information (that of the ground) makes strong computational sense. Such an interpretation is also consistent with theoretical predictions in Tsotsos et al. (1995). It seems logical that a similar principle of active suppression of the task-irrelevant regions (i.e., of the extrastriate visual pathways) may

http://dx.doi.org/10.5772/54835

operate in our non-visual memory task but on a much larger scale, i.e., on a cross-modal instead of intramodal scale. This principle allows task-irrelevant or conflicting pathways to be excluded or 'cut-off' from functional involvement. Thus, the extrastriate suppression surrounding V1 prevents any potential propagation of the entirely non-visual V1 signal through the visual hierarchy pathway, which would be an inappropriate pathway in this case.

4. How does the learning-based reorganization develop? Insights from a congenitally blind case

4.1. Introduction

A congenitally blind novice with no experience with drawing or writing provides an ideal paradigm for investigating the earliest stages of V1 reorganization, and also provides for a critical probe of our amodal sketchpad idea, because congenital blindness eliminates the possibility of any visual mechanisms influencing the neural processing. Specifically, it abolishes not only the bottom-up visual input but also any potential top-down visual processing, such as visual imagery (since congenital blindness eliminates any visual experience on which to base such processing). We were lucky to find such an absolute novice. This individual was well-adapted to operating in the spatial world, including longstanding familiarity with complex tactile manipulations and Braille reading, but had no drawing, writing or even pen-holding experience. Functional MRI was run before and after a week's C-K training in order to test our hypothesis that V1 uses an amodal spatial representation in its operation as the putative memory buffer and to investigate the evolution of brain reorganization as a function of learning to draw.

4.2. Methods: The methods were as described in the general methods section

4.2.1. Subject and training

The congenitally blind subject was a 61-year-old right-handed female, totally blind with no light perception, who lost her vision shortly after birth as a result of rubella (German measles) in her expectant mother, severely and permanently damaging the fetal optic nerves, and also degrading her hearing to some degree. She had not been previously studied by fMRI or behavioral methods of any kind. She is a sophisticated intellect and a fluent Braille reader, with a graduate education and lifetime employment including the professional use of a computer keyboard, and was highly motivated to participate in the study. Nevertheless, despite her Braille fluency, the subject had no experience with writing or drawing, so her training to draw had to start with the basics, such as the proper holding of the pen and key spatial concepts of the representation of 3D structure on a 2D plane. She had relied heavily on active tactile exploration for her whole life, so it was quite surprising that she did not have clear idea of elementary geometric concepts such as a straight line vs a curve, right angles, etc.; she also was unable to reproduce any simple geometric figure or object by drawing. These issues were manifested at all levels of the experimental process – the tactile recognition and memorization phase, the memory recall in drawing, the understanding of spatial relationships, and even the kinesthetic feedback and self-evaluation of her own performance. For example, she could think she had just drawn a straight line when she had actually drawn an almost-closed curve, and so on.

It became clear, however, that these 'negatives' could be turned into significant 'positives' that would allow better tracking of the full evolution of the process of learning to draw. Another advantage was the fact that this subject was an intelligent adult, able both to readily follow instructions and to express back her introspections.

The special C-K technique allowed our subject to learn to draw without using any tactile feedback from the non-drawing (left) hand. As specified in 3.4.1. by eliminating the tactile feedback during drawing, this approach enforces the use of memory signals for guiding the drawing trajectory.

The training procedure was able to inspire and motivate this blind subject to acquire the exciting drawing skill. After only a week, she significantly advanced relative to her starting level, although her capability was still not satisfactory to her. Two months later she came back for two 'refresher' training sessions which she felt brought her up to an adequate skill level. To study the dynamics of the learning process, we ran fMRI before training, as well as after the prolonged period of consolidation.

4.3. Results

4.3.1. Comparative pre/post-training analysis: Increased activation and development of memory specialization

Training-induced enhancement of V1 activation for the memory-guided drawing

The focus of analysis is the occipital region along the calcarine sulcus corresponding to the location of the primary visual cortex, area V1. Comparison of the voxelwise parametric maps in the primary visual cortex during the MemoryDraw task (Fig. 9) reveals dramatic enhancement of the activation (orange coloration) in V1 (green outline) from negligible patchy activation before training (Fig. 9 A) to a massive task-specific activation after training (Fig. 9 B). Remarkably, the extension of post-training activation in V1 approximately corresponds to the spatial extent of that in the blindfolded (normally-sighted) subjects.

Lack of task-specificity before training, but development of MemoryDraw dominance after training

Bar-graphs for the estimated activation in V1 in each hemisphere before training (Fig. 9 C) indicate a lack of task specificity, i.e., similar activation levels for the MD (red) and both control tasks in the left hemisphere, with negligible responses to E/M and MD in the right hemisphere.

In contrast, after training (Fig. 9 D), the MD response (red bars) dominates in both left and right V1. As indicated by the confidence intervals, the following relationships are statistically significant post-training ($p < 0.01$): MD > E/M and MD > S in both hemispheres, and E/M >

http://dx.doi.org/10.5772/54835

S in the left hemisphere. Thus, MD is the task that most powerfully activates V1 bilaterally, showing highly significant % BOLD responses at a low noise level; the E/M task gives much weaker, left-dominant responses; while the motor-control scribbling task, S (which lacks any memory component), is suppressed even in the left hemisphere.

Pre/post comparison of the pattern of response in V1 across tasks shows significant reorganization in this response pattern as a result of training

Pre/post comparison of the pattern of V1 response across three tasks after training (Fig. 9 D) to that before training (Fig. 9 C) shows the following statistically significant ($p < 0.01$) relationships: $E/M_{post} \equiv E/M_{pre,}$ $DM_{post} > DM_{pre}$ and $S_{post} < S_{pre}$ in both the left and the right hemispheres. This analysis implies a significant reorganization in the V1 response pattern as a result of training. Importantly, the V1 response in the memory-guided drawing task (MD) was substantially increased, while the response of the non-memory motor-control task (S) became insignificant. Note that no such increase was present in the area of the left motor cortex (controlling the drawing right hand), which even showed a reduction in response after training, as may be expected when a motor task becomes familiar (e.g., Jenkins et al., 1994). The E/M condition, which used the left hand (controlled by contralateral right motor cortex), developed the expected suppression in the ipsilateral (unused) cortex after training, manifesting the known cross-hemispheric competition in motor cortices (e.g., Nowak et al., 2009).

Bar-graphs show the estimated average activation in each hemisphere (left – LH, right – RH) for Explore/Memorize (E/M, blue), MemoryDraw (MD, red), non-memory drawing Scribble (S, green). The dashed lines indicate 99% confidence interval for the significance of the activation relative to zero. Error bars represent the 99% confidence intervals for assessing significant differences between pairs of activation levels. Pairs of activations are significantly different if they exceed the 99% confidence intervals for both activations. Time-courses show the average time courses of BOLD activity (black lines) for the sequence of three task intervals (white bars). The four dark-grey bars indicate the 20 s rest intervals separating E/M, MD and S tasks.

Top row (A, B): No significant V1 activation before training (in A), but massive, spatiotopic recruitment of V1 in the MemoryDraw task as a result of the CK-training (in B). Second row (C, D): Pre/post training comparison of response amplitudes across all three tasks reveals the development of a clear functional specialization for the MD task after training (in D). Third row (E, F): Response waveform analysis reveals how the immature and non-specific transient 'bursts' before the CK-training (in E) are transformed into a well-developed response waveforms for the memory-drawing MD (in F), but any significant response is lacking for the non-memory drawing S. Bottom row (G, H): In contrast to V1, the left motor area M1 produced well-formed responses for the two right-hand tasks (MD and S) both before (in G), and after training (in H) (but did not respond, and became significant suppressed for the left-hand task E/M).

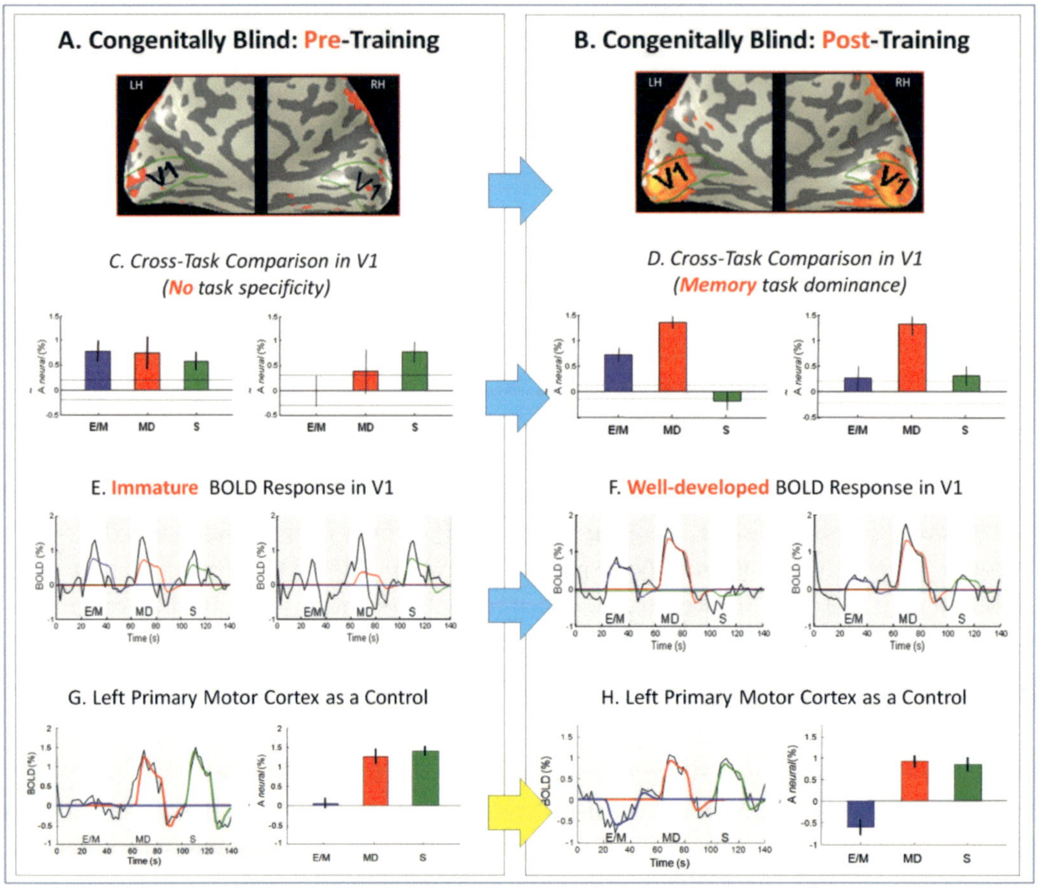

Figure 9. The first three rows present a comparative pre/post-training analysis of the primary visual cortex V1 (green outline), revealing its dramatic training-based functional reorganization, specific to the memory task. The fourth row is a control to show that nonvisual areas, such as the area of the left primary motor cortex that controls the right (drawing) hand, does not significantly change their response characteristics (as opposed to the dramatic pre/post reorganization in V1).

4.3.2. Time course of the BOLD response in V1: Dramatic post-training changes

Before Training: Immature BOLD Response Wave forms in V1 (Fig. 9 E)

The advantage of training a total novice was also manifested by allowing us to capture the very early stages of functional reorganization as expressed in the changes of the time course of the BOLD responses. As seen in Fig. 8 E the average time courses for the sequence of the three task intervals (white bars) have substantial deviations of the waveforms (black lines) from the model prediction fits (color curves). The model takes into account both the task duration and the estimated HRF (see General Methods). The pre-training response waveforms are rudimentary, poorly-developed and noisy, with a prominent transient nature and early offsets long before the end of the 20 s task periods, in spite of the continuous hand movements during the full task period (these continuous hand movements are evident from both the motion-capture records and also from the fully-fledged time course in the motor hand area, Fig. 8 G). Such early offset in V1 implies that the neural response was essentially a brief transient pulse that was immediately withdrawn, suggesting an unsuccessful attempt to activate this area. The subject's self-report and drawing performance were consistent with such undeveloped utilization.

After Training: Well-Developed BOLD Response Waveform in V1 (Fig. 9 F)

Notably, as a result of training, the temporal waveform of the V1 response regularized to become a good match to the linear model prediction (colored curves in Fig. 9 F), with maximum strength for the MemoryDraw task bilaterally. We no longer see the transient early-offset signals. Importantly, however, Scribble, which has no memory involvement, is lacking any significant response.

4.4. Discussion

4.4.1. Support of the amodal spatial 'sketchpad' and its implementation in the primary visual cortex

This study is the first to investigate the temporal evolution of functional reorganization of V1 as a result of a non-visual memory training task. Working with an adult who was an absolute novice has the advantage that such a subject can provide full introspection and produce complex behavioral measures (as opposed to working with difficult-to-communicate-with infants). This investigation thus opens a window to research the developmental evolution of both neural and behavioral changes, benefiting by effectiveness of the Cognitive-Kinesthetic training.

Except for her blindness and reduced hearing, this congenitally blind subject was in a very good physical and mental health, with robust BOLD responses in right-hand motor cortex, M1 (Fig. 9 G, H), so there were no reasons to expect abnormal BOLD responses in V1. Thus, a primary explanation of her V1 response patterns of undifferentiated, transient activation across all the tasks (Fig. 9 C, E) implies the lack of memory-specific role for V1 at that early stage. In particular, the poor response pattern may mean that, despite her lifetime of tactile experience including the professional use of a computer keyboard, the primary visual cortex was not able to serve as an effective 'amodal memory sketchpad' before training. The immature, early-offset waveforms suggest a rudimentary and unsuccessful attempt to activate V1, showing that the six decades of reliance on tactile perception in everyday tasks was evidently insufficient for the development of V1 functionality as demanded by the challenging memory-drawing task.

The rapid, learning-based recruitment of V1 in the MemoryDraw task after blind training (Fig. 9D, F), supports our reconceptualization of the memory 'sketchpad' as an amodal-spatial sketchpad, implemented in the primary visual cortex. Importantly, several parallel measures, such as subject's self-reports on the quality of the memory recall and the objective memory readout (the drawings recorded by our motion-capture system) with the concurrently recorded fMRI activation, all converged in supporting this hypothesis. Importantly, the learning paradigm itself provides causal inferences about the cross-modal changes in V1.

4.4.2. General considerations

Although a mature adult, this congenitally blind individual showed rapid functional reorganization of her brain in the process of learning to draw. It seems particularly surprising to find such reorganization in the primary visual cortex, whose main role is considered to be the early processing modality-specific information from visual input. The implication is that, although this cortical region had never been visually stimulated during the six decades of this blind subject's life, it is still retained sufficient plasticity to be accessible for use when the need arose. Thus, the drawing task was sufficiently demanding to activate functional reorganization that was not instigated by any other task during her life (despite the intensive use of other forms of detailed spatial information such as Braille characters for reading). Further studies are needed to investigate the specific cross-modal mechanisms mediating the V1 reorganization; in general, there is a wide range of theoretical possibilities, such as unmasking of pre-existing connections, synaptic weight changes, or a combination of a number of different mechanisms (e.g., Florence and Kaas, 1995; Jones, 2000; Raineteau and Schwab, 2001; Merabet, 2008; Van Brussel at al., 2011; Likova, 2012a).

Furthermore, the MemoryDraw post-training response map extended to an eccentricity of ~ 10 deg, which is in accordance with that in the blindfolded subjects. Such multidimensional consistency between the congenitally blind and the blindfolded responses is likely not to be accidental but to reflect common mechanisms operating under both short- and long-term visual deprivation. To further explore this finding, a comparative analysis of groups of congenitally and late blind individuals is described in the next section.

5. Is there a universal 'language' of modality-independent space?

5.1. Introduction

The eccentricity overlap of the post-training MD activation in V1 for the congenitally blind (Fig. 9 B) and the blindfolded (Fig. 7 A) appears even more remarkable knowing that what they had in common was that they all memorized the same battery of tactile images, with all images being of the same overall size. Thus, the consistent extent of V1 activation raises a number of fundamental questions, as it is suggestive of i) preservation of the form of topographic organization in V1 despite complete visual deprivation, and moreover, ii) utilization of this topography by cross-modal memory.

Does a topographic operation of the amodal memory sketchpad reflect a universal (i.e., modality-independent) 'language' of spatial representation? This section does not aim to provide complete answers but to raise questions and show their legitimacy based on our results in two additional groups of totally blind people.

http://dx.doi.org/10.5772/54835

5.2. Subjects and methods

A group of 6 congenitally blind and 7 late-onset blind individuals were studied before and after C-K training. All methods were as described in the General Methods section, with delineation of the normal retinotopic and functional organization in the whole-brain averaging by means of the novel training method we have developed. The late-onset blind group had blindness onsets ranging from 3 to 39 years of age. Only patients with total ocular blindness, but not cerebral blindness, were included. (Human vision is well-established to be close to adult level by the age of 3, so the primary requirements for visually processing would have been well developed before the onset of blindness in this late-onset group.)

5.3. Results and discussion

Consistent with the findings from sections 3 and 4, MemoryDraw strongly activated V1 along the calcarine in both blind groups, and remarkably, the activation did not fill in the whole of V1 but extended only about half way posteriorly, covering what in the sighted would be approximately the central 10 deg retinotopic representation. The dashed white lines in Fig. 10 A, B show the 10 deg eccentricity contour, on the basis of our approach to enabling retinotopic mapping in the blind (see Methods). Both the late-onset blind (Fig. 10A) who had full visual experience during early development, and the congenitally blind (Fig. 10B) who did not, show activation extending from the foveal representation all along 10 deg eccentricity contour but not beyond it (to an accuracy of ± 10% of its cortical extent). Thus, despite their total lack of visual experience, even the congenitally blind group adheres to a similar mapping principle as for those with some sighted development.

The fact that eccentricity-robust V1 post-training activation was systematically obtained in all visually-deprived populations is suggestive of topographically organized non-visual processing, and raises an array of questions. How would a topographic organization develo-pin the congenitally blind despite the absence of visual experience to drive it? Is its development entirely genetically predetermined and independent from visual stimulation? How is an already developed retinotopy preserved after the total loss of visual input, as in the case of the late-onset blind subjects? How is a still functioning retinotopy, such as in the blindfolded, rapidly recruited by cross-modal memory? Does it develop by a nature or nurture mechanism, or by an interplay between the two? What is the frame of reference for the implemented metrics?

The current studies leave open questions for future research. A basis for some answers is already emerging. Interestingly, a recent 7-T high-resolution structural MRI study in congenitally blind individuals has established the presence of the stria of Gennari in the human cortex, which is another organizational landmark of V1 (Trampel et al., 2011), indicating that neither the development nor the preservation of this V1 landmark depends on the existence of visual input. Moreover, Cang et al. (2008) demonstrated for the first time that the development of the topographic axes of V1 is genetically controlled. The same group has also shown, however, strong experience-based plasticity in V1 (reviewed in Espinosa and Stryker, 2012). Thus, the plasticity mechanisms in V1 appear to be a complex interplay between genetic and environmental factors, which further tune-up and refine the genetically-determined layout (for a recent review see Maya-Vetencourt and Origlia, 2012).

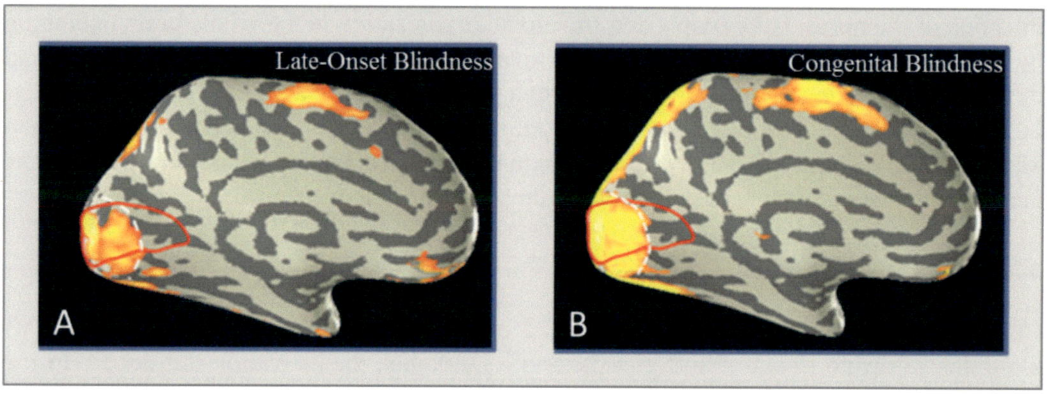

Figure 10. Group data in the blind. (A) The post-training group activation in V1 for the MemoryDraw task in the late-onset blind group expanded to approximately 10 deg eccentricity (dashed white line), similarly to the results in previous sections. (B) Remarkably, even the group data of congenitally blind individuals with no visual experience throughout their life manifested similar eccentricity (dashed white line).

6. Concluding comments

Detailed discussions are included in each of the experimental sections. Here, to sum up, the analyses of the fMRI data from our novel drawing-based memory paradigm and the Cognitive-Kinesthetic training led us to discover a unique pattern of V1 response, cut-off from propagation throughout the visual hierarchy by surrounding deactivation. Consideration of these results, in turn, excludes the most expected hypotheses and suggested a novel straightforward interpretation: That V1 in human implements the visuo-spatial working memory sketchpad concept but does so in an amodal form that transcends the sensory-modality specialization of the visual hierarchy. Based on this reasoning, we generated the prediction that, under the same memory paradigm, V1 will be activated even in the congenitally blind, who had never had any visual stimulation. This prediction was tested in a subsequent experiment, which replicated the freehand memory-drawing paradigm in a congenitally-blind novice. The data from this unique case study confirmed our initial interpretations of i) an amodal working memory role of V1, and ii) a spatiotopic character of the amodal memory representation. Furthermore, they also revealed for the first time the temporal evolution of the BOLD response during this learning-based cross-modal reorganization in the primary visual cortex. The large-scale study in groups of both congenitally and late-onset totally blind participants provided further strong evidence of these discoveries, and also contributed to the formulation of new research questions. The strong training effect implies causal or functionally relevant reorganization of the (non-visual) processing in area

V1. It is also particularly noteworthy that effectiveness of the Cognitive-Kinesthetic training was felt as transformative by the blind individuals in opening a new domain of experience.

Our findings of unusual immature BOLD waveforms in V1 before training might be taken as a warning for neuroimaging studies on plasticity, indicating that finding some unspecified activation is not sufficient as an indicator of functional employment. Additional criteria considering the time-course characteristics of the response have to be analyzed as well. The training-based evolution of cortical response to becoming task-specific and developing fully-fledged BOLD waveforms, eliminates epiphenomenon-based explanations such as those related to excess of excitability or a generic lack of synaptic pruning.

The research reviewed here exemplifies how an innovative learning method, integrated with a complex memory paradigm employing art, can further empower functional neuroimaging to shed light on long-standing puzzles concerning memory and cross-modal plasticity, and lead to re-conceptualization of organizational principles of the brain.

7. Acknowledgements

This research was supported by NSF/SLC Grant to Lora T. Likova. The author thanks Spero Nicholas for his help in data pre-processing and analysis tools, and Christopher W. Tyler for helpful discussions.

Author details

Lora Likova

The Smith-Kettlewell Eye Research Institute, San Francisco, CA, USA

References

[1] Amedi, A., Malach R., Hendler T., Peled S., Zohary E. (2001) Visuo-haptic object-related activation in the ventral visual pathway. Nat Neurosci 4:324–330.

[2] Amedi, A., Jacobson G., Hendler T., Malach R., Zohary E. (2002) Convergence of visual and tactile shape processing in the human lateral occipital complex. Cereb Cortex 12:1202–1212.

[3] Amedi, A., Raz, N., Pianka, P., Malach, R., and Zohary, E. (2003) Early 'visual' cortex activation correlates with superior verbal memory performance in the blind. Nature Neuroscience 6:758–766.

[4] Amedi, A., Floel, A., Knecht, S., Zohary, E., and Cohen, L.G. (2004) Transcranial magnetic stimulation of the occipital pole interferes with verbal processing in blind subjects. Nature Neuroscience 7:1266–1270.

[5] Amedi A., Malach, R., and Pascual-Leone, A. (2005) Negative BOLD Differentiates Visual Imagery and Perception. Neuron, Dec 8 Vol. 48, 859–872.

[6] Amedi, A., Merabet, L.B., Camprodon, J., Bermpohl, F., Fox, S., Ronen, I., Kim, D.S., and Pascual-Leone, A. (2008) Neural and behavioral correlates of drawing in an early blind painter: a case study. Brain Res. 25 1242:252-62.

[7] Baddeley, A.D. (1986) Working Memory. Oxford University Press: New York.

[8] Baddeley, A.D. (2000) The episodic buffer: a new component of working memory? Trends Cognit Sci 4 (11): 417–23.

[9] Baddeley, A.D., and Hitch, G.J. (1974). Working memory. In G Bower (Ed.), The Psychology of Learning and Motivation, Vol. 8, pp. 47-90. San Diego, CA: Academic Press.

[10] Baddeley, A.D. (2003) Working memory: looking back and looking forward. Nature Reviews Neuroscience 4: 829-839.

[11] Block, N. (2003) Tactile sensation via spatial perception. Trends Cogn. Sci. 7, 285-286.

[12] Borowsky, R., Esopenko, C., Cummine, J., and Sarty, G.E. (2007) Neural representations of visual words and objects: a functional MRI study on the modularity of reading and object processing. Brain Topogr. 20:89-96.

[13] Buchsboum, B.R., and D'Esposito, M., (2009) Is there anything special about working memory? In Neuroimaging of Human Memory. Rosler, F., Ranganath, C., Roder, B., Kluwe, R., (Eds.) Oxford Univ. Press: Oxford.

[14] Buechel, C., Price, C., Frackowiak, R.S., and Friston, K. (1998) Different activation patterns in the visual cortex of late and congenitally blind subjects. Brain 121:409–419.

[15] Burton, H., Snyder, A.Z., Conturo, T.E., Akbudak, E., Ollinger, J.M., and Raichle M.E. (2002) Adaptive changes in early and late blind: a fMRI stud of Braille reading. J Neurophysiol 87:589–607.

[16] Burton, H. (2003) Visual cortex activity in early and late blind people. J Neurosci, 23(10):4005–4011.

[17] Burton, H., Sincler, R.J., MacLaren, D.C. (2004) Cortical activity to vibrotactile stimulation: an fMRI study in blind and sighed individuals. Hum Brain Mapp 23:210–228.

[18] Cohen, L.G., Celnik, P., Pascual-Leone, A., Corwell, B., Falz, L., and Dambrosia, J. (1997) Functional relevance of cross-modal plasticity in blind humans. Nature 389:180e–183.

[19] Deibert, E., Kraut M., Kremen S., Hart J.Jr. (1999) Neural pathways in tactil eobject recognition. Neurology. 52:1413–1417.

[20] DeVolder, A.G., Bol, A., Blin, J., Robert, A., Arno, P., and Grandin, C. (1997) Brain energy metabolism in early blind subjects: neural activity in the visual cortex. Brain Res 750:235–244.

[21] Engel, S.A., Glover, G.H., and Wandell, B.A. (1997). Retinotopic organization in human visual cortex and the spatial precision of functional MRI. Cereb. Cortex 7, 181–192.

[22] Florence, S.L., and Kaas, J.H. (1995) Large-scale reorganization at multiple levels of the somatosensory pathway follows therapeutic amputation of the hand in monkeys. 15(12): 8083-95.

[23] Friston, K. J., Jezzard, P., and Turner, R. 1994. Analysis of functional MRI time-series. *Hum.* Brain Map. 1:153–171.

[24] Ganis, G., Thompson, W. L., & Kosslyn, S. M. (2004). Brain areas underlying visual mental imagery and visual perception: An fMRI study. Cognitive Brain Research, 20, 226-241.

[25] Gizewski, E.R., Gasser, T., de Greiff, A., Boehm, A., and Forsting, M. (2003) Cross-modal plasticity for sensory and motor activation patterns in blind subjects. Neuroimage 19(3):968-75.

[26] Goyal, M.S., Hansen, P.J., and Blakemore, C.B. (2006) Tactile perception recruits functionally related visual areas in the late-blind. Neuroreport 17(13):1381-4.

[27] Grossi D and Trojano L. (1999) Constructional apraxia. In Denes G, and Pizzamiglio L (Eds), Handbook of Clinical and Experimental Neuropsychology. Hove: Psychology Press.

[28] Hagen, M.C., Franzen, O., McGlone, F., Essick, G., Dancer, C., Pardo, J.V. (2002) Tactile motion activates the human middle temporal /V5(MT/V5) complex. Eur J Neurosci. 16:957–964.

[29] Hamilton, R.H., Pascual-Leone, A., (1998) Cortical plasticity associated with Braille learning. Trends in Cognitive Sciences, Elsevier, 2(5), 168-174..

[30] Hamilton, R.H, Keenan, J.P., Catala, M., and Pascual-Leone, A. (2000) Alexia for Braille following bilateral occipital stroke in an early blind woman. NeuroReport 11:237–240.

[31] Harrison, S.A., Tong, F. (2009) Decoding reveals the content of visual working memory in early visual cortex. Nature 458:632-5.

[32] Hubel DH, Wiesel TN (1968) Receptive fields and functional architecture of monkey striate cortex. J Physiol. 195:215-43.

[33] Ishai, A., and Sagi, D. (1995) Common mechanisms of visual imagery and perception. Science 268, 1772–1774.

[34] Ishai, A., Ungerleider, L.G., and Haxby, J.V. (2000). Distributed neural systems for the generation of visual images. Neuron 28, 979–990.

[35] Ishai, A., Haxby, J.V., and Ungerleider, L.G. (2002) Visual imagery of famous faces: effects of memory and attention revealed by fMRI. Neuroimage 17, 1729–1741.

[36] Ishai, A. (2009) Retrieving pictures from long-term memory. In Neuroimaging of Human Memory. Rosler, F., Ranganath, C., Roder, B., Kluwe, R., (Eds.) Oxford Univ. Press: New York.

[37] James, T.W., Humphrey, G.K., Gati, J.S., Servos, P., Menon, R.S., and Goodale, M. A. (2002) Haptic study of three-dimensional objects activates extrastriate visual areas. Neuropsychologia 40:1706-1714.

[38] Jenkins, I. H., Brooks, D. J., Nixon, P. D., Frackowiak, R.S.J., Passingham, R. E. (1994) Motor sequence learning: a study with positron emission tomography. J. Neurosci. 14:3775–3790.

[39] Kastner, S., De Weerd, P., Desimone, R., Ungerleider, L.G. (1998) Mechanisms of directed attention in the human extrastriate cortex as revealed by functional MRI. Science 282: 108–111.

[40] Kennedy, J. M. (1993) Drawing and the Blind. New Haven, CT: Yale University Press.

[41] Kennedy, J.M. (2000) Recognizing outline pictures via touch: Alignment theory. In M.A. Heller (Ed.) Touch, Representation and Blindness. Oxford University Press: Oxford (pp. 67-98).

[42] Kennedy, J.M., and Igor, J. (2003) Haptics and projection: Drawings by Tracy, a blind adult. Perception 32(9):1059 –1071.

[43] Kennedy, J.M., and Juricevic, I. (2006) Foreshortening, convergence and drawings from a blind adult. Perception 35(6):847 851.

[44] Kosslyn, S.M., Thompson, W.L., & Alpert, N.M. (1997). Neural systems shared by visual imagery and visual perception: A positron emission tomography study. NeuroImage, 6, 320-334.

[45] Kosslyn, S. M., Thompson, W. L., Kim, I. J., Rauch, S. L., & Alpert, N. M. (1996). Individual differences in cerebral blood flow in area 17 predict the time to evaluate visualized letters. J Cognit Neurosci, 8, 78-82.

[46] Kosslyn, S.M., Pascual-Leone, A., Felician, O., Camposano, S., Keenan, J.P., Thompson, W.L., Ganis, G., Sukel, K.E., and Alpert, N.M. (1999) The role of area 17 in visual imagery: convergent evidence from PET and rTMS. Science 284, 167–170.

[47] Kosslyn, S.M., Thompson, W.L. (2003) When is early visual cortex activated during visual mental imagery? Psychol Bull 129: 723–746, 2003.

[48] Kreiman, G., Koch, C., and Fried, I. (2000) Imagery neurons in the human brain. Nature 408, 357–361.

[49] Lambert, S., Sampaio, E., Mauss, Y., and Scheiber, C. (2004) Blindness and brain plasticity: contribution of mental imagery? An fMRI study. Brain Res. Cogn. Brain Res. 20, 1–11.

[50] Lee, T.S., and Mumford, D. (2003) Hierarchical Bayesian inference in the visual cortex. J Opt Soci Amer, A. 20: 1434-1448.

[51] Likova, L.T., Tyler, C.W., (2008) Occipital network for figure/ground organization. Experimental Brain Research, 2008, 189: 258-67.

[52] Likova, L.T. and Tyler C.W. (2007). Stereomotion processing in the human occipital cortex. NeuroImage 38:293-304.

[53] Likova, L.T. (2010a) Drawing in the blind and the sighted as a probe of cortical reorganization. Human Vision and Electronic Imaging XV. Edited by Rogowitz, Bernice E.; Pappas, Thrasyvoulos N. Proceedings of the SPIE, Volume 7527, 8-14.

[54] Likova, L.T. (2010b) The primary visual cortex as a modality-independent 'screen' for working memory. J Vision, 10(7): 776,

[55] http://www.journalofvision.org/content/10/7/776.

[56] Likova L.T. (2012a) Drawing enhances cross-modal memory plasticity in the human brain: A case study in a totally blind adult. Frontiers in Human Neuroscience.6:44. doi: 10.3389/fnhum.00044.

[57] Likova L.T. (2012b) The spatiotopic 'visual' cortex of the blind. Proc. SPIE 8291, Human Vision and Electronic Imaging XVII, 8291-10L doi:10.1117/12.912257.

[58] Likova L.T. (2013) A cross-modal perspective on the relationships between imagery and working memory. Front. Psychology, 3:561. doi: 10.3389/fpsyg.00561.

[59] Loomis, K.M., and Klatzky R. (2008) Functional equivalence of spatial representations from vision, touch, and hearing: Relevance for sensory substitutions. In M.A. Heller (Ed.) Touch, Representation and Blindness (pp. 67-98), Oxford University Press: New York.

[60] Matteau, I., Kupers, R., Ricciardi, E., Pietrini, P., Ptito, M. (2010) Beyond visual, aural and haptic movement perception: hMT+ is activated by electrotactile motion stimulation of the tongue in sighted and in congenitally blind individuals. Brain Res Bull. Jul 30;82(5-6):264-70.

[61] Mechelli, A., Price, C.J., Friston, K.J., and Ishai, A. (2004) Where bottom-up meets top-down: neuronal interactions during perception and imagery. Cereb. Cortex 14, 1256–1265.

[62] Merabet, L., Rizzo, J., Amedi, A., Somers, D., and Pascual-Leone, A. (2005) What blindness can tell us about seeing again: Merging neuroplasticity and neuroprostheses. Nature Rev. Neurosci 6:71–77.

[63] Merabet, L. (2008) The plastic brain in blind individuals: The cause of disability and the opportunity for rehabilitation. In, Blindness and Brain Plasticity in Navigation and Object Perception. Lawrence Erlbaum Associates, London.

[64] Mumford, D. (1992) On the computational architecture of the neocortex II Biol. Cybern. 66, 241–251.

[65] Nowak DA, Grefkes C, Ameli M, Fink GR (2009) Interhemispheric competition after stroke: brain stimulation to enhance recovery of function of the affected hand. Neurorehabil Neural Repair. 23:641-56.

[66] O'Craven, K.M., and Kanwisher, N. (2000) Mental imagery of faces and places activates corresponding stimulus-specific brain regions. J Cogn Neurosci 12, 1013–1023.

[67] Pascual-Leone, A., Hamilton, R., 2001. The metamodal organization of the brain. Prog. Brain Res. 134, 427-445.

[68] Pascual-Leone, A., Amedi, A., Fregni, F., Merabet, L. (2005) The plastic human brain. Annual Review of Neuroscience, 28: 377-401.

[69] Penfield W, Boldrey E (1937) Somatic motor and sensory representation in the cerebral cortex of man as studied by electrical stimulation. Brain 60:389–443.

[70] Penfield W, Rasmussen T. (1950) The Cerebral Cortex of Man. Macmillan: New York.

[71] Perkel, D.J., Bullier, J., Kennedy, H. (1986) Topography of the afferent connectivity of area 17 in the macaque monkey: a double-labelling study. J Comp Neurol 253:374–402.

[72] Piercy, M., Hecaen H., and Ajuriaguerra, J. (1960) Constructional apraxia associated with unilateral cerebral lesion—Left and right-sided cases compared. Brain 83: 225–242.

[73] Pietrini, P., Furey, M.L., Ricciardi, E., Gobbini, M.I., Wu, W.H., Cohen, L., Guazzelli, M., Haxby, J.V. (2004) Beyond sensory images: Object-based representation in the human ventral pathway. Proc Natl Acad Sci USA 101:5658–5663.

[74] Ponchilla, P. E. (2008) Non-visual sports and Arts: fertile substrates for the growth of knowledge about brain plasticity in people who are blind or have low vision. In Blindness and Brain Plasticity in Navigation and Object Perception. Lawrence Erlbaum Associates, London.

[75] Prather, S.C., Sathian, K. (2002) Mental rotation of tactile stimuli. Brain Res Cogn Brain Res 14: 91–98.

[76] Ptito, M., Fumal A., de Noordhout A.M., Schoenen J., Gjedde A., and Kupers R. (2008) TMS of the occipital cortex induces tactile sensations in the fingers of blind Braille readers. Exp Brain Res. Jan; 184(2):193-200.

[77] Raineteau, O., and Schwab, M.E., (2001) Plasticity of motor systems after incomplete spinal cord injury. Nat Rev Neurosci. 2(4):263-73.

[78] Ranganath, H., (2009) Interrelationships between working memory and long-term memory. In Neuroimaging of Human Memory. Rosler, F., Ranganath, C., Roder, B., Kluwe, R., (Eds.) Oxford Univ. Press: Oxford.

[79] Raz, N., Amedi, A., and Zohary, E. (2005). V1 activation in congenitally blind V1 activation in congenitally blind humans is associated with episodic retrieval. Cereb. Cortex 15, 1459–1468.

[80] Reed, C.L., Shoham, S., Halgren, E. (2004) Neural substrates of tactile object recognition: an fMRI study. Hum Brain Mapp 21:236–246.

[81] Sadato, N., Pascual-Leone, A., Grafman, J., Ibanez, V., Deiber, M.P., Dold, G., Hallett, M. (1996) Activation of the primary visual cortex by Braille reading in blind subjects. Nature 380:526–528.

[82] Schira, M.M., Breakspear, M., Spehar, B., Tyler, C.W., (2010) Modeling magnification and anisotropy in the primate foveal confluence. PLoS, Computational Biology, 6(1):e1000651.

[83] Sereno, M.I., Dale, A.M., Reppas, J.B., Kwong, K.K., Belliveau, J.W., Brady, T.J., Rosen, B.R., and Tootell, R.B. (1995). Borders of multiple visual areas in humans revealed by functional magnetic resonance imaging. Science 26, 889–893.

[84] Slotnick, S.D., Thompson, W.L., Kosslyn, S.M. (2005) Visual mental imagery induces retinotopically organized activation of early visual areas. Cereb Cortex 15:1570–1583.

[85] Sperling G (1963) A model for visual memory tasks. Human Factors, 5: 19-31.

[86] Stoesz, M.R., Zhang, M., Weisser,V.D., Prather, S.C., Mao, H., Sathian, K. (2003) Neural networks active during tactile form perception: common and differential activity during macrospatial and microspatial tasks. Int J Psychophysiol 50: 41–49.

[87] Theoret, H., Merabet, L., Pascual-Leone A. (2004) Behavioral and neuroplastic changes in the blind: Evidence for functionally relevant cross-modal interactions. J Physiol, Paris 98(1-3): 221–233.

[88] Tootell, R.B., Reppas, J.B., Kwong, K.K., Malach, R., Born, R.T., Brady, T.J., Rosen, B.R., Belliveau, J.W. (1995) Functional analysis of human MT and related visual cortical areas using magnetic resonance imaging. J Neurosci 15: 3215–3230.

[89] Trampel, R., Derek, V. M., Ott, R.T. (2011). Do the Congenitally Blind Have a Stria of Gennari? First Intracortical Insights in Vivo. Cereb Cortex, 21(9): 2075-2081.

[90] Trojano L., Grossi D, Flash T. (2009) Cognitive neuroscience of drawing: Contributions of neuropsychological, experimental and neurofunctional studies. *Cortex* , 45(3): 269-277.

[91] Tsotsos, J.K., Culhane, S.M., Kei Wai, W.Y., Lai, Y., Davis, N., NuXo, F. (1995) Modeling visual attention via selective tuning. Artif Intell, 78:507–545.

[92] Tyler, C.W., Likova. L.T., Kontsevich, L.L., Schira, M.M., Wade, A.R. (2005) Enhanced concept of ocipital retinotopy. Current Medical Imaging Reviews 1: 319-329.

[93] Van Boven, R.W., Ingeholm, J.E., Beauchamp, M.S., Bikle, P.C., Ungerleider, L.G. (2005) Tactile form and location processing in the human brain. Proc Natl Acad Sci USA 102:12601–12605.

[94] Uhl, F., Franzen, P., Lindinger, G., Lang, W., Deecke, L. (1991) On the functionality of the visually deprived occipital cortex in early blind persons. Neurosci Lett 124: 256–9.

[95] Uhl, F., Franzen, P., Podreka, I., Steiner, M., Deecke, L. (1993) Increased regional cerebral blood flow in inferior occipital cortex and cerebellum of early blind humans. Neurosci Lett 150: 162–4.

[96] Van Brussel, L., Gerits, A., Arckens, L. (2011) Evidence for cross-modal plasticity in adult mouse visual cortex following monocular enucleation. Cereb. Cortex 21(9): 2133-2146.

[97] Van Essen DC, Maunsell JH (1980) Two-dimensional maps of the cerebral cortex. J Comp Neurol. 191:255-81.

[98] Van Essen DC, Newsome WT, Maunsell JH (1984) The visual field representation in striate cortex of the macaque monkey: asymmetries, anisotropies, and individual variability. Vision Res. 24:429-48.

[99] Weisser, V., Stilla, R., Peltier, S., Hu, X., Sathian, K. (2005) Short-term visual deprivation alters neural processing of tactile form. Exp Brain Res 166:572–582.

[100] Williams, M., Baker, C. I., Op de Beeck, H. P., Shim, W. M. Dang, S., Triantafyllou, C., and Kanwisher, N. (2008) Feedback of visual object information to foveal retinotopic cortex. Nature Neuroscience. 11:1439-45.

[101] Zangaladze, A., Epstein, C.M., Grafton, S.T., Sathian, K (1999). Involvement of visual cortex in tactile discrimination of orientation. Nature 401, 587-590.

[102] Zhang, M., Mariola, E., Stilla, R., Stoesz, M., Mao, H., Hu, X., Sathian, K. (2005) Tactile Discrimination of grating orientation: fMRI activation patterns. Hum Brain Mapp 25:370–377.

Chapter 14

A Systematic Approach to Visual System Rehabilitation — Population Receptive Field Analysis and Real-time Functional Magnetic Resonance Imaging Neurofeedback Methods

T. Dorina Papageorgiou, Amalia Papanikolaou and
Stelios M. Smirnakis

Additional information is available at the end of the chapter

http://dx.doi.org/10.5772/57000

1. Introduction: Is visual rehabilitation inside a cortical scotoma possible, in principle?

Visual information transmission flows from the retinal ganglion cells to the lateral geniculate nucleus and then to the primary visual cortex (V1), the chief cortical relay of visual information to "higher" extrastriate areas. Beyond area V1, visual processing is distributed across multiple interconnected brain areas, the precise role of which and their interactions are not yet, completely understood. To add to the dynamic complexity of the system, feedback from higher areas and modulation by top-down processes, such as attention, are often critical in the formation of visual percepts (Deco and Lee; 2004; Olhausen, 2003; Kastner and Ungerleider, 2000; Mumford, 1994; Hubel and Weisel, 1977).

Impairment of visual function can occur at any point along the visual pathway from the eye to the cortex. We focus our discussion here on lesions of the primary visual cortex (area V1), which result in dense contralateral visual field defects known as "scotomas". Scotomas resulting from area V1 lesions often involve the contralateral half of the visual field resulting in hemianopia, or a contralateral visual field quadrant resulting in quadrantanopia. V1 lesions are the most prevalent injury of the visual cortex, often occurring as a result of posterior cerebral artery (PCA) stroke, hemorrhage, or traumatic brain injury (TBI) (Pambakian and Kennard, 1997; Zhang et al., 2006; Ajina and Kennard, 2012). Twenty to thirty percent of stroke survivors experience visual disability (Taylor, 1997; Gilhotra et al., 2002; Giorgi et al., 2009), while the incidence of significant visual perceptual impairment in TBI victims exceeds 50% in some studies (McKenna et al., 2006; Lew et al., 2007; Elisevich et al., 1984). The loss of visual perception inside a large scotoma can significantly affect the patient's ability to perform daily tasks, navigate in unknown environments, and function independently (Ajina and Kennard, 2012; Rizzo and Robin, 1996; Riggs, et al., 2007). Visual rehabilitation is clearly necessary for the quality of life of these patients. The literature describing visual rehabilitation efforts is extensive and doing justice to it goes well beyond the scope of this chapter. We should mention at the outset that we do not discuss here, the large literature on practicing eye movement strategies or, using prisms to remap the unseen onto the seen part of the visual field. Instead, we focus on novel approaches that aim to enhance perception inside the visual field scotoma.

To date, no established method exists to rehabilitate visual perception in adult patients with lesions of the primary visual cortex. The lack of effective methods for rehabilitation has led to the general perception that the adult visual cortex has decreased capacity to compensate after injury. This engenders diminished hope that successful strategies can be established to promote the recovery of visual perception after cortical injury. This is partly justified, as several attempts claiming to have achieved significant results have failed. The most notable recent example is an effort by Nova Vision claiming that a rehabilitative paradigm based on a "saccade-to-target" task could significantly shrink dense visual field scotomas (Kasten et al.; 1998, 1999; 2000, 2001, 2006; Poggel et al., 2001, 2004, 2006; Sabel et al., 2000, 2004; Werth and Moehrenschlager, 1999; Zihl and VonCramon, 1979; Zihl and Von Carmon, 1985; Jobke et al., 2009). Although early psychophysical training methods and Nova Vision studies were seen as promising (Kerkhoff et al., 1992; Kerkhoff et al., 1994; Julkunen, 2003; Kasten et al., 1995; 1998a, b, 1999; 2000, 2001; Poggel et al., 2001, 2004, 2006; Sabel et al., 2000, 2004; Werth and Moehrenschlager, 1999; Zihl and VonCramon, 1979; Zihl, 1990), later studies implementing rigorous eye movement controls failed to find a reduction of the visual field scotoma in patients with V1 lesions (Reinhard et al., 2005; Horton, 2005a; Horton, 2005b; Pleger et al., 2003).

Although these efforts are disappointing, the rehabilitation of scotomas resulting from V1 injuries is not altogether a cause without hope (Kasten et al., 1998; Huxlin et al., 2009; Schmid et al., 2010; Poggel et al., 2010; Sahraie et al., 2010; Schmid et al., 2009; Alexander and Cowey, 2009). On one hand, patchy injuries to area V1 or its inputs in the optic radiation seem to be amenable to rehabilitation, as training can help recruit and strengthen surviving connections. In support of this, Sabel and colleagues (Kasten et al., 1999) showed that ~74% of patients with partial optic nerve involvement showed significant recovery with training, compared to 29% of patients with post-chiasmatic lesions. This is likely the result of increased recruitment of partially lesioned fiber pathways or islands of residual vision (Fendrich et al., 1992). On the other hand, recent evidence suggests that even when lesions to area V1 or its proximal inputs are dense, it may be possible to some extent to functionally bypass the area of the injury.

First, there exist anatomical pathways that bypass the area of V1 injury (fig. 1). One such pathway projects from the retina to the koniocellular (intercalated) layers of the lateral geniculate nucleus directly, or to the superior colliculus and from there to extrastriate cortex. This pathway originates in the retinal Pγ class of ganglion cells, which comprises ~10% of the total ganglion cell number, is particularly dense near the fovea (Henry and Reid, 2000), and is

http://dx.doi.org/10.5772/57000

known to survive retrograde degeneration following V1 lesions (Cowey and Stoerig, 1989). Another V1-bypassing pathway projects from the retina to the pulvinar directly or via the superior colliculus and from there to the extrastriate visual areas. Notably, although parvocellular and magnocellular projections to the lateral geniculate nucleus and beyond markedly atrophy following striate cortical lesions (Vanburen, 1963; Mihailovic et alo., 1971), superior collicular (Dineen et al., 1982) and pulvinar (Cowey, 1974) projections remain unchanged.

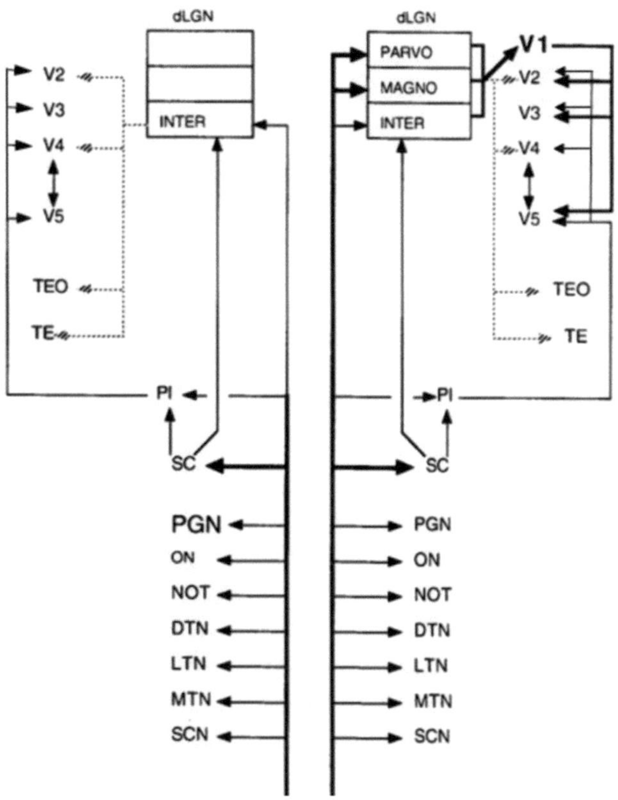

Figure 1. Overview of relevant anatomical visual pathways (Modified with permission from Stoerig and Cowey, 1997): Possible extra-geniculostriate pathways contributing to blindsight behavior, and residual extrastriate cortex activity, adapted from Stoerig and Cowey (Stoerig and Cowey, 1997). This diagram shows known retinofugal inputs, and some of the subsequent projections. On the right, the pathways are shown with intact V1. On the left V1 has been lesioned. Two pathways stand out as potentially mediating the residual activity observed in extrastriate cortex (Rodman et al., 1989; Schmid et al., 2009; Schmid et al., 2010), as well as related aspects of blindsight behavior: **1)** The koniocellular pathway (dotted lines) from the K (intercalated) layers of the thalamus directly to areas V2, V3, V4, V5/MT. This pathway originates in the retinal Pγ class of ganglion cells, comprising ~10% of total ganglion cells, survives retrograde degeneration following V1 lesions, and is particularly dense near the fovea (Cowey and Stoerig, 1989; Henry and Reid, 2000). This pathway receives both direct retinal and superior collicular input. **2)** The projection from the inferior pulvinar to V2, V3, V4, V5/MT, which also receives direct input from the retina, as well as input from the retinotectal (superior colliculus) pathway. PGN: pre-geniculate nucleus, ON: olivary nucleus, NOT: nucleus of optic tract, MTN, LTN, DTN: medial, lateral, dorsal terminal accessory optic nuclei, SCN: suprachiasmatic nucleus, SC: superior colliculus, PI: inferior pulvinar.

Second, these pathways have been shown to be functional under certain conditions. Lesions of area V1 or its post-chiasmatic afferents deprive the extrastriate visual cortex of its main input and result in a dense contralateral visual field scotoma inside which conscious visual perception is thought to be irreversibly lost (Cowey and Stoerig, 1991; Stoerig and Barth, 2001). Remarkably, despite the absence of a conscious visual percept, a capacity to process certain attributes of the visual stimulus persists inside the scotoma, the phenomenon known as "blindsight" (Kluver, 1936; Poppel et al., 1973; Weiskrantz, 1974). The blindsight phenomenon implies that at least some extra-geniculo-striate retinofugal pathways (Cowey, 2010; Schmid et al., 2010; Weiskrantz, 2004; Schoenfeld et al, 2002; Moore et al., 2001; Goebel et al., 2001; Stoerig and Cowey, 1997; Moore et al., 1995; Cowey and Stoerig, 1991; Girard et al., 1991; Pasik and Pasik, 1971) can functionally bypass area V1. This is corroborated by experiments in humans and primates, which have directly demonstrated that extrastriate areas can be modulated by the visual stimulus in the absence of V1 input (Rodman et al., 1989; Baseler et al., 1999; Goebel et al., 2001; Schmid et al., 2009; Schmid et al., 2010). For example, Rodman and Gross demonstrated that area V5/MT can be directly activated through the pathway bypassing area V1 via the superior colliculus (Rodman et al., 1989; Rodman et al., 1990), while Schmid et al. (Schmid et al., 2009; Schmid et al., 2010) demonstrated that early extrastriate areas V2, V3 can be visually modulated by the LGN in the absence of V1 input. Schmid et al. further demonstrated that transiently inactivating LGN in V1 lesioned animals both abolishes visual modulation in areas V2, V3 and returns the monkey's blindsight performance to chance. Unfortunately the V1-bypassing pathways that mediate the blindsight phenomenon are by themselves weak and of limited practical value. The potential of these pathways to induce recovery remains unrealized. This underscores the need to examine the mechanisms underlying the recovery reported in recent studies (Huxlin et al., 2008; Huxlin et al., 2009; Sahraie et al., 2006) in order to understand how to develop effective rehabilitative paradigms. It remains to be examined whether novel neuro-rehabilitative training algorithms may be able to strengthen V1-bypassing pathways enough to derive practical benefit.

Third, training can improve performance inside the scotoma of subjects with area V1 lesions. Behavioral training in healthy subjects can improve visual performance by inducing plastic changes in the physiology of visual networks (Karni and Sagi, 1991; Liu et al., 2000; Yang and Maunsell, 2004; Ahissar and Hochstein, 1997). Perceptual learning is retinotopically specific, suggesting it involves a use-dependent synaptic enhancement induced by pre-and postsynaptic activity (Brown et al., 1988). Studies in humans (Pleger et al., 2003; Taub et al., 2002; Weiller, C. 1998; Lindberg et al., 2003, Takeuchi et al., 2005) and animals (Rudolph et al., 1994; Rudolph and Pasternak, 1999; Rudolph and Delay, 1993, Fabre-Thorpe et al., 1994; Friel et al., 2000; Huxlin and Pasternak, 2004) with V1 lesions, as well as behavioral studies of "blindsight" (Chokron et al., 2008; Stoerig and Cowey, 1997; Sahraie et al., 2006; Overgaard, 2011), suggest that visual performance in the scotoma can also improve with training (Raninen et al., 2007; Henriksson et al., 2007). More recently, following their work on cats (Huxlin et al., 2006), Huxlin et al. trained subjects with V1+ lesions to perform a two-alternative forced choice random dot kinematogram (RDK) direction of motion discrimination task in their blind hemifield (Huxlin et al., 2009). *Remarkably, direction of motion discrimination thresholds recovered from chance to normal at trained locations (Huxlin et al., 2009).* Eye movements were strictly controlled,

http://dx.doi.org/10.5772/57000

and there were no obvious artifacts that could confound the findings. Recovery was retinotopically specific but could be extended by training consecutively adjacent locations that lay progressively deeper inside the scotoma, inducing recovery up to ~20^0 from the scotoma border in one subject. Furthermore, recovery in this task appeared to carry some practical significance, as the subjects' ability to dodge basketballs "thrown" at them from the blind hemifield in a virtual reality environment, improved (Iorizzo et al., 2011). These findings sparked renewed interest in studying visual rehabilitation strategies. This is encouraging but it is necessary to note that visual rehabilitation results appear more variable across the literature (see table 1, in a recent review by Sabel et al. (Sabel et al., 2011), and (Horton, 2005a; Horton, 2005b) for a critical review of the field). Three important questions remain to be answered regarding scotomas resulting from V1 lesions: i) can visual rehabilitative training result in improved visual performance of practical significance? ii) what is the underlying mechanism of recovery? and 3) what is the optimal method for visual rehabilitative training?

In summary, even though visual rehabilitation following area V1+ lesions is a difficult problem, it is not a hopeless endeavor. Anatomical pathways bypassing the area of the lesion exist, and they have been demonstrated to be functional in certain situations. Although early trials have been inconclusive, a recent report by Huxlin et al. (Huxlin et al., 2009), suggests that some recovery is possible, at least in the domain of visual motion perception. Further studies are needed to 1) independently corroborate the results of Huxlin et al. (Huxlin et al., 2009), 2) to understand what visual attributes and what types of lesions are amenable to recovery, and 3) to study the mechanism of recovery. *It is important to note that we do not expect even successful rehabilitation methods to restore vision to pre-lesion levels.* For one, the quality of the restored visual percept will most likely, differ from normal. The reason is that, following V1 lesions, the magnocellular and parvocellular pathways largely degenerate (fig. 1), shifting the balance towards the koniocellular pathway, which is spared. Nevertheless, visual rehabilitation could succeed, by training patients to use the qualitatively and quantitatively different form of visual perception mediated by appropriately strengthened V1-bypassing pathways. The hope is that such an approach will be possible, and may confer considerable practical significance.

2. Using fMRI to functionally characterize and classify cortical lesions and corresponding scotomas

In order to design effective visual rehabilitation strategies for cortical scotomas we have to grapple with the issue of lesion variability. Cortical lesions differ from individual to individual, and this impacts whether or not the resulting scotoma is amenable to rehabilitation. Consequently, some patients show good recovery following visual rehabilitative training (Kasten et al., 1998; Huxlin, 2009) and others no recovery at all (Reinhard et al., 2005; Horton et al., 2005a; Horton et al., 2005b; and our personal observation). It remains unclear what criteria one may use to select patients more likely to recover. Scotomas are mapped using visual field perimetry to determine the part of the visual field where visual perception is impaired. A problem faced in studies of visual rehabilitation is that patients often have heterogeneous lesions, even though the extent and density of their perceptual visual scotomas, measured by perimetry, match. Conversely, the anatomical characterization of the lesion is not always a reliable indication of the properties or, the extent of the resulting scotoma. Consequently, neither visual field perimetry maps nor, purely anatomical information are sufficient indicators of the capacity for rehabilitation.

A measure of the ability of visual stimuli presented inside the scotoma to elicit perceptually sub-threshold activity in spared visual cortex would add valuable information. Functional magnetic resonance imaging (fMRI) can be used to identify which sectors of the visual field scotoma remain able to transmit visual information to spared regions of the visual cortex (fig. 3), downstream from the lesion. This can help to classify functionally different types of lesions that yield similar scotomas, and to identify regions of the scotoma that elicit different patterns of functional activation and may therefore, have different capacity for rehabilitation. The underlying hypothesis is that sectors of the scotoma that can still convey visual information to higher areas, bypassing the cortical lesion, will be more amenable to rehabilitation. Moreover, the extrastriate areas that get activated may reveal clues about the attributes of the visual stimulus that will be more amenable to rehabilitation.

We propose to apply state-of-the-art fMRI methods to characterize voxel by voxel how population receptive fields (Wandell et al., 2007; Dumoulin and Wandell, 2008; Amano et al., 2009; Lee et al., 2013) in spared visual areas are organized to cover the visual field following cortical visual pathway injuries (see figs. 2, and 3) (Baseler et al., 2011). The population receptive field (pRF) of a voxel refers to the region of visual space that elicits a visually-induced modulation of the BOLD (Blood-Oxygen-Level-Dependent) signal in that voxel. Various pRF models have been proposed in the literature (Dumoulin and Wandell, 2008; Zuiderbaan et al., 2012; Lee et al., 2013). The simplest and most commonly used model is a circularly symmetric 2D Gaussian with center (x,y) and radius (σ) (Dumoulin and Wandell, 2008). The BOLD time series predicted by this model is derived by convolving the pRF model with the stimulus sequence and the BOLD hemodynamic response function (HRF; Boynton et al., 1996; Worsley et al., 2002). The pRF's parameters are then estimated by fitting the BOLD signal predicted by the model to the actual BOLD signal measurement obtained from each voxel.

Early methods of retinotopic mapping, such as ring and wedge stimuli (DeYoe et al., 1996; Dougherty et al., 2003; Engel et al., 1994; Sereno et al., 1995), as well as moving bar stimuli (Wandell et al., 2007), which traverse the visual field in different directions can usually provide robust pRF estimates (see fig. 2) rendering them useful for studying cortical reorganization. One limitation of direct-fit pRF estimation methods is that these can result in estimation biases at the scotoma border (Lee et al., 2013). Recently, a 2-step pRF estimation method based on first estimating the pRF topography, thresholding it, and then fitting an appropriate pRF model has been introduced to largely circumvent this problem (Lee et al., 2013). This is the preferred method to use near the scotoma border.

http://dx.doi.org/10.5772/57000

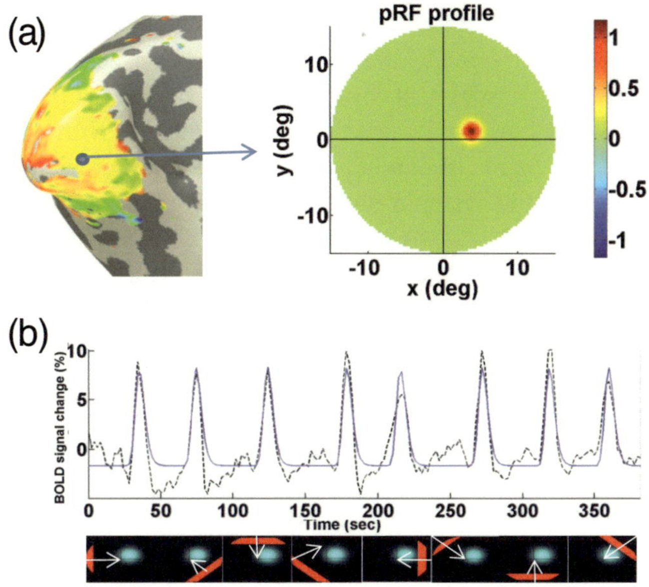

Figure 2. The population receptive field (pRF) model. This model estimates the region of the visual field that elicits a response in a small region (voxel) of the visual cortex. One implementation of the pRF model is a circularly symmetric Gaussian receptive field in visual space whose center and radius are estimated by fitting the BOLD signal responses to the estimated responses elicited by convolving the model with the moving bar stimulus and the hemodynamic response. (**a**) shows the estimated position and size of the pRF in the visual field of a voxel located in V1. (**b**) shows the BOLD time-series (dashed line) and the model prediction (solid line) from the same voxel. The model explains a large amount of variance in the time course data. Below, we illustrate the position and direction of motion of the stimulus bar that elicited the peaks in the BOLD signal.

2.1. Visual field coverage maps can potentially help to guide rehabilitation strategies

Population receptive field analysis can help measure the residual capacity of each area to process visual information following V1+ injuries. Plotting the pRF maps from all voxels in a given area together as a color map reveals the "visual field coverage map" of that area. This represents the part of the visual field that can visually modulate the area. Note that visual field coverage maps of extrastriate areas often overlap with the area of the dense perceptual scotoma, measured by visual field perimetry. This is illustrated in the second panel of figure 3 for area hV5/MT+, an area important for visual motion perception (Zeki et al., 2004; ffytche et al., 2000; Zeki and ffytche et al., 1998). Note that the pRF maps of many area hV5/MT+ voxels lie inside the perceptual scotoma (left upper quadrant in fig. 3). In fact, in this specific case (fig. 3), they cover the whole extent of the scotoma. This implies that hV5/MT+ is activated by visual stimuli presented in the left upper quadrant even though the subject does not perceive these stimuli. This suggests that there is a functional V1-bypassing pathway to area hV5/MT+ that may be able to promote recovery, if harnessed appropriately. Therefore, a promising rehabilitation strategy would be to find ways to strengthen this pathway. It is likely, that it will be easier to rehabilitate visual motion perception inside parts of the scotoma that are covered by the pRF maps of area hV5/MT+. It is also, likely that rehabilitation will be even easier in parts of the scotoma that are also covered by the pRF maps of spared, earlier, visual areas. The third panel of fig. 3 illustrates the visual field coverage map of the spared portion of area V1. Note, that this extends above the horizontal meridian to partly overlap with the area of the scotoma. This defines two regions where visual rehabilitation may be different, according to the above hypothesis. The region of the scotoma indicated by the green arrow is represented in the coverage maps of both area hV5/MT+ and the spared area V1, and is expected to have higher potential for rehabilitation. The region of the scotoma indicated by the blue arrow is represented only in the coverage map of area hV5/MT+ and is expected to present a more difficult challenge for rehabilitation. Visual field coverage maps (Amano et al., 2009) obtained by fMRI are an important adjunct to perimetric maps as they often provide complementary information (personal observation) and will likely be useful in tailoring therapy to appropriate visual field locations.

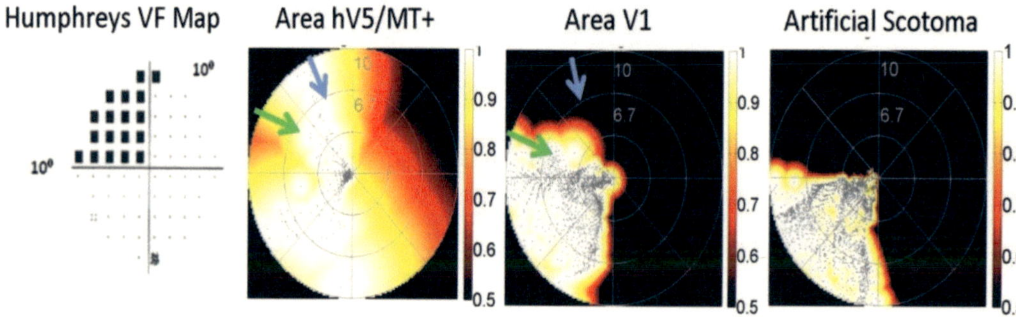

Figure 3. Visual Field Coverage Maps of spared visual areas with significant overlap with the region of the scotoma, define visual field locations that may be more amenable to rehabilitation. *Humphreys VF Map panel:* Humphreys 10-2 visual field map, illustrating that the subject had a dense left upper quadrantanopia. *Area hV5/MT+ panel:* The visual field coverage map (Amano et al., 2009) of the right hV5/MT+ shows visual field locations that evoke significant activity from hV5/MT+ of the lesioned hemisphere. At each visual field location, the highest pRF value of all pRFs that cover this location is plotted. PRF normalization limits the range of values between 0-1. Large values indicate significant visual modulation. Note, that although the subject is blind in the left upper quadrant, the subject's hV5/MT+ responds to stimuli presented in the left upper quadrant. A potential advantage of using visual field coverage maps is that therapy can be individualized to each patient's appropriate visual field locations, which are not necessarily predictable from perimetric maps. *Area V1 Panel:* Visual field coverage map of the *spared* right area V1 extends above the left horizontal meridian into the dense area of the scotoma seen in the Humphreys map. This activity may be induced in orthotopic voxels that survive and are partially active following the V1 lesion (Kasten et al., 1998), or in anatomically ectopic V1 voxels that belong to the upper occipital lobe, which would ordinarily have had receptive fields in the left lower visual field quadrant. Note, that this area does not cover the entire quadrant, as is the case of the coverage map in area hV5/MT+. This mismatch will likely have implications for rehabilitation. *For example, it may be easier to rehabilitate regions of the scotoma where the visual field coverage maps of spared V1 and hV5/MT+ are congruent (green arrow), as opposed to incongruent (blue arrow). Control with AS (artificial scotoma) Panel:* Visual field coverage map from the entire area V1 of a normal subject with an "artificial scotoma" simulating left upper quadrantanopia. By artificial scotoma we mean an area of the stimulus being excluded, in order to simulate the patient's scotoma. Note, that the visual field coverage map in this control case is as expected, i.e. it does not encroach into the left upper quadrant.

2.2. Quantitative pRF measurements provide a useful biomarker for gauging the effect of rehabilitation strategies

Fig. 4 illustrates that pRF measurements in area hV5/MT+ ipsilateral to a chronic V1+ lesion differ from those in the normal hemisphere. Specifically, pRFs in hV5/MT+ voxels of the lesioned hemisphere are smaller on average, and pRF-centers cluster near the vertical meridian (x=0; fig. 4C). Training may further change the pRF topography. Changes in the pRF topography before and after training can be a potentially useful biomarker for evaluating different rehabilitation paradigms before perceptual recovery becomes evident. Applying this approach systematically can help to formulate new hypotheses guiding future rehabilitation attempts. For example, one hypothesis is that following visual motion rehabilitation training the sensitivity to motion stimuli of the human motion selective area complex (hV5/MT+) will increase, and this increase will correlate with behavioral recovery. Alternatively, attentional networks or, other "higher" areas that receive input from hV5/MT+ may reorganize to process visual motion information more effectively. Analyzing pRF maps obtained from visually responsive areas before and after training, will allow us to investigate the above hypotheses and to adopt appropriate rehabilitation strategies.

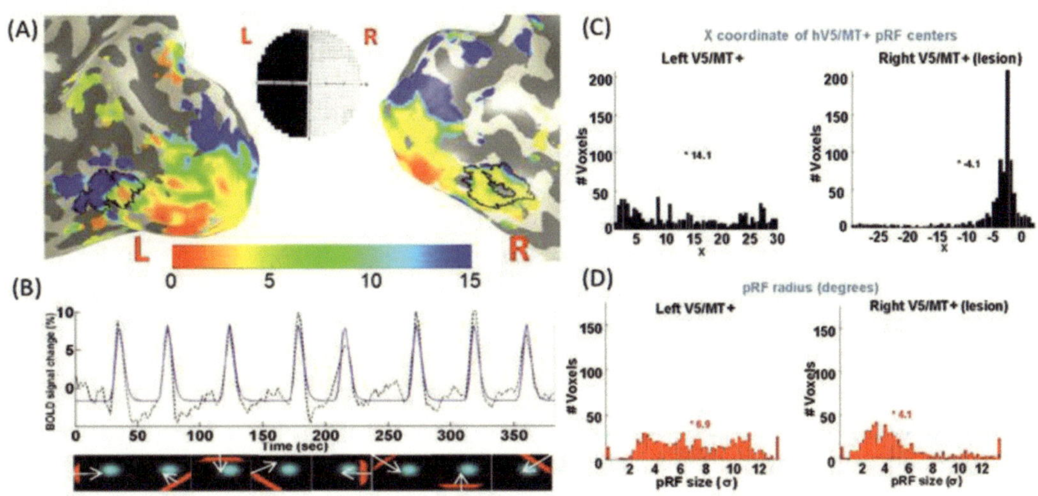

Figure 4. *hV5/MT+* pRF mapping in a hemianopic patient: (A) Inflated occipital lobe of the normal (left) and lesioned (right) hemisphere with hV5/MT+ outlined on the eccentricity map. Mid-inset is a 30-2 Humphrey perimeter showing dense left hemianopia. (B) Illustrates the method used to calculate pRF parameters (same as in fig. 2). The pRF model parameters are optimized to fit the BOLD signal voxel by voxel. Note the close fit between the BOLD time-series (dashed line) and the pRF model prediction (solid line). The position and direction of motion of the stimulus bar that elicited each response is illustrated (bottom). (C) The X coordinate of the pRF centers in hV5/MT+ of the normal (left) hemisphere is evenly distributed while in the lesioned (right) hemisphere it clusters near the vertical meridian (x=0). Note however, that the pRF centers still lie within the scotoma (negative X values). (D): Distribution of pRF radii of hV5/MT+ voxels in the intact (left) and in the lesioned (right) hemisphere. The distribution of pRF radii shifts to smaller values in area hV5/MT+ of the lesioned hemisphere. This may be because ipsilesional hV5/MT+ voxels are driven by a V1-bypassing pathway that drives visual periphery less effectively.

In summary, pRF measurements:

1. Classify subregions of the perimetric scotoma depending on how they are covered by spared visual areas; different regions will likely have different potential for rehabilitation.
2. Allow us to study the mechanism by which rehabilitation strategies improve visual performance.
3. Serve as quantitative biomarkers for evaluating the effects of training before perceptual recovery becomes evident, accelerating the pursuit of a rational strategy for visual rehabilitation.

Other methods of analysis can be applied here, but we do not have the space to do them justice. We mention briefly the promise of recent developments in effective connectivity analysis (Fuji et al., 2009; Stilla et al., 2008; Hinrichs et al., 2006) and diffusion tensor imaging (Wedeen et al., 2012; Yeatman et al., 2012; Van den Stock et al., 2011; Sherbondy et al., 2008; Fields, 2008; Okada et al., 2007; Schoth et al., 2006; Kikuta et al., 2006; Dougherty et al., 2005; Taoka et al., 2005; de Gelder et al., 2005; Reinges et al., 2004; Morris et al., 2001) for studying inter-area pathways that survive post-lesion, and whether they can get stronger by training. Population receptive field and effective connectivity analysis can be used to explore visual system reorganization and recovery following injury, and to generate concrete hypotheses on how to enhance and accelerate recovery of visual function by the application of cutting-edge rehabilitative strategies.

3. How to approach visual rehabilitation?

To date, we have little understanding of how the visual cortex reorganizes after injury, and no proven effective treatment strategies to rehabilitate the recovery of visual perception in the affected portion of the visual field in V1-lesioned patients. Understanding how to manipulate the brain's capacity for plasticity is an important step in the long-term effort to design treatments aiming to enhance the ability of the nervous system to recover after injury. To make progress along this front we need to: i) study the mechanisms by which the adult brain adapts and reorganizes after injury; and ii) devise approaches that will allow us to manipulate the process of reorganization to induce visual recovery.

The network of visual areas can be viewed as a heavily interconnected circuit subject to a series of hierarchy rules. Early areas usually process sensory information initially, by passing it on to higher areas, and in turn, extract "higher" order features and control the flow of information through feedback loops. Increased performance following training can therefore be the result of changes that occur in early areas (Schoups et al., 2001; Yotsumoto et al., 2008; Censor and Sagi, 2009; Karni and Sagi, 2008), or the result of changes that occur in "higher" visual areas and attentional networks (Law and Gold, 2008; Yang and Maunsell, 2004; Lewis et al., 2009). Area V1 injuries, interrupt the cardinal feed-forward pathway but, as discussed above, visually driven information can still activate surviving extrastriate areas through bypassing routes (Cowey, 1974; Dineen et al., 1982; Rodman et al., 1989; Cowey and Stoerig, 1997; Baseler et al.,

http://dx.doi.org/10.5772/57000

1999; Goebel et al., 2001; Schmid et al., 2009; Schmid et al., 2010). The pattern of activity elicited in surviving visual areas interacts with higher "centers" in frontal, parietal and temporal lobes but, in the absence of V1 input, fails to generate a strong visual percept. We suggest here two general, non-mutually exclusive approaches to visual rehabilitation:

"Bottom-up" approach: Visual rehabilitation strengthens V1-bypassing pathways to increase the response elicited in surviving extrastriate areas.

"Top-down" approach: Visual rehabilitation reorganizes higher "centers" to learn to process the changed patterns of activity elicited in extrastriate areas by V1-bypassing inputs.

Non-invasive approaches to visual rehabilitation aim to enhance these pathways by recruiting the mechanisms of plasticity the brain uses for learning. Various behavioral approaches have been used. They usually involve performing a visual task that directs attention to a sub-threshold stimulus, requires a choice, and then provides feedback about correct and incorrect choices (see fig. 5). Although such methods are effective for perceptual learning in general (Yotsumoto et al., 2008; Law and Gold, 2008; Yang and Maunsell, 2004), in the domain of rehabilitation of a dense perceptual scotoma results have been at best variable (Huxlin et al., 2009; Raninen et al., 2007 Sahraie, et al., 2006; Reinhard et al., 2005; Horton et al., 2005a; Horton et al., 2005b; Pleger et al., 2003). The most notable exception has been a recent well-controlled report by Huxlin and co-workers (Huxlin et al., 2009), which demonstrated strong recovery of direction of visual motion perception inside the scotoma of 5 hemianopic subjects (Huxlin et al., 2009; Huxlin, 2006). Two other groups independently, Sahraie et al. and Raninen et al., report that visual sensitivity can improve with training in humans with homonymous scotomas (Sahraie, 2006; Henriksson et al., 2007; Raninen et al., 2007). Encouraging results were also obtained by Pleger and co-workers, who showed that visual cortex reorganization was possible via daily visual stimulation training over a period of 6 months in 3 subjects with partial cortical blindness (Pleger et al., 2003). Although these are encouraging reports, the issue remains far from settled (Horton et al., 2005a; Horton et al., 2005b). Additional studies corroborating recent results and probing the mechanism of recovery are clearly needed to guide the implementation of new, more effective, rehabilitation strategies. In what follows, we discuss the promise and challenges of a new visual rehabilitation approach, which aims to use "real time" fMRI neurofeedback to train subjects to promote plasticity in V1-bypassing pathways relevant to recovery.

4. Introduction to real-time functional magnetic resonance imaging neurofeedback approaches

Less than two decades ago, the real-time functional magnetic resonance (rt-fMRI) method has been introduced (Cox, 1995) in the field of neuro-rehabilitation, which extracts the BOLD (Blood-Oxygen-Level-Dependent) signal from the subject's brain in real-time and uses it to provide feedback to the subject. Since the BOLD signal reflects neural activity this approach is called real-time fMRI neurofeedback (rt-fMRI nFb). Multiple studies have shown that rt-fMRI nFb, can train subjects to modulate the magnitude and spatial extent of the activity elicited in various cortical and subcortical areas (Berman et al., 2011; Bray et al., 2007; Caria et al., 2007; Caria et al., 2010; Chiew et al., 2012; deCharms et al., 2004; deCharms et al., 2005; Frank et al., 2012; Haller et al., 2010; Hamilton et al., 2011; Hinds et al., 2011; Johnson et al., 2012; Johnson et al., 2011; Johnson et al; 2010; Lee et al., 2012; Li et al., 2012; McCaig et al., 2011; Papageorgiou et al., 2009a; Papageorgiou et al., 2009b; Papageorgiou et al., 2013; Posse et al., 2003; Rota et al., 2009; Ruiz et al., 2013; Scharnowski et al., 2012; Shibata et al., 2011; Subramanian et al., 2011; Sulzer et al., 2013; Veit et al., 2012; Weiskopf et al., 2007; Weiskopf et al., 2004; Weiskopf et al., 2003; Yoo and Jolesz, 2002; Yoo et al., 2008; Zotev et al., 2011). The goal of this approach is to train subjects to control the pattern of their brain activity in a way that promotes a desired behavior. It can also be used to boost the neural capacity for learning and plasticity (Shibata et al., 2012; Scharnowski et al., 2012; Shibata et al., 2011; Weiskopf et al., 2004; deCharms et al., 2004; deCharms et al., 2005). If this is applied effectively, it could serve as a useful tool to promote neuro-rehabilitation.

The ability of rt-fMRI nFb to induce a behavioral change was first shown by Weiskopf et al. (Weiskopf et al., 2003) and deCharms et al. (deCharms et al., 2005). In the deCharms et al. study, chronic-pain patients were coached to decrease their pain by learning to control the BOLD signal intensity of the rostral Anterior Cingulate Cortex (rACC), a region known to be involved in pain perception (Apkarian et al., 2005; Peyron et al., 2000). After training, subjects were able to voluntarily increase or decrease the rACC BOLD signal intensity, and this correlated with an increased or decreased level of pain, respectively; i.e., a 50% decrease in the rACC activity of chronic pain subjects corresponded approximately to a 64% decrease in their pain. This effect was specific to rtfMRI nFb training applied to rACC; i.e., no effect was seen after similar training conducted without rtfMRI NFb or, after sham rtfMRI nFb training derived from the activity of another subject's rACC.

Similar results have been obtained in other cortical and subcortical domains (Posse et al., 2003; Lee et al., 2011; Hamilton et al., 2011; Ruiz et al., 2013). A recent study showed that healthy volunteers were able to volitionally regulate the activity of their insula when given rt-fMRI nFb (Lee et al., 2011). Posse and his team (Posse et al., 2003) trained subjects to upregulate their amygdala, an area whose level of activity is associated with sad affect and depression (Wang et al., 2012; Anand et al., 2007; Liu et al., 2011). Amygdala upregulation induced by rt-fMRI feedback was positively correlated with self-ratings of sadness across repeated fMRI sessions (Posse et al., 2003). Conversely, Hamilton et al. used rt-fMRI nFb to train subjects to downregulate subgenual ACC and posterior cingulate cortex, resulting in positive mood induction (Hamilton et al., 2011). Sham rt-fMRI nFb showed no effect. Ruiz et al. showed that schizophrenic patients can be trained by rt-fMRI nFb to voluntarily control their anterior insula bilaterally (Ruiz et al., 2013). The effect of bilateral anterior insula activation is reflected on their ability to recognize face emotion, a known deficit in schizophrenia. These findings collectively, suggest that: 1) rt- fMRI nFb can be used to train subjects to voluntarily control specific areas; 2) changes in the activity of certain areas can be associated with significant behavioral changes; 3) rt-fMRI nFb training can achieve stronger behavioral results than similar training without nFb; and 4) rt-fMRI training can induce plastic changes that can outlast the period of the training inside the magnet and even induce visual perceptual learning (Shibata et al. 2011).

Shibata et al. in a seminal study used rt-fMRI nFb to induce perceptual learning (Shibata et al., 2011). Rt-fMRI nFb methods were used to enable subjects to induce activity patterns in their early visual cortex corresponding to one particular orientation. Initially, the subjects performed an orientation discrimination task and a decoder of area V1/V2 activity was constructed to classify a pattern of the measured fMRI signals into one of three orientations. Once the decoder was constructed, each subject participated in a 5 to 10-day rt-fMRI nFb stage, during which they learned to induce patterns of activity in areas V1/V2 corresponding to the target orientation. During this stage, subjects were instructed to maximize the signal delivered to them via feedback, but were not told how to induce the desired patterns of activity that would result in increased activity. Subjects did not know what was to be learned. Using this strategy they were able to induce visual perceptual learning specific to the target orientation in areas V1/V2. Learning occurred as a function of the subject's ability to elicit the particular pattern of activation corresponding to the target orientation in early visual areas. Remarkably, subjects were able to generate this pattern simply by being given the instruction to maximize feedback, without being aware that the pattern to be elicited had anything to do with orientation. This demonstrated that the rt-fMRI nFb method can be used to induce highly specific activity patterns within a brain region and that repeatedly eliciting the desired pattern of activity is sufficient to induce plasticity in early visual areas. These findings suggest that rt-fMRI nFb training can be used to induce plasticity in a highly selective manner in the visual system.

The studies described above (Ruiz et al., 2013; Linden et al., 2012; Subramanian et al., 2011; Shibata et al., 2011; Lee et al., 2011; Hamilton et al., 2011; deCharms et al., 2004; Posse et al., 2003) show that manipulating the activity in select brain areas can induce plasticity. Modified paradigms can also be designed to increase plasticity along specific pathways, by co-activating input and recipient neuronal populations. For example, a projection from area A to recipient area B in the brain can increase or, decrease in strength, depending on the relative activity between input projections from A and recipient neurons in B. Rt-fMRI nFb does not have the temporal resolution necessary to precisely implement Hebbian mechanisms of plasticity. Nevertheless it can be used to train the subject to voluntarily manipulate the activity level of select neuronal populations in area B while their input from A is presented. The hypothesis is that "top-down" activation of area B enhanced by nFb judiciously paired with "bottom-up" presentation of inputs that activate A, can act together to increase the strength of the projection A->B. Below, we discuss an example outlining how this proposal might work for the rehabilitation of visual motion perception following area V1 lesions (see also fig. 5).

Area hV5/MT+ is associated with global coherent visual motion perception. The goal is to use rt-fMRI nFb methods to strengthen the neural pathways bypassing the V1 lesion and project to area hV5/MT+ in order to improve visual motion perception of random dot kinematogram (RDK) stimuli (see fig. 5). The first step is to train the subject to voluntarily upregulate their hV5/MT+ activity. To do this we ask the subject to practice mental imagery of fully coherent visual motion stimuli in their blind hemifield, moving in the direction of the anticipated stimulus. During mental imagery, the subject receives rt-fMRI nFb proportional to the activity

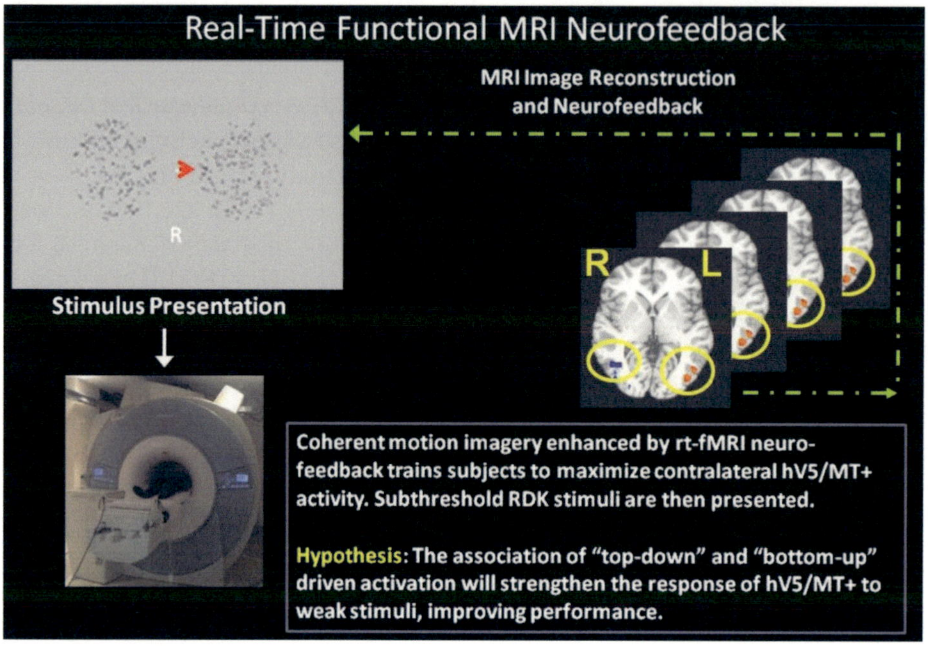

Figure 5. Rt-fMRI nFb paradigm for rehabiliating visual motion perception. The subject lies supine inside the scanner, while s/he is presented with a 0% coherent RDK stimulus. Brain volumes are acquired every 2sec (TR). hV5/MT+ (circled) is selected as our ROI. Turbo Brain Voyager software (Brain Innovation) is used to deliver nFb. The BOLD signal in hV5/MT+ is estimated in real time (every TR), normalized, and "fed" back, via the length of a horizontal arrow at fixation, to train the subject to upregulate area hV5/MT+. Subjects are instructed to attend to the RDK they are cued towards, either right (R) or left (L) and imagine that it is coherent (when in fact it is not). This increases the level of activity in the contralateral hV5/MT+ : (i) superimposing coherent motion via imagery on the right RDK increases left hV5/MT+ activity, which is color-coded in red; (ii) superimposing coherent motion via imagery on the left RDK increases right hV5/MT+ activity, which is color-coded in blue. The subject uses the length of the arrow to determine how successful their strategy is and try harder, if needed. *When the rt-fMRI nFb driven, "top-down," hV5/MT+ activation crosses a threshold, sub-threshold RDK stimuli are presented. The association of "top-down" and "bottom-up" activation will engage Hebbian-like learning mechanisms aiming to strengthen the response of hV5/MT+ to sub-threshold stimuli.* The hypothesis is that once these pathways are strengthened, the presentation of sub-threshold stimuli will elicit enough activity in area hV5/MT+ to improve performance in the direction of motion discrimination task. We note that *hV5/MT+ can be upregulated via rt-fMRI nFb enhanced imagery even when it has lost its V1 input.* Once the subject learns how to upregulate hV5/MT+ inside the magnet over the period of training, we hypothesize that s/he will also be able to transfer this learned voluntary ability during training outside the MRI environment.

in their hV5/MT+ via a visual interface at fixation (red arrow in fig. 5). Subjects are trained to maximize ipsi-lesional hV5/MT+ activity using the imagery task. This nFb mediated, "top-down" increase in hV5/MT+ activity will then be paired with the presentation of visual motion stimuli that are invisible (sub-threshold) to subjects with V1+ lesions. We hypothesize that, by repeatedly presenting sub-threshold visual motion stimuli while hV5/MT+ is activated in a "top-down" fashion by nFb, we will engage Hebbian-like association learning mechanisms (Hebb, 1946; Rebesco and Miller, 2011; Gallistel and Matzel, 2013). These mechanisms will promote plasticity in the surviving, V1-bypassing pathways that get activated by the stimulus presentation and project to area hV5/MT+. In other words, we hypothesize that after nFb training, regions of area hV5/MT+ deprived of V1 input will respond more strongly to visual stimuli presented inside the scotoma, improving performance. If successful, this strategy will induce a "neural bypass" of V1 function with respect to visual motion perception.

One advantage that rt-fMRI nFb methods have over behavioral feedback approaches is that specific brain pathways or areas can be selectively trained. It is then, feasible to train targeted components of the neural circuit that are likely to contribute to recovery. One example is strengthening neural pathways that bypass the region of an injury to promote recovery. Thus far, we have been discussing a univariate rt-fMRI nFb approach, where the nFb provided is proportional to the activity of a specific region of interest (in the above example, area hV5/MT +) or, pattern of activity (Shibata et al., 2011). An alternative to the univariate nFb method is the multivariate classification approach (Papageorgiou et al., 2013; Papageorgiou et al., 2009; LaConte et al., 2007; Mourao-Miranda et al., 2005). In this approach, a classification algorithm, usually, the support vector machine (Vapnik, 1995) is trained on a set of relevant data in order to identify the brain networks that are involved in a specific computation (task). This is akin to the localizer that is used to identify the region of interest in the univariate approach. In fact, in the case of presentation of RDK stimuli, the multivariate SVM approach picks out chiefly area hV5/MT+, as expected (see fig. 6 below). However, in other cases, this approach may reveal different patterns of activity than expected, helping to formulate new hypotheses about the networks that might contribute to rehabilitation. Once the parameters of the classification algorithm are trained, the algorithm can be used to provide neuro-feedback to the subject in separate sessions. This approach may be effective in cases where the pathways that need to be modulated to induce recovery are not known a-priori.

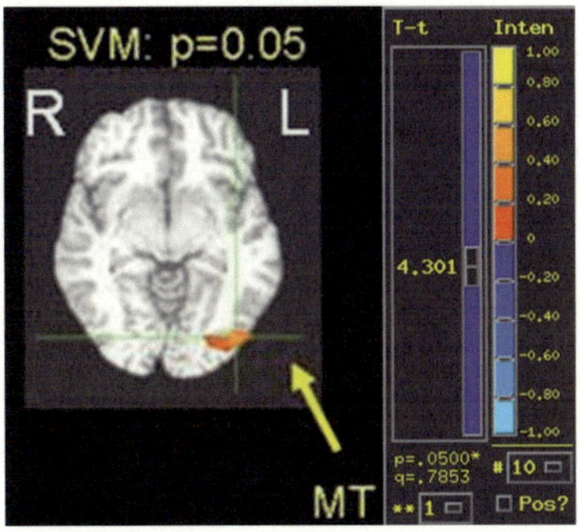

Figure 6. The support vector machine (SVM) algorithm defines area hV5/MT+ when it is trained on coherent motion RDKs. Subjects are fixating while two RDKs are presented simultaneously on symmetric locations in the right (red color-coded activation) and left (blue color-coded activation) visual hemi-fields. The RDK presented in the left visual field was always kept at 0% coherence (no global motion direction), while the RDK in the right visual field alternated between 0% coherence and 100% coherence. A support vector machine algorithm was trained on the whole brain to classify when the coherent RDK (right visual field/red-color coded activation) was presented. The color map indicates the region that was important in classifying the presentation of a 100% coherent vs a 0% coherent RDK in the right visual field. As expected, the area corresponds to area hV5/MT+. Three subjects tested gave consistent results (3dANOVA2 performed in AFNI, p=0.05).

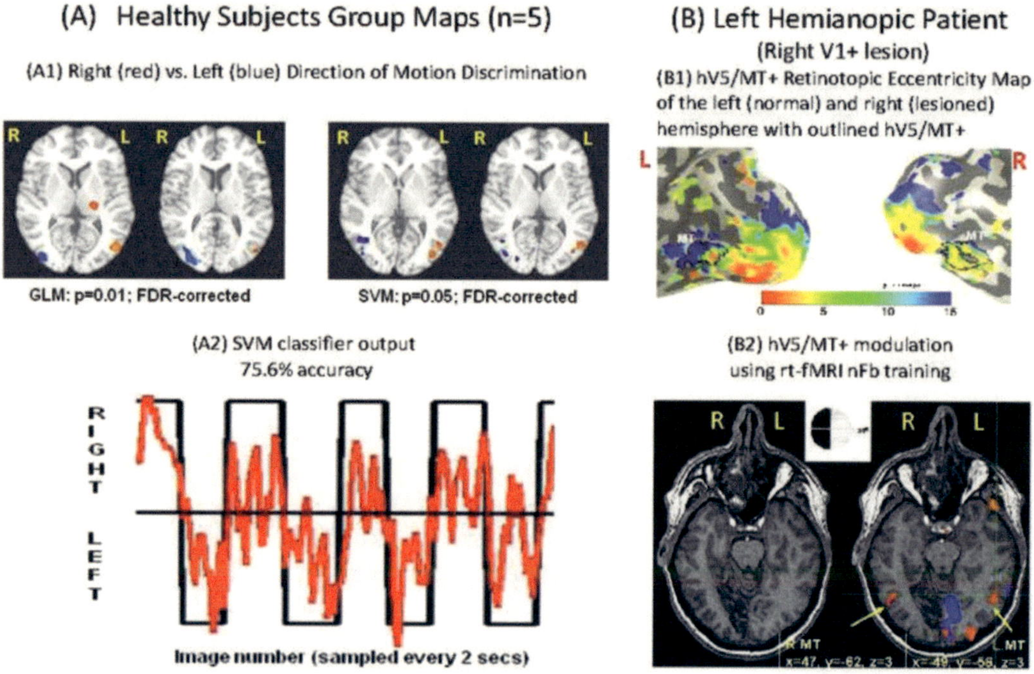

Figure 7. Rt-fMRI nFb can be used to selectively upregulate hV5/MT+ in a top-down manner during an imagery task. In the absence of a coherent moving stimulus, subjects (n=5) were able to use imagery of coherent motion to selectively upregulate hV5/MT+ via rt-fMRI nFb training. (**A**) Non-coherent RDKs were presented symmetrically in the L and the R hemifield and subjects were cued to imagine that the Left or the Right RDK contained coherent motion. The level of contralateral hV5/MT+ activity was delivered via the length of an arrow at the fixation point. (**A1**) Red/blue regions correspond to voxels upregulated when the subject is instructed to imagine coherent motion in the right/left visual field, respectively. Similar results are obtained when using a standard GLM (left panel), versus plotting the weight vector of the support vector machine (SVM) algorithm that classifies whether subjects imagined coherent motion to the left versus the right hemifield (right panel). The regions identified correspond to area hV5/MT+ in both hemispheres (color red, blue). *Therefore,subjects can be trained via rt-fMRI nFb to modulate theirhV5/MT+ activity even in the absence of a coherentlymoving stimulus;* (**A2**) Output (red curve) of a support vector machine classifier indicating the side of the visual field where the subject imagined coherent motion. The correct choice indicated by the black curve; the prediction of the classifier by the red curve. Positive values indicate the subject was instructed to imagine coherent motion on the right, negative on the left. Note that area hV5/MT+ activity can predict the subject's perceptual state. **(B1)** Eccentricity maps of a subject with left hemianopia illustrating that area hV5/MT+ (outlined) of the lesioned (right) hemisphrere can be visually driven. Only the posterior lateral aspect of the two inflated hemispheres is presented. **(B2)** Illustrates that *hV5/MT+ can also be upregulated via rt-fMRI nFb training in a V1+ lesioned patient.* **Left Panel:** Right (R) PCA lesion, resulting in left (L) hemianopia (inset). **Right Panel:** Red color-coded areas represent the activity elicited by the subject's coherent motion imagery in the left (hemianopic) visual field, while blue color-coded area represents non-coherent motion presentation, as generated by GLM. Left hemianopic visual field imagery of coherent motion activated hV5/MT+ bilaterally. Preliminary data for implementing the proposal we outlined above are encouraging as they suggest that ipsilesional hV5/MT+ activity: (1) conveys information about the stimulus **(B1)**, and (2) can be upregulated using rt-fmri nFb imagery in V1+ lesioned patients **(B2)**.

In summary, emerging strategies based on rt-fMRI nFb hold considerable promise, as: 1) they can be used to enhance plasticity in a number of systems (Berman et al., 2011; Bray et al., 2007; Caria et al., 2010; deCharms et al., 2004; Haller et al., 2010; Johnson et al., 2012; Johnston et al., 2010; Lee et al., 2012; Li et al., 2012; McCaig et al., 2011; Papageorgiou et al., 2013; Posse

http://dx.doi.org/10.5772/57000

et al., 2003; Ruiz et al., 2013; Subramanian et al., 2011; Sulzer et al., 2013; Veit al., 2012; Yoo and Jolesz, 2002; Yoo et al., 2008; Zotev et al., 2011) including in the visual system of healthy participants (Scharnowski et al., 2012; Shibata et al., 2011); 2) they may be in some cases superior to normal behavioral methods (deCharms et al., 2005); 3) they can be tailored to induce perceptual learning (plasticity) in a highly specific fashion (Shibata et al., 2011); 4) their multivariate variant can be used to identify pathways relevant to recovery; and 5) induced learning effects are reported to persist outside the magnet, after the end of training (Ruiz et al., 2013; Sulzer et al., 2013). In the long-term, rtfMRI nFb methods promise to induce cortical plasticity that is efficient, robust and specific for each patient. Lessons learned are likely to apply beyond the visual system to disorders of motor function, cognition, speech, language and emotion.

5. Challenges and future considerations using rt-fMRI nFb for visual rehabilitation

Many challenges need to be overcome in order to study the efficacy of rt-fMRI nFb methods in visual rehabilitation. Primarily, we need to develop effective rt-fMRI nFb paradigms in subjects with V1 lesions. The challenge is to implement rt-fMRI nFb training protocols to strengthen specific pathways that are hypothesized to play a role in rehabilitating performance. It is important to understand which pathways are more amenable to rehabilitation and what is the best rt-fMRI nFb paradigm to use. This requires elucidating which factors are "necessary and sufficient variables for learning" (Weiskopf, 2012). The answers to these questions will in general depend on the specifics of the injury that requires rehabilitation, as well as on other factors such as the subject's motivation.

Quantifying the degree of induced reorganization using population receptive field (pRF) methods is complementary to behavioral performance measures and represents a valuable neuroimaging biomarker for studying the mechanism of recovery induced by nFb rehabilitation methods. Information obtained will then, allow us to refine future rehabilitative approaches. Several different pathways may be able to contribute to recovery. For example, in the case of the visual motion rehabilitation example, the focus was on strengthening "bottom-up" pathways to enhance the response of the ipsilesional area hV5/MT+ to the visual motion stimulus. One can hypothesize other strategies that focus instead on reorganizing higher areas, such as frontal eye fields (FEF), supplementary eye fields (SEF) areas involved in the generation of visual motion percepts downstream of hV5/MT+ by "reading out" the weak activity that persists in extrastriate cortex following V1+ lesions. Or, one can focus on strategies that reorganize attentional networks, such as middle frontal gyrus (mFG), intraparietal sulcus (IPS), superior parietal lobule (SPL), and anterior cingulate cortex (ACC) that enhance the weakened responses elicited in surviving areas following V1+ lesions.

There are important technical challenges. The criticism is often mounted that rt-fMRI nFb approaches are impractical because they require large amounts of magnet time. Although this have some truth in it, preliminary studies reveal that time spent inside the magnet is much less than what pure behavioral methods, require. Deciding how many rt-fMRI nFb sessions are needed to induce plasticity is a question that still needs to be answered. Preliminary evidence suggests that as few as 5-10 sessions can be sufficient to induce a strong perceptual learning effect in normal subjects (Shibata et al., 2011), but this will need to be validated in the case of patients with V1 lesions. Healthy participants and patients undergoing rt-fMRI nFb training sessions inside the magnet can learn to voluntarily elicit the desired pattern of activity. Subjects could then, gradually learn to implement this process outside the magnet, transferring their experience from rt-fMRI nFb sessions to ordinary behavioral sessions. We do not have adequate evidence yet, to determine under what conditions training accomplished inside the rt-fMRI environment can be transferred outside the magnet, but there is reason to be hopeful (Ruiz et al., 2013; Sulzer et al., 2013). The clinical applicability of rt-fMRI nFb training will become significantly broader if it becomes feasible to decouple the patient's training sessions from the rt-fMRI nFb environment. Another important question is how long the effects of rt-fMRI nFb training are expected to last and whether this depends on the number of rt-fMRI nFb sessions used for training. Here too, there is reason for optimism given the results of Shibata et al. who managed to induce perceptual learning within 5 sessions of rtfMRI nFb (Shibata et al., 2011).

Many important challenges remain. However, it is now possible to mount a systematic approach to visual rehabilitation using novel rt-fMRI nFb methods guided by pRF analysis of spared visual areas. This approach promises to teach us a lot about the visual system's capacity for plasticity after injury, and offers hope that effective, and robust visual rehabilitation methods, such as the novel rt- fMRI nFb approach will be used in the field of visual neuro-rehabilitation.

6. Conclusion

Rehabilitating visual loss that occurs as a result of primary visual cortex injury is a difficult problem. To date, we have little understanding of the plasticity mechanisms operating in the adult visual system following V1 injury. Consequently, no reliable method exists to effectively rehabilitate V1-lesioned patients who experience loss of visual perception in the contralateral hemifield (Horton, 2005a; Horton, 2005b; Pambakian and Kennard, 1997). Interestingly, recent results have shown that visually driven activity persists in extrastriate cortex following chronic area V1+ lesions (Schmid, 2010; Schmid, 2009; Rodman, 1989; Rodman, 1990; Baseler, 1999), and that it is possible in some cases to rehabilitate visual motion perception (Das, 2010; Huxlin, 2009; Raninen et al., 2007; Henriksson et al., 2007; Sahraie et al., 2010). This confirms the existence of functional pathways that bypass the V1+ lesion, providing direct input to spared extrastriate cortex. Such pathways are generally too weak to result in practical benefit. However, appropriate training strategies may be able to strengthen them sufficiently to induce recovery.

Real-time fMRI neuro-feedback strategies allow subjects to voluntarily modulate activity in certain brain areas or, neural pathways. These methods can be harnessed and used to promote

http://dx.doi.org/10.5772/57000

plasticity (Shibata et al., 2011). For example, we hypothesize that rt-fMRI nFb may be used to strengthen pathways that bypass the region of V1 injury to transmit visual motion information to area hV5/MT+. One paradigm that could, in theory, be used to accomplish this is the following: Subjects are trained by rt-fMRI nFb to voluntarily upregulate their hV5/MT+ activity. Whenever hV5/MT+ activity crosses a pre-set threshold, sub-threshold visual stimuli are presented. Repeated pairing of the "top-down" nFb-driven activation with the "bottom-up" stimulus-driven activation will engage Hebbian-like association learning mechanisms, strengthening the response of hV5/MT+ to visual motion stimuli. Visual motion stimuli that were previously sub-threshold may then, rise above threshold following training, improving performance.

Rehabilitating dense visual field scotomas requires adopting a systematic approach. Changes brought about by new rehabilitation strategies should be mapped and their mechanism studied. We have presented evidence that pRF analysis is an excellent tool for this purpose, quantifying changes and providing rich data for formulating hypotheses about what regions of the visual field may be more amenable to rehabilitation and what pathways contribute to recovery.

We conclude that, even though rt-fMRI nFb methods are currently in their infancy, they hold considerable promise for inducing plasticity in targeted pathways promoting successful rehabilitation. Although here, we have focused on the visual system, principles discussed apply to the neuro-rehabilitation of several other domains of brain function, such as motor control, language, speech (Papageorgiou et al., 2009a; Papageorgiou et al., 2009b; Papageorgiou et al., 2013), emotion and cognition.

Acknowledgements

Part of the work that is presented here was supported by: (i) a McNair Foundation award, a McNair Medical Institute (MMI) award and a Fight for Sight Grant to T.D. Papageorgiou; and (ii) an NEI RO1 (EY019272), DoD (W81XWH-08-2-0146), and an HHMI Early Career Award to S. M. Smirnakis.

Author details

T. Dorina Papageorgiou[1], Amalia Papanikolaou[2] and Stelios M. Smirnakis[1,2]

1 Departments of Neuroscience and Neurology, Baylor College of Medicine, Houston, Texas, USA

2 Departments of MaxPlanck Institute for Biological Cybernetics, Tuebingen, Germany

References

[1] Ahissar, M. and S. Hochstein (1997). "Task difficulty and the specificity of perceptual learning." Nature 387(6631): 401-6.

[2] Ajina, S. and C. Kennard (2012). "Rehabilitation of damage to the visual brain." Rev Neurol (Paris) 168(10): 754-61.

[3] Alexander, I. and A. Cowey (2009). "The cortical basis of global motion detection in blindsight." Exp Brain Res 192(3): 407-11.

[4] Amano, K., B. A. Wandell, et al. (2009). "Visual field maps, population receptive field sizes, and visual field coverage in the human MT+ complex." J Neurophysiol 102(5): 2704-18.

[5] Anand, A., Y. Li, et al. (2007). "Reciprocal effects of antidepressant treatment on activity and connectivity of the mood regulating circuit: an FMRI study." J Neuropsychiatry Clin Neurosci 19(3): 274-82.

[6] Apkarian, A. V., M. C. Bushnell, et al. (2005). "Human brain mechanisms of pain perception and regulation in health and disease." Eur J Pain 9(4): 463-84.

[7] Baseler, H. A., A. Gouws, et al. (2011). "Large-scale remapping of visual cortex is absent in adult humans with macular degeneration." Nat Neurosci 14(5): 649-55.

[8] Baseler, H. A., A. B. Morland, et al. (1999). "Topographic organization of human visual areas in the absence of input from primary cortex." J Neurosci 19(7): 2619-27.

[9] Berman, B. D., S. G. Horovitz, et al. (2012). "Self-modulation of primary motor cortex activity with motor and motor imagery tasks using real-time fMRI-based neurofeedback." Neuroimage 59(2): 917-25.

[10] Boynton, G. M., S. A. Engel, et al. (1996). "Linear systems analysis of functional magnetic resonance imaging in human V1." J Neurosci 16(13): 4207-21.

[11] Bray, S., S. Shimojo, et al. (2007). "Direct instrumental conditioning of neural activity using functional magnetic resonance imaging-derived reward feedback." J Neurosci 27(28): 7498-507.

[12] Brown, T. H., P. F. Chapman, et al. (1988). "Long-term synaptic potentiation." Science 242(4879): 724-8.

[13] Caria, A., R. Sitaram, et al. (2010). "Volitional control of anterior insula activity modulates the response to aversive stimuli. A real-time functional magnetic resonance imaging study." Biol Psychiatry 68(5): 425-32.

[14] Caria, A., R. Veit, et al. (2007). "Regulation of anterior insular cortex activity using real-time fMRI." Neuroimage 35(3): 1238-46.

[15] Censor, N. and D. Sagi (2009). "Global resistance to local perceptual adaptation in texture discrimination." Vision Res 49(21): 2550-6.

[16] Chiew, M., S. M. LaConte, et al. (2012). "Investigation of fMRI neurofeedback of differential primary motor cortex activity using kinesthetic motor imagery." Neuroimage 61(1): 21-31.

[17] Chokron, S., C. Perez, et al. (2008). "From blindsight to sight: cognitive rehabilitation of visual field defects." Restor Neurol Neurosci 26(4-5): 305-20.

[18] Corbetta, M. and G. L. Shulman (2002). "Control of goal-directed and stimulus-driven attention in the brain." Nat Rev Neurosci 3(3): 201-15.

[19] Cowey, A. (2010). "Visual system: how does blindsight arise?" Curr Biol 20(17): R702-4.

[20] Cowey, A. (1974). "Atrophy of retinal ganglion cells after removal of striate cortex in a rhesus monkey." Perception 3(3): 257-60.

[21] Cowey, A. and P. Stoerig (1997). Blindsight in man and monkey. Brain 120(Pt 3): 535-59.

[22] Cowey, A. and P. Stoerig (1989). "Projection patterns of surviving neurons in the dorsal lateral geniculate nucleus following discrete lesions of striate cortex: implications for residual vision." Exp Brain Res 75(3): 631-8.

[23] Cowey, A. and P. Stoerig (1991). "The neurobiology of blindsight." Trends Neurosci 14(4): 140-5.

[24] Cox, R. W. (1996). "AFNI: software for analysis and visualization of functional magnetic resonance neuroimages." Comput Biomed Res 29(3): 162-73.

[25] Das, A. and K. R. Huxlin (2010). "New approaches to visual rehabilitation for cortical blindness: outcomes and putative mechanisms." Neuroscientist 16(4): 374-87.

[26] de Gelder, B., J. S. Morris, et al. (2005). "Unconscious fear influences emotional awareness of faces and voices." Proc Natl Acad Sci U S A 102(51): 18682-7.

[27] deCharms, R. C., K. Christoff, et al. (2004). "Learned regulation of spatially localized brain activation using real-time fMRI." Neuroimage 21(1): 436-43.

[28] deCharms, R. C., F. Maeda, et al. (2005). "Control over brain activation and pain learned by using real-time functional MRI." Proc Natl Acad Sci U S A 102(51): 18626-31.

[29] Deco, G. and T. S. Lee (2004). "The role of early visual cortex in visual integration: a neural model of recurrent interaction." Eur J Neurosci 20(4): 1089-100.

[30] DeYoe, E. A., G. J. Carman, et al. (1996). "Mapping striate and extrastriate visual areas in human cerebral cortex." Proc Natl Acad Sci U S A 93(6): 2382-6.

[31] Dineen, J., A. Hendrickson, et al. (1982). "Alterations of retinal inputs following striate cortex removal in adult monkey." Exp Brain Res 47(3): 446-56.

[32] Dougherty, R. F., M. Ben-Shachar, et al. (2005). "Functional organization of human occipital-callosal fiber tracts." Proc Natl Acad Sci U S A 102(20): 7350-5.

[33] Dougherty RF, V.M. Koch, et al. (2003). "Visual field representations and locations of visual areas V1/2/3 in human visual cortex." J Vis. 3(10):586-98.

[34] Dumoulin, S. O. and B. A. Wandell (2008). "Population receptive field estimates in human visual cortex." Neuroimage 39(2): 647-60.

[35] Elisevich, K. V., R. M. Ford, et al. (1984). "Visual abnormalities with multiple trauma." Surg Neurol 22(6): 565-75.

[36] Engel, S. A., D. E. Rumelhart, et al. (1994). "fMRI of human visual cortex." Nature 369(6481): 525.

[37] Fabre-Thorpe, M., F. Levesque, et al. (1994). "Preservation of pointing accuracy toward moving targets after extensive visual cortical ablations in cats." Cortex 30(4): 585-601.

[38] Fendrich, R., C. M. Wessinger, et al. (1992). "Residual vision in a scotoma: implications for blindsight." Science 258(5087): 1489-91.

[39] ffytche, D.H., A. Howseman, et al. (2000). Human area V5 and motion in the ipsilateral visual field. Eur J Neurosci. 12(8):3015-25.

[40] Fields, R. D. (2008). "White matter in learning, cognition and psychiatric disorders." Trends Neurosci 31(7): 361-70.

[41] Frank, S., S. Lee, et al. (2012). "The obese brain athlete: self-regulation of the anterior insula in adiposity." PLoS One 7(8): e42570.

[42] Friel, K. M., A. A. Heddings, et al. (2000). "Effects of postlesion experience on behavioral recovery and neurophysiologic reorganization after cortical injury in primates." Neurorehabil Neural Repair 14(3): 187-98.

[43] Fujii, T., H. C. Tanabe, et al. (2009). "An investigation of cross-modal plasticity of effective connectivity in the blind by dynamic causal modeling of functional MRI data." Neurosci Res 65(2): 175-86.

[44] Gall, C., B. Steger, et al. (2013). "Evaluation of two treatment outcome prediction models for restoration of visual fields in patients with postchiasmatic visual pathway lesions." Neuropsychologia.

[45] Gallistel, C. R. and L. D. Matzel (2013). "The neuroscience of learning: beyond the Hebbian synapse." Annu Rev Psychol 64: 169-200.

[46] Gilhotra, J. S., P. Mitchell, et al. (2002). "Homonymous visual field defects and stroke in an older population." Stroke 33(10): 2417-20.

[47] Giorgi, R. G., R. L. Woods, et al. (2009). "Clinical and laboratory evaluation of peripheral prism glasses for hemianopia." Optom Vis Sci 86(5): 492-502.

[48] Girard, P., P. A. Salin, et al. (1991). "Visual activity in areas V3a and V3 during reversible inactivation of area V1 in the macaque monkey." J Neurophysiol 66(5): 1493-503.

[49] Goebel, R., L. Muckli, et al. (2001). "Sustained extrastriate cortical activation without visual awareness revealed by fMRI studies of hemianopic patients." Vision Res 41(10-11): 1459-74.

[50] Haller, S., N. Birbaumer, et al. (2010). "Real-time fMRI feedback training may improve chronic tinnitus." Eur Radiol 20(3): 696-703.

[51] Hamilton, J. P., G. H. Glover, et al. (2011). "Modulation of subgenual anterior cingulate cortex activity with real-time neurofeedback." Hum Brain Mapp 32(1): 22-31.

[52] Hebb, D.O. (1949). Distinction in coherent neural network between resting and working brain states. The organization of behavior. New York: Wiley & Sons.

[53] Henriksson, L., A. Raninen, et al. (2007). "Training-induced cortical representation of a hemianopic hemifield." J Neurol Neurosurg Psychiatry 78(1): 74-81.

[54] Henry, S.H. R.C. Reid (2000). The koniocellular pathway in primate vision. Annual Review of Neuroscience 23: 127-153.

[55] Hinds, O., S. Ghosh, et al. (2011). "Computing moment-to-moment BOLD activation for real-time neurofeedback." Neuroimage 54(1): 361-8.

[56] Hinrichs, H., H. J. Heinze, et al. (2006). "Causal visual interactions as revealed by an information theoretic measure and fMRI." Neuroimage 31(3): 1051-60.

[57] Hopfinger, J. B., M. H. Buonocore, et al. (2000). "The neural mechanisms of top-down attentional control." Nat Neurosci 3(3): 284-91.

[58] Horton, J. C. (2005a). "Disappointing results from Nova Vision's visual restoration therapy." Br J Ophthalmol 89(1): 1-2.

[59] Horton, J. C. (2005b). "Vision restoration therapy: confounded by eye movements." Br J Ophthalmol 89(7): 792-4.

[60] Hua, T., P. Bao, et al. (2010). "Perceptual learning improves contrast sensitivity of V1 neurons in cats." Curr Biol 20(10): 887-94.

[61] Hubel, D. H. and Wiesel, T. N. (1977). Ferrier Lecture: Functional Architecture of Macaque Monkey Visual Cortex, Proceedings of the Royal Society of London. Series B, Biological Sciences, Vol. 198, No. 1130 (May 19, 1977), pp. 1-59 Published by: The Royal Society Stable URL: http://www.jstor.org/stable/77245.

[62] Huxlin, K. R., J. M. Williams, et al. (2008a). "A neurochemical signature of visual recovery after extrastriate cortical damage in the adult cat." J Comp Neurol 508(1): 45-61.

[63] Huxlin, K. R. (2008b). "Perceptual plasticity in damaged adult visual systems." Vision Res 48(20): 2154-66.

[64] Huxlin, K. R., T. Martin, et al. (2009). "Perceptual relearning of complex visual motion after V1 damage in humans." J Neurosci 29(13): 3981-91.

[65] Huxlin, K. R. and T. Pasternak (2004). "Training-induced recovery of visual motion perception after extrastriate cortical damage in the adult cat." Cereb Cortex 14(1): 81-90.

[66] Huxlin, K. Plasticity of Visual Perception After Permanent Damage to the Adult Visual Cortex. in "Visual Field restoration Therapy Symposium"— Nanos Annual Meeting. 2006. Tuscon, AZ.

[67] Iorizzo, D. B., M. E. Riley, et al. (2011). "Differential impact of partial cortical blindness on gaze strategies when sitting and walking - an immersive virtual reality study." Vision Res 51(10): 1173-84.

[68] Jobke, S., E. Kasten, et al. (2009). "Vision restoration through extrastriate stimulation in patients with visual field defects: a double-blind and randomized experimental study." Neurorehabil Neural Repair 23(3): 246-55.

[69] Johnson, K. A., K. Hartwell, et al. (2012). "Intermittent "real-time" fMRI feedback is superior to continuous presentation for a motor imagery task: a pilot study." J Neuroimaging 22(1): 58-66.

[70] Johnston, S., D. E. Linden, et al. (2011). "Upregulation of emotion areas through neurofeedback with a focus on positive mood." Cogn Affect Behav Neurosci 11(1): 44-51.

[71] Johnston, S. J., S. G. Boehm, et al. (2010). "Neurofeedback: A promising tool for the self-regulation of emotion networks." Neuroimage 49(1): 1066-72.

[72] Julkunen, L., O. Tenovuo, et al. (2003). "Rehabilitation of chronic post-stroke visual field defect with computer-assisted training: a clinical and neurophysiological study." Restor Neurol Neurosci 21(1-2): 19-28.

[73] Karni, A. and D. Sagi (1991). "Where practice makes perfect in texture discrimination: evidence for primary visual cortex plasticity." Proc Natl Acad Sci U S A 88(11): 4966-70.

[74] Kasten, E., U. Bunzenthal, et al. (2007). "Vision restoration therapy does not benefit from costimulation: A pilot study." J Clin Exp Neuropsychol 29(6): 569-84.

[75] Kasten, E., U. Bunzenthal, et al. (2006). "Visual field recovery after vision restoration therapy (VRT) is independent of eye movements: an eye tracker study." Behav Brain Res 175(1): 18-26.

[76] Kasten, E., E. Muller-Oehring, et al. (2001). "Stability of visual field enlargements following computer-based restitution training -- results of a follow-up." J Clin Exp Neuropsychol 23(3): 297-305.

http://dx.doi.org/10.5772/57000

[77] Kasten, E., D. A. Poggel, et al. (1999). "Restoration of vision II: residual functions and training-induced visual field enlargement in brain-damaged patients." Restor Neurol Neurosci 15(2-3): 273-87.

[78] Kasten, E., D. A. Poggel, et al. (2000). "Computer-based training of stimulus detection improves color and simple pattern recognition in the defective field of hemianopic subjects." J Cogn Neurosci 12(6): 1001-12.

[79] Kasten, E. and B. A. Sabel (1995). "Visual field enlargement after computer training in brain-damaged patients with homonymous deficits: an open pilot trial." Restor Neurol Neurosci 8(3): 113-27.

[80] Kasten, E., S. Wuest, et al. (1998a). "Residual vision in transition zones in patients with cerebral blindness." J Clin Exp Neuropsychol 20(5): 581-98.

[81] Kasten, E., S. Wust, et al. (1998b). "Computer-based training for the treatment of partial blindness." Nat Med 4(9): 1083-7.

[82] Kastner, S. and L. G. Ungerleider (2000). "Mechanisms of visual attention in the human cortex." Annu Rev Neurosci 23: 315-41.

[83] Kerkhoff, G., U. Munssinger, et al. (1992). "Rehabilitation of homonymous scotomata in patients with postgeniculate damage of the visual system: saccadic compensation training." Restor Neurol Neurosci 4(4): 245-54.

[84] Kerkhoff, G., U. Munssinger, et al. (1994). "Neurovisual rehabilitation in cerebral blindness." Arch Neurol 51(5): 474-81.

[85] Kikuta, K., Y. Takagi, et al. (2006). "Early experience with 3-T magnetic resonance tractography in the surgery of cerebral arteriovenous malformations in and around the visual pathway." Neurosurgery 58(2): 331-7; discussion 331-7.

[86] Kluver, H. (1936). An analysis of the effects of the removal of the occipital lobes in monkeys, J Psychol, 2: 49-61.

[87] LaConte, S. M., S. J. Peltier, et al. (2007). "Real-time fMRI using brain-state classification." Hum Brain Mapp 28(10): 1033-44.

[88] Law, C. T. and J. I. Gold (2008). "Neural correlates of perceptual learning in a sensory-motor, but not a sensory, cortical area." Nat Neurosci 11(4): 505-13.

[89] Lee, S., A. Papanikolaou, et al. (2013). "A new method for estimating population receptive field topography in visual cortex." Neuroimage 81: 144-57.

[90] Lee, S., S. Ruiz, et al. (2011). "Detection of cerebral reorganization induced by real-time fMRI feedback training of insula activation: a multivariate investigation." Neurorehabil Neural Repair 25(3): 259-67.

[91] Lew, H. L., J. H. Poole, et al. (2007). "Program development and defining characteristics of returning military in a VA Polytrauma Network Site." J Rehabil Res Dev 44(7): 1027-34.

[92] Lewis, C. M., A. Baldassarre, et al. (2009). "Learning sculpts the spontaneous activity of the resting human brain." Proc Natl Acad Sci U S A 106(41): 17558-63.

[93] Li, X., K. J. Hartwell, et al. (2012). "Volitional reduction of anterior cingulate cortex activity produces decreased cue craving in smoking cessation: a preliminary real-time fMRI study." Addict Biol 18(4): 739-48.

[94] Lindberg, P., H. Forssberg, et al. (2003). "[Rehabilitation after stroke. Imaging techniques show how the cortical reorganization is affected by training]." Lakartidningen 100(51-52): 4289-92.

[95] Linden DE, Habes I, et al. (2012). Real-time self-regulation of emotion networks in patients with depression. PLoS One 7(6):e38115.

[96] Liu, Z. and D. Weinshall (2000). "Mechanisms of generalization in perceptual learning." Vision Res 40(1): 97-109.

[97] Liu X, Zhu XH, Chen W. (2011). Distinction in coherent neural network between resting and working brain states. Brain Connect 1(5):377-88. doi: 10.1089/brain.2011.0044.

[98] Marshall, R. S., J. J. Ferrera, et al. (2008). "Brain activity associated with stimulation therapy of the visual borderzone in hemianopic stroke patients." Neurorehabil Neural Repair 22(2): 136-44.

[99] McCaig, R. G., M. Dixon, et al. (2011). "Improved modulation of rostrolateral prefrontal cortex using real-time fMRI training and meta-cognitive awareness." Neuroimage 55(3): 1298-305.

[100] McKenna, K., D. M. Cooke, et al. (2006). "The incidence of visual perceptual impairment in patients with severe traumatic brain injury." Brain Inj 20(5): 507-18.

[101] Mihailovic LT, Cupic D, Dekleva N. (1971). Changes in the numbers of neurons and glial cells in the lateral geniculate nucleus of the monkey during retrograde cell degeneration. J Comp Neurol 142: 223-9.

[102] Moore, T., H. R. Rodman, et al. (2001). "Direction of motion discrimination after early lesions of striate cortex (V1) of the macaque monkey." Proc Natl Acad Sci U S A 98(1): 325-30.

[103] Moore, T., H. R. Rodman, et al. (1995). "Localization of visual stimuli after striate cortex damage in monkeys: parallels with human blindsight." Proc Natl Acad Sci U S A 92(18): 8215-8.

[104] Morris, J. S., B. DeGelder, et al. (2001). "Differential extrageniculostriate and amygdala responses to presentation of emotional faces in a cortically blind field." Brain 124(Pt 6): 1241-52.

[105] Mourao-Miranda, J., A. L. Bokde, et al. (2005). "Classifying brain states and determining the discriminating activation patterns: Support Vector Machine on functional MRI data." Neuroimage 28(4): 980-95.

[106] Mumford, D. (1994). Neuronal architectures for pattern theoretic problems. In: Large Scale Neuronal Theories of the Brain, Koch C, Davis, JL, eds., MIT Press, pp. 125-152.

[107] Okada, T., Y. Miki, et al. (2007). "Diffusion tensor fiber tractography for arteriovenous malformations: quantitative analyses to evaluate the corticospinal tract and optic radiation." AJNR Am J Neuroradiol 28(6): 1107-13.

[108] Olshausen, B.A. Principles of Image Representation in Visual Cortex. In: The Visual Neurosciences, L.M. Chalupa, J.S. Werner, Eds. MIT Press, 2003: 1603-1615.

[109] Overgaard, M. (2011). "Visual experience and blindsight: a methodological review." Exp Brain Res 209(4): 473-9.

[110] Pambakian, A.L. and C. Kennard (1997). Can visual function be restored in patients with homonymous hemianopia? Br J Ophthalmol (4): p. 324-8.

[111] Papageorgiou, T. D., W. A. Curtis, et al. (2009a). "Neurofeedback of two motor functions using supervised learning-based real-time functional magnetic resonance imaging." Conf Proc IEEE Eng Med Biol Soc 2009: 5377-80.

[112] Papageorgiou, T.D., M. McHenry, J.L. Lisinski, J.P. White, and S.M. LaConte (2009b). Speech rate control using supervised learning-based rea-time fMRI. 15th Annual Meeting of the Organization for Human Brain Mapping, San Franscisco. Neuroimage, 47: p. 97.

[113] Papageorgiou, T.D., LaConte, S.M., Lisinski, J.L., Peng, X., and Smirnakis, S.M, Classification Accuracy and Spatial Networks on Motion Coherence using Real-time fMRI. Hum Brain Mapp, Barcelona, 2010.

[114] Papageorgiou, T. D., J. M. Lisinski, et al. (2013). "Brain-computer interfaces increase whole-brain signal to noise." Proc Natl Acad Sci U S A 110(33): 13630-5.

[115] Pasik, T. and P. Pasik (1971). "The visual world of monkeys deprived of striate cortex: effective stimulus parameters and the importance of the accessory optic system." Vision Res Suppl 3: 419-35.

[116] Pessoa, L., S. Kastner, et al. (2003). "Neuroimaging studies of attention: from modulation of sensory processing to top-down control." J Neurosci 23(10): 3990-8.

[117] Peyron, R., B. Laurent, et al. (2000). "Functional imaging of brain responses to pain. A review and meta-analysis (2000)." Neurophysiol Clin 30(5): 263-88.

[118] Pleger, B., A. F. Foerster, et al. (2003). "Functional magnetic resonance imaging mirrors recovery of visual perception after repetitive tachistoscopic stimulation in patients with partial cortical blindness." Neurosci Lett 335(3): 192-6.

[119] Plow, E. B., S. N. Obretenova, et al. (2011). "Combining visual rehabilitative training and noninvasive brain stimulation to enhance visual function in patients with hemianopia: a comparative case study." PM R 3(9): 825-35.

[120] Poggel, D. A., E. Kasten, et al. (2006). "Improving residual vision by attentional cueing in patients with brain lesions." Brain Res 1097(1): 142-8.

[121] Poggel, D. A., E. Kasten, et al. (2001). "Unusual spontaneous and training induced visual field recovery in a patient with a gunshot lesion." J Neurol Neurosurg Psychiatry 70(2): 236-9.

[122] Poggel, D. A., E. Kasten, et al. (2004). "Attentional cueing improves vision restoration therapy in patients with visual field defects." Neurology 63(11): 2069-76.

[123] Poggel, D. A., I. Mueller, et al. (2010). "Subjective and objective outcome measures of computer-based vision restoration training." NeuroRehabilitation 27(2): 173-87.

[124] Poppel, E., R. Held, et al. (1973). "Leter: Residual visual function after brain wounds involving the central visual pathways in man." Nature 243(5405): 295-6.

[125] Posse, S., D. Fitzgerald, et al. (2003). "Real-time fMRI of temporolimbic regions detects amygdala activation during single-trial self-induced sadness." Neuroimage 18(3): 760-8.

[126] Raninen, A., S. Vanni, et al. (2007). "Temporal sensitivity in a hemianopic visual field can be improved by long-term training using flicker stimulation." J Neurol Neurosurg Psychiatry 78(1): 66-73.

[127] Rebesco, J. M. and L. E. Miller (2011). "Stimulus-driven changes in sensorimotor behavior and neuronal functional connectivity application to brain-machine interfaces and neurorehabilitation." Prog Brain Res 192: 83-102.

[128] Reinges, M. H., T. Krings, et al. (2004). "Functional and diffusion-weighted magnetic resonance imaging for visualization of the postthalamic visual fiber tracts and the visual cortex." Minim Invasive Neurosurg 47(3): 160-4.

[129] Reinhard, J., A. Schreiber, et al. (2005). "Does visual restitution training change absolute homonymous visual field defects? A fundus controlled study." Br J Ophthalmol 89(1): 30-5.

[130] Riggs, R. V., K. Andrews, et al. (2007). "Visual deficit interventions in adult stroke and brain injury: a systematic review." Am J Phys Med Rehabil 86(10): 853-60.

[131] Rizzo, M. and D. A. Robin (1996). "Bilateral effects of unilateral visual cortex lesions in human." Brain 119 (Pt 3): 951-63.

[132] Rodman, H. R., C. G. Gross, et al. (1989). "Afferent basis of visual response properties in area MT of the macaque. I. Effects of striate cortex removal." J Neurosci 9(6): 2033-50.

[133] Rodman, H. R., C. G. Gross, et al. (1990). "Afferent basis of visual response properties in area MT of the macaque. II. Effects of superior colliculus removal." J Neurosci 10(4): 1154-64.

[134] Rota, G., R. Sitaram, et al. (2009). "Self-regulation of regional cortical activity using real-time fMRI: the right inferior frontal gyrus and linguistic processing." Hum Brain Mapp 30(5): 1605-14.

[135] Rudolph, K. and T. Pasternak (1999). "Transient and permanent deficits in motion perception after lesions of cortical areas MT and MST in the macaque monkey." Cereb Cortex 9(1): 90-100.

[136] Rudolph, K. K., V. P. Ferrera, et al. (1994). "A reduction in the number of directionally selective neurons extends the spatial limit for global motion perception." Vision Res 34(24): 3241-51.

[137] Rudolph, T. M. and E. R. Delay (1993). "Recovery of a temporally based visual discrimination after visual cortex lesion in the rat." Behav Brain Res 53(1-2): 189-99.

[138] Ruiz, S., K. Buyukturkoglu, et al. (2013). "Real-time fMRI brain computer interfaces: Self-regulation of single brain regions to networks." Biol Psychol.

[139] Sabel, B. A., P. Henrich-Noack, et al. (2011). "Vision restoration after brain and retina damage: the "residual vision activation theory"." Prog Brain Res 192: 199-262.

[140] Sabel, B. A. and E. Kasten (2000). "Restoration of vision by training of residual functions." Curr Opin Ophthalmol 11(6): 430-6.

[141] Sabel, B. A., S. Kenkel, et al. (2004). "Vision restoration therapy (VRT) efficacy as assessed by comparative perimetric analysis and subjective questionnaires." Restor Neurol Neurosci 22(6): 399-420.

[142] Sahraie, A., M. J. Macleod, et al. (2010). "Improved detection following Neuro-Eye Therapy in patients with post-geniculate brain damage." Exp Brain Res 206(1): 25-34.

[143] Sahraie, A., C. T. Trevethan, et al. (2006). "Increased sensitivity after repeated stimulation of residual spatial channels in blindsight." Proc Natl Acad Sci U S A 103(40): 14971-6.

[144] Schmid, M. C., S. W. Mrowka, et al. (2010). "Blindsight depends on the lateral geniculate nucleus." Nature 466(7304): 373-7.

[145] Schmid, M. C., T. Panagiotaropoulos, et al. (2009). "Visually driven activation in macaque areas V2 and V3 without input from the primary visual cortex." PLoS One 4(5): e5527.

[146] Schoenfeld, M. A., H. J. Heinze, et al. (2002). "Unmasking motion-processing activity in human brain area V5/MT+ mediated by pathways that bypass primary visual cortex." Neuroimage 17(2): 769-79.

[147] Schoth, F., U. Burgel, et al. (2006). "Diffusion tensor imaging in acquired blind humans." Neurosci Lett 398(3): 178-82.

[148] Schoups, A., R. Vogels, et al. (2001). "Practising orientation identification improves orientation coding in V1 neurons." Nature 412(6846): 549-53.

[149] Sereno, M. I., A. M. Dale, et al. (1995). "Borders of multiple visual areas in humans revealed by functional magnetic resonance imaging." Science 268(5212): 889-93.

[150] Scharnowski, F., C. Hutton, et al. (2012). "Improving visual perception through neurofeedback." J Neurosci 32(49): 17830-41.

[151] Sherbondy, A. J., R. F. Dougherty, et al. (2008). "Identifying the human optic radiation using diffusion imaging and fiber tractography." J Vis 8(10): 12 1-11.

[152] Shibata K, L.H. Chang, (2012). Decoding reveals plasticity in V3A as a result of motion perceptual learning. PLoS One. 7(8):e44003.

[153] Shibata, K., T. Watanabe, et al. (2011). "Perceptual learning incepted by decoded fMRI neurofeedback without stimulus presentation." Science 334(6061): 1413-5.

[154] Stilla, R., R. Hanna, et al. (2008). "Neural processing underlying tactile microspatial discrimination in the blind: a functional magnetic resonance imaging study." J Vis 8(10): 13 1-19.

[155] Stoerig, P. and E. Barth (2001). "Low-level phenomenal vision despite unilateral destruction of primary visual cortex." Conscious Cogn 10(4): 574-87.

[156] Subramanian, L., J. V. Hindle, et al. (2011). "Real-time functional magnetic resonance imaging neurofeedback for treatment of Parkinson's disease." J Neurosci 31(45): 16309-17.

[157] Sulzer, J., R. Sitaram, et al. (2013). "Neurofeedback-mediated self-regulation of the dopaminergic midbrain." Neuroimage 83C: 817-825.

[158] Stoerig, P. and A. Cowey (1997). "Blindsight in man and monkey." Brain 120 (Pt 3): 535-59.

[159] Takeuchi, N., T. Chuma, et al. (2005). "Repetitive transcranial magnetic stimulation of contralesional primary motor cortex improves hand function after stroke." Stroke 36(12): 2681-6.

[160] Taoka, T., M. Sakamoto, et al. (2005). "Diffusion tensor imaging in cases with visual field defect after anterior temporal lobectomy." AJNR Am J Neuroradiol 26(4): 797-803.

[161] Taub, E., G. Uswatte, et al. (2002). "New treatments in neurorehabilitation founded on basic research." Nat Rev Neurosci 3(3): 228-36.

[162] Taylor, T. N. (1997). "The medical economics of stroke." Drugs 54 Suppl 3: 51-7; discussion 57-8.

[163] Van den Stock, J., M. Tamietto, et al. (2011). "Cortico-subcortical visual, somatosensory, and motor activations for perceiving dynamic whole-body emotional expressions with and without striate cortex (V1)." Proc Natl Acad Sci U S A 108(39): 16188-93.

[164] Vanburen, J. M. (1963). "Trans-Synaptic Retrograde Degeneration in the Visual System of Primates." J Neurol Neurosurg Psychiatry 26: 402-9.

[165] Vapnik, N. (1995). The nature of statistical learning theory. Springer.

[166] Veit, R., V. Singh, et al. (2012). "Using real-time fMRI to learn voluntary regulation of the anterior insula in the presence of threat-related stimuli." Soc Cogn Affect Neurosci 7(6): 623-34.

[167] Wandell, B. A., S. O. Dumoulin, et al. (2007). "Visual field maps in human cortex." Neuron 56(2): 366-83.

[168] Wang, L., W. Dai, et al. (2012). "Amplitude of low-frequency oscillations in first-episode, treatment-naive patients with major depressive disorder: a resting-state functional MRI study." PLoS One 7(10): e48658.

[169] Wedeen, V. J., D. L. Rosene, et al. (2012). "The geometric structure of the brain fiber pathways." Science 335(6076): 1628-34.

[170] Weiller, C. (1998). "Imaging recovery from stroke." Exp Brain Res 123(1-2): 13-7.

[171] Weiskopf, N. (2012). "Real-time fMRI and its application to neurofeedback." Neuroimage 62(2): 682-92.

[172] Weiskopf, N., R. Sitaram, et al. (2007). "Real-time functional magnetic resonance imaging: methods and applications." Magn Reson Imaging 25(6): 989-1003.

[173] Weiskopf, N., F. Sharnowski, et al., (2004). Self-regulation of local brain activity using real-time functional magnetic resonance imaging (fMRI). J Physiol Paris, 98(4-6): p. 357-73.

[174] Weiskopf, N., R. Veit, et al. (2003). "Physiological self-regulation of regional brain activity using real-time functional magnetic resonance imaging (fMRI): methodology and exemplary data." Neuroimage 19(3): 577-86.

[175] Weiskrantz, L. (2004). "Roots of blindsight." Prog Brain Res 144: 229-41.

[176] Weiskrantz, L., E. K. Warrington, et al. (1974). "Visual capacity in the hemianopic field following a restricted occipital ablation." Brain 97(4): 709-28.

[177] Werth, R. and M. Moehrenschlager (1999). "The development of visual functions in cerebrally blind children during a systematic visual field training." Restor Neurol Neurosci 15(2-3): 229-41.

[178] Worsley, K. J., C. H. Liao, et al. (2002). "A general statistical analysis for fMRI data." Neuroimage 15(1): 1-15.

[179] Yang, T. and J. H. Maunsell (2004). "The effect of perceptual learning on neuronal responses in monkey visual area V4." J Neurosci 24(7): 1617-26.

[180] Yeatman, J. D., R. F. Dougherty, et al. (2012). "Tract profiles of white matter properties: automating fiber-tract quantification." PLoS One 7(11): e49790.

[181] Yoo, S. S. and F. A. Jolesz (2002). "Functional MRI for neurofeedback: feasibility study on a hand motor task." Neuroreport 13(11): 1377-81.

[182] Yoo, S. S., J. H. Lee, et al. (2008). "Neurofeedback fMRI-mediated learning and consolidation of regional brain activation during motor imagery." Int J Imaging Syst Technol 18(1): 69-78.

[183] Yotsumoto, Y., T. Watanabe, et al. (2008). "Different dynamics of performance and brain activation in the time course of perceptual learning." Neuron 57(6): 827-33.

[184] Zeki, S. (2004) Thirty years of a very special visual area, area V5. J. Physiol. 557, 1–2.

[185] Zeki S., and D.H. ffytche (1998). The Riddoch syndrome: insights into the neurobiology of conscious vision. Brain 121(Pt 1):25-45.

[186] Zhang, X., S. Kedar, et al. (2006). "Natural history of homonymous hemianopia." Neurology 66(6): 901-5.

[187] Zihl, J. and D. von Cramon (1979). "Restitution of visual function in patients with cerebral blindness." J Neurol Neurosurg Psychiatry 42(4): 312-22.

[188] Zihl, J. and D. von Cramon (1985). "Visual field recovery from scotoma in patients with postgeniculate damage. A review of 55 cases." Brain 108 (Pt 2): 335-65.

[189] Zotev, V., F. Krueger, et al. (2011). "Self-regulation of amygdala activation using realtime FMRI neurofeedback." PLoS One 6(9): e24522.

[190] Zuiderbaan, W., B. M. Harvey, et al. (2012). "Modeling center-surround configurations in population receptive fields using fMRI." J Vis 12(3): 10.

Chapter 15

Using fMRI to Examine Central Auditory Plasticity

Deborah A. Hall, Cornelis P. Lanting and Douglas E.H. Hartley

Additional information is available at the end of the chapter

http://dx.doi.org/10.5772/29929

1. Introduction

During the last two decades or so, auditory neuroscience has made significant progress in understanding the functional organization of the auditory system in both normal-hearing listeners, patients with sensorineural hearing impairments and/or tinnitus, and following cochlear implantation. Modern brain imaging techniques have made an enormous contribution to that progress by enabling the *in vivo* study of human central auditory function. Since its introduction, magnetic resonance imaging (MRI) has evolved to become one of the most influential methods for investigating non-invasive human brain structure and function. Functional MRI (fMRI) has become the tool of choice for addressing many research questions concerning central auditory plasticity. First, fMRI has good spatial sensitivity and specificity and so can be used to map auditory responses with reasonably fine topographic detail (Logothetis et al., 2001). Second, fMRI is suitable for research use with children, as well as adults. Third, since multiple observations can be made on the same individual, fMRI permits the investigation of longer-term dynamic processes, such as functional plasticity after disease, damage, retraining or therapy. Finally, the need for averaging data across individuals is reduced, further improving the accuracy with which activations can be mapped onto subject-specific brain structure (e.g. Talavage et al., 2004). Despite these advantages, there are significant issues using this technique in certain clinical populations, such as individuals with a cochlear implant (CI). Subsequently, to investigate the effects of CIs on brain reorganization, it is necessary to consider alternative functional imaging methods, such as electroencephalography (e.g. Debener et al., 2008). Moreover, the high level of the scanner acoustic noise can have detrimental effects on central auditory function, patient anxiety and verbal communication between patient and scanner operator (Mechefske et al., 2002). In terms of central auditory function, the scanner acoustic noise can induce increases in the fMRI response of the auditory cortex as well as interfere with the perception of externally presented sound stimuli (Edmister et al., 1999; Hall et al., 1999).

Although- functional imaging of the central auditory pathway has not been widely used in Ear, Nose and Throat or Audiology clinics, there is a growing literature on its potential application. Using insights gained from functional imaging research with normal-hearing subjects, we can now transfer the most efficient protocols to study patients with hearing problems. For example, a choice of methods is available for noise reduction and a range of scanning parameters have been optimised to enhance the detection of brain activity.

This chapter discusses some of the theoretical and technical challenges of auditory fMRI and several of the major themes within this field, notably cortical reorganization underpinning tinnitus and deafness. Also we will consider the effect of CIs on the auditory brain, the difficulties associated with the use of conventional imaging methods in CI populations, and potential application of novel techniques, such as near infrared spectroscopy. Sections not only highlight evidence for neuroplasticity within the central auditory system, but also review what is currently known about the involvement and integration of information from other sensory modalities.

2. Challenges and limitations of auditory fMRI

Rather than directly measuring the stimulus- and task-related electrical activity of neurons, fMRI measures indirect changes in the magnetic properties of the blood. Most common methods make use of the blood-oxygen-level-dependent (BOLD) contrast. This technique is based on the increase in signal intensity caused by an increase in oxygen concentration of blood (Ogawa et al., 1990). The functional image represents the spatial distribution of blood oxygenation levels in the brain and the small fluctuations in these levels over time are correlated with the stimulus input or cognitive task.

To detect the magnetic properties of brain tissue, MR scanners operate using three different types of electromagnetic fields: a very high static field generated by a superconducting magnet, time-varying gradient magnetic fields and pulsed radiofrequency (RF) fields. The latter two fields are much weaker than the first, but all pose a number of unique and considerable technical challenges for conducting auditory fMRI research within this hostile environment.

In the first place, the static and time-varying magnetic fields preclude the use of many types of electronic sound presentation equipment, as well as preventing the safe scanning of patients who are wearing listening devices such as hearing aids or implants. Despite the restriction on the materials that can be used in a scanner, a number of different MR-compatible active headphone driving units have been produced for sound delivery. Several audio systems use electrostatic transducers to ensure high performance. Electrostatic headphones generate sound using a conductive diaphragm placed next to a fixed conducting panel. A high voltage polarises the fixed panel and the audio signal passing through the diaphragm rapidly switches between a positive and a negative signal, attracting or repelling it to the fixed panel and thus vibrating the air. Induced currents in the electronics, caused directly by the time-varying gradient magnetic fields or the RF pulses, are an additional hazard to the electronic devices themselves. Some materials can also absorb the RF energy causing local tissue heating and even burns if in contact with soft tissue. Hence, the digital audio source, electronics and power

http://dx.doi.org/10.5772/29929

supply that drive the sound system are housed outside the RF screened scanner room to avoid electromagnetic interference with MRI. The electrical signal is then transferred from the audio source to the headphones in the RF screened scanner room using either filters through a filter panel or fibre-optic cable through the waveguide. For the same safety reasons, there are restrictions on scanning people who have electronic listening devices. These include hearing aids, cochlear implants and brainstem implants. Official approval for the manufacture of implant devices requires rigorous testing for susceptibility to electromagnetic fields, radiated electromagnetic fields and electrical safety testing (including susceptibility to electrical discharge). Typically however standard listening devices do not meet the more stringent criteria for compatibility with MR scanning. For the patient, risks include movement of the device and localised heating of brain tissue, whereas, for the device, the electronic components may be damaged. Some implant designs *have* been proven to be MR compatible (e.g. Shellock et al., 1993), but they are not routinely supplied in clinical practice. Given that implanted devices cannot be removed without surgical intervention, clinical imaging research of implantees has generally used other brain imaging methods, namely positron emission tomography (Giraud et al., 2001).

The high levels of scanner noise generated by the flexing of the gradient coils in the static magnetic field can reduce audibility and increase listening effort. Scanner noise increases nonlinearly with static magnetic field strength. For example, comparing scanners from 1 to 3 Tesla indicates a difference in sound level of about 10 dB SPL (Price et al., 2001). The sound level measured in the bore of the scanner is typically 99 dB SPL (98 dB(A)). Clearly, exposure to such an intense sound levels without protection is likely to cause a temporary threshold shift in hearing and tinnitus, and could be permanently damaging over a prolonged stay inside the bore of the scanner (Foster et al., 2000). The simplest way to treat the intense noise is to use ear protection in the form of ear defenders and/or ear plugs. Foam ear plugs can compromise the acoustic quality of the experimental sounds delivered to the subject and so circumaural headsets are preferable. Typically transducers are fitted into sound attenuating earmuffs to reduce the ambient noise level at the subject's ears. Considerable attenuation can be achieved in this manner (e.g., 25 dB of attenuation at 600 Hz rising to 40 dB at 4 kHz, Hall et al., 2009). Even with hearing protection, whenever the scanner noise coincides with the presented sound stimulus it produces changes in task performance and increases the attentional demands of the listening task. The frequency range of the scanner noise overlaps with the range that is crucial for speech intelligibility and so speech experiments can be particularly compromised in the noisy environment. For example, Blackman and Hall (2011) reported that speech intelligibility (measured as the proportion of keywords reported correctly) dropped by about 50% for words that were masked by the scanner noise compared to those that were not masked by the noise.

The high levels of ambient scanner noise can interfere with the patterns of brain activity measured during fMRI. The noise not only activates parts of the auditory brain, but also interacts with the patterns of activity evoked by experimental stimuli. Two novel methods have been developed by the Nottingham group to deal with these problems; sparse temporal sampling (Hall et al., 1999) and active noise reduction (Hall et al., 2009). In sparse sampling, a clustered acquisition of brain slices is combined with a long and fixed (e.g. 10 s) inter-scan interval and the scan acquisition is temporally offset from the presentation of the stimulus of interest (typically by 8-10 s). Although the scanner noise resulting from the acquisition still evokes an auditory BOLD response, it does not extend across time to the subsequent scan. Work in the mid-90s demonstrated that when the onset of the scan acquisition is synchronized to the offset of an epoch of the stimulus presentation in a sparse sampling paradigm, the activation associated with the experimental sound can be effectively separated from the activation associated with the scanner sound (Hall et al., 1999). Furthermore, because the scanner sound is temporally offset, it does not produce acoustical masking and does not distract the listener. Sparse sampling is often the scanning protocol of choice for identifying auditory cortical evoked responses in the absence of scanner noise. Active noise reduction achieves significant attenuation of the acoustic energy received at the ears. The system reduces the perceived level of the background noise by producing an output ('cancellation') signal that destructively interferes with the scanner sound (i.e. the incident sound wave) at each ear. Precise control over the frequency and phase of acoustic signals presented to participants is achieved through a digital-signal processing system that delivers sound stimuli via a high-fidelity electrostatic headset. Blackman and Hall (2011) evaluated the benefit of active noise reduction by asking whether sound reduction provided sufficient gains in both audibility and sensitivity to sound-related activity to remove the need for sparse temporal sampling. While both sparse sampling and cancellation were found to improve listening quality, these factors did not provide equivalent benefits for auditory fMRI experiments. Cancellation offered small improvements for spatially specific analyses within Heschl's gyrus and planum polare '(regions that are engaged in auditory perception and speech comprehension)' and for reducing some of the extra-auditory activity associated with the effort required for perceptual discrimination in a noisy environment. Nevertheless, sparse acquisition remained the more robust method for detecting auditory cortical activity by providing proper quiet intervals for stimulus presentation.

3. Central auditory processing: Beyond 3 Tesla fMRI

Brain imaging studies in humans are revealing parallels with the functional organization of the auditory brain discovered in electrophysiological studies in animals: sound-level dependencies as well as topographical mapping of stimulus frequencies are observed. This section provides a foundation to the remaining sections in this chapter by describing the BOLD effect and highlighting experiments that investigate the spatial specificity of this effect as a function of the external magnetic field strength and the parameters of the imaging sequence. We consequently describe fMRI experiments that focus on the functional organization of the auditory system. Specifically, we report an experiment on the topographical mapping of sound frequencies in human auditory cortex and discuss common findings and interpretations.

3.1. Characterizing BOLD signal changes at various field strengths

Research systems typically operate at 3 Tesla, but nowadays MRI systems that operate at magnetic field strengths exceeding this field strength are becoming more widely available.

http://dx.doi.org/10.5772/29929

Nottingham University is the home of the first ultra-high field 7 Tesla MRI scanner for human brain imaging in the UK. The key factor for the development of such high field MRI systems is the improvement in intrinsic signal-to-noise ratio and contrast-to-noise ratio with increasing field strength. A simple motor task, for example, elicits a signal change of approximately 1%, 2.5% and 5% for field strengths of respectively 1.5 Tesla, 3 Tesla and 7 Tesla (Van der Zwaag et al., 2009). Figure 1 shows the results for one of our typical auditory tasks acquired at a magnetic field strength of 7 Tesla. In this short experiment, a sound was alternated with silence in a block design. The signal change was 5.2% indicating that the obtained percent signal change is comparable to that obtained in the above motor task. The resulting increased BOLD effect is of great benefit for fMRI basic research studies and also for clinically focussed studies. It can be exploited to improve the spatial resolution and reduces the number of trials required to demonstrate significant levels of activation. An additional effect is the increased spatial specificity of the BOLD effect with increasing field strength. At higher field strengths the T2*time of blood is relatively short causing its signal, originating mainly from the vasculature, to be attenuated relative to that of tissue at the echo times used for BOLD fMRI (Yacoub et al., 2001). The contribution of venous signal to the BOLD effect is therefore reduced at high field strengths, increasing the spatial specificity of the BOLD effect.

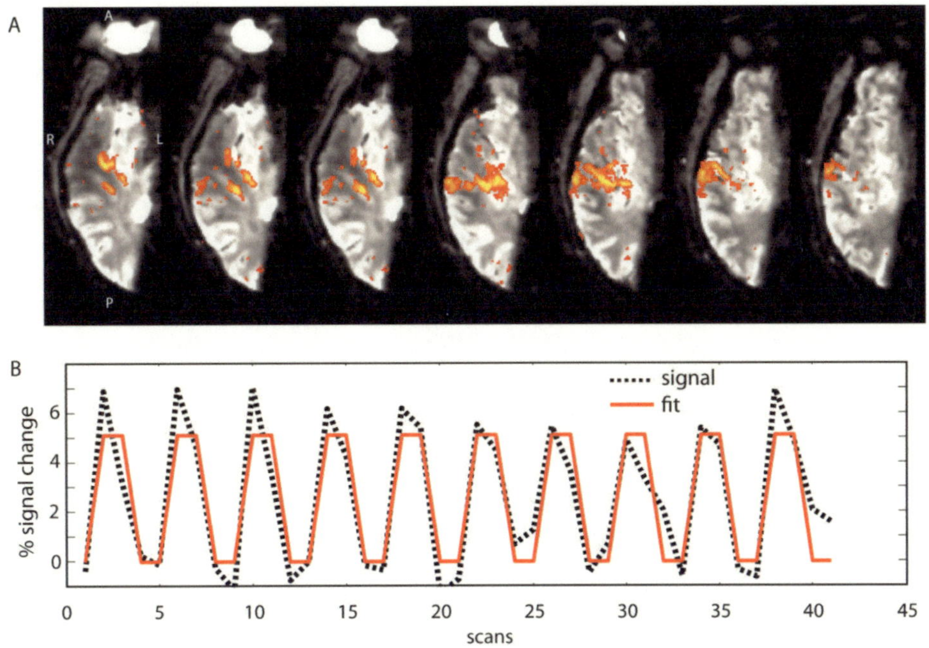

Figure 1. A) Sound-evoked activation maps obtained at 7 Tesla as overlaid on the average EPI image of one subject. The results show the primary auditory cortex, demarcated by the anterior Heschl's sulcus and the posterior Heschl's sulcus, as well as the secondary auditory cortex and higher order areas in the right hemisphere. B) shows the averaged time-course (black) and a linear fit (red) of all the active voxels following a simple block design where blocks of sound were alternated with silence. Abbreviation: A: anterior; P: posterior; R: right, L: left.

3.2. Characterizing the spatial resolution of the BOLD effect

The spatial resolution of the BOLD effect is, compared to most MRI imaging techniques, quite poor. T1-weighted anatomical images can, for example, be easily obtained with an isotropic resolution of well below 1 mm. In contrast, the spatial resolution of the BOLD effect is mostly determined by physiological sources, in the sense that the vascular response to neural activity extends over several millimetres (Logothetis et al., 2001). The BOLD signal originating from small capillaries and surrounding areas is thought to provide good spatial localization for neuronal activation. However, at low and intermediate field strengths, the gradient echo (GE) BOLD effects from large draining veins often contaminate the signal, causing inaccurate localization of the activity and also cause a widening of the point spread function of the BOLD response. GE BOLD is sensitive to extravascular effects around the large vessels while spin-echo (SE) sequences are relatively insensitive to these effects (Duong et al., 2003; Weisskoff et al., 1994). The 180° rephasing pulse in an SE sequence compensates for the constant field heterogeneities to obtain an echo that is weighted in T2 and not in T2*. Spin-echo sequences are therefore more closely related to neuronal activity than GE sequences, although there is still a large contribution of intravascular signal arising from blood. The benefit of high fields (i.e. at field strengths of 3 Tesla and higher) compared to a lower field strength is the reduced T_2 and T_2* of blood relative to that of tissue, reducing the intravascular component even more (Yacoub et al., 2001). However, a major disadvantage with T_2-weighted SE sequences is linked to long inter-scan intervals, resulting in potentially prohibitive acquisition times.

The point spread function describes the response of the MRI scanning to a point source within the brain. Measurements of the point spread function have revealed a full-width at half-maximum (FWHM) of around 2.3 mm tangential to cortical surface of the visual cortex using GE BOLD at 7 Tesla (Shmuel et al., 2007). The FWHM of the point spread function at 7 Tesla is therefore narrower than at 1.5 Tesla (reported at 3.5 mm, Engel et al., 1997) and at 3 Tesla (reported at 3.9 mm, Parkes et al., 2005). Recently, Polimeni et al. (2010) measured the spatial spread of the BOLD response tangential to the cortical surface as a function of cortical depth. Sampling near the pial surface provided the highest signal strength, but it also introduced the most spatial error due to extravascular BOLD effects. Avoiding the surface laminae improved spatial localization considerably, at the cost of sensitivity, implying that optimal spatial resolution in fMRI of the cortex can be achieved using anatomically informed spatial sampling to avoid large pial vessels.

When comparing the point spread function of GE and SE BOLD sequences there is a 13% reduction of the FWHM of the SE sequence, but at a cost of at least 3-fold reduction in contrast to noise ratio (Parkes et al., 2005). fMRI suffers from susceptibility artifacts that are caused by the proximity of air-cavities like the ear canal and nasal cavity causing signal loss and distortion of the images (Ojemann et al., 1997). Many strategies have been proposed to overcome these effects which include the usage of shorter echo times, parallel imaging methods and improved shimming methods. Since static susceptibility artifacts can –at least partly- be refocused using SE techniques, it is less affected by susceptibility artifacts than GE methods and may increase the sensitivity in affected areas (Goense et al., 2008) such as auditory regions in the temporal lobe.

http://dx.doi.org/10.5772/29929

In conclusion, although there is a trade-off between sensitivity and specificity, and the fact that there are still some technical considerations that need to be assessed, most recent imaging literature agree on the advantages of ultra-high field fMRI (i.e. ≥ 7 Tesla), that is, the increased signal-to-noise ratio. This evolution is of special importance for the clinical field in general and the field of auditory research specifically. The next section discusses the application of fMRI to study the functional organization of the auditory cortex.

3.3. Tonotopic organization in the human auditory cortex revealed using fMRI

Many studies have elucidated the functional specialization and hierarchical organization of multiple areas of the auditory cortex in non-human primates using invasive neurophysiological recordings which are obviously not applicable to human studies. A method to map frequency mapping using noninvasive BOLD fMRI in primates is in its early infancy (Petkov, personal communication). At present, there are many applications of BOLD fMRI in experiments on the human auditory system, ranging from the characterization of the effects of sound intensity (Bilecen et al., 2002; Hart et al., 2003), the effects of the side of presentation (Jäncke et al., 2002) to more complex phenomena such as the interaction of scanner noise and sound-evoked responses (Gaab et al., 2007; Hall et al. 1999; Talavage & Edmister 2004) and feature sensitivity and attention modulation (Paltoglou et al., 2011). Mapping one of the most fundamental properties of the auditory system, that of frequency coding, remains elusive and only a few studies have demonstrated a clear topographical arrangement of the response to sound frequency in the auditory pathway. This organization, referred to as 'tonotopy', mirrors the distribution of receptors in the cochlea, with neurons at the base of the cochlea preferentially responding to high frequencies and neurons at the apex responding best to low frequencies (Ehret, 1978). Tonotopy is conveyed up to the level of auditory cortex by fibres in the lemniscal pathway (Figure 3). In the auditory cortex of nonhuman primates, the primary core cortex, consisting of A1, R and RT, exhibit tonotopic gradients that are mirror symmetric to each other (Hackett et al., 2001; Kaas & Hackett 2000; Morel et al., 1993).

Tonotopic organization has been identified in human auditory cortex using a variety of imaging techniques including positron emission tomography (PET) and fMRI. The majority of early studies were based on just two different stimulus frequencies (Bilecen et al., 1998; Lauter et al., 1985; Lockwood et al. 1999; Talavage et al., 2000; Wessinger et al., 1997). From the results a general pattern emerges where high frequencies are represented at the medial end of auditory cortex and low frequencies in the anterior and lateral part of Heschl's gyrus and surrounding areas, mostly as a single low-to-high frequency gradient. More recent studies used more frequencies allowing the identification of more fine-grained frequency gradients. However, the results and interpretations of these studies vary considerably. For example, Langers et al. (2007) reported a single high-to-low gradient extending from posterior medial to anterior lateral auditory areas, similar to earlier studies. Mapping of the left hemisphere at 7 Tesla using surface coils and sparse image acquisition revealed two mirror-symmetric tonotopic maps encompassing three linear gradients of frequency-sensitive BOLD responses in six subjects: (i) one connecting a high-frequency region in Heschl's sulcus with a low-frequency region halfway Heschl's gyrus and (ii) a second connected the same low-frequency region with a high-frequency region near the lateral junction of Heschl's gyrus and the superior temporal gyrus (Formisano et al., 2003). These results suggest that human auditory cortex conforms to the primate model with two mirror-symmetric tonotopic maps in primary auditory core and belt areas that are joined at a common low-frequency boundary (Petkov et al., 2006). Talavage and colleagues reported three consistent gradients, none of which clearly followed the long (medial-to-lateral) axis of Heschl's gyrus (Talavage et al., 2004). Schönwiesner et al. (2002) argued that the observed frequency selectivity does not necessarily represent frequency gradients but instead represent different functional regions within the auditory cortex (Schönwiesner et al., 2002). Based on this assumption, in combination with the models by Kaas et al. (2000), three recent studies investigated tonotopic mapping and related their findings to anatomical models.

Humphries reported two major mirror-symmetric frequency gradients, oriented roughly perpendicular to Heschl's gyrus, rather than parallel as previously assumed (Humphries et al., 2010). They found a single low-frequency area centered on Heschl's gyrus, bordered anteriorly and posteriorly by two high-frequency areas (planum temporale). This seems very similar to previous findings (Woods et al., 2009). Based on the observed mirror-symmetric frequency gradients, both sets of authors proposed boundaries of regions in human auditory cortex homologous to A1 and R regions in non-human primate cortex. Figure 2 shows the schematic outline of auditory fields as based on electrophysiology in primates and the original group data of both Woods et al. (2009) and Humphries et al. (2010) indicating similar tonotopic maps and a close correspondence between these maps and proposed boundaries as compared to primate data.

A different approach was taken by Striem-Amit et al. (2011), who used phase-encoding (travelling wave) fMRI, a technique previously used to map the visual field onto the visual cortex (Engel et al., 1997; Warnking, 2002). They found up to six mirror-symmetric frequency gradients covering most of Heschl's gyrus and extending to the superior temporal gyrus and superior temporal sulcus (Striem-Amit et al., 2011), beyond the areas previously reported by Woods et al. (2009) and Humphries et al. (2010). Such cochleotopic organization was identified in both left and right hemispheres. The maps of Striem-Amit et al. (2011) demonstrate that tonotopy persists well beyond the auditory core and belt.

The discovery of the tonotopic maps and proposed boundaries between auditory cortical fields might help our understanding of these fields and contributes in defining possible connectivity patterns between these fields. These advances ultimately help us understand the functional consequences of e.g., hearing loss and tinnitus which may affect the function or organization of the auditory cortex (Eggermont, 2006; Komiya & Eggermont, 2000). Both conditions are good examples of auditory disorders with an increasing prevalence due to an aging population.

4. Functional magnetic resonance imaging of tinnitus

4.1. A phantom sound?

Tinnitus is an auditory sensation without the presence of an external acoustic stimulus. In 6-20% of the adult population, tinnitus is a chronic condition and severely disrupts quality

http://dx.doi.org/10.5772/29929

of life in 1-3%. Tinnitus is more prevalent in men than in women and its prevalence increases with advancing age (Axelsson & Ringdahl, 1989). Subjective tinnitus is the most common form of the disorder, but its neural origins remain rather elusive. Like any external acoustic percept, tinnitus is associated with neural activity in the central auditory system. Tinnitus-related activity may arise from abnormal cellular mechanisms in neurons of the central auditory system, or may result from aberrant input from the cochlea or non-auditory structures. Subjective tinnitus is often associated with peripheral hearing loss (Eggermont & Roberts, 2004; Nicolas-Puel et al., 2006) and many patients describe tinnitus as a sound in one or both ears. Therefore, it has been thought for many years that the tinnitus-related neural activity must also originate from the cochlea. Clinical observations however indicate the importance of central mechanisms. In patients that underwent sectioning of the eighth cranial nerve as part of retro-cochlear tumor surgery, tinnitus actually arose in 34% of the cases (Berliner et al., 1992). Surgically severing the connections between the cochlea and the brain has also been applied to provide relief of the tinnitus. This procedure was however not successful in 38-85% of cases (Kaltenbach et al., 2005).

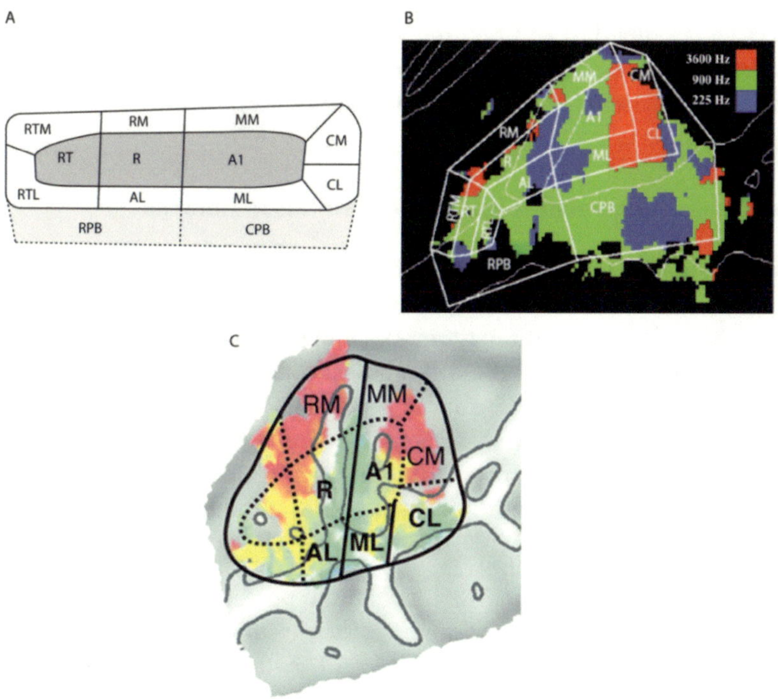

Figure 2. A. Overview of auditory fields in the primate brain showing core areas (A1, R, RT), belt areas (CL, CM, ML, MM, RM, AL, RTL, RTM) and proposed parabelt areas (RPB, CPB); adapted from Kaas et al. (2000). B. The same representation of auditory cortical fields superimposed on tonotopic data in human auditory cortex; adapted from Woods et al., (2009). C. Auditory cortical fields superimposed on tonotopic data; adapted from Humphries et al. (2010). Abbreviations for auditory fields: A1 = primary auditory cortex; R = rostral, T = temporal, M= middle; A = anterior; L = lateral, C=caudal, PB = parabelt.

4.2. Animal studies of tinnitus

Evidence for changes in the firing patterns of neurons in the central auditory system (Figure 3) as a possible substrate of tinnitus is mainly based on animal models. Noise trauma and ototoxic drugs, which are known to cause peripheral hearing loss and tinnitus in humans, result in behavioral responses in animals that seem consistent with the presence of tinnitus. One of these substrates is a change in the spontaneous firing rates of auditory neurons (Eggermont & Roberts, 2004). Since an increased (stimulus-driven) firing rate in auditory neurons typically corresponds to the presence of sound source, an increased spontaneous firing rate could also lead to an auditory percept, i.e., tinnitus. Alternatively, the temporal pattern of spontaneous neural activity could change by increased synchrony of activity across auditory neurons (Noreña & Eggermont 2003; Seki & Eggermont, 2003). The third candidate is a reorganization of the tonotopic map in auditory neurons in the central auditory system (Eggermont 2006; Mühlnickel et al., 1998; Seki & Eggermont, 2003). Although such changes themselves may not directly correspond to tinnitus, they may contribute to abnormal neural activity. For example, cortical reorganization may lead to the over-representation of frequencies corresponding to the frequency at the edge of a high-frequency sloping hearing loss (Eggermont, 2006; Rajan & Irvine, 1998). In other words, auditory-nerve or lower-brainstem neurons that are tuned to an edge frequency could be excessively projected to a region of the auditory cortex. fMRI can be used to identify mechanisms that underlie the generation of tinnitus in humans. Section 4.3 describes the application of fMRI and discusses the outcomes of selected studies.

4.3. Functional magnetic resonance imaging of tinnitus in humans

Although there is much evidence for neural changes based on animal models, there is little corresponding evidence for tinnitus in humans. fMRI is expected to be sensitive to changes in the overall neuronal activity and may reveal changes in neural activity patterns. Yet, given the temporal resolution of typical fMRI studies, changes in synchronous neural activity or spontaneous firing rates may not be apparent from fMRI data. The BOLD signal is a relative measure of tissue properties and so 'activity' cannot be measured directly nor quantified. Instead, fMRI paradigms rely on modulating the hemodynamic response, reflected by neural activity in two (or more) conditions, which allows them to detect differences between states (Ogawa et al., 1990).

Given that tinnitus tends to be a continuous phantom percept, several paradigms have been applied to modulate it in some way. One approach examines sound-evoked responses in tinnitus and control participants, while the second method relies on the ability of a subgroup of people with tinnitus to manipulate their experience by somatic modulation. Examples of include forceful head and neck contractions (Levine, 1999), oral facial movements like jaw clenching or jaw protrusion (Pinchoff et al., 1998), electrical stimulation of the median nerve (Møller et al., 1992) and cutaneous stimulation (Cacace et al., 1999a; 1999b). Other paradigms that could be considered are i) the rapid change of gaze or tonic lateral gaze causing or modulating tinnitus, ii) pharmaceutical interventions that causes a temporal change of the tinnitus (e.g., lidocaine), and, iii) residual inhibition, which refers to the reduc-

http://dx.doi.org/10.5772/29929

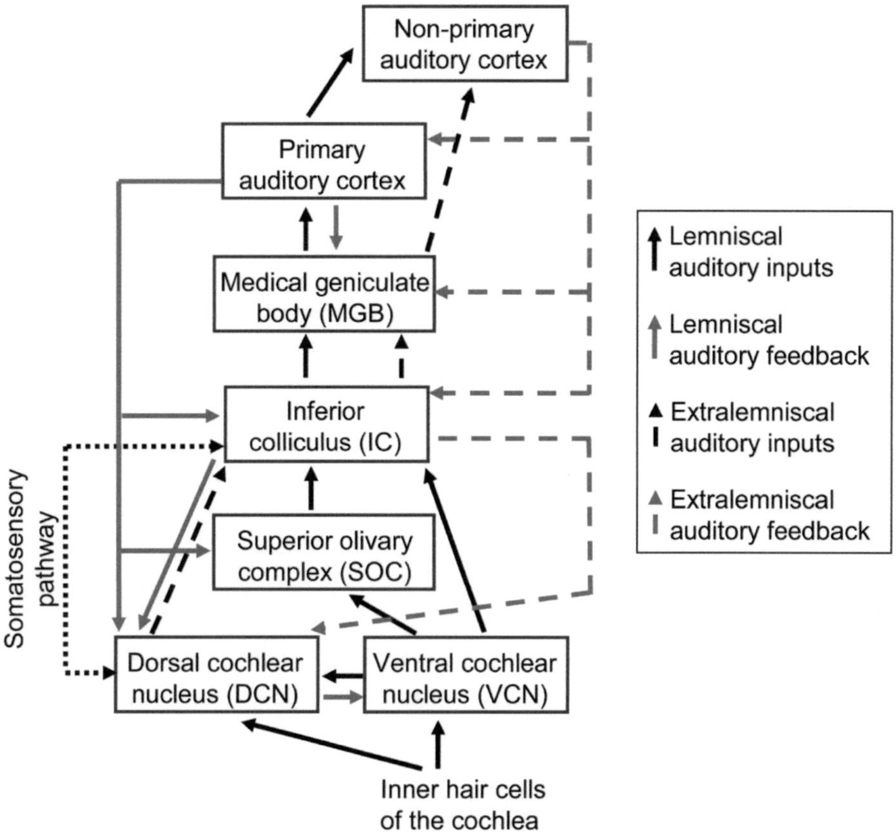

Figure 3. Schematic outline of the auditory system redrawn from Eggermont & Roberts (2004). Inner hairs cells of the cochlea decompose multi-frequency signals into an output that is spatially organised according to frequency (tonotopy). Auditory nerve fibers send inputs to dorsal and ventral cochlear nuclei which both show tonotopic maps. The lemniscal pathway (solid black line) propagates this frequency-specific information throughout all structures. The central auditory system is characterized by strong reciprocal connections (grey lines). The interconnected feedback loops allow the cortex to modulate activity. Changes in dorsal cochlear nucleus, in turn, will directly affect the processing of lemniscal activity at the level of the ventral cochlear nucleus and inferior colliculus. This enables a strong synergy between changes occurring in cortex and brainstem.

tion of the loudness of tinnitus following the offset of an external acoustical stimulus (Roberts, 2007). A review of these paradigms is given by Lanting et al. (2009). Here, we highlight the contribution of fMRI to the current understanding of central mechanisms of tinnitus and explain how interpretations of neuroimaging data may help developing a better understanding of its diverse pathophysiology and of the functional interplay between sensory, cognitive and affective systems.

4.4. Effects of tinnitus on patterns of sound-evoked activity

Compared to non-tinnitus controls, people with unilateral tinnitus initially appeared to show an abnormally small signal change in the inferior colliculus contralateral to the tinnitus percept, but not ipsilateral (Melcher et al., 2000). These findings were interpreted as evidence that tinnitus corresponds with abnormally elevated baseline neural activity on the contralateral side. When an external stimulus is presented, the hemodynamic response reaches saturation, resulting in a reduced difference between the 'sound' and 'silence' conditions. A potential confound to these observed patterns of sound-related activity was the influence of ambient background noise. Hence, in a new experiment, this variable was examined by repeating the study with the noisy helium pump either switched on or off (Melcher et al., 2009). All participants had normal hearing. With the pump off, the tinnitus group showed greater stimulus-evoked activation of the inferior colliculus than the non-tinnitus group. With the pump on, the tinnitus group showed reduced activation compared to controls. This result indicates some degree of auditory response saturation in the brains of people with tinnitus. A possible explanation is that a compensatory mechanism of central auditory gain has increased the response amplitudes close to their maximum limit. Lanting et al. (2008) used a sparse temporal sampling paradigm, minimizing the interaction between the scanner noise, the noise of the pump and auditory brain activity. Stimuli consisted of monaural dynamic rippled broadband noise stimuli at two intensity levels (40 dB and 70 dB SPL). Responses were measured at the level of auditory cortex and inferior colliculus of subjects with unilateral tinnitus and near-normal hearing. These were compared with those of people without tinnitus. This study demonstrated abnormally elevated responses in the inferior colliculus in people with tinnitus subjects compared to controls. Figure 4 shows increased sound-evoked responses, a reduced response lateralization (i.e., stimuli presented to the contralateral and ipsilateral ear gave roughly the same signal change) and a disturbed intensity level dependency in subjects with tinnitus compared to subjects without tinnitus at the level of the inferior colliculus (Lanting et al., 2008).

In summary, the first study by Melcher (Melcher et al., 2000) found *decreased* responses in inferior colliculus on the opposite side to the unilateral tinnitus percept, while the other two studies found overall *increased* responses (Lanting et al., 2008; Melcher et al., 2009). At first sight, these three results appear to be contradictory, but a key methodological difference may account for the discrepancy. While Lanting applied sparse temporal sampling, Melcher scanned continuously. Therefore, Melcher's experiments were performed in high levels of ambient scanner noise which may have caused inferior colliculus to respond excessively to this source of background noise and hence reduce the BOLD response to the experimental sound stimuli. The three papers are thus consistent with the interpretation that the inferior colliculus of people with tinnitus displays a disproportionate response to sound, whether it is ambient sound or experimentally controlled sound.

Although sound-evoked responses appear elevated in people with tinnitus, hyperacusis could be an important alternative explanation. This condition is defined as an abnormal sensitivity to sound and is manifest as a diminished level of sound tolerance possibly elevating sound-evoked responses in the auditory brain. Hyperacusis is often co-morbid with tinnitus (Baguley, 2003). A recent study assessed the neural correlates of hyperacusis, recruiting 13 people with tinnitus and 14 people with no tinnitus (Gu et al., 2010). Hyperacusis was assessed using minimum loudness discomfort levels and a validated questionnaire. Severity of sound level tolerance (i.e. hyperacusis) was correlated with

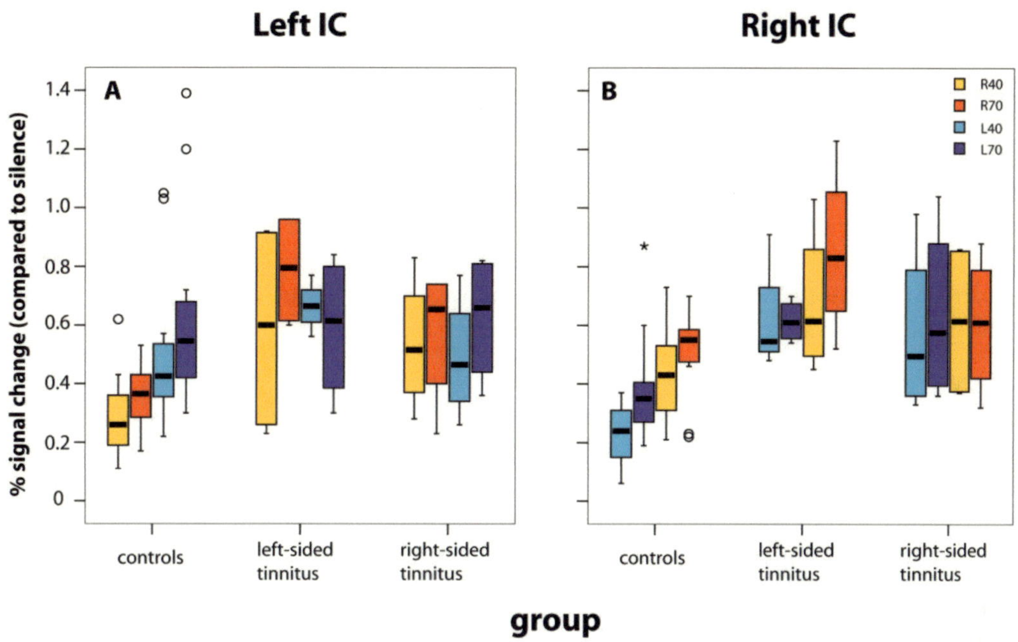

Figure 4. Sound-evoked responses measured in left (**A**) and right (**B**) inferior colliculus (IC). The pattern of evoked responses for the side of presentation (left or right ear) and the sound intensity (40 dB and 70 dB SPL) is different for subjects with unilateral tinnitus as compared to controls. Adapted from Lanting et al. (2008)

elevated activity in the auditory midbrain, thalamus, and primary auditory cortex, despite a fixed sound stimulation level. Sound-evoked activity in the inferior colliculus and medial geniculate body was more associated with hyperacusis than tinnitus, while in primary auditory cortex it was more associated with tinnitus than hyperacusis. Previous findings of increased sound-evoked responses in the inferior colliculus of people with tinnitus (Lanting et al., 2008; Melcher et al., 2009) are most likely to be due to the hypersensitivity to an external sound stimulus than tinnitus per se.

In conclusion, although fMRI has shown general increases in sound-evoked activity in tinnitus, it is more likely to be associated with hyperacusis. Any hypothesized reorganization of tonotopic maps in the auditory cortex has not yet been demonstrated, mostly due to the fact that tonotopic mapping is a relative new field. With the increasing availability of ultra high-field fMRI, this is a promising area for further research.

5. Cross-modal integration in tinnitus

In addition to auditory nerve input, dorsal cochlear nucleus is also innervated by fibers from various parts of the somatosensory system, so this structure is a multi-modal processing station that is considered part of the extralemniscal auditory pathway (Figure 3). Parts of inferior colliculus, medial geniculate body and non-primary auditory cortex are all characterized by sensitivity to somatosensory stimuli (Eggermont & Roberts, 2004). As mentioned in Section 4.3, somatosensory manoeuvers can elicit or modulate tinnitus. Somatic modulation of tinnitus may therefore be considered a special case of multisensory integration—a phenomenon, in which one (sensory) modality influences another. The observation that some people with tinnitus can modulate the loudness or pitch of their tinnitus might be explained by changes in normal multisensory integration. Noise-induced hearing-loss, for example, has been reported to alter the normal somatosensory input (Shore et al., 2008). Somatosensory input to the cochlear nucleus increased after hearing loss, changing the balance of somatosensory and auditory input at the level of the brainstem, which in turn might be the neural correlate of somatosensory modulation of tinnitus.

To investigate this proposed model of cross-modal integration, Lanting et al. (2010) studied two groups: people with tinnitus and matched controls. Tinnitus participants were selected according to their ability to change the psychoacoustical characteristics of their tinnitus by jaw protrusion. Unimodal fMRI responses to sound stimuli and jaw protrusion alone were measured, as well as the response obtained in multimodal condition. To minimize motion artifacts, a sparse sampling method (inter-scan interval = 10 s) was used in which the protrusion started 4 s after the beginning of the acquisition and ended 1.5 s before the next acquisition. Since the jaw was always fully relaxed during scan acquisitions, the task did not degrade data quality. Results showed responses to jaw protrusion throughout the central auditory pathway, in both groups. Auditory responses to jaw protrusion presumably account for the modulation of tinnitus by these manoeuvers. Interestingly, responses in cochlear nucleus and inferior colliculus were greater in people with tinnitus than in controls, suggesting an abnormal auditory-somatic interaction in the patient group. This increased sensitivity to somatosensory inputs may correspond to neuroplastic changes, triggered by pathological change of either the somatosensory or auditory input pathways to dorsal cochlear nucleus (see Figure 3 and also Shore, 2011) that, in turn, lead to compensatory shifts in the balance of excitation and inhibition often attributed to tinnitus.

6. Functional and structural assessments of candidature for cochlear implantation

Hearing loss is the most common sensory deficit. Approximately 1.5% of children and 60% of over-70s have hearing loss and, as the average age of the population increases, the social, economic and psychological impact of hearing loss is likely to become more of a burden. Hearing after mild to moderate deafness can be regained through acoustic amplification with a hearing aid, whilst more severe to profound hearing loss can be partially restored by replacing the non-functioning inner ear with a CI.

A CI is a surgically-implanted auditory prosthesis that provides a sense of hearing to deaf individuals. As part of a preoperative clinical evaluation, computerized-tomography (CT) and MRI are both routinely used to assess CI candidates and to aid pre-operative planning and patient counselling. CT provides excellent structural information about bony detail and is useful for assessing congenital malformations that would affect the course of the facial nerve. MRI on the other hand can visualize soft-tissue with high resolution and is used to evaluate the integrity of the auditory nerve and the patency of the cochlea. Traditional CI candidates were adults with such profound hearing loss they could not benefit from any acoustic hearing. Over the past 25 years CI candidature has been relaxed, to include children and infants in whom auditory deprivation has the most severe developmental consequences. More recently, CIs have been used to treat individuals with some residual acoustic hearing in one or both ears. To date, nearly 200,000 individuals worldwide have been implanted with a CI, approximately half of those being children.

6.1. Prognostic indicators of cochlear implant performance

Although many CI users are able to develop age-appropriate communication and language skills without the need for visual cues, outcomes vary considerably between individuals. A number of factors have been found to correlate with speech outcomes following cochlear implantation (Blamey et al., 1996; Miyamoto et al., 1994). Pre-linguistically deaf adults generally perform more poorly than post-linguistically deaf adults (Busby et al., 1993a; 1993b; Eddington et al., 1978), whilst a long duration of deafness prior to implantation is generally detrimental to performance (O'Donoghue et al., 2000; Sarant et al., 2001). Also, it has been shown that auditory experience can improve clinical performance of cochlear implant users (Blamey et al., 1996; Gantz et al., 1993; Rubinstein et al., 1999). However, individuals with remarkably similar hearing experiences can often have markedly different speech comprehension skills. For example, older pre-linguistically deaf children can occasionally succeed very well in understanding speech, whilst the majority do not (Schramm et al., 2002). Therefore, whilst a number of factors that influence the clinical performance of cochlear implants have been identified, the field currently lacks accurate predictors of cochlear implant performance that can be applied to an individual. Such a prognostic tool, could help direct limited health-care resources towards effective clinical care.

Current national guidelines within the UK suggest that unilateral cochlear implantation is recommended for people with severe to profound hearing loss who do not receive adequate benefit from an acoustic hearing aid. Within this group, simultaneous bilateral cochlear implantation is recommended for i) pre-linguistically deaf children, ii) individuals who are blind, and iii) those who are at risk of cochlear ossification. Therefore, sensitive prognostic tools could be used to mould this crude patient pathway to meet the needs of individual patients, whilst providing realistic indicators of potential outcomes and requirements for rehabilitation.

Behavioural assessments of hearing with a CI are difficult to assess in very young children, as well as individuals with significant physical and/or developmental delay. Therefore, some investigators have focused on developing objective, non-invasive measures of CI function. Electrically-evoked compound action potentials (ECAPs) and electrically-evoked auditory brainstem responses (EABRs) are both routinely used to measure CI function at the level of the auditory nerve and brainstem, respectively. Whilst ECAPs and EABRs are useful clinical tools to ensure that the stimulus levels provided by the CI processor are within the range of perception, there is little evidence to suggest that they correlate well with speech and language outcomes (Miller et al., 2008).

Ultimately, CI users are likely to develop good speech and language skills provided accurate auditory information reaches the auditory cortex. For this reason, attempts have been made to assess functional activation of the auditory cortex with a CI. A number of methods can be used to assess cortical responsiveness to auditory stimulation. However, most of these methods have significant limitations when they are applied to a paediatric CI population.

Earlier in this chapter we discussed the advantages of fMRI in auditory research: a technique that permits the mapping of auditory responses with fine topographic detail, as well multiple observations over time, within both adult and paediatric populations. However, the surgically-implanted magnetic component of a CI device is incompatible with the high magnetic field generated by most MRI scanners. Traditionally, prior to considering MRI, it was necessary to surgically-remove the magnet after cochlear implantation. Recently it has been shown that CI patients can undergo MRI using a low (1.5) Tesla scanner, provided the implanted device is tightly bound before scanning (Crane et al., 2010). Patients in this study received an implant on one side. Although Crane and colleagues observed no magnet displacement whilst visualizing the contralateral- and part of the ipsilateral-internal auditory canal, the artefact surrounding the implanted device (~7x5 cm) precluded functional imaging of the cortex on that side. Similarly, magnetoencephalography (MEG) is limited by the magnetic field associated with current CI devices.

Electroencephalography (EEG) is also used to investigate auditory-evoked cortical potentials. However, the electrical artefact generated by the CI confounds these recordings. Although algorithms have been developed to minimize the effect of CI-generated electrical signals in EEG recordings (Debener et al., 2008), stimulus parameters are generally limited to short-duration sounds such as clicks (Gilley et al., 2006). Responses to longer sounds, including speech and language stimuli, are difficult to extract from the electrical artefact. Functional activation of the auditory cortex can also be assessed using PET that involves the use of ionizing radiation. Due to the cumulative effects of radiation exposure, this technique is not ideal for longitudinal studies, particularly in children.

Sevy et al. (2010) recently demonstrated a relatively novel imaging technique called 'near-infrared spectroscopy' (NIRS) could be used to assess cortical responses to auditory stimulation in children with cochlear implants. NIRS measures relative changes in the tissue contraction, oxy-haemoglobin, deoxy-haemoglobin and total haemoglobin in the cerebral cortex through beaming infra-red light. One of the greatest advantages of this approach is that the optical signals from the auditory cortex are not influenced by the magnetic fields that plague fMRI and MEG recordings. Unlike EEG, NIRS is not affected by CI-generated electrical artifacts. Subjects experience no discomfort with NIRS, as it is a non-invasive test that only requires contacting small optical fibre tips onto the scalp. There are no known risks associated with NIRS. Therefore, unlike PET, it is suitable for multiple measurements in paediatric populations. Although, NIRS provides sufficient resolution to measure evoked responses that are confined to auditory regions of the brain, this spatial resolution (within 1-2 cm of the activated area) is less than that of fMRI. Regardless, compared with alternative techniques, NIRS seems highly suited to assess the responsiveness of the auditory cortex before and after cochlear implantation in paediatric populations.

7. Cross-modal reorganization in the deaf brain

Animal neurophysiology has demonstrated that when sensory neurons lose their input (e.g. vision, touch or hearing) they do not become 'silent'. Instead, they change their responsiveness. In the case of (noise-induced) sensorineural high-frequency hearing loss, neurons previously responding to high frequencies respond to low frequencies (Rajan et al., 1993; Robertson & Irvin, 1989). In the case of complete deafness, neurons become rewired to previously unconnected neurons from another sensory modality (Lomber et al., 2010; Rauschecker et al., 1995).

Cross-modal reorganization refers to the takeover of the processing of sensory information using new cortical pathways. It may be of clinical relevance for hearing-impaired individuals and individuals with a cochlear implant: some people with congenital hearing loss show 'supra-normal' visual abilities (Bavelier et al., 2006; Neville & Lawson, 1987). Furthermore, congenital deafness is known to induce rewiring of inputs within the auditory cortex (Allman et al., 2009; Auer et al., 2007; Finney et al., 2001; 2003; Levänen et al., 1998; Lomber et al., 2010; Nishimura et al., 1999) that plays a vital role in allowing us to process speech and naturally occurring sounds. Recently, Lomber et al. (2010) used cortical cooling techniques in deaf cats to suggest that enhanced visual performance is caused by cross-modal reorganization of the posterior auditory cortex.

Currently, it is not clear whether this rewiring is beneficial to hearing with a cochlear implant (potentially acting as an instructive map to 'train' the reinstated hearing pathways) or detrimental to hearing (by taking away resources from the auditory brain). Regardless, it remains possible that cross-modal reorganization following deafness may prove to be a clinically-useful predictor of performance following cochlear implantation. Lee et al. (2001) used F-18 fluorodeoxyglucose PET to examine resting-state brain activity in pre-lingually deaf individuals. Results indicated an association between hypometabolism in the auditory cortex and the amount of improvement in hearing after cochlear implantation. They speculated that the hypometabolism may be related to cross-modal plasticity and that the individual status of auditory rewiring might be used as an effective predictor of an individual's speech performance following cochlear implantation. More recently, Rouger et al. (2011) demonstrated cross-modal responses in post-lingually deaf adults with cochlear implants engaged in a visual speech reading task using PET brain imaging. Furthermore, they suggested that cross-modal reorganization diminishes in post-lingually deaf adults with CIs with increasing time after implantation (up to at least one year later), towards the levels seen in normal-hearing individuals (Rouger et al., 2011). Doucet et al. (2006) compared cortical evoked potentials between adults with good and poor speech outcomes following cochlear implantation. They found that poor performers exhibited broader, anteriorly-distributed evoked potentials whereas the good performers showed significantly higher amplitudes of the P2 component over the primary visual cortex. Furthermore, Buckley et al. (2006) suggested that cortical responsiveness to visual and auditory evoked potentials accounts for a significant amount of the variability observed in speech perception performance in adults with cochlear implants with pre-lingual, but not post-lingual, hearing loss. Although, approximately half of cochlear implant recipients are children worldwide, currently, little is known about cross-modal brain reorganization during development.

8. Conclusion and future directions

While the results from a number of functional imaging studies in humans provide preliminary insights into plasticity within the central auditory pathway, more information is needed to understand the patterns of neural reorganization and the mechanisms underpinning those changes. This is an exciting time as novel imaging techniques, such as ultra-high field fMRI and NIRS, open new possibilities for investigating the effects of hearing loss, cochlear implantation and tinnitus on the developing and the mature brain. From our review, three key questions capture avenues of future research that are likely to impact on clinical practice: i) What are the separate effects of deafness, hyperacusis, tinnitus and cochlear implantation on central auditory plasticity, including cross-modal reorganization? ii) What are the perceptual consequences of that reorganization, and iii) How does brain reorganization affect and/or predict treatment outcome?

An ultimate goal for functional imaging is to provide the information necessary for evidence-based healthcare, such as targeting tinnitus treatments to specific neural sites of its generation and predicting outcomes following cochlear implantation. Currently, imaging of the central auditory pathway is not used for routine clinical purposes in Ear, Nose and Throat or Audiology clinics. It is our hope that research using imaging techniques may provide valuable insights into mechanisms underlying tinnitus and the variable speech outcomes following cochlear implantation in humans that may tailor our treatment towards the requirements of the individual.

Author details

Deborah A. Hall[1], Cornelis P. Lanting[2] and Douglas E.H. Hartley[3]

1 NIHR National Biomedical Research Unit in Hearing, Nottingham, UK

2 MRC Institute of Hearing Research, Nottingham, UK

3 Department of Otorhinolaryngology, Nottingham University, UK

References

[1] Allman, B.L., Keniston, L.P., & Meredith, M.A. (2009). Adult deafness induces somatosensory conversion of ferret auditory cortex. *Proceedings of the National Academy of Sciences in the United States of America*, 106, 14, pp. 5925-5930.

[2] Auer, E.T., Bernstein, L.E., Sungkarat, W., & Singh, M. (2007). Vibrotactile activation of the auditory cortices in deaf versus hearing adults. *Neuroreport*, 18, 7, pp. 645-648.

[3] Axelsson, A., & Ringdahl, A. (1989). Tinnitus--a study of its prevalence and characteristics. *British Journal of Audiology*, 23, 1, pp. 53-62.

[4] Baguley. D.M. (2003). Hyperacusis. *J R Soc Med*, 96, pp. 582-585.

[5] Baumann, S., Griffiths, T.D., Sun, L., Petkov, C.I., Thiele, A., & Rees, A. (2011). Orthogonal representation of sound dimensions in the primate midbrain. *Nature Neuroscience*, 14, 4, pp. 31-33.

[6] Bavelier, D., Dye, M.W.G., & Hauser, P.C. (2006). Do deaf individuals see better? *Trends in Cognitive Sciences*, 10, 11, pp. 512-518

[7] Berliner, K.I., Shelton, C., Hitselberger, W.E., & Luxford, W.M. (1992). Acoustic tumours: effect of surgical removal on tinnitus. *The American Journal of Otology*, 13, 1, pp. 13- 17.

[8] Bilecen, D., Scheffler, K., Schmid, N., Tschopp, K., & Seelig, J. (1998). Tonotopic organization of the human auditory cortex as detected by BOLD-FMRI. *Hearing Research*, 126, 1- 2, pp. 19-27.

[9] Bilecen, D., Seifritz, E., Scheffler, K., Henning, J., & Schulte, A.C. (2002). Amplitopicity of the human auditory cortex: an fMRI study. *NeuroImage*, 17, 2, pp. 710-718.

[10] Blackman, G., & Hall, D.A. (2011). Reducing the effects of background noise during auditory functional magnetic resonance imaging of speech processing: Qualitative and quantitative comparisons between two image acquisition schemes and noise cancellation. *Journal of Speech, Language and Hearing Research*. 54, pp. 693–704.

[11] Blamey, P., Arndt, P., Bergeron, F., Bredberg, G., Brimacombe, J., Facer, G., Larky, J., Lindstrom, B., Nedzelski, J., Paterson, A., Shipp, D., Staller, S., & Whitford, L., (1996). Factors affecting auditory performance of postlinguistically deaf adults using cochlear implants. *Audiology & Neuro-otology*, 1, 5, pp. 293-306.

[12] Buckley, K.A., & Tobey, E.A. (2011). Cross-modal plasticity and perception in pre- and postlingually deaf cochlear implant users. *Ear and Hearing*, 32, 1, pp. 2-15.

[13] Busby, P.A., Tong Y.C., & Clark G.M. (1993a). Electrode position, repetition rate, and speech-perception by early-deafened and late-deafened cochlear implant patients. *The Journal of the Acoustical Society of America*, 93, 2, pp. 1058-1067.

[14] Busby, P.A., Tong Y.C., & Clark G.M. (1993b). The perception of temporal modulations by cochlear implant patients. *The Journal of the Acoustical Society of America*, 94, 1, pp. 124-131.

[15] Cacace, A.T., Cousins, J.P., Parnes, S.M., Semenoff, D., Holmes, T., McFarland D.J., Davenport, C., Stegbauer, K., & Lovely, T.J. (1999a). Cutaneous-evoked tinnitus. I. Phenomenology, psychophysics and functional imaging. *Audiology & Neuro-otology*, 4, 5, pp. 247-257.

[16] Cacace, A.T., Cousins, J.P., Parnes, S.M., McFarland, D.J., Semenoff, D., Holmes, T., Davenport, C., Stegbauer, K., & Lovely, T.J. (1999b). Cutaneous-evoked tinnitus. II. Review of neuroanatomical, physiological and functional imaging studies. *Audiology & Neuro-otology*, 4, 5, pp. 258-268.

[17] Crane, B. T., Gottschalk, B., Kraut, M., Aygun, N., & Niparko, J. K. (2010). Magnetic resonance imaging at 1.5 T after cochlear implantation. *Otology Neurotology, 31*, 8, 1215-1220.

[18] Debener, S., Hine, J., Bleeck, S., & Eyles, J. (2008). Source localization of auditory evoked potentials after cochlear implantation. *Psychophysiology, 45*, 1, 20-24.

[19] Detre, J.A., & Wang, J. (2002). Technical aspects and utility of fMRI using BOLD and ASL. *Clinical Neurophysiology*, 113, 5, pp. 621-634.

[20] Doucet, M.E., Bergeron, F., Lassonde, M., Ferron, P., & Lepore, F. (2006). Cross-modal reorganization and speech perception in cochlear implant users. *Brain*, 129, pp. 3376-3383.

[21] Duong, T.Q., Yacoub, E., Adriany, G., Hu, X.P., Ugurbil, K., & Kim, S.G. (2003). Microvascular BOLD contribution at 4 and 7 T in the human brain: Gradient-echo and spin-echo fMRI with suppression of blood effects. *Magnetic Resonance in Medicine*, 49, 6, pp. 1019-1027.

[22] Eddington, D.K., Dobelle, W.H., Brackmann, D.E., Mladejovsky, M.G., & Parkin, J.L., (1978). Auditory prostheses research with multiple channel intracochlear stimulation in man. *The Annals of Otology, Rhinology, and Laryngology*, 87, 6.2, pp. 1-39.

[23] Edmister, W.B., Talavage, T.M., Ledden, P.J., & Weisskoff R.M. (1999). Improved auditory cortex imaging using clustered volume acquisitions. *Human Brain Mapping*, 7, pp. 89-97.

[24] Eggermont, J.J. (2006). Cortical tonotopic map reorganisation and its implications for treatment of tinnitus. *Acta Oto-laryngologica. Supplementum*, 556, pp. 9-12.

[25] Eggermont, J.J., & Roberts, L.E. (2004). The neuroscience of tinnitus. *Trends in Neurosciences*, 27, 11, pp. 676-682.

[26] Ehret, G., (1978). Stiffness gradient along the basilar membrane as a basis for spatial frequency analysis within the cochlea. *The Journal of the Acoustical Society of America*, 64, 6, pp. 1723-1726.

[27] Engel, S., Glover, G.H., & Wandell, B. (1997). Retinotopic organization in human visual cortex and the spatial precision of functional MRI. *Cerebral Cortex*, 7, 2, pp. 181-192.

[28] Finney, E.M., Fine, I., & Dobkins, K.R. (2001). Visual stimuli activate cortex in the deaf. *Nature Neuroscience*, 4, 12, pp. 1171-1173.

[29] Finney, E.M., Clementz, B.A., Hickok, G., & Dobkins, K.R. (2003). Visual Stimuli activate auditory cortex in deaf subjects: evidence from MEG. *Neuroreport*, 14, 11, pp. 1425- 1427.

[30] Formisano, E., Kim, D.S., Di Salle, F., Van de Moortele, P, F., Ugurbil, K., & Goebel, R. (2003). Mirror-symmetric tonotopic maps in human primary auditory cortex. *Neuron*, 40, 4, pp. 859-869.

[31] Foster, J.R., Hall, D.A., Summerfield, A.Q., Palmer, A.R., & Bowtell, R.W. (2000). Sound-level measurements and calculations of safe noise dosage during fMRI at 3T. *Journal of Magnetic Resonance Imaging*, 12, pp. 157-163.

[32] Gaab, N., Gabrieli, J.D.E., & Glover, G.H. (2007). Assessing the influence of scanner background noise on auditory processing. I. An fMRI study comparing three experimental designs with varying degrees of scanner noise. *Human Brain Mapping*, 28, 8, pp. 703-720.

[33] Gantz, B.J., Woodworth, G.G., Knutson, J.F., Abbas, P.J., & Tyler, R.S. (1993) Multivariate predictors of success with cochlear implants. *Advances in Oto-Rhino-Laryngology*, 48, pp. 153-167.

[34] Gilley, P. M., Sharma, A., Dorman, M., Finley, C. C., Panch, A. S., & Martin, K. (2006). Minimization of cochlear implant stimulus artifact in cortical auditory evoked potentials. *Clinical Neurophysiology, 117*, 8, 1772-1782.

[35] Giraud A.L., Truy, E., & Frackowiak, R. (2001). Imaging plasticity in cochlear implant patients. *Audiology & Neuro-otology*, 6, 6, pp. 381-393.

[36] Goense, J.B.M., Ku, S.-P., Merkle, H., Tolias, A.S., & Logothetis, N.K. (2008). fMRI of the temporal lobe of the awake monkey at 7 T. *NeuroImage*, 39, 3, pp. 1081-1093.

[37] Gu, J.W., Halpin, C.F., Nam, E.-C., Levine, R., & Melcher, J.R. (2010). Tinnitus, diminished sound-level tolerance, and elevated auditory activity in humans with clinically normal hearing sensitivity. *Journal of Neurophysiology*, 104, 6, pp. 3361-3370.

[38] Hackett, T.A., Preuss, T.M., & Kaas, J.H. (2001). Architectonic identification of the core region in auditory cortex of macaques, chimpanzees, and humans. *Journal of Comparative Neurology*, 222, pp. 197-222.

[39] Hall, D.A., Haggard, M.P., Akeroyd, M.A., Palmer, A.R., Summerfield, A.Q., Elliott, M.R., Gurney, E., & Bowtell, R.W. (1999). Sparse temporal sampling in auditory fMRI. *Human Brain Mapping*, 7, 3, pp. 213-223.

[40] Hall, D.A., Chambers, J., Foster, J., Akeroyd, M.A., Coxon, R., & Palmer, A.R. (2009). Acoustic, psychophysical, and neuroimaging measurements of the effectiveness of active cancellation during auditory functional magnetic resonance imaging. *The Journal of the Acoustical Society of America* 125, 1, pp. 347–359.

[41] Hart, H.C., Hall, D.A., & Palmer, A.R. (2003). The sound-level-dependent growth in the extent of fMRI activation in Heschl's gyrus is different for low- and high-frequency tones. *Hearing Research*, 179, 1-2, pp. 104-112.

[42] Howard, M.A., Volkov, I.O., Abbas, P.J., Damasio, H., Ollendieck, M.C., & Granner, M.A. (1996). A chronic microelectrode investigation of the tonotopic organization of human auditory cortex. *Brain Research*, 724, 2, pp. 260-264.

[43] Humphries, C., Liebenthal, E., & Binder, J.R. (2010). Tonotopic organization of human auditory cortex. *NeuroImage*, 50, 3, pp. 1202-1211.

[44] Jäncke, L., Wüstenberg, T., Schulze, K., & Heinze, H.J. (2002). Asymmetric hemodynamic responses of the human auditory cortex to monaural and binaural stimulation. *Hearing Research*, 170, 1-2, pp. 166-178.

[45] Kaas, J.H., & Hackett, T. (2000). Subdivisions of auditory cortex and processing streams in primates. *Proceedings of the National Academy of Sciences of the United States of America*, 97, 22, pp. 11793-11799.

[46] Kaltenbach, J.A., Zhang, J., & Finlayson, P. (2005). Tinnitus as a plastic phenomenon and its possible neural underpinnings in the dorsal cochlear nucleus. *Hearing Research*, 206, 1-2, pp. 200-226.

[47] Komiya, H., & Eggermont, J.J. (2000). Spontaneous firing activity of cortical neurons in adult cats with reorganized tonotopic map following pure-tone trauma. *Acta Oto-laryngologica*, 120, 6, pp. 750-756.

[48] Langers, D.R.M., Backes, W.H., & Van Dijk, P. (2007). Representation of lateralization and tonotopy in primary versus secondary human auditory cortex. *NeuroImage*, 34, 1, pp. 264-273.

[49] Lanting, C.P., De Kleine, E., Bartels, H., & Van Dijk, P. (2008). Functional imaging of unilateral tinnitus using fMRI. *Acta Oto-laryngologica*, 128, 4, pp. 415-21.

[50] Lanting, C.P., De Kleine, E., & Van Dijk, P. (2009). Neural activity underlying tinnitus generation: results from PET and fMRI. *Hearing Research*, 255, 1-2, pp. 1-13.

[51] Lanting, C.P., De Kleine, E., Eppinga, R.N., & Van Dijk, P (2010). Neural correlates of human somatosensory integration in tinnitus. *Hearing Research*, 267, 1-2, pp. 78-88.

[52] Lauter, J.L., Herscovitch, P., Formby, C., & Raichle, M.E. (1985). Tonotopic organization in human auditory cortex revealed by positron emission tomography. *Hearing Research*, 20, 3, pp. 199-205.

[53] Lee, D. S., Lee, J. S., Oh, S. H., Kim, S. K., Kim, J. W., Chung, J. K., et al. (2001). Cross-modal plasticity and cochlear implants. *Nature, 409*, 6817, 149-150.

[54] Levänen, S., Jousmäki, V., & Hari, R. (1998). Vibration-induced auditory-cortex activation in a congenitally deaf adult. *Current Biology*, 8, 15, pp. 869-872.

[55] Levine, R.A. (1999). Somatic (craniocervical) tinnitus and the dorsal cochlear nucleus hypothesis. *American Journal of Otolaryngology*, 20, 6, pp. 351-362.

[56] Lockwood, A.H. Salvi, R.J., Coad, M.L., Arnold, S.A., Wack, D.S., Murphy, B.W., & Burkard, R.F. (1999). The functional anatomy of the normal human auditory system: responses to 0.5 and 4.0 kHz tones at varied intensities. *Cerebral Cortex*, 9, 1, pp. 65-76.

[57] Logothetis, N.K., Pauls, J., Augath, M., Trinath, T., & Oeltermann, A. (2001). Neurophysiological investigation of the basis of the fMRI signal. *Nature*, 412, 6843, pp. 150-157.

[58] Lomber, S.G., Meredith, M.A., & Kral, A. (2010). Cross-modal plasticity in specific auditory cortices underlies visual compensations in the deaf. *Nature Neuroscience*, 13, 11, pp. 1421-U163.

[59] Lu, H., Golay, X., Pekar, J.J., & Van Zijl, P.C.M. (2003). Functional magnetic resonance imaging based on changes in vascular space occupancy. *Magnetic Resonance in Medicine*, 50, 2, pp. 263-274.

[60] Mechefske, C.M., Geris, R., Gati, J.S., & Rutt, B.K. (2002). Acoustic noise reduction in a 4 T MRI scanner. *Magnetic Resonance Materials in Physics, Biology and Medicine*, 13, 3, pp. 172-176.

[61] Melcher, J.R., Sigalovsky, I.S., Guinan, J.J., & Levine, R.A. (2000). Lateralized tinnitus studied with functional magnetic resonance imaging: abnormal inferior colliculus activation. *Journal of Neurophysiology*, 83, 2, pp. 1058-1072.

[62] Melcher, J.R., Levine, R.A., Bergevin, C., & Norris, B. (2009). The auditory midbrain of people with tinnitus: abnormal sound-evoked activity revisited. *Hearing Research*, 257, 1-2, pp.63-74.

[63] Merzenich, M.M., Knight, P.L., & Roth, G.L. (1975). Representation of cochlea within primary auditory cortex in the cat. *Journal of Neurophysiology*, 38, 2, pp. 231-249.

[64] Miller, J. D., Watson, C. S., Kistler, D. J., Wightman, F. L., & Preminger, J. E. (2008). Preliminary evaluation of the speech perception assessment and training system (SPATS) with hearing-aid and cochlear-implant users. *Proc Meet Acoust, 2*, 1, 1-9.

[65] Miyamoto, R. T., Osberger, M. J., Cunningham, L., Kirk, K. I., Myres, W. A., Robbins, A. M., et al. (1994). Single-channel to multichannel conversions in pediatric cochlear implant recipients. *American Journal of Otology, 15,* 1, 40-45; discussion 45-46.

[66] Morel, A., Garraghty, P.E., & Kaas, J.H., (1993). Tonotopic organization, architectonic fields, and connections of auditory cortex in macaque monkeys. *The Journal of Comparative Neurology*, 335, 3, pp. 437-459.

[67] Møller, A.R., Møller, M.B., & Yokota, M. (1992). Some forms of tinnitus may involve the extralemniscal auditory pathway. *The Laryngoscope*, 102, 10, pp. 1165-1171.

[68] Mühlnickel, W., Elbert, T., Taub, E., & Flor, H. (1998). Reorganisation of auditory cortex in tinnitus. *Proceedings of the National Academy of Sciences of the United States of America*, 95, 17, pp. 10340-10343.

[69] Neville, H. J., & Lawson, D. (1987). Attention to central and peripheral visual space in a movement detection task: an event-related potential and behavioral study. II. Congenitally deaf adults. *Brain Research, 405,* 2, 268-283.

[70] Nicolas-Puel, C., Akbaraly, T., Lloyd, R., Berr, C., Uziel, A., Rebillard, G., & Puel, J-L. (2006). Characteristics of tinnitus in a population of 555 patients: Specificities of tinnitus induced by noise trauma. *The International Tinnitus Journal*, 12, 1, pp. 64-70.

[71] Nishimura, H., Hashikawa, K., Doi, K., Iwaki, T., Watanabe, Y., Kusuoka, H., et al. (1999). Sign language 'heard' in the auditory cortex. *Nature, 397*, 6715, 116.

[72] Noreña, A.J., & Eggermont, J.J. (2003). Changes in spontaneous neural activity immediately after an acoustic trauma: implications for neural correlates of tinnitus. *Hearing Research*, 183, 1-2, pp. 137-153.

[73] O'Donoghue, G.M., Nikolopoulos, T.P., & Archbold, S.M. (2000). Determinants of speech perception in children after chochlear implantation. *Lancet,* 356, 9228, pp. 466-468.

[74] Ogawa, S., Lee, T.M., Kay, A.R., & Tank, D.W. (1990). Brain magnetic resonance imaging with contrast dependent on blood oxygenation. *Proceedings of the National Academy of Sciences of the United States of America*, 87, 24, pp. 9868-9872.

[75] Ojemann, J.G., Akbudak, E., Snyder, A.Z., McKinstry, R.C., Raichle, M.E., & Conturo, T.E. (1997). Anatomic localization and quantitative analysis of gradient refocused echoplanar fMRI susceptibility artifacts. *NeuroImage*, 6, 3, pp. 156-167.

[76] Paltoglou, A.E., Sumner, C.J., & Hall, D.A. (2011). Mapping feature-sensitivity and attentional modulation in human auditory cortex with functional magnetic resonance imaging. *The European Journal of Neuroscience*, 33, 9, pp. 1733-1741.

[77] Parkes, L.M., Schwarzbach, J.V., Bouts, A.A., Deckers, R.H.R., Pullens, P., Kerskens, C.M., & Norris, D.G. (2005). Quantifying the spatial resolution of the gradient echo

and spin echo BOLD response at 3 Tesla. *Magnetic Resonance in Medicine*, 54, 6, pp. 1465-1472.

[78] Petkov, C.I., Kayser, C., Augath, M., & Logothetis, N.K. (2006). Functional imaging reveals numerous fields in the monkey auditory cortex. *PLoS Biology*, 4, 7, pp. e215.

[79] Pinchoff, R.J., Burkard, R.F., Salvi, R.J., Coad, M.L., & Lockwood, A.H. (1998). Modulation of tinnitus by voluntary jaw movements. *The American Journal of Otology*, 19, 6, pp. 785-789.

[80] Polimeni, J.R., Fischl, B., Greve, D.N., & Wald, L.L. (2010). Laminar analysis of 7T BOLD using an imposed spatial activation pattern in human V1. *NeuroImage*, 52, 4, pp. 1334-1346.

[81] Price, D.L., de Wilde, J.P., Papadaki, A.M., Curran, J.S., & Kitney, R.I. (2001). Investigation of acoustic noise on 15 MRI scanners from 0.2 T to 3 T. *Journal of Magnetic Resonance Imaging*, 13, pp. 288-293.

[82] Rajan, R., Irvine, D.R.F., Wise, L.Z., & Heil, P. (1993). Effect of unilateral partial cochlear lesions in adult cats on the representation of lesioned and unlesioned cochleas in primary auditory-cortex. *Journal of Comparative Neurology*, 338, 1, pp. 17-49.

[83] Rajan, R., & Irvine, D.R. (1998) Neuronal responses across cortical field A1 in plasticity induced by peripheral auditory organ damage. *Audiology & Neuro-otology*, 3, 2-3, pp. 123-144.

[84] Rauschecker, J.P. (1995). Compensatory plasticity and sensory substitution in the cerebralcortex. *Trends in Neurosciences*, 18, 1, pp. 36-43

[85] Rauschecker, J.P., & Tian, B. (2000). Mechanisms and streams for processing of "what" and "where" in auditory cortex. *Proceedings of the National Academy of Sciences of the United States of America*, 97, 22, pp. 11800-11806.

[86] Roberts, L.E. (2007). Residual inhibition. *Progress in Brain Research*, 166, pp.487-495.

[87] Robertson, D., & Irvine, D.R.F. (1989). Plasticity of frequency organization in auditory-cortex of guinea-pigs with partial unilateral deafness. *Journal of Comparative Neurology*, 282, 3, pp. 456-471.

[88] Rouger, J., Lagleyre, S., Demonet, J. F., Fraysse, B., Deguine, O., & Barone, P. (2011). Evolution of crossmodal reorganization of the voice area in cochlear-implanted deaf patients. *Human Brain Mapping*, doi: 10.1002/hbm.21331.

[89] Rubinstein, J.T., Parkinson, W.S., Tyler, R.S., & Gantz, B.J. (1999). Residual Speech recognition and cochlear implant performance: Effects of implantation criteria. *American Journal of Otology*, 20, 4, pp. 445-452.

[90] Sarant, J.Z., Blamey, P.J., Dowell, R.C., Clark, G.M., & Gibson, W.P.R. (2001). Variation in speech perception scores among children with cochlear implants. *Ear and Hearing*, 22, 1, pp. 18-28.

[91] Schönwiesner, M., Von Cramon, D.Y., & Rübsamen, R. (2002). Is it tonotopy after all? *NeuroImage*, 17, 3, pp. 1144-11461.

[92] Schramm, D., Fitzpatrick, E., & Seguin, C. (2002). Cochlear implantation for adolescents and adults with prelinguistic deafness. *Otology & Neurotology*, 23, 5, pp. 698-703.

[93] Seki, S., & Eggermont, J. J. (2003). Changes in spontaneous firing rate and neural synchrony in cat primary auditory cortex after localized tone-induced hearing loss. *Hearing Research*, 180, 1-2, pp. 28-38.

[94] Sereda, M., Hall, Deborah., Bosnyak, D.J., Edmondson-Jones, M., Roberts, L.E., Adjamian, P., & Palmer, A.R. (2011). Re-examining the relationship between audiometric profile and tinnitus pitch. *International Journal of Audiology*, 50, 5, pp. 303-312.

[95] Sevy, A.B.G., Bortfeld, H., Huppert, T.J., Beauchamp, M.S., Tonini, R.E., & Oghalai, J.S. (2010). Neuroimaging with near-infrared spectroscopy demonstrates speechevoked activity in the auditory cortex in deaf children following cochlear implantation. *Hearing Research*, 270, 1-2, pp. 39-47.

[96] Shellock, F.G., Morisoli, S., & Kanal, E. (1993). MR procedures and biomedical implants, materials, and devices - 1993 update. *Radiology*, 189, 2, pp. 587-599.

[97] Shmuel, A., Yacoub, E., Chaimow, D., Logothetis, N.K, & Ugurbil, K. (2007). Spatiotemporal point-spread function of fMRI signal in human gray matter at 7 Tesla. *NeuroImage*, 35, 2, pp. 539-552.

[98] Shore, S.E. (2011). Plasticity of somatosensory inputs to the cochlear nucleus - Implications for tinnitus. *Hearing Research*, 18, 281, 1-2, pp. 38-46.

[99] Shore, S.E., Koehler, S., Oldakowski, M., Hughes, L.F., & Syed, S. (2008). Dorsal cochlear nucleus responses to somatosensory stimulation are enhanced after noise-induced hearing loss. *The European Journal of Neuroscience*, 27, 1, pp. 155-168.

[100] Striem-Amit, E., Hertz, U., & Amedi, A. (2011). Extensive cochleotopic mapping of human auditory cortical fields obtained with phase-encoding fMRI. *PloS One*, 6, 3, pp. e17832.

[101] Talavage, T.M., Edmister, W.B., Ledden, P.J., & Weisskoff, R.M. (1999). Quantitative assessment of auditory cortex responses induced by imager acoustic noise. *Human Brain Mapping*, 7, 2, pp. 79-88.

[102] Talavage, T.M, Ledden, P.J, Benson, R.R., Rosen, B.R., & Melcher, J.R. (2000). Frequencydependent responses exhibited by multiple regions in human auditory cortex. *Hearing Research*, 150, 1-2, pp. 225-244.

[103] Talavage, T.M., & Edmister, W.B. (2004). Nonlinearity of fMRI responses in human auditory cortex. *Human Brain Mapping*, 22, 3, pp. 216-228.

[104] Talavage, T.M., Sereno, M.I., Melcher, J.R., Ledden, P.J., Rosen, B.R., & Dale, A.M. (2004). Tonotopic organization in human auditory cortex revealed by progressions of frequency sensitivity. *Journal of Neurophysiology*, 91, 3, pp. 1282-96.

[105] Van der Zwaag, W., Francis, S., Head, K., Peters, A., Gowland, P., Morris, P., & Bowtell, R. (2009). fMRI at 1.5, 3 and 7 T: Characterising BOLD signal changes. *NeuroImage*, 47, 4, pp.1425-1434.

[106] Warnking, J. (2002). fMRI retinotopic mapping—step by step. *NeuroImage*, 17, 4, pp. 1665- 1683.

[107] Weisskoff, R.M., Zuo, C.S., Boxerman, J.L., & Rosen, B.R. (1994). Microscopic susceptibility variation and transverse relaxation: theory and experiment. *Magnetic Resonance in Medicine*, 31, 6, pp. 601-610.

[108] Wessinger, C.M., Buonocore, M.H., Kussmaul, C.L., & Mangun, G.R. (1997). Tonotopy in human auditory cortex examined with functional magnetic resonance imaging. *Human Brain Mapping, 5, 1*, pp.18-25.

[109] Woods, D.L, Stecker, G.C., Rinne, T., Herron, T.J., Cate, A.D., Yund, E.W., Liao, I., & Kang, X. (2009). Functional maps of human auditory cortex: effects of acoustic features and attention. *PloS ONE, 4, 4*, pp. e5183.

[110] Yacoub, E., Shmuelm A., Pfeuffer, J., Van de Moortele, P.F., Adriany, G., Andersen, P., Vaughan, J.T., Merkle, H., Ugurbil, K., & Hu, X.P. (2001). Imaging brain function in humans at 7 Tesla. Magnetic Resonance in Medicine, 45, 4, pp. 588-594.

Section 3

Advances in Motor Cortical Network Applications

Chapter 16

The Neurofunctional Architecture of Motor Imagery

Aymeric Guillot, Franck Di Rienzo and Christian Collet

Additional information is available at the end of the chapter

http://dx.doi.org/10.5772/30961

1. Introduction

The remarkable ability of the human brain to mentally simulate sensations and actions has become a major focus of research during the last two decades. Such mental simulation process is called mental imagery and resembles perception, but without the immediate actual corresponding sensory input. Typically, mental imagery is often referred to as the experience of seeing with the mind's eye, hearing with the mind's ear. In other words, mental images can be generated from any sensorial system. Imagery experience is usually understood as evocation, copy, or reconstruction of actual perceptual experience from the long term memory systems, which is first recalled and then internally reproduced within working memory. At other times, participants may also anticipate possible or forthcoming events through imagery as a consequence of what they are expecting. A key feature of this simulation process is that it gives the ability to mentally rehearse motor acts without overt body movement, which is known as the motor imagery (MI) experience. Practically, MI is more often used when the whole human body or body segments are involved, hence requiring participants to imagine the body as the generator of acting forces and not only the effects of these forces on the external world. In this latter case, the information from the body is mainly used to build the mental image of the movement, e.g. tactile or any kind of proprioceptive data, from the muscles, the joints or the inner ear. In the wealth of motor learning literature, it is now well-established that MI is a valuable complement to physical practice to enhance cognitive and motor performance (Driskell et al., 1994; Guillot and Collet, 2008), as well as to promote motor recovery (Sharma, Pomeroy, and Baron, 2006; de Vries and Mulder, 2007). Interestingly, different types of MI can be described and easily combined. While people commonly report using visual imagery (either through a first-person or third-person perspective), auditory imagery, and kinaesthetic imagery (Kosslyn et al., 1990), each perceptual modality can be accompanied by imagery. On the other hand, and with reference to perception, mental imagery combines several cues from different sensorial systems, making the imagery a global sensory-motor experience. Investigating and assessing MI remains difficult due to its concealed nature.

Early on, MI has attracted attention from cognitive neuroscientists. In particular, understanding the neural correlates of goal-directed action, whether executed or imagined, has been an important purpose of cognitive brain research since the advent of neuroimaging techniques such as functional Magnetic Resonance Imaging (fMRI). Providing physiological recordings that correlate to the MI experience was therefore a major issue of the past two decades. In short, but not only, neuroimaging research demonstrated that MI engages motor systems, and that the cerebral plasticity due to actual practice also occurred during MI as well. These findings help to explain why MI can improve actual performance, and further contribute to motor memory consolidation. The present chapter was designed to provide an overview of the neural correlates of MI, with a threefold aim: *i)* to explain why recording human brain functioning with fMRI during MI is of particular interest and provides an objective measurement of imagery processes, *ii)* to give a clear picture of the neural networks mediating MI and actual execution of the same movement, while considering the issue of brain plasticity in the light of individual differences in imagery ability, motor expertise, the different forms of imagery, and the level of motor learning, and *iii)* to consider the recent fMRI methodological advances, e.g. using real-time fMRI, as well as recording changes in functional and effective connectivity in the neural networks activated during MI.

2. Using fMRI to assess brain activity during motor imagery

The most common aim of fMRI data analysis is to investigate the neural processes mediating higher cognitive functions, and detect correlations between brain activations and the task the participants have to perform during the scan. Since the early 1980s, fMRI has come to dominate the MI-related brain mapping research. During typical fMRI image acquisitions, participants are asked to alternatively perform a movement, either physically or mentally. These experimental conditions are also compared with a rest/control condition.

There are two main types of experimental design in fMRI: the block design and the event-related design. The block design paradigm consists of several epochs representing experimental conditions, during which stimuli are presented and actual execution or MI are required from participants, which are interleaved with time blocks of rest. Event-related designs rather associate brain processes with discrete events, with stimulus events being randomly presented one at a time, and being separated by an inter-stimulus interval of a pre-determined length. Habituation effects can be reduced, and stimuli are usually presented either tens of seconds apart or at a faster rate (e.g., every second). It could be noted that event-related designs are more sensitive to the details of the hemodynamic response model use. Figure 1 provides a schematic illustration of these two experimental designs when comparing the neural networks mediating MI and actual execution of the same movement.

There is not a universal procedure yet to acquire and process functional data and many researchers independently developed their methods and analyses, using different software. Practically, the most commonly technique used to map brain activity is based on the Blood Oxygenation Level Dependent (BOLD) effect, measuring regional differences in hemoglobin oxygenation. Therefore, a considerable amount of experimental studies investigated BOLD changes during MI. They primarily aimed at comparing the neuronal activations recorded

http://dx.doi.org/10.5772/30961

during MI and actual practice of the same movement. For example, Grèzes and Decety (2001) provided evidence of a functional equivalence between intending, simulating, observing, and performing an action, as shown by the great deal of overlap among the corresponding neural substrates. Interestingly, brain activations recorded during MI do however not exactly match those of actual movement execution (Lotze and Halsband, 2006). Some differences also appear when comparing different types of MI or with regard to the individual features of the sample of participants. Put simply, fMRI allows the measurement of different neuronal activity in participants with high vs. low MI ability (Guillot et al., 2008), as well as differences in the patterns of cerebral activity related to the degree of expertise of the participants with the motor task (Lotze et al., 2003; Ross et al., 2003; Milton et al., 2007, 2008). Based on such evidence, fMRI offers a way to evidence the MI experience (Guillot et al., 2008). Finally, recent fMRI data suggested that different types of imagery elicit different patterns of neural activation (Solodkin et al., 2004; Jackson et al., 2006; Guillot et al., 2009; Lorey et al., 2009), while others nicely demonstrated, through positron emission tomography (PET), that the cerebral plasticity occurring after physical practice was reflected during MI (Lafleur et al., 2002; Jackson et al., 2003), hence reinforcing the principle of functional equivalence between MI and physical performance of the same task. Before we consider these issues and review the main related findings in the next section, an important methodological question remains to be addressed. Specifically, what are the additional benefits provided by fMRI to the study of MI compared to other neuroimaging and electrophysiological techniques, and, in turn, what are the limitations of fMRI recordings?

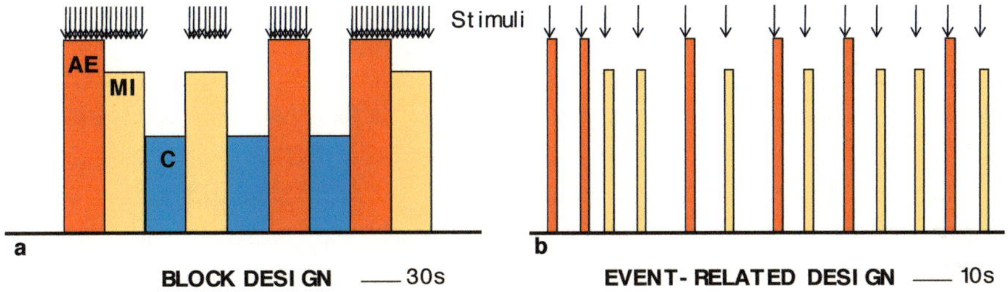

Figure 1. Schematic representation of a block design (a) and an event-related design (b) fMRI paradigm. In the block design paradigm, sustained periods of 20–30s are used for actual execution (AE), motor imagery (MI), and control (C) conditions during which several trials are performed successively (vertical arrows). Before each block of practice, brief periods of time (2-3s) are necessary to provide participants some instructions related to the forthcoming condition. Brief 'rest' periods can also be included. In the event-related design, each experimental condition is presented at random (orange and red vertical bars and vertical arrows).

A first advantage of fMRI, compared to PET, is its non-invasive nature. There is no radiation exposure, and no need for injection of radioactive materials with fMRI. Another strength is its high spatial resolution allowing the measurement of brain activity from deep structures. Unlike electrophysiological techniques such as electroencephalography (EEG) or even magnetoencephalography (MEG) which are somewhat biased towards the cortical surface, fMRI records signal from all regions of the brain, including subcortical structures. This is of particular importance when pinpointing the neural correlates of MI as activation of cerebellum and basal ganglia is commonly reported. As every technique, however, fMRI has also some methodological disadvantages or weaknesses for the study of MI (e.g., Dietrich, 2008). A first limit is related to the causality between brain activation and the task the participants perform in the scanner. Although the baseline activity is subtracted from that recorded during the task and the result is averaged for all participants, the causality between activated structures and characteristics of the task is not straightforward. In other words, some brain structures could remain activated although this activity is not directly related to the task. Such argument for the causality however also holds for other brain imaging techniques. Second, the temporal resolution, of about several seconds must also be considered, as fMRI cannot track the rapid temporal dynamics of the functioning brain compared to high temporal resolution methods, e.g. EEG or MEG. Neuromodulatory effects of some cognitive functions such as attention and memory are also likely to affect the spatiotemporal resolution of the signal as they can affect large masses of cells, and potentially induce larger changes in the fMRI signal than the sensory signals themselves (Logothetis, 2008). Fourth, regarding data interpretation, advanced statistical methods are required to identify active voxels, while modifications of the brain activity might be due either to an excitatory or inhibitory influence. Finally, the position in the scanner and the limitation of body movements make it difficult to investigate complex motor skills. Indeed, the participant is lying in supine position in the scanner, and then performs both actual execution and mental representation of the same movement. However, in such situation, only simple movement of fingers, hands, feet, or tongue, or more complex movements of limited amplitude can actually be studied.

As reported by Logothetis (2008), and despite the shortcomings mentioned above, fMRI remains probably the best current available technique for providing insights into brain function and formulating interesting testable hypotheses. Logothetis also stated that some limitations of fMRI are directly due to inappropriate experimental protocols that ignore the circuitry and complex functional organization of the human brain. This is a critical aspect that must be considered before drawing general conclusions and when comparing results across experimental studies. In particular, a specific attention should be paid to the control condition from which MI-related brain activity is reported. In fMRI data analyses, images of activity from experimental and control conditions are subtracted to determine significant peaks of activation. Hence, a difference which does not reach the significant threshold does not necessarily reflect a lack of activity, but can also be due to similar brain activations during both experimental and control tasks. In other words, if the control task is not thoroughly determined, the remaining activity of some brain regions might bias the pattern of activity observed during the experimental condition, i.e. MI. Inspection of the MI literature reveals great differences across studies. Among others, participants were asked either to listen for auditory stimuli without engaging in MI (perceptual control condition), having rest, projecting into a restful state, fixating a static stimulus on a screen, or even imagining a 'neutral' static motor task. Obviously, such differences might significantly influence the pattern of activity recorded when contrasting the experimental and the control conditions. This is likely to render the functional brain map, hence hindering the conclusions that one can reach. Another example is the fact that MI can be performed either with open or closed eyes. Marx et al. (2003) suggested that eye closure is likely to improve imagination, but in some circumstances, MI paradigms might require participants to open their eyes to fixate a screen and/or wait for instructions. As some activation or deactivation of cortical regions could be undetected with eyes open, however, MI could ideally be performed with closed eyes as it was asked for the rest condition (Marx et al., 2004).

3. Neural correlates of motor imagery

MI research provided convincing evidence that there is a functional equivalence between MI and motor performance (e.g., Jeannerod, 1994; Grèzes and Decety, 2001; Holmes and Collins, 2001). It is therefore not surprising that movement execution and MI reveal a high overlap of active brain regions, i.e. imagery draws on almost the same neural network that is used in actual perception and motor control (Murphy et al., 2008). Considerable evidence in support of the functional equivalence theory comes from fMRI studies which demonstrated that both cortical and subcortical brain areas are activated during MI. All studies showed that secondary motor areas are recruited during MI, but that the primary motor cortex might also be part of the network involved in imagined actions. Cerebellar and basal ganglia activations are also consistently reported during MI (Gerardin et al., 2000; Munzert et al., 2009). Finally, apart from motor-related areas, other regions such as parietal lobules are activated during mental simulation of actions (Gao et al., 2011). Taken together, these data support that MI and motor performance share the same neural substrate (Figure 2), but interestingly there are also some differences within the pattern of activity in these areas, as well as with regards to the characteristics of the participants or the content of the MI task.

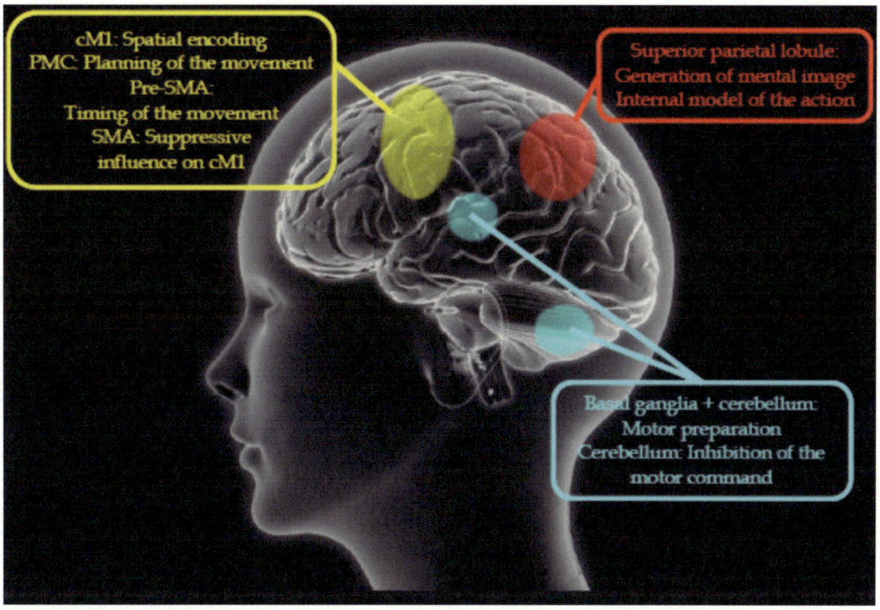

Figure 2. Schematization of brain activity during MI and possible roles of motor-related regions.

3.1. Contribution of the primary motor cortex

Since the pioneering PET study by Roland et al. (1980), who did not find a significant activation of the contralateral primary motor cortex (cM1) during MI, the question of the contribution of cM1 during imagined actions has attracted considerable attention and remains controversial (for reviews, see Lotze and Halsband, 2006; Sharma et al., 2008; Munzert et al., 2009; Lotze and Zentgraf, 2010). For example, Binkofski et al. (2000), Gerardin et al. (2000), Kuhtz-Bushbeck et al. (2003) and Hanakawa et al. (2003, 2008) failed to find peaks of activation in cM1, while Dechent et al. (2004) reported fleeting involvement. In contrast, several studies reported significant activation in cM1 during MI (Leonardo et al., 1995; Sabbah et al., 1995; Porro et al. 1996, 2000; Roth et al., 1996; Lotze et al., 1999; Ehrsson et al. 2003; Nair et al., 2003; Solodkin et al. 2004; Guillot et al., 2008, 2009). In a seminal paper, Ehrsson et al. (2003) even showed that the content of MI was reflected in the pattern of motor cortical activation. Accordingly, they found that MI of hand, foot and tongue movements specifically activated the corresponding hand, foot and tongue sections of cM1. Transcranial magnetic studies also revealed cM1 activation during MI (for review, see Stinear et al., 2010). Finally, and as previously reviewed by Munzert et al. (2009), similar cM1 activation was found during MI of movements with the impaired limb to that during motor execution with the unimpaired limb in amputees (Ersland et al., 1996). The same result was found in healthy controls during motor execution of the same movement and in patients with spinal cord injury (Alkadhi et al., 2005; Cramer et al., 2005). The fact that these patients do not need to inhibit the movement might have influenced the nature of the MI process (Munzert et al., 2009). A recent MEG study however revealed similar patterns of activity in cM1 during MI and actual practice of the same movement in a patient with spinal cord injury, while a loss of power was observed in the matched healthy control participant (Di Rienzo et al., 2011). These latter data support fMRI findings and further confirm both the contribution of cM1 during MI and the weakened inhibitory processes during MI in patients with spinal cord injury. Cerebral activations governing the inhibition of action during MI have not been yet fully explored. This would probably offer fruitful fMRI investigations in the future.

With regards to cM1 activation, Sharma et al. (2006) suggested that discrepancies in results may be due, at least partially, to methodological differences and difficulties in monitoring compliance with MI instructions. Definitely, the advent of fMRI techniques to investigate the contribution of cM1 during MI brought significant data into the debate. Furthermore, Sharma and colleagues argued that previous experimental studies did not specifically address whether the subdivisions of cM1 were differentially involved during MI. Furthermore, the primary motor cortex can be subdivided into an anterior component, thought to be executive in nature, while its posterior part would be involved in cognitive tasks or non-executive functions (Sanes and Donoghue, 2000). Accordingly, Sharma et al. (2008) reported that the cluster distribution in the anterior part of cM1 was significantly reduced during MI compared to the physical execution, while that of the posterior part was similar. Therefore, they proposed that the role of cM1 and its subdivisions may be related to spatial encoding. Despite these results, determining how MI participates to the activation of cM1 remains difficult. Typically, cM1 activation during MI is usually smaller compared to that during overt motor execution, and is not systematically observed in all participants (Munzert et al., 2009). It was also demonstrated that cM1 was increasingly involved along with the complexity of movements (Kuhtz-Buschbeck et al., 2003). As well, activation of cM1 might be differentially influenced by MI instructions, MI ability and motor expertise (Lotze and Zentgraf, 2010).

Taken together, and despite some controversial data, fMRI studies support that cM1 is actually activated during MI but more weakly than during actual movement. Such activation is not essential for imagery and the neurons are not in the same location than those active during movement execution (Lotze and Halsband, 2006). Moreover, cM1 might not only have an execution function for the motor system (Lotze and Zentgraf, 2010), and it is particularly preparation for MI that may impact significantly cM1 activation (Johnson et al., 2002; Zang et al., 2003; Munzert et al., 2009).

3.2. Activation of the other motor-related regions

Unlike the contribution of cM1, there is a general consensus regarding the activation of secondary motor areas of the cerebral cortex during MI. Accordingly, fMRI studies provided strong evidence that motor-related areas such as the ventral and dorsal parts of the premotor cortex (PMC), the supplementary motor area (SMA) and the pre-SMA, are active during MI of both simple and complex movements as well as subcortical areas such as the cerebellum and the putamen. Overall, it was shown that MI might activate a subset of the areas required for movement execution, albeit to a lesser extent (e.g., Macuga and Frey, 2011). Conversely, other activations could also be more robust in MI than during motor execution (e.g., Gerardin et al., 2000) thus leading to a partial overlapping of the networks mediating MI and actual execution of the same movement.

The activation of the SMA is now well-established during MI (Guillot et al., 2008; 2009; Hanakawa et al., 2003; Lotze et al., 1999; Munzert et al., 2009; Olsson et al., 2008; Solodkin et al., 2004). As neurons in the SMA are integrated in the functional loop controlling motor actions planning, it is therefore quite reasonable that similar preparatory aspects of the movement are engaged during MI. Kasess et al. (2008) further demonstrated that the SMA may inhibit activity of cM1 to prevent movement execution during MI. Interestingly, some authors also reported that the overlap of SMA activity during MI and motor execution is partial, and that the pre-SMA might rather play a specific role during imagined movements. Hence, the pattern of activity in this region could be higher during MI than during actual practice of the same task (Gerardin et al., 2000). Pre-SMA is known as being involved in acquiring new sequences, planning spatio-temporal actions, and updating motor plans for subsequent temporally ordered movements. As suggested by Malouin et al. (2003), the pre-SMA might thus be activated during MI to provide proper movement sequencing and timing. Taken together, MI would activate both the SMA and pre-SMA, not only for the preparation of the movement, but also for its suppressive influence on cM1 thus preventing movement execution.

Overlapping activity during MI and motor execution was also found in the premotor cortex (PMC-Guillot et al., 2008; Lotze and Halsband, 2006; Munzert et al., 2009), with potential more robust activity in the ventral part of the PMC during MI (Gerardin et al., 2000). The dorsal part of the PMC is involved in the preparation and control of movement, while the ventral part may play a crucial role in the planning of movements. Neurons with mirror properties which are active during action imitation and recognition were found in this region neighbouring Brodmann areas 44/45 (Rizzolatti et al., 1998). This may partially explain why more robust activity was reported in this area during MI than during motor execution.

Several studies also reported that MI and motor execution activate subcortical regions such as the basal ganglia and the cerebellum (e.g., Gerardin et al., 2000; Guillot et al., 2008; Lotze and Halsband, 2006; Munzert et al., 2009; Nair et al., 2003). These structures are known to strongly contribute to motor learning and motor preparation, and are closely connected to cortical motor-related areas. More specifically, the cerebellum contributes to predict movement outcomes as well as to correct the movement on the basis of sensori-motor feedback. Evidence that the cerebellum participates in the internal representation of movement, and therefore to MI, also comes from studies showing that cerebellar stroke disrupts MI (Battaglia et al., 2006; Gonzalez et al., 2005). In a seminal paper, Lotze et al. (1999) even postulated that the posterior cerebellum might play a crucial role in the inhibition of the motor command during MI (see also Lotze and Halsband, 2006). Likewise, basal ganglia play a substantial role in the storage of learned motor sequences, as well as in motor preparation (Alexander and Crutcher, 1990). While there is a general consensus regarding the activation of the basal ganglia during MI, few studies were devised to determine the specific role of this region. Li (2004) demonstrated that lesion of the putamen is likely to impair MI. More generally, however, studies on the effect of basal ganglia dysfunction on MI revealed inconsistent results (Heremans et al., 2011), with possible, but not systematic, MI impairment in patients with Parkinson's disease. Taken together, and even if experimental studies still have to understand in greater details how the cerebellum and the basal ganglia specifically contribute to the MI process, there is no doubt about their activation during imagined movements. Interestingly, distinct contributions of the cortico-striatal and cortico-cerebellar anatomical systems have been proposed in the motor learning literature (Doyon and Ungerleider, 2002; Doyon and Benali, 2005). Although functional interactions between these anatomical systems are essential at the beginning of the learning process, there is ample evidence that the cerebellum plays a weaker role when the sequence is well learned and has reached asymptotic performance (for review, see Doyon and Benali, 2005), while basal ganglia remain active during the later stages of motor learning. By comparing the pattern of cerebral activations in 13 skilled and 15 unskilled imagers during both physical execution and MI sequence of finger movements, Guillot et al. (2008) showed that poor imagers not only need to recruit the cortico-striatal system, but also activate the cortico-cerebellar system as well. Findings also suggested that compared to poor imagers, good imagers would have a more efficient recruitment of movement engrams. Although this remains a working hypothesis awaiting experimental investigation, the authors concluded that the pattern of cerebral activation recorded during MI in poor imagers might improve and evolve close to that observed in good imagers after MI training. This might suggest that the expected changes in subcortical brain activations during MI would reflect those elicited by the process of actual motor learning (see also Lafleur et al., 2002; Jackson et al., 2003).

http://dx.doi.org/10.5772/30961

3.3. Activation of parietal regions

Activation of parietal areas including possibly the somatosensory cortex, but more certainly the inferior and superior parietal lobules and the precuneus, is frequently reported during MI (Binkofski et al., 2000; Gerardin et al., 2000; Guillot et al., 2009; Hanakawa et al., 2003; Lotze et al., 1999; Munzert et al., 2009; Nair et al., 2003). More importantly, the parietal cortex, and perhaps predominantly the left parietal cortex, could play a critical role in the formation of mental images. Experimental studies in patients with lesions located in the superior region of the parietal cortex showed that the temporal congruence between actual and MI times, known as being a reliable measure of MI quality (Guillot and Collet, 2005; Guillot et al., 2011; Malouin et al., 2008), was affected (Malouin et al., 2004; Sabate et al., 2007; Sirigu et al., 1996). Sirigu et al. (1996) postulated that the parietal cortex might set up an internal model of the forthcoming movement making the participants to predict the expected movement outcome. After parietal damage, coding for the spatial properties of the movement to be imagined and predicting the temporal feature of the movement might therefore be more difficult, hence reflecting the critical role of the posterior parietal cortex in the use of MI.

Interestingly, bilateral parietal lesions resulted in a complete unawareness of movement execution during imagery (Schwoebel et al., 2002). In fact, the patient exhibited hand movements during MI of the same body segments, but explicitly denied that they occurred. These data strongly support that parietal areas are critically involved in the generation and guidance of imagined movements. More importantly, they further suggest that the patient failed to inhibit the motor consequences of MI, and therefore, that parietal areas may also play a critical role in the inhibitory processes of the motor command during MI. Finally, and in conjunction with previous fMRI data (Gerardin et al., 2000), these findings suggest a functional dissociation between cortical areas underlying MI and motor execution. The primary sensory area would be more activated during motor execution, due to sensory feedback processes, whereas MI might elicit greater activity both in the inferior and superior parietal cortices, especially in the left hemisphere.

3.4. The effect of motor imagery content

A limited number of fMRI studies examined the influence of MI content on brain activations, and most specifically whether visual imagery and kinaesthetic imagery recruit different neural networks. In a first study looking at this issue, Binkofski et al. (2000) reported that the anterior part of the intraparietal sulcus was more active during kinaesthetic imagery of finger movements, while visual imagery yielded stronger activation in the posterior part. Bilateral activations were also reported during kinaesthetic imagery in the opercular portion of the ventral PMC. In a seminal paper, Solodkin et al. (2004) later investigated the neural networks mediating physical execution, visual imagery and kinaesthetic imagery of hand movements. Although these two types of imagery shared similar neural substrates, including the connection from the superior parietal lobule to the SMA, the main difference was found in the inputs from the superior parietal lobule and the SMA to cM1, which were opposite to those observed during motor execution. In the same line of research, Guillot et al. (2009) examined whether the same group of participants with high imagery abilities recruited similar or distinct brain activations during visual imagery and kinaesthetic imagery of complex hand movements. Visual activated predominantly the visual pathways including the occipital regions and the precuneus, whereas the pattern of kinaesthetic imagery involved mainly motor-associated structures and the inferior parietal lobule (Figure 3). These data support the hypothesis that visual imagery of a motor sequence refers to the visual properties of visual perception, while kinaesthetic imagery includes in greater extent motor simulation processes related to the form and timing of actual movements (Michelon et al. 2006). Although visual and kinaesthetic imagery shared common brain structures, these data provide strong evidence that the patterns of neural activity mediating the ability to perform a specific MI type are partially different. Practically, it might suggest that participants are able to favor one sensory modality to form mental images, although MI is a multimodal and multidimensional construct.

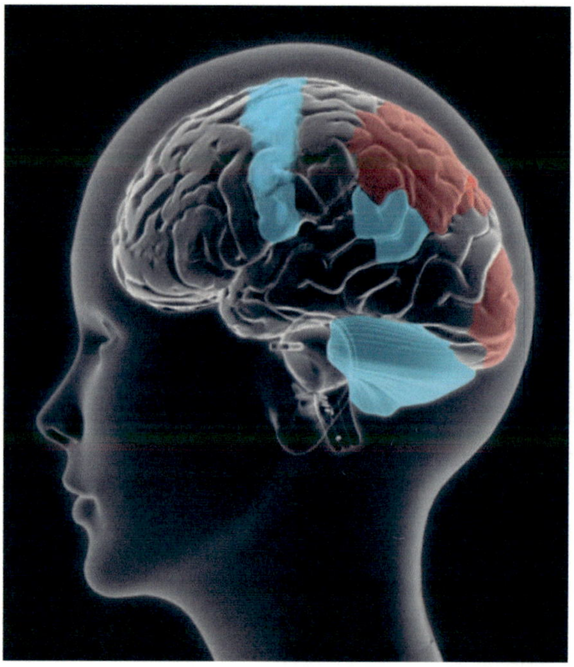

Figure 3. Schematization of brain activations during visual and kinaesthetic imagery. During kinaesthetic imagery (blue), the pattern of activity involves motor-related regions mainly, including cM1, PMC, SMA, cerebellum and basal ganglia (hidden from the view), as well as the inferior parietal cortex. In contrast, more activity is observed during visual imagery (red) in occipital regions and superior parietal lobule, including the precuneus.

Similar data have been reported in regards to the visual imagery perspectives, that is, by comparing first and third person imagery perspectives. The first study looking at this issue, was conducted by Ruby and Decety (2001) with PET scan. They investigated the neural networks mediating MI when the participants were asked either to mentally simulate an action or to imagine someone else (*i.e.,* the experimenter) performing the same movement. They found that the inferior parietal lobule and the somatosensory cortex were more activated during the first-person than the third-person perspective, while increased activity was observed in the precuneus and during the third-person perspective. Then, fMRI experiments were designed to examine in greater details whether using either the first or third-person imagery perspective lead to motor-areas-related changes. Jackson et al. (2006) explored the neural circuits involved in imitation and perspective-taking, and showed that first-person perspective was more tightly coupled to the sensory-motor system than the third-person perspective, which requires additional visuospatial transformation. More recently, Lorey et al. (2009) also found stronger activation of motor and motor-related brain structures, like the inferior parietal lobe, when imagining a movement through a first-person compared to a third-person perspective. They argued that proprioceptive information on actual body posture would be more relevant for internal visual imagery users, which reveals that bodily information is mainly integrated into the image of one's own body movement.

More generally, these results provide clear evidence that both imagery types (e.g. visual vs. kinaesthetic imagery) and perspectives (first-person vs. third-person imagery) are partially mediated through separate neural systems, which may therefore contribute differently during the process of motor learning and neurological rehabilitation. This assumption is supported by a remarkable fMRI study demonstrating that MI engages systematically the organized sections of cM1 in a somatotopic manner, i.e. that the content of the mental images is reflected in the pattern of motor cortical activation (Ehrsson et al., 2003).

3.5. Neural mechanisms for motor expertise

The fact that MI efficacy depends on the individual ability to form accurate mental images is now well-established. Although not systematic, this capacity to mentally simulate forthcoming actions may be influenced by the individual level of expertise in the corresponding motor task. While the concepts of motor expertise and MI ability measurements have been considered early on, researchers are able, for a short time, to assess the content of MI objectively using thorough procedures, in particular since the advent of functional brain mapping studies (Guillot et al., 2010). Taken overall, fMRI data strongly support the existence of distinct neural mechanisms of expertise in MI, as a function of the individual skill level. For instance, lower cortical activation was recorded in professional violinists as compared to amateurs (Lotze et al., 2003). Similar differences were observed in the neural networks mediating MI in novice and expert athletes (Milton et al., 2008). By comparing brain activations of six golfers of various handicaps during MI of a golf swing, Ross et al. (2003) found decreased activations of the SMA and cerebellum as a function of golf skill level, i.e. an inverse relationship between brain activity and skill level. Also, golf swing MI yielded few peaks of activation in the basal ganglia and cingulate gyri across all skill levels. Milton et al. (2007) finally reported that the posterior cingulate, the amygdala-forebrain complex, and the basal ganglia were activated in novices but not in elite golfers during motor planning of a golf swing movement, hence confirming that the neural networks controlling both motor planning and MI are influenced by the individual skill level. As a whole, changes in cerebral activations confirmed previous investigations showing that levels of expertise during the motor learning process are supported by different neural networks (Doyon and Benali, 2005).

Analogously, researchers investigated the neural networks mediating musical experience, training onset, and training stages. Langheim et al. (2002) first showed in participants who imagined playing a musical selection with their instrument that an associative network including the superior parietal lobule, the inferior frontal gyrus and the bilateral lateral cerebellum was activated. This network would be particularly activated during the coordination of the complex spatial and timing components of musical performance. As mentioned above, Lotze et al. (2003) compared the patterns of brain activation during auditory imagery in experienced and novice musicians, with the professionals reporting frequent use of imagery with high vividness. Interestingly, experienced musicians recruited very few cerebral areas, while amateurs manifested a widely distributed activation map. In the professional group, however, more activation was observed in the SMA, the PMC, the superior parietal lobule and the cerebellum. Finally, Kleber et al. (2007) reported an increased activation in the fronto-parietal regions during imagined singing, suggesting increased involvement of working memory processes during imagery.

To summarize, fMRI data provided strong evidence that partially separate neural networks are activated during MI of both motor and musical performance in regards to the individual differences in the level of expertise. These findings therefore strongly support the hypothesis of distinct neural mechanisms for expertise in imagery, independently of the imagery type, with the network integrating the superior parietal lobule mediating the imagery activity of highly experienced people.

4. fMRI advances in the study of motor imagery

Going beyond the usual comparison of brain activation patterns that are associated with MI and motor execution of the same movement is the next step in this field of research. To do so, researchers may take advantage of the recent methodological fMRI developments. Both the use of real time fMRI and the recording of changes in functional and effective connectivity in the neural networks activated during MI are of particular interest. Also, reporting more systematically negative BOLD responses might contribute to expand our knowledge on neuronal inhibition during MI.

It is now possible to image human brain functioning in real time with fMRI (Esposito et al., 2003; Weiskopf et al., 2003; for an extensive review, see deCharms, 2008). This technique provides a reconstruction of the raw data obtained with the brain scan, while the scan is going on. There are several exciting research perspectives for considering the contribution of real time fMRI dedicated to MI study. This technique may first be useful during the MI learning process, especially when the pattern of activation during mental simulation is not the one expected. Indeed, real time images are expected to provide participants some objective information related to the vividness of MI, since they have been educated to gain some familiarity with the neuroanatomy before neurofeedback fMRI sessions. Receiving a feedback from brain activation in predetermined regions of interest seems possible. An average 2-5s time lapse usually remains necessary, due to the physiological delay of the Hemodynamic

http://dx.doi.org/10.5772/30961

response (a more simple feedback can also be used, e.g. a simple score using a Likert-type scale). A well-known study supporting the use of real time fMRI was published by Yoo and Jolesz (2002), who used visual feedback of brain activation to guide participants to adjust their motor task performance and to achieve the desired modulation of cortical activity. For example, during simple hand movements, participants spontaneously involved more muscle groups and increased tapping frequency along fMRI sessions. The biofeedback given to participants with regard to the activated neural network helped them to modulate their cortical activation. A significant illustration of the strength of this methodology in the field of MI was offered by deCharms et al. (2004) during imagery of a manual action task. In this study, participants received feedback about the activation level in the somatomotor cortex with a simple virtual reality interface. The results showed that they enhanced the fMRI level of activation driven by MI in the somatomotor cortex through the course of training. Moreover, the activation of this region after MI training was as robust as that recorded during actual practice. In a more recent study, Yoo et al. (2008) showed that real time fMRI might help individuals to learn how to increase region-specific cortical activity associated with a MI task. Practically, the level of increased activation in motor areas was consolidated after the 2-week self-practice period. Regarding the method for presentation of neurofeedback (intermittent presentation or continuous presentation) and the nature of the neurofeedback being presented to participants during a MI task (true or false neurofeedback regarding brain activations), Johnson et al. (2010) further reported that the intermittent presentation of feedback was more effective than the continuous presentation in promoting self-modulation of brain activity. Accordingly, regular interruptions in neurofeedback presentation allowed central processing and integration of the information conveyed in the feedback regarding brain activations. The authors also reported that false feedback resulted in irrelevant brain activations with regards to the regions of interest targeted by experimental instructions. Finally, Xie et al. (2011) supported the effectiveness of delivering neurofeedback during MI using real time fMRI, and further provided evidence that the SMA was controllable by participants. These data strongly support that real time fMRI is a valuable technique to investigate whether participants are able to use a cognitive strategy to control a target brain region in real time. In the field of neurorehabilitation, for example, a similar approach could be used to learn how controlling pain by learning to control the brain regions that mediate pain perception (deCharms et al., 2005). Indeed, there are multiple therapeutic applications of real time fMRI but it is too early to predict success or failure and new experimental results are awaited.

Real time fMRI might also be particularly useful in developing brain-computer interfaces (deCharms, 2008). A brain-computer interface is a novel communication system that translates human thoughts or intentions into a control signal without using any muscle activity (for review, see Pfurtscheller and Neuper, 2010). To date, many brain-computer interface systems have used MI tasks to modulate sensorimotor EEG activity taken to operate and control an external device. Pfurtscheller and Neuper (2010) nicely demonstrated how brain-computer interface systems using MI can contribute to help patients with various motor impairments and paralysis. Despite the temporal limitations of fMRI, using fMRI data for brain-computer interface remains plausible as many cognitive processes change slowly, over seconds or minutes (deCharms, 2008). Practically, the method is appealing although future experimental studies are necessary to determine its feasibility and effectiveness, such as during motor recovery of patients with motor impairment. Real time fMRI might also be used to explore the state of consciousness and communicate with patients in a persistent vegetative state. Accordingly, Owen et al. (2006) detected awareness in such a patient following instructions to mentally imagine moving around a house and playing tennis. Brain activations were observed in the parahippocampal gyrus, the posterior parietal lobe and the lateral PMC during MI of walking, and in the SMA when imagining playing tennis. These data provide evidence that real time fMRI might be used along with MI to actually communicate with people and/or patient who are physically or conventionally unable to interact with their environment (deCharms, 2008).

Creating functional connectivity maps of distinct spatial distributions of temporally correlated brain regions is another methodological advanced tool offered by fMRI. Functional and effective connectivity can be used to examine interactions among brain regions. Practically, these techniques go beyond the usual activation maps obtained through peak-detection methods, by looking respectively at temporal correlation between the time course of activation of two regions, and the influence of one neuronal population over another (Doyon and Benali, 2005; Friston and Büchel, 2003; Friston et al., 1995). Multiple innovative data-driven methods have been proposed to investigate the changes observed in cerebral networks over time, or to study functional and effective connectivity. Some reliable examples are the structural equation modeling and the dynamic causal modeling. In the field of MI, few researchers examined the inter-relationships among brain areas selectively activated along different experimental conditions, that is, the effective connectivity between network components. Solodkin et al. (2004) explored the effective connectivity of the neural networks mediating motor execution, visual imagery, and kinaesthetic imagery of a finger to thumb opposition task. Their results provided evidence that the networks underlying these behaviors were almost different, despite the extensive overlap between motor execution and kinaesthetic imagery. In particular, inputs from SMA and lateral PMC to cM1, which were facilitated during motor execution, exhibited the opposite activity during kinaesthetic imagery, suggesting a physiological mechanism whereby the system prevents overt movements. A second study looking at the effective connectivity between SMA and cM1 suggested that the lack of activation in cM1 during MI might result from a suppressive influence of the SMA (Kasess et al., 2008). More recently, Gao et al. (2008) and Chen et al. (2009) reported that, in right-handed participants, more brain regions showed effective connections to the SMA during right-hand MI than during left-hand MI, but that the strength of the casual influence was stronger during left-hand MI. Furthermore, they found forward and backward effective connectivity between the SMA and the bilateral dorsal PMC, the contralateral primary and somatosensory cortex, and cM1. A last study by the same group of researchers confirmed these findings, and further showed that motor execution has some increased causal connections because of additional processes for the overt behavior stage (Gao et al., 2011). Taken together, these experimental studies highlight the advantages of studying functional and/or effective connectivity through fMRI to expand our understanding of the neural underpinnings of MI.

http://dx.doi.org/10.5772/30961

Reporting an elevation in the fMRI BOLD signal, namely positive BOLD, has become a common routine for mapping neural activity in the human brain. Aside from positive BOLD signal changes, several studies have also observed negative BOLD responses. The negative BOLD signal is a physiological process correlated with a corresponding decrease in cerebral blood flow, oxygen consumption, and neuronal activity (for review, see Shmuel et al., 2002). Practically, negative BOLD signal might reflect less activation for a given task as compared to a baseline condition. The negative BOLD response might be caused by a reduction in cerebral blood flow and is associated with decrease in the rate of oxygen consumption. From a functional viewpoint, the negative BOLD response is further considered to carry information content that is stimulus-specific (Bressler et al., 2007). Its neurobiological mechanisms are yet not well understood, and much of the debate has centered on whether its source is primarily vascular or neuronal. The neuronal origin might reflect a suppression of the local neuronal activity and/or a reduction in the afferent input, whereas the vascular origin refers to a reduction in cerebral blood flow to the less demanding regions due to the increase in flow to the demanding areas, without a necessary decrease in neuronal activity in the negative regions (Schmuel et al., 2002). Until recently, very few experimental studies were designed to analyze negative BOLD responses during either visual or MI tasks. An interesting and innovative paper by Amedi et al. (2005) investigated the pattern of brain deactivation during visual imagery and compared it to the neural correlates of visual perception. While they found that visual imagery and visual perception share similar neural substrate, these two conditions yielded different brain-deactivation profiles as shown by negative BOLD responses. Of particular interest is that the authors reported a robust deactivation in early auditory areas as well as a selective deactivation in the somatosensory cortex during visual imagery, hence supporting that visual imagery is associated with deactivation of non-visual sensory processing. Based on these data, they stated that what they named 'pure visual imagery' might be characterized by an isolated activation of visual cortical areas with concurrent deactivation of sensory inputs that may potentially disrupt the image created by the mind's eye. They also considered that deactivation could be the consequence of filtering out irrelevant stimuli. The correlation between the level of deactivation and the vividness of visual imagery might support the hypothesis that participants with high imagery ability are able to shut down or disconnect the 'useless' cortices. To the best of our knowledge, unfortunately, there are no experimental studies specifically designed to look for similar results in MI paradigms *per se,* since most experimental work looking at the negative BOLD signal primarily focused on stimulus-based sensorial tasks rather than cognitive processes such as MI. Interestingly, it has been suggested that the prolonged negative BOLD signal reflects an inhibition or suppression of neuronal firing patterns (Raichle, 1998; Kastrup et al., 2008). Looking more systematically for negative BOLD responses might therefore be of particular interest to investigate the inhibitory processes of the motor command during imagined movements. Indeed, motor command inhibition mechanisms during MI remains rather unclear since no direct fMRI brain signal is known to characterize neural inhibitory processes, and inhibition during MI has been primarily studied from brain activation signals. The negative BOLD response might therefore be an interesting methodological complement to extend our knowledge of motor inhibition.

5. Conclusion

The data reviewed in this chapter strongly support that recording human brain functioning with fMRI during MI provides an objective measurement of the neural networks underlying MI processes. Although they are not totally overlapping, it is now well-established that the neural substrates mediating MI and actual execution of the same task are quite similar. Spurred by recent fMRI methodological advances, the next step of MI research will certainly contribute to understand in greater details the neural correlates of imagined movements, as well as the inhibition of the motor command. MI studies could therefore benefit from the use of real time fMRI, effective connectivity, and also negative BOLD. This latter technique might be of particular interest in investigating the inhibitory processes of the motor command during MI. Cerebral regions which are inhibited during mental operations or human behavior are certainly of the same scientific interest than those which are activated.

Author details

Aymeric Guillot*, Franck Di Rienzo and Christian Collet

Centre de Recherche et d'Innovation sur le Sport, Performance Motrice, Mentale et du Matériel (P3M), Université de Lyon, Université Claude Bernard Lyon1, France

References

[1] Alexander, G.E., and Crutcher, M.D. (1990). Functional architecture of basal ganglia circuits: neural substrates of parallel processing. *Trends in Neuroscience, 13*, 266-271.

[2] Alkadhi, H., Brugger, P., Boendermaker, S.H., Crelier, G., Curt, A., Hepp-Reymond, M.-C., and Kollias, S.S. (2005). What disconnection tells about motor imagery: evidence from paraplegic patients. *Cerebral Cortex, 15*, 131–140.

[3] Amedi, A., Malach, R., and Pascual-Leone, A. (2005). Negative BOLD Differentiates Visual Imagery and Perception. *Neuron, 48*, 859-872.

[4] Battaglia, F., Quartarone, A., Ghilardi, M.F., Dattola, R., Bagnato, S., Rizzo, V., Morgante, L., and Girlanda, P. (2006). Unilateral cerebellar stroke disrupts movement preparation and motor imagery. *Clinical Neurophysiology, 117*, 1009–1016.

[5] Bressler, D., Spotswood, N., and Whitney, D. (2007). Negative BOLD fMRI Response in the Visual Cortex Carries Precise Stimulus-Specific Information. *Plos One, 2*, e410.

[6] Binkofski, F., Amunts, K., Stephan, K.M., Posse, S., Schormann, T., Freund, H.J., Zilles, K., and Seitz, R.J. (2000). Broca's region subserves imagery of motion: a combined cytoarchitectonic and fMRI study. *Human Brain Mapping, 11*, 273-285.

[7] Chen, H., Yang, Q., Liao, W., Gong, Q., and Shen, S. (2009). Evaluation of the effective connectivity of supplementary motor areas during motor imagery using Granger causality mapping. *NeuroImage, 47*, 1844-1853.

[8] Cramer, S.C., Lastra, L., Lacourse, M., and Cohen, M.J. (2005). Brain motor system function after chronic, complete spinal cord injury. *Brain, 128*, 2941–2950.

[9] De Vries, S., and Mulder, T. (2007). Motor imagery and stroke rehabilitation: a critical discussion. *Journal of Rehabilitation Medicine, 39*, 5–13.

[10] deCharms, R.C. (2008). Applications of real-time fMRI. *Nature Review Neuroscience, 9*, 720-729.

[11] deCharms, R.C., Christoff, K., Glover, G.H., Pauly, J.M., Whitfield, S., and Gabrieli J.D. (2004). Learned regulation of spatially localized brain activation using real-time fMRI. *NeuroImage, 21*, 436-443.

[12] deCharms, R.C., Maeda, F., Glover, G.H., Ludlow, D., Pauly, J.M., Soneji, D., Gabrieli, J.D., and Mackey, S.C. (2005). Control over brain activation and pain learned by using real-time functional MRI. *PNAS, 102*, 18626-18631.

[13] Dechent, P., Merboldt, K.D., and Frahm, J. (2004). is the human primary motor cortex involved in motor imagery? *Cognitive Brain Research, 19*, 138-144.

[14] Dietrich, A. (2008). Imaging the imagination: the trouble with motor imagery. *Methods, 45*, 319-324.

[15] Di Rienzo, F., Guillot, A., Delpuech, C., Rode, G., Grangeon, M., and Collet C. (2011). Inhibition of the motor command during motor imagery: a MEG study with a tetraplegic patient. 17th Annual Meeting of the Organization on Human Brain Mapping, June 26-30, Quebec City, Canada, p. 40.

[16] Doyon, J., and Benali, H. (2005). Reorganization and plasticity in the adult brain during learning of motor skills. *Current Opinion in Neurobiology, 25*, 161-167,.

[17] Doyon, J., and Ungerleider, L.G. (2002). Functional anatomy of motor skill learning. In: *Neuropsychology of Memory*, Squire L.R and Schacter D.L. (Ed.), pp. 225-38, Guilford Press.

[18] Driskell, J. E., Copper, C., and Moran, A. (1994). Does mental practice enhance performance? *Journal of Applied Psychology, 79*, 481–492.

[19] Ehrsson, H.H., Geyer, S., and Naito, E. (2003). Imagery of voluntary movement of fingers, toes and tongue activates corresponding body-part-specific motor representations. *Journal of Neurophysiology, 90*, 3304-3316.

[20] Ersland, L., Rosen, G., Lundervold, A., Smievoll, A.I., Tillung, T., Sundberg, H., and Hugdahl, K. (1996). Phantom limb imaginary fingertapping causes primary motor cortex activation: an fMRI study. *Neuroreport,8*, 207–210.

[21] Esposito, F., Seifritz, E., Formisano, E., Morrone, R., Scarabino, T., Tedeschi, G., Cirillo, S., Goebel, R., and Di Salle, F. (2003). Real-time independent component analysis of fMRI time-series. *NeuroImage, 20,* 2209-2224.

[22] Friston, K.J., and Buchel, C. (2003). Functional connectivity. In: *Human Brain Function,* Frackowiak, R.S.J. et al., Eds), Academic Press (2nd Ed.).

[23] Friston, K., Ungerleider, L., Jezzard, P., and Turner, R., (1995). Characterizing modulatory interactions between V1 and V2 in human cortex with fMRI. *Human Brain Mapping, 2,* 211-224.

[24] Gao, Q., Chen, H., and Gong, Q. (2008). Evaluation of the effective connectivity of the dominant primary motor cortex during bimanual movement using Granger causality. *Neuroscience Letters, 443,* 1-6.

[25] Gao, Q., Duan, X., and Chen, H. (2011). Evaluation of effective connectivity of motor areas during motor imagery and execution using conditional Granger causality. *NeuroImage, 54,* 1280-1288.

[26] Gerardin, E., Sirigu, A., Lehericy, S., Poline, J.B., Gaymard, B., Marsault, C., Agid, Y., and Le Bihan, D. (2000). Partially overlapping neural networks for real and imagined hand movements. *Cerebral Cortex, 10,* 1093-1104.

[27] Gonzáles, B., Rodríguez, M., Ramirez, C., and Sabaté, M. (2005). Disturbance of motor imagery after cerebellar stroke. *Behavioral Neuroscience, 119,* 622–626.

[28] Grezes, J., and Decety, J. (2001). Functional anatomy of execution, mental simulation, observation, and verb generation of actions: a meta-analysis. *Human Brain Mapping, 12,* 1-19.

[29] Guillot, A., and Collet, C. (2005). Duration of mentally simulated movement: a review. *Journal of Motor Behavior, 37,* 10-20.

[30] Guillot, A., and Collet, C. (2008). Construction of the motor imagery integrative model in sport: a review and theoretical investigation of motor imagery use. *International Review of Sport and Exercice Psychology, 1,* 31–44.

[31] Guillot, A., Collet, C., Nguyen, V.A., Malouin, F., Richards, C., and Doyon, J. (2008). Functional neuroanatomical networks associated with expertise in motor imagery ability. *NeuroImage, 41,* 1471-1483.

[32] Guillot A., Collet C., Nguyen VA., Malouin F., Richards C., and Doyon J. (2009). Brain activity during visual versus kinesthetic imagery: an fMRI study. *Human Brain Mapping, 30,* 2157-2172.

[33] Guillot A., Louis M., and Collet C. (2010). Neurophysiological substrates of motor imagery ability. In: *The neurophysiological foundations of mental and motor imagery,* Guillot A. and Collet C. (Eds.) pp. 109-124, Oxford University Press.

[34] Guillot, A., Hoyek, N., Louis, M., and Collet, C. (2011). Understanding the timing of motor imagery: recent findings and future directions. *International Review of Sport and Exercice Psychology, in press.*

[35] Hanakawa, T., Immisch, I., Toma, K., Dimyan, M.A., Van Gelderen, P., and Hallett, M. (2003). Functional properties of brain areas associated with motor execution and imagery. *Journal of Neurophysiology, 89*, 989-1002.

[36] Hanakawa, T., Dimyan, M.A., and Hallett, M. (2008). Motor planning, imagery, and execution in the distributed motor network: a time-course study with functional MRI. *Cerebral Cortex, 18*, 2775–2788.

[37] Heremans, E., Feys, P., Nieuwboer, A., Vercruysse, S., Vandenberghe, W., Sharma, N., Helsen, W. (2011). Motor imagery ability in patients with early-and mid-stage Parkinson disease. *Neurorehabilitation and Neural Repair, 25*, 168-177.

[38] Holmes, P.S., and Collins, D.J. (2001). The PETTLEP approach to motor imagery: A functional equivalence model for sport psychologists. *Journal of Applied Sport Psychology, 13*, 60-83.

[39] Jackson, P.L., Meltzoff, A.L., and Decety, J. (2006). Neural circuits involved in imitation and perspective-taking. *NeuroImage, 31*, 429-439.

[40] Jackson, P.L., Lafleur, M.F., Malouin, F., Richards, C.L., and Doyon, J. (2003). Functional cerebral reorganization following motor sequence learning through mental practice with motor imagery. *NeuroImage, 20*, 1171-1180.

[41] Jeannerod, M. (1994). The representing brain: neural correlates of motor intention and imagery. *Behavioral and Brain Sciences, 17*, 187-202, 1994.

[42] Johnson, S.H., Rotte,M., Grafton, S.T., Hinrichs, H., Gazzaniga,M.S., and Heinze, H.J. (2002). Selective activation of a parietofrontal circuit during implicitly imagined prehension. *NeuroImage, 17*, 1693–1704.

[43] Johnson, K.A., Hartwell, K., Lematty, T., Borckardt, J., Morgan, P.S., Govindarajan, K., Brady, K., and George, M.S. (2010). Intermittent "Real-time" fMRI Feedback Is Superior to Continuous Presentation for a Motor Imagery Task: A Pilot Study. *Journal of Neuroimaging, in press*, doi: 10.1111/j.1552-6569.2010.00529.

[44] Kasess, C.H., Windischberger, C., Cunnington, R., Lanzenberger, R., Pezawas, L., and Moser, E. (2008). The suppressive influence of SMA on M1 in motor imagery revealed by fMRI and dynamic causal modeling. *NeuroImage, 40*, 828-837.

[45] Kastrup, A., Baudewig, J., Schnaudigel, S., Huonker, R., Becker, L., Sohns, J.M., Dechent, P., Klingner, C., and Wittec, O.W. (2008). Behavioral correlates of negative BOLD signal changes in the primary somatosensory cortex. *Neuroimage, 41*, 1341-1371.

[46] Kleber, B., Birbaumer, N., Veit, R., Trevorrow, T., and Lotze, M. (2007). Overt and imagined singing of an Italian aria. *NeuroImage*, 36, 889-900.

[47] Kosslyn, S.M., Segar, C., Pani, J., and Hillger, L.A. (1990). When is imagery used in everyday life? A diary study. *Journal of Mental Imagery, 14*, 131–52.

[48] Kuhtz-Buschbeck, J.P., Mahnkopf, C., Holzknecht, C., Siebner, H., Ulmer, S., and Jansen, O. (2003). Effector-independent representations of simple and complex imagined finger movements: a combined fMRI and TMS study. *European Journal of Neuroscience, 18*, 3375-3387.

[49] Lafleur, M.F., Jackson, P.L., Malouin, F., Richards, C.L., Evans, A.C., and Doyon, J. (2002). Motor learning produces parallel dynamic functional changes during the execution and imagination of sequential foot movements. *NeuroImage, 2*, 142-57.

[50] Langheim, F.J.P., Callicott, J.H., Mattay, V.S., Duyn, J.H., and Weinberger, D.R. (2002). Cortical systems associated with covert music rehearsal. *NeuroImage, 16,* 901-908, 2002.

[51] Leonardo, M., Fieldman, J., Sadato, N., Campbell, G., Ibanez, V., Cohen, L., Deiber, M.-P., Jezzard, P., Pons, T., Turner, R., Le Bihan, D., and Hallett, M. (1995). A functional magnetic resonance imaging study of cortical regions associated with motor task execution and motor ideation in humans. *Human Brain Mapping, 3*, 83–92.

[52] Li, C.R. (2000). Impairment of motor imagery in putamen lesions in humans. *Neuroscience Letters*, 287, 13-16.

[53] Logothetis, N. K. (2008). What we can do and what we cannot do with fMRI. *Nature,* 453, 869-878.

[54] Lorey, B., Bischoff, M., Pilgramm, S., Stark, R., Munzert, J., and Zentgraf, K. (2009). The embodied nature of motor imagery: the influence of posture and perspective. *Experimental Brain Research, 194,* 233-243.

[55] Lotze, M., and Halsband, U. (2006). Motor imagery. *Journal of Physiology (Paris),* 99, 386-395.

[56] Lotze, L., Montoya, P., Erb, M., Hulsmann, E., Flor, H., Klause, U., Birbaumer, N., and Grodd, W. (1999). Activation of cortical and cerebellar motor areas during executed and imagined hand movements: an fMRI study. *Journal of Cognitive Neuroscience, 11,* 491-501.

[57] Lotze, M., Scheler, G., Tan, H.R.M., Braun, C., and Birbaumer, N. (2003). The musician's brain: functional imaging of amateurs and professionals during performance and imagery. *NeuroImage, 20*, 1817-1829, 2003.

[58] Lotze, M. and Zentgraf, K. (2010). Contribution of the primary motor cortex to motor imagery. In: *The neurophysiological foundations of mental and motor imagery*, Guillot A. and Collet C. (Eds.), pp. 31-46, Oxford University Press.

[59] Macuga, K.L., and Frey, S.H. (2011). Neural representations involved in observed, imagined, and imitated actions are dissociable and hierarchically organized. *NeuroImage, in press.*

[60] Malouin, F., Belleville, S., Richards, C. L., Desrosiers, J., and Doyon, J. (2004). Working memory and mental practice outcomes after stroke. *Archives of Physical Medicine and Rehabilitation, 85*, 177–183.

[61] Malouin, F., Richards, C., Durand, A., and Doyon, J. (2008). Reliability of mental chronometry for assessing motor imagery ability after stroke. *Archives of Physical Medicine and Rehabilitation, 89*, 311-319.

[62] Malouin, F., Richards, C.L., Jackson, P.L., Dumas, F., and Doyon, J. (2003). Brain activations during motor imagery of locomotor-related tasks: a PET study. *Human Brain Mapping, 19*, 47–62.

[63] Marx, E., Deutschländer, A., Stephan, T., Dieterich, M., Wiesmann, M., and Brandt, T. (2004). Eyes open and eyes closed as rest conditions: Impact on brain activation patterns. *NeuroImage, 21*, 1818-1824.

[64] Marx, E., Stephan, T., Nolte, A., Deutschländer, A., Seelos, K.C., Dieterich, M., and Brandt, T. (2003). Eye closure in darkness animates sensory systems. *NeuroImage, 19*, 924–934.

[65] Michelon, P., Vettel, J.M., and Zacks, J.M. (2006). Lateral somatotopic organization during imagined and prepared movements. *Journal of Neurophysiology, 95*, 811-822.

[66] Milton, J., Solodkin, A., Hlustîk, P., and Small, S.L. (2007).The mind of expert motor performance is cool and focused. *NeuroImage, 35*, 804-813.

[67] Milton, J., Small, S.L., and Solodkin, A. (2008). Imaging motor imagery: methodological issues related to expertise. *Methods, 45*, 336-341.

[68] Munzert, J., Lorey, B., and Zentraf, K. (2009). Cognitive motor processes: The role of motor imagery in the study of motor representations. *Brain Research Reviews, 60*, 306-326.

[69] Murphy, S., Nordin, S.M., and Cumming, J. (2008). Imagery in sport, exercise and dance. In: *Advances in sport psychology*, Horn T.S. (Eds.), pp. 306–315, Champagne, IL: Human Kinetics.

[70] Nair, D.G., Purcott, K.L., Fuchs, A., Steinberg, F., and Kelso, J.A.K. (2003). Cortical and cerebellar activity of the human brain during imagined and executed unimanual and bimanual action sequences: a functional MRI study. *Cognitive Brain Research, 15*, 250-260.

[71] Olsson, C.J., Jonsson, B., and Nyberg, L. (2008). Internal imagery training in active high jumpers. *Scandinavian Journal of Psychology, 49*, 133–140.

[72] Owen, A.M., Coleman, M.R., Boly, M., Davis, M.H., Laureys, S., and Pickard, J.D. (2006). Detecting Awareness in the Vegetative State. *Science, 313*, 1402.

[73] Pfurtscheller, G., and Neuper, K. (2010). EEG-based brain–computer communication. In: *The neurophysiological foundations of mental and motor imagery*, Guillot A. and Collet C. (Eds.), pp. 203-212, Oxford University Press.

[74] Porro, C.A., Francescato, M.P., Cettolo, V., Diamond, M.E., Baraldi, P., Zuiani, C., Bazzocchi, M., and Di Prampero, P.E. (1996). Primary motor and sensory cortex activation during motor performance and motor imagery: a functional magnetic resonance imaging study. *Journal of Neuroscience, 16*, 7688-7698.

[75] Porro, C.A., Cettolo, V., Francescato, M.P., and Baraldi, P. (2000). Ipsilateral involvement of primary motor cortex during motor imagery. *European Journal of Neuroscience, 12*, 3059-3063.

[76] Raichle, M.E. (1998). Behind the scenes of functional brain imaging: a historical and physiological perspective. *Proceedings of the National Academy of Sciences (USA), 95*, 765–772.

[77] Rizzolatti, G., Luppino, G., and Matelli, M. (1998). The organization of the cortical motor system: new concepts. *Electroencephalographic and Clinical Neurophysiology, 106*, 283–296.

[78] Roland, P.E., Larsen, B., Lassen, N.A., and Skinhoj, E. (1980). Supplementary motor area and other cortical areas in organisation of voluntary movements in man. *Journal of Neurophysiology, 43*,118-136.

[79] Ross, J.S., Tkach, J., Ruggieri, P.M., Lieber, M., and Lapresto, E. (2003). The mind's eye: Functional MR imaging of golf motor imagery. *American Journal of Neuroradiology, 24*, 1036-1044.

[80] Roth, M., Decety, J., Raybaudi, M., Massarelli, R., Delon-Martin, C., Segebarth, C.M., Morand, S., Gemignani, A., Decorps, M., and Jeannerod, M. (1996). Possible involvement of primary motor cortex in mentally simulated movement: A functional magnetic resonance imagery study. *Neuroreport, 7*, 1280-1284.

[81] Ruby, P., and Decety, J. (2001). Effect of subjective perspective taking during simulation of action: a PET investigation of agency. *Nature Neuroscience, 4*, 546-50.

[82] Sabate, M., Gonzalez, B., and Rodriguez, M. (2007). Adapting movement planning to motor impairments: The motor-scanning system. *Neuropsychologia, 45*, 378-386.

[83] Sabbah, P., Simond, G., Levrier, O., Habib M, Trabaud V, Murayama N, Mazoyer BM, Briant JF, Raybaud C, and Salamon G. (1995). Functional magnetic resonance imaging at 1.5T during sensory motor and cognitive tasks. *European Neurology, 35*, 131–135.

[84] Sanes, J.N., and Donoghue, J.P. (2000). Plasticity and primary motor cortex. *Annual Review of Neuroscience, 23*, 393-415.

[85] Schmuel, A., Yacoub, E., Pfeuffer, J., Van de Moortele, P.F., Adriany, G., Hu, X., and Ugurbil, K. (2002). Sustained negative BOLD, blood flow and oxygen consumption response and its coupling to the positive response in the human brain. *Neuron, 36*, 1195-1210.

[86] Schwoebel, J., Boronat, C.B., and Coslett, H.B. (2002). The man who executed "imagined" movements: evidence for dissociable components of the body schema. *Brain and Cognition, 50*, 1–16.

[87] Sharma, M., Pomeroy, V.M., and Baron, J.C. (2006). Motor imagery: a backdoor to the motor system after stroke? *Stroke, 37*, 1941–1952.

[88] Sharma, M., Jones, P.S., Carpenter, T.A., and Baron, J.C. (2008). Mapping the involvement of BA 4a and 4p during motor imagery. *NeuroImage, 41*, 92–99.

[89] Sirigu, A., Duhamel, J.R., Cohen, L., Pillon, B., Dubois, B., and Agid, Y. (1996). The mental representation of hand movements after parietal cortex damage. *Science, 273,* 1564-1568.

[90] Solodkin, A., Hlustik, P., Chen, E.E., and Small, S.L. (2004). Fine modulation in network activation during motor execution and motor imagery. *Cerebral Cortex, 14*, 1246-1255.

[91] Stinear, C.M. (2010). Corticospinal facilitation during motor imagery. In: *The neurophysiological foundations of mental and motor imagery*, Guillot A. and Collet C. (Eds.), pp. 47-61, Oxford University Press.

[92] Weiskopf, N., Veit, R., Erb, M., Mathiak, K., Grodd, W., Goebel, R., and Birbaumer, N. (2003). Physiological self-regulation of regional brain activity using real-time functional magnetic resonance imaging (fMRI): methodology and exemplary data. *NeuroImage, 19*, 577-86.

[93] Xie, B., Ma, X., Yao, L., Long, Z., and Zhao, X. (2011). Real time fMRI data analysis using region of interest selection based on fast ICA. *Progress in Biomedical Optics and Imaging – Proceedings of SPIE, 7965*, article 79651s.

[94] Yoo, S.S, and Jolesz, F.A. (2002). Functional MRI for neurofeedback: feasability study on a hand motor task. *Neuroreport, 13*, 1377-81.

[95] Yoo, S.S., Lee, J.H., O'Leary, H., Panych, L.P., and Jolesz, F.A. (2008). Neurofeedback fMRI-mediated learning and consolidation of regional brain activation during motor imagery. *International Journal of Imaging Systems and Tecnhology, 18*, 69-78.

[96] Zang, Y., Jia, F., Weng, X., Li, E., Cui, S., Wang, Y., Hazeltine, E., and Ivry, R. (2003). Functional organization of the primary motor cortex characterized by event-related fMRI during movement preparation and execution. *Neuroscience Letters, 337*, 69–72.

Chapter 17

Feedback Regulation of Limb Position Characterized Using FMRI

Aaron J. Suminski and Robert A. Scheidt

Additional information is available at the end of the chapter

http://dx.doi.org/10.5772/30740

1. Introduction

Functional magnetic resonance imaging (FMRI) is a powerful tool for exploring the neural basis of sensory control of movement and it has profitably been used to study simple actions like finger-tapping (Rao et al., 1996), compensation for visual feedback distortions during movement (Imamizu et al., 2000) and regulation of isometric force (Peck et al., 2001; Kawato et al., 2003; Vaillancourt et al., 2003). However, use of FMRI to image neural mechanisms contributing to more complex motor tasks like stabilization of limb position in the face of environmental perturbation has been limited because such tasks require mechanically-active, MRI-compatible devices capable to perturb the limb in a controlled manner. Here, we present a case study on the development and validation of a MRI-compatible device specifically for use in studying sensorimotor control in the presence of environmental perturbations. We then demonstrate the device in a functional imaging study of limb posture regulation wherein healthy human subjects stabilized their wrists against predictable and unpredictable loads. In particular, we sought to understand how the brain uses somatosensory information to adjust behavioral strategies for load compensation.

We anticipated that multiple strategies might be used to stabilize the limb against perturbation, and that distinct neural mechanisms would implement these strategies. We therefore hypothesized that at least two distinct neural mechanisms contribute to the stabilization of wrist position in the presence of persistent environmental perturbations. A first mechanism likely mediates the online control of endpoint position (not joint torque) via feedback control. Feedback control attempts to adjust motor commands to cancel deviations of the limb from its desired state. Thus we expect regions contributing to feedback control of wrist position to show increased FMRI BOLD response in the presence of positioning errors, independent of joint torque magnitude. Moreover, temporal variations in the BOLD response in these regions should correlate with variations in wrist position on a moment-by-moment basis. A second mechanism likely monitors performance over a longer timeframe than an online feedback controller. This mechanism would initiate discrete, conditional, corrective actions when feedback control fails to eliminate persistent errors. Regions contributing to this higher-order evaluation of performance should demonstrate BOLD responses that reflect changes in positioning errors with a longer temporal integration period than that used for moment-by-moment feedback control.

Portions of this chapter have appeared previously in separate publications (Suminski et al., 2007b; Suminski et al., 2007a).

2. Development of a MRI-compatible manipulandum

Devices intended for use in MRI environments must: 1) satisfy noise tolerance and size limitations imposed by MR scanner technologies, and 2) be constructed of MR compatible materials (Schenck, 1996; Chinzei et al.). The large static magnetic field generated by the scanner precludes use of ferromagnetic materials that would otherwise be attracted into the scanner bore, compromising safety of both the research subject and the scanner. It is also essential that all actuators and sensors in the device be impervious to rapidly switching imaging gradients and that device operation does not disturb the homogeneity of the magnetic field, which would lead to image distortion. Finally, the device must have a small form factor, capable to fit inside the scanner bore without causing subject discomfort. To date, a small number of such robotic devices have been developed for use in neuroscience research or rehabilitation applications (Suminski et al., 2002; Ganesh et al.,2004; Diedrichsen et al., 2005; Flueckiger et al., 2005; Khanicheh et al., 2005; Gassert et al., 2006; Khanicheh et al., 2008; Yu et al., 2008). In this section we describe the design, performance characteristics and validation of a novel, MR compatible, 1 degree-of-freedom (DOF), pneumatically actuated robot for motor control research (Fig. 1). Our aim was to create a device able to both monitor and perturb wrist motion during FMRI, and to demonstrate device safety and efficacy as a tool for the study of complex motor behavior in human subjects.

2.1. Device design and performance characteristics

We used a pneumatic actuator to exert computer-controlled torques about the wrist because this type of actuator can be MR compatible, small, light-weight and back-driveable. A single bellows-type actuator was enclosed within a curved volume. This actuator transmits force from compressed air to a wall rigidly attached to the device's handle. Pressurizing the actuator generates an extensor torque about the subject's right wrist whereas pulling a vacuum within the actuator imparts a flexor torque. Air pressure within the actuator is sensed by a pressure transducer (26PC Series, Honeywell International, Inc., Morristown, NJ), amplified (x25) and low-pass filtered (20Hz cutoff frequency) in hardware. Joint angle is sensed with an optical encoder (HEDM-6540, Agilent Technologies, Inc., Palo Alto, CA) located on the underside of the device. The device monitors wrist position to within 0.05° and wrist torque to within 0.001

Nm. Only the manipulandum, pressure transducer, optical encoder and necessary instrumentation are located within the MR environment; all other control components are located in the scanner control room. The manipulandum can accommodate both right- and left-handed individuals, providing 80° range of motion at the wrist (40° flexion to 40° extension). Pressure within the actuator is regulated by a Proportion Air QB3 electro-pneumatic pressure valve (Proportion-Air Inc., McCordsville, IN). Wrist angle and actuator pressure data are acquired at a rate of 1000 samples per second. Commands to the pressure valve are generated at the same rate.

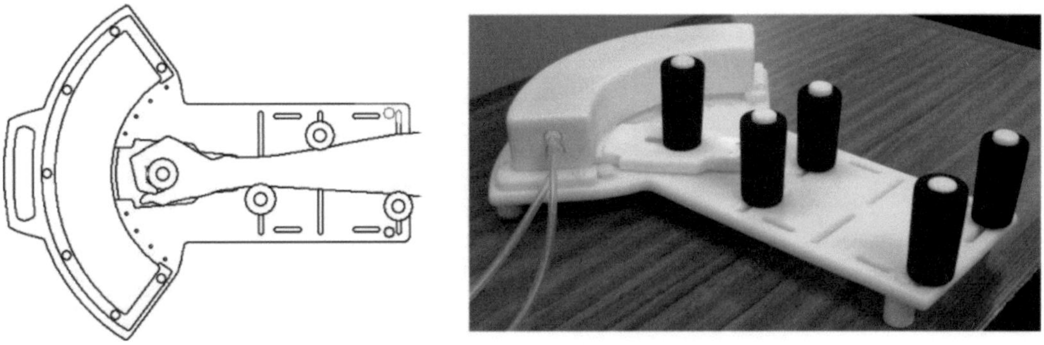

Figure 1. Schematic representation and a photograph of the 1 DOF pneumatic manipulandum.

We implemented a proportional-integral-derivative (PID) controller to improve the ability of the device to quickly and accurately regulate pressure within the pneumatic actuator.

$$P_C(t) = K_P e(t) + K_I \int e(t)dt + K_D \dot{e}(t) \tag{1}$$

where $P_C(t)$ is the commanded actuator pressure in units of psi, e(t) is the difference between the measured and desired actuator pressure in units of psi, K_P is the proportional gain, K_I is the integral gain, and K_D is the derivative gain. Ziegler-Nichols tuning rules were used to tune the controller (Ziegler and Nichols, 1942), yielding the following gain values: $K_P = 3.3$, $K_I = 14$, and $K_D = 0.055$. Under PID control, step response rise times (10% to 90% steady state) were 77ms and 90ms for 1 and 2 PSI step changes, respectively, with modest overshoot (19%; Fig. 2A). Due to time required to transmit air from the MRI control room to the actuator, we observed an average time delay (command onset to 10% steady state) of 62ms for both step responses.

We identified the bandwidth of the closed-loop system by assessing the system's ability to track changes in commanded actuator pressure having a 1 PSI peak-to-peak 'chirp' profile sweeping from 0 to 5Hz. The device is able to track commanded pressure changes within ±15% of the peak pressure up to 1.6 Hz (Fig. 2B). These frequency response characteristics allow the robot to apply torsional spring-like loads about the wrist. By way of demonstration, we commanded the robot to apply two separate position-dependent loads (0.075 and 0.15 Nm/°)

and estimated the realized spring constants obtained during 25° flexion/extension movements performed by a representative human subject. The estimated stiffness of the two spring-like loads, obtained by fitting a linear model to the joint torque vs. joint angle data, were 0.059 and 0.134 Nm/° respectively, yielding an average error of 16%. In both cases, the torque-angle relationships were linear, with regression r^2 values exceeding 96% and 99% for the 0.075 and 0.15 Nm/° loads, respectively (Fig. 2C).

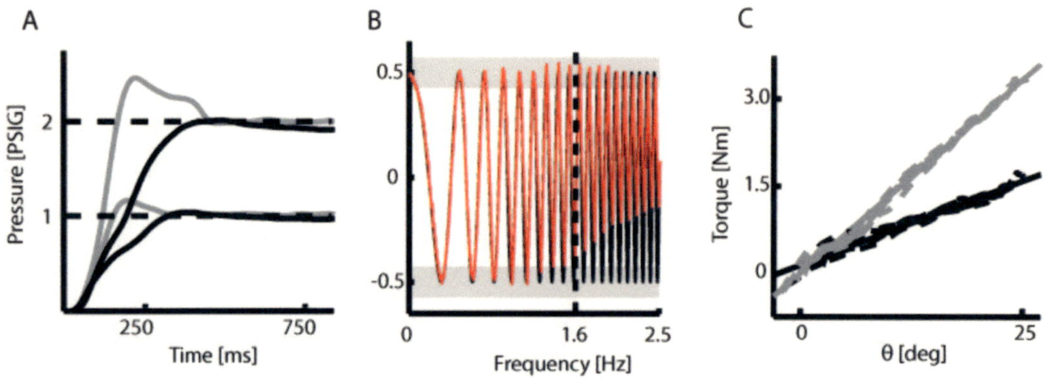

Figure 2. (A) Responses of the robot to 1 and 2 PSI step changes in pressure under open loop control (i.e. no pressure feedback to controller; black lines) and closed loop PID control (i.e. real-time feedback of actual actuator pressure to the controller; gray lines). (B) The frequency response of the system under closed loop control (red line) was identified by assessing the system's ability to track a desired actuator pressure signal sweeping from 0 to 5Hz (black line). The upper bound on the system's bandwidth was defined to be the frequency at which the controller was unable to regulate the actuator pressure within 15% of the peak commanded pressure (horizontal gray bars). (C) Spring constants for two simulated loads (0.075 and 0.15 Nm/°; dashed black and gray lines, respectively) estimated by fitting a line to the joint torque and wrist angle data (solid lines). Figure adapted from Suminski et al., 2007a.

2.2. MR-compatibility testing

We validated the simultaneous acquisition of manipulandum data and FMRI images by scanning a spherical head phantom both with and without the robotic device in a 3.0T GE Excite HD MR scanner (General Electric Healthcare, Milwaukee, WI). The phantom (Fig. 3A, P; GE Model #: 2359877) was supported within a split transmit/receive quadrature head coil (Fig. 3A, HC; GE Model #: 2376114). A gradient echo, echo planar imaging (EPI) pulse sequence (29 contiguous sagittal slices; echo time (TE) = 25ms, inter-scan period (TR) = 2s, flip angle = 77°, field of view (FOV) = 24cm, 64 x 64 matrix; 3.75x3.75x6 mm spatial resolution) was used to verify: 1) that operation of the robot during scanning does not induce significant artifacts in functional images, and 2) that the robot can measure pressure and joint angle without contamination from gradient switching noise during EPI.

Validation testing used a blocked experimental design (Duration = 270s). During "Motion" states, the computer cycled the device's handle through a sinusoidal trajectory (0.25 cycles per second) whereas the device remained motionless during "No Motion" states (50% duty cycle; period = 60s). Raw, complex k-space data (I and Q channels) were collected to allow analysis

http://dx.doi.org/10.5772/30740

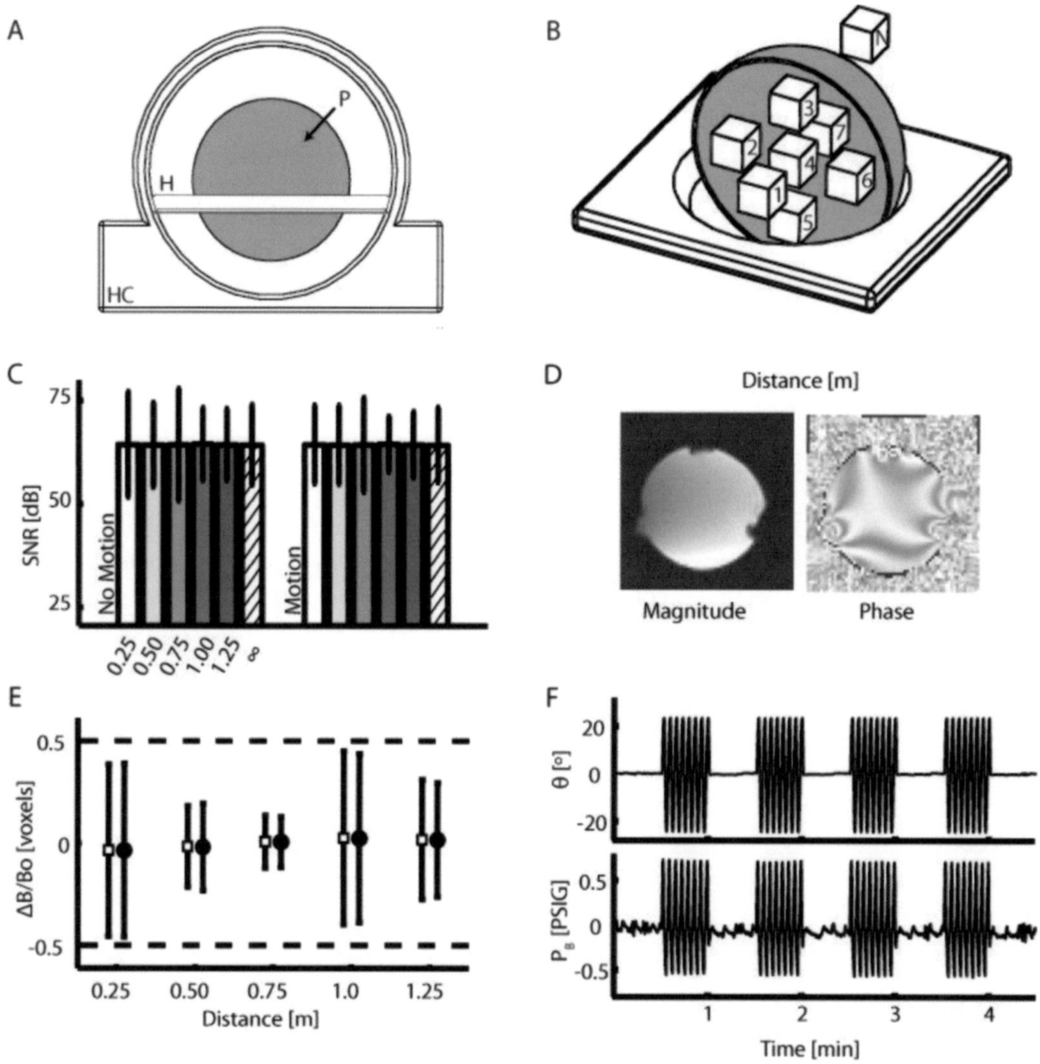

Figure 3. (A) Set-up of the head coil (HC), phantom holder (H), and phantom (P). (B) Holder and phantom cut-away showing details of ROIs used to calculate signal-to-noise ratio (SNR) and field homogeneity. (C) SNR for ROI 3 vs. device operating distance measured from the center of the imaging volume. Error bars represent 95% confidence intervals about the mean SNR at each distance. (D) Sagittal slice of magnitude and phase images of the phantom. Susceptibility artifact in the lower portion of the images was caused by the phantom holder, and was seen in images both with and without the manipulandum in the scanner. (E) Estimate of field homogeneity in ROI 3 from phase data collected in the "No Motion" (open squares) and "Motion" (filled circles) states. We were interested in the distribution of homogeneity values, so error bars in this panel represent ± 2 SD about the mean and thus 95% of the data lie within these bounds. (F) Representative wrist angle and bellows pressure time series collected when the device was 0.50 m from the imaging volume. Angle and pressure measurements are not adversely effected by scanner operation. Figure adapted from Suminski et al., 2007a.

of both magnitude and phase MR images. We quantified the effects of simultaneous operation of the robot and scanner during "Motion" and "No Motion" states by imaging the phantom with the robot at 6 distances from the center of the imaging volume (0.25m, 0.50m, 0.75m, 1.0m, and 1.25m) as well as in a control condition with the robot operating in the scanner control room (∞). The phantom was sampled using 7 equal-volume regions of interest (ROI) distributed within its spherical boundary to test whether the robot induced amplitude and phase anisotropies during scanning.

We computed three measures to assess compatibility of the robot and MR scanner; two evaluated MR signal quality during robot operation in the "Motion" and "No Motion" states and the third evaluated the effects of echo planar imaging on measurements of handle position and actuator pressure. First, we calculated the signal to noise ratio (SNR) within each ROI for each robot-distance condition using the magnitude images:

$$SNR_{ROI} = \frac{\mu_{ROI}}{0.665 * \sigma_{noise}} \tag{2}$$

μ_{ROI} is the time series average within a given ROI, and σ_{noise} is an estimate of noise obtained by calculating the standard deviation of the time series in the magnitude images in an identically sized ROI located outside the phantom (Fig. 3B; ROI "N"). The factor 0.665 corrected for changes in the statistical distribution of σ_{noise} caused by calculating the magnitude image from the original complex MR data (Haacke et al., 1999). Values of SNR varied across the seven ROIs but were insensitive to the robot's distance within each ROI (eg. Fig 3C). Three-way ANOVA found main effects of both ROI location ($p < 0.0005$) and robot distance ($p < 0.0005$) but no effect of robot motion state ($p = 0.929$). Comparison of SNR at each of the five distances relative to the control condition (∞) revealed a small but significant 0.64 dB and 0.90 dB *increase* in SNR at 1.0 m and 1.25 m as compared to control ($p < 0.05$), but no change in SNR for the other distances ($p > 0.7$). Thus, we found no systematic degradation of functional MR SNR as a function of robot distance from the imaging volume.

Second, we used the phase images to quantify changes in magnetic field homogeneity induced by robot operation within the scanner suite. We computed the average change in the static magnetic field for each ROI (ΔB_{ROI}):

$$\Delta B_{ROI} = \frac{\phi_{ROI}}{-\gamma \, T_E} \tag{3}$$

where, ϕ_{ROI} is the average change in each ROI's phase time series with respect to baseline (i.e. ∞), γ is the gyromagnetic ratio, and T_E is the echo time of the EPI sequence (Haacke et al., 1999). We then normalized ΔB_{ROI} to the magnitude of the static magnetic field (B_0 = 3.0T) yielding a unit-less quantity corresponding to the homogeneity of the magnetic field ($\Delta B/B_0$) in parts per million. This normalization process allows comparison of the field homogeneity and the bandwidth/voxel (39Hz or 0.32 ppm) of the EPI sequence. If introduction of the robot into the scanning environment disturbs the field homogeneity by > 0.16 ppm (i.e. ½ voxel), the actual and measured voxel locations would be inconsistent causing inaccuracy in the resultant images. As shown for a representative ROI (Fig. 3B), field inhomogeneity induced by the robot was well below ½ voxel at each distance in the "No Motion" and "Motion" conditions. One-sided t-tests rejected the hypothesis that the field distortion exceeded ½ voxel in either condition ($p < 0.0005$ in each case). Thus, image quality was not compromised by operating the robotic device within the MR scanner.

Finally, we quantified the effects of echo planar imaging on robot operation by calculating SNR for the actuator pressure (SNR_P) and wrist angle (SNR_A) signals while the computer drove the robot's handle through a sinusoidal trajectory:

$$SNR = 20*\log_{10}\left(\frac{RMS_{Signal} - RMS_{Noise}}{RMS_{Noise}}\right) \tag{4}$$

Root mean squared (RMS) values of actuator pressure and joint angle were calculated during "Motion" and "No Motion" states to approximate signal and noise respectively. Neither joint angle nor pressure SNR varied systematically as a function of robot distance from the center of the scanner bore. Individual two-sample t-tests found no difference in SNR_A or SNR_P ($p > 0.705$) when compared to baseline measures obtained when the robot was operated outside the scanning environment (∞).

In conclusion, we have implemented a pneumatically actuated manipulandum that applies controlled joint torques and measures joint angle at the wrist. This device neither degrades fMRI signal quality nor is itself compromised by rapidly switching imaging gradients.

2.3. Comparison of the device with other MR-compatible devices

In the last decade, several robotic devices have been developed for use during MR scanning (see Gassert et al., 2008 for review). The 1 DOF manipulandum we developed compares favorably to other MR-compatible devices used in neuroscience research or rehabilitation applications. For example, the device developed by Hidler et al. (Hidler et al., 2006) only monitors the torque/force generated by a subject whereas our device can simulate dynamic environments by generating controlled torques about the wrist. Other MR-compatible devices can apply dynamic loads using Lorentz coils (Riener et al., 2005), ultrasonic motors (Flueckiger et al., 2005), electrorheological fluids (Khanicheh et al., 2008) or hydrostatic pistons (Gassert et al., 2006). However, in contrast to the device presented by Riener and colleagues, our device does not degrade image quality when operated less than 1m from the scanner's isocenter. Because the devices presented by Flueckinger (Flueckiger et al., 2005) and Gassert (Gassert et al., 2006) are not backdriveable, they can not simulate realistic dynamic loads during movements requiring rapid changes in direction whereas our device clearly can do so. Recently, Yu, et al. compared 1 DOF MR-compatible devices containing hydraulic and pneumatic actuators (Yu et al., 2008) and concluded that pneumatic actuation was favorable for fast, force controlled applications, whereas hydraulic actuation was best for applications requiring accurate position control (Yu et al., 2008). And while the 2 DOF device presented by Diedrichsen (Diedrichsen et al., 2005) offers the ability to perturb planar reaching movements of the arm, perturbation of proximal limb segments can lead to considerable head motion that must be accounted for during analysis of fMRI data (Diedrichsen and Shadmehr, 2005). In contrast, our current design limits motion to the wrist, which may lead to fewer head motion artifacts in the fMRI dataset.

3. Limb position regulation with proprioceptive feedback

For a first demonstration of the robot's utility, we examined how the brain uses proprioceptive feedback of limb position for the moment-by-moment (i.e. on-line) feedback stabilization of wrist position during a compensatory tracking task. Limb stabilization is important because meaningful interaction with the world frequently requires stabilization of hand-held items (eg. holding a young child's hand) and/or movement of such objects between stabilized positions or "postures" (eg. turning the steering wheel of a car). At any moment, task performance may be compromised due to environmental perturbations (eg. the car hitting a pothole) requiring corrective action to maintain desired performance.

The central nervous system can employ three strategies acting on different timescales to compensate for errors arising during stabilization. First, it may utilize feedback regulation of joint position via segmental (Sinkjaer and Hayashi, 1989) and transcortical ("long loop") reflex pathways (Evarts and Tanji, 1976; Strick, 1978; Evarts and Fromm, 1981) to minimize errors. Alternatively, subjects may increase the impedance of the limb via voluntary co-activation (Milner and Cloutier, 1993) of muscles whose actions oppose one another (i.e. antagonist muscles). Finally, subjects may generate discrete, feedforward, corrective movements when feedback mechanisms and impedance regulation fail to adequately reduce perceived errors (Haaland and Harrington, 1989; Fagg et al., 1998). These strategies are not mutually exclusive, but are complementary in two ways. First, they reduce performance errors over different timescales ranging from the short-latency mechanical responses of antagonist coactivation and reflex action to the reduction of persistent errors by discrete adjustment of behavioral goals. Second, they provide the flexibility in motor output needed to respond to task-dependent tradeoffs between accuracy and muscular effort, thus providing the behavioral basis for optimality in human motor control (Todorov and Jordan, 2002; Scott, 2004). Much is yet unknown about the neuromuscular control of limb position (i.e. posture stabilization), including which aspects of environmental perturbation are compensated on a moment-by-moment basis, and what performance criteria might cause a subject to generate a discrete corrective movement during stabilization.

Ten healthy right-handed volunteers (5 female) participated in two experimental sessions performed on separate days. They performed identical wrist stabilization tasks both days. In one session, subjects stabilized the wrist against robotic perturbation while inside a mock MR scanner. This allowed recording of electromyographic (EMG) data from task-relevant muscles. In the other session, subjects performed the experiment while undergoing FMRI scanning in a 1.5T General Electric Signa scanner equipped with a 3-axis local gradient head coil and an elliptical endcapped quadrature radiofrequency head coil. In both sessions, subjects rested supine in the scanner with their head constrained by foam padding to reduce head motion. With arms at their sides, subjects grasped the robot handle with their right hand. The handle and wrist axes of rotation were aligned and the frame of the device was secured to both the subject's forearm and the inner wall of the scanner bore for support.

3.1. Experimental procedure

Both sessions consisted of a blocked experimental design that alternated between periods of rest and active wrist stabilization. Each stabilization trial was conducted in 5 phases (Fig. 4). During the 30 s prior to stabilization (phase 1), the subject was instructed to relax while the robot held the hand in a comfortable resting posture of 40° flexion (θ_r). 3 s prior to the start of stabilization (phase 2), the robot moved the relaxed hand to the target posture (20° flexion) and held it there until the onset of the stabilization period. The purpose of this phase was to provide a salient haptic cue of the desired wrist angle about which subjects were to stabilize. During the 30 s stabilization period (phase 3), subjects were instructed to hold their wrist steady at the target angle during two experimental conditions in which the device was programmed to apply either a predictable, constant extensor torque about the wrist (*CT*, mean = 1.2 Nm) or unpredictable, pseudo-random extensor torques comprised of band-limited Gaussian "white" noise with a high-frequency cutoff of 1.6 Hz (*RT*; 1.2±1.1 Nm; mean ± SD). At the end of the stabilization period, subjects were instructed to relax while the robot moved the passive hand to its resting position at 40° flexion (phase 4), after which resting EMG continued to be monitored for 3 s (phase 5). Direct view of the wrist was precluded and subjects received no visual feedback of hand motion during stabilization phase 3. Instead, subjects viewed a stationary fixation target that was back-projected onto a screen located at their feet and was visible using prism glasses. The fixation target moved in concert with the hand during passive movement phases 2 and 4 and thus provided an implicit visual representation of the desired wrist angle during stabilization phase 3.

During each 3-minute imaging run, *CT* and *RT* stabilization trials were each presented one time in pseudo-random order, along with 30 s periods of inactivity (rest) preceding and following stabilization. Each subject performed 10 of these runs in each session. Whole-brain images were acquired using a single-shot, blipped gradient-echo echo-planar pulse sequence (19 contiguous sagittal 7-mm slices, TE = 40 ms, TR = 2.5 s, 90° flip angle, FOV = 24 cm, 64x64 matrix, 3.75-mm in-plane resolution). 72 whole-brain images were acquired in each run. BOLD contrast was used to image the hemodynamic related changes in the brain occurring during the two stabilization tasks. Before functional imaging, high-resolution 3D spoiled gradient recalled at steady-state T1-weighted anatomic images were collected for anatomic localization and co-registration (TE=5 ms, TR=24 ms, 40° flip angle, slice thickness=1.2 mm, FOV=24 cm, 256x192 matrix).

3.2. Behavioural correlates of stabilization

We found that that wrist torque perturbations elicited changes in wrist angle (θ) and wrist angular velocity ($d\theta/dt$) despite instruction to hold the hand steady. We quantified kinematic performance using the root mean square (RMS) stabilization error. Not unexpectedly, subjects

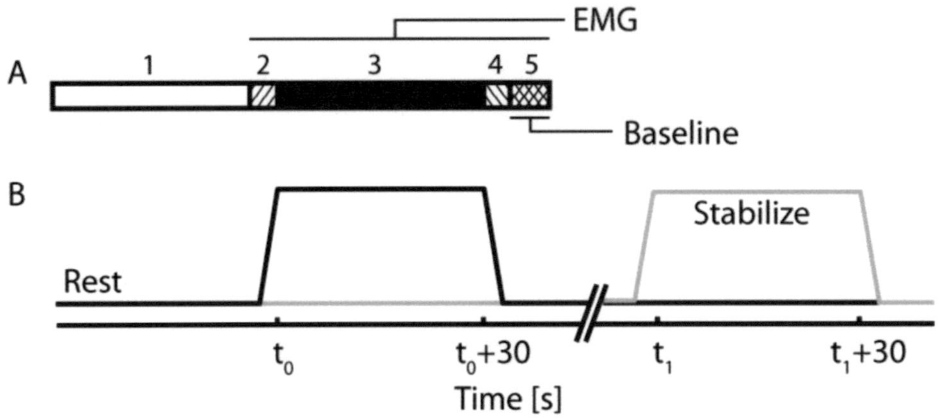

Figure 4. (A) Single stabilization trials were conducted in 5 phases. See text for details. (B) During each imaging run, subjects stabilized their wrist against constant (gray line) or pseudo-random (black line) extensor torque perturbations. Figure adapted from Suminski et al., 2007b.

were less able to maintain steady hand posture while perturbed by band-limited pseudo-random torques than by constant torques. Paired t-tests found greater errors in *RT* vs. *CT* trials ($p <= 0.0005$, both sessions). We also observed positional drift over time (a significant slope in the joint angle time series over the final 20 seconds a trial) in 91% of *RT* trials (88 of 96) and 75% of *CT* trials (72 of 95). The absolute magnitude of this drift was 0.155 ± 0.104 °/s and 0.017 ± 0.028 °/s in the *RT* and *CT* cases respectively.

Elevated Coactivity: We quantified coordination between muscles in the arm by estimating the amount of antagonist muscle co-activity at the wrist, elbow and shoulder joints using a coactivity measure also known as 'wasted contraction' (Thoroughman and Shadmehr, 1999) [*C(t)*]. We considered the anterior and postierior deltoid as shoulder antagonists, the biceps and triceps as elbow antagonists, and flexor carpi radialis and extensor carpi radialis as wrist antagonists. ANOVA and post-hoc Dunnet's t-tests found significant elevations in muscle co-activity values (Fig. 5A) during phase 3 ***RT*** stabilization at each joint ($p < 0.0005$) and during phase 3 ***CT*** stabilization at the wrist ($p = 0.03$). No significant EMG activity was observed at any joint during passive movement (phases 2 and 4) when compared to rest.

Feedback Regulation: We evaluated the contributions of reflex-mediated responses to wrist stabilization by analyzing the cross-correlation between wrist angular velocity and measured EMG responses in the ***RT*** condition. For wrist flexor FCR, we consistently observed increased EMG activity lagging wrist extension by 59.0 ± 44.2 ms (Fig. 5B-d). The timing of this load-dependent activity was within the range of delays expected for transcortical reflex compensation for muscle stretch (Evarts and Vaughn, 1978; Strick, 1978; Matthews, 1981). For wrist extensor ECR, we observed decreased EMG activity lagging wrist extension by 40.3 ± 23.7 ms (Fig. 5B-c). The sign and latency of these EMG changes are consistent with an unloading response mediated by spinal circuits (Sinkjaer et al., 2000). We also observed an increase in ECR activity with lag of 484.8± 118.3 ms (Fig. 5B-e) in 5 out of 10 subjects. This later response is consistent with a strategy of voluntary co-activation about the wrist since no contemporaneous decrease in flexor activity was observed.

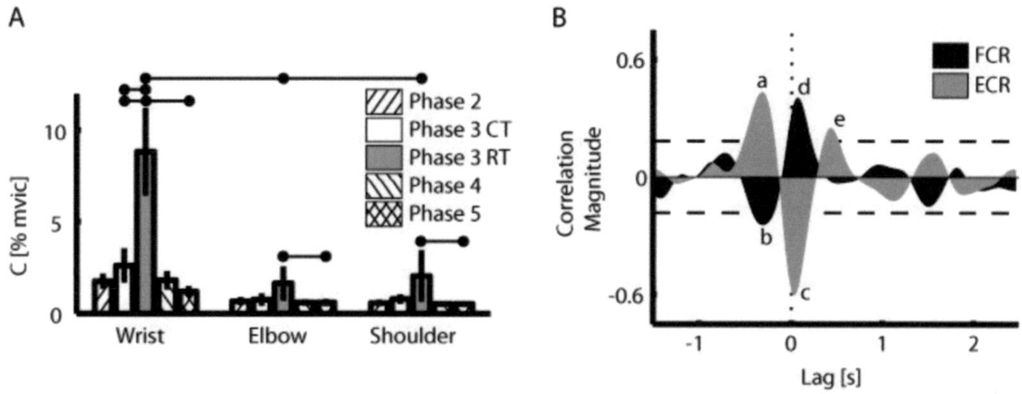

Figure 5. (A) Population average of antagonist muscle coactivity. Horizontal bars indicate significant differences between stabilization conditions ($p < 0.05$). Hatched bars: co-activity during phases 2, 4, and 5. White and gray bars: co-activity during periods of active stabilization against *CT* and *RT* respectively. Error bars indicate ± 2 SEM. (B) Cross-correlation of wrist angular velocity ($d\theta/dt$) and EMG from flexor carpi radialis (*FCR;* black) and extensor carpi radiallis (*ECR*; gray) for a representative subject. Significant correlations were observed with EMG both lagging/leading changes in $d\theta/dt$. Horizontal dashed lines: 95% confidence interval about zero correlation. Figure adapted from Suminski et al., 2007b.

Discrete Corrections: Consistent with expectation, we observed many rapid movements that did not appear to be a direct mechanical consequence of moment-by-moment fluctuations in the robot's commanded wrist torque. To identify the onset of discrete corrective movements, we first computed the average wrist angle trajectory for each subject in each condition. We then removed the trial-averaged trajectory from every individual trial to obtain a 'corrected' wrist angle time series. We considered as discrete corrections only those motions wherein the angular velocity of corrected wrist angle trajectories exceeded 5°/s if they occurred >1 second after the start of stabilization. Discrete corrective movements occurred in 95% of ***RT*** trials (91 of 96) and 19% of ***CT*** trials (18 of 95). The majority of corrective movements were directed appropriately to reduce positioning errors (56.1% and 80% of movements in the ***RT*** and ***CT*** cases, respectively).

In summary, the behavioral and electromyographic data revealed that subjects compensated for environmental perturbations using a combination of three readily identifiable strategies. Subjects modulated limb impedance via co-activation of agonist/antagonist muscle pairs spanning the wrist, elbow and shoulder. Correlation analysis found that subjects also invoke both spinal and supraspinal reflexes to compensate for the perturbations. Finally, subjects generated discrete corrective movements to reduce performance errors that likely accumulated due to the lack of visual feedback during stabilization (cf. Wann and Ibrahim, 1992). We next sought to characterize the neural mechanisms contributing to each of these strategies during experiments conducted within the MR scanner.

3.3. Neural correlates of stabilization

Functional images were generated and analyzed within the Analysis of Functional NeuroImages (AFNI) software package (Cox, 1996). The first three images in each run were discarded to allow for equilibration of the magnetic field. Individual run time series were then concatenated and aligned in three-dimensional space using an interactive, linear, least squares method. Voxel-wise multiple linear regression was used to determine the amount of FMRI signal contrast between the two task conditions (***CT*** and ***RT*** stabilization) and the resting baseline. The resulting functional images for ***RT*** and ***CT*** were interpolated to obtain a volumetric grid having $1mm^3$ voxel volumes, coregistered, and then converted into the Talairach stereotaxic coordinate space (Talairach and Tournoux, 1988). To facilitate group analysis, the Talairached functional images were spatially blurred using a 4-mm Gaussian full-width half-maximum filter to compensate for inter-subject anatomical variability. In all across-subject analyses, a activation volume and thresholding technique was used to correct for multiple comparisons in the group analysis. Appropriate activation volume and individual voxel p-value thresholds ($p=0.005$) were estimated by performing 5000 iterations in a Monte-Carlo simulation using the AlphaSim tool within AFNI (Cox, 1996). The location of activated regions in group statistical parametric maps was obtained using the Talairach atlas (Talairach and Tournoux, 1988) for cerebral activations and the Schmahmann atlas (Schmahmann, 2000) for activations in the cerebellum and its nuclei. Cortical activations were visualized using CARET (Van Essen et al., 2001); http://brainmap.wustl.edu/caret).

Changes in BOLD signal intensity (relative to rest) correlated with periods of ***CT*** or ***RT*** stabilization in many brain regions that contribute to control of the upper extremity (Table 1; Fig. 6A). We visualized the time series of BOLD activations for each ROI during stabilization against each load type for each subject. Two patterns of activation became evident. The first, an example of which is shown for an ROI spanning left primary motor and somatosensory cortices (Fig. 6B; M1/S1), was characterized by increases in percentage BOLD signal intensity change (PSC) throughout the ***RT*** stabilization periods and during the passive hand movements preceding and following ***CT*** stabilization. Motion sensitivity during active stabilization and passive movement might be expected from brain regions concerned with the estimation of - and on-line feedback correction for - performance errors (Marsden et al., 1972; Lee and Tatton, 1975; Evarts and Vaughn, 1978; Marsden et al., 1978; Strick, 1978; Horne and Butler, 1995; Scott, 2004). This idea is based on the assumption that passive movement of the wrist induces a discrepancy (error) between actual limb position and that expected given the recent history of motor output. The second pattern, an example is shown for the right middle frontal gyrus (Fig. 6B; BA 46), was characterized by a parallel increase in PSC during both stabilization conditions with no sensitivity to passive movement. This pattern might be expected from regions (i.e. prefrontal cortex, anterior cingulate cortex, rostral aspects of dorsal premotor and SMA cortices, and inferior aspects of posterior parietal cortex) involved in supervisory aspects of control such as evaluating the success of an ongoing task (Carter et al., 1998) as well as the

http://dx.doi.org/10.5772/30740

resetting of behavioral goals when ongoing performance is deficient (Jahanshahi et al., 1995; Winstein et al., 1997; Jueptner and Weiller, 1998; Sakai et al., 1999). As shown for an ROI in the cerebellum (Fig 6B, CBLM), some regions exhibited an intermediate pattern (i.e. some sensitivity to stabilization type but little sensitivity to passive movement). To quantify these observations, we defined a 2-element feature vector $\Psi = \{\bar{S},\ \Delta_M\}$ summarizing the average increase in PSC induced by passive movement in phase 2 ($\bar{S}$) and the difference in PSC between ***RT*** and ***CT*** trials during stabilization (Δ_M). The $\bar{S}$ feature was the average of 4 consecutive TR samples from each of the ***RT*** and ***CT*** trials, beginning 5 seconds (i.e. 2 TR) prior to the start of the stabilization period. The Δ_M feature was computed as the difference between the average ***RT*** and ***CT*** PSC time series during the middle 20 seconds (8 TR) of stabilization.

Using these features, we performed a k-means cluster analysis to identify ROIs demonstrating common patterns of sensitivity to trial type and/or passive movement. Two distinct groups were identified: ROIs that were sensitive to both trial type and passive movement (Cluster I; red regions in Fig. 6A), and those demonstrating increased BOLD activation over the duration of the trial with little sensitivity to passive movement (Cluster II; blue regions in Fig. 6A). Group separation and membership was visualized by plotting the coordinates of each ROI on the axes defined by the features of Ψ (Fig. 6C; Cluster I = red circles; Cluster II = blue squares). Regions showing significant sensitivity to passive movement ($\bar{S} \neq 0$; $p < 0.05$) are indicated by filled symbols. These include left M1/S1, SPL/IPL (BA 5 and 7), and SMA/CCZ. The left M1/S1, SPL/IPL (BA 5 and 7), and right STG (BA 39) were also significantly modulated by task type ($\Delta_M > 0$; $p < 0.05$). ROIs with membership in Cluster I were characterized by both a high sensitivity to trial type (being more active during RT trials) and a robust response to passive movement of the wrist. Such sensitivity to positioning errors is necessary for regions directly involved in the feedback control of movement kinematics (i.e. limb position, velocity and acceleration). Indeed, responsivity to passive movement has been considered a hallmark of brain regions participating in the optimal feedback control of movement (Scott, 2004). In contrast, Cluster II ROIs exhibited generally elevated BOLD activation during stabilization with little sensitivity to trial type or passive movement. This pattern of activity is expected from regions involved in supervisory aspects of control rather than in the moment-by-moment cancellation of position errors.

We performed a second BOLD regression analysis to identify brain regions involved in generating the discrete corrective movements observed behaviorally. We limited our investigation to the ***CT*** stabilization blocks because corrections were unambiguous and most often directed to decrease kinematic error. In particular, we wished to know whether the regions generating corrective movements might be a subset of the networks previously described or different networks. For this event-related analysis, we created input reference functions for each subject indicating the onset times of discrete corrective movements during ***CT*** stabilization. These time series had a value of 1 during each sampling instant (TR) wherein a discrete corrective movement occurred and 0 otherwise. Additional reference time series were included to mitigate effects of head motion and to model the average activity used to maintain limb position during CT stabilization. Regressors were convolved with a γ-variate function and the resulting functional images were pre-processed as described above. Voxel-wise, one-sided t-

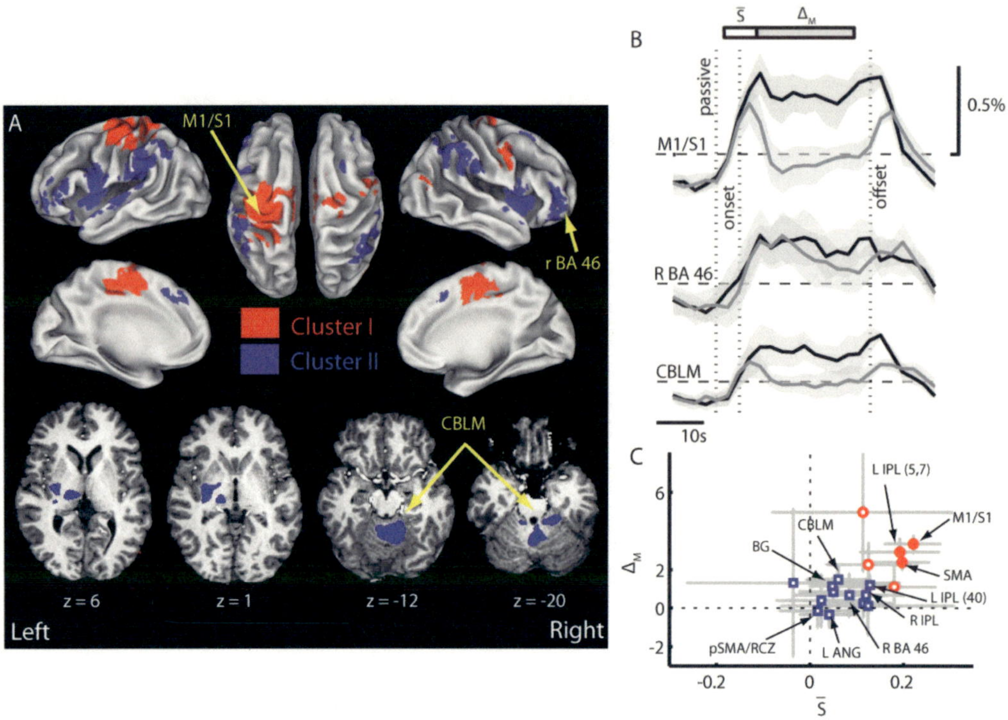

Figure 6. (A) Population maps showing significantly activated ROIs during periods of random ***RT*** or constant ***CT*** stabilization ($p < 0.05$; corrected for multiple comparisons). Cluster I ROIs are shown in red while Cluster II ROIs are shown in blue. Upper panels: activations mapped onto inflated representations of the cerebral hemispheres. Lower panels: activations in the basal ganglia, thalamus, cerebellar cortex (lobule IV-VI), and right dentate nucleus. (B) Percent signal change (PSC) of BOLD time series from ***RT*** (black) and ***CT*** (gray) stabilization periods. Shaded regions: ±2 SEM. Scale bars above the PSC plots indicate the time intervals used to calculate the components of $\Psi = \{\bar{S}, \Delta_M\}$. (C) Group membership for 18 ROIs was visualized by plotting the coordinates of each ROI within the plane defined by the features of Ψ (red: Cluster I; blue: Cluster II). Filled symbols: region where PSC differs significantly from resting baseline. Figure reproduced with permission from Suminski et al., 2007b.

tests found specific regions in Cluster I and Cluster II to be modulated discrete movement corrections. Interestingly, BOLD signal intensity *increased* with respect to baseline in a subset of regions in Cluster II, while a subset of regions in Cluster I exhibited BOLD signal *decreases* (Fig. 7; red and blue areas, respectively). These results highlight the manner in which Cluster I and II regions work together to maintain the position of the limb during stabilization against a deterministic perturbation: During CT trials, discrete corrective movements induce a persistent decrease in the magnitude of kinematic errors, thus reducing the need for feedback regulation of position and likely contributing to the inhibition of Cluster I ROIs. Conversely, we found the BOLD activity increased in a subset of Cluster II ROIs at the specific times when discrete movements were performed. Such behavior would be expected from brain regions involved in the resetting of behavioral goals (reference wrist angle) when performance is deficient.

	Tailarach Coordinates					
	Hem	**X**	**Y**	**Z**	**Vol**	**Mean T**
Cluster I						
Precentral Gyrus (BA 4,6)	L	-30.1	-19.4	51.3	14166	5.884
Medial Frontal Gyrus (BA 6)	B	-0.5	-11.6	48	7856	5.0334
Cingulate Gyrus (BA 24,31)						
Inferior/Superior Parietal Lobule (BA 5,7,40)	L	-30.3	-38.3	56.6	2630	4.7958
Precentral Gyrus (BA 6)	R	49.6	-2.4	39	957	4.5288
Precentral Gyrus (BA 6)	R	22.3	-15.4	61.2	866	4.2839
Middle Temporal Gyrus (BA 39)	R	52.6	-68.9	15.9	596	4.3556
Cluster II						
Cortical						
Insula (BA 13)	L	-47.5	0.2	8.3	8102	4.8657
Superior Temporal Gyrus (BA 22,41)						
Inferior Parietal Lobule (BA 40)	L	-50.9	-31.5	24.9	7477	4.9995
Insula (BA 13)	R	42.7	5.6	4.8	5633	5.0229
Superior Temporal Gyrus (BA 22)						
Inferior Parietal Lobule (BA 40)	R	53.4	-40.6	36.4	5125	4.3849
Supramarginal Gyrus						
Middle/Inferior Frontal Gyrus (BA 10,46)	R	39.2	38.5	3.5	2668	4.5259
Middle/Inferior Frontal Gyrus (BA 10,45,46)	L	-42	29.8	12.5	1334	4.3385
Inferior Parietal Lobule (BA 40)	L	-49.1	-53.3	39.9	1079	4.1923
Angular Gyrus						
Medial Frontal Gyrus (BA 6,8)	B	-3	25.4	41.5	1033	4.156
Cingulate Gyrus (BA 32)						
Superior Frontal Gyrus (BA 8)						
Superior Frontal Gyrus (BA 6)	R	22.1	13.3	51.6	687	4.5912
Middle Frontal Gyrus (BA 10)	L	-38.2	49.2	3.4	582	4.2741
Subcortical and Cerebellum						
Cerebellar Cortex Lobule IV, V, VI	B	5.9	-49.4	-14	9571	5.2885
Dentate	R					
Basal Ganglia and Thalamus	L	-19.7	-17	10.1	6684	5.025
L = Left; R = Right; B = Bilateral; BA = Broadman's Area						

Table 1. Brain regions exhibiting significant task-related activation during ***RT*** or ***CT*** trials.

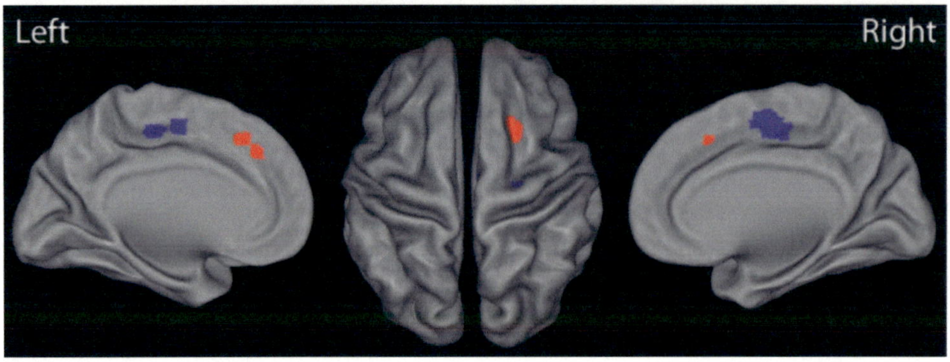

Figure 7. Activation maps showing regions where BOLD signal correlated with the generation of discrete corrective movements ($p < 0.05$, corrected for multiple comparisons). Increased BOLD activiy was observed in a subset of Cluster II regions [red: right pre-PMd, left medial frontal gyrus (BA 8) and rostral cingulate]. BOLD activaions decreased relative to resting state activity in Cluster I regions (blue: bilateral SMA and right PMd).

We performed a final set of multiple linear regression analyses to identify brain regions explicitly involved in the moment-by-moment and long-term evaluation and correction for kinematic and/or kinetic performance errors. We wished to know whether the activations identified in the preceding analyses were related to compensation for kinematic errors, generation of wrist torques, or both. Here, we performed four separate regressions using input reference functions corresponding to the magnitude of RMS wrist angle errors and RMS torque both on a trial-by-trial and TR-by-TR basis ($\boldsymbol{Error}_{Trial}$, $\boldsymbol{Error}_{TR}$, $\boldsymbol{Torque}_{Trial}$, and $\boldsymbol{Torque}_{TR}$). Each function was created from error or pressure data measured during the corresponding stabilization run within the MR-scanner. The value at each (TR) sampling instant for $\boldsymbol{Error}_{Trial}$ and $\boldsymbol{Torque}_{Trial}$ was set equal to the RMS error (or torque) value computed throughout the corresponding stabilization period (30 s integration window). The value at each (TR) sampling instant for $\boldsymbol{Error}_{TR}$ and $\boldsymbol{Torque}_{TR}$ (reference functions that varied on a moment-by-moment basis) was set equal to the RMS error (torque) value computed during that TR sampling period (2.5 s integration window). In all cases, the reference time series were convolved with a γ-variate function to model the temporal filtering properties of the hemodynamic response. Additional reference time series included head motion parameters and functions representing periods of ***RT*** and ***CT*** stabilization. The resulting functional images were processed as described above. Voxel-wise, one-sided t-tests were used to identify regions where BOLD signal covaried with $\boldsymbol{Error}_{Trial}$, $\boldsymbol{Error}_{TR}$, $\boldsymbol{Torque}_{Trial}$, or $\boldsymbol{Torque}_{TR}$.

Regions demonstrating increased sensitivity to errors that change on a moment-by-moment timescale ($\boldsymbol{Error}_{TR}$; Fig. 8 red) were predominantly members of Cluster I, while regions sensitive to kinematic errors on a much longer timescale ($\boldsymbol{Error}_{Trial}$; Fig. 8 blue) were more likely to be members of Cluster II (see also Table 2). Activation of the left IPL in the $\boldsymbol{Error}_{Trial}$ contrast was located more posterior to those activated regions of IPL in the $\boldsymbol{Error}_{TR}$ contrast. There were no overlapping activations in the $\boldsymbol{Error}_{Trial}$ and $\boldsymbol{Error}_{TR}$ contrasts. Importantly, no regions were found to be active in contrasts examining differences between either $\boldsymbol{Torque}_{Trial}$ or $\boldsymbol{Torque}_{TR}$ and resting baseline.

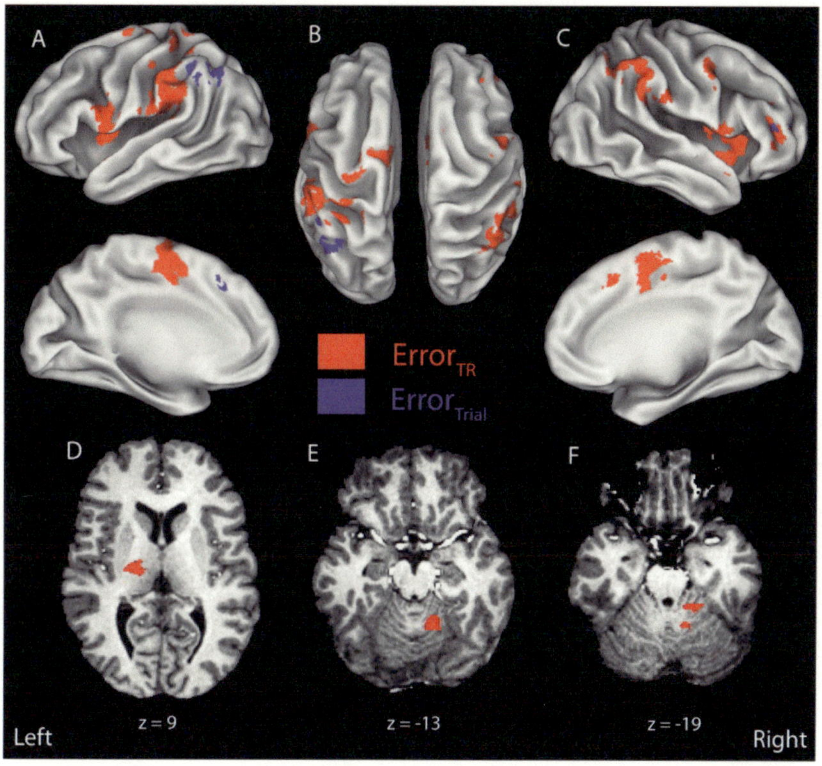

Figure 8. Activation maps showing regions wherein the BOLD response correlated with kinematic performance errors (i.e. state estimation errors) on both a moment-by-moment (***Error***$_{TR}$, red regions) and a trial-by-trial basis (***Error***$_{Trial}$, blue regions) ($p < 0.05$, corrected for multiple comparisons). Figure reproduced with permission from Suminski et al., 2007b.

		Tailarach Coordinates				
	Hem	**X**	**Y**	**Z**	**Vol**	**Mean T**
ErrorTR						
Cortical						
Inferior Parietal Lobule (BA 40)	L	51.3	29.9	28.4	3599	4.7941
Supramarginal Gyrus						
Postcentral Gyrus (BA 2)						
Insula (BA 13)						
Medial Frontal Gyrus (BA 6)	B	1.5	4.9	50	3056	4.7047
Superior Frontal Gyurs (BA 6)	L					
Cingulate Gyrus (BA 24)	R					

	Tailarach Coordinates					
	Hem	**X**	**Y**	**Z**	**Vol**	**Mean T**
Insula (BA 13)	R	-41	-8.3	2.9	2245	4.7498
Superior Temporal Gyrus (BA 22)						
Insula (BA 13)	L	47	-0.5	11.8	1986	4.9146
Superior Temporal Gyrus (BA 22)						
Inferior Parietal Lobule (BA 40)	R	-50.4	42	38.7	1706	4.2648
Supramarginal Gyrus						
Inferior Parietal Lobule (BA 5,40)	L	35.6	38.4	53.3	761	4.2931
Inferior Frontal Gyrus (BA 46)	R	-35.7	-34.5	14.9	756	5.0422
Middle Frontal Gyurs (BA 10)						
Precentral Gyrus (BA 6)	R	-46.3	2.3	43	247	4.7634
Precentral Gyrus (BA 44)	R	-53.7	-5.6	9.8	180	4.0774
Cingulate Gyrus (BA 32)	R	-8.6	-17.8	43.7	152	5.0572
Precentral Gyrus (BA 4)	L	28.5	21.6	63.2	115	3.9996
Subcortical and Cerebellum						
Basal Ganglia and Thalamus	L	18.8	14.4	7.1	1036	4.4593
Putamen						
Globus Pallidus (Medial/Lateral)						
Ventral Posterior Lateral						
Ventral Lateral						
Cerebellar Cortex Lobule IV, V, VI	R	-15.4	48.5	-16.1	871	4.2192
Dentate						
ErrorTrial						
Inferior Parietal Lobule (BA 40)	L	41.5	54.3	36.6	556	4.302
Supramarginal Gyrus						
Medial Frontal Gyrus (BA 8)	L	6.9	-20.3	43.6	45	3.9183
L = Left; R = Right; B = Bilateral; BA = Broadman's Area						

Table 2. Brain regions activated in the $Error_{TR}$ and $Error_{Trial}$ contrasts.

4. Summary and future directions

This chapter has described the development and validation of a novel, 1 DOF pneumatically actuated manipulandum. We demonstrated that the device was: 1) capable of generating computer controlled perturbations of movement and 2) compatible with the MR scanner such that performance of neither the device nor image quality was affected by robot operation. We then demonstrated device utility in a study of wrist posture stabilization against environmental perturbation. We provided behavioral evidence that subjects invoke three complementary compensatory responses when stabilizing the wrist in the absence of ongoing visual feedback of task performance. These compensatory responses include feedback regulation via spinal and long-loop reflexes, impedance modulation via antagonist muscle co-activation and feedforward, discrete corrective movements. Analysis of functional neuroimages obtained from the same subjects performing the same tasks revealed two distinct networks that were differentially excited by the task. The first (Cluster I) included a cerebello-thalamo-cortical network previously implicated in the online computation and feedback correction of errors (Marsden et al., 1972; Lee and Tatton, 1975; Evarts and Vaughn, 1978; Marsden et al., 1978; Strick, 1978; Horne and Butler, 1995). BOLD signal changes within these regions were correlated with moment-by-moment fluctuations in state estimation errors (Fig. 7, red regions; $\boldsymbol{Error}_{TR}$, 2.5 s integration period). That is, BOLD activity in these regions was elevated mainly during random torque perturbations and during passive movement of the hand, despite the absence of significant muscle activation during passive manipulation in the behavioral experiments and the production of the same average wrist torques in both tasks. A second network (Cluster II) exhibited similarly elevated BOLD activity during performance of both stabilization tasks. Brain regions demonstrating this pattern include the prefrontal cortex, rostral aspects of dorsal premotor and SMA cortices, and inferior aspects of posterior parietal cortex. These brain regions have previously been reported to contribute to the planning and execution of internally-generated, discrete motor actions (Jahanshahi et al., 1995; Winstein et al., 1997; Jueptner and Weiller, 1998; Sakai et al., 1999). Consistent with those reports, we found that BOLD activity increased within rostral PMd, SMA and cingulate cortex during time periods wherein subjects generated discrete corrective movements. Although discrete corrections were transient events, they gave rise to long-lasting performance improvements in most cases and so, they likely represent a resetting of the reference wrist angle when online (moment-by-moment) feedback control failed to satisfy subjective performance constraints. To do so, however, subjects first needed to determine when error had grown sufficiently large to warrant correction and then plan the direction and magnitude of corrective action. This type of supervisory role likely was supported by Cluster II ROIs demonstrating high sensitivity to state estimation errors over a longer, 30 s timeframe (Fig. 7, blue regions).

These results highlight the importance of both postural and movement trajectory control mechanisms in peripheral limb stabilization and suggest a possible neural basis for the distinct postural and trajectory (movement) control mechanisms recently isolated during point-to-point arm movements and movement sequences (Ghez et al., 2007; Scheidt and Ghez, 2007). Additional studies are needed to better understand how the brain combines the different control processes to minimize performance errors and how the brain uses information from multiple sensory feedback modalities to optimize limb stabilization and movement control. Such work will be greatly facilitated by the use of mechanically-active tools that apply physical perturbations to the limb while subjects undergo concurrent functional MR imaging.

Acknowledgements

This work was supported by NSF BES0238442, NIH NCRR M01-RR00058, 2 P01 MH51358, R01 HD053727 and by the Alvin W. and Marion Birnschein Foundation. We thank Kristine Mosier and Steve Rao for constructive feedback on an earlier version of this manuscript, Jeff Goldstein for crafting the manipulandum, Dr. N. Bansal for helpful suggestions regarding the statistical handling of the data, as well as Vanai Roopchansingh, Sally Durgerian, Matthew Verber and the MCW MRI technicians for assistance during FMRI data collection and analysis.

Author details

Aaron J. Suminski[1,2] and Robert A. Scheidt[1,3,4]

1 Biomedical Engineering, Marquette University, Milwaukee, WI, USA

2 Organismal Biology and Anatomy, University of Chicago, Chicago, IL, USA

3 Neurology, Medical College of Wisconsin, Milwaukee, WI, USA

4 Physical Medicine and Rehabilitation, Feinberg School of Medicine, Northwestern University, Chicago, IL, USA

References

[1] Carter CS, Braver TS, Barch DM, Botvinick MM, Noll D, Cohen JD (1998) Anterior cingulate cortex, error detection, and the online monitoring of performance. Science 280:747-749.

[2] Chinzei K, Kikinis R, Jolesz FA (1999) MR compatibility of mechatronic devices: design criteria. In: Medical Image Computing and Computer-Assisted Intervention Meeting - MICCAI'99., pp 1020-1030. Cambridge, UK.

[3] Cox RW (1996) AFNI: software for analysis and visualization of functional magnetic resonance neuroimages. Comput Biomed Res 29:162-173.

[4] Diedrichsen J, Shadmehr R (2005) Detecting and adjusting for artifacts in fMRI time series data. Neuroimage 27:624-634.

[5] Diedrichsen J, Hashambhoy Y, Rane T, Shadmehr R (2005) Neural correlates of reach errors. J Neurosci 25:9919-9931.

[6] Evarts EV, Tanji J (1976) Reflex and intended responses in motor cortex pyramidal tract neurons of monkey. J Neurophysiol 39:1069-1080.

[7] Evarts EV, Vaughn WJ (1978) Intended Arm Movements in Response to Externally Produced Arm Displacements in Man. In: Cerebral Motor Control in Man: Long Loop Mechanisms (Desmedt JE, ed), pp 178-192. Basel ; New York: Karger.

[8] Evarts EV, Fromm C (1981) Transcortical reflexes and servo control of movement. Can J Physiol Pharmacol 59:757-775.

[9] Fagg AH, Barto AG, Houk JC (1998) Learning to reach via corrective movements. In: Tenth Yale Workshop on Adaptive and Learning Systems, pp pp. 179-185. New Haven, CT.

[10] Flueckiger M, Bullo M, Chapuis D, Gassert R, Perriard Y (2005) fMRI compatible haptic interface actuated with traveling wave ultrasonic motor. In: IEEE Industry Applications Society Fortieth Annual Meeting pp 2075-2082 2073. Kowloon, Hong Kong.

[11] Ganesh G, Gassert R, Burdet E, Bleuler H (2004) Dynamics and control of an MRI compatible master-slave system with hydrostatic transmission. In: IEEE International Conference on Robotics and Automation, pp 1288-1294. New Orleans, LA.

[12] Gassert R, Burdet E, Chinzei K (2008) Opportunities and challenges in MR-compatible robotics: reviewing the history, mechatronic components, and future directions of this technology. IEEE Eng Med Biol Mag 27:15-22.

[13] Gassert R, Moser R, Burdet E, Bleuler H (2006) MRI/fMRI-compatible robotic system with force feedback for interaction with human motion. IEEE/ASME Transactions on Mechatronics 11:216-224.

[14] Ghez C, Scheidt R, Heijink H (2007) Different learned coordinate frames for planning trajectories and final positions in reaching. J Neurophysiol 98:3614-3626.

[15] Haacke EM, Brown RW, Thompson MR, Venkatesan R (1999) Magnetic resonance imaging : physical principles and sequence design. New York: Wiley.

[16] Haaland KY, Harrington DL (1989) Hemispheric control of the initial and corrective components of aiming movements. Neuropsychologia 27:961-969.

[17] Hidler J, Hodics T, Xu B, Dobkin B, Cohen LG (2006) MR compatible force sensing system for real-time monitoring of wrist moments during fMRI testing. J Neurosci Methods 155:300-307.

[18] Horne MK, Butler EG (1995) The role of the cerebello-thalamo-cortical pathway in skilled movement. Prog Neurobiol 46:199-213.

[19] Imamizu H, Miyauchi S, Tamada T, Sasaki Y, Takino R, Putz B, Yoshioka T, Kawato M (2000) Human cerebellar activity reflecting an acquired internal model of a new tool.[comment]. Nature 403:192-195.

[20] Jahanshahi M, Jenkins IH, Brown RG, Marsden CD, Passingham RE, Brooks DJ (1995) Self-initiated versus externally triggered movements. I. An investigation using

measurement of regional cerebral blood flow with PET and movement-related potentials in normal and Parkinson's disease subjects. Brain 118 (Pt 4):913-933.

[21] Jueptner M, Weiller C (1998) A review of differences between basal ganglia and cerebellar control of movements as revealed by functional imaging studies. Brain 121 (Pt 8):1437-1449.

[22] Kawato M, Kuroda T, Imamizu H, Nakano E, Miyauchi S, Yoshioka T (2003) Internal forward models in the cerebellum: fMRI study on grip force and load force coupling. Progress in Brain Research 142:171-188.

[23] Khanicheh A, Mintzopoulos D, Weinberg B, Tzika AA, Mavroidis C (2008) MR_CHIROD v.2: magnetic resonance compatible smart hand rehabilitation device for brain imaging. IEEE Trans Neural Syst Rehabil Eng 16:91-98.

[24] Khanicheh A, Muto A, Triantafyllou C, Weinberg B, Astrakas L, Tzika A, Mavroidis C (2005) MR compatible ERF driven hand rehabilitation device. In: IEEE 9th International Conference on Rehabilitation Robotics, pp 7-12. Chicago, IL.

[25] Lee RG, Tatton WG (1975) Motor responses to sudden limb displacements in primates with specific CNS lesions and in human patients with motor system disorders. Can J Neurol Sci 2:285-293.

[26] Marsden CD, Merton PA, Morton HB (1972) Servo action in human voluntary movement. Nature 238:140-143.

[27] Marsden CD, Merton PA, Morton HB, Adam JER, Hallett M (1978) Automatic and Voluntary Responses to Muscle Stretch in Man. In: Cerebral Motor Control in Man: Long Loop Mechanisms (Desmedt JE, ed), pp 167-177. Basel ; New York: Karger.

[28] Matthews PBC (1981) Muscle spindles their messages and their fusimotor supply. In: Sect, 1: The Nervous System (Brookhart JM, Mountcastle VB, Brooks VB, eds), pp 189-228. Bethesda: Am. Phisiol. Soc.

[29] Milner TE, Cloutier C (1993) Compensation for mechanically unstable loading in voluntary wrist movement. Experimental Brain Research 94:522-532.

[30] Peck KK, Sunderland A, Peters AM, Butterworth S, Clark P, Gowland PA (2001) Cerebral activation during a simple force production task: changes in the time course of the haemodynamic response. Neuroreport 12:2813-2816.

[31] Rao SM, Bandettini PA, Binder JR, Bobholz JA, Hammeke TA, Stein EA, Hyde JS (1996) Relationship between finger movement rate and functional magnetic resonance signal change in human primary motor cortex. J Cereb Blood Flow Metab 16:1250-1254.

[32] Riener R, Villgrattner T, Kleiser R, Nef T, Kollias S (2005) fMRI-Compatible Electromagnetic Haptic Interface. In: 27th Annual International Conference of the IEEE-EMBS, pp 7024-7027. Shanghi, China.

[33] Sakai K, Hikosaka O, Miyauchi S, Sasaki Y, Fujimaki N, Putz B (1999) Presupplementary motor area activation during sequence learning reflects visuo-motor association. J Neurosci 19:RC1.

[34] Scheidt RA, Ghez C (2007) Separate adaptive mechanisms for controlling trajectory and final position in reaching. J Neurophysiol 98:3600-3613.

[35] Schenck JF (1996) The role of magnetic susceptibility in magnetic resonance imaging: MRI magnetic compatibility of the first and second kinds. Medical Physics 23:815-850.

[36] Schmahmann JD (2000) MRI atlas of the human cerebellum. In. San Diego: Academic Press.

[37] Scott SH (2004) Optimal feedback control and the neural basis of volitional motor control. Nat Rev Neurosci 5:532-546.

[38] Sinkjaer T, Hayashi R (1989) Regulation of wrist stiffness by the stretch reflex. J Biomech 22:1133-1140.

[39] Sinkjaer T, Andersen JB, Ladouceur M, Christensen LO, Nielsen JB (2000) Major role for sensory feedback in soleus EMG activity in the stance phase of walking in man. J Physiol 523 Pt 3:817-827.

[40] Strick PL (1978) Cerebellar Involvement in 'Volitional' Muscle Responses to Load Changes. In: Cerebral Motor Control in Man: Long Loop Mechanisms (Desmedt JE, ed), pp 85-93. Basel ; New York: Karger.

[41] Suminski AJ, Ropella KM, Scheidt RA (2002) A pneumatically actuated manipulandum for neuromotor control research. In: IEEE EMBS Society Meeting, pp 2347-2348 vol.2343. Houston, TX.

[42] Suminski AJ, Zimbelman JL, Scheidt RA (2007a) Design and validation of a MR-compatible pneumatic manipulandum. J Neurosci Methods 163:255-266.

[43] Suminski AJ, Rao SM, Mosier KM, Scheidt RA (2007b) Neural and electromyographic correlates of wrist posture control. J Neurophysiol 97:1527-1545.

[44] Talairach J, Tournoux P (1988) Co-planar stereotaxic atlas of the human brain. Stuttgart/New York: Thieme.

[45] Thoroughman KA, Shadmehr R (1999) Electromyographic correlates of learning an internal model of reaching movements. J Neurosci 19:8573-8588.

[46] Todorov E, Jordan MI (2002) Optimal feedback control as a theory of motor coordination. Nat Neurosci 5:1226-1235.

[47] Vaillancourt DE, Thulborn KR, Corcos DM (2003) Neural basis for the processes that underlie visually guided and internally guided force control in humans. J Neurophysiol 90:3330-3340.

[48] Van Essen DC, Drury HA, Dickson J, Harwell J, Hanlon D, Anderson CH (2001) An integrated software suite for surface-based analyses of cerebral cortex. J Am Med Inform Assoc 8:443-459.

[49] Wann JP, Ibrahim SF (1992) Does limb proprioception drift? Exp Brain Res 91:162-166.

[50] Winstein CJ, Grafton ST, Pohl PS (1997) Motor task difficulty and brain activity: investigation of goal-directed reciprocal aiming using positron emission tomography. J Neurophysiol 77:1581-1594.

[51] Yu N, Hollnagel C, Blickenstorfer A, Kollias SS, Riener R (2008) Comparison of MRI-Compatible Mechatronic Systems With Hydrodynamic and Pneumatic Actuation. Mechatronics, IEEE/ASME Transactions on 13:268-277.

[52] Ziegler JG, Nichols NB (1942) Optimum settings for automatic controllers. ASME Trans 64:759-768.

Chapter 18

Active and Passive fMRI for Presurgical Mapping of Motor and Language Cortex

Victoria A.L. Mosher, Einat Liebenthal and Bradley G. Goodyear

Additional information is available at the end of the chapter

http://dx.doi.org/10.5772/30844

1. Introduction

1.1. Overview of fMRI and its use in presurgical mapping

Functional magnetic resonance imaging (fMRI) using the blood-oxygenation level dependent (BOLD) contrast measures local changes in the concentration of paramagnetic deoxyhemoglobin associated with an increase in blood flow to active regions of the brain (Bandettini, Wong, Hinks, Tikofsky, & Hyde, 1992; Kwong, et al., 1992; Ogawa, Lee, Kay, & Tank, 1990). Together with structural MRI, BOLD fMRI is increasingly being used for presurgical functional mapping of epilepsy and brain tumor patients (Binder, et al., 1996; Haberg, Kvistad, Unsgard, & Haraldseth, 2004; Hall & Truwit, 2008; Krings, et al., 2001; Lee, et al., 1999; Van Westen, Skagerberg, Olsrud, Fransson, & Larsson, 2005; Liebenthal, 2011). Presurgical fMRI provides patient-specific functional information that may facilitate maximal tumor or epileptic tissue surgical resection, with minimal damage to surrounding gray and white matter structures that support vital sensory and cognitive functions. Presurgical mapping of brain regions and their spatial relationship to the lesion can in some cases help predict possible deficits in sensory, motor or cognitive functions due to surgery or due to continued lesion growth (Haberg, Kvistad, Unsgard, & Haraldseth, 2004). In this way, presurgical mapping can guide the decision for course of intervention (extent of resection, alternative treatment, etc.) and the decision to conduct additional intra-operative functional mapping (e.g., intra-cortical stimulation-ICS). Finally, presurgical mapping is also valuable for planning the pathway of surgical approach.

The main advantages of fMRI in particular for clinical applications are its high spatial resolution, non-invasive nature, relative safety, and the good access to MRI scanners in most medical centers. Compared to functional mapping using ICS, or the intracarotid sodium amobarbital procedure (Wada test) to assess hemispheric language and memory dominance, fMRI is far less unpleasant to the patient and it is also less expensive (Medina et al., 2004). Unlike ICS, fMRI is performed prior to the surgery, without the time constraints and limitations on patient cognitive ability imposed by the surgery. The Wada test, named after Canadian neurologist and epileptologist Juhn Atsushi Wada, was developed 50 years ago to assess the hemispheric dominance of language and thereby the risk of language decline in patients undergoing brain surgery (Wada & Rasmussen, 1960). The procedure involves some risk to the patient, however, and it is unpleasant, consisting of intracarotid injection of a barbiturate to one brain hemisphere, during which time the other hemisphere is examined for language and memory functions.

One concern sometimes raised with respect to fMRI presurgical mapping is that it identifies neuronal networks involved in a specific function, including regions that may actually be expendable (Desmond & Annabel Chen, 2002). The problem is that, surgical resection of expendable areas of a functional network identified with fMRI should not cause a post-surgical functional decline, and it would therefore be safe and desirable for achieving the surgery goals to resect them. In contrast, ICS is thought to indicate strictly regions that are essential for a function, and should therefore be preserved during surgery in order to prevent functional damage. However it is important to understand that, as with fMRI, a similar concern actually applies to ICS in that regions that are remote from the stimulated region but connected to it may also be affected by the stimulation and 'erroneously' considered essential.

Another difficulty relates to the high inter-and intra-subject variability in the extent of activation foci observed with fMRI even with constant scanning parameters and consistent task performance (Liu, Zhang, Brown, Yue, 2004). This problem usually arises because the extent of active areas is determined by a statistical significance threshold that depends on the quality and quantity of the acquired signal, irrespective of the physiological boundaries of functional fields. In an effort to address this issue, recent work examined the spatial and temporal reproducibility of the relative amplitude of the fMRI signal and found it to be more stable than the absolute amplitude (Voyvodic, 2006). It was further determined empirically that thresholding individual activation maps at 40% of the most active voxels resulted in highly reproducible activity in the sensorimotor hand region during a finger tapping task, both within and across subjects (Voyvodic, Petrella, Friedman, 2009). Thus, relative amplitude maps can be used to improve the reliability and anatomical specificity of functional MR mapping.

Despite the differences between the techniques, there is generally very good concordance between fMRI and ICS mapping of somatosensory and motor functions (Bartos, et al., 2009; Hirsch, et al., 2000; Holodny, et al., 2000; Jack, et al., 1994; Krishnan, et al., 2004; Lehericy, Duffau, et al., 2000; Pujol, et al., 1996; Yetkin, et al., 1997). Nevertheless, in language mapping, reported levels of task sensitivity and specificity are variable and strongly dependent on the particular task that is used (Bookheimer, 2007). High sensitivity (i.e., fMRI activates all or most sites identified by ICS) but low specificity (i.e., fMRI activates additional sites not identified by ICS) has been reported for auditory tasks and in particular word generation to auditory cues (FitzGerald, et al., 1997). Higher sensitivity can generally be achieved with the conjunction

http://dx.doi.org/10.5772/30844

of several fMRI language tasks, but specificity remains low (Pouratian, Bookheimer, Rex, Martin, & Toga, 2002; Rutten, Ramsey, van Rijen, Noordmans, & van Veelen, 2002).

Perhaps the most serious concern with regard to fMRI presurgical mapping (and functional mapping in general) is that the predictive value of this technique for patient outcome has not yet systematically been established in different patient types and using specific activation paradigms. Several studies in small cohorts of brain tumor patients suggest that fMRI mapping can be predictive of surgical outcome. A distance of 10 mm or more between tumor and functional cortex has been suggested to carry reduced risk for postsurgical neurological deficits (Haberg, et al., 2004; Krishnan, et al., 2004). In epilepsy, fMRI of language has been demonstrated to be strongly correlated with Wada language asymmetry (Binder, et al., 1996; Lehericy, Cohen, et al., 2000; Liegeois, et al., 2002) and to predict language (Sabsevitz, et al., 2003) and verbal memory (Binder et al., 2008) outcome after left anterior temporal lobe resection. Despite these recent advances, more work is required to establish guidelines for the use of fMRI in presurgical mapping.

1.2. Selection of fMRI paradigms

Factors related to the choice of fMRI activation paradigm (active or passive experimental task, control task, level of task performance, and the number of trials per task condition) have a critical impact on the resulting functional maps and their interpretation, particularly for mapping cognitive functions such as language and memory (Binder, Swanson, Hammeke, & Sabsevitz, 2008; Bookheimer, 2007). In terms of motor paradigms, simple hand/finger movements are usually sufficient to get a reliable map of motor cortex activity. These types of paradigms will be presented later in the chapter when we compare task-based fMRI with passive resting-state paradigms.

In terms of language tasks, active tasks requiring an overt response are generally considered more reliable because they allow monitoring of the patient's performance, and are thought to engage targeted brain regions more strongly (Lee, Jack, & Riederer, 1998). An active control task rather than passive rest (i.e., no task) is recommended as a baseline for contrasting with an active language task, because activation in the control condition can account for general executive, attentional and motor processes that are nonspecific to the tested cognitive function (Binder, Swanson, et al., 2008; Demonet, et al., 1992). In addition, activation during passive rest can be dominated by continuous processing of linguistically-mediated thoughts (whether day dreaming or intense mental processing), thereby effectively reducing the differential activation of language networks between the active and control tasks (McKiernan, D'Angelo, Kaufman, & Binder, 2006; Stark & Squire, 2001).

Finally, the task demands should be adjusted according to the patient's level of performance. For example, shorter and/or easier versions of the tasks may be needed for young children who tend to have a shorter attention span and a different knowledge base. As will be elaborated in the second part of this chapter, an interesting alternative approach to active task paradigms is passive resting-state fMRI. The resting-state paradigm is independent of patient performance, is time efficient, and can therefore serve as an alternative for patients who cannot comply with active task demands.

1.3. Comparative evaluation of three fMRI language tasks

To demonstrate the impact of the choice of paradigm on resulting fMRI maps, we compared three language tasks in terms of sensitivity and specificity in regions of interest for language. These tasks are: an auditory semantic word decision task (Binder, et al., 1997), an auditory definition naming task, and a silent word generation task.

The semantic task consists of a dual-criteria go/no-go semantic word decision (SWD) to auditorily presented animal names (i.e., press the keypad if "this animal is found in North America and used by people"), contrasted with a dual-criteria go/no-go auditory decision task (press the keypad if "this tone sequence contains two high-pitch tones"). Sixteen blocks of 8 trials alternating between the tasks are presented in two separate runs. This task has been shown to reliably activate regions in left inferior and middle frontal gyri, left middle temporal gyrus, left fusiform gyrus, and left angular gyrus, involved in phonological and lexico-semantic functions (Binder, et al., 1997). The degree of lateralization of these activated areas has been shown to correspond with Wada language asymmetry in individual epilepsy patients (Binder, et al., 1996) and predicts language and verbal memory outcome after left temporal lobectomy (Binder, et al., 2008; Sabsevitz, et al., 2003).

The auditory definition (or responsive) naming task (DNT) consists of overtly naming (speaking the name of) an object, animal, famous place, or famous person, in response to an auditory description (for example, "a horse with stripes" or "what the king wears on his head") (Hamberger & Seidel, 2003). This task is contrasted with an auditory control task consisting of speaking the number of siren sounds embedded in a sequence of white noise. This task uses an event-related design with 128 naming trials and 128 control trials, presented intermixed in 4 separate runs. Relative to the SWD, the DNT places additional demands on the language system because it involves word retrieval and overt speech. This task is expected to engage stronger, or more extensive activation within language areas, particularly in the left inferior frontal, middle temporal and parietal cortex (Okada & Hickok, 2006; Tomaszewki Farias, Harrington, Broomand, & Seyal, 2005).

The silent word generation (SWG) task consists of covertly finding words that start with certain letters, each displayed on a screen for 5 seconds. The control task consists of visually fixating on letters from foreign alphabets. We compared the SWG task with the SWD and DNT tasks because it is simple to implement, and therefore often used for fMRI language mapping in clinical settings. Nevertheless, a significant drawback of this task is that there is no monitoring of task performance. In addition, the task engages primarily phonological processing, but imposes fewer demands on the semantic system.

The reliability of each of the three fMRI language tasks was examined in eleven right handed neurologically-healthy volunteers, using sensitivity and specificity measures described previously (Hirsch, et al., 2000). Target regions of interest (ROIs) for language (inferior frontal gyrus, IFG; middle temporal gyrus, MTG; and angular gyrus, AG) were defined using a digital atlas based on cytoarchitectonic and macro anatomical labels (Eickhoff, et al., 2005). An adaptive thresholding scheme, consisting of mapping the activation in each subject as a percentage of local excitation, was used as a means to reduce the variability in the extent of activation between subjects (Voyvodic, 2006). The amplitude threshold each ROI was set individually at 40% of the peak positive activation within that ROI. This level of thresholding was selected because it was shown to yield highly reproducible and specific individual motor activation maps in a large cohort of subjects (Voyvodic, et al., 2009). Individual maps were then grouped using a random effects model to account for intersubject variability and capture effects in the whole population. Sensitivity measures across the group for each task were computed based on the individual mean value of the t-statistic in all voxels exceeding the threshold in the target ROIs. Specificity was measured across subjects as the proportion of the above-threshold positive activation volume (computed as the mean t-value weighted by the number of voxels exceeding the threshold) in the target ROI relative to the whole brain. Lateralization indices (LI) were computed for each of the ROIs in the language tasks because of the clinical relevance of language dominance as a predictor of post surgical neurological deficit, and because quantitative measures of laterality show better within-subject reproducibility than quantitative measures of the volume of activation (Harrington, Buonocore, & Farias, 2006). The LI was computed as the ratio of the difference and the sum of the activation in each left and right region of interest (LI=[QLHQRH]/[QLH+QRH]), where the activation in the left and right hemispheres (QLH and QRH, respectively) was represented by the sensitivity or specificity measures described above. In this formulation, LI varies between 1 and-1, with positive values reflecting left hemisphere dominance and negative values reflecting right hemisphere dominance. LI measures incorporating both the magnitude and extent of activation were selected here because they have been shown to be more robust and reproducible (Fernandez et al., 2003; Jansen et al., 2006). Note that the size of the contralateral ROIs defined in the atlas were similar though not identical, but this did not bias the LI comparison between tasks because any ROI asymmetries were consistent across tasks.

The group maps of fMRI activation in the language ROIs for the three language tasks are shown in Figure 1, and the corresponding sensitivity and specificity measures in each ROI and task are shown in Figures 2 and 3. Activation was observed in all ROIs for all tasks. The LI based on the sensitivity measure (Figure 2) was similar for all tasks in IFG. However, it was larger for SWD and DNT in AG, with a similar trend in MTG. LI was largest for DNT when all three regions were combined. The LI based on the specificity measure (Figure 3) was larger for SWG in IFG, and larger for SWD and DNT in AG, with no differences observed in MTG. The LI was largest for DNT when all three regions were combined.

Figure 4 shows an example of presurgical fMRI language maps using the SWD and SWG tasks in a patient with right frontotemporal low-grade glioma. Activation for both the SWD and SWG tasks was left lateralized. Only the SWD task activated the angular gyrus. The SWG task produced stronger activation in IFG and MFG. Figure 5 shows an example of fMRI language maps in a neurologically healthy left handed volunteer. In this subject, the three maps consistently indicate right hemisphere dominance for language. As in previous examples, only the SWD and DNT tasks activated the angular gyrus. The SWG task activated specifically the frontal cortex.

Taken together, these results suggest that an important advantage of the semantic word decision (SWD) and definition naming (DNT) tasks over the silent word generation (SWG) task, is that the former tasks activate the angular gyrus (AG) with high sensitivity and specificity. The AG activation can be used to compute more robust lateralization indices. In contrast, the SWG task activates the inferior frontal gyrus (IFG) with high specificity, such that lateralization patterns in this area are more robust with the SWG compared to the SWD task. Thus, an optimal set of paradigms for presurgical language mapping would consist of a combination of the SWG task with the SWD or DNT tasks.

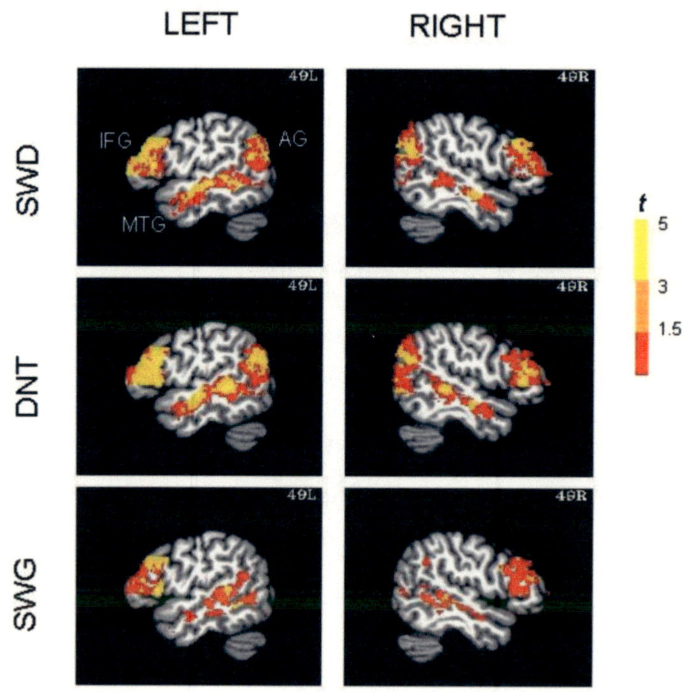

Figure 1. FMRI activation within the three language ROIs for a group of right handed, neurologically healthy adults (n=11), during auditory semantic word decision (SWD), auditory definition naming (DNT), and silent word generation (SWG) tasks. The individual activation maps in each task were thresholded at 40% of the maximum activation in each ROI, projected into standard stereotaxic space (Talairach & Tournoux, 1988), and combined across subjects using random effects analysis. The fMRI activity in the three language ROIs is displayed on sagittal slices of a template brain (Colin N27), at 49 mm to the left (49L, left panels) and to the right (49R, right panels) of the midsagittal plane. In this map, t=2.57 corresponds to p<0.01 (orange), and t=3.3 corresponds to p<0.001 (yellow).

http://dx.doi.org/10.5772/30844

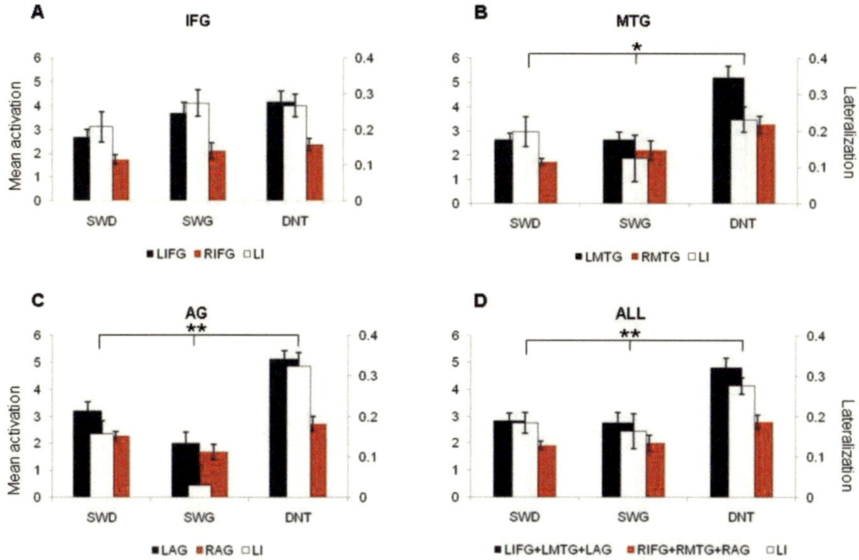

Figure 2. The sensitivity measure (expressed as mean *t*-value of supra threshold voxels) and corresponding lateralization index (LI) in the regions of interest for language (A. left and right inferior frontal gyri, B. left and right middle temporal gyri, C. left and right angular gyri – LIFG, RIFG, LMTG, RMTG, LAG, RAG, respectively), during semantic word decision (SWD), silent word generation (SWG), and definition naming (DNT). $*p<0.1$; $**p < 0.05$.

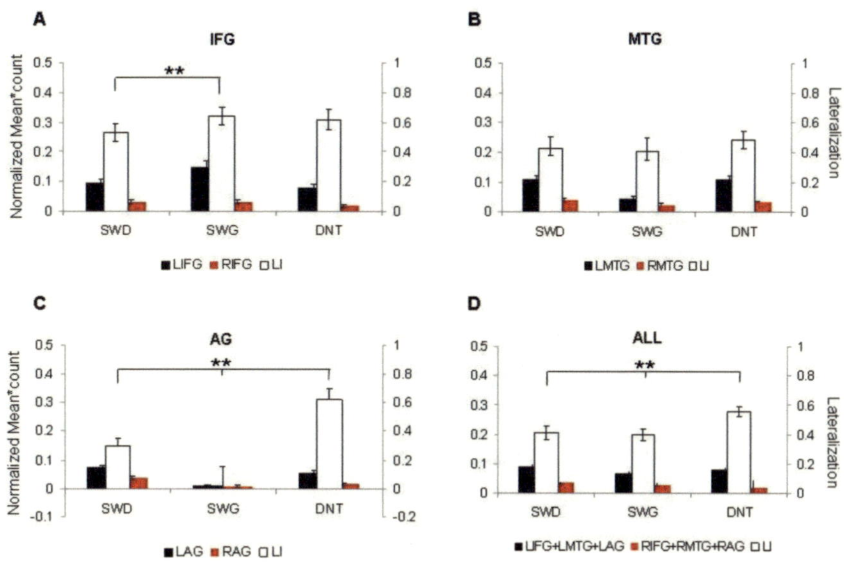

Figure 3. The specificity measure (expressed as mean *t*-value of supra threshold voxels weighted by voxel count in each ROI, normalized to the whole brain) and corresponding LI in each language task. Abbreviations as in Figure 2. $**p < 0.05$.

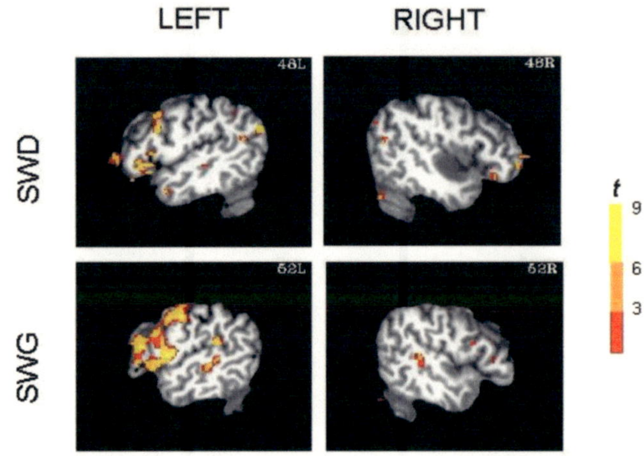

Figure 4. Presurgical fMRI language maps for the semantic word decision (SWD) and silent word generation (SWG) tasks, in a patient with right frontotemporal low-grade glioma (37 year old, right handed, female).

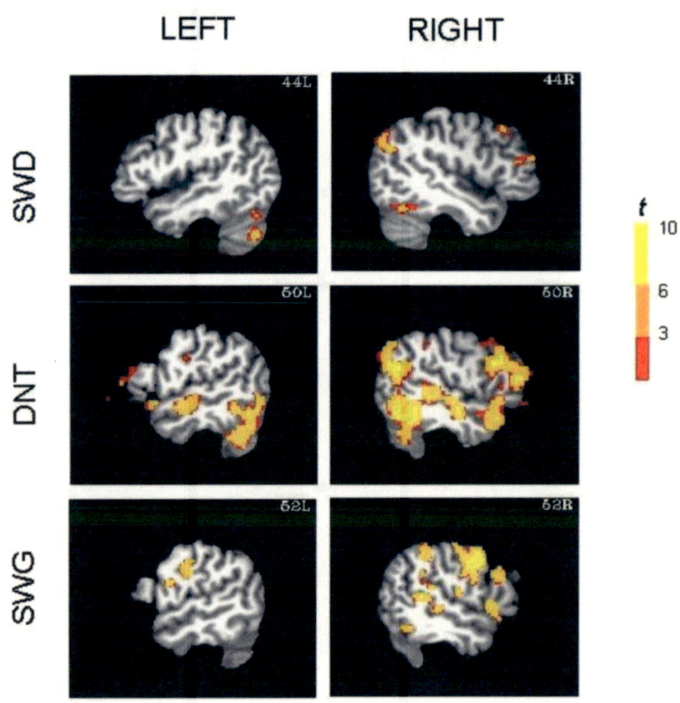

Figure 5. FMRI language maps in the three language tasks, in a neurologically healthy, left handed volunteer (Edinburgh Handedness Inventory score=-0.8). In these maps, the activation was thresholded at 40% of the maximal activation in the entire brain. The three maps are consistent with one another in suggesting right hemisphere dominance for language.

http://dx.doi.org/10.5772/30844

2. The origins of resting-state fMRI

As with many major scientific advances, the use of fMRI to investigate the resting state of the human brain was discovered essentially by accident. In 1995, Biswal and colleagues (Biswal, Yetkin, Haughton, & Hyde, 1995) initially meant to isolate physiological noise in fMRI data collected from subjects during rest; however, it was observed that a prominent low-frequency signal (< 0.1 Hz) remained after removing the physiological noise. To investigate this further, high frequencies were filtered from all voxels in the brain, and the signal from the left primary sensorimotor cortex was compared to the signal from all other voxels in the brain. A high level of correlation was observed with the right primary sensorimotor cortex and other motor areas. It was thus hypothesized that spontaneous low-frequency fluctuations in the BOLD signal were the result of functional connections between these brain regions.

With fMRI itself still in its infancy and no clear explanation of how the resting-state signal was generated, there remained some controversy as to whether the resting-state signal was neurologically significant. One study found that the strength of left/right inter-hemispheric correlations during the resting state decreased when subjects held their breath, and returned to normal levels once the subjects began breathing again (Biswal, Hudetz, Yetkin, Haughton, & Hyde, 1997). This study provided initial evidence that spontaneous low-frequency fluctuations were indeed blood-oxygenation dependent, like the BOLD signal. It was later observed using voltage sensitive dyes that neurons in the visual cortex exhibit patterns of spontaneous activity similar to the activation patterns elicited in response to stimuli of a specific visual orientation angle (Kenet, Bibitchkov, Tsodyks, Grinvald, & Arieli, 2003). This was one of the first studies to provide evidence that the resting-state signal had a physiological basis. In recent years, a number of studies have demonstrated a link between physiological, electrophysiological, and BOLD processes and the resting-state signal (Laufs, et al., 2003; Leopold, Murayama, & Logothetis, 2003; Mantini, Perrucci, Del Gratta, Romani, & Corbetta, 2007).

Through the validation of the resting-state phenomenon, resting-state fMRI has become a valuable tool to investigate how the brain functions at rest and how brain networks at rest are organized; however, the physiological mechanisms underlying the resting-state signal are still under investigation. Higher field scanners (3 Tesla and above) with improved signal-to-noise ratios, improved imaging sequences and head coils, and a better understanding of fMRI in general, have also contributed to the ubiquity of resting-state research. Many resting-state studies have been published since the first study, investigating a multitude of disease and normal brain states (e.g., Lowe, 2010; van den Heuvel & Hulshoff Pol, 2010). Since resting-state fMRI is still a relatively new concept, appropriate methods of collecting and analyzing resting-state data are still under development. With these advances in methodology, resting-state fMRI will continue to be an invaluable research tool for investigations of normal and pathological human brain function. However, as we will discuss in the remainder of this chapter, resting-state fMRI has the added potential of becoming a powerful diagnostic and assessment tool for individual clinical cases (Freilich & Gaillard, 2010; Sakoglu, et al., 2011).

3. Resting-state fMRI: The basics

3.1. The resting-state signal

The resting-state fMRI signal resides within a narrow frequency range of 0.01 to 0.1 Hz (De Luca, et al., 2006; Damoiseaux, et al., 2006). The similarity of these low-frequency fluctuations between spatially distinct brain regions implies a functional connection. The degree of similarity, which quantifies the strength of the functional connection, is called *functional connectivity*, and the collection of brain regions identified as connected constitute a *resting-state network*. These resting-state networks are consistent with our knowledge of structural connections via the connective white matter tracts. In a study of patients with multiple sclerosis, a significant correlation was found between the integrity of the white matter tracts (i.e., structural connectivity) within the motor pathway and functional connectivity of the motor network (Lowe, et al., 2008). This suggests that when white matter cannot functional properly, functional connectivity is diminished or even lost in the specific resting-state network. Even in healthy control subjects, the strength of functional connections within resting-state networks seems to be correlated with the strength of anatomical connections in the same area, as measured by diffusion tensor imaging (Skudlarski, et al., 2008). These studies have helped to establish a link between our knowledge of the anatomical connections in the brain and the resting-state fMRI signal.

3.2. The procedure

Acquiring resting-state fMRI data is relatively simple because of its non-demanding nature in terms of effort required from the subject. The subject is asked to remain still inside the MR scanner for a period of five to ten minutes, think of nothing in particular, and stare at a fixation cross or keep their eyes closed. It has been demonstrated that when the eyes are open, the functional connections between the thalamus and visual cortex are weakened as opposed to when the eyes are closed (Zou, et al., 2009). Conversely, it has been shown that the functional connections are stronger within the "default mode" network (consisting of the posterior cingulate cortex, precuneus, medial prefrontal cortex, ventral anterior cingulate cortex and lateral parietal cortex) when subjects have their eyes open (Yan, et al., 2009). Taken together, these studies suggest that the benefit of having subjects stare at a fixation cross or keep their eyes closed differs between resting-state networks, and must be taken into account when a study is planned.

Once the data has been collected, it is subjected to a pre-processing pipeline the same as that for task fMRI: motion correction, spatial and temporal smoothing, and spatial normalization to a standard template (i.e., the Montreal Neurological Institute brain or the Talairach atlas). In terms of the analysis of resting-state data, there are two broad categories: model-dependent or model-independent (Joel, Caffo, van Zijl, & Pekar, 2011; Margulies, et al., 2010). A model-dependent analysis assumes *a priori* knowledge of the approximate location of a brain region within the resting-state network under investigation. Using this knowledge, a seed region (consisting of a voxel or voxels) is typically defined either anatomically or by an apposite task. The average time course of this seed region is extracted and compared to the time course of every other voxel in the brain, usually by some variant of the General Linear Model including nuisance signals from cerebral spinal fluid, white matter, the average signal from the whole brain and physiological recordings of cardiac and respiratory signals (Glover, Li, & Ress, 2000; Murphy, Birn, Handwerker, Jones, & Bandettini, 2009). The voxels whose time courses are strongly correlated to the seed region are considered to be functionally connected to the seed region, and constitute a whole-brain map of the resting-state network. In contrast, a model-independent analysis is performed when there is no specific hypothesis about a brain region and its inclusion in a resting-state network. One common example of this approach is independent component analysis (ICA) (Hu, Yang, LaConte, & Weng, 2008). The ICA algorithm identifies groupings of spatial clusters (i.e., components) that are maximally independent and whose BOLD signal fluctuations are similar; however, knowledge of brain networks is still required to select and group the components to form a final map of the desired resting-state network. Whether the approach to resting-state connectivity analysis is ICA or a seed region, there is good agreement as to the number and location of these networks in the healthy human brain across studies (Damoiseaux, et al., 2006; De Luca, Beckmann, De Stefano, Matthews, & Smith, 2006; Kiviniemi, et al., 2009).

The ease of data collection from the subject's perspective is just one of the many reasons why resting-state is becoming a popular brain imaging method. As we will discuss in the next few sections, the resting-state method is being applied to a number of neurological and psychiatric disease states, which could eventually advance resting-state fMRI towards actual clinical use.

3.3. Current uses of resting-state fMRI

Because there is no effort required by the subject during imaging, resting-state fMRI is quickly becoming a popular research tool. The majority of resting-state investigations are conducted in tandem with task-related fMRI, to gain complementary information about the brain's functional networks. For some patient populations, however, task-based fMRI may not be possible because patients are non-compliant. Resting-state fMRI may permit the investigation of patients who would otherwise not be studied using fMRI. Thus, resting-state fMRI may be a good candidate for translation into the clinical realm (Freilich & Gaillard, 2010; Sakoglu, et al., 2011). A few of the current uses and techniques of resting-state fMRI will be outlined in this section.

To better understand the nature and mechanisms of functional connectivity, current studies combine resting-state fMRI with diffusion tensor imaging (DTI), an MR technique that provides an anatomical map of white matter tract integrity and quantifies structural connectivity (Skudlarski, et al., 2008). This combination of techniques permits a direct investigation of the relationship between structural and functional connectivity, since white matter tracts may provide the means for distant brain areas to remain functionally connected through correlated fluctuations in the BOLD signal. Studies have shown a strong correlation between structural and functional connectivity in the brain on a whole-brain scale (Skudlarski, et al., 2008) and for individual functional networks (Greicius, Supekar, Menon, & Dougherty, 2009). Hence, these types of studies provide insight into how structure and function are directly related in the human brain.

Resting-state fMRI is also being used to investigate a number of brain diseases and disorders. Disrupted resting-state networks have been identified in neurological disorders such as Alzheimer's disease (Jiang, et al., 2006) and psychopathologies such as schizophrenia (Liu, et al., 2007). One group has even used resting-state fMRI to distinguish patients with major depressive disorder from healthy controls (Craddock, Holtzheimer, Hu, & Mayberg, 2009), demonstrating that resting-state fMRI has the potential for individual patient diagnosis. Studies have also demonstrated changes in resting-state connectivity over the course of mild cognitive impairment (Wang, et al., 2011) and recovery after stroke (Park, et al., 2011). Hence, there is potential for resting-state fMRI in the clinical assessment of disease progression and recovery, as well as in the assessment of treatment strategies.

3.4. Advantages of resting-state fMRI

Resting-state fMRI has a number of advantages over task-based fMRI and other imaging modalities. Unlike PET, fMRI in general is noninvasive and unlike EEG, deep-brain activity can be more easily detected. In comparison to task-based fMRI, there are little or no practice effects associated with multiple resting-state sessions. The main advantage of resting-state fMRI, however, is that it requires no effort from the patient. This is especially helpful when studying patients with physical or psychological deficits who cannot perform or understand the required task. For example, most task-based fMRI studies of stroke involving the upper extremities require patients to perform repetitive finger or hand movements (Buma, Lindeman, Ramsey, & Kwakkel, 2010; Calautti & Baron, 2003). For many stroke patients, it is difficult to move a single finger or hand, and often other fingers or body parts move in unison (i.e. mirror movements) (Nelles, et al., 1998). As a result, it is difficult to obtain a reliable anatomical map of brain activity in response to the desired movement. Typically, these patients, who make up a large proportion of stroke patients, are simply not studied, creating a significant patient selection bias in research studies. As a result, this poses severe limits on the translation of task-based fMRI into clinical use. Since resting-state fMRI does not require any movement or action from the patient, this problem can be avoided. All patients, even the most severely impaired, can undergo a resting-state study, thus increasing the likelihood of gaining significant insight into the disease.

3.5. Disadvantages of resting-state fMRI

Despite continual advances in resting-state fMRI, there are still some drawbacks to the method. The most significant problem is that the physiological basis of the resting-state signal is still not completely understood. As discussed above, a number of studies support the notion that the resting-state BOLD signal has a neuronal basis (e.g., Lowe, et al., 2008; Skudlarski, et al., 2008). However, because of the inherent difficulty associated with recording neuronal activity from multiple spatially distinct sites, no studies have explicitly established the link between spontaneous fluctuations of the BOLD signal associated with the resting-state signal and neural activity.

Another problem is that although subjects are asked to lie still and not think of anything in particular, it is likely that periods of rest actually possess frequent states of unrest, such as

http://dx.doi.org/10.5772/30844

transient cognitive processes and inner monologues. These states interfere with the accurate determination of resting-state networks, especially those that involve the frontal cortices (Squire & Stark, 2001; McKiernan, et al., 2006). Indeed, it is impossible to achieve a state of "language rest". However, as we will demonstrate later in this chapter, this disadvantage can actually be exploited for the purposes of presurgical mapping of language processing.

Unless a seed region for resting-state analysis can be defined easily using an anatomical atlas, like motor cortex (Yousry, et al., 1995), an appropriate task-based experiment designed to activate a particular area of the brain is still required. This becomes more difficult if resting-state maps of higher cognitive functions are desired. A recent study, however, has demonstrated that ROIs can be defined using the resting-state data by selecting the voxels that exhibit the greatest inter-voxel cross-correlation (Goodyear & Golestani, 2011), thus removing the need for the task-based scan.

Of course, there are several disadvantages to fMRI in general. Compared to other brain imaging techniques such as EEG, fMRI possesses relatively low temporal resolution. This is because the hemodynamic response takes approximately two seconds to initiate, 6 to 8 seconds to reach its peak, and tens of seconds to return to baseline. In addition, because the BOLD signal is dependent on blood oxygenation and blood flow, there are many factors that can influence the generated signal such as age, brain pathology, and medications (Haller & Bartsch, 2009).

4. Resting-state fMRI for presurgical mapping

The downside to using any functional MRI technique for presurgical mapping, however, is that the brain shifts following tissue resection due to gravity, brain swelling and cerebrospinal fluid drainage (Nimsky, et al., 2000). This complicates the interpretation of functional localization post-surgically. Inaccuracy in brain function localization associated with this brain shift can be avoided using intraoperative MRI to assist in anatomical registration of fMRI data (Nimsky, Ganslandt, Hastreiter, & Fahlbusch, 2001). However, brain shift remains a problem for centers that do not have access to intraoperative MR scanners.

As noted above, resting-state fMRI circumvents the limitations associated with task-based fMRI: no task is necessary to isolate a specific brain region, it can be used on any patient regardless of their level of impairment, and multiple maps can be created from one short scan. This technique may also provide an easy means to track plasticity or recruitment of brain function that may occur following surgery. One advantage specific to epilepsy research is that resting-state fMRI can be used to study epileptogenic networks and foci, potentially assisting presurgical plannning (Guye, et al., 2010; Moeller, et al., 2011).

To date, very few studies have investigated resting-state fMRI for the purposes of presurgical mapping. Therefore, its feasibility and effectiveness as a presurgical tool is yet to be determined. For resting-state fMRI to become a clinically useful tool, it must provide reliable information on a single subject level. One study demonstrated good concordance between resting-state fMRI and ICS of patients with sensorimotor tumors, in terms of the identification of sensorimotor cortex (Zhang, et al., 2009). Another study demonstrated similar findings in patients with tumors or epileptic foci near the motor cortex. The patients that underwent surgery following fMRI mapping showed no new neurological deficits post-surgery (Liu, et al., 2009). Together, these studies provide the necessary framework to help move presurgical planning using resting-state fMRI from research into practice. However, before clinical application can be realized, studies are needed to validate resting-state techniques for presurgical mapping in terms of the brain regions that are identified and the dependence statistical threshold on deciding if brain regions are indeed significantly connected.

5. Clinical cases of resting-state fMRI

As an example of a typical case for presurgical motor fMRI, a patient with a right-hemisphere frontal lobe glioma in proximity to the motor cortex was referred to our centre for clinical fMRI. The clinical question was whether the patient exhibited a normal pattern of predominantly right-hemisphere motor activity in response to left hand movements and whether this activity was in close proximity or abutting the glioma. The motor task consisted of sequential finger tapping at a self-regulated pace of approximately 2 Hz. This task is performed easily by most patients and is effective in reliably activating sensorimotor regions, including contralateral primary somatosensory, premotor, and supplementary motor regions and the ipsilateral cerebellum (Gountouna et al., 2009; Yoo et al., 2005). The task was designed as twelve 30-sec blocks alternating between three visually cued conditions, right hand, left hand, and rest. The patient also underwent a 7-minute resting-state scan while staring at a fixation cross.

The results of the task-based fMRI and resting-state fMRI are shown in Figure 6. In response to left hand movement task, a typical pattern of brain activity was observed. A predominance of right hemisphere motor activity was observed within motor and sensorimotor cortices, like that for healthy control subjects. Bilateral premotor activity, however, was also present, and frontal activity associated with the cognitive components of the task was also observed. The conclusions drawn from this clinical fMRI study were that the patient exhibited a normal pattern of motor and sensorimotor activity, with an atypical distribution of bilateral premotor activity, possibly the result of functional compensation in response to the impinging glioma. It was advised that any resection should attempt to avoid the premotor regions lateral to the glioma.

A contralateral pattern of motor activity is associated with strong interhemispheric resting-state correlations in healthy subjects (Damoiseaux, et al., 2006). This was also the case for this patient (Figure 6). Strong intra-and inter-hemispheric correlations were observed, as well as strong correlations with the supplementary motor area (SMA) along the midline. The SMA is associated with the coordination of bimanual and complex movements (Sadato, Yonekura, Waki, Yamada, & Ishii, 1997; Wu, Wang, Hallett, Li, & Chan). Activity within the SMA was not observed for the simple hand movement task. In this case, the conclusions were that the interhemispheric resting-state connections are intact and that care must be taken during

http://dx.doi.org/10.5772/30844

resection to avoid the SMA along the medial edge of the glioma. This clinical case is a clear example where task-based fMRI and resting-state fMRI provide complementary and clinically relevant information for the purposes of presurgical planning.

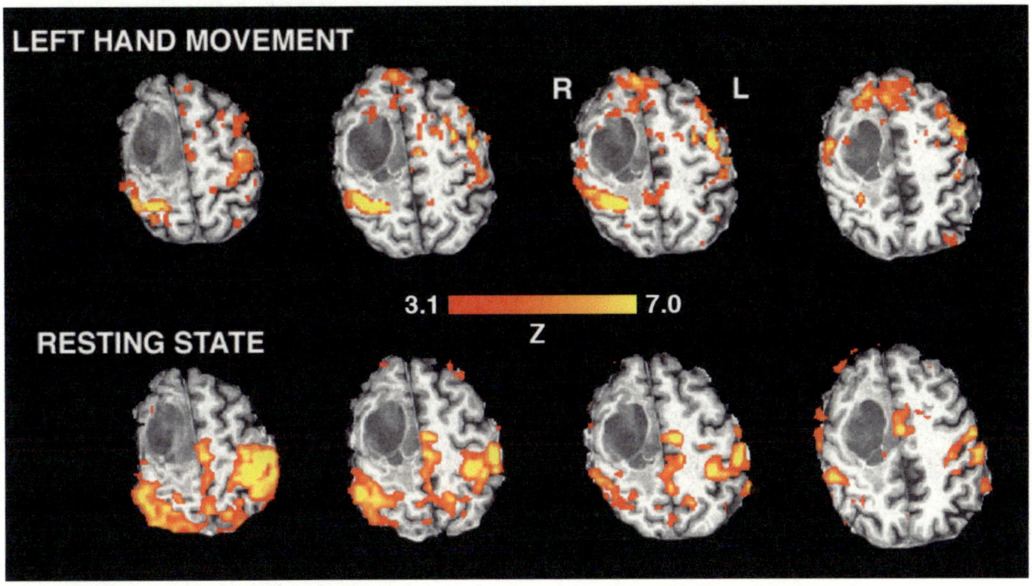

Figure 6. Motor task fMRI and resting-state fMRI data from a presurgical patient with a right frontal lobe glioma (clusters corrected to a significance level of α=0.05, using Gaussian Random Fields theory). For the motor task (left hand movement), a typical predominance of brain activity was observed within the right motor and sensorimotor cortices. Some prefrontal and frontal activity was also observed, likely associated with the cognitive components of the task (e.g., self-pacing of movement). In this case, it was concluded that motor activity is not in close proximity to the glioma. Resting-state fMRI using the right motor cortex as the seed region revealed strong correlations with left motor cortex as well as with supplementary motor cortex along the midline in close proximity to the glioma. Red-to-yellow indicates the increasing relative strength of brain activity (for the task) or resting-state correlation (both expressed as a statistical Z-score).

As examples of typical cases for presurgical language fMRI, two patients with left temporal lobe epilepsy were referred to our centre for clinical fMRI. The clinical question was whether the patient exhibited a normal pattern of predominantly left-hemisphere inferior frontal gyrus (i.e., Broca's area) activity in response to language processing. The language task consisted of the SWD task described earlier. The patients also underwent a 5-minute resting-state scan while staring at a fixation cross. Patients were not instructed to try to suppress any internal dialogue, but rather to just let their mind wander.

The results of the task-based fMRI and resting-state fMRI for the first patient are shown in Figure 7. In response to the language tasks, bilateral activity was observed within the inferior frontal gyrus (i.e., Broca's area). It was concluded from the task-related fMRI study that there was an atypical bilateral representation of language processing in this patient. A number of other brain regions were also activated by these tasks, which involved significant cognitive components. Of interest, however, was the lack of left-hemisphere Wernicke's area activity in response to these tasks.

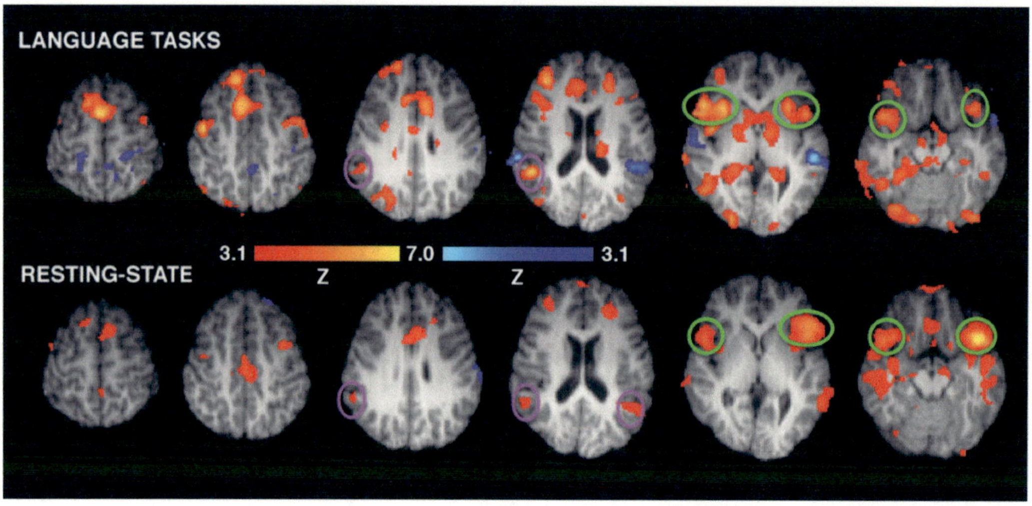

Figure 7. Language task fMRI and resting-state fMRI data from a presurgical patient with left temporal lobe epilepsy (clusters corrected to a significance level of α=0.05, using Gaussian Random Fields theory). For the language tasks (verb generation and lexical retrieval), significant bilateral activity within the inferior frontal gyri (green circles) was observed. Other cognitive areas associated with the task were also observed, including the anterior cingulate, caudate, and prefrontal cortex. Language processing was represented bilaterally. Resting-state fMRI using the left inferior frontal gyrus as the seed region demonstrated strong correlations with right inferior frontal gyri as well as bilateral Wernicke's area (purple circles). In this case, the conclusion was also a bilateral representation of language processing. Red-to-yellow indicates the increasing relative strength of brain activity (for the task) or resting-state correlation. Blue-to-light blue indicates the increasing relative strength of negative brain activity (for the task) or negative resting-state correlation (both expressed as a statistical Z-score).

The resting-state fMRI map revealed strong similarities to the language task fMRI; strong inter-hemispheric correlations were observed between the inferior frontal gyri, as well as between left and right Wernicke's areas. These identified regions suggest that the patient was indeed not in a resting-state for language, and in fact these results further suggest that internal dialogue during rest may be sufficient to determine the contribution of each hemisphere to language processing.

The results of the task-based fMRI and resting-state fMRI for the second patient are shown in Figure 8. In response to the language tasks, activity was observed predominantly within the left inferior frontal gyrus. It was concluded from the task-related fMRI study that there was a typical left-hemisphere dominant representation of language processing in this patient. A number of other brain regions were also activated by these tasks, which involved significant cognitive components.

As with the other clinical case, the resting-state fMRI map was similar to the language task fMRI map; weak or absent inter-hemispheric correlations were observed between the inferior frontal gyri, as well as between left and right Wernicke's areas. These results also suggest that

http://dx.doi.org/10.5772/30844

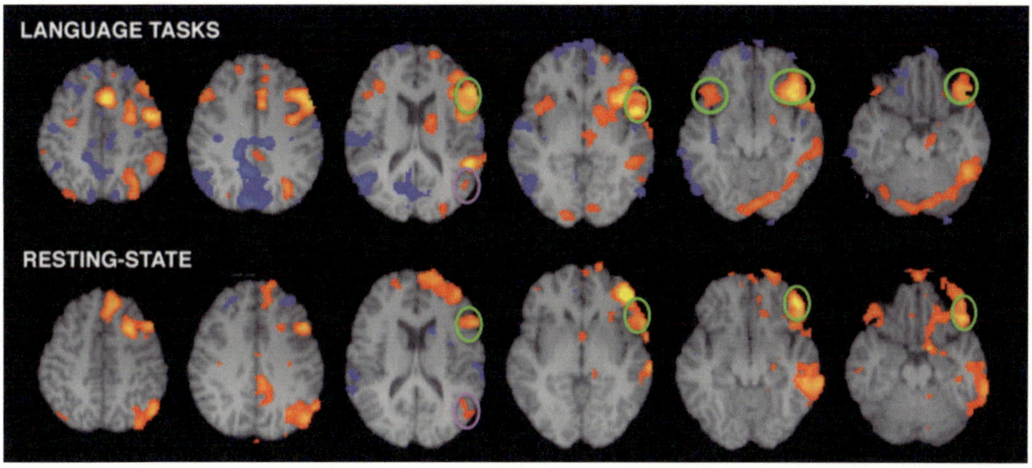

Figure 8. Language task fMRI and resting-state fMRI data from another presurgical patient with left temporal lobe epilepsy. For the language tasks (verb generation and lexical retrieval), significant left hemisphere dominant activity within the inferior frontal gyrus (green circles) was observed. Other cognitive areas associated with the task were also observed. In this case, the conclusion was a left hemisphere dominant representation of language processing. Resting-state fMRI using left inferior frontal gyrus as the seed region demonstrated strong correlations only within hemisphere, including left Wernicke's area (purple circles). In this case, the conclusion was also a left hemisphere dominant representation of language processing. Red-to-yellow indicates the increasing relative strength of brain activity (for the task) or resting-state correlation. Blue-to-light blue indicates the increasing relative strength of negative brain activity (for the task) or negative resting-state correlation.

internal dialogue may be sufficient to determine the contribution of each hemisphere to language processing.

6. Conclusions

In task-based fMRI, task selection has a critical impact on the resulting maps and their interpretation. In this chapter, we have shown that an important advantage of the semantic word decision and definition naming tasks over the commonly used silent word generation task is that the former tasks activate the angular gyrus with high sensitivity and specificity, allowing for more reliable estimation of hemispheric dominance in semantic processing. In contrast, the silent word generation task activates the inferior frontal gyrus with high specificity, such that estimation of lateralization patterns in this area is more robust with this task. Thus, a comprehensive set of paradigms for presurgical language mapping would consist of a combination of tasks targeting both phonological and semantic processing.

Resting-state fMRI provides a potential alternative to task-based fMRI in cases where patients are unable to comply with task instructions. As we have demonstrated in this chapter, resting-state fMRI of the motor network of individual patients provides important clinically relevant information complementary to task-based fMRI related to motor activity. We have also demonstrated that resting-state fMRI of the language network can also elucidate the laterality of language processing.

In summary, active or passive fMRI can provide a comprehensive delineation (and perhaps overestimation) of somatomotor and language regions surrounding a lesion, and this information can assist in assessing the risks of a surgery and the potential benefit of intrasurgical functional mapping. If the functional regions identified by fMRI are found to be distant from the lesion, this can facilitate a decision for radical tumor resection. If fMRI activation associated with vital sensory, motor or cognitive functions is demonstrated near the radiological tumor boundary and partial resection is suggested, ICS mapping can be used to identify the precise resection margin during the surgery itself. In this case, the presurgical information can be useful in speeding up the ICS procedure and in guiding the extent of the craniotomy (Bartos, et al., 2009; Rutten, et al., 2002).

Author details

Victoria A.L. Mosher[1], Einat Liebenthal[1,3] and Bradley G. Goodyear[1]

1 University of Calgary, Canada

University of Manitoba, Canada

2 Medical College of Wisconsin, USA

References

[1] Acharya, J. N., Dinner, D. S. (1997). Use of the intracarotid amobarbital procedure in the evaluation of memory. *J Clin Neurophysiol, 14*(4), 311-325.

[2] Bandettini, P. A., Wong, E. C., Hinks, R. S., Tikofsky, R. S., & Hyde, J. S. (1992). Time course EPI of human brain function during task activation. *Magn Reson Med, 25*(2), 390-397.

[3] Bartos, R., Jech, R., Vymazal, J., Petrovicky, P., Vachata, P., Hejcl, A., et al. (2009). Validity of primary motor area localization with fMRI versus electric cortical stimulation: a comparative study. *Acta Neurochir (Wien), 151*(9), 1071-1080.

[4] Binder, J. R., Frost, J. A., Hammeke, T. A., Cox, R. W., Rao, S. M., & Prieto, T. (1997). Human brain language areas identified by functional magnetic resonance imaging. *J Neurosci, 17*(1), 353-362.

[5] Binder, J. R., Sabsevitz, D. S., Swanson, S. J., Hammeke, T. A., Raghavan, M., & Mueller, W. M. (2008). Use of preoperative functional MRI to predict verbal memory decline after temporal lobe epilepsy surgery. *Epilepsia, 49*(8), 1377-1394.

[6] Binder, J. R., Swanson, S. J., Hammeke, T. A., Morris, G. L., Mueller, W. M., Fischer, M., et al. (1996). Determination of language dominance using functional MRI: a comparison with the Wada test. *Neurology, 46*(4), 978-984.

[7] Binder, J. R., Swanson, S. J., Hammeke, T. A., & Sabsevitz, D. S. (2008). A comparison of five fMRI protocols for mapping speech comprehension systems. *Epilepsia, 49*(12), 1980-1997.

[8] Binder, J. R. (2011). Functional MRI is a valid noninvasive alternative to Wada testing. *Epilepsy Behav, 20*(2), 214-222.

[9] Biswal, B., Hudetz, A. G., Yetkin, F. Z., Haughton, V. M., & Hyde, J. S. (1997). Hypercapnia reversibly suppresses low-frequency fluctuations in the human motor cortex during rest using echo-planar MRI. *J Cereb Blood Flow Metab, 17*(3), 301-308.

[10] Biswal, B., Yetkin, F. Z., Haughton, V. M., & Hyde, J. S. (1995). Functional connectivity in the motor cortex of resting human brain using echo-planar MRI. *Magn Reson Med, 34*(4), 537-541.

[11] Bookheimer, S. (2007). Pre-surgical language mapping with functional magnetic resonance imaging. *Neuropsychology Review, 17*(2), 145-155.

[12] Buma, F. E., Lindeman, E., Ramsey, N. F., & Kwakkel, G. (2010). Functional neuroimaging studies of early upper limb recovery after stroke: a systematic review of the literature. *Neurorehabil Neural Repair, 24*(7), 589-608.

[13] Calautti, C., & Baron, J. C. (2003). Functional neuroimaging studies of motor recovery after stroke in adults: a review. *Stroke, 34*(6), 1553-1566.

[14] Craddock, R. C., Holtzheimer, P. E., Hu, X. P. P., & Mayberg, H. S. (2009). Disease State Prediction From Resting State Functional Connectivity. *Magnetic Resonance in Medicine, 62*(6), 1619-1628.

[15] Damoiseaux, J. S., Rombouts, S. A., Barkhof, F., Scheltens, P., Stam, C. J., Smith, S. M., et al. (2006). Consistent resting-state networks across healthy subjects. *Proc Natl Acad Sci U S A, 103*(37), 13848-13853.

[16] De Luca, M., Beckmann, C. F., De Stefano, N., Matthews, P. M., & Smith, S. M. (2006). fMRI resting state networks define distinct modes of long-distance interactions in the human brain. *Neuroimage, 29*(4), 1359-1367.

[17] De Luca, M., Beckmann, C. F., De Stefano, N., Matthews, P. M., & Smith, S. M. (2006). fMRI resting state networks define distinct modes of long-distance interactions in the human brain. *Neuroimage, 29*(4), 1359-1367.

[18] Demonet, J. F., Chollet, F., Ramsay, S., Cardebat, D., Nespoulous, J. L., Wise, R., et al. (1992). The anatomy of phonological and semantic processing in normal subjects. *Brain, 115 (Pt 6)*, 1753-1768.

[19] Desmond, J. E., & Annabel Chen, S. H. (2002). Ethical issues in the clinical application of fMRI: factors affecting the validity and interpretation of activations. *Brain Cogn, 50*(3), 482-497.

[20] Eickhoff, S. B., Stephan, K. E., Mohlberg, H., Grefkes, C., Fink, G. R., Amunts, K., et al. (2005). A new SPM toolbox for combining probabilistic cytoarchitectonic maps and functional imaging data. *Neuroimage, 25*(4), 1325-1335.

[21] Fernandez, G., Specht, K., Weis, S., Tendolkar, I., Reuber, M., Fell, J. (2003). Intrasubject reproducibility of presurgical language lateralization and mapping using fMRI. *Neurology, 60*, 969-975.

[22] FitzGerald, D. B., Cosgrove, G. R., Ronner, S., Jiang, H., Buchbinder, B. R., Belliveau, J. W., et al. (1997). Location of language in the cortex: a comparison between functional MR imaging and electrocortical stimulation. *AJNR Am J Neuroradiol, 18*(8), 1529-1539.

[23] Freilich, E. R., Gaillard, W. D. (2010). Utility of functional MRI in pediatric neurology. *Curr Neurol Neurosci Rep, 10*(1), 40-46.

[24] Glover, G. H., Li, T. Q., & Ress, D. (2000). Image-based method for retrospective correction of physiological motion effects in fMRI: RETROICOR. *Magn Reson Med, 44*(1), 162-167.

[25] Goodyear, B. G., & Golestani, A. M. (2011). Regions of interest for resting-state fMRI analysis determined by inter-voxel cross-correlation. *Neuroimage, 56*(1), 246-251.

[26] Greicius, M. D., Supekar, K., Menon, V., & Dougherty, R. F. (2009). Resting-State Functional Connectivity Reflects Structural Connectivity in the Default Mode Network. *Cerebral Cortex, 19*(1), 72-78.

[27] Guye, M., Bettus, G., Bartolomei, F., Confort-Gouny, S., Guedj, E., Chauvel, P., et al. (2010). Role of resting state functional connectivity MRI in presurgical investigation of mesial temporal lobe epilepsy. *Journal of Neurology Neurosurgery and Psychiatry, 81*(10), 1147-1154.

[28] Haberg, A., Kvistad, K. A., Unsgard, G., & Haraldseth, O. (2004). Preoperative blood oxygen level-dependent functional magnetic resonance imaging in patients with primary brain tumors: Clinical application and outcome. *Neurosurgery, 54*(4), 902-914.

[29] Hall, W. A., & Truwit, C. L. (2008). Intraoperative MR-guided neurosurgery. *J Magn Reson Imaging, 27*(2), 368-375.

[30] Haller, S., & Bartsch, A. J. (2009). Pitfalls in FMRI. *Eur Radiol, 19*(11), 2689-2706.

[31] Hamberger, M. J., & Seidel, W. T. (2003). Auditory and visual naming tests: normative and patient data for accuracy, response time, and tip-of-the-tongue. *J Int Neuropsychol Soc, 9*(3), 479-489.

[32] Harrington, G. S., Buonocore, M. H., & Farias, S. T. (2006). Intrasubject reproducibility of functional MR imaging activation in language tasks. *AJNR Am J Neuroradiol, 27*(4), 938-944.

[33] Hirsch, J., Ruge, M. I., Kim, K. H., Correa, D. D., Victor, J. D., Relkin, N. R., et al. (2000). An integrated functional magnetic resonance imaging procedure for preoperative mapping of cortical areas associated with tactile, motor, language, and visual functions. *Neurosurgery, 47*(3), 711-721; discussion 721-712.

[34] Holodny, A. I., Schulder, M., Liu, W. C., Wolko, J., Maldjian, J. A., & Kalnin, A. J. (2000). The effect of brain tumors on BOLD functional MR imaging activation in the adjacent motor cortex: implications for image-guided neurosurgery. *AJNR Am J Neuroradiol, 21*(8), 1415-1422.

[35] Hu, X. P., Yang, Z., LaConte, S., & Weng, X. C. (2008). Ranking and averaging independent component analysis by reproducibility (RAICAR). *Human Brain Mapping, 29*(6), 711-725.

[36] Jack, C. R., Jr., Thompson, R. M., Butts, R. K., Sharbrough, F. W., Kelly, P. J., Hanson, D. P., et al. (1994). Sensory motor cortex: correlation of presurgical mapping with functional MR imaging and invasive cortical mapping. *Radiology, 190*(1), 85-92.

[37] Jansen, A., Deppe, M., Schwindt, W., Mohammadi, S., Sehlmeyer, C., Knecht, S. (2006). Interhemispheric dissociation of language regions in a healthy subject. *Arch Neurol, 63*, 1344-1346.

[38] Jiang, T. Z., Wang, L., Zang, Y. F., He, Y., Liang, M., Zhang, X. Q., et al. (2006). Changes in hippocampal connectivity in the early stages of Alzheimer's disease: Evidence from resting state fMRI. *Neuroimage, 31*(2), 496-504.

[39] Joel, S. E., Caffo, B. S., van Zijl, P. C., & Pekar, J. J. (2011). On the relationship between seed-based and ICA-based measures of functional connectivity. *Magn Reson Med.*

[40] Kenet, T., Bibitchkov, D., Tsodyks, M., Grinvald, A., & Arieli, A. (2003). Spontaneously emerging cortical representations of visual attributes. *Nature, 425*(6961), 954-956.

[41] Kiviniemi, V., Starck, T., Remes, J., Long, X., Nikkinen, J., Haapea, M., et al. (2009). Functional segmentation of the brain cortex using high model order group PICA. *Hum Brain Mapp, 30*(12), 3865-3886.

[42] Krings, T., Reinges, M. H., Erberich, S., Kemeny, S., Rohde, V., Spetzger, U., et al. (2001). Functional MRI for presurgical planning: problems, artefacts, and solution strategies. *J Neurol Neurosurg Psychiatry, 70*(6), 749-760.

[43] Krishnan, R., Raabe, A., Hattingen, E., Szelenyi, A., Yahya, H., Hermann, E., et al. (2004). Functional magnetic resonance imaging-integrated neuronavigation: correlation between lesion-to-motor cortex distance and outcome. *Neurosurgery, 55*(4), 904-914; discusssion 914-905.

[44] Kwong, K. K., Belliveau, J. W., Chesler, D. A., Goldberg, I. E., Weisskoff, R. M., Poncelet, B. P., et al. (1992). Dynamic magnetic resonance imaging of human brain activity during primary sensory stimulation. *Proc Natl Acad Sci U S A, 89*(12), 5675-5679.

[45] Laufs, H., Krakow, K., Sterzer, P., Eger, E., Beyerle, A., Salek-Haddadi, A., et al. (2003). Electroencephalographic signatures of attentional and cognitive default modes in spontaneous brain activity fluctuations at rest. *Proc Natl Acad Sci U S A, 100*(19), 11053-11058.

[46] Lee, C. C., Jack, C. R., Jr., & Riederer, S. J. (1998). Mapping of the central sulcus with functional MR: active versus passive activation tasks. *AJNR Am J Neuroradiol, 19*(5), 847-852.

[47] Lee, C. C., Ward, H. A., Sharbrough, F. W., Meyer, F. B., Marsh, W. R., Raffel, C., et al. (1999). Assessment of functional MR imaging in neurosurgical planning. *AJNR Am J Neuroradiol, 20*(8), 1511-1519.

[48] Lehericy, S., Cohen, L., Bazin, B., Samson, S., Giacomini, E., Rougetet, R., et al. (2000). Functional MR evaluation of temporal and frontal language dominance compared with the Wada test. *Neurology, 54*(8), 1625-1633.

[49] Lehericy, S., Duffau, H., Cornu, P., Capelle, L., Pidoux, B., Carpentier, A., et al. (2000). Correspondence between functional magnetic resonance imaging somatotopy and individual brain anatomy of the central region: comparison with intraoperative stimulation in patients with brain tumors. *J Neurosurg, 92*(4), 589-598.

[50] Leopold, D. A., Murayama, Y., & Logothetis, N. K. (2003). Very slow activity fluctuations in monkey visual cortex: implications for functional brain imaging. *Cereb Cortex, 13*(4), 422-433.

[51] Liebenthal, E. (2011). Introduction to Brain Imaging. In: Biomechanics of the brain (Miller K, ed), pp 41-68. Springer Science+Business Media, LLC.

[52] Liegeois, F., Connelly, A., Salmond, C. H., Gadian, D. G., Vargha-Khadem, F., & Baldeweg, T. (2002). A direct test for lateralization of language activation using fMRI: comparison with invasive assessments in children with epilepsy. *Neuroimage, 17*(4), 1861-1867.

[53] Liu, H., Buckner, R. L., Talukdar, T., Tanaka, N., Madsen, J. R., & Stufflebeam, S. M. (2009). Task-free presurgical mapping using functional magnetic resonance imaging intrinsic activity. *J Neurosurg, 111*(4), 746-754.

[54] Liu, Z. N., Zhou, Y., Liang, M., Tian, L. X., Wang, K., Hao, Y. H., et al. (2007). Functional disintegration in paranoid schizophrenia using resting-state fMRI. *Schizophrenia Research, 97*(1-3), 194-205.

[55] Lowe, M. J. (2010). A historical perspective on the evolution of resting-state functional connectivity with MRI. *MAGMA, 23*(5-6), 279-288.

[56] Lowe, M. J., Beall, E. B., Sakaie, K. E., Koenig, K. A., Stone, L., Marrie, R. A., et al. (2008). Resting state sensorimotor functional connectivity in multiple sclerosis inversely correlates with transcallosal motor pathway transverse diffusivity. *Hum Brain Mapp, 29*(7), 818-827.

[57] Mantini, D., Perrucci, M. G., Del Gratta, C., Romani, G. L., & Corbetta, M. (2007). Electrophysiological signatures of resting state networks in the human brain. *Proc Natl Acad Sci U S A, 104*(32), 13170-13175.

[58] Margulies, D. S., Bottger, J., Long, X., Lv, Y., Kelly, C., Schafer, A., et al. (2010). Resting developments: a review of fMRI post-processing methodologies for spontaneous brain activity. *MAGMA, 23*(5-6), 289-307.

[59] McKiernan, K. A., D'Angelo, B. R., Kaufman, J. N., & Binder, J. R. (2006). Interrupting the "stream of consciousness": an fMRI investigation. *Neuroimage, 29*(4), 1185-1191.

[60] Medina, L., Aguirre, E., Bernal, B., Altman, N. R. (2004). Functional MR inaging versus Wada test for evaulation of language laterlaization: cost analysis. *Radiology, 230*, 49-54.

[61] Moeller, F., Maneshi, M., Pittau, F., Gholipour, T., Bellec, P., Dubeau, F., et al. (2011). Functional connectivity in patients with idiopathic generalized epilepsy. *Epilepsia, 52*(3), 515-522.

[62] Murphy, K., Birn, R. M., Handwerker, D. A., Jones, T. B., & Bandettini, P. A. (2009). The impact of global signal regression on resting state correlations: are anti-correlated networks introduced? *Neuroimage, 44*(3), 893-905.

[63] Nelles, G., Cramer, S. C., Schaechter, J. D., Kaplan, J. D., Finklestein, S. P. (1998). Quantitative assessment of mirroe movements after stroke. *Stroke, 29*(6), 1187-1187.

[64] Nimsky, C., Ganslandt, O., Cerny, S., Hastreiter, P., Greiner, G., & Fahlbusch, R. (2000). Quantification of, visualization of, and compensation for brain shift using intraoperative magnetic resonance imaging. *Neurosurgery, 47*(5), 1070-1079.

[65] Nimsky, C., Ganslandt, O., Hastreiter, P., & Fahlbusch, R. (2001). Intraoperative compensation for brain shift. *Surgical Neurology, 56*(6), 357-364.

[66] Ogawa, S., Lee, T. M., Kay, A. R., & Tank, D. W. (1990). Brain magnetic resonance imaging with contrast dependent on blood oxygenation. *Proc Natl Acad Sci U S A, 87*(24), 9868-9872.

[67] Okada, K., & Hickok, G. (2006). Identification of lexical-phonological networks in the superior temporal sulcus using functional magnetic resonance imaging. *Neuroreport, 17*(12), 1293-1296.

[68] Park, C. H., Chang, W. H., Ohn, S. H., Kim, S. T., Bang, O. Y., Pascual-Leone, A., et al. (2011). Longitudinal changes of resting-state functional connectivity during motor recovery after stroke. *Stroke, 42*(5), 1357-1362.

[69] Pouratian, N., Bookheimer, S. Y., Rex, D. E., Martin, N. A., & Toga, A. W. (2002). Utility of preoperative functional magnetic resonance imaging for identifying language cortices in patients with vascular malformations. *J Neurosurg, 97*(1), 21-32.

[70] Pujol, J., Conesa, G., Deus, J., Vendrell, P., Isamat, F., Zannoli, G., et al. (1996). Presurgical identification of the primary sensorimotor cortex by functional magnetic resonance imaging. *J Neurosurg, 84*(1), 7-13.

[71] Rutten, G. J., Ramsey, N. F., van Rijen, P. C., Noordmans, H. J., & van Veelen, C. W. (2002). Development of a functional magnetic resonance imaging protocol for intraoperative localization of critical temporoparietal language areas. *Ann Neurol, 51*(3), 350-360.

[72] Sabsevitz, D. S., Swanson, S. J., Hammeke, T. A., Spanaki, M. V., Possing, E. T., Morris, G. L., 3rd, et al. (2003). Use of preoperative functional neuroimaging to predict language deficits from epilepsy surgery. *Neurology, 60*(11), 1788-1792.

[73] Sadato, N., Yonekura, Y., Waki, A., Yamada, H., & Ishii, Y. (1997). Role of the supplementary motor area and the right premotor cortex in the coordination of bimanual finger movements. *J Neurosci, 17*(24), 9667-9674.

[74] Sakoglu, U., Upadhyay, J., Chin, C. L., Chandran, P., Baker, S. J., Cole, T. B., Fox, G. B., Day, M., Luo, F. (2011). *Biochem Pharmacol, 81*(12), 1374-1387.

[75] Skudlarski, P., Jagannathan, K., Calhoun, V. D., Hampson, M., Skudlarska, B. A., & Pearlson, G. (2008). Measuring brain connectivity: Diffusion tensor imaging validates resting state temporal correlations. *Neuroimage, 43*(3), 554-561.

[76] Squire, L. R., & Stark, C. E. L. (2001). When zero is not zero: The problem of ambiguous baseline conditions in fMRI. *Proceedings of the National Academy of Sciences of the United States of America, 98*(22), 12760-12765.

[77] Stark, C. E., & Squire, L. R. (2001). When zero is not zero: the problem of ambiguous baseline conditions in fMRI. *Proc Natl Acad Sci U S A, 98*(22), 12760-12766.

[78] Talairach, J., Tournoux, P. (1998). Co-planar stereotaxic atals of the human brain. New York: Thieme Medical Publishers,

[79] Tomaszewki Farias, S., Harrington, G., Broomand, C., & Seyal, M. (2005). Differences in functional MR imaging activation patterns associated with confrontation naming and responsive naming. *AJNR Am J Neuroradiol, 26*(10), 2492-2499.

[80] van den Heuvel, M. P., & Hulshoff Pol, H. E. (2010). Exploring the brain network: a review on resting-state fMRI functional connectivity. *Eur Neuropsychopharmacol, 20*(8), 519-534.

[81] Van Westen, D., Skagerberg, G., Olsrud, J., Fransson, P., & Larsson, E. M. (2005). Functional magnetic resonance imaging at 3T as a clinical tool in patients with intracranial tumors. *Acta Radiol, 46*(6), 599-609.

[82] Voyvodic, J. T. (2006). Activation mapping as a percentage of local excitation: fMRI stability within scans, between scans and across field strengths. *Magn Reson Imaging, 24*(9), 1249-1261.

[83] Voyvodic, J. T., Petrella, J. R., Friedman, A. H. (2009). fMRI activation mapping as a percentage of local excitation: consistent presurgical motor maps without threshold adjustment. *J Magn Reson Imaging, 29*, 751-759.

[84] Wada, J., Rasmussen, T. (1960). Intracarotid injection of Sodium Amytal for the lateralization of cerebral speech dominance. Experimental and clinical observations. *J Neurosurg, 17*, 266-282.

[85] Wang, Z., Liang, P., Jia, X., Qi, Z., Yu, L., Yang, Y., et al. (2011). Baseline and longitudinal patterns of hippocampal connectivity in mild cognitive impairment: Evidence from resting state fMRI. *J Neurol Sci, 309*, 79-85.

[86] Wu, T., Wang, L., Hallett, M., Li, K., & Chan, P. Neural correlates of bimanual antiphase and in-phase movements in Parkinson's disease. *Brain, 133*(Pt 8), 2394-2409.

[87] Yan, C., Liu, D., He, Y., Zou, Q., Zhu, C., Zuo, X., et al. (2009). Spontaneous brain activity in the default mode network is sensitive to different resting-state conditions with limited cognitive load. *PLoS One, 4*(5), e5743.

[88] Yetkin, F. Z., Mueller, W. M., Morris, G. L., McAuliffe, T. L., Ulmer, J. L., Cox, R. W., et al. (1997). Functional MR activation correlated with intraoperative cortical mapping. *AJNR Am J Neuroradiol, 18*(7), 1311-1315.

[89] Yousry, T. A., Schmid, U. D., Jassoy, A. G., Schmidt, D., Eisner, W. E., Reulen, H. J., et al. (1995). Topography of the cortical motor hand area: prospective study with functional MR imaging and direct motor mapping at surgery. *Radiology, 195*(1), 23-29.

[90] Zhang, D. Y., Johnston, J. M., Fox, M. D., Leuthardt, E. C., Grubb, R. L., Chicoine, M. R., et al. (2009). Preoperative Sensorimotor Mapping in Brain Tumor Patients Using Spontaneous Fluctuations in Neuronal Activity Imaged with Functional Magnetic Resonance Imaging: Initial Experience. *Neurosurgery, 65*(6), 226-236.

[91] Zou, Q., Long, X., Zuo, X., Yan, C., Zhu, C., Yang, Y., et al. (2009). Functional connectivity between the thalamus and visual cortex under eyes closed and eyes open conditions: a resting-state fMRI study. Hum Brain Mapp, 30(9), 3066-3078.

Section 4

Brain Neuroimaging Applications in Disease Processes

Chapter 19

Functional MRI in Alzheimer's Disease

Julia Kivistö, Hilkka Soininen and Maija Pihlajamäki

Additional information is available at the end of the chapter

http://dx.doi.org/10.5772/47912

1. Introduction

Alzheimer's disease (AD) is the most common form of dementia affecting millions of people worldwide. AD results in progressive brain atrophy, memory loss and widespread neurologic deterioration. The first AD-related neuropathological changes appear in the medial temporal lobe (MTL) memory structures already years prior to the manifestation of clinical dementia. Atrophy of the MTL structures as revealed by structural magnetic resonance imaging (MRI) is nowadays considered to be a valid diagnostic marker at the mild cognitive impairment stage, although structural imaging findings may be somewhat nonspecific. "Mild cognitive impairment" (MCI) is in fact one of the recent concepts to describe the possible prodromal state of AD, that is a stage between healthy aging and full-blown clinical AD, for example in terms of neuropathological, imaging and cognitive changes.

Despite extensive research, the fundamental neural basis of memory impairment characteristic to early AD is still largely unknown. Particularly, the relationship between β-amyloid (Aβ) pathology and alterations in memory function remains to be fully elucidated. During recent years, clinical functional MRI (fMRI) has provided tools to investigate the neural underpinnings of AD-related cognitive alterations and thus novel insights into the pathognomonic changes in the MTL structures and related whole-brain memory networks. The ultimate clinical goal of fMRI research is to develop means to reliably define alterations in brain function related to the earliest symptoms of AD before development of significant irreversible structural damage. Since the MTL memory structures are known to be the site of early neuropathological alterations (e.g., neurofibrillary tangles) in AD, previous fMRI studies have largely focused on investigating this particular region of the brain. FMRI during tasks probing episodic memory encoding, which is the cognitive function most characteristically impaired in early AD, is of particular interest.

In this chapter, we will summarize previous studies demonstrating changes in task-related fMRI activity, primarily focusing on memory tasks, as well as studies investigating resting state fMRI findings in clinical AD patients compared to healthy elderly individuals. In a nutshell, fMRI studies in AD patients have demonstrated hypoactivation of the MTL structures during memory task performance, whereas studies in elderly individuals with MCI have reported both increased and decreased MTL responses depending on the severity of the cognitive impairment and underlying structural atrophy. Additionally, recent fMRI findings in MCI and AD patients are beginning to reveal functional abnormalities between the MTL and posteromedial regions such as posterior cingulate and precuneal cortices. In addition to MCI and clinical AD, we will also review recent advances in our understanding of the neuroimaging correlates of cognitively intact elderly subjects at increased risk to develop AD in terms of carrying the apolipoprotein E ε4 (*APOE* ε4) allele.

The long asymptomatic or minimally symptomatic phase of AD provides a potential period for early therapeutic interventions to slow down – and perhaps ultimately prevent – the progression to clinical dementia. Large-scale worldwide multimodal imaging studies on reliable predictors of AD are on-going. There is great hope that imaging of the MTL memory structures and related whole-brain networks would facilitate early diagnosis of AD and other dementias as well as improve treatment options of these devastating diseases in the near future.

2. Alzheimer's disease

AD was originally described in 1907 by the German physician Alois Alzheimer (Alzheimer, 1907; Maurer et al., 1997). Today it is the most common form of dementia in the elderly (Bookmayer et al., 1998). AD is a progressive neurodegenerative syndrome which typically begins with insidious impairment of episodic memory (*i.e.*, memory for past personal experiences in a particular spatial and temporal context).

The most common form of AD is often termed sporadic or late-onset AD as opposed to the relatively rare early-onset forms of the disease (Tanzi & Bertram, 2001). For late-onset AD, the main known genetic risk factor is the *APOE* ε4 allele in chromosome 19 (Bertram et al., 2007; Saunders et al., 1993). Neuropathologically, the disease is characterized by the accumulation of extracellular deposits of Aβ plaques, intracellular neurofibrillary tangles (NFTs) consisting of hyperphosphorylated tau protein, and brain atrophy with regional synaptic, neuronal, and axonal loss (Braak & Braak, 1991). Interestingly, as opposed to the NFT pathology, at the early stages of the disease, Aβ accumulation is often modest within the MTL memory structures but more pronounced, for example, in the posteromedial cortices of the brain. Presentation and clinical course of the AD syndrome can, however, be very variable. Accordingly, AD can be heterogeneous in terms of genetic background (Bertram et al., 2007), response to treatment (Kaduszkiewicz et al., 2005) as well as neuropathological and neuroradiological patterns (Henry-Feugeas, 2007; Jagust et al., 2008; Jellinger, 2002).

http://dx.doi.org/10.5772/47912

The diagnosis of AD relies on clinical judgement. Perhaps the most widely used criteria for defining AD were developed by the National Institute of Neurological and Communicative Disorders and Stroke / Alzheimer's Disease and Related Disorders Association (NINCDS-ADRDA; McKhann et al., 1984). A recent proposal for new research criteria for AD suggests that various biomarkers could be used as supportive features in the diagnostics to improve the specific identification of AD as early as possible (Dubois et al., 2007; Dubois et al., 2010). According to this suggestion, the diagnosis of AD requires meeting the core criterion of significant episodic memory impairment together with at least one or more of the supportive biomarker criteria. Hippocampal atrophy as revealed by structural MRI is one of the most widely documented supportive biomarkers of AD, in addition to abnormal cerebrospinal fluid (CSF) Aβ and tau findings, and a specific pattern of temporoparietal hypometabolism as indicated by [^{18}F]fluorodeoxyglucose positron emission tomography (FDG-PET). Currently, a diagnosis of *definite* AD can, however, only be done by *post mortem* neuropathological evaluation.

The concept of MCI (mild cognitive impairment) refers to subjects with cognitive impairment beyond that expected for their age and education but who are not demented (Petersen et al., 1999; Petersen et al., 2001; Petersen et al., 2004; Petersen et al., 2009; Winblad et al., 2004). During recent years, MCI has had a number of definitions. Diagnostic criteria for the amnestic subtype of MCI widely used during recent years are as follows: 1) memory complaint, preferably corroborated by an informant; 2) objective memory impairment; 3) normal general cognitive function; 4) intact activities of daily living; and 5) not demented (Petersen et al., 2001). Persons who present with amnestic MCI have an increased risk of developing clinical AD with an annual conversion rate of 12 – 15 %, in contrast to 1 – 2 % risk of conversion in healthy elderly individuals (Petersen et al., 1999). Not all subjects with MCI progress to dementia / AD, but a high number of MCI subjects remain stable or may even revert back to normal during follow-up (Ganguli et al., 2004; Gauthier et al., 2006; Larrieu et al., 2002; Petersen, 2004). As noted above, however, a lively discussion on revisions of the MCI / AD diagnostic criteria is on-going among researchers and clinicians.

3. Methodological basis for fMRI in AD

During the past fifteen years, fMRI – together with complementary imaging modalities and sophisticated data analysis methods – has proved to be a very useful tool in investigating the neural basis of intact human memory and other higher cognitive functions. Activation of the hippocampal and parahippocampal regions of the MTL during successful memory encoding has been demonstrated in several fMRI studies in healthy young subjects (Brewer et al., 1998; Pihlajamäki et al., 2003; Sperling et al., 2003b; Stern et al., 1996; Wagner et al., 1998;). These human fMRI findings strongly support the notion that the MTL structures are critical for encoding new events into long-term memory (Eichenbaum, 2000; Mesulam, 1998; Squire & Zola-Morgan, 1991).

Clinical fMRI research into the pathophysiology of age-associated neurodegenerative diseases has become established more recently. Neuroimaging tools such as fMRI provide *in vivo* methods to investigate the integrity of the resting human brain as well as mapping neural networks supporting higher cognitive functions (*e.g.*, memory). fMRI is non-invasive, radiation-free and offers a combination of good spatial and reasonable temporal resolution. Nowadays, the most widely used fMRI technique to measure hemodynamic changes related to underlying cellular activity is based on imaging of the endogenous blood-oxygen-level-dependent (BOLD) contrast (Kwong et al., 1992; Ogawa et al., 1992). In a nutshell, the relative decrease in the amount of deoxygenated hemoglobin enhances the MRI signal locally in brain areas activated during a particular cognitive task. In addition to observed increases in BOLD signal in "activated" brain areas, it has recently been shown that negative BOLD responses are also related to underlying neural activity and originate in decreases in neuronal activity below spontaneous activity in "deactivated" brain regions (Logothetis et al., 2001; Shmuel et al., 2006).

Typically, fMRI experiments compare the BOLD signal during one cognitive condition (*e.g.*, encoding novel pictures) to a control task (*e.g.*, viewing familiar pictures) or to a passive baseline condition (*e.g.*, visual fixation on a cross-hair). This can be done in a "block design" paradigm, in which stimuli of each cognitive condition are grouped together in blocks lasting 20–40 s, or in "event-related" paradigms, in which single stimuli from several different conditions are interspersed.

In addition to functional activation studies, there has recently been considerable interest in studying the baseline activity, or the "default mode" activity, of the resting human brain using FDG-PET and various fMRI techniques (Buckner et al., 2005; Gusnard & Raichle, 2001). It is very interesting that the same brain areas which show high default mode activity and predilection for task-induced fMRI deactivation responses have also demonstrated the earliest hypometabolic changes in AD in previous FDG-PET studies as well as early accumulation of Aβ pathology in recent molecular PET studies using a tracer called [11C]Pittsburgh Combound B, or PIB (Buckner et al., 2005; Cavanna & Trimble, 2006; Klunk et al., 2004; Minoshima et al., 1997; Nestor et al., 2003).

Taken together, fMRI based on BOLD contrast offers a unique, safe and widely available technique for the study of intact human cognition as well as alterations in neuronal function related to healthy aging and dysfunction related to neurodegenerative diseases such as AD.

4. FMRI activation studies

4.1. FMRI activation studies in AD and MCI patients

The hallmark of early AD is the inability to form new enduring episodic memories. At the same time, mild AD patients typically present with neuropathological changes such as synaptic alterations, selective neuronal loss and neurofibrillary tangles in the MTL structures (Braak & Braak 1991; Gomez-Isla et al., 1997; Hyman et al., 1984; Kordower et al., 2001; Scheff et al., 2006; Scheff et al., 2007). In addition to the critical role of the MTL, successful memory formation is thought to require a carefully synchronized interplay between the

http://dx.doi.org/10.5772/47912

MTL and large-scale neural networks (Buckner et al., 2005; Eichenbaum, 2000; Lavenex & Amaral, 2000; Mesulam, 1998; Squire & Zola-Morgan, 1991; Suzuki, 2007; Tulving & Markowitsch, 1998).

Given the prominence of MTL pathology and structural atrophy in early AD, the pioneering fMRI studies on AD focused on investigating alterations in hippocampal activation during various episodic memory tasks (Kato et al., 2001; Machulda et al., 2003; Rombouts et al., 2000; Small et al., 1999; Sperling et al., 2003a). To date, there are several fMRI studies, which have consistently reported diminished or absent MTL activation in AD compared to healthy elderly controls (Fig. 1), during encoding numerous different types of novel stimuli such as faces, face-name pairs, line-drawings, scenes, and geometric shapes (Dickerson et al., 2005; Golby et al., 2005; Grön & Riepe, 2004; Hämäläinen et al., 2007; Kato et al., 2001; Machulda et al., 2003; Pariente et al., 2005; Remy et al., 2005; Rombouts et al., 2000; Small et al., 1999; Sperling et al., 2003a).

Subjects with amnestic MCI (Petersen et al., 2001; Petersen et al., 2009) are an important group to investigate, as they are at increased risk for developing dementia, AD in particular. Consonant with the notion of clinical heterogeneity, results of fMRI studies in MCI subjects relative to controls and AD patients have been variable, findings of hippocampal activation ranging from hyperactivation during encoding (Dickerson et al., 2004; Dickerson et al., 2005; Hämäläinen et al., 2007; Kircher et al., 2007; Woodard et al., 2009; Yassa et al., 2010) to hypoactivation both during encoding and retrieval tasks (Johnson et al., 2006a; Machulda et al., 2003; Mandzia et al., 2009; Petrella et al., 2006). In addition to the heterogeneity of the MCI population, some of the diversity of fMRI results in MCI subjects may be explained by differences in subject selection criteria and the level of clinical severity and underlying MTL atrophy, fMRI paradigms and their difficulty as well as functional imaging and data analysis methods. The mechanistic underpinnings of the observed MTL hyperactivation still remain unclear. In addition to pathological changes in cellular, synaptic or neurotransmitter activity, for example, multiple non-neural factors (such as resting hypoperfusion and metabolism) may also confound the interpretation of BOLD fMRI results in MCI and AD. Typically MCI subjects with significantly impaired memory have, however, similar to AD patients, shown decreased hippocampal activity compared with controls (Machulda et al., 2003; Petrella et al., 2006).

Interestingly, there is also evidence of increased MTL activity in AD patients during specific contrasts, primarily involving the brain response to repetitive stimuli. Golby et al. (2005) reported impaired fMRI repetition suppression paralleled by more MTL activation in AD patients than in older controls during processing of repeated scenes. Another recent study provided evidence that the normal suppression of MTL activity to repeated face-name pairs as compared to visual fixation is impaired in AD (Pihlajamäki et al., 2008). Similar findings have been reported in individuals with amnestic MCI (Johnson et al., 2004). Failure of the hippocampus and surrounding MTL cortices to discriminate familiar from novel information at encoding has also been related to both poor associative recognition memory and poor performance in neuropsychological tests of episodic memory across a range of age and cognitive impairment (Pihlajamäki et al., 2011).

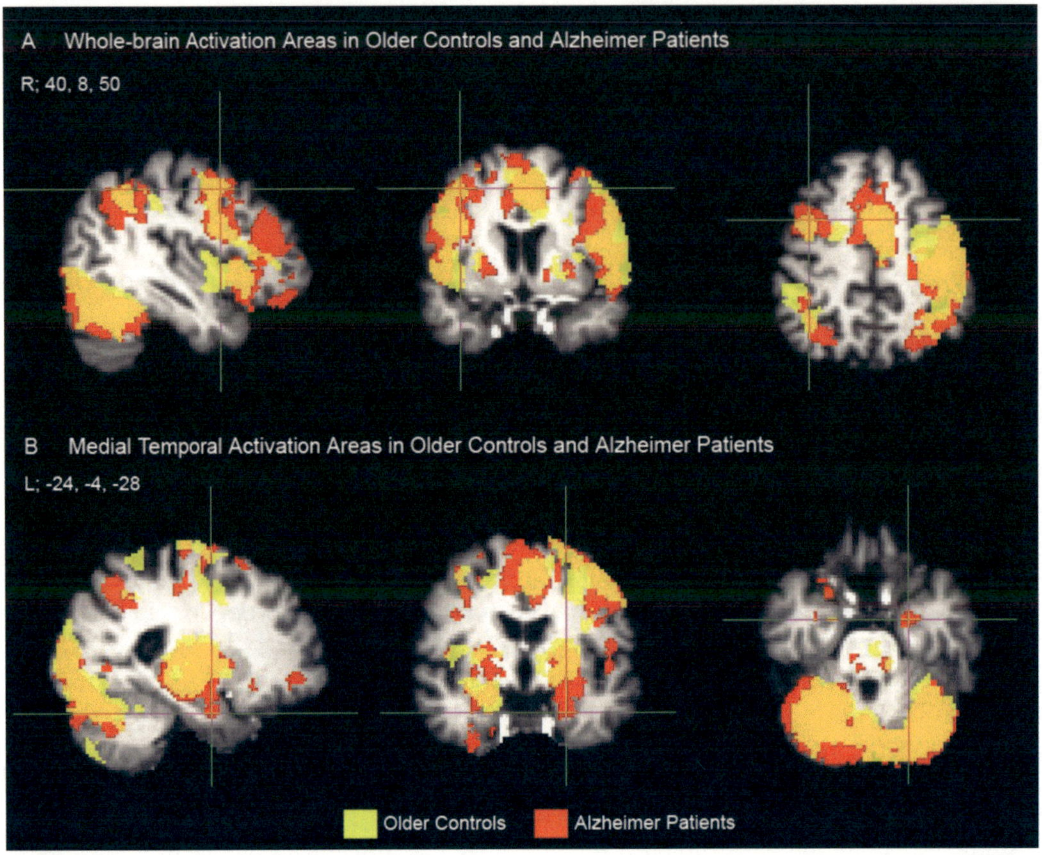

Figure 1. Increased fMRI activity in neocortical (A) and medial temporal (B) brain areas during processing of repeated face-name stimuli in patients with Alzheimer's disease (in red) relative to healthy older control subjects (in yellow). A: Crosshair is located in the right (R) prefrontal cortex, MNI coordinate: 40, 8, 50; B: Crosshair is located in the left (L) anterior hippocampus, MNI coordinate: -24, -4, -28.

As mentioned above, not only the hippocampus but also whole-brain neural networks interconnected with the MTL, are critical for higher cognitive functions such as episodic memory formation (Buckner et al., 2005; Eichenbaum, 2000; Mesulam, 1998). It can be hypothesized that – as opposed to focal changes in the MTL only – multiple nodes within these networks and their mutual interconnectivity are affected at the earliest stage of AD (Scheff et al., 2006; Selkoe, 2002). A recent meta-analysis (Schwindt & Black, 2009) of both fMRI and FDG-PET memory activation studies of AD identified several cortical regions as being more likely to show greater encoding-related activation in controls than in AD patients, including the ventrolateral prefrontal, precuneal, cingulate and lingual cortices. On the other hand, in addition to consistent findings of decreased MTL activation during novel encoding memory tasks, several groups have found evidence of increased fMRI or PET activation in neocortical brain regions, such as frontal and parietal cortices in mild AD patients compared to controls (Grady et al., 2003; Sperling et al., 2003a; Pariente et al., 2005; Celone et al., 2006). This may represent a compensatory process in the setting of MTL dysfunction.

In summary, previous fMRI studies in AD patients, compared to healthy elderly individuals, have reported decreased MTL activity during processing of novel versus repeated information (Dickerson et al., 2005; Golby et al., 2005; Rombouts et al., 2000; Sperling et al., 2003b). These findings of decreased hippocampal activity when comparing novel to repeated stimuli in AD or MCI patients are, in fact, likely to be explained at least to some degree by failure of hippocampal repetition suppression to repeatedly presented stimuli as reviewed above (Golby et al., 2005; Johnson et al., 2004; Pihlajamäki et al., 2008; Pihlajamäki et al., 2011). There is also converging evidence that AD patients show increases in brain activity to compensate for cognitive difficulties in brain regions such as frontal areas that are pathologically less affected than the MTL at the early stages of the disease (Braak & Braak, 1991; Johnson & Albert, 2000; Lehtovirta et al., 1996). Future studies combining multiple imaging modalities such as various structural and functional MRI techniques and PIB-PET imaging (Klunk et al., 2004) are likely to expand our knowledge of the relationships between cognitive impairment, neuropathological changes and alterations in functional imaging patterns.

4.2. FMRI activation studies in *APOE* ε4 carriers

Several experiments in cognitively intact elderly control subjects have demonstrated that elderly individuals are able to activate their hippocampus during successful associative encoding largely to the same degree as young subjects (Miller et al., 2008a; Rand-Giovannetti et al., 2006; Sperling et al., 2003a; Sperling et al., 2003b), although age-related alterations in fMRI activity during normal aging have also been reported (Cabeza et al., 2004; Daselaar et al., 2006a, Daselaar et al., 2006b; Dennis et al., 2007). It has been suggested that age-related changes in memory performance may primarily be due to alterations in cortical regions or in the connectivity between the MTL and neocortical regions.

Similarly to MCI subjects, results of fMRI studies comparing activation in *APOE* ε4 carriers at risk for AD versus their non-carrier counterparts have been diverse. Increased hippocampal and cortical activation has been reported during tasks such as encoding novel and repeated pictures or face-name pairs, encoding and retrieval of word-pairs, a letter fluency task, and an auditory verbal *n*-back working memory task, and has usually been interpreted to reflect compensatory neural mechanisms (Bondi et al., 2005; Bookheimer et al., 2000; Burggren et al., 2002; Dickerson et al., 2005; Fleisher et al., 2005; Han et al., 2007; Pihlajamäki & Sperling, 2009; Smith et al., 2002; Wishart et al., 2006). At the same time, several studies have demonstrated reduced functional brain activity in the MTL and other brain areas in cognitively normal ε4 carriers (Borghesani et al., 2007; Lind et al., 2006; Smith et al., 1999; Trivedi et al., 2006). Several of the above mentioned studies have carefully matched the study groups regarding age, gender and cognitive performance. It is difficult to draw firm conclusions of the ε4 effects on BOLD fMRI activation pattern. In the most recent large-scale fMRI studies, more complex patterns of alterations in brain activation differentially affected by *APOE* ε4 and family history of AD have been suggested (Bassett et al., 2006; Johnson et al., 2006). Longitudinal fMRI testing of subjects at genetic risk for AD would likely be informative, optimally in combination with metabolic FDG- and molecular PIB-PET imaging to improve our understanding of the temporal sequence of events early in the course of prodromal AD.

4.3. FMRI in prediction of cognitive decline

Since some of the MCI subjects will remain stable and some will progress to dementia over time, great interest has been focused on attempts to identify the features predicting future conversion from MCI to clinical AD. As reviewed above, previous cross-sectional fMRI studies in subjects with MCI / prodromal AD have reported variable results, ranging from MTL hypoactivation to hyperactivation compared to cognitively normal elderly individuals. It has been hypothesized that subjects in early phases of prodromal AD may present a short period of paradoxical hippocampal hyperactivation, which is then followed by loss of hippocampal activation along with progressive cognitive decline. FMRI studies with clinical follow-up data on MCI subjects have reported that increased MTL activity at baseline in MCI compared to elderly control subjects may indicate higher likelihood of subsequent cognitive decline (Dickerson et al., 2004; Miller et al., 2008b). Similar findings of a temporary period with abnormally enhanced MTL activity during a preclinical stage of AD have been reported in *APOE* ε4 carriers relative to non-carriers (Bookheimer et al., 2000).

Recently, one longitudinal fMRI study (O'Brien et al., 2010) demonstrated both the highest hippocampal activation at baseline and the greatest loss of hippocampal activation during follow-up in cognitively impaired subjects with the most rapid decline during the follow-up period. The authors concluded that cognitive decline is associated with loss of hippocampal activation and suggested that fMRI may prove valuable in tracking very early progression of brain dysfunction on the trajectory towards clinical AD but prior to the point of irreversible neuronal loss and significant macroscopic atrophy. Thus, it seems that there may be a temporary phase of abnormal MTL hyperactivity along the course of MCI to clinical AD, which may in turn be an indicator of compensatory neural mechanisms recruited in MCI subjects in order to keep memory performance close to the level of cognitively normal elderly subjects. Around the conversion from MCI to clinical AD, the ability to compensate for the MTL pathology is lost, which is then seen as poor task performance and disrupted hippocampal fMRI activity. In other words, hippocampal hyperactivity observed in some previous fMRI studies during the progression of MCI to clinical AD may be a compensatory phenomenon, but it may also be a harbinger of impending hippocampal failure.

5. FMRI resting state studies in AD, MCI and *APOE* ε4 carriers

In addition to task-related fMRI activation studies primarily focusing on the MTL function, recent functional imaging studies have demonstrated AD-related alterations in the so called brain "default mode", or resting state activity. The default mode of the human brain was originally identified by its consistent activity increases during passive task states as compared to a wide range of goal-directed activation tasks (Buckner et al., 2008; Gusnard & Raichle, 2001; Mazoyer et al., 2001; Raichle et al., 2001; Shulman et al., 1997). Regions of the

http://dx.doi.org/10.5772/47912

default network show high resting glucose metabolism and blood flow relative to other brain regions as well as coordinated low frequency fluctuations in states of relative rest (Buckner et al., 2008; Minoshima et al., 1997; Raichle et al., 2001; Shulman et al., 1997). Anatomically, the key default mode regions consist of the posteromedial and lateral parietal regions as well as midline and lateral frontal regions. Interestingly, the same default mode regions which are upregulated at rest appear to be suppressed during various cognitive activities, including intentional encoding of new memories (Pihlajamäki et al., 2008; Rombouts et al., 2005a; Shulman et al., 1997). Deactivation of key nodes of the default mode network, in coordination with hippocampal activation, seems in fact to be a prerequisite for successful memory encoding (Daselaar et al., 2004; Miller et al., 2008a; Weissman et al., 2006).

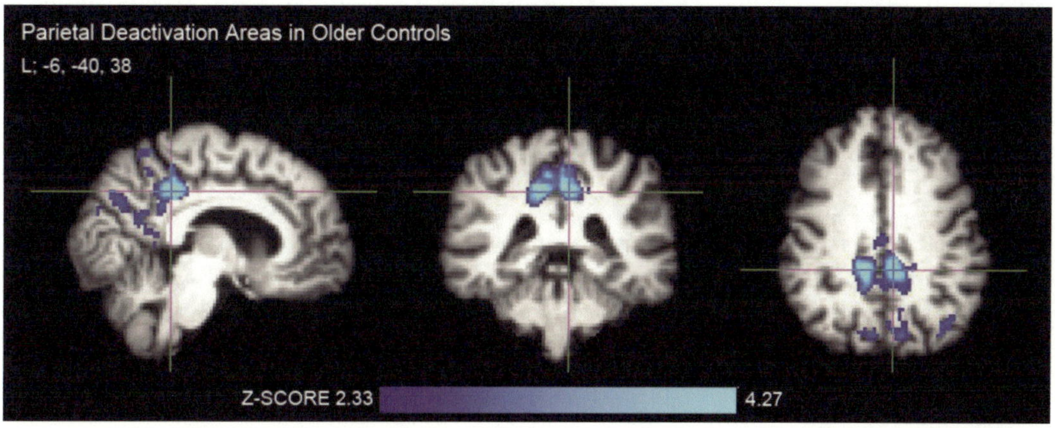

Figure 2. FMRI deactivation areas during processing of repeated face-name stimuli in healthy older subjects (in blue). Crosshair is located in the left (L) posterior cingulated cortex, MNI coordinate: -6, -40. 38.

Previous FDG-PET studies in clinical AD patients and older individuals at risk for AD have revealed hypometabolism of the posteromedial and other association cortical regions (Herholz et al. 2002; Minoshima et al. 1997; Mosconi et al. 2008b; Rapoport 1991; Reiman et al., 1996). In recent fMRI studies of AD, corresponding brain default mode regions have been found to demonstrate an abnormal fMRI task-induced deactivation pattern (Greicius et al., 2004; Lustig et al., 2003; Petrella et al., 2007a; Petrella et al., 2007b; Pihlajamäki et al., 2008; Rombouts et al., 2005a; Rombouts et al., 2005b). Predilection of the posteromedial core regions of the default network to demonstrate task-induced deactivation (*i.e.*, relative decreases in the BOLD fMRI signal) during paradigms requiring attention to external stimuli has been consistently demonstrated in both young and old healthy subjects (Fig. 2) using a multitude of cognitive stimuli and both fMRI and PET imaging modalities (Buckner et al., 2005, 2008; Cavanna & Trimble, 2006; Daselaar et al., 2004; Fransson & Marrelec, 2008; Gusnard et al., 2001; Mazoyer et al., 2001; Miller et al., 2008a; Otten & Rugg, 2001; Raichle et al., 2001; Shulman et al., 1997). In other words, the relative decreases in fMRI signal normally observed in the default mode regions in healthy subjects performing a cognitive task are not seen in AD patients, or may even be reversed to a paradoxical activation response (Grady et al., 2006; Lustig et al., 2003; Miller et al., 2008a; Otten & Rugg, 2001).

Recent fMRI studies have also revealed alterations in the deactivation pattern in elderly individuals at risk for AD by virtue of their *APOE* e4 genotype or evidence of MCI. The pattern of fMRI task-induced deactivation seems to be progressively disrupted along the continuum from normal aging to MCI and to clinical AD and more impaired in ε4 carriers than in non-carriers (Persson et al., 2008; Pihlajamäki et al., 2009). It is likely that the pathologically affected MCI or AD brain (Selkoe, 2002; Tanzi, 2005) is no longer capable of "turning off" the default mode activity during focused cognitive processing as it should in order to optimally recruit other networks – such as the hippocampal memory network or the frontoparietal attentional network – for task performance (Eichenbaum, 2000; Mesulam, 1998; Miller et al., 2008a; Weissman et al., 2006). This finding is consonant with recent studies in healthy young and elderly adults demonstrating that the ability to suspend default mode activity during goal-directed cognitive tasks, *i.e.* to reallocate neurocognitive resources to those brain regions optimal for the task performance, may be critical for successful cognitive performance (Daselaar et al., 2004; Grady et al., 2006; Miller et al., 2008a; Otten & Rugg, 2001; Weissman et al., 2006). Posteromedial cortical regions of the default mode network not only overlap topographically with the pattern of FDG-PET hypometabolism but also with the distribution of the fibrillar Aβ deposition in AD (Buckner et al., 2005). Recent PIB-PET studies, the tracer PIB labeling Aβ plaques (Edison et al., 2007; Klunk et al., 2004), have suggested that the posteromedial cortical areas of high default mode activity may in fact be among the earliest sites of Aβ pathology in AD (Mintun et al., 2006). Thus, functional (as well as structural) imaging studies focusing on this region are pathobiologically very relevant when searching for potential early markers of prodromal AD.

In MCI and AD patients, alterations observed in the fMRI deactivation pattern of the posteromedial cortices (Greicius et al., 2004; Lustig et al., 2003; Petrella et al., 2007a; Petrella et al., 2007b; Pihlajamäki et al., 2008; Rombouts et al., 2005a; Rombouts et al., 2005b) may also reflect remote effects of the MTL pathology and atrophy. The MTL, which is thought to be responsible for the episodic memory deficits observed in amnestic MCI and AD and is known to present the earliest neurofibrillary changes and neuronal and synapse loss, is strongly interconnected to the posteromedial nodes of the default network (Braak & Braak, 1991; Gomez-Isla et al., 1997; Hyman et al., 1984; Insausti et al., 1987; Kordower et al., 2001; Leichnetz, 2001; Suzuki & Amaral, 1994). It is also possible that local structural atrophy of the underlying posteromedial brain regions may explain some of the findings of altered fMRI deactivation (Greicius et al., 2004; Lustig et al., 2003; Petrella et al., 2007a; Petrella et al., 2007b; Pihlajamäki et al., 2008; Rombouts et al., 2005a; Rombouts et al., 2005b).

A more recent imaging approach, that is the functional connectivity MRI (fcMRI), identifies brain systems via intrinsic functional (activity) correlations (Damoiseaux et al., 2006; Fox & Raichle, 2007; Greicius et al., 2003; Greicius & Menon, 2004). Recent fcMRI studies in healthy young subjects have demonstrated the consistency of the resting state networks in the human brain and have also corroborated the findings of altered task-induced deactivation in MCI and AD relative to controls (Celone et al., 2006; Sorg et al., 2007; Supekar et al., 2008;

Wang et al., 2007; Zhou et al., 2008). As an example, functional connectivity between the posteromedial and MTL cortices has been reported to be impaired even in MCI subjects relative to healthy elderly controls (Sorg et al., 2007; Zhou et al., 2008). Resting state fcMRI between the MTL and posteromedial cortices has also been demonstrated to reflect underlying structural connectivity as revealed by diffusion tensor imaging. Future studies investigating the relations between, for example, the fMRI task-induced activation / deactivation, resting fcMRI and PIB-PET amyloid imaging findings will further expand our understanding of the role of the Aβ pathology and impaired default mode network function in the pathogenesis and cognitive symptomatology of AD.

6. Conclusions and future directions of fMRI in AD

There have been a number of promising clinically relevant imaging studies targeting brain functional alterations in AD, MCI and subjects at-risk for AD relative to normal aging. Functional imaging during memory paradigms has shown evidence of specific alterations in the MTL and related whole-brain memory networks that may be able to differentiate the process of very early AD from normal aging. The greatest potential of functional imaging most likely lies in the study of very early stages of dementias, at the point of emerging neuronal dysfunction without significant macroscopic brain atrophy. In the context of early diagnostics of AD, the most interesting and challenging target group to be investigated continues to be the elderly subjects with subtle memory impairment as these subjects would still have preserved brain function and thus scope for therapeutic interventions. Recent revisions of the criteria for AD and MCI strongly emphasize the use of imaging biomarkers in future clinical diagnostics of these disorders. Structural MRI evaluation of the hippocampus is already widely used as a supportive biomarker for AD diagnosis.

The past two decades have seen remarkable advances in our understanding of the pathophysiology of neurodegenerative dementias. As reviewed above, fMRI has many potential advantages in studying patients with cognitive impairment. FMRI can be acquired on a standard clinical magnet during the same session as structural imaging. Because it is non-invasive and subjects are not exposed to radiation, fMRI can be safely repeated many times over the course of longitudinal studies. Perhaps the greatest potential advantage of fMRI is that we can image patients with memory disorders while they are attempting to do the type of cognitive process that is causing them difficulty in their daily living. The use of event-related designs enables investigation of the hemodynamic correlates of specific behavioral events, such as successful long-term memory formation.

There are, however, several challenges in performing fMRI studies in patients with neurodegenerative dementias. It is likely that fMRI will remain quite problematic in examining patients with more severe cognitive impairment. High-field fMRI with optimized imaging parameters can offer spatial resolution as high as in the order of 1 mm, or even less. This is, however, currently not realistic with demented patients as the technique is sensitive to head motion. Inherently, the signal-to-noise ratio of BOLD signal changes between activation and baseline conditions is low, which necessitates repeated measurements and thus leads to relatively long scanning sessions. There is a need for continued technical advances, such as real-time motion correction and high-speed acquisition, to fully realize the potential of this technology in dementia research. Also, if the patients are not able to adequately perform the cognitive task, one of the major advantages of fMRI activation studies is lost. Differences in task performance between patient and control groups complicate data interpretation, as the ability to perform the task may greatly influence the pattern and degree of observed fMRI activity. Resting state fMRI can, however, be performed with less co-operative subjects and is thus better applicable to imaging more severely impaired patients.

It is important to remember that BOLD fMRI is an indirect measure of neuronal activity. The BOLD fMRI signal, and neurovascular coupling linking cellular activity to hemodynamic changes, is likely to undergo changes during healthy aging and during AD-related pathological processes. Some of the changes that may occur even in healthy elderly subjects include, for example, increased atherosclerosis. In AD, the presence of Aβ in the cerebral vasculature, together with altered neurotransmitter activity, impairs synaptic, neuronal and glial function, and may thus lead to attenuated BOLD response. Both increased and decreased BOLD fMRI responses have, however, been reported in MCI and AD and MCI compared to elderly controls, which does not support the view of attenuation of the BOLD signal solely due to vascular reasons (Golby et al., 2005; Grady et al., 2003; Sperling et al., 2003a). The alterations in BOLD activity reported in AD also appear to be quite regionally specific and dependent on the nature of the cognitive task, thus making it relatively unlikely that the changes observed in fMRI studies represent global pathophysiological alterations in neurovascular coupling.

In terms of using fMRI in longitudinal or pharmacological studies, it is critical to complete further validation experiments. The reproducibility of BOLD signal changes within young healthy individuals during memory encoding tasks across separate days is reported to be reasonable (Sperling et al., 2002; Harrington et al., 2006). However, reproducibility of task-related or resting state fMRI activity in older and cognitively impaired subjects has not yet been well established. More longitudinal functional imaging studies are needed to track the evolution of alterations in the fMRI activation / deactivation pattern over the course of the cognitive continuum from healthy aging to AD. It is also important to evaluate the contribution of structural atrophy to changes observed with functional imaging. A combination of structural MRI, fMRI and other functional and molecular imaging techniques such as PIB-PET may eventually serve as a valuable method for the *in vivo* detection of AD prior to clinical dementia, at the point when disease modifying therapies would likely be most efficacious.

In summary, despite technical challenges, there have been a number of promising fMRI studies in elderly individuals with prodromal AD. Neuroimaging, and in particular BOLD fMRI has produced invaluable information and will likely enable even deeper understanding of the human brain function both in health and disease in the future. Carefully designed future studies using multimodal imaging are hoped to yield us new tools that aid in the early identification of subjects likely to develop dementia. Further longitudinal studies are needed to track the evolution of brain functional alterations over the course of the cognitive continuum from healthy aging to clinical dementias such as AD, and perhaps also the pharmacological efficacy for novel disease-modifying therapies.

Acknowledgements

The original work from our laboratory described in this chapter was supported by Health Research Council of the Academy of Finland, grant #121038 (HS) and grants #108188 and #214050 (MP), and Kuopio University Hospital EVO grants #477311 and #5772720. We express special gratitude to the subjects who participated in this study.

Author details

Julia Kivistö[1], Hilkka Soininen[1,2] and Maija Pihlajamäki[1,2]

1 Department of Neurology, Kuopio University Hospital, Kuopio, Finland

2 Unit of Neurology, Institute of Clinical Medicine, School of Medicine, University of Eastern Finland, Kuopio, Finland

References

[1] Alzheimer, A. (1907). Über eine eigenartige Erkrankung der Hirnrinde. Allgem Z Psychiatr Psych-Gerisch Med, , 64, 146-148.

[2] Bassett, S. S., Yousem, D. M., Cristinzio, C., Kusevic, I., Yassa, M. A., Caffo, B. S., & Zeger, S. L. (2006). Familiar risk for Alzheimer's disease alters fMRI activation patterns. Brain, , 129, 1229-1239.

[3] Bertram, L., Mcqueen, M. B., Mullin, K., Blacker, D., & Tanzi, R. E. (2007). Systematic metaanalyses of Alzheimer's disease genetic association studies: the AlzGene database. Nat Genet, , 39, 17-23.

[4] Bondi, M. W., Houston, W. S., Eyler, L. T., & Brown, G. G. (2005). fMRI evidence of compensatory mechanisms in older adults at genetic risk for Alzheimer disease. *Neurology*, 64, 501-508.

[5] Bookheimer, S. Y., Strojwas, M. H., Cohen, M. S., Saunders, A. M., Pericak-vance, M. A., Mazziotta, J. C., & Small, G. W. (2000). Patterns of brain activation in people at risk for Alzheimer's disease. *N Engl J Med.*, 343, 450-456.

[6] Borghesani, P. R., Johnson, L. C., Shelton, A. L., Peskind, E. R., Aylward, E. H., Schellenberg, G. D., et al. (2007). Altered medial temporal lobe responses during visuospatial encoding in healty APOE*4 carriers. Neurobiol Aging, , 82, 239-259.

[7] Braak, H., & Braak, E. (1991). Neuropathological stageing of Alzheimer-related changes. *Acta Neuropathol (Berl),*, 82, 239-259.

[8] Brewer, J. B., Zhao, Z., Desmond, J. E., Glover, G. H., & Gabrieli, J. D. (1998). Making memories: brain activity that predicts how well visual experience will be remembered. *Science*, 281, 1185-1187.

[9] Brookmayer, R. Gray,.S & Kawas, C. ((1998). Projections of Alzheimer's disease in the United States and the public health impact of delaying disease onset. Am J Public Health, , 88, 1337-1342.

[10] Buckner, R. L. (2004). Memory and executive function in aging and AD: multiple factors that cause decline and reserve factors that compensate. *Neuron*, 44, 195-208.

[11] Buckner, R. L., Snyder, A. Z., & Shannon, B. J. LaRossa, G., Sachs, R., Fotenos, A. F., Sheline, Y.I., Klunk, W.E., Mathis, C.A., Morris, J.C. & Mintun, M.A. ((2005). Molecular, structural, and functional characterization of Alzheimer's disease: Evidence for a relationship between default activity, amyloid, and memory. Journal of Neuroscience, , 25, 7709-7717.

[12] Buckner, R. L., Andrews-hanna, J. R., & Schacter, D. L. (2008). The brain's default network: Anatomy, function and relevance to disease. *Annals of the New York Academy of Sciences*, 1124, 1-38.

[13] Burggren, A. C., Small, G. W., Sabb, F. W., & Bookheimer, S. Y. (2002). Specificity of brain activation patterns in people at genetic risk for Alzheimer disease. *Am J Geriatr Psychiatry,*, 10, 44-51.

[14] Cavanna, A. E., & Trimble, M. R. (2006). The precuneus: A review of its functional anatomy and behavioural correlates. *Brain*, 129, 564-583.

[15] Cabeza, R., Daselaar, S. M., Dolcos, F., Prince, S. E., Budde, M., & Nyberg, L. (2004). Task-independent and task-specific age effects on brain activity during working memory, visual attention and episodic retrieval. *Cereb Cortex,*, 14, 364-375.

[16] Celone, K. A., Calhoun, V. D., Dickerson, B. C., Atri, A., Chua, E. F., Miller, S., Depeau, K., Rentz, D. M., Selkoe, D. J., Blacker, D., Albert, M. S., & Sperling, R. A. (2006). Alterations in memory networks in mild cognitive impairment and Alzheimer's disease: an independent component analysis. *J. Neurosci.*, 26, 10222-10231.

[17] Damoiseaux, J. S., Rombouts, S. A., Barkhof, F., Scheltens, P., Stam, C. J., Smith, S. M., & Beckmann, C. F. (2006). Consistent resting-state networks across healthy subjects. *Proc Natl Acad Sci U S A*, 103, 13848-13853.

[18] Daselaar, S. M., Prince, S. E., & Cabeza, R. (2004). When less means more: deactivations during encoding that predict subsequent memory. *NeuroImage*, 23, 921-927.

[19] Daselaar, S. M., Fleck, M. S., Dobbins, I. G., Madden, D. J., & Cadeza, R. (2006a). Effects of healty aging on hippocampal and rhinal memory functions: an event-related study. Cereb Cortex, , 16, 1771-1782.

[20] Daselaar, S. M., Fleck, M. S., & Cebeza, R. (2006b). Triple dissociation in the medial temporal lobes: recollection, familiarity, and novelty. *J Neurophysiol,*, 96, 1902-1911.

[21] Dennis, N. A., Kim, H., & Cadeza, R. (2007). Effects of aging on true and false memory formation: an fMRI study. *Neuropsychologia*, 45, 74-79.

[22] Dickerson, B. C., Salat, D. H., Bates, J. F., Atiya, M., Killiany, R. J., Greve, D. N., et al. (2004). Medial temporal lobe function and structure in mild cognitive impairment. Ann Neurol, , 56, 27-35.

[23] Dickerson, B. C., Salat, D. H., Greve, D. N., Chua, E. F., Rand-giovannetti, E., Rentz, D. M., Bertram, L., Mullin, K., Tanzi, R. E., Blacker, D., Albert, M. S., & Sperling, R. A. (2005). Increased hippocampal activation in mild cognitive impairment compared to normal aging and AD. *Neurology*, 65, 404-411.

[24] Dubois, B., Feldman, H. H., Jacova, C., Dekosky, S. T., Barberger-gateau, P., Cummings, J., Delacourte, A., Galasko, D., Gauthier, S., Jicha, G., Meguro, K., Brien, O., Pasquier, J., Robert, F., Rossor, P., Salloway, M., Stern, S., Visser, Y., & Scheltens, P. J. P. ((2007). Research criteria for the diagnosis of Alzheimer's disease: revising the NINCDS-ADRDA criteria. Lancet Neurol, , 6, 734-746.

[25] Dubois, B., Feldman, H. H., Jacova, C., Cummings, J. L., Dekosky, S. T., Barberger-gateau, P., Delacourte, A., Frisoni, G., Fox, N. C., Galasko, D., Gauthier, S., Hampel, H., Jicha, G., Meguro, K., Brien, O., Pasquier, J., Robert, F., Rossor, P., Salloway, M., Sarazin, S., De Souza, M., Stern, L. C., Visser, Y., & Scheltens, P. J. P. ((2010). Revising the definition of Alzheimer's disease: a new lexicon. Lancet Neurol, , 9, 1118-1127.

[26] Edison, P. E., Archer, H. A., Hinz, R., Hammers, A., Pavese, N., Tai, Y., et al. (2007). Amyloid, hypometabolism, and cognition in Alzheimer disease: an [11C]PIB and [18F]FDG PET study, Neurology , 68, 501-508.

[27] Eichenbaum, H. system for declarative memory. Nat Neurosci , 1, 41-50.

[28] Fleisher, A. S., Houston, W. S., Eyler, L. T., Frye, S., Jenkins, C., Thal, L. J., & Bondi, M. W. (2005). Identification of Alzheimer disease risk by fMRI. Arch Neurol, , 62, 1881-1888.

[29] Fox, M. D., & Raichle, M. E. (2007). Spontaneous fluctuations in brain activity observed with functional magnetic resonance imaging,. *Nat Rev Neurosci*, 8, 700-711.

[30] Fransson, P., & Marrelec, G. (2008). The precuneus/posterior cingulate cortex plays a pivotal role in the default mode network: Evidence from a partial correlation network analysis. *Neuroimage*, 42, 1178-1184.

[31] Ganguli, M., Dodge, H. H., Shen, C., & Dekosky, S. T. (2004). Mild cognitive impairment amnestic type: an epidemiologic study. *Neurology*, 63, 115-121.

[32] Gauthier, S., Reisberg, B., Zaudig, M., Petersen, R. C., Ritchie, K., Broich, K., Belleville, S., Brodaty, H., Bennett, D., Chertkow, H., Cummings, J. L., De Leon, M., Feldman, H., Ganguli, M., Hampel, H., Scheltens, P., Tierney, M. C., Whitehouse, P., & Winblad, B. (2006). Mild cognitive impairment. Lancet, , 367, 1262-1270.

[33] Golby, A., Silverberg, G., Race, E., Gabrieli, S., Shea, O., Knierim, J., Stebbins, K., & Gabrieli, G. J. ((2005). Memory encoding in Alzheimer's disease: an fMRI study of explicit and implicit memory. Brain, 128, 773-787.

[34] Gomez IslaT., Hollister, R., West, H., Mui, S., Growdon, J.H., Petersen, R.C., Parisi, J.E. & Hyman, B.T. ((1997). Neuronal loss correlates with but exceeds neurofibrillary tangles in Alzheimer's disease. Ann Neurol, , 41, 17-24.

[35] Grady, C. L., Mcintosh, A. R., Beig, S., Keightley, M. L., Burian, H., & Black, S. E. (2003). Evidence from functional neuroimaging of a compensatory prefrontal network in Alzheimer's disease. *J Neurosci,*, 23, 986-993.

[36] Grady, C. L., Springer, M. V., Hongwanishkul, D., Mcintosh, A. R., & Winocur, G. (2006). Age related changes in brain activity across the adult lifespan. J Cogn Neurosci, , 18, 227-241.

[37] Greicius, M. D., Krasnow, B., Reiss, A. L., & Menon, V. (2003). Functional connectivity in the resting brain: a network analysis of the default mode hypothesis. Proc Natl Acad Sci U S A, , 100, 253-258.

[38] Greicius, M. D., & Menon, V. (2004). Default-mode activity during a passive sensory task: uncoupled from deactivation but impacting activation,. J Cogn Neurosci, , 16, 1484-1492.

[39] Greicius, M. D., Srivastava, G., Reiss, A. L., & Menon, V. (2004). Default-mode network activity distinguishes Alzheimer's disease from healthy aging: Evidence from functional MRI. Proc Natl Acad Sci USA, , 101, 4637-4642.

[40] Grön, G., & Riepe, M. W. (2004). Neural basis for the cognitive continuum in episodic memory from health to Alzheimer's disease. Am J Geriatr Psychiatry, , 12, 648-652.

[41] Gusnard, D. A., & Raichle, M. E. (2001). Searching for a baseline: Functional imaging and the resting human brain. Nat Rev Neurosci, , 2, 685-694.

[42] Harrington, G. S. Tomaszewski Farias, S., Buonocore, M. H., & Yonelinas, A. The intersubject and intrasubject reproducibility of FMRI activation during three encoding tasks: Implications for clinical applications. Neuroradiology, 2006.

[43] Henry-feugeas, M. C. (2007). MRI of the `Alzheimer syndrome`. *J Neuroradiol,*, 34, 220-227.

[44] Hyman, B. T., Van Hoesen, G. W., Damasio, A. R., & Barnes, C. L. (1984). Alzheimer's disease: cell-specific pathology isolates the hippocampal formation. Science, 225, 1168-1170.

[45] Hämäläinen, A., Pihlajamäki, M., Tanila, H., Hänninen, T., Niskanen, E., Tervo, S., et al. (2007). Increased fMRI responses during encoding in mild cognitive impairment. *Neurobiol Aging,*, 28, 1889-1903.

[46] Insausti, R., Amaral, D. G., & Cowan, W. M. (1987). The entorhinal cortex of the monkey: II. Cortical afferents,. J Comp Neurol, , 264, 356-395.

[47] Jagust, W. J., Zheng, L., Harvey, D. J., Mack, W. J., Vinters, H. V., Weiner, M. W., Ellis, W. G., Zarow, C., Mungas, D., Reed, B. R., Kramer, J. H., Schuff, N., Decarli, C., & Chui, H. C. (2008). Neuropathological basis of magnetic resonance images in aging and dementia. *Ann Neurol,*, 63, 72-80.

[48] Jellinger, K. A. (2002). Alzheimer disease and cerebrovascular pathology: an update. *J Neural Transm,*, 109, 813-836.

[49] Johnson, K. A., & Albert, M. S. (2000). Perfusion abnormalities in prodromal AD. *Neurobiol Aging,*, 21, 289-299.

[50] Johnson, S. C., Baxter, L. C., Susskind-wilder, L., Connor, D. J., Sabbagh, M. N., & Caselli, R. J. (2004). Hippocampal adaptation to face repetition in healthy elderly and mild cognitive impairment. *Neuropsychologia*, 42, 980-989.

[51] Johnson, S. C., Schmitz, T. W., Trivedi, M. A., Ries, M. L., Torgerson, B. M., Carlsson, C. M., et al. (2006). The influence of Alzheimer disease family history and apolipoprotein E epsilon4 on mesial temporal lobe activation. *J Neurosci,*, 26, 6069-6076.

[52] Kaduszkiewicz, H., Zimmermann, T., & Beck-bornholdt, H. P. Van Den Bussche, H. (2005). Cholinesterase inhibitors for patients with Alzheimer's disease: systematic review of randomised clinical trials. BMJ, , 331, 321-327.

[53] Kato, T., Knopman, D., & Liu, H. (2001). Dissociation of regional activation in mild AD during visual encoding: a functional MRI study. *Neurology*, 57, 812-816.

[54] Kircher, T. T., Weis, S., Freyman, K., Erb, M., Jessen, F., Grodd, W., Heun, R., & Leube, D. T. (2007). Hippocampal activation in MCI patients is necessary for successful memory encoding. J Neurol Neurosurgery Psychiatry, , 78, 812-818.

[55] Klunk, W. E., Engler, H., Nordberg, A., Wang, Y., Blomqvist, G., Holt, D. P., Bergström, M., Savitcheva, I., Huang, G. F., Estrada, S., Ausén, B., Debnath, M. L., Barletta, J., Price, J. C., Sandell, J., Lopresti, B. J., Wall, A., Koivisto, P., Antoni, G., Mathis, C. A., & Långström, B. (2004). Imaging brain amyloid in Alzheimer's disease with Pittsburgh Compound-B. Ann Neurol, , 55, 306-319.

[56] Kordower, J. H., Chu, Y., Stebbins, G. T., Dekosky, S. T., Cochran, E. J., Bennett, D., & Mufson, E. J. (2001). Loss and atrophy of layer II entorhinal cortex neurons in elderly people with mild cognitive impairment. Ann Neurol, , 49, 202-213.

[57] Kwong, K. K., Belliveau, J. W., Chesler, D. A., Goldberg, I. E., Weisskoff, R. M., Poncelet, B. P., Kennedy, D. N., Hoppel, B. E., Cohen, M. S., Turner, R., et al. (1992). Dynamic magnetic resonance imaging of human brain activity during primary sensory stimulation. Proc Natl Acad Sci USA, , 89, 5675-5679.

[58] Larrieu, S., Letenneur, L., Orgogozo, J. M., Fabrigoule, C., & Amieva, H. Le Carret, N., Barberger-Gateau, P. & Dartigues, J.F. ((2002). Incidence and outcome of mild cognitive impairment in a population-based prospective cohort. Neurology, 59, 1594-1599.

[59] Lavenex, P., & Amaral, D. G. (2000). Hippocampal-neocortical interaction: a hierarchy of associativity. *Hippocampus*, 10, 420-430.

[60] Lind, J., Persson, J., Ingvar, M., Larsson, A., Curts, M., Van Broeckhoven, C., Adolfsson, R., Bäckman, L., Nilsson, L. G., Petersson, K. M., & Nyberg, L. (2006). Reduced functional brain activity response in cognitively intact apolipoprotein E epsilon 4 carriers. Brain, , 129, 1240-1248.

[61] Lehtovirta, M., Soininen, H., Laakso, M. P., Partanen, K., Helisalmi, S., Mannermaa, A., Ryynänen, M., Kuikka, J., Hartikainen, P., & Riekkinen, P. J. Sr. ((1996). SPECT and MRI analysis in Alzheimer's disease: relation to apolipoprotein E epsilon 4 allele. J Neurol Neurosurg Psychiatry, , 60, 644-649.

[62] Leichnetz, G. R. (2001). Connections of the medial posterior parietal cortex (area 7m) in the monkey,. Anat Rec, , 263, 215-236.

[63] Logothetis, N. K., Pauls, J., Augath, M., Trinath, T., & Oeltermann, A. (2001). Neurophysiological investigation of the basis of the fMRI signal. *Nature*, 412, 150-157.

[64] Lustig, C., Snyder, A. Z., Bhakta, M., Brien, O., Mcavoy, K. C., Raichle, M., Morris, M. E., & Buckner, J. C. R.L. ((2003). Functional deactivations: change with age and dementia of the Alzheimer type. Proc Natl Acad Sci USA, , 100, 14504-14509.

[65] Machulda, M. M., Ward, H. A., Borowski, B., Gunter, J. L., Cha, R. H., Brien, O., Petersen, P. C., Boeve, R. C., Knopman, B. F., Tang-wai, D., Ivnik, D. F., Smith, R. J., Tangalos, G. E., Jack, E. G., & Jr, C. R. (2003). Comparison of memory fMRI response among normal, MCI and Alzheimer's patients. *Neurology*, 61, 500-506.

[66] Mandzia, J. L., Mcandrews, M. P., Grady, C. L., Graham, S. J., & Black, S. E. (2009). Neural correlates of incidental memory in mild cognitive impairment: an fMRI study. *Neurobiol Aging,*, 30, 717-730.

[67] Maurer, K., Volk, S., & Gerbaldo, H. and Alzheimer's disease. Lancet, , 349, 1546-1549.

[68] Mazoyer, B., Zago, L., Mellet, E., Bricogne, S., Etard, O., Houde, O., Crivello, F., Joliot, M., Petit, L., & Tzourio-mazoyer, N. (2001). Cortical networks for working memory and executive functions sustain the conscious resting state in man. Brain Res Bull, , 54, 287-298.

[69] Mckhann, G., Drachman, D., Folstein, M., Katzman, R., Price, D., & Stadlan, E. M. (1984). Clinical diagnosis of Alzheimer's disease: report of the NINCDS-ADRDA Work Group under the auspices of Department of Health and Human Services Task Force on Alzheimer's Disease. *Neurology*, 34, 939-944.

[70] Mesulam, M. M. (1998). From Sensation to cognition. *Brain,*, 121, 1013-1052.

[71] Miller, S. L., Celone, K., Depeau, K., Diamond, E., Dickerson, B. C., Rentz, D., Pihlajamäki, M., & Sperling, R. A. memory impairment associated with loss of parietal deactivation but preserved hippocampal activation. Proc Natl Acad Sci USA, , 105, 2181-2186.

[72] Miller, S. L., Fenstermacher, E., Bates, J., Blacker, D., Sperling, R., & Dickerson, B. C. (2008b). Hippocampal activation in adults with mild cognitive impairment predicts subsequent cognitive decline. *J Neurol Neurosurg Psychiatry,*, 79, 630-635.

[73] Minoshima, S., Giordani, B., Berent, S., Frey, K. A., Foster, N. L., & Kuhl, D. E. (1997). Metabolic reduction in the posterior cingulate cortex in very early Alzheimer's disease. *Ann Neurol,*, 42, 85-94.

[74] Mintun, M. A., Larossa, G. N., Sheline, Y. I., Dence, C. S., Lee, S. Y., Mach, R. H., et al. (2006). [11C]PIB in a nondemented population: potential antecedent marker of Alzheimer disease. *Neurology*, 67, 446-452.

[75] Nestor, P. J., Fryer, T. D., Smielewski, P., & Hodges, J. R. (2003). Limbic hypometabolism in Alzheimer's disease and mild cognitive impairment. *Ann Neurol,*, 54, 343-351.

[76] Brien, O, Keefe, J.L., O, & La, K.M. . & Sperling, R.A. (2010). Longitudinal fMRI study in elderly reveals loss of hippocampal activation with clinical decline. Neurology, Vol. 71, pp. 1969-76.

[77] Otten, L. J., & Rugg, M. D. (2001). When more means less: neural activity related to unsuccessful memory encoding. Curr Biol, , 11, 1528-1530.

[78] Ogawa, S., Tank, D. W., Menon, R., Ellermann, J. M., Kim, S. G., Merkle, H., & Ugurbil, K. (1992). Intrinsic signal changes accompanying sensory stimulation: Functional brain mapping with magnetic resonance imaging. Proc Natl Acad Sci USA, , 89, 5951-5955.

[79] Pariente, J., Cole, S., Henson, R., Clare, L., Kennedy, A., Rossor, M., Cipoloti, L., Puel, M., Demonet, J. F., Chollet, F., & Frackowiak, R. S. (2005). Alzheimer's patients engage an alternative network during a memory task. Ann Neurol, , 58, 870-879.

[80] Persson, J., Lind, J., Larsson, A., Ingvar, M., Sleegers, K., Van Broeckhoven, C., Adolfsson, R., Nilsson, L. G., & Nyberg, L. (2008). Altered deactivation in individuals with genetic risk for Alzheimer's disease. *Neuropsychologia*, 46, 1679-1687.

[81] Petersen, R. C., Smith, G. E., Waring, S. C., Ivnik, R. J., Tangalos, E. G., & Kokmen, E. (1999). Mild cognitive impairment: clinical characterization and outcome. *Arch Neurol,*, 56, 303-308.

[82] Petersen, R. C., Doody, R., Kurz, A., Mohs, R. C., Morris, J. C., Rabins, P. V., Ritchie, K., Rossor, M., Thal, L., & Winblad, B. (2001). Current concepts in mild cognitive impairment. *Arch Neurol,*, 58, 1985-1992.

[83] Petersen, R. C. (2004). Mild cognitive impairment as a diagnostic entity. *J Intern Med,*, 256, 183-194.

[84] Petersen, R. C., Roberts, R. O., Knopman, D. S., Boeve, B. F., Geda, Y. E., Ivnik, R. J., Smith, G. E., & Jack, C. R. Jr. ((2009). Mild Cognitive impairment: Ten years later. Arch Neurol, , 66, 1447-1455.

[85] Petrella, J. R., Krishnan, S., Slavian, M. J., Tran, T. T., Murty, L., & Doraiswamy, P. M. (2006). Mild cognitive impairment: evaluation with 4-T functional MR imaging. *Radiology*, 240, 177-186.

[86] Petrella, J. R., Prince, S. E., Wang, L., Hellegers, C., & Doraiswamy, P. M. (2007a). Prognostic value of posteromedial cortex deactivation in mild cognitive impairment. PLoS ONE, e1104., 2

[87] Petrella, J. R., Wang, L., Krishnan, S., Slavin, M. J., Prince, S. E., Tran, T. T., & Doraiswamy, P. M. (2007b). Cortical deactivation in mild cognitive impairment: high-field-strength functional MR imaging. *Radiology*, 245, 224-235.

[88] Pihlajamäki, M., Tanila, H., Hänninen, T., Könönen, M., Mikkonen, M., Jalkanen, V., Partanen, K., Aronen, H. J., & Soininen, H. (2003). Encoding of novel picture pairs activates the perirhinal cortex: an fMRI study. *Hippocampus*, 13, 67-80.

[89] Pihlajamäki, M., Depeau, K. M., Blacker, D., & Sperling, R. A. (2008). Impaired medial temporal repetition suppression is related to failure of parietal deactivation in Alzheimer's disease. Am J Geriatr Psychiatry, , 16, 283-292.

[90] Pihlajamäki, M., Keefe, O., Bertram, K., Tanzi, L., Dickerson, R. E., Blacker, B. C., Albert, D., & Sperling, M. S. R.A. ((2010). Evidence of altered posteromedial cortical fMRI activity in subjects at risk for Alzheimer disease. Alzheimer Disease and Associated Disorders, 24, 28-36.

[91] Pihlajamäki, M., & Sperling, R. A. (2009). Functional MRI assessment of task-induced deactivation of the default mode network in Alzheimer's disease and at-risk older individuals. Behavioral Neurology, , 21, 77-91.

[92] Pihlajamäki, M., Keefe, O., Brien, K. , O., Blacker, J., & Sperling, D. R.A. ((2011). Failure of repetition suppression and memory encoding in aging and Alzheimer's disease. Brain Imaging Behav, , 5, 36-44.

[93] Raichle, M. E. MacLeod, A.M., Snyder, A.Z., Powers, W.J. Gusnard, D.A. & Shulman, G.I. ((2001). A default mode of brain function. Proc Natl Acad Sci U S A , 98, 676-682.

[94] Remy, F., Mirrashed, F., Cambell, B., & Richter, W. (2005). Verbal episodic memory impairment in Alzheimer's disease: a combined structural and functional MRI study. *Neuroimage*, 25, 253-266.

[95] Rombouts, S. A., Barkhof, F., Veltman, D. J., Machielsen, W. C., Witter, M. P., Bierlaagh, M. A., Lazeron, R. H., Valk, J., & Scheltens, P. (2000). Functional MR imaging in Alzheimer's disease during memory encoding. *Am J Neuroradiol,*, 21, 1869-1875.

[96] Rombouts, S. A., Barkhof, F., Goekoop, R., Stam, C. J., & Scheltens, P. (2005a). Altered resting state networks in mild cognitive impairment and mild Alzheimer's disease: An fMRI study. Hum Brain Mapp, , 26, 231-239.

[97] Rombouts, S. A., Goekoop, R., Stam, C. J., Barkhof, F., & Scheltens, P. (2005b). Delayed rather than decreased BOLD response as a marker for early Alzheimer's disease. *Neuroimage*, 26, 1078-1085.

[98] Saunders, A. M., Strittmatter, W. J., Schmechel, D., George-hyslop, P. H., Pericak-vance, M. A., Joo, S. H., Rosi, B. L., & Gusella, J. F. Crapper-MacLachlan, D.R., Alberts, M.J., et al. ((1993). Association of apolipoprotein E allele epsilon 4 with late-onset familial and sporadic Alzheimer's disease. Neurology, 43, 1467-1472.

[99] Scheff, S. W., Price, D. A., Schmitt, F. A., & Mufson, E. J. (2006). Hippocampal synaptic loss in early Alzheimer's disease and mild cognitive impairment. Neurobiol Aging, , 27, 1372-1384.

[100] Scheff, S. W., Price, D. A., Schmitt, F. A., Dekosky, S. T., & Mufson, E. J. (2007). Synaptic alterations in CA1 in mild Alzheimer disease and mild cognitive impairment. *Neurology*, 68, 1501-1508.

[101] Schwindt, G. C., & Black, S. E. (2009). Functional imaging studies of episodic memory in Alzheimer's disease: A quantative meta-analysis. Neuroimage, , 45, 181-190.

[102] Selkoe, D. J. (2002). Alzheimer's disease is a synaptic failure, Science, 298, 789-791.

[103] Shmuel, A., Augath, M., Oeltermann, A., & Logothetis, N. K. (2006). Negative functional MRI response correlates with decreases in neuronal activity in monkey visual area V1. *Nat Neurosci,*, 9, 569-577.

[104] Shulman, G., Fiez, J. A., Corbetta, M., Buckner, R. L., Miezin, F. M., & Raichle, M. E. (1997). Common blood flow changes across visual tasks: II. Decreases in cerebral cortex. J Cogn Neurosci, , 9, 648-663.

[105] Small, S. A., & Perera, G. M. DeLaPaz, R., Mayeux, R. & Stern, Y. ((1999). Differential regional dysfunction of the hippocampal formation among elderly with memory decline and Alzheimer's disease. Ann Neurol, , 45, 466-472.

[106] Smith, C. D., Andersen, A. H., Kryscio, R. J., Schmitt, F. A., Kindy, M. S., Blonder, L. X., & Acison, M. J. (1999). Altered brain activation in cognitively intact individuals at high risk for Alzheimer's disease. *Neurology*, 53, 1391-1396.

[107] Smith, C. D., Andersen, A. H., Kryscio, R. J., Schmitt, F. A., Kindy, M. S., Blonder, L. X., & Acison, M. J. (2002). Women at risk for AD show increased parietal actication during a fluent task. Neurology, , 58, 1197-1202.

[108] Sorg, C., Riedl, V., Muhlau, M., Calhoun, V. D., Eichele, T., Läer, L., Drzezga, A., Förstl, H., Kurz, A., Zimmer, C., & Wohlschläger, A. M. (2007). Selective changes of resting-state networks in individuals at risk for Alzheimer's disease. *Proc Natl Acad Sci USA*,, 104, 18760-18765.

[109] Sperling, R., Greve, D., Dale, A., Killiany, R., Holmes, J., Rosas, H. D., Cocchiarella, A., Firth, P., Rosen, B., Lake, S., Lange, N., Routledge, C., & Albert, M. (2002). Functional MRI detection of pharmacologically induced memory impairment. *Proc Natl Acad Sci USA.*, 99, 455-460.

[110] Sperling, R., Bates, J., Chua, E., Cocchiarella, A., Schacter, D. L., Rosen, B., Schacter, D. L., & Albert, M. S. (2003a). fMRI studies of associative encoding in young and elderly controls and mild AD patients. J Neurol Neurosurg Psychiatry, , 74, 44-50.

[111] Sperling, R., Chua, E., Cocchiarella, A., Rand-giovannetti, E., & Poldrack, R. Schacter, D.L & Albert, M. ((2003b). Putting names to faces: successful encoding of associative memories activates the anterior hippocampal formation. Neuroimage, 20, 1400-1410.

[112] Squire, L. R., & Zola-morgan, S. (1991). The medial temporal lobe memo. ry system. Science, , 11, 1380-1386.

[113] Stern, C. E., Corkin, S., Gonzalez, R. G., Guimaraes, A. G., Baker, J. R., Jennings, P. J., Carr, C. A., Sugiura, R. M., Vedantham, V., & Rosen, B. R. (1996). The hippocampal formation participates in novel picture encoding: evidence from functional magnetic resonance imaging. *Proc Natl Acad Sci USA*,, 93, 8660-8665.

[114] Supekar, K., Menon, V., Rubin, D., Musen, M., & Greicius, M. D. (2008). Network analysis of intrinsic functional brain connectivity in Alzheimer's disease. PLoS Comput Biol, e1000100., 4

[115] Suzuki, W. A., & Amaral, D. G. (1994). Perirhinal and parahippocampal cortices of the macaque monkey: cortical afferents. J Comp Neurol, , 350, 497-533.

[116] Suzuki, W. A. (2007). Making new memories: The role of the hippocampus in new associative learning. *Ann N Y Acad Sci*,, 1097, 1-11.

[117] Tanzi, R. E., & Bertram, L. (2001). New frontiers in Alzheimer's disease genetics. *Neuron*, 32, 181-184.

[118] Trivedi, M. A., Schmitz, T. W., Ries, M. L., Torgerson, B. M., Sager, M. A., Hermann, B. P., Asthana, S., & Johnson, S. C. (2006). Reduced hippocampal activation during episodic encoding in middle-aged individuals at genetic risk of Alzheimer's disease: a cross-sectional study. *BMC Med,*, 4, 1 EOF.

[119] Tulving, E., & Markowitsch, H. J. (1998). Episodic and declarative memory: role of the hippocampus. *Tulving, E. & Markowitsch, H.J. (1998). Episodic and declarative memory: role of the hippocampus. Hippocampus, Vol. 8, pp. 198-204.*, 8, 198-204.

[120] Wagner, A. D., Schacter, D. L., Rotte, M., Koutstaal, W., Maril, A., Dale, A. M., Rosen, B. R., & Buckner, R. L. (1998). Building memories: remembering and forgetting of verbal experiences as predicted by brain activity. *Science*, 281, 1188-1191.

[121] Wang, K., Liang, M., Wang, L., Tian, L., Zhang, X., Li, K., & Jiang, T. (2007). Altered functional connectivity in early Alzheimer's disease: A resting-state fMRI study. Hum Brain Mapp, , 28, 967-978.

[122] Weissman, D. H., Roberts, K. C., Visscher, K. M., & Woldorff, M. G. (2006). The neural bases of momentary lapses in attention. Nat Neurosci, , 9, 971-978.

[123] Winblad, B., Palmer, K., Kivipelto, M., Jelic, V., Fratiglioni, L., Wahlund, L. O., Nordberg, A., Bäckman, L., Albert, M., Almkvist, O., Arai, H., Basun, H., Blennow, K., De Leon, M., Decarli, C., Erkinjuntti, T., Giacobini, E., Graff, C., Hardy, J., Jack, C., Jorm, A., Ritchie, K., Van Duijn, C., Visser, P., & Petersen, R. C. (2004). Mild cognitive impairment--beyond controversies, towards a consensus: report of the International Working Group on Mild Cognitive Impairment. J Intern Med, , 256, 240-246.

[124] Wishart, H. A., Saykin, A. J., Rabin, L. A., Santulli, R. B., Flashman, L. A., Guerin, S. J., Mamourian, A. C., Belloni, D. R., Rhodes, C. H., & Mcallister, T. W. (2006). Increased brain activation during working memory in cognitively intact adults with the APOE epsilon4 allele. *Am J Psychiatry,*, 163, 1603-1610.

[125] Woodard, J. L., Seidenberg, M., Nielson, K. A., Antuono, P., Guidotti, L., Durgerian, S., Zhang, Q., Lancaster, M., Hantke, N., Butts, A., & Rao, S. M. (2009). Semantic memory activation in amnestic mild cognitive impairment. Brain, , 132, 2068-2078.

[126] Yassa, M. A., Stark, S. M., Bakker, A., Albert, M. S., Gallagher, M., & Stark, C. E. (2010). High-resolution structural and functional MRI of hippocampal CA3 and dentate gyrus in patients with amnestic Mild Cognitive Impairment. *Neuroimage*, 51, 1242-1252.

[127] Zhou, Z. Dougherty, Jr. J.H, Hubner, K.F., Bai, B., Cannon R.L. & Hutson R.K. ((2008). Abnormal connectivity in the posterior cingulate and hippocampus in early Alzheimer's disease and mild cognitive impairment. Alzheimers Dement, , 4, 265-270.

Chapter 20

Structural and Functional Magnetic Resonance Imaging in Hepatic Encephalopathy

Long Jiang Zhang, Guang Ming Lu and Hui Mao

Additional information is available at the end of the chapter

http://dx.doi.org/10.5772/29563

1. Introduction

Hepatic encephalopathy (HE) is a neuropsychiatric syndrome that develops in patients with severe liver diseases and/or portal-systemic shunting. HE is characterized by a wide spectrum of clinical manifestations, ranging from alterations of psychometric performance to stupor and coma (Rovira et al., 2008; Cordoba J., 2011). Several non-invasive neuroimaging techniques, such as particularly magnetic resonance imaging (MRI) and magnetic resonance spectroscopy (MRS), are used for the diagnosis and prognosis of HE. These MR techniques can identify and measure the abnormal accumulation and increase of metabolite, such as glutamine and glutamate (Glx) as a result of HE in the central nervous system (CNS). Under normal circumstances, these substances are efficiently metabolized by the liver. This chapter will review the pathophysiology of HE and its conventional MRI, MRS and functional MRI findings.

2. Pathophysiology of hepatic encephalopathy

Various hypotheses have been proposed to explain the complex neuropsychiatric syndrome seen in HE. Imbalance between inhibitory and excitatory neurotransmission and hyperammonemia is the primary and most widely accepted hypothesis for HE (Rovira et al., 2008; Butterth et al., 2003; Cordoba J., 2011; Cordoba & Minguczb., 2008). Downregulation of glutamate receptors and an increase in inhibitory neurotransmission result in clinical manifestations of HE. In addition, patients with liver failure or portal-systemic shunt surgery have elevated levels of circulating ammonia, which enters the brain through the blood-brain barrier, and increases the ammonia concentration in the cerebral blood up to four fold (normally in the order of two) in liver failure. Hyperammonemia disease leads to profound astrocyte changes, including astrocyte swelling in acute HE and Alzheimer type II astrocyte changes in chronic HE (Matsusue et al., 2005).

3. Clinical features of hepatic encephalopathy

The Working Party at the 11th World Congresses of Gastroenterology, held in Vienna in 1998, recommended the nomenclature and types of HE (Table 1) (Ferenci et al., 2002). The West Haven criteria for semi-quantitative grading of HE are summarized in Table 2. HE can be classified in three main groups: episodic, chronic, and minimal, based on duration and characteristics of the clinical manifestations. Episodic HE is characterized by the development of an impaired mental state, neuromuscular abnormalities, asterixis (tremor, jerking movement of the wrist), fetor hepaticus (metcarpans pass into the lungs resulting in the presence of ammonia and ketones in the breath), and hyperventilation, which develops during a short period of time and fluctuates in severity. Chronic HE can be further classified into subgroups: relapsing HE and persistent HE. Relapsing HE manifests itself as frequent episodes of acute HE, while persistent HE refers to manifestations that do not reverse despite adequate treatment. Characteristic manifestations of severe persistent HE are dementia, Parkinsonism, or myelopathy in combination with neurologic involvement, such as ataxia, gait abnormalities, tremor (Rovira et al., 2008). Minimal HE refers to those patients HE type with cirrhosis or portal-systemic shunts who have subtly abnormal cognitive and/or neurophysiologic functions. These abnormalities cannot be detected by the standard clinical examination and can only be determined by a detailed assessment of the patient's history and a comprehensive neurologic evaluation of cognitive performance and motor functions. The neuropsychological features of the minimal HE type include abnormalities in executive functions, particularly in selective attention and psychomotor speed (Amodio et al., 2004). However, other abnormalities, such as memory impairments, are also seen (Amodio et al., 2004). A complete psychometric assessment by a neuropsychologist is the best way to determine the extent of the attention-related cognitive impairment of a HE patient. A large number of neuropsychologic tests, such as the number connection test (NCT), the line-tracing, and inhibitory control test have been developed and are applied to describe cognitive abnormalities in patients without any clinical evidence of HE (Ferenci et al., 2002; Amodio et al., 2010). Testing across various neuropsychological domains is probably the optimal approach in order to identify cognitive and motor system abnormalities, such as attention and fine motion control. A standardized test battery, including the NCT type A and NCT type B, the line-tracing, the serial-dotting, and the digit-symbol tests (PSE-Syndrome-Test) has a high specificity for HE as compared with other metabolic encephalopathies (Ferenci et al., 2002). Changes in EEG/evoked responses and neuroimaging findings are non-specific and may not be able to provide sufficient information for the diagnosis of minimal HE.

HE type	Nomenclature	Subcategory	Subdivisions
A	Encephalopathy associated with acute liver failure		
B	Encephalopathy associated with portal-systemic bypass and no intrinsic hepatocellular disease		
C	Encephalopathy associated with cirrhosis and portal cirrhosis and portal hypertension/ or portal-systemic shunts	Episodic HE	Precipitated, spontaneous, recurrent
		Persistent HE	Mild, severe, treatment-dependent
		Minimal HE	

Table 1. The nomenclature of HE

Grade	Criteria
1	Trivial lack of awareness, euphoria or anxiety, shortened attention span, impaired performance of addition
2	Lethargy or apathy, minimal disorientation for time or place, subtle personality change, inappropriate behaviour, impaired performance of subtraction
3	Somnolence to semi stupor, but responsive to verbal stimuli; confusion; gross disorientation
4	Coma (unresponsive to verbal or noxious stimuli)

Table 2. West Haven criteria for semiquantitative grading of HE

4. Conventional MRI findings

In MRI exam of HE, bilateral hyperintensities at basal ganglia in T1 weighted images but without the corresponding abnormal T2 weighted signal intensity is a typical imaging feature of HE (Figure 1A).This imaging characteristic is attributed to hypermanganesemia (Gover et al., 2006; Mcphail & Taylor-Robinson., 2010).

The observed hyperintensities in bilateral basal ganglia on T1 weighted images can be reduced, or even disappear, after liver transplantation (Figure 1B and 1C) (Naegele et al., 2000; Cordoba et al., 2001; Long et al., 2009). However, no correlation between signal intensities on T2 weighted images in basal ganglia and the clinical encephalopathy or neuropsychological test performance was found, after a number of studies (Spahr et al., 2002; Thuluvath et al., 1995). Although mild brain oedema has been reported in the studies using other advanced MR sequences, no mild abnormalities can be detected in the conventional T1 and T2 weighted images, in addition to basal ganglia hyperintensity. Some studies reported high signal intensity at the centrum semiovale or corticospinal tract (along the hemispheric white matter in or around the corticospinal tract) on fast fluid attenuated inversion recovery (FLAIR) T2-

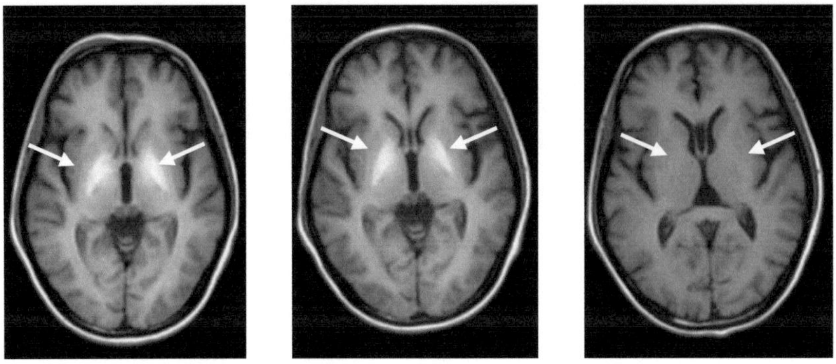

A B C

Figure 1. T1 weighted spin echo images of a selected axial brain section from a patient before and after receiving liver transplantation A) An axial T1 weighted image shows the symmetrical areas with high intensity in bilateral basal ganglia (arrows) before liver transplantation; B) 1 week after liver transplantation; and C) 4 months after liver transplantation, noticing that the symmetrical high intensity in bilateral basal ganglia shown in A, B, has disappeared.

weighted images (Figure 2) (Mínguez et al., 2007; Rovira et al., 2008). This is strikingly similar to signalintensity abnormalities noted in cases of amyotrophic lateral sclerosis, a neurodegenerative disease that affects motor neurons. The progressive normalization of the abnormal high signal-intensity in patients with cirrhosis can be found after the successful liver transplantation or effective treatment. (Mínguez et al., 2007; Rovira et al., 2008). Therefore, observation of FLAIR white matter hyperintensities without additional imaging sequences, such as T2 weighted images cannot be used to diagnose HE.

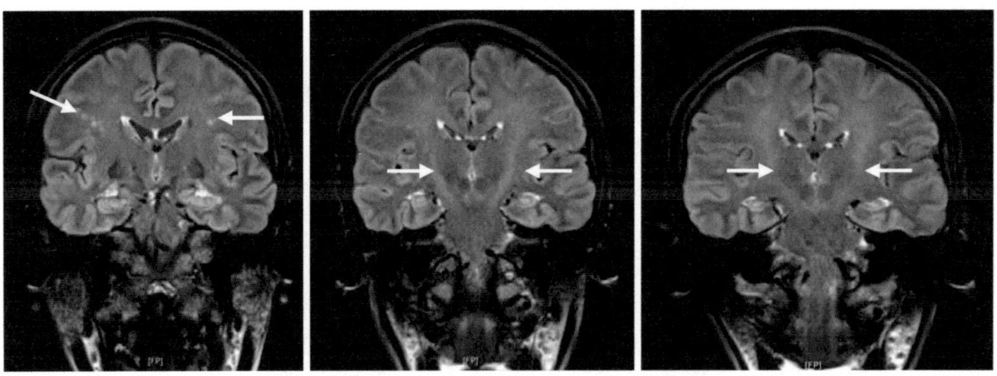

A B C

Figure 2. Coronal FLAIR images of a cirrhotic patient A to C. Coronal FLAIR images show the areas with abnormal high signal intensity in or around the corticospinal tract (arrows).

5. Magnetic transfer imaging (MTI)

Magnetic transfer imaging (MTI) is a singular MRI technique that has been shown to be useful in the diagnosis of HE. It is mainly based on the interaction (cross-relaxation) between protons in a relatively free environment and those in which motion is restricted (immobile water). Exchange of the saturated magnetization with free water reduces the signal intensity observed in the subsequent MR images. The degree of the signal-intensity loss depends on the attenuation of the macromolecules in a given tissue. MTI allows the measurement of magnetic transfer ratios (MTR), which reflect brain parenchymal changes that may not be visible using the standard MR techniques (Mcphail & Taylor-Robinson., 2010; Grover et al., 2006). Lower MTR can be the result of pathologies that alter the structural integrity and the relative macromolecular-water composition of brain parenchyma, such as in multiple sclerosis plaques (Rovira et al., 1999) or in end-stage cirrhosis (Rovira et al., 2001). Severe MTR decrease may be attributed to the demyelination and axonal loss, but a less severe decrease can be the result of inflammation and moderate demyelination.

All studies using MTI in HE found reduced MTR values in the examined brain regions of HE patients compared to those of controls (Miese et al., 2006; Miese et al., 2009; Poveda et al., 2010). Observed MTR decrease is mild (approximately 10%) and is not accompanied by significant abnormalities on conventional T1-and T2-weighted images, compared with the other metabolic and neurodegenerative diseases, such as experimental autoimmune encephalomyelitis, toxic demyelination, progressive multifocal leukoencephalopathy, human immunodeficiency virus encephalitis, and multiple sclerosis. The MTR reduction seems to have a temporal process in HE, an early involvement of basal ganglia and white matter has been shown (Miese et al., 2009). The MTR decrease almost returns to normal values after liver transplantation, thus supporting the hypothesis that the reduced MTR values reflect mild reversible brain edema.

6. Diffusion weighted imaging and diffusion tensor imaging

Diffusion weighted imaging (DWI) is an MRI method that generates brain images based on the image contrast that is dependent on the mobility of H2O molecules in the different tissue compartments (Thoeny & De Keyzer, 2011). Water molecule motion follows the principles of Brownian motion (Provenzale et al., 2006). In a container of water, molecules undergo free, thermally agitated diffusion, which is also called isotropic (means equal in all directions). However, limited diffusion is observed since the movement of water molecules is restricted by their interactions with cell membranes and macromolecules, causing the directional specific motion of water molecules in environments, which is called anisotropic diffusion (Figure 3). DWI derives its image contrast on the basis of differences in the mobility of protons (primarily associated with water) between tissues (Thoeny & De Keyzer., 2011). Apparent diffusion coefficient (ADC) is a measurement obtained from DWI that provides estimation of water motion. DWI has a potential to locate the compartment where the increase of mobility of water molecules is more prominent. Therefore, it can be applied to differentiate cytotoxic oedema from vasogenic oedema in HE patients (Lodi et al., 2004; Mies et al., 2006; Sugimoto et al., 2008). Brain oedema refers to excess water accumulation in the intracellular and extracellular spaces of the brain. In cytotoxic oedema, there is shift of water from extracellular to intracellular compartments producing cellular swelling expressing low ADC, while in vasogenic oedema the reverse happens, producing increased ADC (Schaefer et al., 2000). Recent studies on the application of mean diffusivity measurements within normal-appearing white matter of chronic cirrhotic patients have shown significant increase in brain water diffusivity, which is more pronounced with increased severity of HE (Lodi et al., 2004). These diffusivity values correlated with neuropsychological impairment (Kumar et al., 2008) and increased venous ammonia (Lodi et al., 2004). These findings suggest an accumulation of water in the extracellular compartment and a hypothesis of cortical astrocytic swelling is not supported, but rather increased interstitial fluid or chronic demyelination (Lodi et al., 2004; Sugimoto et al., 2008; Miese et al., 2006) due to glutamine accumulation is cited as the cause of diffuse brain oedema in chronic liver failure (Lodi et al., 2004; Sugimoto et al., 2008; Miese et al., 2006).

On the other hand, the morphology and structure of tissues vary in different organs and pathologic states. Water motion occurs preferentially in some directions in certain tissues due to the presence of obstacles that limit molecular movement in some directions. Diffusion-tensor imaging (DTI) can be used to obtain anisotropy information about water diffusion in tissues when specific diffusion weighted gradients in at least six directions are applied. The most commonly used invariant indices for the measurement of anisotropic diffusions by DTI are; the relative anisotropy (RA), the fractional anisotropy (FA), and the volume ratio (VR) indices (Figure 4A). In brain imaging applications, these indices provide a quantitative measurement of the changes of white matter integrity in different brain regions that are affected by HE. Using DTI together with advanced fibre-tracking algorithms based on the obtained diffusion tensors, white matter tractography can be derived from DTI data. Therefore, it is possible to noninvasively construct 3D trajectories of neural tracts, allowing the modelling of white matter neural connectivity (Figure 4B) (Mukherjee et al., 2008a; Mukherjee et al., 2008).

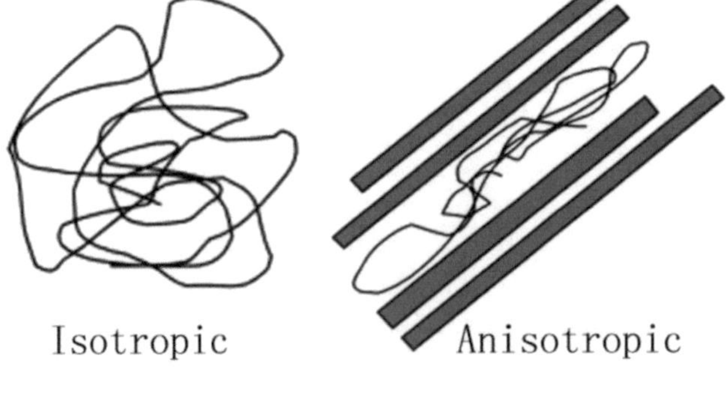

Figure 3. Schematic illustrations describe the isotropic diffusion and anisotropic restricted diffusion

http://dx.doi.org/10.5772/29563

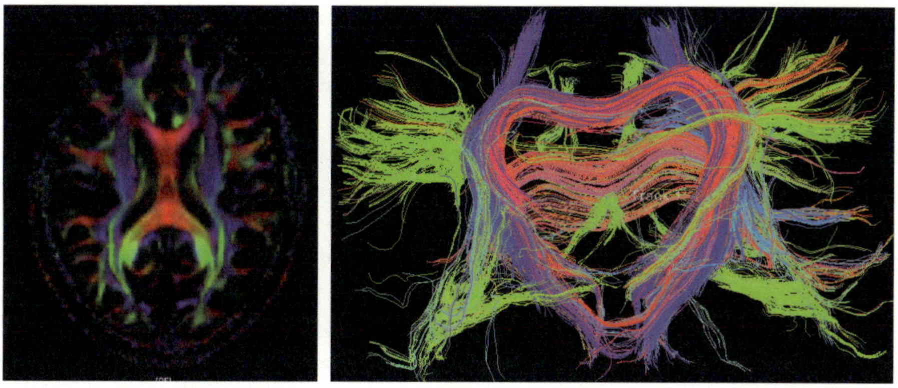

A B

Figure 4. Presentations of diffusion tensor images of a cirrhotic patient acquired at a 3T MR scanner A. FA image, B. tractography of the brain derived from diffusion tensor imaging

FA, an index reflecting the white matter integrity, shows no significant changes in patients with chronic HE compared with normal subjects, indicating an absence of microstructural damage in these patients (Kale et al., 2006). Recently, a study by Chavarria et al. (Chavarria et al., 2010) into a rat model of acute liver failure provided experimental evidence to support the cytotoxic origin of brain oedema. Their results suggested that the metabolism of ammonia in astrocytes induces an increase of glutamine and lactate, which may mediate cortical cellular swelling. Saksena et al. (Saksena et al., 2007) also found decreased mean diffusivity values in patients with acute liver failure, suggesting an increase in the intracellular brain water content. Other studies also found decreased ADC value and high cortical signal intensity in patients with acute HE (McKinney et al., 2010; Toru et al., 2011). Therefore, it appears that two different types of brain oedema, intracellular in acute forms and probably interstitial in chronic forms, may exist in liver failure.

7. Magnetic Resonance Spectroscopy (MRS)

MRS is a non-invasive analytical method for analysing brain metabolites, often accompanied by MRI in neuroradiology practices. MR spectroscopic analysis offers a tool to interrogate the molecular process within body tissue and fluids in vivo and in vitro. In clinical application, MRS is capable of identifying and measuring the individual chemicals in the brain regions localized and sampled from MR images. 1H MRS can provide information about brain metabolites such as choline (Cho), creatine (Cr), N-acetyl aspartate (NAA), glutamine and glutamate (Glx), as well as osmolytes, such as myoinositol (mI) and taurine (Ross et al., 1994; Kreis R et al., 1992; Cordoba et al., 2002; Mcphail & Taylor-Robinson. 2010). NAA is a derivative of an amino acid found mostly in neurons; therefore, it is commonly considered as a measure of neuronal density. The peak at 3.19-3.24 ppm of MR spectra obtained in vivo represents a total Cho, including soluble membrane phospholipids: phosphorylcholine (PCho), glycerophosphocholine (GPC) and a relatively small amount of free choline. PCho is involved in the synthesis of the insoluble membrane phospholipids (Vance DE,1996), while GPC is a product of membrane degradation and free choline is involved in synthesis of the neurotransmitter, acetylcholine, as well as of membrane synthesis (Michel et al., 2006). An increase in the total Cho peak is associated with an increase in membrane breakdown or turnover, myelination or inflammation (Astley et al., 2009). The peak of Cr spectrum is assigned at 3.02 ppm. This peak represents a combination of molecules containing creatine and phosphocreatine. When energy supply is insufficient from the Kreb's cycle, ATP is produced from adenosine diphosphate (ADP) instead of glucose. This reaction is buffered by the phosphocreatine–creatine (PCr-Cr) system, where PCr donates a phosphate group to become Cr. It is assumed that the Cr peak reflects energy use. Glutamine and glutamate (Glx) are amino acids. Glutamate is the most abundant amino acid in the brain and is released by approximately 90% of excitatory neurons. The spectral peaks of these molecules are often grouped together as Glx in the spectra obtained at the low field, because their overlap makes it hard to resolve them separately. Two compounds become increasingly separated in the higher magnetic fields; therefore, with the increasing clinical availability of stronger magnets, there is growing importance and interests in studying these compounds. Myo-inositol (mI), a simple sugar-alcohol compound, and precursor for inositol lipid synthesis, is a putative glial cell marker because it is located primarily in glial cells and not in neurons. Astrocytes, microglia, and macrophages have been shown to increase their levels of the Na+/mI cotransporter (SMIT) in response to stress or injury and an accumulation of mI, suggesting a role of this compound in inflammation and a marker for astrogliosis and microglia activity (Gupta et al., 2010).

1H MRS has been extensively used in HE studies and has consensus findings on the intracellular metabolite changes in HE. Typical 1H MRS findings of HE include: lower Cho/Cr and mI/Cr and higher Glx/Cr in all examined brain regions compared with controls (Figure 5) (Taraow et al., 2003; Weissenborn et al., 2004; Stewart et al., 2005; Mechtcheriakov et al., 2005; Miese et al., 2006; Weissenborn et al.,2007; Verma et al., 2008). Ammonia plays an important role in the series of metabolite changes in patients with cirrhosis. The increase in brain glutamine during HE increases intracellular osmolality. To maintain osmotic equilibrium, the astrocytes lose osmolytes such as mI and Cho, and a large amount of water enters the astrocyte and results in cellular swelling and impairment of cellular metabolism, further influencing neuronal and astrocyte function and the interaction between them. MRS measurable metabolite changes have been shown to correlate with improved psychometric performance (Binesh et al., 2006) following treatment (Haseler et al., 1998) and following liver transplantation (Naegele et al., 2000; Córdoba et al., 2001; Long et al., 2009). Recent pilot studies on in vitro 1H MRS of human blood or urine in acute liver failure have demonstrated significant metabolite abnormalities between survivors and nonsurvivors (Saxena et al., 2006). They found glutamine in serum and the urine glutamine: creatinine ratio was higher in non-surviving patients compared with surviving patients [serum glutamine, 3.08 (1.68-7.11) vs. 0.56 (0.34-0.99) mM, median and range; P=0.0001 and urine glutamine:creatinine ratio, 1.72 (0.24-7.76) vs. 0.39 (0.1-0.84), P=0.1], and urine urea:creatinine ratio was higher in surviving patients compared with non-surviving patients [10.83 (0.2-22.6) vs. 2.09 (0.96-4.0), P=0.002]. Their study indicated MRS had a potential use for clinical decision making about the need for advanced therapeutic intervention, such as artificial liver support or emergency liver transplantation in acute liver failure. Of these available MRS measurements, mI seems to be a more sensitive biomarker than Cho to detect HE. mI levels were negatively related to the Child-Pugh score and degree of HE (Lee et al., 1999; Zhang et al., 2010). However, values in a normal range have not been established and the diagnostic accuracy of 1H-MRS remains uncertain.

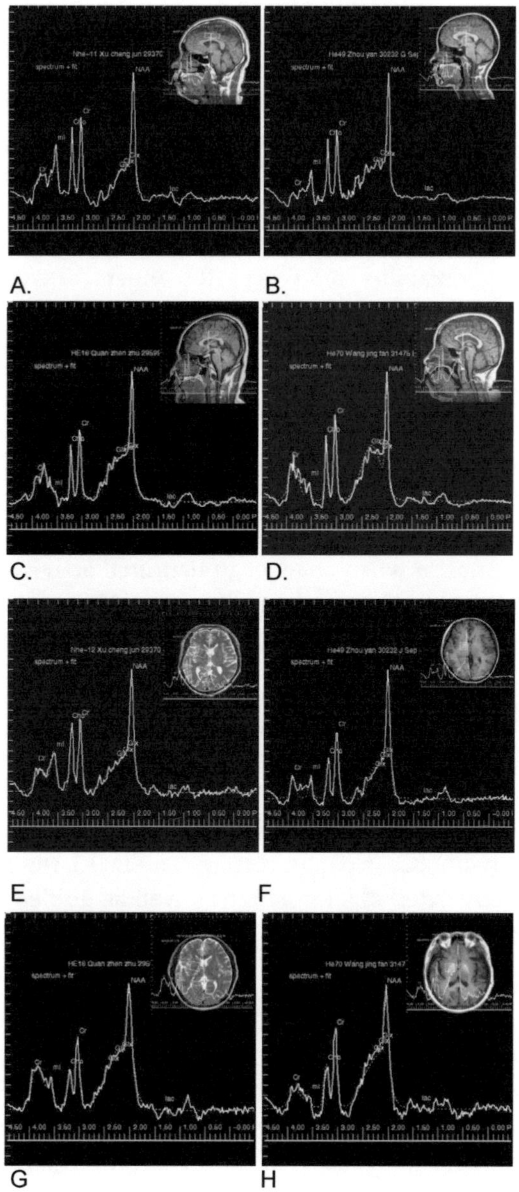

Figure 5. A set of brain MR spectra of HE patients and normal control A) A MRS spectrum obtained from the anterior cingulate cortex (ACC) of a healthy adult. B) A MRS spectrum of the anterior cingulate cortex from a 25-year-old fe-

male cirrhotic patient. C) MRS of the anterior cingulate cortex of a 44-year-old minimal hepatic encephalopathy (MHE) patient shows decreased ml and increased Glx. D) MRS of anterior cingulate cortex of a 70-year-old HE patient shows decreased ml and Cho and increased Glx. E) MRS of right basal ganglia of the same subject as in A). F) MRS of right basal ganglia in the same patient as in B) shows decreased ml and Cho. G) MRS of right basal ganglia of the same patient as in (C) shows decreased ml and Cho and increased Glx. H) MRS of right basal ganglia of the same patient in (D) shows decreased ml and Cho and increased Glx. The spectra were collected using single voxel PRESS sequence at 1.5T scanner.

In clinical practice, accurately differentiating some metabolites, such as Glx and taurine, can be challenging with the clinical MR scanners currently available. Two-dimensional MR spectroscopy may enhance the spectral resolution to distinguish metabolites based on Jcoupling patterns of individual molecules. On the other hand, more sophisticated statistical techniques, such as principal components analysis (PCA), enable assessment of the variation in individual metabolites based on the data obtained from MRS measurements (Barba et al., 2008; Singhal et al., 2010). These techniques can provide quantitative information of metabolic processes and changes in the diseased tissue, and therefore can be helpful in the understanding of HE pathogenesis. Wide availability of 3 Tesla MR scanners, as well as short TE acquisition, can improve the spectral resolution and thus, allow the differentiation of Glx from NAA (Sawara et al., 2004).

8. Blood oxygen level dependent functional MRI

Functional magnetic resonance imaging (fMRI) detects blood oxygen level-dependent (BOLD) changes in MRI signal that arise when changes in neuronal activity occur in a region of the brain after responding to a stimulus or performing a specific task (Ogawa et al., 1992). The BOLD signal is the ratio of oxy-haemoglobin to deoxy-haemoglobin during brain activation (Gore JC., 2003).

According to experimental paradigms, fMRI can be classified to task-related and resting state fMRI. In task-related fMRI design, a subject is placed in the magnet of an MRI machine, where various different kinds of stimulus are administered in a controlled fashion. Two main experimental paradigms are commonly used in task-related fMRI studies: block design and event-related paradigms (Gore JC., 2003). In resting state fMRI, the participants are required to rest in the absence of any external stimulation as well as guided to avoid reiterating any goal-directed thought pattern in their heads (Damoiseaux et al., 2006).

Many behavioural studies have claimed the existence of attention alterations in cirrhotic patients without overt HE (Zhang et al., 2007a; Randolph et al., 2009; Bajaj et al., 2009). There are only a few fMRI studies into HE. Zafiris et al. (Zafiris et al., 2004) first analysed the pathologically impaired neural mechanisms of cirrhotic patients using fMRI. Nine subjects with minimal HE (MHE) and 10 controls underwent scanning while they indicated the apparent transition from a steady light to the onset of a flicker light. Judgment-related BOLD activation was decreased in MHE patients compared to controls in the right inferior parietal cortex (IPL). Impaired neural interaction between the IPL and the parietooccipital cortex (Poc), the intraparietal sulcus, the anterior cingulate cortex (ACC), the right

http://dx.doi.org/10.5772/29563

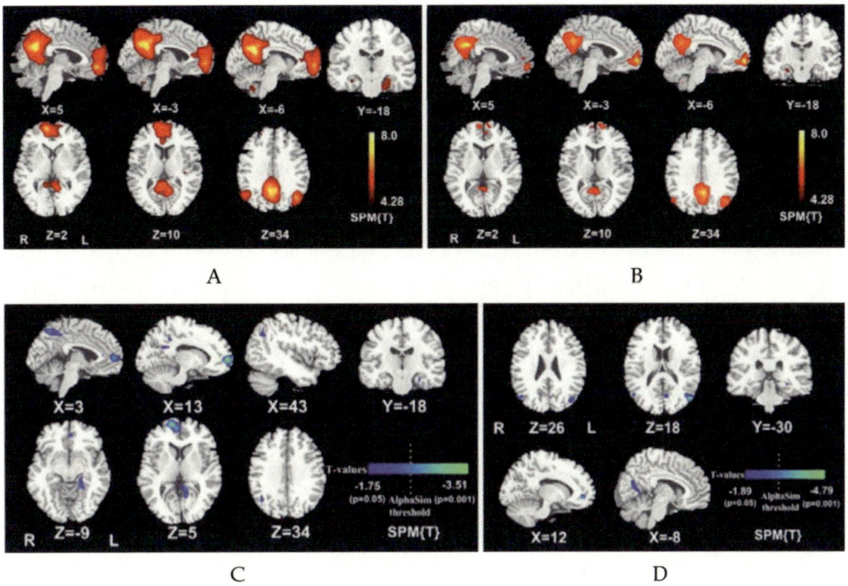

Figure 6. Default mode network changes of HE patients A) DMN of the controls consists of bilateral precuneus/posterior cingulate cortex, medial prefrontal cortex, anterior cingulate cortex, angular gyri, and temporal pole ($P < 0.05$, FDR corrected). B) DMN of HE patients consists of bilateral precuneus/posterior cingulate cortex, medial prefrontal cortex, anterior cingulate cortex, angular gyri, and temporal pole ($P < 0.05$, FDR corrected). Patterns in the HE patients are similar but with a reduced size compared to the ones of normal subjects in (A). C) Differences between the DMN of the HE patients and the controls are noted in the left posterior cingulate cortex and bilateral precuneus, right angular gyrus, bilateral middle frontal cortex, and left parahippocampus ($P<0.05$ for all, uncorrected) and in the right middle frontal gyrus and left posterior cingulate cortex ($P<0.05$ corrected, combined height threshold $P < 0.01$ and a minimum cluster size of 24 voxels). D) Statistical t-maps of the vein blood ammonia against z-scores in ICA in the HE group. The right middle frontal gyrus, left posterior cingulate cortex, left parahippocampus, bilateral angular gyri ($P<0.05$ for all, uncorrected) and the left angular gyrus colour-coded bright blue ($P<0.05$ corrected, combined height threshold $P<0.01$ and a minimum cluster size of 24 voxels) had a negative correlation with venous blood ammonia. From Zhang et al., Brain default-mode network abnormalities in hepatic encephalopathy: a resting-state functional MRI study. Hum Brain Mapp, 2011, DOI:10.1002/hbm.21295, with permission

prefrontal cortex (PFC), the medial temporal lobe, and the extrastriate cortex V5 and an enhanced coupling between IPL and the postcentral cortex were found in MHE patients. This study demonstrated an impaired and compensatory neural mechanism during visual judgment in the earliest stages of HE. Zhang et al. (Zhang et al., 2007a) used a Chinese character Stroop task as the target stimulus to investigate the neural mechanism of the cognitive control impairment in cirrhotic patients using fMRI. This study showed that simple tasks (incongruous word-reading tasks) increased the activity of the precentral and postcentral gyri in cirrhotic patients, and harder tasks (incongruous colour-naming tasks) decrease the activity of bilateral SFG, MFG, IFG, medial wall frontal gyrus, anterior cingulate cortex (ACC), temporal cortex, and parietal cortex. This study provides suggests that the abnormal anterior cingulate cortex–prefrontal cortex–parietal "fusiform" cortex circuit could be the neural substrarte responsible for this impaired cognitive control. However, the task-induced fMRI is better suited to the patients with low-grade or minimal HE who can perform the test tasks (Mcphail and Taylor-Robinson, 2010).

Recently, increasing attention has been given to detecting brain activities during resting state (Greicius et al., 2003; Greicius et al., 2004; Zhang et al., 2007b; Zhang et al., 2011). Most reported resting-state network components include; the default mode network (DMN), the sensorimotor component, the executive control component, up to three visual components, two lateralized fronto-parietal components, the auditory component and the temporoparietal component (Rosazza & Minati, 2011). As already reported, these resting-state networks consist of anatomically distinct, but functionally connected regions, which display a high level of correlated BOLD signal activity. Nowadays the DMN is the most studied network in comparison to task-induced brain networks. DMN includes the medial prefrontal cortex, rostral anterior cingulate, posterior cingulated cortex (PCC), and precuneus (Greicius et al., 2003; Greicius et al., 2004). The DMN is described by increased activity during rest, while decreased activation suppressed during cognitively demanding tasks, such as visual and auditory attention, language processing, memory, and motor activities. Abnormal default function network has been known to relate to Alzheimer disease (AD), autism, attention deficit hyperactivity disorder, schizophrenia, epilepsy, and other diseases (et al., 2005; Rombouts et al., 2005; Kennedy et al., 2006; Liang M, et al., 2006; Tian et al., 2006). In a study by Zhang et al. (Zhang et al., 2007), the DMN of the cirrhotic patients was investigated using a block-design in which a modified Chinese Stroop task was used as the target stimulus.

In this study, a subtraction method was used to study the resting-state network in patients with hepatic cirrhosis. The functional data suggest that cirrhotic patients may have a deactivated DMN. The absence of deactivation in the PCC and precuneus may be a sensitive, rather than specific, marker in patients with hepatic cirrhosis. Recently, in a resting state fMRI study (Zhang et al., 2011), a significantly reduced functional connectivity in the right middle frontal gyrus, left precuneus, and left PCC in the HE patients was observed when compared to the controls. A negative correlation was shown between left angular gyrus and left PCC activation and venous blood ammonia levels, suggesting this can be an important biomarker for HE (Figure 6).

9. Conclusion

In conclusion, HE appears as diffuse mild brain oedema, and is associated with cognition functional changes such as attention, and functional brain area changes, such as default mode brain areas. HE-associated functional and physiological abnormalities have been demonstrated by a variety of MR techniques. MRI offers a range of capabilities for investigating HE in clinical diagnosis. The recent development of fMRI and MRS expands the applications of the study of HE with the capability of assessing functional changes in brain affected by HE. These advances opens up the opportunities to better understand the pathopsychological mechanisms of HE.

http://dx.doi.org/10.5772/29563

Acknowledgements

We express our thanks for the grants from Natural Scientific Foundation of China (30700194 and 81230032 to L.Z.) and Natural Scientific Foundation of Jiangsu Province of China (BK2007572 to L. Z.).

Author details

Long Jiang Zhang[1], Guang Ming Lu[2] and Hui Mao[2]

1 Department of Medical Imaging, Nanjing University College of Clinical Medicine, Nanjing, China

2 Department of Radiology, Emory University School of Medicine, Atlanta, Georgia, USA

References

[1] Amodio P, Montagnese S, Gatta A, Morgan MY (2004) Characteristics of minimal hepatic encephalopathy. Metab Brain Dis 19:253-26

[2] Amodio P, Ridola L, Schiff S, et al. (2010) Improving the inhibitory control task to detect minimal hepatic encephalopathy. Gastroenterology 139:510-51

[3] Astley SJ, Richards T, Aylward EH, et al. (2009) Magnetic resonance spectroscopy outcomes from a comprehensive magnetic resonance study of children with fetal alcohol spectrum disorders. Magn Reson Imaging 27:760-77

[4] Bajaj JS, Wade JB, Sanyal AJ (2009) Spectrum of neurocognitive impairment in cirrhosis Implications for the assessment of hepatic encephalopathy. Hepatology 50:2014-202

[5] Barba I, Chatauret N, Garcia-Dorado D, Cordoba J (2008) A 1H nuclear magnetic resonancebased metabonomic approach for grading hepatic encephalopathy and monitoring the effects of therapeutic hypothermia in rats. Liver Int 28:1141–114

[6] Binesh N, Huda A, Thomas MA, et al. (2006) Hepatic encephalopathy: a neurochemical, neuroanatomical, and neuropsychological study. J Appl Clin Med Phys 7:86-9

[7] Butterworth RF (2003) Pathogenesis of hepatic encephalopathy: new insights from neuroimaging and molecular studies. J Hepatol 39:278-28

[8] Chavarria L, Oria M, Romero-Gimenez J, Alonso J, Lope-Piedrafita S, Cordoba J (2010) Diffusion tensor imaging supports the cytotoxic origin of brain edema in a rat model of acute liver failure. Gastroenterology 138:1566-157

[9] Córdoba J (2011) New assessment of hepatic encephalopathy. J Hepatol 54:1030-104

[10] Córdoba J, Alonso J, Rovira A, et al. (2001) The development of low-grade cerebral edema in cirrhosis is supported by the evolution of (1)H-magnetic resonance abnormalities after liver transplantation. J Hepatol 35:598-60

[11] Córdoba J, Mínguez B (2008) Hepatic encephalopathy. Semin Liver Dis 28:70-8

[12] Córdoba J, Sanpedro F, Alonso J, Rovira A (2002) 1H magnetic resonance in the study of hepatic encephalopathy in humans. Metab Brain Dis 17:415-42

[13] Damoiseaux JS, Rombouts SA, Barkhof F, et al. (2006) Consistent resting-state networks across healthy subjects. Proc Natl Acad Sci U S A 103:13848-1385

[14] Dupont S, Duron E, Samson S, et al. (2010) Functional MR imaging or Wada test: which is the better predictor of individual postoperative memory outcome? Radiology 255:128-13

[15] Ferenci P, Lockwood A, Mullen K, Tarter R, Weissenborn K, Blei AT (2002) Hepatic encephalopathy--definition, nomenclature, diagnosis, and quantification: final report of the working party at the 11th World Congresses of Gastroenterology, Vienna, 1998. Hepatology 35:716-72

[16] Gore JC (2003) Principles and practice of functional MRI of the human brain. J Clin Invest 112:4–

[17] Gotman J, Grova C, Bagshaw A, Kobayashi E, Aghakhani Y, Dubeau F (2005) Generalized epileptic discharges show thalamocortical activation and suspension of the default state of the brain. Proc Natl Acad Sci U S A 102:15236–1524

[18] Greicius MD, Krasnow B, Reiss AL, Menon V (2003) Functional connectivity in the resting brain: a network analysis of the default mode hypothesis. Proc Natl Acad Sci U S A 100:253-25

[19] Greicius MD, Srivastava G, Reiss AL, Menon V (2004) Default-mode network activity distinguishes Alzheimer's disease from healthy aging: evidence from functional MRI. Proc Natl Acad Sci U S A 101:4637-464

[20] Grover VP, Dresner MA, Forton DM, et al. (2006) Current and future applications of magnetic resonance imaging and spectroscopy of the brain in hepatic encephalopathy. World J Gastroenterol 12:2969-297

[21] Gupta RK, Yadav SK, Rangan M, et al. (2010) Serum proinflammatory cytokines correlate with diffusion tensor imaging derived metrics and (1)H-MR spectroscopy in patients with acute liver failure. Metab Brain Dis 25:355-36

[22] Haseler LJ, Sibbitt WL Jr, Mojtahedzadeh HN, Reddy S, Agarwal VP, McCarthy DM (1998) Proton MR spectroscopic measurement of neurometabolites in hepatic encephalopathy during oral lactulose therapy. AJNR Am J Neuroradiol 19:1681- 168

[23] Kale RA, Gupta RK, Saraswat VA, et al. (2006) Demonstration of interstitial cerebral edema with diffusion tensor MR imaging in type C hepatic encephalopathy. Hepatology 43:698-70

[24] Kennedy DP, Redcay E, Courchesne E (2006) Failing to deactivate: resting functional abnormalities in autism. Proc Natl Acad Sci U S A 103:8275–828

[25] Kreis R, Ross BD, Farrow NA, Ackerman Z (1992) Metabolic disorders of the brain in chronic hepatic encephalopathy detected with H-1 MR spectroscopy. Radiology 182:19-2

[26] Kumar R, Gupta RK, Elderkin-Thompson V, et al. (2008) Voxel-based diffusion tensor magnetic resonance imaging evaluation of low-grade hepatic encephalopathy. J Magn Reson Imaging 27:1061-106

[27] Lee JH, Seo DW, Lee YS, et al. (1999) Proton magnetic resonance spectroscopy (1H-MRS) findings for the brain in patients with liver cirrhosis reflect the hepatic functional reserve. Am J Gastroenterol 94:2206-221

[28] Liang M, Zhou Y, Jiang T, et al. (2006) Widespread functional disconnectivity in schizophrenia with resting-state functional magnetic resonance imaging. Neuroreport 17:209–21

[29] Lodi R, Tonon C, Stracciari A, et al. (2004) Diffusion MRI shows increased water apparent diffusion coefficient in the brains of cirrhotics. Neurology 62:762-76

[30] Long LL, Li XR, Huang ZK, Jiang YM, Fu SX, Zheng W (2009) Relationship between changes in brain MRI and (1)H-MRS, severity of chronic liver damage, and recovery after liver transplantation. Exp Biol Med (Maywood) 234:1075-108

[31] Matsusue E, Kinoshita T, Ohama E, Ogawa T (2005) Cerebral cortical and white matter lesions in chronic hepatic encephalopathy: MR-pathologic correlations. AJNR Am J Neuroradiol 26:347-35

[32] McKinney AM, Lohman BD, Sarikaya B, et al. (2010) Acute hepatic encephalopathy: diffusion-weighted and fluid-attenuated inversion recovery findings, and correlation with plasma ammonia level and clinical outcome. AJNR Am J Neuroradiol 31:1471-147

[33] McPhail MJ, Taylor-Robinson SD (2010) The role of magnetic resonance imaging and spectroscopy in hepatic encephalopathy. Metab Brain Dis 25:65-7

[34] Mechtcheriakov S, Schocke M, Kugener A, et al. (2005) Chemical shift magnetic resonance spectroscopy of cingulate grey matter in patients with minimal hepatic encephalopathy. Neuroradiology 47:27-3

[35] Michel V, Yuan Z, Ramsubir S, Bakovic M (2006) Choline transport for phospholipid synthesis. Exp Biol Med (Maywood) 231:490-50

[36] Miese F, Kircheis G, Wittsack HJ, et al. (2006) 1H-MR spectroscopy, magnetization transfer, and diffusion-weighted imaging in alcoholic and nonalcoholic patients with cirrhosis with hepatic encephalopathy. AJNR Am J Neuroradiol 27:1019-102

[37] Miese FR, Wittsack HJ, Kircheis G, et al. (2009) Voxel-based analyses of magnetization transfer imaging of the brain in hepatic encephalopathy. World J Gastroenterol 15:5157-516

[38] Mínguez B, Rovira A, Alonso J, Córdoba J (2007) Decrease in the volume of white matter lesions with improvement of hepatic encephalopathy. AJNR Am J Neuroradiol 28:1499-50

[39] Mukherjee P, Berman JI, Chung SW, Hess CP, Henry RG (2008) Diffusion tensor MR imaging and fiber tractography: theoretic underpinnings. AJNR Am J Neuroradiol 29:632-64

[40] Mukherjee P, Chung SW, Berman JI, Hess CP, Henry RG (2008) Diffusion tensor MR imaging and fiber tractography: technical considerations. AJNR Am J Neuroradiol 29:843-85

[41] Naegele T, Grodd W, Viebahn R, et al. (2000) MR imaging and (1)H spectroscopy of brain metabolites in hepatic encephalopathy: time-course of renormalization after liver transplantation. Radiology 216:683-69

[42] Ogawa S, Tank DW, Menon R, et al. (1992) Intrinsic signal changes accompanying sensory stimulation: functional brain mapping with magnetic resonance imaging. Proc Natl Acad Sci U S A 89:5951–595

[43] Poveda MJ, Bernabeu A, Concepción L, et al. (2010) Brain edema dynamics in patients with overt hepatic encephalopathy A magnetic resonance imaging study. Neuroimage 52:481-48

[44] Provenzale JM, Mukundan S, Barboriak DP (2006) Diffusion-weighted and perfusion MR imaging for brain tumor characterization and assessment of treatment response. Radiology 239:632-64

[45] Randolph C, Hilsabeck R, Kato A, et al. (2009) Neuropsychological assessment of hepatic encephalopathy: ISHEN practice guidelines. Liver Int 29:629-63

[46] Rombouts SA, Barkhof F, Goekoop R, Stam CJ, Scheltens P (2005) Altered resting state networks in mild cognitive impairment and mild Alzheimer's disease: an fMRI study. Hum Brain Mapp 26:231–23

[47] Rosazza C, Minati L (2011) Resting-state brain networks: literature review and clinical applications. Neurol Sci 2011; DOI 10.1007/s10072-011-0636-y

[48] Ross BD, Jacobson S, Villamil F, et al. (1994) Subclinical hepatic encephalopathy: proton MR spectroscopic abnormalities. Radiology 193:457-46

[49] Rovira A, Alonso J, Cordoba J (2008) MR imaging findings in hepatic encephalopathy. AJNR Am J Neroradiol 29:1612–162

[50] Rovira A, Alonso J, Cucurella G, et al. (1999) Evolution of multiple sclerosis lesions on serial contrast-enhanced T1-weighted and magnetization-transfer MR images. AJNR Am J Neuroradiol 20:1939– 4

[51] Rovira A, Grive′ E, Pedraza S, et al. (2001) Magnetization transfer ratio values and proton MR spectroscopy of normal-appearing cerebral white matter in patients with liver cirrhosis. AJNR Am J Neuroradiol 22:1137– 114

[52] Rovira A, Mínguez B, Aymerich FX, et al. (2007) Decreased white matter lesion volume and improved cognitive function after liver transplantation. Hepatology 46:1485-149

[53] Saksena S, Rai V, Saraswat VA, et al. (2008) Cerebral diffusion tensor imaging and in vivo proton magnetic resonance spectroscopy in patients with fulminant hepatic failure. J Gastroenterol Hepatol 23:e111-11

[54] Sawara K, Kato A, Yoshioka Y, Suzuki K (2004) Brain glutamine and glutamate levels in patients with liver cirrhosis: assessed by 3.0-T MRS. Hepatol Res 30:18-2

[55] Saxena V, Gupta A, Nagana Gowda GA, Saxena R, Yachha SK, Khetrapal CL (2006) 1H NMR spectroscopy for the prediction of therapeutic outcome in patients with fulminant hepatic failure. NMR Biomed 19:521–52

[56] Schaefer PW, Grant PE, Gonzalez RG (2000) Diffusion-weighted MR imaging of the brain. Radiology 217:331-34

[57] Singhal A, Nagarajan R, Hinkin CH, et al. (2010) Two-dimensional MR spectroscopy of minimal hepatic encephalopathy and neuropsychological correlates in vivo. J Magn Reson Imaging 32:35-4

[58] Spahr L, Burkhard PR, Grötzsch H, Hadengue A (2002) Clinical significance of basal ganglia alterations at brain MRI and 1H MRS in cirrhosis and role in the pathogenesis of hepatic encephalopathy. Metab Brain Dis 17:399-41

[59] Stewart CA, Reivich M, Lucey MR, Gores GJ (2005) Neuroimaging in hepatic encephalopathy. Clin Gastroenterol Hepatol 3:197-20

[60] Sugimoto R, Iwasa M, Maeda M, et al. (2008) Value of the apparent diffusion coefficient for quantification of low-grade hepatic encephalopathy. Am J Gastroenterol 103:1413- 142

[61] Tarasów E, Panasiuk A, Siergiejczyk L, et al. (2003) MR and 1H MR spectroscopy of the brain in patients with liver cirrhosis and early stages of hepatic encephalopathy. Hepatogastroenterology 50:2149-215

[62] Thoeny HC, De Keyzer F (2011) Diffusion-weighted MR imaging of native and transplanted kidneys. Radiology 259:25-3

[63] Thuluvath PJ, Edwin D, Yue NC, deVilliers C, Hochman S, Klein A (1995) Increased signals seen in globus pallidus in T1-weighted magnetic resonance imaging in cirrhotics are not suggestive of chronic hepatic encephalopathy. Hepatology 21:440-44

[64] Tian L, Jiang T, Wang Y, et al. (2006) Altered resting-state functional connectivity patterns of anterior cingulate cortex in adolescents with attention deficit hyperactivity disorder. Neurosci Letters 400:39–4

[65] Toru S, Matumura K, Kawaguchi R, Kobayashi T, Irie T (2011) Widespread cortical lesions on diffusion-weighted imaging in acute portal systemic shunt encephalopathy caused by primary biliary cirrhosis. AJNR Am J Neuroradiol 32:E55-5

[66] Vance DE: Phospholipid biosynthesis. In: Biochemistry of Lipids, Lipoproteins and Membranes. Vance DE and Vance J (eds.). Amsterdam, Elsevier Science Publishers B. V., pp. 153-181, 199

[67] Verma A, Saraswat VA, Radha Krishna Y, Nath K, Thomas MA, Gupta RK (2008). In vivo 1H magnetic resonance spectroscopy-derived metabolite variations between acuteon- chronic liver failure and acute liver failure. Liver Int 28:1095-110

[68] Weissenborn K, Ahl B, Fischer-Wasels D, et al. (2007) Correlations between magnetic resonance spectroscopy alterations and cerebral ammonia and glucose metabolism in cirrhotic patients with and without hepatic encephalopathy. Gut 56:1736-174

[69] Weissenborn K, Bokemeyer M, Ahl B, et al. (2004) Functional imaging of the brain in patients with liver cirrhosis. Metab Brain Dis 19:269-28

[70] Zafiris O, Kircheis G, Rood HA, Boers F, Haussinger D, Zilles K (2004) Neural mechanism underlying impaired visual judgement in the dysmetabolic brain: an fMRI study. Neuroimage 22:541–55

[71] Zhang L, Qi R, Wu SY, et al. (2011) Brain default-mode network abnormalities in hepatic encephalopathy: a resting-state functional MRI study. Human Brain Mapping DOI: 10.1002/hbm.2129

[72] Zhang LJ, Yang G, Yin J, Liu Y, Qi J (2007a) Neural mechanism of cognitive control impairment in patients with hepatic cirrhosis: a functional magnetic resonance imaging study. Acta Radiol 48:577-58

[73] Zhang LJ, Yang G, Yin J, Liu Y, Qi J (2007b) Abnormal default-mode network activation in cirrhotic patients: a functional magnetic resonance imaging study. Acta Radiol 48:781-78

[74] Zhang LJ, Yin JZ, Qi J, Lu GM (2010) Metabolic changes of anterior cingulate cortex in patients with hepatic cirrhosis: a magnetic resonance spectroscopy study. Hepatol Res 40:777-78

Chapter 21

Application of Diffusion- and Perfusion- Weighted Imaging in Acute Ischemic Stroke

Vincent Lai

Additional information is available at the end of the chapter

http://dx.doi.org/10.5772/30857

1. Introduction

Stroke is one leading cause of morbidity and mortality worldwide, with acute ischemia constituting approximately 80% of all cases (Srinivasan et al. 2006). In the past imaging was primarily used to exclude hemorrhage and other causes. Yet diagnostic classification and clinical prognosis are major challenges in acute stroke. This is of particular importance since the approval of intravenous tissue plasminogen activator therapy for acute ischemic stroke within a narrow therapeutic window of 4.5 hours (Hacke et al. 2008; Wahlgren et al. 2008). It then becomes clear that advance imaging techniques would be much needed to identify this subgroup of patient who could be beneficial for thrombolytic therapy.

Computed tomography (CT) scan has been the traditional method for acute stroke imaging with excellent sensitivity for acute intracerebral hemorrhage, but demonstrates limited sensitivity for detection of ischemic infarction, with a reported sensitivity of 40-60% only within 6 hours of onset, and a negative predictive valve of 17% (von Kummer et al. 2001). Magnetic resonance (MR) imaging has no doubt superior spatial and contrast resolution as compared with CT, achieving higher sensitivity and accuracy. It has even emerged over last few years to become as sensitive as CT in depiction of acute intracranial hemorrhage (Fiebach et al. 2004; Kidwell et al. 2004). Perfusion-weighted imaging (PWI) and diffusion-weight imaging (DWI) had played an influencing role since their introduction in late 1980s. Application DWI and PWI offer complex information regarding infarction and perfusion deficits, significantly improving the diagnostic yield. In a recent meta-analysis, Schellinger et al. (2010) found that the ability to correctly diagnose acute ischemic stroke on MR was significantly higher than CT, demonstrating an odds ratio of 25 ($p<0.0001$; confidence interval: 8-79). It also brings forward the concept of salvageable brain tissue as indicated by penumbra which is a rather dynamic entity that exists within a narrow range of perfusion pressures, carrying significant implication on therapy selection and prediction of clinical outcome. The goal is to salvage the ischemic but viable penumbral tissue from progression into infarction.

2. Diffusion-weighted imaging

In DWI, the random motion or diffusion pulse of water molecules in a tissue compartment through a magnetic field gradient causes phase shifting leading to intravoxel dephasing and hence loss of signal intensity (Poustchi-Amin et al. 2001). In region of acute infarct or irreversible ischemia, the decreased diffusion of water molecules due to presence of cytotoxic edema (neuronal and glial cell swelling) from intracellular water accumulation and disruption of membrane ionic homeostasis will manifest as an increase in signal intensity, reflecting a decrease in apparent diffusion coefficient (ADC). ADC is "apparent" only because the actual measurement cannot be obtained by DWI due to presence of unmeasured variables that influence the rate of diffusion, such as temperature of the tissue and actual route of the diffusing molecule. Images are acquired sequentially with the axis of diffusion sensitization in three orthogonal planes, with their average yielding an image with minimal anisotropy-related hyperintensity. As a result, infracted tissue appears bright in comparison with normal bran tissue. This has significant impact in helping to identify area of acute ischemia from chronic ischemic, and hence stratify treatment plan.

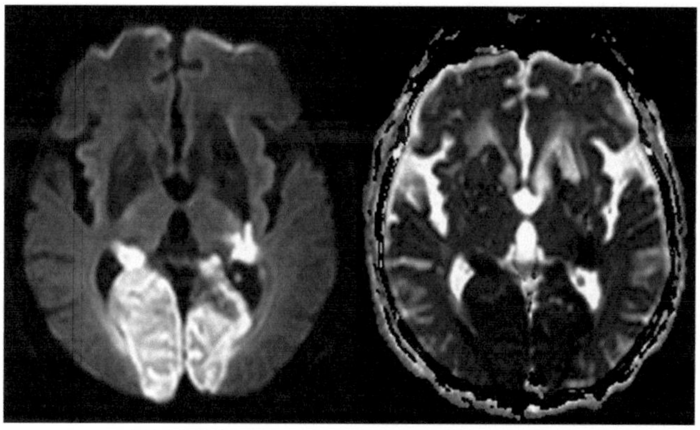

(a) DWI (b) ADC

Figure 1. a) DWI shows hyperintense signal along bilateral posterior cerebral arterial territories. b) corresponding ADC map demonstrates hypointense signal.

In human, restricted diffusion showing increased DWI signal with reduced ADC value (Figure 1) have been observed as early as 30 minutes after onset of ischemia (Srinivasan et al. 2006). ADC continues to decrease for 3-5 days, reaching its minimum value of about 50% between 24-48 hours after symptoms onset before returning to baseline at approximately 1-4 weeks, so called pseudonormalization. At this stage, mildly hyperintense DWI signal with normalized ADC value will be seen. ADC increases afterwards with value higher than normal value associated with variable DWI signal (from T2 effect) for upto months, presumably related to increased quantity of extracellular water. It is important to note that DWI signal can remain hyperintense for a long period (initially due to restricted diffusion but latter due to T2 shine-through effect) and cannot reliably estimate infarct age alone, while reduced ADC value almost always point to acute infarction. Yet, tissue with very low ADC value may recover and no single threshold can distinguish salvageable from unsalvageable tissue (Fiehler et al. 2002). Therefore we must interpret DWI with ADC maps together cautiously.

DWI-ADC carries high sensitivity (88-100%) and specificity (86-100%) in depiction of acute ischemia (Beaulieu et al. 1999; Gonzalez et al. 1999; Lövblad et al. 1997, 1998). However, with increasing cases of DWI-negative stroke reported and bearing in mind the existence of patient selection bias in the reported series, the true sensitivity probably lies between 80%-90% in general (Schellinger et al. 2010). In a prospective randomized study comparing between DWI and CT scan in evaluation of acute ischemic stroke, Fiebach et al. (2002) reported that the sensitivity of infarct detection was significantly better based on DWI (91%) than on CT (61%), achieving good interrater homogeneity and better accuracy even for readers with limited experience. This is due to relative insensitivity of CT for ischemia beyond 10ml per 100/min. as compared with DWI (Barber et al. 1999). False negative is unusual but is generally more common in posterior fossa stroke (Oppenheim et al. 2000) due to the presence of susceptibility artifacts from basal skull. Risk of hemorrhagic transformation is one feared complication and frequently asked question, especially when thrombolysis is contemplated. It had been postulated that a DWI lesion volume > $100cm^3$ was highly associated with hemorrhagic transformation, in particular when there was no associated DWI-PWI mismatch, hence contraindication to thrombolysis. However this would require further study for confirmation. Mixed results had been shown regarding the accuracy of baseline DWI in determination of final infarct volume. While it was widely thought that it would underestimate the final size of infarction (Yamada et al. 2002), Warach et al. (2000) in his double-blinded placebo-controlled study involving patients suffering from anterior-circulation stroke syndromes, found that baseline DWI lesion volume significantly correlated with baseline National Institute of Health Stroke Scale (NIHSS) scores and follow-up lesion volume at 12 weeks. Therefore, it is well established that DWI is accurate and superior to CT for diagnosis of acute ischemic stroke and can probably predict baseline clinical stroke severity as well as final lesion volume.

3. Perfusion-weighted imaging

PWI provides information on hypoperfused area and momentary hemodynamic state of the brain tissue before structural brain tissue damage, therefore to identify areas of reversible ischemia early before it progresses to permanent infarction. It can be performed by exogenous method (dynamic or steady state susceptibility contrast-enhanced MR imaging by injection of MR contrast) or by endogenous method (arterial spin labeling by magnetically tagging water protons in arterial blood).

3.1. Mechanism

Dynamic susceptibility contrast-enhanced MR perfusion imaging involves rapid imaging with gradient echo or spin echo echo-planar imaging during the first passage of a paramagnetic contrast agent through the entire brain. 10 to 20 brain slices are usually repeatedly imaged immediately before and for 70 to 80 seconds after contrast injection, with 20 to 40 images per slice obtained. It makes use of the tissue signal changes caused by susceptibility (T2*) effects to generate signal drop (shortening of T2 and T2*) and creates a hemodynamic time-signal intensity curve for each voxel, which in turns depends on cerebral blood volume and cerebral blood flow (Seevinck et al. 2010). It is through the dynamic monitoring of these signal changes in perfused tissue that hemodynamic parameters can be estimated. In contrary, steady state susceptibility contrast-enhanced MR perfusion allows estimation of blood volume, micro-vessel density and vessel size from the contrast-induced signal changes. Spin echo and gradient echo transverse relaxation rates are measured before and after contrast injection, typically with use of ultrasmall superparamagnetic iron oxide particles (longer half-life and higher suscept-ibility difference) from which parameters are derived (Wu EX et al. 2004). This method however is mainly of use in research level only.

Quantification depends on measuring the signal intensity change within tissues which itself is the susceptibility effect of the contrast agent. The ram images are processed on a pixel-by-pixel basis to generate perfusion maps of different parameters. Time-to-peak (TTP) can be readily obtained by simple data processing from the tissue response curve. Cerebral blood flow (CBF), cerebral blood volume (CBV) and mean transit time (MTT) are time dependent parameters requiring utilization of computational deconvolution methods (such as singular value decomposition, circular value decomposition, Fourier transformation) with knowledge of arterial input function for quantification. The arterial input function is defined under visual control by drawing 5-10 intravascular voxels within the proximal & distal segments of the concerned artery of the unaffected hemisphere (Zaro-Weber et al. 2010a). Point to note is that these are relative values only due to variable confounding factors which will be discussed in later section.

Arterial spin labeling is a non-invasive method exploiting the spins of endogenous water protons to measure perfusion without the need of gadolinium. An inversion radiofrequency pulse is used to tag & invert the spin polarity of the flowing arterial protons with respect to stationary arterial proton, at a level proximal to the imaging slab (Deibler et al. 2008). The tagged spins then enter the imaging plane and exchange with tissue, allowing pair-wise subtraction to yield the perfusion map after acquisition of multiple runs for signal intensity averaging. The detected MR signals would be analyzed and perfusion parameters can be obtained. However, it has a low signal-to-noise ratio requiring longer imaging time.

3.2. Image analysis

Perfusion imaging is the method of choice for demonstrating ischemic brain tissue in hyper-acute stage of stroke. The central portion with maximal perfusion abnormality will progress irreversibly to infarction while the outer area with only slight perfusion abnormality may recover (Hossmann 1994; Rohl et al. 2001). No single threshold has thus far been able to reliably discriminate salvageable from unsalvageable tissue definitely (Wu O et al. 2006) since absolute quantification of cerebral perfusion is not really possible and perfusion thresholds change with time (Butcher et al. 2003). Interpretation is based on the concept that as CBF falls, CBV is initially unchanged or increased (from vasodilatation due to autoregulation) in conjunction with prolonged MTT and TTP. If such autoregulatory mechanism is exhausted, CBV will then decrease and CBF continues to fall to a level of irreversible ischemia with eventual infarction. Therefore a mismatch between MTT/TTP and CBV or CBF map indicates presence of potentially salvageable tissue (Figure 2).

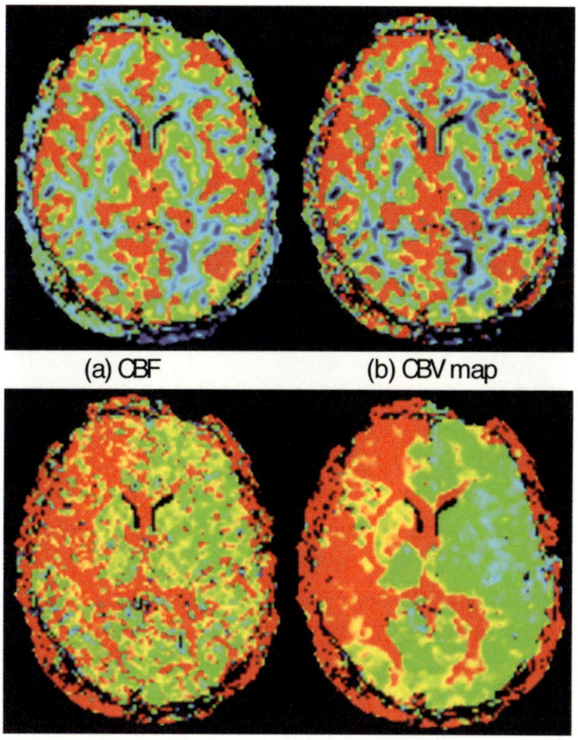

Figure 2. Patient with right internal carotid artery (ICA) occlusion shows mild reduced a) CBF & b) CBV; but marked elevated c) MTT & d) TTP along right ICA territory.

3.3. Parameters threshold

Depending on different processing methods, differing results regarding the reliability and threshold of each parameter resulted. MTT or TTP delay relative to the contralateral hemisphere has been used in preference to CBF & CBV because of the more conspicuous changes and better correlation with initial stroke severity and change in infarct size, hence final functional outcomes in individual patient (Kane et al. 2007; Muir et al. 2006). In the contrary, pre-treatment lesions on DWI and ADC maps were non-discriminative.

3.3.1. *Time-to-peak (TTP)*

Restrepo et al. (2004) found that TTP map was most useful in demonstrating perfusion changes, showing good signal-to-noise ratio facilitating identification of lesion boundary, rendering it a parametric surrogate marker for efficacy of acute stroke treatment. A perfusion delay of 6 seconds relative to the non-affected hemisphere on TTP maps acquired before the initiation of stroke treatment helped predict the lesion size on T2-weighted images on day 8 irrespective of the treatment regimen (Kidwell et al. 2002; Nighoghossian et al. 2003; Seitz et al. 2005). Neumann-Haefelin et al. (1999) reported a TTP delay of >6 seconds best predicted final infarct volume while a TTP delay >4 seconds correlated best with acute stage neurologic deficit, signifying the true penumbra lies between 4 to 6 seconds. Kajimoto et al. (2003) depicted same result with a TTP delay of >4 seconds (sensitivity 79%, specificity 96%) in his study evaluating patients with chronic stroke. Sobesky et al. (2004 & 2005) validated these threshold values in patients with acute stroke, and demonstrated matched results when comparing with PET-derived data, also showing a TTP delay threshold of >4 seconds with sensitivity of 84% & specificity of 77%, and a penumbral volume threshold of >6 seconds. Takasawa et al. (2008) later reported a TTP delay threshold of >4.8 seconds which is also similar, as supported more recently, by Zaro-Weber et al. (2010a) showing a TTP delay threshold of >4.2 seconds best predicted penumbral flow, with a sensitivity of 91% and specificity of 82%. These consistent findings strengthen the TTP delay threshold to lie between >4-5 seconds.

3.3.2. *Mean transit time (MTT)*

Grandin et al. (2001 & 2003) reported a MTT delay of 8.1 seconds best predicted infarct growth whereas a MTT delay of 5.3 seconds defined the hypoperfused area that did not progress to infarction, to indicate oligemic tissue. Yamada et al (2002) supported the accuracy of MTT maps showing a sensitivity of 85-94% and positive predictive value of 54-61% for final infarct volume. Such results were in coherent with other studies, with Parsons et al. (2002) depicting a >6 seconds threshold, Bristow et al. (2005) depicting a >7seconds threshold, Takasawa et al. (2008) depicting a> 6-7 seconds threshold, and more recently, Zaro-Weber (2010a) demonstrating a >5.3 seconds threshold with a sensitivity of 88% & specificity of 78%. Therefore, we can conclude that a MTT delay between 5.3 -8.1 seconds denotes the true penumbra.

3.3.3. *Cerebral blood flow (CBF)*

CBF has found to be a good indicator of penumbral flow. Early study by Liu et al. (2000) revealed a cut off value of 48% of the unaffected hemisphere for depiction of penumbral flow. From positron emission tomography (PET) scan, it was found that neurological dysfunction occurred after CBF dropped below 20mL/100g of brain tissue per minute while infarction occured shortly after CBF dropped below 10mL/100g of brain tissue per minute (Latchaw et al. 2003). Zaro-Weber et al. confirmed in two of his recent trials (2009 & 2010a) that threshold values of <20mL/100g/min (sensitivity 76%, specificity 96%) and <21.7mL/100g/min (sensitivity 89%, specificity 87%) respectively were of best estimate of penumbral flow, adding to its consistency and strength in validation. Therefore, a CBF between 10 -20mL/100g of brain tissue per minute becomes the intermediate zone and represents penumbral flow.

3.3.4. Cerebral blood volume (CBV)

CBV shows the least correlation with penumbral flow with limited studies validating its penumbral threshold level in past literatures, especially in comparison with PET. Liu et al. (2000) depicted the threshold level to be 87% of the unaffected hemisphere while Hatazawa et al. (1999) found the threshold level to be 85% in their studies comparing with single photon emission CT (SPECT). Zaro-Weber et al. (2010a) recently, in the only study comparing with PET, found the penumbral threshold flow at <1.5mL/100g with a sensitivity of 82% and specificity of 79%.

4. Diffusion-weighted — Perfusion-weighted imaging

Ischemia is a complex and multifactorial process. Its clinical severity on initial presentation is a major determinant of clinical outcome, correlating strongly with the extent of the lesion. Proportion of patients with diffusion/perfusion mismatch and its size will depend on the severity of stroke and time delay to imaging. Based on the heterogeneity of the ADC and PWI values in depiction of ischemic lesion (Bandera et al. 2006), as well as occasional recovery of DWI-abnormal tissue (Davis & Donnan 2005), it becomes clear that it is unlikely any single DWI or PWI parameters or threshold values will reliably predict the degree of ischemia or infarct. Combination of diffusion- and perfusion-weighted imaging has revolutionized acute stroke imaging, allowing early & accurate delineation of the actual ischemic lesion or penumbra. Diffusion-perfusion mismatch is a quantitative and reliable measure with important pathophysiologic & therapeutic implications, becoming the surrogate of ischemic penumbra that provides valuable information for neurointervention planning, especially within the crucial therapeutic window. The Desmoteplase for Acute Ischemic Stroke and Dose Escalation of Desmoteplase for Acute Ischemic Stroke studies demonstrated that patients with PWI-DWI mismatchs who were treated with desmoteplase experienced higher rate of reperfusion and better clinical outcome (Furlan et al. 2006; Hacke et al. 2005). Earlier studies by Parsons et al. (2001 & 2002) demonstrated greater recanalization, reperfusion and penumbral salvage in a group of patients with DWI-PWI mismatch compared with the control group in acute stroke patients treated with intravenous tissue plasminogen activator. Luby & Warach (2007) also found that mismatch volumes can provide highly reliable and consistent results with improved sensitivity. Early PWI lesions are typically larger than early DWI lesions. Finding of normal diffusion-weighted images and altered perfusion parameters (mismatch) is a simple and practical mean to define the tissue at risk, which correlates with poor clinical outcome (Derex et al. 2004; Parsons et al. 2002). If untreated or reperfusion does not occur, progressive enlargement of the DWI lesion over time is expected, signifying progressive infarction of the initial PWI lesion. This can be rapid in patient with poor collateral circulation. Conversely, a pattern of DWI lesion equaling the volume of PWI lesion indicates absence of penumbra or no salvageable tissue.

5. Reliability and pitfalls

The reliability of mismatch volume and percentage of measurements had been discussed though not extensively investigated. Obtaining precise parameters is one major problem as only the "relative" but not absolute perfusion changes can be measured due to presence of significant confounding factors, hence nowadays we mostly rely on the relative maps of perfusion. So, exactly how precise can the penumbral threshold be quantitatively estimated by CBF, CBV, TTP and MTT? What are the overall accuracy, reproducibility and reliability?

Firstly, perfusion imaging is sensitive to vascular delays (tracer arrival delay and tissue transit time) and dispersion effects (Calamante et al. 2002), especially in the derivation of the nondeconvoluted TTP map. Any minor physiologic change such as heart rate and cardiac output can potentially change the mismatch zone. Lack of standardization between MR machines, measurement technique or associated errors from pulse sequence parameters, image processing and analytic software are other major concerns. Measurement of arterial input function is particularly problematic with lack of standardized protocol in measurement of signal intensity around the artery. Besides, the tissue signal intensity does not demonstrate a linear relationship to tissue contrast concentration (Latchaw et al. 2003). Past studies attempted to optimize perfusion maps by venous output function (Knutsson et al. 2007), separated measurement of CBV (Newman et al. 2003), advanced mathematical correction (van Osch et al. 2003), partial volume artifact correction, and bulk-blood correction (Zaharchuk et al. 2009) but were found technically difficult requiring further validation.

We have to understand that considerable individual variation regarding the penumbral flow exists and absolute quantification of the perfusion maps is difficult. So far, perfusion maps had been mainly validated with respect to infarct delineation on follow-up MR imaging but with uncertainty of perfusion changes during the time interval in between (Zaro-Weber et al. 2010a). Improved validation had been made by direct comparison with PET for in vivo perfusion measurement in normal subjects and chronic cerebrovascular dusease, but with only few in acute stroke setting (Grandin et al. 2005; Ibaraki et al. 2007; Lin et al. 2001; Ostergaard et al. 1998; Tanaka et al. 2006), raising uncertainty that such values may not precisely demonstrate the exact tissue at greatest risk of infarction, hence leading to misleading & unreliable interpretation (Keir & Wardlaw 2000). Despite so, PET-based calibration of perfusion imaging demonstrated superior result with best correction of the absolute perfusion values. Nevertheless, different post-processing algorithms differ, producing variations in lesion size and measured parameters. Indeed, ideal technique of perfusion imaging and its post-processing have not yet established. Zaro-Weber et al. (2010b) found that the variation of penumbral flow for each patient was high and hence with great error, emphasizing the need for an individual correction procedure or calibration with independent cohort. In a later study, he also found that the mean flow rate on contralateral hemisphere well explained the variability of individual perfusion thresholds such that it could be used to yield the best estimate of penumbral threshold. Based on these, a PET-validated look-up table was derived to calibrate the best individual penumbral threshold from the values of the contralateral hemispheric reference, allowing a much improved volumetric congruence. Linear regression analysis demonstrated best agreement for TTP and good agreement for CBF and MMT while that for CBV was less pronounced. Despite this seemingly advantage, further validation (especially in an independent cohort) will be needed before its implementation in clinical use. In the meantime, standardization in both the data processing methods and software is also much needed. A consensus on which parameter is best to be used for prediction should also be made.

Arterial spin labeling has additional limitation regarding its own method, in that assumptions on tagging efficiency, delay time to imaging and flow quantification are based on normal population. Whether the same setting can be translated to various disease states has yet to be tested properly (Deibler et al. 2008). Besides, it has a relatively low signal-to-noise ratio, further adding to uncertainty in its accuracy. Therefore dynamic contrast enhanced susceptibility MR imaging technique is still the preferred method for quantitative perfusion imaging.

6. Transient ischemic attack

Transient ischemic attack should not be taken lightly as it carries a subsequent risk of ischemic stroke of 10% within 7 days and 30% within 90 days of symptoms onset, in particularly during the first 48 hours (Johnston et al. 2000; Mlynash et al. 2009). Subsequent risk of myocardial infarction, arrhythmia and congestive heart failure are increased. Hence appropriate prevention strategy is important in reducing the risk.

Establishment of the diagnosis of transient ischemic attack has been a major challenge since most patient present normal neurologically with resolution of the clinical deficit at the time of consultation and imaging, resulting in negative finding on conventional MR brain imaging. DWI & PWI can provide objective measurement of brain ischemia, even better than angiography as lesions are mostly small and may reflect occlusion of distal intracranial vessels that is angiographically occulted. Past studies demonstrated variable lesion depiction rate from 16 to 67% on DWI in patients with TIA (Crisostomo et al. 2003; Kastrup et al. 2002; Kidwell et al. 1999; Krol et al. 2005; Redgrave et al. 2007; Restrepo et al. 2004; Rovira et al. 2002) but it had been argued that these patients should be labeled as ischemic stroke with infarction rather, regardless of the duration of symptoms with a clinical implication of potential poorer clinical outcome. This was still much debatable as report had described complete resolution of both DWI and PWI abnormalities on follow-up images (Kidwell et al. 1999; Neumann-Haefelin et al. 2000; Rovira et al. 2002). Initial studies had also found that approximately one-thirds of the DWI lesions would normalized to a certain degree if they were imaged within a 3 hour time window (Fiehler et al. 2002; Kidwell et al. 2000). PWI should substantiate the yield given its ability to reveal hypoperfusion accompanying or preceding development of DWI abnormalities. Nevertheless, the occurrence of patients with positive PWI and negative DWI was variable, reported to be 3 to 33% (Krol et al. 2005; Mlynash et al. 2009; Restrepo et al. 2004). Ma et al. (2009) explained in his study model that the classical mismatch pattern would fragment with time due to a number of more random events such as variable ages & composition of the clots, random locations of vessel occlusion with consequent differences in vulnerability of tissue, and variable degree of collateralization from neighboring vessels.

Therefore, the yield is expected to be >50% if such patients are scanned urgently after symptom onset. It is useful in helping to identify a subset of patients with diffusion-perfusion mismatch who would require close observation, and possible further investigation such as angiography assessment which would then have a high yield (>50%) for detection of potentially relevant lesion, hence to assist in subsequent management. It helps to differentiate those that who are likely to improve or reverse spontaneously from those who are likely to progress.

7. Clinical impact

There is no true imaging gold standard in acute stroke. DWI has a well established role in depiction of acute infarction while it is also clear that perfusion imaging can provide information that cannot be depicted in other techniques, thus seemingly superior. After all, is there a role for perfusion imaging in affecting clinical decision making and hence, influencing patient outcomes? Previously, we have already discussed that perfusion-diffusion mismatch correlates with poor clinical outcome and that treatment of large perfusion-diffusion mismatch can result in improved clinical outcome. Such can lead to the attracting prospect of expanding the narrow therapeutic time window by selecting eligible patients via DWI-PWI mismatch, as demonstrated in various series & trials (Albers et al. 2006; Davis et al. 2008; Furlan et al. 2006; Hacke et al. 2005 & 2009; Parsons et al. 2002; Schellinger et al. 2007), but with lack of proper prospective study or strong evidence-based validation so far to support its role in alteration of patient outcomes. Beside, the exact role of perfusion imaging was often vague and its benefits in decision making were usually under-evaluated or retrospectively compared with clinical parameters only (Provenzale et al. 2008). Further, it is not yet established that patients without mismatch will be unresponsive to thrombolysis, since DWI lesions can potentially resolve as discussed previously. More importantly, nowadays in most stroke centers, decisions for administration of thrombolytic agents in patients suffering from acute ischemic stroke are often made without prior perfusion imaging, nor even diffusion imaging. Decisions are routinely made on clinical ground in particular for those presenting at <4.5 hours, raising doubts over its added value. This is perhaps the main reason for lack of literature addressing the influencing role of perfusion imaging in decision making of thrombolysis. In addition, the feasibility and practicality of MR imaging in acute stroke assessment are always in doubt. Cleary, time is an important factor. The whole process of perfusion imaging together with subsequent data analysis can be time-consuming especially in less experience hands. After all, routine use of PWI to assess acute ischemic stroke is open to dispute. Until we have a compelling randomized prospective blinded study assessing the efficacy of perfusion imaging in decision making in assessment of acute ischemic stroke, perhaps the most convincing role at present is prediction of final infarct volume which demonstrates strong correlation, and hence helping to predict the clinical outcome in individual patient. On the other hand, it can also be of great use in monitoring of evolution of ischemia and subsequent therapeutic response. Despite these seemingly limitations and arguments on methodical issues, many believe that the accumulated evidence of the importance of DWI-PWI mismatch is already good enough to establish a role in daily practice if allow.

8. Conclusion

With advances in imaging technology, there is substantial evidence and data to support the feasibility & superiority of MR imaging in acute ischemic stroke. PWI is complementary to DWI in acute stroke assessment. DWI-PWI mismatch provides a good, rapid & reliable estimation of the penumbra, helping to select suitable candidate for acute thrombolytic therapy, especially for patients presenting with unknown onset time or beyond the 4.5 hours therapeutic window. PWI lesion can resolve rapidly after successful thrombolysis or reperfusion, leading to improved clinical outcome. Only then will we know for every individual patient, at certain time after stroke, with particular clinical features and parameters on DWI/PWI, which will be the best clinical management.

Author details

Vincent Lai

The University of Hong Kong, Hong Kong

References

[1] Albers GW, Thijs VN, Wechsler L, Kemp S, Schlaug G, Skalabrin E, Bammer R, Kakuda W, Lansberg MG, Shuaib A, Coplin W, Hamilton S, Moseley M & Marks MP; DEFUSE Investigators. (2006). Magnetic resonance imaging profiles predict clinical response to early reperfusion: the diffusion and perfusion imaging evaluation for understanding stroke evolution (DEFUSE) study. *Ann Neurol*, Vol.60, No.5, pp. 508-517

[2] Bandera E, Botteri M, Minelli C, Sutton A, Abrams KR & Latronico N. (2006). Cerebral blood flow threshold of ischemic penumbra and infarct core in acute ischemic stroke. A systemic review. *Stroke*, Vol.37, No.5, pp. 1334-1339

[3] Barber PA, Darby DG, Desmond PM, Gerraty RP, Yang Q, Li T, Jolley D, Donnan GA, Tress BM & Davis SM. (1999). Identification of major ischemic change. Diffusion-weighted imaging versus computed tomography. *Stroke*, Vol.30, No.10, pp. 2059-2065

[4] Beaulieu C, de Crespigny A, Tong DC, Moseley ME, Albers GW & Marks MP. (1999). Longitudinal magnetic resonance imaging study of perfusion and diffusion in stroke: evolution of lesion volume and correlation with clinical outcome. *Ann Neurol*, Vol.46, No.4, pp. 568-578

[5] Bristow MS, Simon JE, Brown RA, Eliasziw M, Hill MD, Coutts SB, Frayne R, Demchuk AM & Mitchell JR. (2005). MR perfusion and diffusion in acute ischemic stroke:

human gray and white matter have different thresholds for infarction. *J Cereb Blood Flow Metab,* Vol.25, No.10, pp. 1280-1287

[6] Butcher K, Parsons M, Baird T, Barber A, Donnan G, Desmond P, Tress B & Davis S. (2003). Perfusion thresholds in acute stroke thrombolysis. *Stroke,* Vol.34, No.9, pp. 2159-2164

[7] Calamante F, Gadian DG & Connelly A. (2002). Quantification of perfusion using bolus tracking magnetic resonance imaging in stroke: assumptions, limitations, and potential implications for clinical use. *Stroke,* Vol.33, No.4, pp. 1146-1151

[8] Crisostomo RA, Garcia MM & Tong DC. (2003). Detection of diffusion-weighted MRI abnormalities in patients with transient ischemic attack: correlation with clinical characteristics. *Stroke,* Vol.34, No.4, pp. 932-937

[9] Davis SM & Donnan GA. (2005). Using mismatch on MRI to select thrombolytic responders: an attractive hypothesis awaiting confirmation. *Stroke,* Vol.36, No.5, pp. 1106-1107

[10] Davis SM, Donnan GA, Parsons MW, Levi C, Butcher KS, Peeters A, Barber PA, Bladin C, De Silva DA, Byrnes G, Chalk JB, Fink JN, Kimber TE, Schultz D, Hand PJ, Hankey G, Muir K, Gerraty R, Tress BM & Desmond PM; EPITHET investigators. (2008). Effects of alteplase beyond 3 h after stroke in the Echoplanar Imaging Thrombolytic Evaluation Trial (EPITHET): a placebo-controlled randomized trial. *Lancet Neurol,* Vol.7, No.4, pp. 299-309

[11] Deibler AR, Pollock JM, Kraft RA, Tan H, Burdette JH & Maldjian JA. (2008). Arterial spina-labeling in routine clinical practice, part 1: Technique and artifacts. *AJNR Am J Neuroradiol,* Vol.29, No.7, pp. 1228-1234

[12] Derex L, Nighoghossian N, Hermier M, Adeleine P, Berthezene Y, Philippeau F, Honnorat J, Froment JC & Trouillas P. (2004). Influence of pretreatment MRI parameters on clinical outcome, recanalization and infarct size in 49 stroke patients treated by intravenous tissue plasminogen activator. *J Neurol Sci,* Vol.225, No.1-2, pp. 3-9

[13] Fiebach JB, Schellinger PD, Jansen O, Meyer M, Wilde P, Bender J, Schramm P, Jüttler E, Oehler J, Hartmann M, Hähnel S, Knauth M, Hacke W & Sartor K. (2002). CT and diffusion-weighted MR imaging results in higher accuracy and lower interrater variability in the diagnosis of hyperacute ischemic stroke. *Stroke,* Vol.33, No.9, pp. 2206-2210

[14] Fiebach JB, Schellinger PD, Gass A, Kucinski T, Siebler M, Villringer A, Olkers P, Hirsch JG, Heilnd S, Wilde P, Jansen O, Röther J, Hacke W & Sartor K. (2004). Stroke magnetic resonance imaging is accurate in hyperacute intracerebral hemorrhage: a multicenter study on the validity of stroke imaging. *Stroke,* Vol.35, No.2, pp. 502-506

[15] Fiehler J, Foth M, Kucinski T, Knab R, von Bezold M, Weiller C, Zeumer H & Röther J. (2002). Severe ADC decreases do not predict irreversible tissue damage in humans. *Stroke,* Vol.33, No.1, pp. 79-86

[16] Furlan AJ, Eyding D, Albers GW, Al-Rawi Y, Lees KR, Rowley HA, Sachara C, Soehngen M, Warach S & Hacke W; DEDAS Investigators. (2006). Dose Escalation of Desmoteplase for Acute Ischemic Stroke (DEDAS): evidence of safety and efficacy 3 to 9 hours after stroke onset. *Stroke,* Vol.37, No.5, pp. 1227-1231

[17] Gonzalez RG, Schaefer PW, Buonanno FS, Schwamm LH, Budzik RF, Rordorf G, Wang B, Sorensen AG & Koroshetz WJ. (1999). Diffusion-weighted MR imaging: diagnostic accuracy in patients imaged within 6 hours of stroke symptom onset. *Radiology,* Vol.210, No.1, pp. 155-162

[18] Grandin CB, Duprez TP, Smith AM, Mataigne F, Peeters A, Oppenheim C & Cosnard G. (2001). Usefulness of magnetic resonance-derived quantitative measurements of cerebral blood flow and volume prediction of infarct growth in hyperacute stroke. *Stroke,* Vol.32, No.5, pp. 1147-1153

[19] Grandin CB. (2003). Assessment of brain perfusion with MRI: methodology and application to acute stroke. *Neuroradiology,* Vol.45, No.11, pp. 755-766

[20] Grandin CB, Bol A, Smith AM, Michel C & Cosnard G. (2005). Absolute CBF and CBV measurements by MRI bolus tracking before and after acetazolamide challenge: repeatability and comparison with PET in humans. *Neuroimage,* Vol.26, No.2, pp. 525-535

[21] Hacke W, Albers G, Al-Rawi Y, Boqousslavsky J, Davalos A, Eliasziw M, Fischer M, Furlan A, Kaste M, Lees KR, Soehngen M & Warach S; DIAS Study Group. (2005). The Desmoteplase in Acute Ischemic Stroke Trial (DIAS): a phase II MRI-based 9-hour window acute stroke thrombolysis trial with intravenous desmoteplase. *Stroke,* Vol.36, No.1, pp. 66-73

[22] Hacke W, Kaste M, Bluhmki E, Brozman M, Dávalos A, Guidetti D, Larrue V, Lees KR, Medeghri Z, Machnig T, Schneider D, von Kummer R, Wahlgren N, Toni D & ECASS investigators. (2008). Thrombolysis with alteplase 3 to 4.5 hours after acute ischemic stroke. *N Engl J Med,* Vol.359, No.13, pp. 1317-1329

[23] Hacke W, Furlan AJ, Al-Rawi Y, Davalos A, Fiebach JB, Gruber F, Kaste M, Lipka LJ, Pedraza S, Ringleb PA, Rowley HA, Schneider D, Schwamm LH, Leal JS, Söhngen M, Teal PA, Wilhelm-Ogunbiyi K, Wintermark M & Warach S. (2009). Intravenous desmoteplase in patients with acute ischemic stroke selected by MRI perfusion-diffusion weighted imaging or perfusion CT (DIAS-2): a prospective, randomized, double-blind, placebo-controlled study. *Lancet Neurol,* Vol.8, No.2, pp. 141-150

[24] Hatazawa J, Shimosegawa E, Toyoshima H, Ardekani BA, Suzuki A, Okudera T & Miura Y. (1999). Cerbral blood volume in acute brain infarction : a combined study

with dynamic susceptibility contrast MRI and 99MTC-HMPAO-SPECT. *Stroke,* Vol. 30, No.4, pp. 800-806

[25] Hossmann KA. (1994) Viability thresholds and the penumbra of focal ischemia. *Ann Neurol,* Vol.36, No.4, pp. 557-565

[26] Ibaraki M, Ito H, Shimoseqawa E, Toyoshima H, Ishigame K, Takahashi K, Kanno I & Miura S. (2007). Cerebral vascular mean transit time in healthy humans: a comparative study with PET and dynamic susceptibility contrast-enhanced MRI. *J Cereb Blood Flow Metab,* Vol.27, No.2, pp. 404-413

[27] Johnston SC, Gress DR, Browner WS & Sidney S. (2000). Short-term prognosis after emergency-department diagnosis of TIA. *JAMA,* Vol.284, No.22, pp. 2901-2906

[28] Kajimoto K, Moriwaki H, Yamada N, Hayashida K, Kobayashi J, Miyashita K & Naitomi H. (2003). Cerebral hemodynamic evaluation using perfusion-weighted magnetic resonance imaging : comparison with positron emision tomography values in chronic occlusive carotid disease. *Stroke,* Vol.34, No.7, pp. 1662-1666

[29] Kane I, Carpenter T, Chappell F, Rivers C, Armitage P, Sandercock P & Wardlaw J. (2007). Comparison of 10 different magnetic resonance imaging processing methods in acute ischemic stroke: Effect on lesion size, proportion of patients with diffusion/ perfusion match, clinical scores, and radiologic outcomes. *Stroke,* Vol.38, No.12, pp. 3158-3164

[30] Kastrup A, Schulz JB, Mader I, Dichqans J & Küker W. (2002). Diffusion-weighted MRI in patients with symptomatic internal carotid artery disease. *J Neurol,* Vol.249, No.9, pp. 1168-1174

[31] Keir SL & Wardlaw JM. (2000). Systemic review of diffusion and perfusion imaging in acute ischemic stroke. *Stroke,* Vol.31, No.11, pp. 2723-2731

[32] Kidwell CS, Alger JR, Di Salle F, Starkman S, Villablanca P, Bentson J & Saver JL. (1999). Diffusion MRI in patients with transient ischemic attacks. *Stroke,* Vol.30, No.6, pp. 1174-1180

[33] Kidwell CS, Saver JL, Mattiello J, Starkman S, Vinuela F, Duckwiler G, Gobin YP, Jahan R, Vespa P, Kalafut M & Alger JR. (2000). Thrombolytic reversal of acute human cerebrak ischemic injury shown by diffusion/perfusion magnetic resonance imaging. *Ann Neurol,* Vol.47, No.4, pp. 462-469

[34] Kidwell CS, Saver JL, Starkman S, Duckwiler G, Jahan R, Vespa P, Villablanca JP, Liebeskind DS, Gobin YP, Vinuela F & Alger JR. (2002). Late secondary ischemic injury in patients receiving intraarterial thrombolysis. *Ann Neurol,* Vol.52, No.6, pp. 698-703

[35] Kidwell CS, Chalela JA, Saver JL, Starkman S, Hill MD, Demchuk AM, Butman JA, Patronas N, Alger JR, Latour LL, Luby ML, Baird AE, Leary MC, Tremwel M, Ovbiagele B, Fredieu A, Suzuki S, Villablanca JP, Davis S, Dunn B, Todd JW, Ezzeddine

MA, Haymore J, Lynch JK, Davis L & Warach S. (2004). Comparison of MRI and CT for detection of acute intracerebral hemorrhage. *JAMA*, Vol.292, No.15, pp. 1823-1830

[36] Knutsson L, Borjesson S, Larsson EM, Risberg J, Gustafson L, Passant U, Stahlberg F & Wirestam R. (2007). Absolute quantification of cerebral blood flow in normal volunteers : correlation betwen Xe-133 SPECT and dynamic susceptibility contrast MRI. *J Magn Reson Imaging*, Vol.26, No.4, pp. 913-920

[37] Krol AL, Coutts SB, Simon JE, Hill MD, Sohn CH & Demchuk AM; VISION Study Group. (2005). Perfusion MRI abnormalities in speech or motor transient ischemic attack patients. *Stroke*, Vol.36, No.11, pp. 2487-2489

[38] Latchaw RE, Yonas H, Hunter GJ, Yuh WT, Ueda T, Sorensen AG, Sunshine JL, Biller J, Wechsler L, Higashida R & Hademenos G. (2003). Guidelines and recommendations for perfusion imaging in cerebral ischemia: a scientific statement for healthcare professionals by the writing group on perfusion imaging, from the Council on Cardiovascular Radiology of the American Heart Association. *Stroke*, Vol.34, No.4, pp. 1084–1104

[39] Lin W, Celik A, Derdeyn C, An H, Lee Y, Videen T, Ostergaard L & Powers WJ. (2001). Quantitative measurements of cerebral blood flow in patients with unilateral carotid artery occlusion: a PET and MR study. *J Magn Reson Imaging*, Vol.14, No.6, pp. 659-667

[40] Liu Y, Karonen JO, Vanninen RL, Ostergaard L, Roivainen R, Nuutinen J, Perkio J, Kononen M, Hamalainen A, Vanninen EJ, Soimakallio S, Kuikka JT & Aronen HJ. (2000). Cerebral hemodynamics in human acute ischemic stroke : a study with diffusion- and perfusion-weighted magnetic resonance imaging an SPECT. *J Cereb Blood Flow Metab*, Vol.20, No.6, pp. 910-920

[41] Lövblad KO, Baird AE, Schlaug G, Benfield A, Siewert B, Voetsch B, Conner A, Burzynski C, Edelman RR & Warach S. (1997). Ischemic lesion volumes in acute stroke by diffusion-weighted magnetic resonance imaging correlate with clinical outcome. *Ann Neurol*, Vol.42, No.2, pp. 164-170

[42] Lövblad KO, Laubach HJ, Baird AE, Curtin F, Schlaug G, Edelman RR & Warach S. (1998). Clinical experience with diffusion-weighted MR in patients with acute stroke. *AJNR Am J Neuroradiol*, Vol.19, No.6, pp. 1061-1066

[43] Luby M & Warach S. (2007). Reliability of MR perfusion-weighted and diffusion-weighted imaging mismatch measurement methods. *AJNR Am J Neuroradiol*, Vol.28, No.9, pp. 1674-1678

[44] Ma H, Zavala JA, Teoh H, Churilov L, Gunawan M, Ly J, Wright P, Phan T, Arakawa S, Davis SM & Donnan GA. (2009). Fragmentation of the classical magnetic resonance mismatch "penumbral" pattern with time. *Stroke*, Vol.40, No.12, pp. 3752-3757

[45] Mlynash M, Olivot JM, Tong DC, Lansberg MG, Eyngorn I, Kemp S, Moseley ME & Albers GW. (2009). Yield of combined perfusion and diffusion MR imaging in hemispheric TIA. *Neurology,* Vol.72, No.13, pp. 1127-1133

[46] Muir KW, Buchan A, von Kummer R, Rother J & Baron JC. (2006). Imaging of acute stroke. *Lancet Neurol,* Vol.5, No.9, pp. 755-768

[47] Newman GC, Delucia-Deranja E, Tudorica A, Hospod FE & Patlak CS. (2003). Cerebral blood volume measurements by T*2-weighted MRI and contrast infusion. *Magn Reson Med,* Vol.50, No.4, pp. 844-855

[48] Neumann-Haefelin T, Wittsack HJ, Wenserski F, Siebler M, Seitz RJ, Modder U & Freund HJ. (1999). Diffusion- and perfusion-weighted MRI: the DWI/PWI mismatch region in acute stroke. *Stroke,* Vol.30, No.8, pp. 1591-1597

[49] Neumann-Haefelin T, Wittsack HJ, Wenserski F, Li TQ, Moseley ME, Siebler M & Freund HJ. (2000). Diffusion- and perfusion-weighted MRI in a patient with a prolonged reversible ischemic neurological deficit. *Neuroradiology,* Vol.42, No.6, pp. 444-447

[50] Nighoghossian N, Hermier M, Adeleine P, Derex L, Dugor JF, Philippeau F, Ylmaz H, Honnorat J, Dardel P, Berthezène Y, Froment JC & Trouillas P. (2003). Baseline magnetic resonance imaging parameters and stroke outcome in patients treated by intravenous tissue plasminogen activator. *Stroke,* Vol.34, No.2, pp. 458-463

[51] Oppenheim C, Logak M, Dormont D, Lehericy S, Manai R, Samson Y, Marsault C & Rancurel G. (2000). Diagnosis of acute ischemic stroke with fluid-attenuated inversion recovery and diffusion-weighted sequences. *Neuroradiology,* Vol.42, No.8, pp. 602-607

[52] Ostergaard L, Johannsen P, Host-Poulsen P, Vestergaard-Poulsen P, Asboe H, Gee AD, Hansen SB, Cold GE, Gjedde A & Gyldensted C. (1998). Cerebral blood flow measurements by magnetic resonance imaging bolus tracking: comparison with [^{15}O]H_2O positron emission tomography in humans. *J Cereb Blood Flow Metab,* Vol.18, No.9, pp. 935-940

[53] Parsons MW, Yang Q, Barber PA, Darby DG, Desmond PM, Gerraty RP, Tress BM & Davis SM. (2001). Perfusion magnetic resonance imaging maps in hyperacute stroke: relative cerebral blood flow most accurately identifies tissue destined to infarct. *Stroke,* Vol.32, No.7, pp. 1581-1587

[54] Parsons MW, Barber PA, Chalk J, Darby DG, Rose S, Desmond PM, Gerraty RP, Tress BM, Wright PM, Donnan GA & Davis SM. (2002). Diffusion- and perfusion-weighted MRI response to thrombolysis in stroke. *Ann Neurol,* Vol.51, No.1, pp. 28-37

[55] Poustchi-Amin M, Mirowitz SA, Brown JJ, McKinstry RC & Li T. (2001) Principles and applications of echo-planar imaging: a review for the general radiologist. *Radiographics,* Vol.21, No.3, pp. 767-779

[56] Provenzale JM, Shah K, Patel U & McCrory DC. (2008). Systemic review of CT and MR perfusion imaging for assessment of acute cerebrovascular disease. *AJNR Am J Neuroradiol,* Vol.29, No.8, pp. 1476-1482

[57] Redgrave JN, Coutts SB, Schulz UG, Briley D & Rothwell PM. (2007). Systemic review of associations between the presence of acute ischemic lesions on diffusion-weighted imaging and clinical predictors of early stroke risk after transient ischemic attack. *Stroke,* Vol.38, No.5, pp. 1482-1488

[58] Restrepo L, Jacobs MA, Barker PB & Wityk RJ. (2004). Assessment of transient ischemic attack with diffusion- and perfusion-weighted imaging. *AJNR Am J Neuroradiol,* Vol.25, No.10, pp. 1645-1652

[59] Røhl L, Ostergaard L, Simonsen CZ, Vestergaard-Poulsen P, Andersen G, Sakoh M, Le Bihan D & Gyldensted C. (2001) Viability thresholds of ischemic penumbra of hyperacute stroke defined by perfusion-weighted MEI and apparent diffusion coefficient. *Stroke,* Vol.32, No.5, pp. 1140-1146

[60] Rovira A, Rovira-Gols A, Pedraza S, Grivé E, Molina C & Alvarez-Sabin J. (2002). Diffusion-weighted MR imaging in the acute phase of transient ischemic attacks. *AJNR Am J Neuroradiol,* Vol.23, No.1, pp. 77-83

[61] Schellinger PD, Thomalla G, Fiehler JF, Köhrmann M, Molina CA, Neumann-Haefelin T, Ribo M, Singer OC, Zaro-Weber O & Sobesky J. (2007). MRI-based and CT-based thrombolytic therapy in acute stroke within and beyond established time windows: an analysis of 1210 patients. *Stroke*, Vol.38, No.10, pp. 2640-2645

[62] Schellinger PD, Bryan RN, Caplan LR, Detre JA, Edelman RR, Jaigobin C, Kidwell CS, Mohr JP, Sloan M, Sorensen AG & Warach S. (2010). Evidence-based guideline: The role of diffusion and perfusion MRI for the diagnosis of acute ischemic stroke: report of the Therapeutics and Technology Assessment Subcommittee of the American Academy of Neurology. *Neurology,* Vol.75, No.2, pp. 177-185

[63] Seevinck PR, Deddens LH & Dijkhuizen RM. (2010). Magnetic resonance imaging of brain angiogenesis after stroke. *Angiogenesis,* Vol.13, No.2, pp. 101-111

[64] Seitz RJ, Meisel S, Weller P, Junghans U, Wittsack HJ & Siebler M. (2005). Initial ischemic event: perfusion-weighted MR imaging and apparent diffusion coefficient for stroke evaluation. *Radiology*, Vol.237, No.3, pp. 1020-1028

[65] Sobesky J, Zaro Weber O, Lehnhardt FG, Hesselmann V, Thiel A, Dohmen C, Jacobs A, Neveling M & Heiss WD. (2004). Which time-to-peak threshold best identifies penumbral flow? A comparison of perfusion-weighted magnetic resonance imaging and positron emission tomography in acute ischemic stroke. *Stroke,* Vol.35, No.12, pp. 2843-2847

[66] Sobesky J, Zaro Weber O, Lehnhardt FG, Hesselmann V, Neveling M, Jacobs A & Heiss WD. (2005). Does the mismatch match the penumbra? Magnetic resonance

imaging and positron emission tomography in early ischemic stroke. *Stroke,* Vol.36, No.5, pp. 980-985

[67] Srinivasan A, Goyal M, Azri FA & Lum C. (2006). State-of-the-art imaging of acute stroke. *Radiographics,* Vol.26, Suppl 1, pp. s75-s95

[68] Takasawa M, Jones PS, Guadagno JV, Christensen S, Fryer TD, Harding S, Gillard JH, Williams GB, Aigbirhio FI, Warburton WA, Ostergaard L & Baron JC. (2008). How reliable is perfusion MR in acute stroke?: validation and determination of the penumbra threshold against quantitative PET. *Stroke,* Vol.39, No.3, pp. 870-877

[69] Tanaka Y, Nariai T, Nagaoka T, Akimoto H, Ishiwata K, Ishii K, Matsushima Y & Ohno K. (2006). Quantitative evaluation of cerebral hemodynamics in patients with moyamoya disease by dynamic susceptibility contrast magnetic resonance imaging-comparison with positron emission tomography. *J Cereb Blood Flow Metab,* Vol.26, No.2, pp. 291-300

[70] Tissue plasminogen activator for acute ischemic stroke. The National Institute of Neurological Disorders and Stroke rt-PA Stroke Study Group. (1995). *N Engl J Med,* Vol.333, No. 24, pp. 1581-1587

[71] van Osch MJ, Vonken EJ, Viergever MA, van der Grond J & Bakker CJ. (2003). Measuring the arterial input function with gradient echo sequences. *Magn Reson Med,* Vol. 49, No.6, pp. 1067-1076

[72] von Kummer R, Bourquain H, Bastianello S, Bozzao L, Manelfe C, Meier D & Hacke W. (2001). Early prediction of irreversible brain damage after ischemic stroke at CT. *Radiology,* Vol.219, No.1, pp. 95-100.

[73] Wahlgren N, Ahmed N, Dávalos A, Hacke W, Millán M, Muir K, Roine RO, Toni D, Lees KR & SITS investigators. (2008). Thrombolysis with alteplase 3-4.5 h after acute ischemic stroke (SITS-ISTR) : an observational study. *Lancet,* Vol.372, No.9646, pp. 1303-1309

[74] Warach S, Pettigrew LC, Dashe JF, Pullicino P, Lefkowitz DM, Sabounjian L, Harnett K, Schwiderski U & Gammans R. (2000). Effect of citicoline on ischemic lesions as measured by diffusion-weighted magnetic resonance imaging. Citicoline 010 Investigators. *Ann Neurol,* Vol.48, No.5, pp. 713-722

[75] Wu EX, Tang H & Jensen JH. (2004). Applications of ultrasmall superparamagnetic iron oxide contrast agents in the MR study of animal models. *NMR Bomed,* Vol.17, No.7, pp. 478-483

[76] Wu O, Christensen S, Hjort N, Dijkhuizen RM, Kucinski T, Fiehler J, Thomalla G, Röther J & Ostergaard L. (2006). Characterizing physiological heterogeneity of infarction risk in acute human ischemic stroke using MRI. *Brain,* Vol.129, No.9, pp. 2384-2393

[77] Yamada K, Wu O, Gonzalez RG, Bakker D, Ostergaard L, Copen WA, Weisskoff RM, Rosen BR, Yagi K, Nishimura T & Sorensen AG. (2002). Magnetic resonance perfu-

sion-weighted imaging of acute cerebral infarction: effect of the calculation methods and underlying vasculopathy. *Stroke*, Vol.33, No.1, pp. 87-94

[78] Zaharchuk G, Bammer R, Straka M, Newbould RD, Rosenberg J, Olivot JM, Mlynash M, Lansberg MG, Schwartz NE, Marks MM, Albers GW & Moseley ME. (2009). Improving dynamic susceptibility contrast MRI measurement of quantitative cerebral blood flow using corrections for partial volume and nonlinear contrast relaxivity : a Xenon computed tomographic comparative study. *J Magn Reson Imaging,* Vol.30, No. 4, pp. 743-752

[79] Zaro-Weber O, Moeller-Hartmann W, Heiss WD & Sobesky J. (2009). The performance of MRI-based cerebral blood flow measurements in acute and subacute stroke compared with 15O-water positron emission tomography : identification of penumbral flow. *Stroke,* Vol.40, No.7, pp. 2413-2421

[80] Zaro-Weber O, Moeller-Hartmann W, Heiss WD & Sobesky J. (2010a). MRI perfusion maps in acute stroke validated with 15o-water positron emission tomography. *Stroke,* Vol.41, No.3, pp. 443-449

[81] Zaro-Weber O, Moeller-Hartmann W, Heiss WD & Sobesky J. (2010b). A simple positron emission tomography-based calibration for perfusion-weighted magnetic resonance maps to optimize penumbral flow detection in acute stroke. *Stroke,* Vol.41, No. 9, pp. 1939-1945

Section 5

Neuroimaging of Decision Making and Social Valuation

The Brain is not "as-if" — Taking Stock of the Neuroscientific Approach on Decision Making

Kirsten G. Volz and Gerd Gigerenzer

Additional information is available at the end of the chapter

http://dx.doi.org/10.5772/31393

1. Introduction

How do we make decisions? This question has long engaged researchers from various disciplines, including philosophy, psychology, economics, and cognitive neuroscience. Recently, a new discipline – neuroeconomics – emerged, devoted to addressing exactly this question by means of interdisciplinary endeavors. Broadly stated, neuroeconomics sets out "to promote interdisciplinary collaborations on the topics lying at the intersection of the brain and decision sciences in the hope of advancing both theory and research in decision making" (Society for Neuroeconomics, www.neuroeconomics.org), so that it can be used "for discriminating among standard economic models" (Maskin, 2008, p. 1788) and thus inform economic theory.

Neuroeconomics evolved as a field in the late 1990s, an offspring of behavioral economics and cognitive neuroscience. Behavioral economics is itself a relatively new field, just two decades old, and is commonly traced back to Herbert Simon's concept of bounded rationality (Simon, 1989). All three fields share the same goal of making theories of decision making more realistic by using insights from psychology or cognitive neuroscience. But what may be unrealistic about classical theories of decision making? It appears that there is little evidence that humans make decisions by calculating the expected utility (EU) or any of its variants outside the world of choices between gambles, if at all (e.g., Kahneman & Tversky, 2000; Lopes, 1981; Rubinstein, 1988; Russo & Dosher, 1983). Daniel Friedman and Shyam Sunder (2011) reviewed the literature on risky choice from 1950 up to 2010:

Most theories of risky choice postulate that a decision maker maximizes the expectation of a Bernoulli (or utility or similar) function. We tour 60 years of empirical search and conclude that no such functions have yet been found that are useful for out-of-sample predictions. Nor do we find practical applications of Bernoulli functions in major risk-based industries such as finance, insurance and gambling. (p. 1)

The important methodological concept is "out-of-sample prediction." There is plenty of evidence that EU theory and its variants, such as prospect theory, can *fit* their parameters to data after the fact, but the real test is in *prediction*. "Out of sample" means that a utility function is fitted to a randomly chosen part of the data – say one half – and the fitted function is then tested on the other part (known as *cross-validation*). If the fitted utility function captures the actual process of decision making, then that function should predict the same person's or group's choices. Friedman and Sunder (2011) claim having found no evidence of this in the 60 years of research they reviewed, neither for individual utility functions nor for group utility functions, where groups are defined by demography, gender, and the like.

Neoclassical economists have, in fact, never proposed that people actually weight utilities with probabilities and carry out computations to maximize expected utility. On the contrary, they insist that their theory says nothing about processes either at the level of cognitive psychology and/or its neural implementation. Following Milton Friedman's (1953) "as-if" methodology of "positive economics," neoclassical economists, like the behaviorist B. F. Skinner, saw value only in predicting behavior, not in modeling cognitive processes. EU theory, prospect theory, equity aversion theory, and other utility theories are "as-if" models that are deliberately *not* meant to model the cognitive processes or give information on the neural implementation. The theoretical direction that much of cognitive neuroscience on decision making has taken, however, is to apply the as-if framework as a theory for how the brain works. For example, Glimcher and his colleagues suggested that "the neoclassical/revealed preference framework might prove a useful theoretical tool for neuroscience" (Glimcher et al., 2009, p. 7). Subsequently, the concepts of 'expected value' and 'expected utility' found their ways into neuroeconomics, and there is broad consensus that neuroeconomics is about studying "the computations that the brain carries out in order to make value-based decisions, as well as the neural implementation of those computations" (Rangel et al., 2008, p. 545).

In this chapter, we will provide a positive alternative to the main path pursued by the cognitive neuroscience of decision making. One of the authors (GG) has spent decades on opening the "black box" and investigating the cognitive processes of decision making in humans (Gigerenzer, 2008; Gigerenzer et al., 1999, 2011), including expert groups of doctors, judges, and managers (Gigerenzer & Gray, 2011). Consistent with D. Friedman and Sunder (2011), Gigerenzer and colleagues found little evidence that utility functions predict (as opposed to fit) laypersons' and experts' behavior. Moreover, none of the professional groups investigated uses EU maximization in practice, while there is evidence that they rely strongly on heuristics (Gigerenzer et al., 2011). Given these findings and the overall aim of unraveling human decision making processes, we argue that the neuroscience of decision making would benefit from the following principles so as to approach a psychologically valid account of human decision making (in the brain):

1. *Study the neural correlates of process theories, not of as-if theories*. As explained above, the aim of as-if models, such as EU theory and its modifications is to explain behavior on an aggregate level by explicitly ignoring the underlying cognitive processes. Examples are

the explanation of cabdrivers' labor supply decisions (Crawford & Meng, 2010), farmers' decision behavior given new Common Agricultural Policy (Serrao & Coelho, 2004), or individuals' risky choice behavior (Kahneman & Tversky, 1979). Cognitive neuroscience, however, is explicitly dedicated to the process level, i.e., cognitive neuroscientists want to understand "*how* [emphasis added] the brain enables mind" (Gazzaniga, 2000, p. xii). That is, the object of interest is "the cognitive rules that people follow and the knowledge representations that those rules operate on" (Gazzaniga, 2000, p. 6). Accordingly, the two disciplines work at different levels of description, paying – by definition – no or only little attention to how descriptions at one specific level are related to descriptions at other levels (e.g., Craver, 2007; Marr, 1982). Thus, testing for the neural correlates of EU theory or its variants appears to us as if researchers would interpret an as-if model as a cognitive process model. One solution to this inconsistency would be to test for the neural correlates of process models of decision making, such as heuristics and their adaptive use (Gigerenzer et al., 2011; Payne et al., 1992, 1993; Volz et al. 2006, 2010). For instance, research on investigating how the brain encodes magnitude and probability (e.g., Tobler et al., 2005; Preuschoff et al., 2006, 2008) or how risk is coded (Christopoulos et al., 2009) does not need the EU framework of weighting and integration (of all utilities and probabilities). Magnitude and probability are also components of heuristic models, such as the priority heuristic (Brandstätter et al. 2006; see below). Researchers testing for the neural correlates of an as-if model – such as EU and its variants – however, need to explain why they reinterpret these models in a way they were explicitly not meant to be interpreted.

2. *Study decision making in uncertain ("large") worlds, not only certain ("small") worlds.* The large majority of neuroscientific studies on decision making examines behavior predominantly in situations where everything is known for certain, including all alternatives, consequences, and their probabilities (a so-called small world). A prototypical example is the gambling paradigm. Variants are probabilistic learning-or probabilistic tracking paradigms, in which events are exclusively governed by a specific probability function and individuals are encouraged to estimate the probability of reward. Tasks of the latter sort might be more or less difficult (e.g., depending on whether encountering a volatile or a stable environment, cp. Beherns et al., 2007), but individuals know on which variable(s) to concentrate, i.e., they know about the structure of the task.

To study decision making in gambling and similar small worlds can be an interesting subject. Yet as a study of cognitive processes, it is unclear what this approach can tell us about decision making under uncertainty, that is, situations where not all alternatives, consequences, and probabilities are known for sure. Recent research indicates that neither the normative nor the descriptive results can be automatically generalized. First, what is rational in small worlds is *not* generally rational in uncertain worlds (Gigerenzer et al., 2011). On the contrary, models that are optimal in a small world are typically only second-best when uncertainty is introduced, as in out-of-sample prediction or out-of-population prediction (Czerlinski et al., 1999; Gigerenzer & Brighton, 2009). One reason is that optimization models tend to overfit. Second, small worlds and large worlds may require entirely different skills and strategies. For example, whereas it might suffice to calculate the expected value in a lottery, this skill will not be sufficient for deciding whether or not to be vaccinated against swine flu, which share to buy, or whom to marry. Decision making under risk (small worlds) and under uncertainty (large worlds) do require different skills: statistical thinking and heuristic thinking. It is not possible to extrapolate from small to large worlds except for the rare case in which the former approximates the latter. Savage, known as the father of Bayesian decision theory, has drawn a clear line between the study of small and large worlds:

Jones is faced with the decision whether to buy a certain sedan for a thousand dollars, a certain convertible also for a thousand dollars, or to buy neither and continue carless. The simplest analysis, and the one generally assumed, is that Jones is deciding between three definite and sure enjoyments, that of the sedan, the convertible, or the thousand dollars. Chance and uncertainty are considered to have nothing to do with the situation. This simple analysis may well be appropriate in some contexts; however, it is not difficult to recognize that Jones must in fact take account of many uncertain future possibilities in actually making his choice. The relative fragility of the convertible will be compensated only if Jones's hope to arrange a long vacation in a warm and scenic part of the country actually materializes; Jones would not buy a car at all if he thought it is likely that he would immediately be faced by a financial emergency arising out of the sickness of himself or of some member of his family; he would be glad to put the money on a car, or almost any durable goods, if he feared extensive inflation." (Savage, 1954, p. 83).

Note that studying decision making under uncertainty (as opposed to risk) does not require bringing the complexity of the world into the scanner. It only requires studying tasks where not all alternatives, consequences, and probabilities are known for sure. Well-known examples of large world paradigms are the city task by Goldstein and Gigerenzer (2002), in which individuals are presented with pairs of cities (e.g., Portland – Virginia Beach) and have to infer which city in each pair had the larger population, the fever task by Persson & Rieskamp (2009), and Bröder et al.'s (2010) tasks in which individuals have to judge which of two patients (each with a specific combination of symptoms) had reached the more advanced and dangerous stage.

Implement predictive and competitive tests as methodological standards. The practice of using good fits to support theories has been very popular, in psychology as well as in behavioral economics in the past. However, a good fit by itself is not an adequate test of a theory (Roberts & Pashler, 2000; Roberts & Sternberg, 1993; Wexler, 1978). Fitting means that the data is already known, and dependent on the number and kind of adjustable parameters, one can always achieve a good fit. Accordingly, the practice is now changing and researchers now determine how the theory constrains possible outcomes (i.e., how well it predicts), how actual outcomes agree with those constraints, and whether plausible alternative outcomes would have been inconsistent with the theory (Ahn et al., 2008; Roberts & Pashler, 2000; Bröder et al., 2010; Persson & Rieskamp, 2009). Besides fitting, a second limitation of many tests is that only one theory is being tested, and one does not know whether other theories would predict behavior better. Examples for competitive tests are found in Brandstätter et al., (2006) on models predicting choice between lotteries (small worlds), and in Dhami (2003) on British magistrates' bail decisions (uncertain worlds). Following up on this progression, we would like to push forward

http://dx.doi.org/10.5772/31393

the implementation of predictive and competitive tests also in the cognitive neuroscience of decision making, which only now started to be pursued (e.g., Christopoulos et al., 2009; Venkatraman et al., 2009). Obviously, we specifically would like to suggest incorporating heuristic models in competitive testing on the neural level since this hasn't been pursued intensively yet. Exceptions are from our own research: In an fMRI study on the neural correlates of the Recognition Heuristic (RH)[1], we predicted the neuronal activation patterns that different theories for RH-based decisional processes made. Results revealed that processes underlying RH-based decisions go beyond simply choosing the recognized alternative and are rather distinguished by judgments about the ecological rationality of the RH (Volz et al., 2006).

The three issues are linked. We begin our discussion of them by introducing the basic concepts of modern decision theory and methodology.

2. As-If models and process models

It is often worthwhile to ask where ideas come from. That helps to understand why we ask the questions we ask. Expected value theory has its origin in the social gambling behavior of aristocrats. Its year of birth is commonly dated 1654, when the Chevalier de Méré asked the mathematicians Pierre Fermat and Blaise Pascal for advice in gambling. Their exchange of letters is seen as the beginning of probability and decision theory (for the psychological interpretation of the classical theory, and decline, see Daston, 1988; Gigerenzer et al., 1989). The resulting theory of rational choice is known today as expected value (EV) theory:

$$\mathrm{EV}(x) = \Sigma p_i x_i \tag{1}$$

where p_i is the probability and x_i the (monetary) value of each outcome (i=1, …, n) of a given alternative x. For example, being offered the choice between two alternatives, Gamble A: 490.000€ with p=1.0 and Gamble B: 1 million with p=.5 and 0€ with p=.5, which would you choose? According to EV theory, you should choose the risky Gamble B because it maximizes EV. However, the choices of reasonable people did not always conform to theory. This discrepancy received the greatest attention in the St. Petersburg problem, which led Daniel Bernoulli (1738/1954) to modify EV theory into what is now called expected utility (EU) theory:

$$\mathrm{EU}(x) = \Sigma p_i u_i \tag{2}$$

As one can see, the basic idea is the same: Instead of choosing the alternative that maximizes EV, one now chooses the alternative that maximizes EU. Bernoulli assumed that u=log(x), modeling diminishing returns, but other functions have been proposed, including nonmono-

1 The RH can be stated as follows: "If one of two objects is recognized and the other is not, then infer that the recognized object has the higher value with respect to the criterion." (Goldstein &Gigerenzer, 2002, p. 76).

tonic functions (e.g., Friedman & Savage, 1948). In the second half of the 20th century, systematic deviations from EU theory were shown in experimental studies, including the Allais paradox and the fourfold pattern (Kahneman & Tversky, 2000; Lopes, 1994). Simon (1989) and Selten (2001) proposed a radical break from EU theory, based on empirical psychological principles, such as the study of adaptive heuristics (Gigerenzer et al., 1999; Leland, 1994; Payne, Bettman, & Johnson, 1993; Tversky, 1972). However, most of the economic community retained the EU framework but modified it by adding more adjustable parameters, as in prospect theory (Kahneman & Tversky, 1979):

$$V(x)=\sum_{i=0}^{n}\pi_i^{+}v(x_i)+\sum_{i=-m}^{0}\pi_i^{-}v(x) \quad (3)$$

where π are the decision weights for positive and negative domains (gains and losses). The added flexibility – five adjustable parameters – fitted the data better than EU theory, as long as the additional parameter values were suitably chosen. Given that the form of the relation between u and x is left unspecified, one has to estimate it for each person or group. Yet, as mentioned above, the fitted utility functions of EU and its modifications have not been shown to be useful in out-of-sample prediction of behavior (Friedman & Sunder, 2011). The problem that Bernoulli functions lack predictive power is rarely mentioned in the neuroscience literature.

Prospect theory and similar attempts to make models of decision making more realistic led to "the paradoxical result that the models became even less psychologically plausible" (Gigerenzer & Selten, 2001, p. 5). For example, try to compute the subjective values of risky prospects following prospect theory:

An individual chooses among two or more lotteries according to the following procedure. First, transform the probabilities of all outcomes associated with a particular lottery using a nonlinear probability-transformation function. Then transform the outcomes associated with that lottery (i.e., all elements of its support). Third, multiply the transformed probabilities and corresponding transformed lottery outcomes, and sum these products to arrive at the subjective value associated with this particular lottery. Repeat these steps for all remaining lotteries in the choice set. Finally, choose the lottery with the largest subjective value, computed according to the method above. (Berg & Gigerenzer, 2010, pp. 4-5)

Yet, as explained above, prospect theory should not be mistaken as a cognitive process model. On the contrary, economists since Samuelson (1937) and Friedman (1953) have thought of EU and its variants as *as-if-models*, not as theories that describe the cognitive or neural processes, and neoclassical economists treat the mind as a black box. Economic theory is about the prediction of behavior, not of process. This is the main reason why many economists see little value in neuroeconomics (for economic research) (e.g., Gul & Pesendorfer, 2008; Harrison, 2008; Marchionni & Vromen, 2010; Rubinstein, 2008).

Hitherto, most of the neuroscience of decision making set out to investigate the neural correlates of as-if models, assuming that the mind engages in some form of utility calculations. A great number of neuroeconomic studies aimed at identifying "the brain activations coding the key decision parameters of expected value (magnitude and probability)." (Tobler et al., 2007, p. 1621). Fox and Poldrack (2009, p. 165) claim that there "has been substantial progress in understanding the neural correlates of prospect theory." Others present the assumptions of EU theory as true with regard to the cognitive processes weighting and integrating, but did neither proof whether individuals indeed engaged in these cognitive processes nor elaborated on the issue of as-if models vs. process models. "Decision makers integrate the various dimensions of an option into a single measure of its idiosyncratic subjective value and then choose the option that is most valuable. Comparisons between different kinds of options rely on this abstract measure of subjective value, a kind of "common currency" for choice" (Kable & Glimcher, 2009, p. 734).

The assumption that choice is always based on a weighted integration of all relevant aspects is the most widely held belief across disciplines when it comes to decision making processes; irrespective of whether decisions entail choosing between risky prospects, such as in gambling situations, or between uncertain options, such as a suitable capital investment, a proper university degree, or a life partner: "When deciding between different options, individuals are guided by the expected (mean) value of the different outcomes and by the associated degrees of uncertainty." (Tobler et al., 2007, p. 1621).

These claims are somewhat astounding, at least to us. The experimental literature indicates that full integration (compensation) of the various attributes is the exception rather than the rule (e.g., Gigerenzer & Gaissmaier, 2011; Hauser, 2011; Marewski et al., 2010). Often, humans rely on heuristics that are noncompensatory, such as one good reason or lexicographic decisions. A classic review of 45 studies in which the process of decision making was analyzed by means of eye movement, mouse lab, and other process tracking studies concluded (Ford et al., 1989):

The results firmly demonstrate that non-compensatory strategies were the dominant mode used by decision makers. Compensatory strategies were typically used only when the number of alternatives and dimensions were small or after a number of alternatives have been eliminated from consideration. (p. 75)

Influential neuroeconomic papers, including those cited above, yet, seem to focus on the EU framework and hence seem to accept its basic assumptions as a realistic description of cognitive and neural processes. These assumptions about the nature of decision making thus include (Katsikopoulos & Gigerenzer, 2008):

1. *Independent evaluations:* Every option has a value that is measured by a single number (options are not evaluated relative to other options).

2. *Exhaustive search:* The value of an option is calculated by using all available information (for gambles, the probabilities and values for all possible outcomes).

3. *Trade-offs:* To calculate an option's value, low values on one attribute (e.g., a value) can be compensated by high values on another attribute (e.g., a probability).

4. *Optimization.* The final choice is based on the maximization of utility or some other function of value.

These four assumptions are maintained through almost all modifications of EU theory; what is modified in reaction to inconsistent data are the functions of the probabilities and values, as illustrated in Equations 2 and 3. Most important, the optimization assumption requires studying a small world in which all alternatives, consequences, and probabilities are known, as in a lottery. We will turn to the consequences of this assumption below.

What would a process model look like? A process model would not take the four assumptions as axiomatic, but consider the empirical evidence. In brief, evidence shows that the first three assumptions sometimes hold, typically in small worlds, but are most of the time not descriptively correct (e.g., Bröder & Gaissmaier, 2007; Hauser, 2011). Specifically, there is little evidence that individuals use the same strategy for each decision; instead it implements an "adaptive toolbox" with multiple strategies, called heuristics (Bröder, Persson? Gigerenzer & Selten 2001; Honda et al., 2011; Payne et al., 1993). One interesting result is that whereas modifications of EU such as the Allais paradox grow increasingly complex in order to represent deviating behavior in the EU framework, simple heuristics that are based on evidence on cognitive processes can dispense with this added complexity. One illustration of a process model is the priority heuristic, which applies to lotteries (Brandstätter et al., 2006). Like many heuristics, it has three building blocks: a search rule, a stopping rule, and a decision rule.[2] The priority heuristic consists of the following steps:

Search rule: Go through reasons in the following order: Minimum gain, probability of minimum gain, maximum gain.

Stopping rule: Stop examination if the minimum gains differ by 1/10 (or more) of the maximum gain; otherwise, stop examination if probabilities differ by 1/10 (or more) of the probability scale.

Decision rule: Choose the gamble with the more attractive gain (probability).

The term "attractive" refers to the gamble with the higher (minimum or maximum) gain and the lower probability of the minimum gain. The priority heuristic models difficult choices (same or similar expected values of the alternatives) and nontrivial choices (no alternatives dominates the other, weakly or strongly). The search, stopping, and decision rules were derived from psychological studies (Brandstätter et al., 2006, 2008).

Consider one of the problems, known as the Allais paradox, which shows an inconsistency between the actual observed choices and the predictions of EU theory, or more specifically the transgression of the common consequence effect (Allais 1953, p. 527):

2 Note that EU theory has no search and stopping rules, because it assumes full information or exhaustive search.

A:	100 million	for sure
B:	500 million	$p = .10$
	100 million	$p = .89$
	0	$p = .01$

C:	100 million	$p = .11$
	0	$p = .89$
D:	500 million	$p = .10$
	0	$p = .90$

Most people prefer *A* over *B*, but *D* over *C*, violating EU theory (because *C* and *D* are obtained from *A* and *B* by eliminating a.89 probability to win 100 million from both gambles). EU does not predict whether *A* or *B* will be chosen; it simply makes predictions of the type "if *A* is chosen over *B*, then it follows that *C* is chosen over *D*." The priority heuristic, in contrast, makes stronger predictions: It predicts whether *A* or *B* will be chosen, and whether *C* or *D* will be chosen. Consider the choice between *A* and *B*. The maximum payoff is 500 million, and therefore the aspiration level is 50 million; 100 million and 0 represent the minimum gains of the choice problem. Because the difference (100 million) exceeds the aspiration level of 50 million, the minimum gain of 100 million is considered good enough and people are predicted to select gamble *A*. That is, the heuristic predicts the majority choice correctly.

In the second choice problem, the minimum gains (0 and 0) do not differ. Hence, the probabilities of the minimum gains are attended to, p=.89 and.90, a difference that does not meet the aspiration level. Thus, the higher maximum gain (500 million versus 100 million) decides the choice, and the prediction is that people will select gamble *D*. Again, this prediction is consistent with the choice of the majority. Together, the pair of predictions amount to the Allais paradox.

This simple three-step process generates not only the Allais paradox, but several other so-called anomalies as well. It simultaneously implies common consequence effects, common ratio effects, reflection effects, and the fourfold pattern of risk attitude (Katsikopoulos & Gigerenzer, 2008). Note that the priority heuristic logically *implies* these behaviors, whereas prospect theory is only *consistent* with these.

The priority heuristic is an example of a process model that assumes dependent rather than independent evaluation: The two alternatives are compared, one by one, on the attributes, rather than each alternative having a utility independent of the other. Second, it assumes limited rather than exhaustive search, that is, it does not always use all of the information but employs a stopping rule. Third, it makes no trade-offs between the attributes: For instance, the first attribute alone can determine the decision. Finally, it is not based on the computation of a maximum or minimum of a function including weighting and summing, but is a model of bounded rationality in the sense of Simon.

In sum, EU theory and its modifications have been proposed as as-if models in neoclassical economics precisely because there is little evidence that they describe the psychological or neuronal processes underlying a decision. Behavioral economics largely retained the as-if framework, adding realism in the form of one or a few adjustable parameters and continuing to produce as-if models (Berg & Gigerenzer, 2010). Given this fact and the case that cognitive neuroscience investigations are by definition foremost interested in process models, we wonder why most of neuroeconomic research attempted to locate the computations involved in as-if theories in the brain and did not turn to distinct process models of decision making. Thus, an alternative is to study the neural correspondents of heuristic processes, such as search rules, stopping rules, and decision rules (e.g., Volz et al., 2006, 2010).

3. The brain is adapted to the uncertainty of large worlds, not to the certainty of lotteries and small worlds

EU and its modifications all involve optimization. Optimization means finding the maximum or minimum of a function and proving that it is the best choice. Optimization requires a world in which all alternatives, consequences, and probability distributions are known for certain (otherwise one cannot know what the optimal alternative is). The father of modern Bayesian decision theory, L. J. Savage (1954), distinguished between small worlds in which these conditions are met and where Bayesian theory is optimal, and large worlds in which these conditions are not met and where it would be "ridiculous" to use Bayesian decision theory (Binmore, 2008). The prototype for a small-world problem is the lottery. Many of the standard neuroscientific tasks on choice and decision making have been modeled after the lottery: two-armed bandit tasks, intertemporal choice tasks, and the ultimatum game, among others. For example, in the financial decision-making task by De Martino and colleagues (2006), participants had the choice between a sure option, i.e., how much money can be won on a specific trial (v=£20 with p=1.0) and a gamble option, i.e., the exact specification of the probabilities of winning or losing a specific amount of money (v=£50 with p=.37 and v=£0 with p=.63). Likewise, outcomes are certain in intertemporal choice paradigms, where, for instance, participants have to choose between smaller monetary rewards delivered immediately (€13.83 today) and larger monetary rewards delivered only later (€17.29 in 4 weeks) (e.g., Albrecht et al., 2011; McClure et al., 2004).

Without any disparagement intended, we note that these conditions rarely apply to real-world decisions. In the real world, part of the relevant information is missing or has to be estimated from small samples and the future is naturally unpredictable, so that optimization is out of reach. Such decisions include which capital investment to favor, which used car to buy, or whether to take a specific cancer check-up. Moreover, even if all alternatives, consequences, and probabilities are known, computing the best alternative in such situations can be computationally intractable. For example, computing Bayes' rule is NP-hard, that is, it becomes computationally intractable for brain and computers alike when large numbers of attributes are involved. Even in the simplest case with binary predictors only and a binary criterion, the number of leaves (exits) of the full decision tree increases exponentially (2^n) with the number n of predictors. In contrast, in a fast-and-frugal tree, which is a heuristic model of Bayesian-type decision making, the number of leaves increases with n+1 only (Martignon et al., 2011).

http://dx.doi.org/10.5772/31393

3.1. Cognitive processes in large worlds differ from those in small worlds

The crucial question is, what can we learn about the brain if we focus experimental tasks on small worlds? Can we generalize the findings from small world studies to large worlds? Consider first the sub-question: Can we generalize the norms? To begin with, note that the very reason to construct small worlds is that EU theory or some other optimization model can be used to determine the best alternative. But the best strategy in a small world is not necessarily the best one in a large world. By way of illustration, two examples may suffice. Consider financial investment once again. A normative theory of how to allocate a sum of money to *N* assets is Markowitz's Nobel prize-winning mean-variance model. Like all optimizing theories, it assumes perfect knowledge about the relevant parameters. Is the theory optimal in the real world of financial investment, where parameter values are not known for certain but need to be estimated? The answer is no. De Miguel and colleagues (2009) showed that the simple 1/*N* heuristic (allocate your money equally to *N* assets) performs better in out-of-sample prediction. Note that the heuristic achieves better performance because it ignores part of the information, which makes it robust, while the mean-variance portfolio tries to integrate all information to estimate the weights and suffers from overfitting.

Second, consider sequential decision making where cues need to be ordered in a way that improves inductive inferences. The optimal ordering can be determined easily in a small world, and it leads to more accurate inferences than when cues are ordered simply, as with the take-the-best heuristic. Yet, in out-of-sample prediction, the optimal ordering is no longer best, and the simple ordering leads to more accurate inferences (Gigerenzer & Brighton, 2009; Martignon & Hoffrage, 1999). These results illustrate *less-is-more effects*, that is, situations where less computation and information enables more accurate inferences. They also explain why a certain degree of memory and capacity limitations can be beneficial (e.g., Goldstein & Gigerenzer, 2002; Hertwig & Todd, 2003). The study of ecological rationality answers the question of the worlds in which a given heuristic is successful, relative to other strategies. For instance, to allocate one's money to *N*=50 assets, an estimated 500 years of stock data would be needed for the mean-variance portfolio to finally exceed the performance of the 1/*N* heuristic (DeMiguel et al, 2009).

In sum, good decision making in large worlds requires different cognitive processes than those studied in small worlds. Studying a world of certainty may very likely miss the cognitive processes in the brain that deal with uncertainty (such as limited search and aspiration levels), thus extrapolation is simply not feasible. Furthermore, one cannot extend claims about a person's cognitive processes based on a rational or irrational behavior in a small world to corresponding claims for the same person in a large world.

3.2. Bounded rationality is the study of cognition in "Large worlds"

In behavioral economics, the term "bounded rationality" is correctly attributed to H. A. Simon but incorrectly identified with a program that is not his: the study of deviations from optimality. For instance, Daniel Kahneman (2003, p. 1449) noted: "Our research attempted to obtain a map of bounded rationality, by exploring the systematic biases that separate the beliefs that people have and the choices they make from the optimal beliefs and choices assumed in rational-agent models." In this research, it is assumed that the conditions for rational models (i.e., for a small world) hold and can thus define optimal reasoning. Simon (1989), however, asked a fundamentally different question, leading to a different research program.

Simon's question: "How do human beings reason when the conditions for rationality postulated by the model of neoclassical economics are not met?" (p. 377)

The study of behavior in lotteries – and other small worlds where the conditions of rationality are met – does not address Simon's question. In large worlds, people cannot optimize, but need to satisfice by relying on heuristics from the brain's adaptive toolbox. The term 'satisficing', originally introduced by Simon, is the Northumbrian term for 'satisfying' and is used to refer to decision strategies that ignore part of the available information and involve little computation. For example, when having to decide quickly whether an incoming patient should be treated as low-risk or high-risk in an emergency room, medical decisions implement fast and frugal heuristics (cp. Gigerenzer et al., 1999). We believe that as long as research continues to focus on small worlds, we will learn little about the brain's decision processes that people use *outside the laboratory*, i.e., in uncertain decision situations, which structurally differ from small world problems.

Why are lotteries studied so extensively in decision research and neuroeconomics and used to define rational choice? This appears particularly surprising given that we do not come across gambling situations very frequently in the real world, nor that most of our daily decision situations structurally resemble gambling situations. There are probably three reasons for this tradition. The first is the historical origin of aristocrats asking mathematicians for advice on gambling problems – not on whom to marry, what job offer to accept, or how to invest money wisely. The second reason is the fondness for optimization models. And the third is the misinterpretation of Simon's concept of bounded rationality as constrained optimization, or the deviation of human judgment from optimization.

In sum, much of cognitive neuroscience studies decision making in small worlds, where optimization is defined. This focus imposes limits on the understanding of brain processes, both normatively and descriptively. Normative behavior in a small world is not necessarily so in large worlds, where robustness counts. Furthermore, cognitive processes underlying bounded rationality, such as heuristic search, stopping rules, and aspiration levels, have little chance of being "detected" when one is looking for the neural correlates of EU theory and other as-if theories.

4. Methodology

4.1. Prediction versus fitting

In their methodological critique of behavioral economics, Binmore and Shaked (2010) recall the importance of distinguishing between data fitting and prediction:

The scientific gold standard is prediction. It is perfectly acceptable to propose a theory that fits existing experimental data and then to use the data to calibrate the parameters of the model.

http://dx.doi.org/10.5772/31393

But, before using the theory in applied work, the vital next step is to state the proposed domain of application of the theory and to make specific predictions that can be tested with the data that was used neither in formulating the theory nor in calibrating its parameters. (p. 89)

Whereas the dependent variable in behavioral decision research is the actual choice behavior of individuals, groups, or the entire economy, the primary dependent variable in almost all neuroeconomic studies to date is the BOLD signal change. Accordingly, the gold standard in neuroscience is the prediction of the BOLD response as a function of the manipulation of the independent variable(s). Neuroeconomic studies thus aim at predicting the BOLD signal changes as a function of choice behavior, or, in the words of Glimcher and colleagues (2009, p. 7): "Neuroscientists [...] were interested in describing the algorithmic mechanisms of choice. Their goal was to describe the neurobiological hardware that supported choice behavior in situations ranging from perceptual decision making to the expression of more complicated preferences." Yet the authors also noted that "what they [the neuroscientists] lacked was an overarching theoretical framework for placing their neural measurements into context." This lacuna was filled by Glimcher and colleagues' (2009) suggestion "that the neoclassical/revealed preference framework might prove a useful theoretical tool for neuroscience. What followed was the rapid introduction to the neuroscientific literature of such concepts as expected value and expected utility" (p. 7).

This statement reveals three issues that we propose to characterize the state of the art in current neuroeconomic research:

1. Cognitive neuroscientific and neuroeconomic research on decision making primarily pursues the neoclassical as-if approach while so far neglecting process theories, such as the program on fast-and-frugal heuristics (cp. Gigerenzer, Hertwig, & Pachur, 2011; for exceptions, see Volz et al., 2006, 2010).

2. The predominant focus on concepts such as expected value and expected utility has led to the generally accepted assumption in the cognitive neuroscience literature on decision making that the choice process comprises a first evaluation stage in which values for each option are determined and a subsequent decision phase in which the individual chooses on the basis of this common currency. Accordingly, these two stages have been investigated separately, and a considerable number of studies have focused on the neurobiological implementation of the anticipation/valuation phase, which is – by definition – characterized by a lack of choice behavior. As noted above, there is no compelling empirical evidence supporting the assumption that decision makers go through the stages of the expected value or utility model (e.g., Ford et al., 1989; Friedman & Sunder, 2011; Venkatraman et al., 2009). Even in gambling situation, individuals seem to process information in a way that is incompatible with the EU framework, as revealed by imaging studies (Venkatraman et al., 2009), verbal protocol studies (Cokely & Kelley, 2009), eye tracking studies (Russo & Dosher, 1983), and information board studies (Mann & Ball, 1994; Payne & Braunstein, 1978).

3. The presentation of data on the neurobiology of value-based decision making, which are considered to "yield new insights into behavior" (Kable & Glimcher, 2009, p. 734), may

make researchers forget that neuroimaging does *not* allow "to look into the head" objectively or "watch the brain while deciding". Rather, by setting up an experiment, researchers are driven by their theoretical assumptions, which subsequently lead to very specific experimental manipulations, such as manipulating the probability and the value of choice options, as well as to very specific analyses, such as conducting an analysis "to identify candidate regions whose activity correlated with a linear model of EV." (Knutson et al., 2005; p. 4807). Thus, fMRI data themselves do not tell us what is going on in the brain. The data are always interpreted in light of a theory; more important, the very choice of a theory determines a study's problems and experimental set-up. For instance, optimization theory, such as EU and Bayes, confines studies to small worlds, where *the* best outcome is computationally tractable.

Returning to the issue of the degree to which neuroeconomic studies use prediction, we would like to recapitulate that the scientific method is hypothesis-driven. In other words, (neuroeconomic) researchers *first* hypothesize about the alleged brain areas being crucially involved in the respective cognitive process(es) and *then* conduct significance testing. Hypothesis-driven research needs to be distinguished from so-called "see-and-tell" studies or "fishing expeditions." The latter are appropriate for finding hypotheses, but should not be presented as testing hypotheses by performing significant tests after the fact. The practice of (mis)presenting a hypothesis-finding study as a hypothesis-testing study is possibly due to disregard for the value of hypotheses finding, high degrees of freedom in data collection and post hoc analysis, and the practice of null-hypothesis significance testing (NHST) (Loftus, 1996; Simmons et al., 2011; Vul et al., 2009). This is not to say that we take issue with research that describes the data of an experiment, which may lead to specific testable predications; we solely disapprove of presenting significant results after the fact, that is, selling ad hoc interpretations as genuine experimental hypotheses testing.

The very majority of neuroeconomic studies use prediction. That is, as a function of the experimental design, more or less elaborated anatomical hypotheses are made. For example, in the realm of value-based decision making, it has been predicted and shown that the ventral striatum encodes the subjective value of both primary (e.g., pleasant sweet taste [O'Doherty et al., 2002], naturalistic food aromas [Bragulat et al., 2010], or erotic stimuli [Sescousse et al., 2010]) and secondary rewards (e.g., money [Delgado et al., 2000; Elliott et al., 2000; Knutson et al., 2000]); that the ventral striatum reflects the anticipation and expectation of (monetary) rewards (Breiter et al., 2001; Knutson et al., 2000, 2003, 2005; Preuschoff et al., 2006;); and that the ventral striatum is directly related to an individual's actual behavioral preferences (Kable & Glimcher, 2007; O'Doherty et al., 2006;). A similar story seems to hold for the medial prefrontal cortex (mPFC) (e.g., Elliott et al., 2003; Lebreton et al., 2009; Plassmann et al., 2007; Tom et al., 2007; for a review see Rushworth et al., 2011). That is, the mPFC seems to reflect the "some aspect of both expected reward value prior to the making of a choice and the received reward value after a choice is made" (Rushworth et al., 2011, p. 1056). Contemporary research suggests that the mPFC valuation signal is an automatic one reflecting the value of an option even in the absence of any choice (Lebreton et al., 2009; Smith et al., 2010). Further endeavors to differentiate the roles of the mPFC and the ventral striatum suggested that the latter specifically represents anticipated gain magnitude, whereas the mPFC also represents anticipated gain probability; it was also suggested that the mPFC integrates the two components of expected value (Knutson et al., 2005; Knutson & Wimmer, 2007; but see also Behrens et al., 2007; Rushworth & Behrens, 2008, for the suggestion that anterior cingulate cortex activity encodes the integrated value of actions).

In the realm of intertemporal choice, it has been predicted and shown that decisions involving immediately available rewards draw specifically on limbic structures, whereas intertemporal choices per se, irrespective of delay, draw on the lateral prefrontal cortex and associated structures (Albrecht et al., 2010; McClure et al., 2004; Mitchell et al., 2011). Moreover, by taking into account the actual choice behavior of subjects, it has been predicted and shown that activity in the ventral striatum, the mPFC, and within the posterior cingulate cortex "tracks the revealed subjective value of delayed monetary rewards" (Kable & Glimcher, 2007, p. 1625). In the realm of social decision making, activation within neural structures that are involved in emotional processing, such as the insula and amygdala, have been predicted and shown to be activated for a number of so-called decision anomalies such as rejecting unfair offers in the UG (Sanfey et al., 2003; Tabibnia et al., 2008), reciprocating trust (van den Bos et al., 2009), displaying inequity aversion (Haruno & Frith, 2010), or third-party punishment decisions (Buckholtz et al., 2008); for an overview, see Rilling & Sanfey, 2011). Amygdala or insula activation has generally been considered to provide important affective biases to social decisions.

All these predictions were made in small world settings, where all alternatives, consequences, and probability distributions were known or could be learned from experience. That is, participants had to value specific items such as food, money, or specific abstract symbols representing money, or had to choose between (two) risky outcomes. As outlined before, these results are very interesting by itself and can help in determining the cognitive processes involved in different decision making tasks. For example, individuals using the priority heuristic (PH) when making decisions between risky gambles, by definition, first go through the reasons in the order of minimum gain, probability of minimum gain, maximum gain and they stop their examination if the minimum gain differs by 1/10 (or more) of the maximum gain; otherwise they stop the examination if probabilities differ by 1/10 (or more) of the probability scale. Given the formal description of the PH, individuals applying the PH simply compare the numbers of the two gambles and thus the corresponding anatomical hypothesis would predict activation within a parietal network, which has been shown to support number comparison processes (Deheaene & Cohen, 1995; Pinel et al., 2001). Yet, if activation within striatal, dorsal anterior cingulate cortex, or inferior frontal cortex would elicit, i.e., within the network suggested to reflect processes involved in EU calculations, this finding would raise issues.

4.2. Competitive tests of models versus testing only one

Given the abundance of evidence on the neurobiology of choice behavior, "for a neuroeconomist, then, these studies constitute overwhelming evidence that a value system exists and can be functionally localized" (Glimcher, 2009, p. 514). This may certainly apply to small world settings. Yet, achieving a good fit of the expected value model/neoclassical framework to the neuroscientific observations does not necessarily mean that a good model or *the* best model is found. This is because "all models are wrong, but some of them predict better than others and may lead to novel questions." (Gigerenzer & Gaissmaier, 2011). Thus, although EV models and their variants may capture the algorithmic mechanisms of choice behavior for small-world problems, a convincing argument would require tests of several models to determine how well the model of choice performs against competing ones. This procedure differs from the widespread practice of null hypothesis testing. Since neuroeconomics seeks an algorithmic description of the mechanisms for choice behavior and the cognitive strategies underlying economic and everyday decisions, implementing comparative tests of formal models seems indispensable. Only by pursuing this strategy, we consider, can one accomplish the ambitious aim of describing the mechanisms of how decisions are made (in the brain).

To our knowledge, hitherto there are only a handful of imaging studies that use competitive testing on the neural level, for example in risky decision making (Basten et al., 2010; Christopolous et al., 2010; Hsu et al., 2009; Venkatraman et al., 2009) and we are not aware of imaging studies that tested whether competing models such as heuristic models (e.g., priority heuristic for risky gambles) fit to the neural data. We would be eager to learn whether the activation in a specific area may also or better be explained by a competing (heuristic) model of choice.

One reason for this may be that in the implemented small-world problems, the very experimental set-up primes one specific strategy, namely EV calculations. Experimenters may have thereby induced the decision strategy in which they were interested. For example, in gambling paradigms the only information that is presented concerns probabilities and incentives (e.g., Behrens et al., 2007; Cooper & Knutson, 2008; De Martino et al., 2006; Elliott et al., 2000; Hsu et al., 2009; Knutson et al., 2005; Shiv et al., 2005; Preuschoff et al., 2006, 2008; Tom et al., 2007; Yacubian et al., 2006). Nevertheless, different decision strategies are also conceivable in gambling paradigms, such as probability matching, the priority heuristic, guessing, win-stay/lose-shift and others whose correlates could be investigated and compared (for probability matching in monkeys, see Morris et al., 2006; Niv et al., 2006). Indeed, Venkatraman and colleagues (2009) when investigating risky choice behavior on the neural level, found neural predictors of strategic variability in decision making: "Choices that maximized gains or minimized losses were predicted by fMRI activation in ventromedial prefrontal cortex or anterior insula, respectively. However, choices that followed a simplifying strategy (i.e., attending to overall probability of winning) were associated with activation in parietal and lateral prefrontal cortices." (p. 593). Most interestingly, activation within the mPFC predicted individual variability in strategic preferences through differential functional connectivity with parietal and insular cortex. In our view, studies like Venkatraman's et al. are very promising and should be extended to large world decision situations.

With these exceptions, the focus hitherto has been on the investigation of the anticipation/expectation phase, in which participants are presented with a specific gamble, bet, or other well-defined option and are encouraged to choose advantageously. For example, in the probabilistic monetary incentive delay task by Knutson and colleagues, participants in the anticipation phase were presented with an icon indicating exactly the incentive that could be obtained on that specific trial and its probability. After having seen the trial specification, participants awaited the feedback, which was probabilistically determined, sometimes even independent of participants' choice. The fMRI analyses in studies using such small world settings are generally stimulus-dependent, that is, regressors are generated on the basis of measures built into the task, such as magnitude of gain/loss, probability of gain/loss, valence or salience of events, trial-specific EVs, or the like. Accordingly, analyses reveal activation in areas responsible for processing specific task characteristics.

Recently, further progress has been made in showing – for risky decisions– that in combining information from the three areas that are particularly responsive to changes in magnitude (striatum), objective risk coding (doral anterior cingulate cortex (dACC)), and risk aversion (inferior frontal gyrus (IFG)), "these BOLD responses were informative enough to allow an ideal observer to detect the overt choice: a risky choice was more probably when striatal and cingulate activity was higher, whereas increased BOLD signals from IFG correlated with increased probability of a safe choice." (Christopoulos et al., 2009, p. 12581). In other words, the BOLD responses from these areas combined essentially decoded the behavioral choice.

Such findings are very interesting by itself and especially give information on how the brain (in small world settings) encodes and processes specific task characteristics and combines them to produce overt choice behavior; yet, it remains an open and interesting issue whether these processes will manifest themselves in large world decision situations alike. In other words, the description of the neurobiology of value-based decisions in the small world does not necessarily imply these processes and activations in large-world decisions, yet this remains to be seen. To clarify this problem, we recommend imaging studies on large-world decisions that determine and compare the neural correlates of different decision strategies, including those modeled by the neoclassical approach and the heuristic approach.

Another recent development, which we think should be pursued and extended, is model-based fMRI, "which involves the application of computational models in the design and analysis of neuroimaging experiments" (O'Doherty et al., 2007, p. 35). The standard operating procedure is to "first fit the computational model to subjects' actual behavior to find specific values for the free parameters in the model []. Once the best-fitting model parameters have been found, then the different model components can be regressed against the fMRI data [] to identify areas where the model-predicted time series show significant correlations with the actual changes in blood oxygenation level-dependent (BOLD) signal over time." (O'Doherty et al., 2007, p. 37). This model-based approach has been used to investigate questions of how the brain integrates cost and benefits during decision making (Basten et al., 2010), how it trades off between amount and delay during intertemporal decisions (Kable & Glimcher, 2007), how different experiences (volatility) are weighted in guiding future actions (Behrens et al., 2007), and whether the brain encodes a nonlinear probability function when evaluating risky choices as predicted by prospect theory (Hsu et al., 2009; Boorman & Sallet, 2009). Accordingly, in the various studies, the authors fitted several functions to the subjects' actual choices (e.g., various nonlinear probability-weighting functions or discount functions) and used the best fitting ones for the fMRI analyses. Results were interpreted to show that "the brain, thus, weighs costs against benefits by combining neural benefit and cost signals into a single, difference-based neural representation of net value" (Basten et al., 2010, p. 21767); that "the neural tradeoffs between amount and delay that are captured by the neurometric discount functions match the behavioral tradeoffs between these variables that are captured by the psychometric discount functions" (Kable & Glimcher, 2007, p. 1631); that "humans repeatedly combine prior and subsequent information as data accumulate over time," as reflected by ACC activity; and "that activity in the striatum during valuation of monetary gambles are nonlinear in probabilities in the pattern predicted by prospect theory, suggesting that probability distortion is reflected at the level of the reward encoding process." (Hsu et al., 2009, p. 2231). Despite these impressive results, it should be kept in mind that only different forms of neoclassical models were applied and compared, that is, competitive tests were carried out only within the same model class of as-if models. Moreover, these models were fit to the behavioral data (for problems with data fitting, see above), and the results cannot conclusively demonstrate that the respective activated regions implement the process specified within the model. Yet, in our view, it is a step in the right direction and should definitely be pursued and extended.

5. Taking stock: As-If models, small worlds and comparative tests

5.1. As-If models and neuroeconomics

In reviewing the neuroeconomic literature it appears that the overarching theoretical framework is the neoclassical as-if theory. The consensus that "the brain must perform multiple value computations to make sound choices" (Hare et al., 2008, p. 5623) just as in the as-if theory seems too strong and questionable given that hitherto only few studies investigated competing process models.

5.2. Small worlds and neuroeconomics

The majority of cognitive neuroscience studies on decision making use variants of small-world problems, prototypical gambling tasks, in which all alternatives, consequences and their probabilities are known. In nongambling decision situations, such as intertemporal choice paradigms, decisions are not even risky, but have certain outcomes. Examples are: "Do you prefer $24 now or $33 in 4 weeks?" or "How much are you willing to pay for a candy bar?" (cp. Hare et al., 2008; Kable & Glimcher, 2007; Lebreton et al., 2009; McClure et al., 2004; Plassmann et al., 2007). Moreover, in several other decision making and conditioning tasks, participants had extensive experience with the stimuli and the associated rewards, so that choice might simply have entailed assessing the likelihoods of rewards, as for example in probability-tracking tasks and their variants (cp., Behrens et al., 2007; Huettel et al., 2005). Subsuming conditioning under decision making is a questionable practice that leads to every response, conscious or unconscious, being called decision making, so that the term becomes meaningless. Accordingly, decisions such as whether to select surgery or radiation therapy for a tumor, to invest retirement savings in the stock market or treasury bills, to pursue a university degree, or to accept a new job – all of which were mentioned as prototypical for decision making (examples taken from Rustichini, 2009; Tom et al., 2007; Trepel et al., 2005) – have almost nothing in common with the experimental investigations of decision making processes in neuroeconomic studies (so far).

At this point, we return to Savage's distinction between small worlds and large worlds. Savage, the father of modern Bayesian decision theory, considered Bayesian rationality as normative for small worlds, holding the opinion that a decision maker should engage in subjective EU maximization when making decisions under risk in such a world. But Savage also wrote that it would be "ridiculous" to use Bayesian rationality outside small worlds. Instead, he emphasized that rational models cannot be automatically assumed to provide the correct answer in real-life decision situations. Large worlds need a different kind of rationality, and the study of heuristic decision making is one way to model rationality under uncertainty.

With that said, we agree with Glimcher (2009) that it might be the case that "dopamine neurons lead to the direct computation of SV [subjective value] under some conditions" (p. 514), whereby "some conditions" should be translated to "small-world problems". Although we believe that none of the proposed subjective EU maximization strategies is computationally feasible in large worlds, those who make the claim that dopamine neurons compute EU in small worlds need to show this by performing competitive tests with models that do not involve optimization.

A promising course of research, we think, is based on modeling cognitive decision making processes by means of alternative models, including models of heuristics, and testing and comparing their respective neural correlates.

Given that we share neuroscientists' overall aim of understanding how decisions are made, our main concern, however, is to promote the use of uncertain large-world paradigms in the neuroeconomic literature. For one, it is indispensable to incorporate heuristics into the theoretical portfolio, since previous research on large-world problems has shown that laypeople and experts (e.g., doctors, judges, managers) use heuristics in an ecologically rational way, that is, when the heuristics are adapted to the structure of the environment (Dhami, 2003; Gigerenzer & Gaissmaier, 2011; Gigerenzer et al., 2011; Wübben & Wangenheim, 2008). This research indicated that individuals, especially under time pressure and information overload, switch to simple noncompensatory strategies that use few cues rather than engaging in exhaustive compensatory strategies such as EU maximization.

5.3. Predictive and competitive tests and neuroeconomics

The inventory for this issue is elaborated in section 4. In short, neuroeconomics does a good job of predicting BOLD signal changes as a function of the manipulation of the independent variable(s), but we hold that neuroeconomic studies have a large developmental potential for competitive tests, especially with regard to heuristic models. As outlined above, our second concern is to promote predictive and competitive tests as methodological standards in neuroeconomics (where it is feasible). For example, cognitive neuroscience investigations shall compare different decision strategies that individuals actually used (cp. Venkatraman et al., 2009). Examples of general methodological approaches to determine which decision rules people (very likely) follow are the use of outcome-based strategy classification such as the maximum-likelihood strategy classification (Bröder & Schiffer, 2003; Bröder, 2010) or ACT-R (Anderson, 2007), cognitive process analysis (if it is useful) (Witteman & van Geenen, 2010), or a multiple measure strategy classification combining outcomes and decision times (Bergert & Nosofsky, 2006; Bröder & Gaissmaier, 2007; Glöckner, 2010; Rieskamp & Otto, 2006).

6. Conclusion

Neuroeconomics or the study of the neurobiology of decision making set out with ambitious aims, namely, "revealing the neurobiological mechanisms by which decisions are made" (Glimcher & Rustichini, 2004, p. 447) and "providing a biologically sound theory of how humans make decisions that can be applied in both the natural and the social sciences." (Rangel et al., 2008, p. 545). Back in 2005, Camerer, Loewenstein, and Prelec outlined two types of contributions that they suggested neuroscience could make to economics, or more broadly to decision sciences, namely, an incremental and a radical approach. The incremental approach was formulated as "neuroscience adds variables to conventional accounts of decision making or suggests specific functional forms to replace "as if" assumptions that have never been well supported empirically." (Camerer et al., 2005, p. 10).

We highly appreciate this statement, yet, as the previous literature review showed, in most neuroeconomic studies, the as-if models were not replaced (e.g., by models of heuristic decision making) but instead retained as descriptions of how the brain works.

The radical approach, on the other hand, is about neuroscience informing economic theory. According to Camerer and colleagues (2005), neuroscience "points to an entirely new set of constructs to underlie economic decision making. The standard economic theory of constrained utility maximization is most naturally interpreted either as the result of learning based on consumption experiences (which is of little help when prices, income, and opportunity sets change), or careful deliberation – a balancing of the costs and benefits of different options […]." (p.10). In this passage, the authors appear to be in line with us. First, they acknowledge that for most of our real-life decisions we cannot rely on an extensive learning history and thus cannot define priors (as necessary for Bayesian calculations), and second, they admit that the decision process as prescribed in (neo-)classical decision theory for actual flesh-and-blood human beings is simply intractable: "The variables that enter into the formulation of the decision problem are precisely the variables that should affect the decision if the person had unlimited time and computing ability." (p. 10). In this way, Camerer and colleagues seem to concur with Savage that small-world and large-world decision situations differ fundamentally and that the investigation of small-world experiments alone may produce unrealistic data having no relevance for understanding how the brain deals with the real world. Yet, as we have shown in this contribution, the study of optimization in small worlds is still the norm of present-day neuroeconomics.

We find the advent of neuroeconomics promising and hope that it will continue on its original course so that we can arrive at psychologically valid accounts of human decision making. This, in our view, is only possible if the neuroscience of decision making forsakes as-if theories,

http://dx.doi.org/10.5772/31393

investigates decision making under uncertainty (large worlds) rather than only under risk (small worlds), and implements competitive testing that includes heuristic models.

Author details

Kirsten G. Volz[1] and Gerd Gigerenzer[2]

1 Werner Reichardt Centre for Integrative Neuroscience, Tuebingen, Germany

2 Max Planck Institute for Human Development, Berlin, Germany

References

[1] Ahn, W.-Y., Busemeyer, J.R., Wagenmakers, E.-J., & Stout, J.C. (2008). Comparison of decision learning models using the generalization criterion method. *Cognitive Science*, Vol. 32, pp. 1376-1402

[2] Albrecht et al. (2011). What is for me is not for you: Brain correlates of intertemporal choice for self and other. *Social Cognitive and Affective Neuroscience*, Vol. 6, pp. 218-225

[3] Allais, M. (1953). Le comportement de l'homme rationnel devant le risque: Critique des postulates et axioms de l'école américaine. *Econometrica*, Vol. 21, pp. 503-546

[4] Anderson, J.R. (2007). Using brain imaging to guide the development of a cognitive architecture, In: *Integrated models of cognitive systems*, W.D. Gray, (Ed.), pp. 49-62, Oxford University Press, New York

[5] Basten, U., Biele, G., Heekeren, H.R. & Fiebach, C.J. (2010). How the brain integrates costs and benefits during decision making, *Proceedings of the National Academy of Sciences*, Vol. 107, pp. 21767-21772

[6] Behrens, T.E.J., Wollrich, M.W., Walton, M.E. & Rushworth, M.F.S. (2007). Learning the value of information in an uncertain world. *Nature Reviews Neuroscience*, Vol. 10, pp. 1214-1221

[7] Berg, N. & Gigerenzer, G. (2010). As-if behavioral economics: Neoclassical economics in disguise? *History of Economic Ideas*, Vol. 18, pp. 133-166

[8] Bergert, F.B. & Nosofsky, R.M. (2007). A response-time approach to comparing generalized rational and take-the-best models of decision making. *Journal of Experimental Psychology: Learning, Memory, and Cognition*, Vol. 331, pp. 107-129

[9] Bernoulli, D. (1738/1954). Specimen theoriae novae de mensura sortis. *Comentarii academiae scientarum imperialis Petropolitanae*, Vol. 5, pp. 175-192. (Engl. Transl. by L.

Sommer, Exposition of a new theory on the measurement of risk. *Econometrica,* Vol. 22, pp. 23-26

[10] Binmore, K. & Shaked, A. (2010). Experimental economics: Where next? *Journal of Economic Behavior & Organization,* Vol. 73, pp. 87-100

[11] Binmore, K. (2008). Behavioral economics and its applications. *The Economic Journal,* Vol. 118, pp. F248-F251

[12] Boorman, E. D. & Sallet, J. (2009). Mean-variance or prospect theory? The nature of value representations in the human brain. *The Journal of Neuroscience,* Vol. 29, pp. 7945-7947

[13] Bragulat, V., Dzemidzic, M., Bruno, C., Cox, C.A., Talavage, T., Considine, R.V. & Kareken, D.A. (2010). Food-related odor probes of brain reward circuits during hunger: A pilot fMRI study. *Obesity,* Vol. 18, pp. 1566-1571

[14] Brandstätter, E., Kühberger, A. & Schneider, F. (2002). A cognitive-emotional account of the shape of the probability weighting function. *Journal of Behavioral Decision Making,* Vol. 15, pp. 79-100

[15] Brandstätter, E., Gigerenzer, G. & Hertwig, R. (2006). The priority heuristic: Making choices without trade-offs. *Psycholocial Review,* Vol. 113, pp. 409-432

[16] Brandstätter, E., Gigerenzer, G. & Hertwig, R. (2008). Risky choice with heuristics: Reply to Birnbaum (2008), Johnson, Schulte-Mecklenbeck, and Willemsen (2008) and Rieger and Wang (2008). *Psychological Review,* Vol. 115, pp. 281-290

[17] Breiter, H.C., Aharon, I., Kahnemann, D., Dale, A. & Shizgal, P. (2001). Functional imaging of neural responses to expectancy and experience of monetary gains and losses. *Neuron,* Vol. 30, pp. 619-639

[18] Bröder, A. & Gaissmaier, W. (2007). Sequential processing of cues in memory-based multi-based multi-attribute decisions. *Psychonomic Bulletin and Review,* Vol. 45, pp. 895-900

[19] Bröder, A., Newell, B.R., & Platzer, C. (2010). Cue integration vs. Exemplar-based reasoning in multi-attribute decisions from memory: A matter of cue representation. *Judgment and Decision Making,* Vol. 5, pp. 326-338

[20] Bröder, A. & Schiffer, S. (2003). Bayesian strategy assessment in multi-attribute decision making. *Journal of Behavioral Decision Making,* Vol. 16, pp. 193-213

[21] Bröder, A. (2010). Outcome-based strategy classification, In: *Foundations for Tracing Intuition: Challenges and Methods,* A. Glöckner & C. Witteman, (eds.), pp. 61–82, Psychology Press, 978-1-84872-019-0, New York

[22] Buckholtz, J.W., Asplund, C.L., Dux, P.E., Zald, D.H., Gore, J.C., Jones, O.D. & Marois, R. (2008). The neural correlates of third-party punishment. *Neuron,* Vol. 60, pp. 930-940

[23] Camerer, C., Loewenstein, G., & Prelec, D. (2005). Neuroeconomics: How neuroscience can inform economics. *Journal of Economic Literature*, Vol. XLIII, pp. 9-64

[24] Christopoulos, G.I., Tobler, P.N., Bossaerts, P., Dolan, R.J., & Schultz, W. (2009). Neural correlates of value, risk, and risk aversion contributing to decision making under risk. *Journal of Neuroscience*, Vol. 29, pp. 12574-12583

[25] Cokely, E.T. & Kelley, C.M. (2009). Cognitive abilities and superior decision making under risk: A protocol analysis and process model evaluation. *Judgment and Decision Making*, Vol. 4, pp. 20-33

[26] Cooper , J.C. & Knutson, B. (2008). Valence and salience contribute to nucleus accumbens activation. *NeuroImage*, Vol. 39, pp. 538-547

[27] Craver, C.F. (2007). Explaining the brain: Mechanisms and the mosaic unity of neuroscience. Oxford: Oxford University Press.

[28] Crawford, V.P. & Meng, J. (2010). New York City cabdrivers' labor supply revisited: reference-dependent preferences with rational-expectations targets for hours and income. Available at SSRN: http://ssrn.com/abstract=1851878

[29] Czerlinski, J., Gigerenzer, G., & Goldstein, D.G. (1999). How good are simple heuristics? In: *Simple heuristics that make us smart*, G. Gigerenzer, P.M. Todd, and the ABC research group (eds.), pp. 97-118, New York: Oxford University Press

[30] Daston, L.J. (1988). *Classical Probability in the Enlightenment*, Princeton Univ. Press, Princeton, NJ

[31] Dhami, M.K. (2003). Psychological models of professional decision making. *Psychological Science*, Vol. 14, pp. 175-180

[32] De Martino, B., Kumaran, D., Seymour, B. & Dolan, R.J. (2006). Frames, biases, and rational decision-making in the human brain. *Science*, Vol. 313, pp. 684-687

[33] Dehaene, S. & Cohen, L. (1995). Towards an anatomical and functional model of number processing. *Mathematical Cognition*, Vol. 1, pp. 83-120

[34] Delgado. M.R., Nystrom, L.E., Fissell, C., Noll, D.C. & Fiez, J.A. (2000). Tracking the hemodynamic responses to reward and punishment in the striatum. *Journal of Neurophysiology*, Vol. 84, pp. 3072-3077

[35] DeMiguel, V., Garlappi, L., Nogales, F.J. & Uppal, R. (2009). A generalized approach to portfolio optimization: Improving performance by constraining portfolio norms. *Management Science*, Vol. 55, pp. 798-812

[36] Dhami, M. (2003). Psychological models of professional decision making. *Psychological Science*, Vol. 14, pp. 175-180

[37] Elliot, R., Friston, K.J. & Dolan, R.J. (2000). Dissociable neural responses in human reward systems. *The Journal of Neuroscience*, Vol. 20, pp. 6159-6165

[38] Elliot, R., Newman, J.L., Longe, O.A. & Deakin, J.F.W. (2003). Differential response patterns in the striatum and orbitofrontal cortex to financial reward in humans: A parametric functional magnetic resonance imaging study. *The Journal of Neuroscience,* Vol. 23, pp. 303-307

[39] Ford, J.K., Schmitt, N., Schlechtman, S.L., Hults, B.H. & Doherty, M.L. (1989). Process tracing methods: contributions, problems and neglected research questions. *Organizational Behavior and Human Decision Processes,* Vol. 43, pp. 55-117

[40] Fox, C.R. & Poldrack, R.A. (2009). Prospect theory and the brain, In: *Neuroeconomics: Decision Making and the Brain,* P.W. Glimcher, C. Camerer, E. Fehr, & R.A. Poldrack, (eds.), pp. 145-173, New York, NY: Academic Press, 978-0-12-374176-9

[41] Friedman, D. & Sunder, S. (2011). Risky curves: From unobservable utility to observable opportunity sets, *Cowles Foundation Discussion Paper No. 1819,* Cowles Foundation for Research in Economics, Yale University, New Haven CT

[42] Friedman, M. & Savage, L.J. (1948). The utility analysis of choices involving risk. *Journal of Political Economy,* Vol. 56, pp. 279-304

[43] Friedman, M. (1953). *Essays in Positive Economics,* Chicago Press, Chicago

[44] Friedmann, D. & Sunder, S. (2011). Risky curves: From unobservable utility to observable opportunity sets. *Cowles Foundation Discussion Paper No. 1819,* Cowles Foundation for Research in Economics, Yale University, New Haven CT

[45] Gazzaniga, M.S. (2000). Cognitive Neuroscience: A reader. Malden, MA: Blackwell Publishers

[46] Gigerenzer, G. (2008). Why heuristics work. *Perspectives on Psychological Science,* Vol. 3, pp. 20-29

[47] Gigerenzer, G. & Brighton, H. (2009). Homo heuristics: Why biased minds make better inferences. *Topics in Cognitive Science,* Vol. 1, pp. 107-143

[48] Gigerenzer, G. & Gaissmaier, W. (2011). Heuristic decision making. *Annual Review Psychology,* Vol. 62, pp. 451-482

[49] Gigerenzer, G. & Gray J.A.M. (2011). Launching the century of the patient, In: *Better Doctors, Better Patients, Better Decisions: Envisioning Health Care 2020,* G. Gigerenzer & J.A.M. Gray, (eds.), pp. 3-29, MIT Press, 978-0-262-01603-2, Cambridge MA

[50] Gigerenzer, G. & Selten, R. (2001). Rethinking rationality. In: *Bounded rationality: The adaptive toolbox.* , G. Gigerenzer & R. Selten, (eds.), pp. 1-12, MIT Press, Cambridge MA

[51] Gigerenzer, G., Hertwig, R. & Pachur, T. (eds.) (2011). *Heuristics: The Foundations of Adaptive Behavior,* Oxford University Press, New York

[52] Gigerenzer, G., Swijtink, Z., Porter, T., Daston, L., Beatty, J. & Krüger, L. (1989). *The Empire of Chance. How Probability Changed Science and Everyday Life,* Cambridge University Press, Cambridge

[53] Gigerenzer, G., Todd, P.M. & the ABC Research Group (1999). *Simple Heuristics that Make Us Smart,* Oxford University Press, New York

[54] Glimcher, P.W. (2009). Choice: Towards a standard back-pocket model. In: *Neuroeconomics: Decision Making and the Brain,* P.W. Glimcher, C. Camerer, E. Fehr, & R.A. Poldrack, (eds.), pp. 503-521, New York, NY: Academic Press, 978-0-12-374176-9

[55] Glimcher, P.W. & Rustichini, A. (2004). Neuroeconomics: The consilience of brain and decision. *Science,* Vol. 306, pp. 447-452

[56] Glimcher, P.W., Camerer, C., Fehr, E., & Poldrack, R.A. (2009). Introduction: A brief history of neuroeconomics, In: *Neuroeconomics: Decision Making and the Brain,* P.W. Glimcher, C. Camerer, E. Fehr, & R.A. Poldrack, (eds.), pp. 1-12, New York, NY: Academic Press, 978-0-12-374176-9

[57] Glöckner, A. (2010). Multiple measure strategy classification: Outcomes, decision times and confidence ratings, In: *Foundations for Tracing Intuition: Challenges and Methods,* A. Glöckner & C. Witteman, (eds.), pp. 83-105, Psychology Press, 978-1-84872-019-0, New York

[58] Goldstein, D.G. & Gigerenzer, G. (2002). Models of ecological rationality: The recognition heuristic. *Psychological Review,* Vol. 109, pp. 75-90

[59] Gul, F. & Pesendorfer, W. (2008). The case for mindless economics, In: *The Foundations of Positive and Normative Economics: A Handbook,* A. Caplin & A. Schotter, (eds.), pp. 3-39, Oxford, Oxford University Press

[60] Hare, T.A., O'Doherty, J., Camerer, C.F., Schultz, W. & Rangel, A. (2008). Dissociating the role of the orbitofrontal cortex and the striatum in the computation of goal values and prediction errors. *The Journal of Neuroscience,* Vol. 28, pp. 5623-5630

[61] Harrison, G.W. (2008). Neuroeconomics: A critical reconsideration. *Economics and Philosophy,* Vol. 24, pp. 303-344

[62] Haruno, M. & Frith, C.D. (2010). Activity in the amygdala elicited by unfair divisions predicts social value orientation. *Nature Neuroscience,* Vol. 13, pp. 160-161

[63] Hauser, J. (2011). A marketing science perspective on recognition-based heuristics (and the fast-and-frugal paradigm). *Judgment and Decision Making,* Vol. 6, pp. 396-408

[64] Hertwig, R. & Todd, P.M. (2003). More is not always better: The benefits of cognitive limits, In: *Thinking: Psychological Perspectives on Reasoning, Judgment and Decision making,* D. Hardman & L. Macchi, pp. 213-233, John Wiley & Sons Ltd, 0-471-49457-7, Chichester, United Kingdom

[65] Honda, H., Abe, K., Matsuka, T., & Yamagishi, K. (2011). The role of familiarity in binary choice inferences. *Memory and Cognition*, Vol. 39, pp. 851-863

[66] Hsu, M., Krajbich, I., Zhao, C. & Camerer, C.F. (2009). Neural response to reward anticipation under risk is nonlinear in probabilities. *The Journal of Neuroscience,* Vol. 29, pp. 2231-2237

[67] Huettel, S.A., Song, A.W. & McCarthy, G. (2005). Decisions under uncertainty: Probabilistic context influences activation of prefrontal and parietal cortices. *The Journal of Neuroscience,* Vol. 25, pp. 3304-3311

[68] Kable, J.W. & Glimcher, P.W. (2007). The neural correlates of subjective value during intertemporal choice. *Nature Reviews Neuroscience,* Vol. 10, pp. 1625-1633

[69] Kable, J.W. & Glimcher, P.W. (2009). The neurobiology of decision: Consensus and controversy. *Neuron*, Vol.63, pp. 733-745

[70] Kahneman, D. & Tversky, A. (1979). Prospect theory: An analysis of decision under risk. *Econometrica,* Vol. 47, pp. 163-291

[71] Kahneman, D. & Tversky, A. (eds.) (2000). *Choices, Values and Frames,* Cambridge University Press, Cambridge

[72] Kahneman, D. (2003). Maps of bounded rationality: Psychology for behavioral economics. *The American Economic Review,* Vol. 93, pp. 1449-1475

[73] Katsikopoulos, K.V. & Gigerenzer, G. (2008). One reason decision-making: Modeling violations of expected utility theory. *Journal of Risk and Uncertainity,* Vol. 37, pp. 35-56

[74] Knutson, B. & Wimmer, G.E. (2007). Splitting the difference: How does the brain code reward episodes? *Annals of the New York Academy of Sciences,* Vol. 1104, pp. 54-69

[75] Knutson, B., Fong, G.W., Bennett, S.M., Adams, C.M. & Hommer, D. (2003). A region of mesial prefrontal cortex tracks monetarily rewarding outcomes: Characterization with rapid event-related FMRI. *NeuroImage,* Vol. 18, pp. 263-272

[76] Knutson, B., Taylor, J., Kaufman, M., Peterson, R. & Glover, G. (2005). Distributed neural representation of expected value. *The Journal of Neuroscience,* Vol. 25, pp. 4806-4812

[77] Knutson, B., Westdorp, A., Kaiser, E. & Hommer, D. (2000). fMRI visualization of brain activity during a monetary incentive delay task. *NeuroImage,* Vol. 12, pp. 20-27

[78] Lebreton, M., Jorge, S., Michel, V., Thirion, B. & Pessiglione, M. (2009). An automatic valuation system in the human brain. Evidence from functional neuroimaging. *Neuron*, Vol. 64, pp. 431-439

[79] Leland, H.E. (1994). Corporate dept value, bond convenants, and optimal capital structure. *The Journal of Finance,* Vol. 49, pp. 1213-1252

[80] Lopes, L.L. (1981). Decision making in the short run. *Journal of Experimental Psychology: Human Learning and Memory,* Vol. 7, pp. 377-385

[81] Lopes, L.L. (1994). Psychology and economics. Perspectives on risk, cooperation, and the marketplace. *Annual Review of Psychology,* Vol. 45, pp. 197-227

[82] Loftus, G.R. (1996). Psychology will be a much better science when we change the way we analyze data. *Current Directions in Psychological Science,* Vol. 5, pp. 161-171

[83] Mann, L. & Ball, C. (1994). The relationship between search strategy and risky choice. *Australian Journal of Psychology,* Vol. 46, pp. 131-136

[84] Marchionni, C. & Vromen, J. (2010). Neuroeconomics: hype or hope? *Journal of Economic Methodology,* Vol. 17, pp. 103-106

[85] Marewski, J.N., Gaissmaier, W. & Gigerenzer, G. (2010). Good judgments do not require complex cognition. *Cognitive Processing,* Vol. 11, pp. 103-121

[86] Martignon, L. & Hoffrage, U. (1999). Why does one-reason decision making work? A case study in ecological rationality, In: *Simple heuristics that make us smart,* G. Gigerenzer, P.M. Todd & the ABC Research Group, pp. 119– 140, Oxford University Press, New York

[87] Martignon, L., Vitouch, O., Takezawa, M. & Forster, M. R. (2011). Naive and yet enlightened: From natural frequencies to fast and frugal decision trees, In: *Heuristics: The Foundations of Adaptive Behavior,* G. Gigerenzer, R. Hertwig & T. Pachur (eds.), pp. 136-150, Oxford University Press, Oxford

[88] Marr, D. (1982). Vision. San Francisco: W.H. Freeman

[89] Maskin, E. (2008). Can neural data improve economics? *Science,* Vol. 321, pp. 1788-1789

[90] McClure, S.M., Laibson, D.I., Loewenstein, G. & Cohen, J.D. (2004). Separate neural systems value immediate and delayed monetary rewards. *Science,* Vol. 306, pp. 503-507

[91] Mitchell, J.P., Schirmer, J., Ames, D.L. & Gilbert, D.T. (2011). Medial prefrontal cortex predicts intertemporal choice. *Journal of Cognitive Neuroscience,* Vol. 23, pp. 857-866

[92] Morris, G., Nevet, A., Arkadir, D., Vaadia, E. & Bergman, H. (2006) Midbrain dopamine neurons encode decisions for future action. *Nature Neuroscience,* Vol. 9, pp. 1057-1063

[93] Niv, Y., Daw, N.D. & Dayan, P. (2006). Choice values. *Nature Neuroscience,* Vol. 9, pp. 987-988

[94] O'Doherty, J.P., Hampton, A. & Kim, H. (2007). Model-based fMRI and its application to reward learning and decision making. *Annals of the New York Academy of Sciences,* Vol., 1104, pp. 35-53

[95] O'Doherty, J.P., Buchanan, T.W., Seymour, B. & Dolan, R.J. (2006). Predictive neural coding of reward preference involves dissociable responses in human ventral midbrain and ventral striatum. *Neuron,* Vol. 49, pp. 157-166

[96] O'Doherty, J.P., Deichmann, R., Critchley, H.D. & Dolan, R.J. (2002). Neural responses during anticipation of a primary taste reward. *Neuron,* Vol. 33, pp. 815-826

[97] Payne, J.W., Bettman, J.R., Coupey, E.,& Johnson, E.J. (1992). A constructive process view of decision making – Multiple strategies in judgment and choice. *Acta Psychologica,* Vol. 80, pp. 107-141

[98] Payne, J.W., Bettman, J.R. & Johnson, E.J. (1993). *The Adaptive Decision Maker,* Cambridge University Press, Cambridge

[99] Payne, J.W. & Braunstein, M.L. (1978). Risky choice: An examination of information acquisition behavior. *Memory and Cognition,* Vol. 6, pp. 554-561

[100] Persson, M., & Rieskamp, J. (2009). Inferences from memory: Strategy- and exemplar-based judgment models compared. *Acta Psychologica,* Vol. 130, pp. 25-37

[101] Pinel, P., Dehaene, S., Riviere, D., & LeBihan, D. (2001). Modulation of parietal activation by semantic distance in a number comparison task. *Neuroimage,* Vol. 14, pp. 1013-1026

[102] Plassmann, H., O'Doherty, J. & Rangel, A. (2007). Orbitofrontal cortex encodes willingness to pay in everyday economic transactions. *The Journal of Neuroscience,* Vol. 27, pp. 9984-9988

[103] Preuschoff, K., Bossaerts, P. & Quartz, S.R. (2006). Neural differentiation of expected reward and risk in human subcortical structures. *Neuron,* Vol. 51, pp. 381-390

[104] Preuschoff, K., Quartz, S.R. & Bossaerts, P. (2008). Human insula activation reflects risk prediction errors as well as risk. *The Journal of Neuroscience,* Vol. 28, pp. 2745-2752

[105] Rangel, A., Camerer, C. & Montague, P.R. (2008). A framework for studying the neurobiology of value-based decision making. *Nature Reviews Neuroscience,* Vol. 9, pp. 545-556

[106] Rieskamp, J. & Otto, P.E. (2006). SSL: A theory of how people learn to select strategies. *Journal of Experimental Psychology: General,* Vol. 135, pp. 207-236

[107] Rilling, J.K. & Sanfey, A.G. (2011). The neuroscience of social decision-making. *Annual Review of Psychology,* Vol. 62, pp. 23-48

[108] Roberts, S. & Pashler, H. (2000). How persuasive is a good fit? A comment on theory testing. *Psychological Review,* Vol. 107, pp. 358-367

[109] Roberts, S. & Sternberg, S. (1993).The meaning of additive reaction-time effects: Tests of three alternatives. In: Attention and Performance XIV: Synergies in experimental

psychology, artificial intelligence, and cognitive neuroscience – A Silver Jubilee, D.E. Meyer & S. Kornblum (eds.), pp. 611-653, Cambridge, MA: MIT Press

[110] Rubinstein, A. (1988). Similarity and decision-making under risk (is there a utility theory resolution tot he Allais Paradox?). *Journal of Economic Theory*, Vol. 46, pp. 145-153

[111] Rubinstein, A. (2008). Comments on neuroeconomics. *Economics and Philosophy*, Vol. 24, pp. 485-494

[112] Rushworth, M.F.S. & Behrens, T.E.J. (2008). Choice, uncertainty and value in prefrontal and cingulate cortex. *Nature Neuroscience*, Vol. 11, pp. 389-397

[113] Rushworth, M.F.S., Noonan, M.P., Boorman, E.D., Walton, M.E. & Behrens, T.E. (2011). Frontal cortex and reward-guided learning and decision-making. *Neuron*, Vol. 70, pp. 1054-1069

[114] Russo, J.E., & Dosher, B.A. (1983). Strategies for multiattribute binary choice. Journal of Experimental Psychology: *Learning, Memory, and Cognition*, Vol. 9, pp. 676-696

[115] Rustichini, A. (2009). Neuroeconomics: What have we found, and what should we search for. *Current Opinion in Neurobiology*, Vol. 19, pp. 672-677

[116] Samuelson, P.A. (1937). A note on the measurement of utility. *The Review of Economic Studies*, Vol. 4, pp. 155-161

[117] Sanfey, A.G., Rilling, J.K., Aronson, J.A., Nystrom, L.E. & Cohen, J.D. (2003). The neural basis of economic decision-making in the ultimatum game. *Science*, Vol. 300, pp. 1755-1758

[118] Savage, L.J. (1954). *Foundations of Statistics.* Wiley, New York

[119] Selten, R. (2001) What is bounded rationality? In: *Bounded Rationality: The Adaptive Toolbox*, G. Gigerenzer & R. Selten (eds.), pp. 13-36, MIT Press, Cambridge MA

[120] Serrao, A. & Coelho, L. (2004). Cumulative Prospect Theory : A study oft he farmers' decision behavior in the Alentejo dryland region of Protugal. Presentation at the American Agricultural Economics Association Annual Meeting, Denver, Colorado, August 1-4, 2004.

[121] Sescousse, G., Redouté, J. & Dreher, J.-C. (2010). The architecture of reward value coding in the human orbitofrontal cortex. *The Journal of Neuroscience*, Vol. 30, pp. 13095-13104

[122] Shiv, B., Loewenstein, G., Bechara, A., Damasion, H., & Damasio, A.R. (2005). Investment behavior and the negative side of emotion. *Psychological Science*, Vol. 16, pp. 435-439

[123] Simon, H.A. (1989). The scientist as a problem solver. In: *Complex Information Processing: The Impact of Herbert A. Simon*, D. Klar & K. Kotovsky (eds.), pp. 373-398, Erlbaum, Hillsdale NJ

[124] Simmons, J.P., Nelson, L.D. & Simonsohn, U. (2011). False-positive psychology: Undisclosed flexibility in data collection and analysis allows presenting anything as significant. *Psychological Science*, in press

[125] Smith, D.V., Hayden, B.Y., Truong, T.K., Song, A.W., Platt, M.L., & Huettel, S.A. (2010). Distinct value signals in anterior and posterior ventromedial prefrontal cortex. *Journal of Neuroscience*, Vol. 30, 2490-2495

[126] Tabibnia, G., Satpute, A.B. & Lieberman, M.D. (2008). The sunny side of fairness: Preference for fairness activates reward circuitry (and disregarding unfairness activates self-control circuitry). *Psychological Science*, Vol. 19, pp. 339-347

[127] Tobler, P.N., Fiorillo, C.D., & Schultz, W. (2005). Adaptive coding of reward value by dopamine neurons. *Science*, Vol. 307, pp. 1642-1645

[128] Tobler, P.N., O`Doherty, J.P., Dolan, R.J. & Schultz, W. (2007). Reward value coding distinct from risk attitude-related uncertainty coding in human reward systems. *Journal of Neurophysiology*, Vol. 97, pp. 1621-1632

[129] Tom, S.M., Fox, C.R., Trepel, C., & Podrack, R.A. (2007). The neural basis of loss aversion in decision-making under risk. *Science*, Vol. 315, pp. 515-518

[130] Trepel, C., Fox, C.R. & Poldrack, R.A. (2005). Prospect theory on the brain? Toward a cognitive neuroscience of decision under risk. *Cognitive Brain Research*, Vol. 23, pp. 34-50

[131] Tversky, A. (1972). Elimination by aspects: A theory of choice. *Psychological Review*, Vol. 79, pp. 281-299

[132] Van den Bos, W., van Dijk, E., Westenberg, M., Rombouts, S.A., & Crone, E.A. (2009). What motivates repayment? Neural correlates of reciprocity in the Trust Game. *Social Cognitive and Affective Neuroscience*, Vol. 4, pp. 294-304

[133] Venkatraman, V., Payne, J.W., Bettman, J.R., Luce, M.F., & Huettel, S.A. (2009). Separate neural mechanisms underlie choice and strategic preferences in risky decision making. *Neuron*, Vol. 62, pp. 593-602

[134] Vul, E., Harris, C., Winkielman, P. & Pashler, H. (2009). Puzzling high correlations in fMRI studies of emotion, personality, and social cognition. *Perspectives on Psychological Science*, Vol. 4, pp. 274-2980

[135] Volz, K.G., Schooler, L.J. & von Cramon, D.Y. (2010). It just felt right: The neural correlates of the fluency heuristic. *Consciousness and Cognition*, Vol. 19, pp. 829-837

[136] Volz K.G., Schooler, L.J., Schubotz, R.I., Raab, M., Gigerenzer, G. & von Cramon, D.Y. (2006). Why you think Milan is larger than Modena: Neural correlates of the recognition heuristic. *Journal of Cognitive Neuroscience*, Vol. 18, pp. 1924-1936

[137] Wexler, K. (1978). A review of John R. Andereson's Language, Memory, and Thought. *Cognition*, Vol. 6, pp. 327-351

[138] Witteman, C. & van Geenen, E. (2010). Cognitive process analysis, In: *Foundations for Tracing Intuition: Challenges and Methods*, A. Glöckner & C. Witteman, (eds.), pp. 45-60, Psychology Press, 978-1-84872-019-0, New York

[139] Wübben, M. & von Wangenheim, F. (2008). Instant customer base analysis: Managerial heuristics often "get it right". *Journal of Marketing*, Vol. 72, pp. 82-93

[140] Yacubian, J., Gläscher, J., Schroeder, K., Sommer,T., Braus, D.F. & Büchel, C. (2006). Dissociable systems for gain- and loss-related value predictions and errors of prediction in the human brain. The Journal of Neuroscience, Vol. 26, pp. 9530-9537

Chapter 23

Using fMRI to Study Valuation and Choice

P. Read Montague, Ann H. Harvey and Ulrich Kirk

Additional information is available at the end of the chapter

http://dx.doi.org/10.5772/47993

1. Introduction

The ability to make decisions relies on brain mechanisms designed to value our environment and elicit appropriate actions based on those values. These mechanisms allow an agent to predict the value of a potential action both immediately and into the future, and then execute the chosen action. Because the ability to find food or choose a mate directly impacts the survival of a species, it is easy to see how these goal-seeking behaviors would acquire reward value in the brain. But what is the internal value of a piece of art, a label on a bottle, an idea, or a social gesture by another person? Humans use resources (both energy and money) to acquire these types of abstract rewards, and they affect decision-making behavior in a manner similar to primary rewards such as food, water, and sex. In fact, abstract rewards can be powerful motivators: pursuit of these rewards can even cause humans to forego basic needs to acquire them. In this chapter we review experiments in the field of neuroimaging that explore how value is constructed in the human brain across a variety of domains. We then focus on a series of experiments conducted to probe the brain responses underlying preference decisions for art, and how these preferences can be altered by external variables in the environment. These experiments combine neuroscience, psychology, and economics to probe the underlying neurobiology of valuation and choice behavior in humans.

2. The study of valuation and choice in the human brain

Like all mobile creatures, a variety of human decisions requires the ability to differentially value choices and their potential outcomes. Many of these decisions relates to the acquisition of food, water, and sex, choices that allow us to survive as a species. In the learning literature, these types of appetitive stimuli are known as primary rewards. Early work on reward processing in alert monkeys demonstrated that midbrain dopaminergic neurons track the delivery of such primary rewards (Romo and Schultz, 1990). Since that time, a large body of ongoing research in this area has provided evidence that dopaminergic target structures are a key component in the brain's reward circuit, and they subsequently respond to an array of primary rewards (Schultz et al, 1997, Breiter et al, 1997, Berns et al, 2001).

However, we make many decisions in our everyday life that have little or no connection to such primary rewards. We make decisions about which brand of paper towels to buy, what to wear, how to spend our afternoon, etc. These types of decisions have components beyond simply meeting our basic needs, and there is often no clear best answer. And yet we routinely make preference judgments and choices for these types of stimuli. Furthermore, we get satisfaction from making these decisions, even in the absence of receiving any primary reward for our efforts. Neuroimaging experiments over the last several years have shed light on the neural mechanisms associated with valuation of abstract rewards. One area that has been shown repeatedly to be activated by diverse rewarding stimuli is the ventromedial prefrontal cortex (VMPFC). This brain region has been shown in several neuroimaging studies to be active for a variety of primary and abstract rewards including sports cars (Erk et al, 2002), cola preference (McClure et al, 2004), pleasant odors (Rolls et al, 2003), wine price (Plassmann et al, 2008), facial attractiveness (O'Doherty et al 2003) and money (Knutson et al, 2003). Mounting evidence in these studies and others supports the idea that this region is involved in converting the value of these diverse stimuli to a common scale for action selection (Montague and Berns, 2002).

2.1. Current research on valuation and choice

The overall goal of this research is to use fMRI to understand how value is constructed for abstract rewards in the human brain. Are value mechanisms for abstract rewards plugging into the same mechanisms dedicated to primary rewards? And if valuation is dynamic, how do context changes modulate responses in these brain areas and subsequently alter the value of an abstract reward? Such a mechanism for comparing value across a set of choices must fit into the existing data on valuation mechanisms for primary rewards, but also extend to the broader context of choices that do not involve primary reward feedback. The series of experiments herein elucidate the neural correlates of valuation for abstract rewards and begin to parameterize ways in which the signal can be modulated by external influences including familiarity and social gestures.

2.2. Using art to study valuation

In this chapter we summarize a series of experiments to study subjective decision-making by looking at how humans value visual art. Art represents an interesting category of rewards because they maintain their value even when separated from a primary reinforcer. Despite the fact that that value is a complex variable that depends largely on a person's current internal and external states, humans are incredibly efficient at assigning value to abstract rewards and making appropriate decisions. For our experiments, we assume that expressed behavioral preference for a piece of art represents its internal value. Expressed preference is a common tool used in marketing, psychological, behavioral economics, and imaging experiments to assess the value of rewards. While this is just one example of a subjective decision, it is well-

http://dx.doi.org/10.5772/47993

suited to study using fMRI for several reasons. The first reason is simply a practical one: visual art is an easy stimulus to deliver to subjects while inside an MRI scanner. Secondly, people can fairly easily evaluate their preference for art. Even subjects with very little exposure to art can find pieces that they enjoy and pieces that they dislike without having a working knowledge of the art or artist. Third, painting preferences are very unique to the individual. For a given set of paintings, the art that you like/dislike will be different from the art that I like/dislike. This makes the stimulus set more robust for analysis, since we will be unlikely to accidentally bias our analysis toward certain features of the stimuli.

In the first set of experiments, we examined neural responses to abstract rewards using images of paintings as stimuli. Human subjects viewed paintings passively in an fMRI experiment, and were asked to rate their subjective preference and familiarity for every painting after the scanning session (Figure 1). 80 paintings were chosen to represent a wide array of abstract and representational art from the Western tradition.

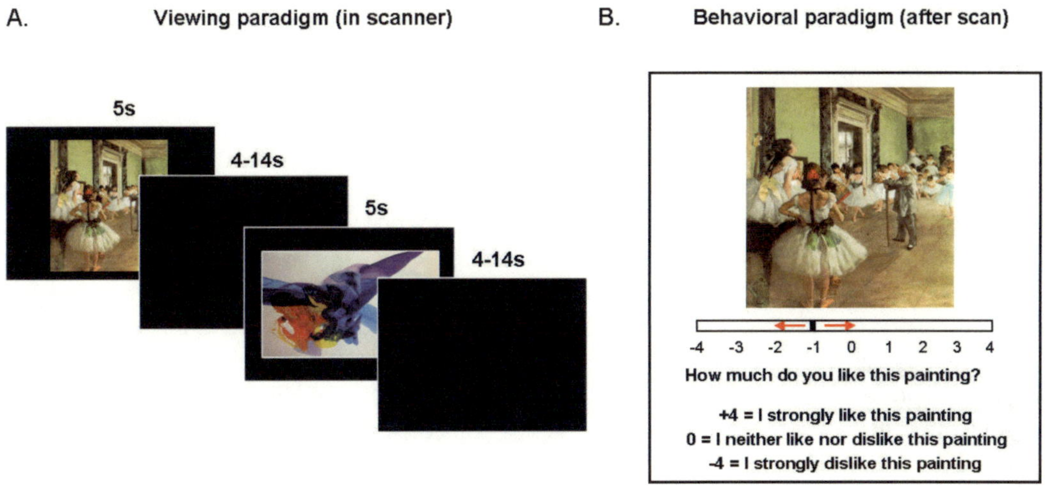

Figure 1. Experiment paradigm for (A) passive viewing of paintings during fMRI task and (B) behavioral ratings after fMRI session

We first examined the behavioral relationship between preference and familiarity for the paintings (Figure 2A). When we compared the subjective ratings for the two parameters for each painting, we found a striking, positive correlation (Figure 3.2A, $y = 1.04x + 0.086$, $R^2 = 0.976$). Preference varied linearly with familiarity, although of course it is not known if these scales themselves are linear with respect to internal states of preference or familiarity. BOLD responses in several brain regions were found to vary significantly with subjective preference and with familiarity (Figure 2B). Considerable overlap was found to exist between the two groups, including activation in the ventromedial prefrontal cortex, the bilateral caudate, retrosplenial cortex, and bilateral parahippocampal gyrus, further supporting the relationship between these parameters found behaviorally.

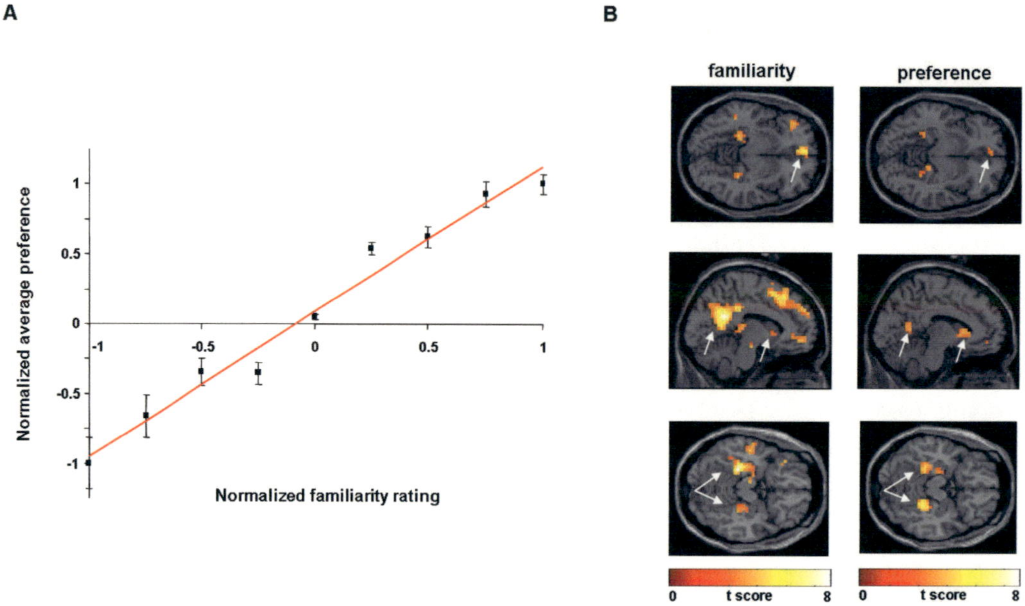

Figure 2. Behavioral correlation between average preference and familiarity (A) and BOLD responses during passive viewing of paintings that correlate with preference and familiarity (B).

The results of the art viewing experiment provide evidence that BOLD responses correlated with painting preference overlap with brain regions known to process a variety of rewarding stimuli. We also demonstrate a strong behavioral and neural correlation between preference and familiarity, suggesting that familiarity is a key factor in the valuation of abstract rewards. These results demonstrate that BOLD responses in several overlapping brain regions encode expressed preference and familiarity for artwork. The region of the vmpfc overlaps with a region previously found to encode preference for colas (McClure et al 2004). This work provides evidence that preferences for abstract stimuli such as art can generate reward responses in the brain similar to primary rewards. The second important finding is that familiarity for these stimuli is encoded in similar brain regions, and may play an important role in determining preferences for art. Notably, we located these neural signatures while subjects passively viewed the artwork in the scanner and were not asked to evaluate the paintings at the time of the viewing. The fact that neural responses during passive viewing correlate with behavioral ratings after the presentation suggest that our brains are doing continuous valuation of stimuli in our environment even in the absence of explicit decisions.

2.3. The influence of favors on valuation and decision-making

Subjective decisions may include simple aesthetic choices, but many of our most important decisions are subtle and complex value choices, as in the case of medical decisions and prescribing behavior in the medical and biomedical device industries. Every year, physicians prescribe millions of dollars worth of prescription drugs to treat their patients' illnesses. Often

http://dx.doi.org/10.5772/47993

there are comparable drug treatment strategies for a given disease, with slight variations in side effects, cost, etc. How does a physician decide between drug A and drug B? Perhaps in her experience she has found that drug A is more easily tolerated by her patients, or the cost of drug A is lower. But we know that there are other factors at work as well. Each year the pharmaceutical industry directs approximately 90% of its $21 billion marketing budget toward physicians (Brennan et al 2006). This includes sponsored events, lunches, gifts, conferences, and other physician-industry interactions. The effect of such gifts on physician decision-making behavior is unmistakable. In several reports on such issues, while physicians routinely deny that gifts could influence their prescribing behavior, and yet receiving samples, industry-paid meals, funding for travel and paid speaking engagements from a pharmaceutical company for a certain drug have all been found to result in more favorable viewpoints and increased prescribing of the particular drug (Wazana, 2000). The fact that physicians are largely unaware of the potential biasing effects of these gestures suggests that external influences are acting in ways that subvert cognitive awareness or control. While biases can be self-serving (Dana and Loewenstein, 2003), they may also be rooted in biological mechanisms that subvert cognitive control. For example, social gestures from a sender to a receiver are known to bias decisions made by the recipient, especially when the decision relates to the sender – a feature of social exchange known as reciprocity. Recent experiments have begun to shed light on the neural mechanisms of reciprocity-eliciting gestures (King-Casas et al, 2005, Rilling et al, 2002, Li et al, 2009, van den Bos et al, 2009); however, almost nothing is known about the influence of an 'open-loop' favor where there is no possibility for reciprocating interactions between the sender and the receiver.

We next investigated how external social value can manipulate value judgments for abstract rewards. In the same way that familiarity and preference were tightly linked, perhaps preference judgments depend on other contextual factors. A particularly salient real-world example is the use of monetary favors to influence preferences. We hypothesized that social favors can bias subjective preferences, even for objects unrelated to the favor itself. We have addressed this hypothesis by employing a collection of monetary favor and preference experiments. A monetary favor is sent from an agent to a subject and we tested the influence of this favor on subjects' preference for art, a domain where there is no objectively 'correct' answer and no economic relationship exists between the favor and the preference judgment.

During functional brain imaging sessions, subjects passively viewed images of artwork paired with company logos; one of the companies sponsored the subjects' participation in the experiment (Figure 3). After scanning, the subjects provided preference ratings for all the paintings.

The behavioral results showed that preference for paintings paired with the sponsoring company's logo was higher than those paired with the non-sponsor logo (Harvey et al 2010). The functional imaging results revealed that activity in the ventromedial prefrontal cortex tracked the sponsorship category of the painting (Figure 4, Harvey et al 2010). This imaging study is the first to indicate that a monetary favor transfers value to a proxy and influences judgment of objects in spatial proximity, with corresponding changes in neural activity as measured by fMRI. Furthermore, the experiment was designed with no reciprocal interaction

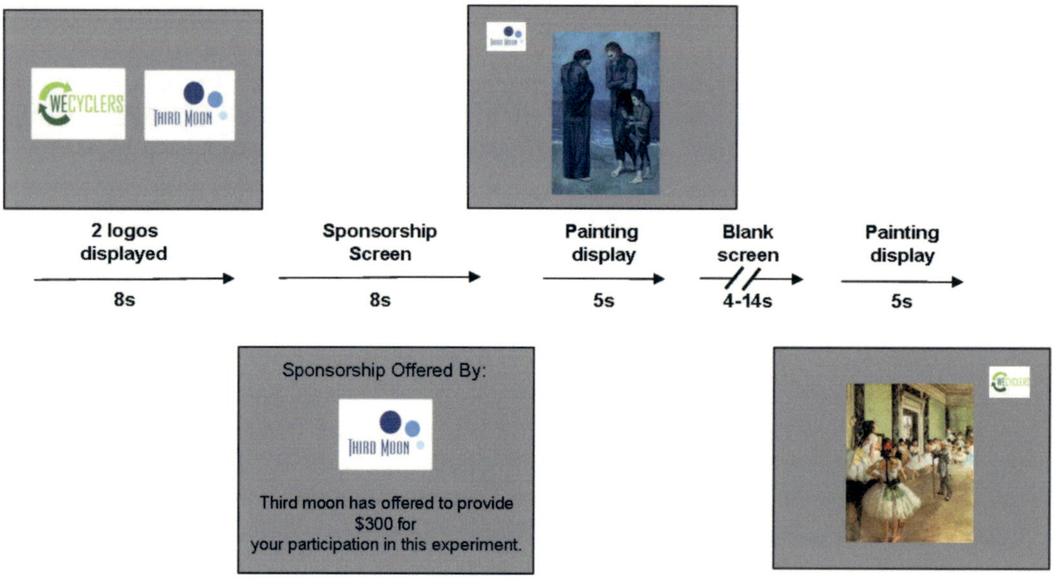

Figure 3. Experiment paradigm for the favor task (Harvey et al. 2010). Subjects are shown 2 logos, and one of the companies represented by the logo is indicated as the sponsor company. Subjects then viewed a series of paintings paired with either the sponsor company logo, or another non-sponsor logo. After the MRI session subjects rated each of the paintings based on their preference.

between the company and the receiver, suggesting that the favor is capable of influencing behavior even when the receiver has no explicit stake in the outcome of the interaction.

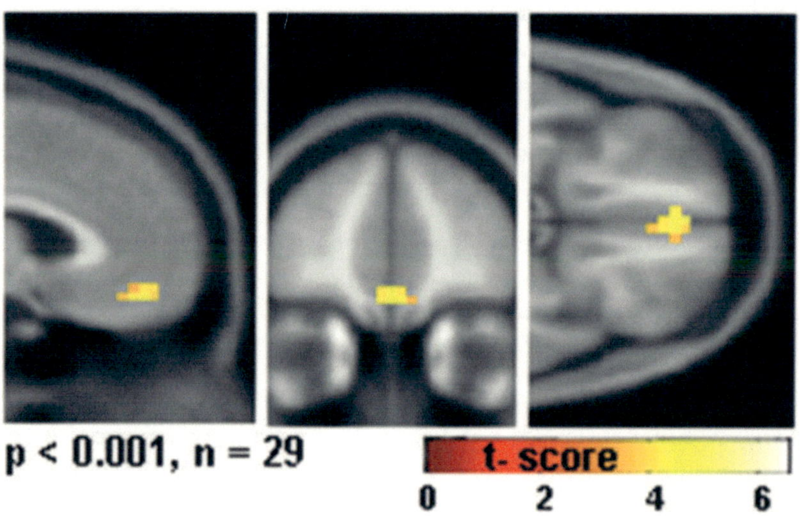

Figure 4. Region of the VMPFC that differentially responds to sponsor and non-sponsor paintings in the favor task (Harvey et al 2010). The voxels displayed show BOLD activity for linearly increasing painting preference.

http://dx.doi.org/10.5772/47993

We hypothesized that social favors can bias subjective preferences, even for objects unrelated to the favor itself. We addressed this hypothesis by employing a collection of monetary favor and revealed preference experiments. Using an art-viewing paradigm possessing no objectively correct answer for preferring one piece of art over another in conjunction with fMRI, we show that sponsorship of the experiment by a company endows the company's logo with the capacity to bias revealed preference for art displayed next to the logo. These sponsorship effects do not possess a special collection of brain responses, but instead modulate responses in neural networks normally activated by a wide range of preference judgments. The results raise the important possibility that monetary favors bias judgments by acting through existing valuation mechanisms, and individuals may therefore have difficulty detecting the gesture's influence over their subjective preferences even for objects seemingly unrelated to the favor.

2.4. Mitigation of the effects of favors using domain expertise

In the last series of experiments we examined whether the biasing effects of the social gesture on valuation judgments could be mitigated. In the case of medical professionals, biases from monetary gifts or other favors from pharmaceutical companies is thought to be mitigated through a variety of mechanisms. These include disclosing potential conflicts of interest, not accepting large gifts or favors, and oversight from the institution regarding biases in prescribing behavior. In addition, the fact that medical professionals have expertise in their domain of decision-making is taken as an argument that they should be more objective in their judgment than the average person making a similar decision. To test this directly, we recruited a group of participants with expertise in the domain of art to perform the favor task outlined in the previous section. The hypothesis was that these subjects have training in assessing art that may insulate them against the biasing gesture of the sponsoring company. And if so, there should be neurobiological correlates to the difference in behavior between art experts and non-experts.

Art experts with a minimum of 5 years of experience working in a visual art-related field were asked to undergo an fMRI scan while they viewed images of artwork, paired either with a sponsor company logo or another non-sponsor company logo. After the fMRI session, each subject rated their preference for the art that was displayed in the scanner. The setup was identical to the previously discussed experiments, except that student art was used rather than well known paintings to eliminate the possibility that differences in results between art experts and non-experts was due solely to familiarity with the paintings.

In the behavioral results, there was no difference in preference for the sponsor paintings compared to the non-sponsor paintings, in contrast to the control subjects who showed a significantly higher preference for sponsor paintings. We hypothesized that if the experts did not show an effect of the sponsorship on preference ratings for paintings, BOLD activity in the VMPFC should not be sensitive to the sponsorship in the expert group as originally seen in the control subjects whose behavior reflected the bias toward preferring sponsor paintings. This was indeed the case: while both the controls and art experts had activity in the VMPFC that correlated linearly with painting preference for all paintings, the experts did not show differential activity in this region for sponsor compared to non-sponsor paintings.

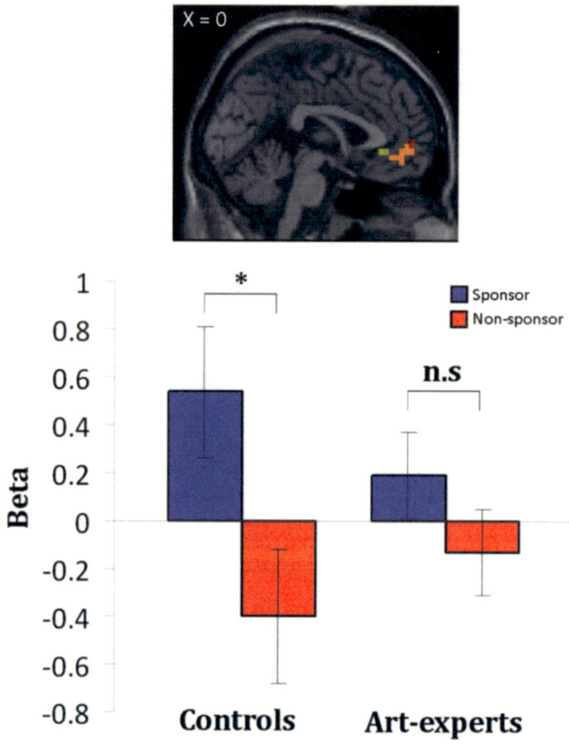

Figure 5. Activity in the VMPFC (top panel) increases linearly for painting preference in both control subjects and art experts. However, while control subjects showed differential activity in this area for sponsor compared to non-sponsor paintings (bottom panel), the VMPFC of the art experts did not discriminate between the two types of paintings (Kirk et al. 2011).

The next natural question based on the behavioral results was: if art experts do not show an effect of sponsorship on their behavior or activity in the ventromedial prefrontal cortex, which region of the brain might be involved in the mitigation of this effect? To explore this further, we examined activity in the dorsolateral prefrontal cortex (DLPFC), a region known to be involved in executive control (Wagner et al. 2011) and modulation of valuation in other experiments (Hare et al. 2009, Li et al. 2011).

We generated a contrast to look for differences in BOLD response between art experts and controls in the DLPFC for linearly increases preferences for paintings. This region was significantly more active in a linear preference regressor for art experts than controls (Figure 6). No other brain region showed differential activity for this contrast.

We also found that activity in this region of the DLPFC was functionally connected to the VMPFC, and the coupling of these two regions was stronger during presentation of sponsor paintings than during presentation of non-sponsor paintings. Thus the DLPFC may be a region through which modulation of activity in the VMPFC may occur. Furthermore, we repeated the same functional connectivity analysis in a group of controls that were selected because

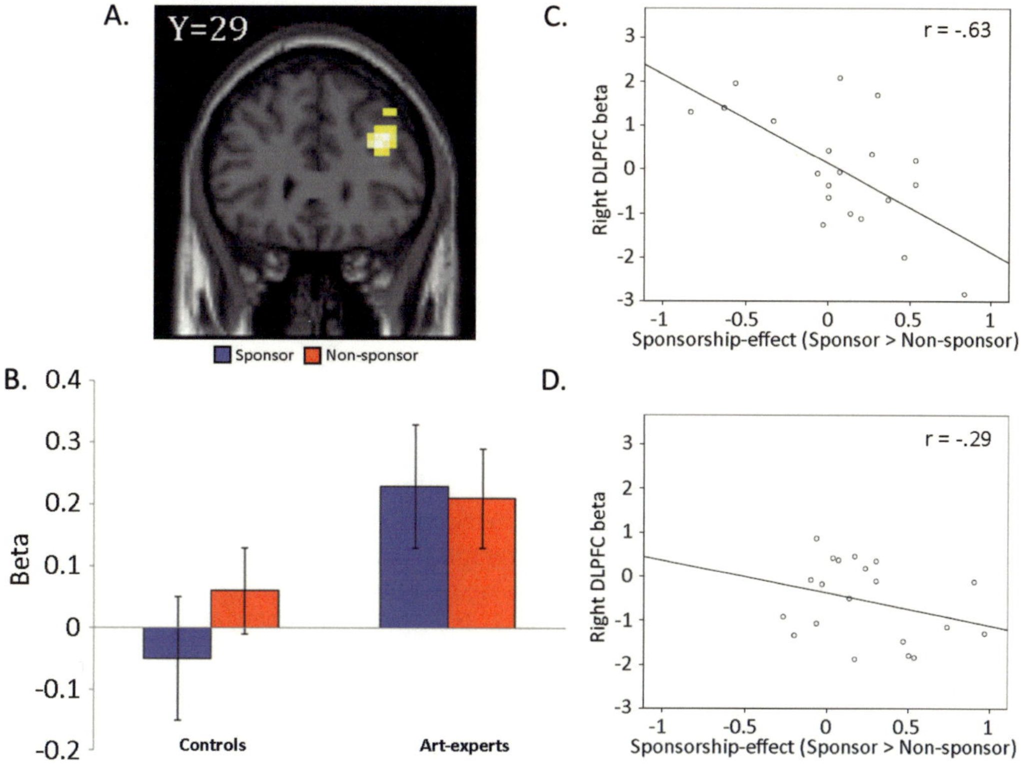

Figure 6. (A) BOLD responses in the DLPFC are greater for linear correlations with preference in the art experts compared to control subjects. (B) Beta responses in this region for sponsor and non-sponsor paintings in both controls and art-experts (from Kirk et al. 2011).

they did not show a behavioral sponsorship bias. In these participants, we replicated the functional connectivity result between the VMPFC and the DLPFC (Kirk et al 2011). The result indicates that the modulation of the VMPFC by DLPFC may be a more general mechanism by which a person could be insulated by biasing maneuvers, rather than being specific to the individuals with specific domain expertise such as the art experts used in this study.

2.5. Future research

In future experiments, we plan to further explore the neurobiology underlying human valuation and behavioral bias, and the ways in which values may be altered with external influences, training, or intervention. There are several lines of questions that lend themselves to the types of experimental manipulations described in this chapter. For example, it may be possible to teach people to train their responses either behaviorally or with real-time feedback during fMRI sessions to become less sensitive to bias. This could be extremely useful for professions where expertise and decision-making should be immune to external influence. One particular profession would be physicians and prescribing behavior: if it is possible to assess bias and make changes to their behavior, such intervention may be a much better option than current methods of disclosing conflicts of interest as a method of mitigating bias. Additionally, it is possible that there exist subtypes of people within the population that may be more or less susceptible to bias. Paradigms such as these could be used to parametrically evaluate susceptibility and the effects of intervention. And finally, if a person is biased in one domain are they necessarily biased in another domain as well? Perhaps art experts are not only insulated from bias within the domain of art viewing but across multiple domains that require evaluation without being susceptible to bias. Future research in the areas described will hopefully lead to the discovery of strategies to improve decision-making behavior and knowledge of potential bias across a range of domains.

3. Conclusion

We have used fMRI in conjunction with behavioral paradigms to investigate the neural correlates of subjective decision-making. Our metric for subjective decisions was behavioral preference for visual art. While this is just one example of a subjective decision, it is well-suited to study using fMRI for several reasons. The first reason is simply a practical one: visual art is an easy stimulus to deliver to subjects while inside an MRI scanner. Secondly, people can fairly easily evaluate their preference for art. Even subjects with very little exposure to art can find pieces that they enjoy and pieces that they dislike without having a working knowledge of the art or artist. Third, painting preferences are very unique to the individual. For a given set of paintings, the art that you like/dislike will be different from the art that I like/dislike. This makes the stimulus set more robust for analysis, since we will be unlikely to accidentally bias our analysis toward certain features of the stimuli. In the first experiment, visual art was passively presented to subjects while they underwent an fMRI scan. We showed that subjects' stated preference for the paintings, made *after* the scanning session was complete, correlated linearly with the neural responses in several brain regions. The regions with increased BOLD activity included the ventral striatum and the ventromedial prefrontal cortex, regions that are activated by a wide range of rewarding stimuli. In our experiments, however, subjects were not asked to make judgments and were unaware that they would be rating the paintings after the scanner session. This manipulation provides evidence that the brain is likely making continuous valuation judgments even without any explicit query. The results provide further support to the idea that the mechanisms that encode value for primary rewards largely overlap with those for subjective assessments of stimuli with little or no link to such rewards.

In the second set of experiments, we were interested in whether or not external social influences could alter preferences for art. Previous studies have shown that other external factors can alter perception and preferences, with corresponding changes in neural responses. People perceive art as more beautiful when they believe it to be created by an artist compared to a computer (Kirk et al, 2008), the taste of wine is more pleasant (Plassmann et al, 2008) and the perceived efficacy of analgesics is improved if the product is believed to be more expensive (Waber et al 2008). Rather than price or perceived quality, we designed an experiment to test the effect of a monetary favor on subjects' preference judgments. Favors are a particular type

http://dx.doi.org/10.5772/47993

of open-loop social gesture whereby an agent provides a gift with no explicit expectation of reciprocity. Recent neurobiological experiments have sought to explore the neural underpinnings of reciprocity (King-Casas et al, 2005, Rilling et al, 2002), but to date no fMRI experiments have addressed the influence of monetary favors on decision-making. In our experiment we used subjective preferences for artwork as a tool for measuring the influence of a monetary favor. When a company paid the subject for their participation in the experiment, the visual juxtaposition of the "sponsor" logo made individuals more likely to like those paintings as compared to paintings paired with another, non-sponsor logo. We ran the experiment at three levels of monetary compensation; while there was a trend toward preferring sponsor paintings in all conditions, only the $300 effect was significant. Despite the fact that it took the largest sum of money to influence preferences in this task, it is unlikely that small favors have no biasing effect on behavior. This task, compared to real-world examples of biasing gestures, was relatively mild. The subjects had no option to pay back the company; the gesture was a single-shot interaction rather than a repeated encounter, as is typical of most social gestures. Furthermore, there was absolutely *no incentive* for subjects to change their preference ratings based on the sponsorship; they were paid exactly the same amount no matter how they answered. The fact that such a mild gesture influences preference judgments completely unrelated to the favor suggests that favors are plugging into reciprocity mechanisms without the individual being aware of the effect. The neural results showed that, rather than a global effect of sponsorship on any brain region, the influence of the sponsor adds linearly to brain responses that more generally encode painting preference.

And finally, we examined the ways in which valuation bias from a favor can be mitigated with domain expertise. We chose art experts for the study because of their experience with the type of stimuli used in the study. We hypothesized that expertise within a domain is one way in which an individual might be insulated by biasing maneuvers such as a monetary favor or other social gesture. While a monetary favor was capable of biasing behavior for a group of control subjects in the previous experiments, the art experts in this study did not show a behavioral preference for sponsor compared to non-sponsor paintings. In the fMRI data, we showed that the VMPFC, a region of the brain known to respond during valuation experiments, was responsive to increasing painting preference for both controls and art experts. However, this region was not differentially sensitive to sponsorship in the art experts, suggesting that the biasing maneuver of the monetary favor was somehow mitigated in this group. We found a candidate region of the dorsolateral prefrontal cortex that may play a role in the mitigation of bias. This region was functionally connected to the VMPFC in the art experts, and also in a group of controls that did not show the sponsorship effect. The results suggest that domain expertise may be one way in which individuals may insulate themselves from bias, but other strategies may be involved for control subjects who show similar effects.

Acknowledgements

We would like to thank the members of the Human Neuroimaging Laboratory for assistance in conducting the studies discussed in this chapter. In particular we thank the MRI technologists who assisted in collecting the fMRI data, the software development team that assisted in coding the task stimuli and development of the Hyperscan technology and NEMO experiment environment. We thank Ron Fisher for early work and collaboration on the art experiment paradigm. We would also like to thank George Denfield for co-author contributions to the published experiments discussed in the chapter (Harvey et al 2010).

Author details

P. Read Montague, Ann H. Harvey and Ulrich Kirk

Virginia Tech Carilion Research Institute, Virginia, USA

References

[1] Berns, G.S., McClure, S.M., Pagnoni, G., Montague, P.R. (2001). Predictability modulates human brain response to reward. J Neurosci. 21, 2793-8.

[2] Breiter, H.C., Gollub, R.L., Weisskoff, R.M., Kennedy, D.N., Makris, N., Berke, J.D., Goodman, J.M., Kantor, H.L., Gastfriend, D.R., Riorden, J.P., Mathew, R.T., Rosen, B.R., Hyman, S.E. (1997). Acute effects of cocaine on human brain activity and emotion. Neuron 19, 591-611.

[3] Breiter, H.C., Aharon, I., Kahneman, D., Dale, A., Shizgal, P. (2001). Functional imaging of neural responses to expectancy and experience of monetary gains and losses. Neuron 30, 619-39.

[4] Brennan, TA, Rothman, DJ, Blank, L, Blumenthal, D, Chimonas, SC, Cohen, JJ, Goldman, J, Kassirer, JP, Kimball, H, Naughton, J, Smelser, N. (2006). Health industry practices that create conflicts of interest. JAMA 295, 429-433.

[5] van den Bos, W, van Dijk, E, Westenberg, M, Rombouts, S A R B, & Crone, E A (2009) What motivates repayment? Neural correlates of reciprocity in the Trust Game. Soc Cogn Affect Neurosci 4:294-304.

[6] Dana, J., Loewenstein, G. (2003). A social science perspective on gifts to physicians from industry. JAMA 290, 252-255.

[7] Erk, S., Spitzer, M., Wunderlich, A.P., Galley, L., Walter, H. (2002). Cultural objects modulate reward circuitry. NeuroReport 13, 2499-2503.

[8] Hare TA, Camerer CF, Rangel A (2009) Self-control in decision-making involves modulation of the VMPFC valuation system. Science 324:646-648.

[9] Harvey, A.H., Kirk, U., Denfield, G.H., Montague, P.R. (2010). Monetary favors and their influence on neural responses and revealed preference. J Neurosci 30, 9597-9602.

[10] King-Casas, B., Tomlin, D.A., Anen, C., Camerer, C.F., Quartz, S.R., Montague, P.R. (2005). Getting to know you: reputation and trust in a two-person economic exchange. Science 308, 78-83.

[11] Kirk, U, Skov, M, Hulme, O, Christensen, MS, & Zeki, S (2009) Modulation of aesthetic value by semantic context: an fMRI study. Neuroimage 44:1125-1132.

[12] Kirk, U., Harvey, A.H., Montague, P.R. (2011). Domain expertise insulates against judgment bias by monetary favors through a modulation of ventromedial prefrontal cortex. PNAS 108, 10332-10336.

[13] Li, J, Xiao, E, Houser, D, & Montague, PR (2009) Neural responses to sanction threats in two-party economic exchange. Proc Natl Acad Sci U S A 106:16835-16840.

[14] Li J, Delgado MR, Phelps EA (2011) How instructed knowledge modulates the neural systems of reward learning. Proc Natl Acad Sci U S A. 108:55-60.

[15] Knutson, B., Fong, G.W., Bennett, S.M., Adams, C.M., Hommer, D. (2003). A region of mesial prefrontal cortex tracks monetarily rewarding outcomes: characterization with rapid event-related fMRI. Neuroimage 18: 263-272.

[16] McClure, S.M., Li, J., Tomlin, D., Cypert, K.S., Montague, L.M., Montague, R.M. (2004). Neural correlates of behavioral preference for culturally familiar drinks. Neuron 44, 379-387.

[17] Montague, P.R., Berns, G.S. (2002). Neural Economics and the biological substrates of valuation. Neuron 36, 265-284.

[18] O'Doherty, J, Winston, J, Critchley, H, Perrett, D, Burt, DM, & Dolan, RJ (2003) Beauty in a smile: the role of medial orbitofrontal cortex in facial attractiveness. Neuropsychologia 41:147-155.

[19] Rilling, J.K., Gutman, D.A., Zeh, T.R., Pagnoni, G., Berns, G.S., Kilts, C.D. (2002). A neural basis for social cooperation. Neuron 35, 395-405.

[20] Rolls, E.T., Kringelbach, M.L., de Araujo, I.E. (2003). Different representations of pleasant and unpleasant odours in the human brain. Eur J Neurosci 18, 695-703.

[21] Romo, R, Schultz, W (1990). Dopamine neurons of the monkey midbrain: contingencies of responses to active touch during self-initiated arm movements. J. Neurophysiology 63(3):592-606.

[22] Schultz W, Dayan P, Montague PR. (1997). A neural substrate of prediction and reward. Science 14, 1593-9.

[23] Schultz, W, Romo, R (1990). Dopamine neurons of the monkey midbrain: contingencies of responses to stimuli eliciting immediate behavioral reactions. J. Neurophysiology 63, 607-624.

[24] Wagner AD, Maril A, Bjork RA, Schacter DL (2001) Prefrontal contributions to executive control: fMRI evidence for functional distinctions within lateral prefrontal cortex. Neuroimage 14(6):1337-1347.

[25] Wazana, A. (2000). Physicians and the pharmaceutical industry: is a gift ever just a gift? JAMA 283, 373-380.

Chapter 24

Social Pain and the Brain: How Insights from Neuroimaging Advance the Study of Social Rejection and Variants of Normal

Richard S. Pond, Jr., Stephanie B. Richman,
David S. Chester and C. Nathan DeWall

Additional information is available at the end of the chapter

http://dx.doi.org/10.5772/31141

1. Introduction

Imagine a time when a close loved one has given you the "silent treatment". When describing how this experience made you feel, terms such as "hurt", "pained", and "broken-hearted" may have come to mind. Most everyone at one time or another has experienced the pain of social rejection, whether it was in the form of unrequited love or in the form of punishment, such as when a social click ostracizes an outcast. Over the past decade, social psychologists have conducted a great deal of experimental research to uncover the detrimental effects of social rejection. However, social psychologists are just beginning to understand the neural basis of rejection. In the current chapter, we will review the neuroscientific research on social rejection, and we will discuss the implications from neuroimaging studies that advance theory and research in the area. Specifically, we will highlight how neuroimaging has been used to uncover the neural similarities between physical and social pain.

We will begin this chapter with a brief introduction to social rejection research. Specifically, we will review research that explores the cognitive, emotional, and behavioral consequences of rejection. We will illustrate the effects of rejection within the domains of: antisocial and prosocial behavior, self-regulation, self-defeating behaviors, and intelligent thought. We will end this section with a discussion of the research that suggests an overlap between physical and social pain.

Next we will introduce the neuroscientific approach to the study of social rejection. Specifically, we will review research that has used functional magnetic resonance imaging to help uncover the similarities between social and physical pain. In this section we will focus on activation in the dorsal anterior cingulate cortex and the anterior insula, as they are associated with the affective component of physical and social pain. We will then discuss how acetaminophen, an over-the-counter medication for treating physical pain, reduces activity in these neural regions among rejected people.

Last, we will discuss how neuroimaging has helped social psychologists identify those individuals who are most vulnerable to social rejection. We will review research on the personality characteristics that modulate the neural responses of rejection. Specifically, we will discuss how one's level of attachment anxiety, self-esteem, and emotion differentiation (i.e., aptitude for using discrete emotion categories to capture one's felt experience) either heighten or reduce activity in the neural regions associated with the distress of social rejection.

2. Why is there social pain?

The desire for social connection is among the most basic of human motivations. This desire is so strong that is has become known as a need, specifically the "need to belong" (Baumeister & Leary, 1995). All people in all cultures, to at least some degree, have an innate need to form and maintain interpersonal relationships. This need most likely developed over the course of evolutionary history, as social animals like humans have always depended on others for survival. In ancient times, groups provided a variety of advantages to their members (see Axelrod & Hamilton, 1981; Barash, 1977; Buss, 1990, 1991; Hogan et al., 1985; Moreland, 1987). Such advantages include providing mates, sharing resources, and helping to care for offspring. Tasks necessary for survival in ancient times, such as hunting large animals or keeping vigilance against predators, were best accomplished by group cooperation. Even today, people still remain dependent on other people for their survival. Most of us do not grow our own food, sew our own clothes, or build our own houses, just to name a few. Natural selection thus favors those who are motivated to be included, as such people are more likely to survive and reproduce.

In order to ensure that people continued associating with others, they required a system that motivates quick responses to signs of exclusion and punishes those who do not avoid it (MacDonald & Leary, 2005). Such a system would motivate people to identify social acceptance and seek out interpersonal relationships. Over time, humans have developed such a system. The "need to belong" creates in people a fundamental need for positive, enduring relationships with others as well as aversive reactions to a lack of social connection (Baumeister & Leary, 1995). The joy people experience after satisfying their need to belong in a group setting, as well as the consequences they face after their state of belongingness is thwarted, should work as a motivating factor to avoid social exclusion and seek out interpersonal relationships.

3. Responses to social pain

If the need to belong is not met or if it is thwarted, people suffer a host of deleterious physical and psychological consequences. Social exclusion thwarts the need to belong, because it is directly contrary to the desired state of social acceptance. Thus, as we present below, exclusion is extremely aversive and people's bodies and minds react accordingly. Exclusion leads people to feel social pain, similar to how being injured leads people to feel physical pain. Additional consequences of social exclusion include impairments in cognitive functioning, increased aggressive behavior, self-defeating behavior, and self-regulatory deficits. Social exclusion can lead to positive behaviors as well, but only if such behaviors have a chance of promoting social acceptance. Each of these consequences to social exclusion will be presented and discussed in the sections below.

3.1. Cognitive responses

Social exclusion reduces cognitive performance. Participants who were told they would end up alone later in life (future alone condition), compared to those who were told they would have lots of friends (future belonging condition) or those who were told they would be accident-prone (misfortune control), attempted fewer problems and answered fewer questions correctly on the General Mental Abilities Test (Baumeister et al., 2002: for a description of the GMAT see Janda, 1996 and Janda et al., 1995). Participants in the future alone condition also performed worse on difficult GRE questions about a passage they had read, compared to those in the other two conditions. Finally, future alone participants were no different in their ability to correctly recall nonsense syllables but did show relatively poor performance on analytical GRE questions, compared to participants in the two control conditions (Baumeister et al., 2002). Social exclusion leads to cognitive deficits that specifically impair logic and reasoning ability, although not simple recall. These findings speak to the view that exclusion leads to a deficit in controlled processes and executive functions, potentially because of the need to devote one's self-regulatory resources to stifling emotional distress brought about by social exclusion.

3.2. Impact on self-control

As suggested above, social exclusion may impact one's use of self-regulatory resources. Indeed, social exclusion does deplete people's self-regulatory energy. For example, participants who were told they would end up alone later in life, compared to those who were told that they would have lots of friends or those who were told that they would be accident-prone, were less able to make themselves drink a healthy, but bad-tasting beverage (Baumeister et al., 2005). Participants who were excluded by being told no one in a group wanted to work with them ate more cookies in a taste-testing exercise than those who were told everyone in a group wanted to work with them (Baumeister et al., 2005). Excluded participants also persisted less on a frustrating task and performed worse on a dichotic listening task compared to non-excluded participants (Baumeister et al., 2005). Self-regulation is critical in overcoming one's impulses. Decreased ability to eat healthy foods despite their taste, as well as overcoming the desire to eat unhealthy foods and ignoring distractions, are prime examples of self-regulatory failure. Thus, these studies indicate that participants who have just experienced social exclusion are relatively more unwilling or unable to self-regulate effectively.

Perhaps due to these self-regulatory deficits, socially excluded people engage in a variety of self-defeating behaviors. Excluded participants, compared to non-excluded participants, were more likely to choose to participate in a relatively riskier lottery (i.e. one that had an overall lower net gain and in which losing also included listening to aversive noise; Twenge et al., 2002). Excluded participants were also more likely than their non-excluded counterparts to choose a variety of unhealthy over healthy behaviors. These included choosing to eat unhealthier foods, reading entertainment magazines instead of receiving feedback about their health, and opting to receive a resting versus running pulse measure (Twenge et al., 2002). Finally, excluded participants were more likely to procrastinate rather than prepare for an upcoming test, compared to non-excluded participants (Twenge et al., 2002). Following social exclusion, people are more likely to engage in behaviors that are contrary to their own interests.

3.3. Behavioral responses: Aggression versus altruism

Not only is social exclusion associated with self-defeating behavior, it is also associated with retaliation. Social exclusion is robustly associated with aggressive behavior (see Leary & Quinlivan, 2006 for a review). Excluded individuals, compared to their non-excluded counterparts, give more damagingly negative job candidate evaluations, make strangers listen to annoying tape recordings, blast strangers with more intense and prolonged noise, and dole out large amounts of hot sauce to people who express a strong dislike for spicy food. Excluded people act more aggressively not only toward their rejecters, but also towards members of a similar social group and even towards people completely unassociated with the rejecters (DeWall et al., 2010). The link between rejection and aggression also extends beyond the laboratory. Feeling rejected is one of the most common precipitating factors associated with domestic violence, in which men murder their wives (Barnard et al., 1982; Crawford & Gartner, 1992). Feeling rejected is also associated with aggression among women towards their husbands (Downey et al., 1998). Finally, there is some evidence that suggests school shootings are associated with social rejection (Leary et al., 2003). Across a variety of different situations both inside and outside a laboratory setting, rejection leads to aggression.

It does not always lead to aggression, however. When the target is the rejecter or when there is no hope for social reconnection, rejected people will act aggressively. In this case, there is no reason for excluded people to overcome their aggressive impulses. However, when it is possible to regain social acceptance, excluded people are motivated to act in a way that will regain their acceptance (see DeWall & Richman, 2011 for a review). Excluded participants, compared to non-excluded participants, awarded more money to a partner based on their partner's average drawing, even though doing so meant they were less likely to win the money back (Maner et al., 2007). This effect only held up if excluded participants expected to meet their partner (thus having the possibility of social reconnection.) If they did not, they had no motivation to behave prosocially and assigned less money to their partner than non-excluded participants did. Participants will behave prosocially in a group setting as well, if doing so can lead to social acceptance. Socially excluded participants, particularly those who were highly sensitive to rejection, ingratiated themselves by contributing more money to a group task than

http://dx.doi.org/10.5772/31141

non-excluded participants (Romero-Canyas et al., 2010). Socially excluded people will behave more prosocially if doing so can buy their acceptance.

3.4. Emotional responses

Thus far we know how social exclusion affects people both cognitively and behaviorally. Excluded people seem to experience a deficit in controlled processes and executive function as indicated by decreased cognitive performance and self-regulation as well as increased self-defeating behavior and aggression. One hypothesis for this effect is that exclusion causes emotional distress, which requires the use of cognitive resources to reduce its impact. Given what we know about the importance of belongingness (e.g. Baumeister & Leary, 1995), social exclusion should be a distressing and aversive experience. Thus after social exclusion, we would expect participants to report greater negative emotional states than accepted participants. However, empirical research shows that this is not the case. Socially excluded people often report emotional states that do not significantly differ from participants in acceptance or control conditions (Baumeister et al., 2002; Gardner et al., 2000; Twenge et al., 2001; Twenge & Campbell, 2003; Twenge et al., 2002; Zadro et al., 2004). Why are excluded participants numb to negative emotional distress?

One reason socially excluded people report numbness to negative emotional states may be that exclusion leads to a defensive state of cognitive deconstruction (DeWall & Baumeister, 2006). The deconstructed state (Baumeister, 1990) is characterized by emotional numbness, an altered perception of time, thoughts of meaninglessness, lethargy, and avoidance of self-focused attention. Socially excluded people show all of these behaviors (Twenge et al., 2003), suggesting that exclusion may bring about the deconstructed state. This deconstructed state may also explain the deficits in controlled processes as well as the increases in aggressive behavior exhibited by people who have been socially excluded. It may offer excluded people a temporary reprieve from feeling the intense pain or distress that can accompany threats to belongingness. However, after the reprieve, the social pain returns.

4. Overlap between social and physical pain

As we have hinted thus far, social and physical pain have much in common. Just as people experiencing social pain suffer deficits in self-regulation and executive functioning, so too do people experiencing physical pain suffer these same consequences. The similarities between the two types of pain extend beyond their consequences. Social and physical pain have numerous other psychological and physiological similarities. For this reason, we propose that severe social pain impacts the body in a similar way as physical injury.

4.1. Consequences of physical pain

The consequences of experiencing physical pain are very similar to those of experiencing social pain. People who are in physical pain, similar to people in social pain, experience deficits in cognitive functioning and self-regulation as well as engage in a variety of self-defeating behaviors (for a review, see Solberg Nes et al., 2009). For example, just as people in social pain experience cognitive deficits that impair their logic and reasoning ability, so too do people in physical pain. In particular, chronic pain patients perform relatively worse on tasks that measure working and recognition memory, free recall, verbal fluency, and vocabulary (Landro et al., 1997; Park et al., 2001). Physical pain decreases cognitive functioning on a variety of tasks.

Additionally, similar to participants who are experiencing social pain, participants experiencing physical pain engage in more self-defeating behavior. For example, passive (e.g. relying on doctors, avoiding activities) instead of active (e.g. problem-solving, aiming to control pain) strategies of coping are less likely to help chronic pain patients (Callahan, 2000; Ferrando et al., 2004; Keefe et al., 1989; Snow-Turek et al., 1996; Zautra, 1999). Despite this knowledge, however, chronic pain patients are more likely to attempt to manage their pain through passive strategies (Callahan, 2000; Zautra, 1999). They are also less likely to engage in physical activity (Brown & Nicassio, 1987; Epker & Gatchel, 2000), another behavior that is encouraged and beneficial to the patients' recovery (Abel et al., 2005; Burckardt, 2002; Carlson et al., 2001; Nichols & Glenn, 1992; Smith et al., 2006). Thus, people experiencing chronic physical pain, just like people experiencing social pain, engage in behaviors that are contrary to their own interests.

4.2. Psychological similarities

Physical and social pain share several psychological and physiological similarities in addition to their shared consequences. For example, physical pain shares a linguistic similarity with social pain. People describing social pain use metaphors to physical pain such as "broken hearted," "emotionally scarred," or "crushed" (MacDonald & Leary, 2005). In the English language, not only do people describe social pain with reference to physical pain, but there is literally no way to describe it without making that reference (Leary & Springer, 2001). This linguistic link can be found across a wide variety of languages and cultures, including German, Hebrew, Mandarin, and Inuktitut, as well as at least 10 others. (MacDonald & Leary, 2005). Although the linguistic similarity between social and physical pain does not give direct proof of an overlap between their mechanisms, it does show that people think of social and physical pain in extremely similar ways. If social and physical pain share a common psychological or physiological basis, there should be evidence of a crossover between the two types of pain. Specifically, more extreme physical pain should be associated with increased sensitivity to social pain and vice versa. Indeed, the similarity goes beyond mere metaphor.

Personality traits related to a fear of social pain are also associated with pain tolerance, giving further evidence to a relationship between physical and social pain. Introverts, people who are overall less social and more afraid of rejection, have a lower pain tolerance than extraverts, people who are overall more social and less afraid of rejection (Phillips & Gatchel, 2000). This research demonstrates that people who have a lower tolerance for social pain also have a lower tolerance for physical pain. The connection between fear of social pain and pain tolerance is known to work in the opposite direction as well. Increased physical pain, such as chronic pain sufferers experience over time, is associated with behaviors indicative of a fear of social pain.

Chronic pain sufferers become more introverted, socially anxious, and avoidant of social situations the longer their chronic pain continues (Phillips & Gatchel, 2000; Sharp & Harvey, 2001). As people continue to experience physical pain, they become increasingly averse to social pain. These findings show a clear link between physical and social pain. They support the idea that both physical and social pain are managed by the same psychological and physiological systems.

4.3. Physiological similarities

Evidence for physiological overlap in physical and social pain stems from the early work of Panksepp and colleagues (Herman & Panksepp, 1978; Panksepp et al., 1978; Panksepp et al, 1978). They noticed the many similarities between the two types of pain and proposed an evolutionary explanation. They hypothesized that the link between social and physical pain exists because evolution piggybacked these neurological systems on top of each other. Animals needed to adapt to increasing social interaction and instead of creating an entirely new system for doing so, evolution used an already existing one. It piggybacked responses to social pain onto the existing systems hard-wired to respond to physical pain. The result of these shared systems is that social events will activate the body's pain response system and possibly have repercussions on how it registers physical pain. This explanation would account for the similarities we have discussed so far, but what evidence do we have for it?

Psychobiological research supports this evolutionary theory and indicates at least two such systems that respond to both physical and social pain. The periaqueductal gray (PAG) brain structures, which receive input from the body's injury detection system (nociceptive system), and anterior cingulated cortex (ACC) are involved in both the detection of physical pain and in animal bonding behavior (Craig & Dostrovsky, 1999). Activation of the PAG elicits separation distress cries from rats (Panksepp, 1998) and lesions to this area lead to reduced separation distress cries (Wiedenmayer et al., 2000). Young rats detect separation from their mother, an occurrence that is socially painful, by means of the same system that detects physical injury. Similarly, lesioning the cingulate eliminates separation distress in hamsters and squirrel monkeys (Maclean & Newman, 1988; Murphy et al., 1981). Administration of oxytocin and opiods such as morphine, used to diminish physical pain, also lead to reduced social pain in the form of reduced separation distress cries in rats (Carden et al., 1996; Carden & Hofer, 1990; Insel & Winslow, 1991). Pharmacological physical pain relief soothes distressed young rats that have been separated from their mothers. These findings indicate that the PAG and ACC are systems that respond to both physical and social pain.

Additional evidence for the overlap between physical and social pain mechanisms stems from research showing that physical pain can be alleviated by social support. Social support, which alleviates social pain, also alleviates physical pain across a variety of situations. Increased social support is associated with reduced chronic pain (Phillips & Gatchel, 2000), labor pain (Klaus et al., 1986; Niven, 1985), cardiac pain (Chalmers et al., 1995; Cogan & Spinnato, 1988), and postoperative pain (Lidderdale & Walsh, 1998). The link between physical pain and social support has also been demonstrated experimentally, which showed that participants had a higher pain threshold, as measured by the cold pressor task (i.e., participants immersed one hand into a container of ice cold water and kept it there until they could no longer tolerate it) when social support was applied before the experience (Brown et al., 2003). Social support can buffer both physical and social pain. This suggests that social and physical pain share many of the same psychological and physiological mechanisms.

4.4. Social pain causes physical numbness

Experimental evidence gives further support to the shared mechanisms between the physical and social pain systems. While social support alleviates physical pain, social pain results in physical numbness. That is, socially painful events can have an analgesic effect (i.e., decreased sensitivity to physical pain). As we mentioned earlier, humans experience emotional numbness following social exclusion in the lab (e.g. Baumeister et al., 2002). A similar phenomena is also found among nonhuman animals. Isolation, a phenomenon that is associated with social pain, produces reduced sensitivity to physical pain in rat pups (Kehoe & Blass, 1986a, 1986b; Naranjo & Fuentes, 1985; Spear et al., 1985), mice (Konecka & Sroczynska, 1990), cows (Rushen et al., 1999), and chicks (Sufka & Hughes, 1990; Sufka & Weed, 1994).

Recent research shows that exclusion also produces insensitivity to physical and emotional pain in people. Participants who were told they would end up alone later in life, compared to those who were told they would have lots of friends or those who were told they would be accident-prone, showed significantly higher pain thresholds as evidenced by resisting greater amounts of pressure from a pressure algometer (DeWall & Baumeister, 2006). Exclusion also affected emotional forecasting, simulating emotional responses to possible future events. Excluded participants, compared to those in the two control conditions, predicted relatively neutral emotional reactions to their college football game winning and losing against a rival team (DeWall & Baumeister, 2006). Excluded participants also exhibited significantly less empathy for a person experiencing social pain compared to participants in either control condition. Importantly, increased pain threshold and tolerance scores were a significant predictor of empathy scores. Thus, the increased threshold and tolerance for physical pain following exclusion was related to lack of empathic concern for others. These results were replicated for feeling empathy toward a person in physical pain as well (DeWall & Baumeister, 2006). Next we will introduce the neuroscientific approach to the study of social rejection, reviewing research that has used functional magnetic resonance imaging to help uncover the neural similarities between social and physical pain.

5. Neuroscience and social rejection

While the use of a biologically-based approach to understand social-psychological processes can be dated to Ancient Greece and Galen's four humours, neuroscience has only been substantially used to investigate this topic in the past decade or so (Lieberman, 2007). In this time, we have seen a diminished viewpoint among scholars that looking at the brain to understand the complex world of human social dynamics is an irredeemably reductionist endeavor, as well as the emergence of the field of social neuroscience (a.k.a. social cognitive neuroscience) replete with its own dedicated organizations, conferences and journals (see

http://dx.doi.org/10.5772/31141

Harmon-Jones & Winkielman, 2007). As testament to this exponential growth, the number of PsycINFO database hits that return from a search of the term *social neuroscience* have multiplied by an order of more than 16 from 2000 to 2007 (Harris, 2009). Functional magnetic resonance imaging (hereafter fMRI) has been at the forefront of the charge in this neuroscientific revolution of social psychology, providing an unobtrusive avenue through which to elucidate the human mind by measuring the hemodynamic functions of the brain.

As social neuroscience has burgeoned over the past 10-15 years, so has research on social rejection. As we have shown so far in this chapter, conventional methodologies such as overt behavior and self-report have dominated this area of inquiry and have revealed a dizzying amount of information about the powerful and dynamic nature of social rejection. However, as social phenomena are increasingly placed under the lens of investigators using neuroimaging techniques, the literature on social rejection has become infused with a wealth of brain research, though often from researchers without that goal in mind. For instance, Roy Baumeister and Mark Leary's Belongingness Hypothesis (1995), which posits that humans are consummately social beings that require social bonds in the similarly urgent way that they require food and water, was unintentionally substantiated by cognitive neuroscientists who found that the default mode network's regions (i.e., dorsomedial prefrontal cortex, medial prefrontal cortex, ventromedial prefrontal cortex, precuneus, the tempoparietal junction, fusiform gyrus, and temporal poles) are strikingly similar to those involved in social cognition (for a review see Lieberman, 2010). For instance, these areas are crucial for making sense of other people's minds (i.e., mentalizing), for empathizing with others and making moral judgments, as well as for reflecting and forming knowledge about the self (Lieberman, 2010). Thus, when people are not doing other tasks, they engage in thoughts about themselves and their social relationships. Only a remarkably social species would have evolved a brain which, given a lack of external preoccupation, 'prefers' to engage in socially-relevant mental activity. Additional fMRI studies revealed that activation in the ventral striatum, a brain region associated with the feeling of reward, was associated with positive social feedback about oneself (Izuma et al., 2008) and altruistic helping of others (Moll et al., 2006). Demonstrating such a pervasive social-preoccupation and prosocial-orientation of the human brain implicates social rejection as one of the largest threats to our well-being.

Aside from the neuroscience studies that have incidentally advanced the field of social rejection research, relatively few studies have focused explicitly on the topic. Importantly, these studies have garnered a colossal degree of interest (and controversy) from the scientific community and the general populace. For example, the first published fMRI study on social rejection (Eisenberger et al., 2003) has been cited over 400 times in less than eight years. Later in this chapter, we will go into detail about several of these experiments, but beforehand we would like to discuss the features of functional neuroimaging that have made them so impactful.

6. Advantages of functional neuroimaging in social rejection research

Given that social rejection research has flourished with the use of conventional measurements such as self-report, overt behavior, and response-time, why should any researcher opt to use neuroimaging techniques in this line of investigation (or any other for that matter)? We argue that functional neuroimaging, fMRI in particular, is a necessary toolbox to fully understand complex social processes, given a laundry list of beneficial attributes that are not available through conventional methodologies. As a most evident benefit, neuroimaging data is not confounded by the psychometric imprecision that is often commensurate with self-report and certain behavioral measures such as response bias and introspective inaccuracy. Additionally, neuroimaging techniques allow researchers to meaningfully distinguish and associate psychological processes on the virtue of whether they are associated with similar or dissimilar neural regions (Lieberman, 2007). Neuroimaging alone can determine when two psychological processes reflect *different* neural mechanisms despite being experientially *similar*, or conversely, when two psychological processes reflect *similar* neural mechanisms despite being perceived as *different*. As an example of the latter capability of fMRI, the medial prefrontal cortex activates similarly to judgments that focus on the self and other individuals who have close relationships with us (Mitchell et al., 2006). By associating or disassociating such mental states, we can further our understanding of the dynamic interactions in the mind and how the brain's structure influences the expression of our thoughts, feelings and behavior. Furthermore, by describing psychological mechanisms in terms of their neural correlates, neuroimaging replaces the often abstract and subjective definitions produced by conventional methods with more objective delineations. Instead of referring to an arbitrarily-defined cognitive sub-mechanism (e.g., the visuospatial sketchpad), we can discuss the neuroanatomical region(s) that process is associated with, its size and position in the brain, the connections that region has to other brain areas and the effects which those connections have (i.e., inhibitory, excitatory). By doing so, psychologists and neuroscientists alike ground models of the mind in concrete parameters that are readily translatable to other fields such as medical and biological sciences. Not only can neuroimaging data help with advances in medical and pharmacological treatments, but can also serve as the basis for understanding behavioral processes, as a result of the novel hypotheses generated via the fMRI technology, which elucidates physiological underpinnings. These powerful advantages of functional neuroimaging are the forces driving the productivity and quality of fMRI research on social rejection.

7. Exemplary fMRI studies on social rejection

Now we will focus on four of the aforementioned advantages of adopting neuroimaging methodologies. Each advantage will be accompanied by an fMRI study on social rejection that exemplifies it. It is our hope that in this section we will demonstrate that social rejection research is a line of inquiry that has substantially benefited from the use of fMRI techniques and that these material gains argue for the necessity of functional neuroimaging to understand complex social phenomena. Summary information for each study can be found below in Figure 1 and in Table 1.

http://dx.doi.org/10.5772/31141

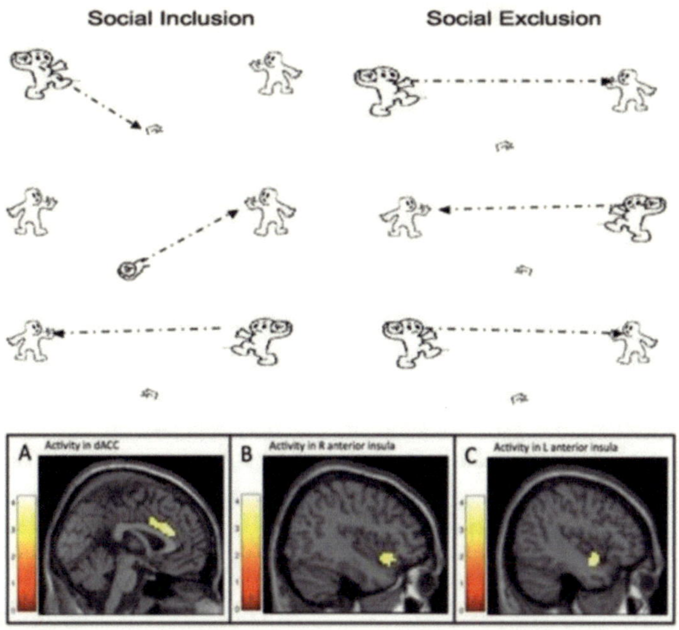

Figure 1. The top represents example stimuli from the Cyberball game used in fMRI research on social rejection. Participants typically play the inclusion round during the first scan and the exclusion round during the second scan. The bottom shows areas in the brain that are associated with the affective component of pain (i.e., dACC and anterior insula) which also become activated during exclusion (vs. inclusion) in the Cyberball paradigm.

Study:	Paradigm:	Analysis of Interest:	Regions of Interest:
Eisenberger et al., 2003	Cyberball paradigm: 1st scan = included 50% of time 2nd scan = excluded after 7 throws	Whole-brain analysis: Exclusion vs. Inclusion	dACC, RVPFC
Eisenberger et al., 2007	Cyberball paradigm: 1st scan = included 50% of time 2nd scan = excluded after 7 throws	Region of interest regressions with: Daily social support, Cortisol reactivity, Distress	dACC, DSFG
DeWall et al., 2010	Cyberball paradigm: 1st scan = included 33% of time 2nd scan = excluded after 3 throws	Region of interest regression with: Tylenol condition	dACC, anterior insula, amygdala
Masten et al., 2011	Cyberball paradigm: 1st scan = included 33% of time 2nd scan = excluded after 10 throws	Region of interest regression with: Observer ratings of distress, Attributions of discrimination, Self-reported distress	dACC, anterior insula

Table 1. Summary of the studies presented in Section 7. Neural activity of interest is during exclusion vs. inclusion.

7.1. Associating seemingly-distinct psychological processes based on shared neural substrates: Eisenberger et al., 2003

Naomi Eisenberger and her colleagues (2003) were not only the first researchers to report fMRI findings on social rejection, but their results confirmed a striking hypothesis that social injury elicits a nearly identical neural response as harm to one's *soma*. To achieve these results, they adapted a computerized, ball-tossing task (Cyberball: Williams et al., 2000) to the fMRI environment that immersed participants in experiences of social acceptance and rejection. Participants were told that they would play a virtual ball-tossing game inside the scanner, via the Internet, with two other participants who were also in scanners. In reality, computer programs represented the other players. During the first round of the game, participants were included (i.e., received a ball toss from one of the virtual players) throughout the round. However, during the second round, the virtual players stopped throwing the participant the ball after he or she had received three throws. Participants were excluded for the remainder of the game and watched as the two virtual players continued without them. After the scanning procedures, participants reported the degree of social distress (a facet of social pain) they experienced due to the rejection manipulation. Traditionally, the neurological mechanisms of pain have been separated into three components: 1. the sensory component, 2. the affective-motivational component, and 3. the evaluative component (Melzack & Casey, 1968). In their study, Eisenberger and colleagues (2003) discovered neural activation that was specific to social rejection in three brain regions previously associated with the affective component of physical pain (e.g., Bush et al., 2000; Carter et al., 2000; Foltz & White, 1968; Lieberman et al., 2007; Price, 2000; Rainville et al., 1997). One of these regions was the dorsal portion of the anterior cingulate cortex (hereafter dACC), a key region for the affective component of pain (see Apkarian et al., 2005 for a review), which is often conceptualized as an 'alarm system' that monitors the external environment for elements that deviate from the ideal (Bush et al., 2000; Carter et al., 2000). It then responds by eliciting feelings of distress that motivate the individual to repair or assuage the discrepancy (see Eisenberger & Lieberman, 2004). The second region was the anterior portion of the insula, a cortical structure previously associated with both negative affect (e.g., Lane et al., 1997) and visceral pain (e.g., Aziz et al., 2000). Last, this study reported activation in two voxel-clusters of the right-ventral prefrontal cortex (hereafter RVPFC), an area of the neocortex which functions to regulate aversive experiences (e.g., Petrovic & Ingvar, 2002). As predicted, activation in the dACC predicted *greater* levels of social distress, while conversely, the RVPFC predicted *lesser* amounts of social distress. DACC activation also mediated the inhibitory relationship between the RVPFC and social distress, suggesting that the RVPFC reduces subjective distress to social threat by inhibiting the dACC response to it. Taken together, these results are astoundingly analogous to those from physical pain research (e.g., Foltze & White, 1968; Kong et al., 2006; Petrovic et al., 2005; Rainville et al., 1997), establishing the functional similarity between physical and social pain, which has great implications for the understanding and treatment of various psychopathologies, such as post-traumatic stress disorder (PTSD; Felmingham et al., 2007) and anxiety (Simmons et al., 2008). Perhaps most importantly, these findings suggest that maintaining social connections should be conceptualized as a fundamental human need, to the same degree as physical safety, since evolution has given the two equal standing as evidenced by the same neural underpinnings.

The Eisenberger and colleagues (2003) study is exemplary in another related way, in that it largely resolved a decades-long line of research that was previously relegated to harmful experiments on nonhuman animals. Various researchers have suspected the possible social/physical pain overlap that we have discussed here in detail, but since functional neuroimaging techniques were not yet available, their research has taken such forms as measuring the distress vocalizations of socially-isolated bird chicks (Panksepp et al., 1978) and ablating the cingulate cortex of squirrel monkeys (MacLean & Newman, 1988). Thanks to fMRI, it is no longer necessary for a great deal of research questions to be performed on nonhuman animals, who contribute results that are not always readily translatable to humankind and who are often harmed in the process.

7.2. Translating psychological processes to their biological mechanisms: Eisenberger et al., 2007

While research on social rejection and isolation had established that a prolonged lack of interpersonal connections is extremely deleterious to physical and mental health (e.g., House et al., 1988; Berkman & Syme, 2007), the biological mechanisms through which this process occurred was largely unknown. Capitalizing on the unique ability of neuroimaging to lay bare the brain's response to social rejection in a format which could then be statistically assessed in relation to other biological processes, Naomi Eisenberger and her colleagues (2007) pitted two popular hypotheses of how social connections improve health outcomes against one another. In these hypotheses, which are not necessarily mutually exclusive, social bonds:

a. lessen the extent to which events are perceived as threatening via reduced activation in brain regions associated with HPA activation (namely the dACC, amygdala and insula).

and/or

b. increase the coping resources an individual can apply to a threat via greater activation of 'top-down' brain regions associated with self-regulation (namely the ventrolateral- and medial-prefrontal cortices).

Both of these postulated mechanisms would down-regulate the hypothalamic-pituitary-adrenal axis (hereafter HPA axis) response to threats, the activation of which releases the stress hormone, cortisol, into the bloodstream (i.e., cortisol reactivity). Since cortisol suppresses the immune system, both of the aforementioned explanations would ultimately protect individuals by preventing the reduced immune-function that is commensurate with the physiological stress response. To test these competing theories, participants had endogenous cortisol measurements taken before and after a stress-inducing task, with increases in cortisol over the course of the stressful task indicating a larger degree of cortisol reactivity. Self-reported social distress was recorded from each participant immediately after the stress task as well. Over the following 10 days, participants reported the degree of social support they felt at various times throughout that period. At the end of the 10 days, participants had fMRI scans taken of their brain while they performed a variant of the ball-tossing task described in the Eisenberger et al., 2003 study which manipulated experiences of social acceptance and rejection. The authors of the study planned to assess the degree to which neural responses to social rejection were associated with social support, cortisol reactivity during the stress task, and self-reported social distress. The results yielded no support for the hypothesis that social support increases activation in brain regions that would facilitate coping with stressful events, since no neural regions were activated above statistical threshold in the acceptance condition. However, regression analyses on the rejection condition revealed there was substantial evidence for the hypothesis that social support improves physical health by reducing activation in brain regions associated with social distress/pain. Specifically, activation during social rejection in both an *a priori* region-of-interest analysis of the dACC and Brodmann's Area 8 of the dorsal superior frontal gyrus (hereafter DSFG) were both *negatively* correlated with social support and *positively* correlated with both social distress and cortisol reactivity. Rejection-specific activation in both the dACC and DSFG mediated the inhibitory relationship between social support and cortisol reactivity, which implicates these two regions as the mechanism through which social support reduces the physiological stress response. Moreover, rejection-specific activation in the hypothalamus mediated the relationships between the dACC and DSFG and cortisol reactivity. This suggests that the dACC and DSFG influence stress responses by modulating the activity of the HPA axis. Summarizing these findings, this study supports the claim that when people encounter a social threat (e.g., public speaking, social rejection), their everyday levels of social support reduce activation in both the dACC and DSFG. Reduced activation in the dACC and DSFG subsequently reduces HPA axis activation, cortisol release and suppression of the immune system. Aside from the clear implications these findings have for treatments and therapies, this study is a prime example of the process through which functional neuroimaging allows for a more objective delineation of a behavioral response into its physiological components.

7.3. Generating novel hypotheses based on physiological principles: DeWall et al., 2010

Recent work from our lab stands as a clear example of the ability of fMRI to produce novel psychological hypotheses based purely on physiological knowledge of neural correlates. Since previous research had established the overlap between brain regions involved in social and physical pain (e.g., Eisenberger et al., 2003), DeWall and colleagues (2010) tested whether a popular physical-pain-reliever, acetaminophen, would have similar analgesic effects for social pain. In the first study, twice-daily ingestion of acetaminophen (compared to placebo) reduced self-reported social pain over a time period of three weeks. To assess whether these self-reports translated to diminished pain responses in the brain, participants who had either taken acetaminophen or placebo twice-daily for three weeks (same as study 1) were placed in an fMRI scanner and then socially-accepted and -rejected by a similar version of the computerized, ball-tossing task used in the two previous studies.

Participants who had taken acetaminophen showed reduced activation (as compared to those who took placebo) during rejection (as compared to acceptance) in both brain regions previously associated with social pain, the dACC and anterior insula. Participants who took acetaminophen also showed lesser activation of the amygdala during social-rejection, a brain region involved in producing 'fight-or-flight' responses. By understanding the physiological similarities between social and physical pain, a novel hypothesis was vetted about a psychological process with powerful implications for our understanding and treatment of social rejection.

7.4. Covert measurement of social processes that avoid self-report biases: Masten et al., 2011

Much of the phenomena that interest social psychologists are sensitive to issues of self-presentation and demand characteristics. Thus, psychologists favor measurement tools that avoid or are at least robust to these self-report biases. For example, psychologists interested in studying prejudice may opt for measuring implicit attitudes of their participants, because people may not respond truthfully on self-report questionnaires if holding racist attitudes is looked down upon socially. The fMRI scanner represents a powerful tool that can be used to assess such attitudes in a way that is not subject to self-report biases. A recent rejection study conducted by Masten and colleagues (2011) is an excellent example for how neuroimaging can be utilized to investigate potentially sensitive topics. Specifically, they investigated the neural correlates of negative social treatment associated with racial discrimination.

In their study, African American participants encountered two white (1 male, 1 female) confederates during the informed consent procedure. The participants were told that they would be playing a virtual ball-tossing game (i.e., Cyberball) with the two confederates, while they were each inside fMRI scanners. In reality, participants played the virtual ball-tossing game with a preprogrammed computer. During the initial scan, participants were included by their virtual partners, who threw the ball to the participant one third of the time. However, during the second scan, participants were excluded from their partners after receiving ten throws. After the scanning session, participants completed self-report measures of distress and discriminatory attributions. Observers also rated the participants' level of distress during videotaped interviews, in which the participants discussed their feelings about being excluded during the game.

Using a whole-brain analyses approach, Masten and colleagues (2011) found that exclusion (vs. inclusion) increased activity in the anterior insula, and rACC, and decreased activity in the VLPFC, which was consistent with previous research (Eisenberger et al., 2003, 2007). Yet, in region of interest analyses, self-reported distress was not related to activation in these areas during exclusion, as previous research would suggest. However, observer-rated distress was related to increased activity in the dACC and anterior insula and decreased activity in the VLPFC and ALPFC, which is consistent with previous research. These findings highlight the robustness of fMRI procedures to self-report biases. Last, the authors found that the more participants attributed the exclusion experience to racism, the less activation they experienced in the dACC during exclusion. This last finding suggests that attributing negative social events to discrimination provides a protective function, which allows the individual to better cope with the distress.

8. Individual difference factors that moderate neural responses to social rejection

In the previous sections of this chapter, we described how most everyone desires at least some level of social acceptance and that experiences of rejection can have very powerful physical and psychological consequences. We described how the use of fMRI methodologies have helped social psychologists identify the shared neural substrates of physical and social pain, and how this tool has helped psychologists generate and test novel hypotheses about the physiological processes that underlie social pain. In this final section, we will discuss recent research from our lab, which utilized fMRI procedures to identify those individuals who are most vulnerable to social rejection. Specifically, we will concentrate on two individual difference factors that modulate neural responses to rejection, namely one's level of attachment anxiety and one's capacity for identifying and describing emotional experiences. People who are high in attachment anxiety possess an intense desire for intimacy and are highly sensitive to the potential for rejection (Fraley & Shaver, 2000). Thus, they may exhibit heightened neural responses to rejection, compared to individuals who are less anxiously attached. Similarly, people who have a low capacity for identifying and describing their emotional experiences may show heightened neural responses to rejection, because they respond more negatively to stressful situations than people who have a greater capacity for describing their emotions (Barrett et al, 2004; Kashdan et al., 2010; Tugade et al., 2004). Table 2 below presents a brief summary of the two studies that will be described in more detail in sections 8.1 and 8.2

Study:	Paradigm:	Analysis of Interest:	Regions of Interest:
DeWall et al., 2011a	Cyberball paradigm: 1st scan = included 33% of time 2nd scan = excluded after 3 throws	Region of interest regressions with: Anxious attachment, Avoidant attachment	dACC, anterior insula
DeWall et al., 2011b	Cyberball paradigm: 1st scan = included 33% of time 2nd scan = excluded after 3 throws	Region of interest regressions with: Emotional differentiation X Self-esteem	dACC, anterior insula

Table 2. Summary of studies presented in Section 8. Neural activity of interest is during exclusion vs. inclusion trials.

8.1. Neural responses to rejection depend on attachment style

Prior work has shown that belongingness threats can cause a variety of negative emotional, cognitive, and behavioral outcomes (e.g., Baumeister et al., 2002; DeWall et al., 2009; Twenge et al., 2001, 2003), and that these threats activate some of the same neural substrates as those underlying physical pain (Eisenberger et al., 2003). However, people vary a great deal in how they experience and maintain social relationships. For example, some people may have a high need for social acceptance and are especially attuned to signs of rejection, whereas others may be uncomfortable with close relationships. Such differences in attachment style may have direct implications for neural responses to experiences of rejection.

DeWall and colleagues (2011) tested whether individual differences in anxious and avoidant attachment styles moderated neural responses to social rejection. They predicted that people with high attachment anxiety (depicted as having a great desire for closeness and vigilant to rejection cues; see Fraley & Shaver, 2000) would show heightened neural activity in those regions associated with the processing of social rejection (i.e., dACC and anterior insula). On the other hand, they predicted that those individuals who demonstrated a high level of avoidant attachment (depicted as being uncomfortable with closeness and uses regulatory strategies to minimize attention to attachment-related events and information) would show dampened activity within those same neural regions. In their study, participants completed the Attachment Style Questionnaire (ASQ; Feeney et al., 1994) in the lab, and then returned three weeks later to complete a virtual ball-tossing game in an fMRI scanner. Following Eisenberger and colleagues' (2003, 2007) and DeWall and colleagues' (2010) procedures (i.e., the Cyberball paradigm), participants believed that they were playing the virtual ball-tossing game with two other participants already in fMRI scanners. In reality, computer programs represented the other players. During the first scan, the virtual players regularly tossed the ball to the participants. However, during the second scan, the virtual players excluded those participants who received three throws. After playing the ball-tossing game, participants completed a measure of social distress.

As predicted by DeWall and colleagues (2011), participants with a high level of attachment anxiety exhibited more activity in the dACC and anterior insula while experiencing social exclusion (vs. social inclusion). These results suggest that anxious attachment is associated with greater negative responses to belongingness threats. On the other hand, a high level of avoidant attachment was related to less activation within the dACC and anterior insula during social exclusion (vs. social inclusion). These results suggest that people who reflect the avoidant attachment style detach from attachment-relevant situations, thereby providing a buffer to the negative effects of potential rejection. These findings support and extend prior work on the pain of social rejection. Moreover, they demonstrate how neuroimaging methodologies can be used to allow social psychologists a better understanding of how personality characteristics moderate physiological responses to social situations.

8.2. Self-esteem and level of emotional differentiation interact to predict neural responses to rejection

Attachment style represents just one factor that can increase or decrease one's vulnerability to social threats. However, there are surely many other personality characteristics that can intensify or buffer the negative effects of rejection. Self-esteem is one potential factor that has garnered recent attention from social psychologists. People who generally perceive that others reject them tend to have low self-esteem. In contrast, people who generally perceive acceptance from others tend to have high self-esteem (Leary et al., 1995). Using a similar virtual ball-tossing paradigm as the studies above, Onoda and colleagues (2010) have provided some of the first evidence from neuroimaging that people with low self-esteem experience more distress during rejection than people high in self-esteem. They showed that people with low self-esteem exhibit greater activation within the dACC during a simulated rejection experience (vs. inclusion). Similarly, using a social evaluative task, Somerville and colleagues (2010) demonstrated that people with low self-esteem exhibited greater activation in a more ventral area of the anterior cingulate cortex after receiving negative social feedback (versus positive social feedback).

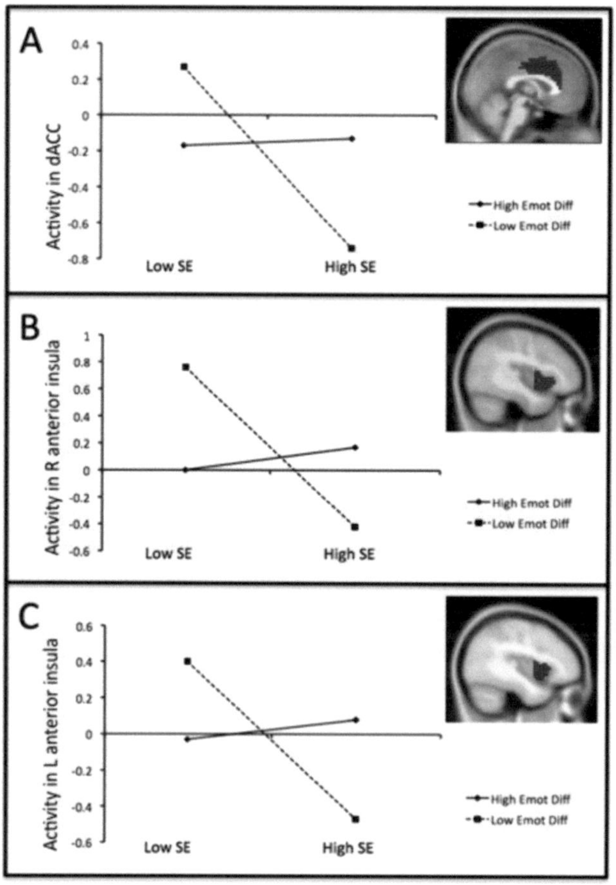

Figure 2. Neural activity during exclusion (vs. inclusion) within the dACC (Panel A) and anterior insula (Panels B & C) as a function of self-esteem and emotion differentiation (based on a priori anatomically-defined ROI analyses; significance was defined as $p < .05$). The functionally-defined ROI's are depicted in blue.

Following the work of Onoda and colleagues (2010) and Somerville and colleagues (2010), DeWall and coauthors (2011) explored whether one's capacity for identifying and describing emotional experiences (i.e., emotional differentiation) moderated the effects of self-esteem on neural responses to social rejection. People who are better at differentiating their emotional experiences respond more positively to stress and depend less on maladaptive coping strategies, such as abusing alcohol (Barrett et al, 2004; Kashdan et al., 2010; Tugade et al., 2004). On the other hand, people who are less able to identify and differentiate their emotional experiences react more negatively to stress, possess more negative attitudes, and are more likely to use those maladaptive coping strategies. Thus, DeWall and colleagues (2011) predicted that greater emotional differentiation should be linked to psychological resilience to social rejection experiences. However, among individuals who exhibit a lower capacity for emotional differentiation, those with low self-esteem will be most vulnerable to experiences of rejection, which will be evidenced in heightened responses in the dACC and anterior insula.

In DeWall and colleagues' (2011) study, participants completed daily diary measures of self-esteem and negative emotion for a three-week assessment period. After the diary portion of the study, participants completed the virtual ball-tossing game (i.e., the Cyberball paradigm) with two ostensible partners while in an fMRI scanner (first scan = inclusion round, second scan = exclusion round). Last, participants completed a measure of social distress. All neuroimaging data were preprocessed and analyzed using Statistical Parametric Mapping (SPM8; Wellcome Department of Cognitive Neurology, Institute of Neurology, London, UK). As with the study described in section 8.1, anatomically-defined region of interest (ROI) analyses were implemented based on a priori hypotheses regarding the involvement of the dACC and anterior insula in processing social rejection. Differential activity in each ROI during exclusion vs inclusion was examined, as well as how this activity related to individuals' self-esteem, emotion differentiation, and the interaction between self-esteem and emotion differentiation (significance was defined as $p < .05$). Parameter estimates of activity during exclusion vs inclusion were entered as dependent variables in hierarchical multiple regression analyses, with self-esteem and emotion differentiation scores entered as predictors in the first step and their interaction entered as a predictor in the second step. Finally, supplemental whole-brain analyses were implemented, thresholded at $p < .005$ and 20 voxels for a priori defined regions of interest, while all other regions were examined at a threshold corrected for false discovery rate.

As predicted, regression analyses showed that among low emotional differentiators, lower self-esteem was strongly associated with greater activation in the dACC and the anterior insula during exclusion (vs. inclusion). However, high emotional differentiation bore no relation to activity in these neural regions during exclusion (vs. inclusion), regardless of selfesteem level (see Figure 2). These findings extend prior neuroimaging research linking low self-esteem to increased vulnerability to social rejection, by demonstrating that one's level of emotional differentiation may reinforce or buffer this relationship. Additionally, this research is an example of how neuroimaging can be utilized to explore the physiology of psychological resilience and expand upon prior conclusions drawn from behavioral and/or self-report methodologies.

9. Criticisms of the Cyberball paradigm

For the most part, the fMRI literature on social rejection is nested within the Cyberball paradigm (Williams et al., 2000), in which participants believe that they are playing a ball-tossing computer game over the Internet with two other participants who are, ostensibly, also in MRI scanners. However, the nature of the rejection experience during Cyberball is not without its limitations. For example, in order for the rejection manipulation to maintain believability and ecological validity, all participants are included by their virtual partners during the first fMRI scan and excluded by those same partners during the second scan. The blocks must be presented in this order to ensure that all participants have the same expectations from one scan to the next. That is, if the blocks were presented randomly, then those participants who encountered exclusion during the first scan would naturally expect exclusion on the second scan. Inclusion during the second scan might appear unrealistic.

Although the presentation of inclusion blocks and exclusion blocks are not usually randomized in the Cyberball paradigm, follow-up fMRI research has replicated the patterns of brain activation commonly reported with Cyberball, using other manipulations of rejection. For example, participants show increased activity within the dorsal ACC while viewing rejection-themed images as opposed to acceptance-themed images or abstract paintings of positive and negative valence (Kross et al., 2007). Similarly, people show increased dACC activity while viewing pictures with disapproving facial expressions (Burklund et al., 2007). Moreover, Sebastian and colleagues (2010) found that activation within the amygdala and subgenual ACC increased when participants viewed social threat words (e.g., pathetic), as opposed to neutral words, during a rejection-themed emotional Stroop task. They also found an Age x Valence interaction, such that adults who viewed rejection-themed words, in contrast to acceptance or neutral control words, exhibited decreased activity within the right VLPFC. However, this activity was not present in adolescents who viewed the rejection-themed words. This finding suggests that maturation of the right VLPFC, an area associated with affect regulation, continues during adolescence. Last, a recent study conducted by Bolling and colleagues (2011) actually modified the Cyberball paradigm, such that participants played several blocks of the game, which alternated between inclusion and exclusion. They found that participants' self-reported distress after Cyberball was comparable to previous research and did not decline or become less upsetting over time. Additionally, the authors replicated previous research by showing, in whole-brain and region of interest analyses, that brain activity within the ventral anterior cingulate cortex (vACC), the posterior cingulate cortex (PCC), and the right anterior insula increased during exclusion (vs. inclusion). Each of these studies give support for the validity of the Cyberball paradigm for fMRI research.

Another criticism of the Cyberball paradigm for fMRI research is that the manipulation may not be strong enough to mimic the sensation of physical pain, nor is it especially prolonged. Thus, experiences of intense social pain that are drawn out over time may prove to be more beneficial in examining the neural similarities between physical and social pain. For example, a recent study by Kross and colleagues (2011) shows that the overlap between social and physical pain may be more extensive than previously understood. They argued that neural activation during social pain should not just mirror the affective component of physical pain (e.g., activation within the dACC and anterior insula), but if the socially painful experience is extreme enough then it should also mirror the sensory component of physical pain (e.g., activation within the secondary somato-sensory cortex and the dorsal posterior insula). In their study, they recruited participants who have recently experienced an unwanted romantic break-up. Those participants then performed two counterbalanced tasks during an fMRI scanning session: 1. a Social Rejection task, and 2. a Physical Pain task. During the rejection-trials of the social rejection task, participants viewed a headshot of their ex-partner and thought about their specific break-up. During friend-trials of the social rejection task, participants viewed a headshot of a close friend of the same sex as the ex-partner and thought of a positive experience with that friend. During the physical pain task, participants experienced painfully hot thermal stimulation to an area on their left arm during hot trials and non-noxious thermal stimulation to the same area during warm trials. In a whole-brain conjunction analyses and region of interest analyses, the authors found that the brain activation during social rejection overlapped with the brain activation during physical pain in the areas associated with the affective component of pain (i.e., dACC and anterior insula), replicating previous research (e.g., Eisenberger et al., 2003). Moreover, the brain activation during social rejection also overlapped with the brain activation during physical pain in areas associated with the sensory component of pain (i.e., the thalamus and the secondary somatosensory cortex). Thus, one of the implications of this research is that to fully understand the overlap between physical and social pain, researchers must employ stronger manipulations of rejection than what the Cyberball paradigm might provide.

10. Future directions

There are many possible directions for future neuroscientific research concerning social pain. One avenue that will provide fruitful exploration concerns how the brain regulates the immune system in response to physical and social threats. The current literature supports the idea that the neural mechanisms for social and physical pain overlap, but how far does this overlap extend and what implications does this overlap have for people in terms of health and disease? A recent study from Slavich and colleagues (2010) provides initial evidence that individual differences in neural responses to social threat may make one more or less susceptible to disease. In their study, participants performed the Trier Social Stress Test (TSST; Kirschbaum et al., 1993), in which they prepared and delivered an unrehearsed speech and performed mental arithmetic in front of a panel of socially rejecting raters. Participants then provided a saliva sample, which was assayed for two markers of inflammatory response: 1. a soluble receptor for tumor necrosis factor-α (sTNFαRII) and 2. interleukin-6 (IL-6). At a second laboratory session, a subsample of these participants played the Cyberball game, in which they experienced exclusion, during an fMRI scanning session. The authors found that the TSST increased levels of sTNFαRII and IL-6, compared to baseline levels. Once more, in region of interest analyses, increased activity within the dACC and anterior insula during exclusion (vs. inclusion) was associated with increased levels of sTNFαRII after the TSST, but not IL-6. These findings give initial evidence that the neural mechanisms underlying social pain are associated with susceptibility to inflammatory responses in socially stressful situations, which has implications for the link between stress and disease.

Another future direction for fruitful research centers around the emerging area of genomic imaging, in which investigators are able to model the neural responses towards a stimulus among individuals genetically predisposed for various characteristics or not. For example, Way and colleagues (2009) conducted initial research that applied this method to the study of social rejection. They were interested in whether variation in the μ-opioid receptor gene (a site where morphine acts) was associated with individual differences in sensitivity to social rejection. In their study, participants completed a rejection sensitivity questionnaire and provided a saliva sample that was assayed for the A118G polymorphism (a measure of the μ-opioid receptor gene). A subsample of participants then completed the Cyberball game, in which they experienced exclusion, during a second laboratory session. The investigators expected that participants with the G allele on the A118G polymorphism would be more sensitive to rejection, because the G allele is related to the reduced potency of opiates. As expected, participants who were G allele carriers reported more sensitivity to rejection. Additionally, in region of interest analyses, activity in the dACC and anterior insula during exclusion (vs. inclusion) was greater for G allele carriers than A allele homozygotes. Moreover, activity in the dACC during exclusion (vs. inclusion) significantly mediated the association between individual variation in the A118G polymorphism and individual differences in rejection sensitivity. These findings provide initial evidence for the genetic overlap between physical and social pain, however much more work still needs to be done in this area.

11. Conclusion

Humans are social creatures, and the need to belong is a powerful motivator for sustenance of social integration. Over the past decade, social psychologists have worked hard to understand the consequences of rejection at the cognitive, emotional, and behavioral levels. However, we are just beginning to understand the neural basis underlying social rejection. Recently, social psychologists have been able to apply functional magnetic resonance imaging to this research area. Doing so, we have gained a better understanding of the neural similarities between physical and social pain. Specifically, psychologists have identified the dorsal anterior cingulate cortex and the anterior insula (areas associated with the affective component of physical pain) as particularly important in the processing of social rejection. More recently, researchers have employed fMRI procedures to extend upon this work. Through neuroimaging, psychologists have shown that an over-the-counter medication, acetaminophen, reduces feelings of rejection by dampening activity in the dACC and anterior insula during rejection experiences. Neuroimaging has also been useful in helping researchers identify those individuals who are most vulnerable to social rejection, exclusion, and ostracism.

As technology becomes more advanced and user-friendly, the researcher's toolkit will expand to incorporate new scientific approaches. Functional brain imaging can be applied in a variety of ways to the study of human behavior. This chapter is just one example of an area in which neuroimaging procedures can be easily adapted. Although we don't fully understand the neural basis of social pain, psychologists have been able to gain ground in this area because of the availability of fMRI technology.

The brain is an immensely complex organ. In order to gain a better understanding of the behaviors that it produces, we need to make full use of the research tools at our disposal.

http://dx.doi.org/10.5772/31141

Functional brain imaging can offer valuable insights to the social sciences. These insights lead to better questions, the generation of novel hypotheses, and more elegant methods for testing those hypotheses.

Author details

Richard S. Pond, Jr., Stephanie B. Richman, David S. Chester and C. Nathan DeWall

University of Kentucky, United States of America

References

[1] Abell, J.E., Hootman, J.M., Zack, M.M., Moriarty, D., Helmick, C.G. (2005). Physical activityand health related quality of life among people with arthritis. *Journal of Epidemiology and Community Health*, Vol. 59, No.5, (May 2005), pp. 380–385, ISSN 1470-2738.

[2] Apkarian, A.V., Bushnell, M.C., Treede, R., & Zubieta, J. (2005). Human brain mechanisms of pain perception and regulation in health and disease. *European Journal of Pain*, Vol.9, No.4, (August 2005), pp. 463–484, ISSN 1090-3801.

[3] Axelrod, R., & Hamilton, W.D. (1981). The evolution of cooperation. *Science.* Vol. 211,No.4489, (March 1981), pp. 1390-1396, ISSN 0036-8075.

[4] Aziz, Q., Schnitzler, A., & Enck, P. (2000). Functional neuroimaging of visceral sensations. *Journal of Clinical Neurophysiology*, Vol.17, No.6, (November 2000), pp. 604-612, ISSN 0736-0258.

[5] Barash. D.P. (1977). *Sociobiology and behavior*, Elsevier, ISBN 0444990364, New York.

[6] Barnard, G.W., Vera, H., Vera, M.I., & Newman, G. (1982). Til death do us part: A study of spouse murder. *Bulletin of the American Academy of Psychiatry and the Law*, Vol. 10, No.4, (December 1982), pp. 271-280, ISSN 7165803.

[7] Barrett, L.F., Tugade, M.M., & Engle, R.W. (2004). Individual differences in working memory capacity and dual-process theories of the mind. *Psychological Bulletin*, Vol. 130, No.4, (July 2004), pp.553-573, ISSN 00332909.

[8] Baumeister, R.F. (1990). Suicide as escape from the self. Psychological Review, Vol. 97, No.1,(January 1990), pp. 90-113, ISSN 1939-1471.

[9] Baumeister, R.F., & Leary, M.R. (1995). The need to belong: Desire for interpersonal attachments as a fundamental human motivation. *Psychological Bulletin*, Vol.117, No. 3, (May 1995), pp. 497-529, ISSN 0033-2909.

[10] Baumeister, R.F., Twenge, J.M., & Nuss, C.K. (2002). Effects of social exclusion on cognitive processes: Anticipated aloneness reduces intelligent thought. *Journal of Personality and Social Psychology,* Vol.83, No.4, (October 2002), pp. 817-827, ISSN 1939-1315.

[11] Baumeister, R.F., DeWall, C.N., Ciarocco, N.J., & Twenge, J.M. (2005). Social exclusion impairs self-regulation. *Journal of Personality and Social Psychology,* Vol.88, No.4 (April 2005), pp. 589-604, ISSN 1939-1315.

[12] Berkman, L.F., & Syme, S.L. (1979). Social networks, host resistance, and mortality: A nine-year follow-up study of Alameda county residents. *American Journal of Epidemiology*, Vol.109, No.2, (February 1979), pp. 186-204, ISSN 0002-9262.

[13] Bolling, D.Z., Pitskel, N.B., Deen, B., Crowley, M.J., McPartland, J.C., Mayes, L.C., & Pelphrey, K.A. (2011). Dissociable brain mechanisms for processing social exclusion and rule violation. *Neuroimage,* Vol.54, No.3, (February 2011), pp. 2462-2471, ISSN 1095-9572.

[14] Brown, G., & Nicassio, P. (1987). Development of a questionnaire for the assessment of active and passive coping strategies in chronic pain patients. *Pain.* Vol.31, No.1, (October 1987), pp. 53–64, ISSN 0304-3959.

[15] Brown, J.L., Sheffield, D., Leary, M.R., & Robinson, M.E. (2003). Social support and experimental pain. *Psychosomatic Medicine,* Vol.65, No.2, (March-April 2003), pp.276–283, ISSN 1534-7796.

[16] Buckley, K., Winkel, R., & Leary, M.R. (2004). Reactions to acceptance and rejection: Effects of level and sequence of relational evaluation. *Journal of Experimental Social Psychology,* Vol.40, No.1, (January 2004), pp.14-28, ISSN 0022-1031.

[17] Burckardt, C.S. (2002). Nonpharmacologic management strategies in fibromyalgia. *Rheumatic Disease Clinics of North America,* Vol.28, No.2, (May 2002), pp. 291–304, ISSN 0889-857X.

[18] Burklund, L.J., Eisenberger, N.I., & Lieberman, M.D. (2007). The face of rejection: Rejection sensitivity moderates dorsal anterior cingulated activity to disapproving facial expressions. *Social Neuroscience,* Vol.2, No.3–4, (July 2007), pp. 238–253, ISSN 1747-0919.

[19] Bush, G., Luu, P., & Posner, M.I. (2000). Cognitive and emotional influences in anterior cingulated cortex. *Trends in Cognitive Sciences,* Vol.4, No.6, (June 2000), pp. 215–222, ISSN 1879-307X.

[20] Buss, D.M. (1990). The evolution of anxiety and social exclusion. *Journal of Social and Clinical Psychology,* Vol.9, No.2, (June 1990), pp. 196-210, ISSN 0736-7236.

[21] Buss, D.M. (1991). Evolutionary personality psychology. *Annual Review of Psychology,* Vol. 42, pp. 459-491, ISSN 0-8243-0242-7.

[22] Callahan, C.D. (2000). Stress, coping, and personality hardiness in patients with temporomandibular disorders. *Rehabilitative Psychology,* Vol.45, No.1, (February 2000), pp. 38–48, ISSN 1939-1544.

[23] Carden, S.E., & Hofer, M.A. (1990). Socially mediated reduction of isolation distress in rat pups is blocked by naltrexone but not by Ro 15–1788. *Behavioral Neuroscience,* Vol. 104, No.3, (June 1990), pp. 457–463, ISSN 1939-0084.

[24] Carden, S.E., Hernandez, N., & Hofer, M.A. (1996). The isolation and companion comfort responses of 7- and 3-day-old rat pups are modulated by drugs active at the opioid receptor. *Behavioral Neuroscience,* Vol.110, No.2, (April 1996), pp. 324–330, ISSN 1939-0084.

[25] Carlson, C.R., Bertrand, P.M., Ehrlich, A.D., Maxwell, A.W., & Burton, R.G. (2001). Physical self-regulation training for the management of temporomandibular disorders. *Journal of Orofacial Pain,* Vol.15, No.1, (Winter 2001), pp. 47–55, ISSN 11889647.

[26] Carter, C.S., MacDonald, III, A.M., Botvinick, M.M., Ross, L.L., Stenger, V.A., Noll, D., & Cohen, J. (2000). Parsing executive processes: Strategic vs. evaluative functions of the anterior cingulate cortex. *Proceedings of the National Academy of Sciences,* Vol.97, No.4, (February 2000), pp. 1944 – 1948, ISSN 0027-8424.

[27] Chalmers, B., Wolman, W.L., Nikodem, V.C., Gulmezoglu, A.M., & Hofmeyer, G.J. (1995). Companionship in labour: Do the personality characteristics of labour supporters influence their effectiveness? *Curationis,* Vol.18, No.4, (December 1995), pp. 77–80, ISSN 8697543.

[28] Cogan, R., & Spinnato, J.A. (1988). Social support during premature labor: Effects on labor and the newborn. *Journal of Psychosomatic Obstetrics and Gynecology,* Vol.8, No.3, (June 1988), pp. 209–216, 1743-8942.

[29] Craig, A.D., & Dostrovsky, J.O. (1999). Medulla to thalamus. In: *Textbook of pain,* P. Wall & R. Melzack (Eds.), pp. 183–214, Churchill Livingstone, ISBN 0443062528, New York.

[30] Crawford, M., & Gartner, R. (1992). *Women killing: Intimate femicide in Ontario, 1974-1990.* Women We Honour Action Committee, Toronto.

[31] DeWall, C.N., & Baumeister, R.F. (2006). Alone but feeling no pain: Effects of social exclusion on physical pain tolerance and pain threshold, affective forecasting, and interpersonal empathy. *Journal of Personality and Social Psychology,* Vol.91, No.1, (July 2006), pp. 1-15, ISSN 1939-1315.

[32] DeWall, C.N., MacDonald, G., Webster, G.D., Masten, C.L., Baumeister, R.F., Powell, C., Combs, D., Schurtz, D.R., Stillman, T.F., Tice, D.M., & Eisenberger, N.I. (2010). Acetaminophen reduces social pain: Behavioral and neural evidence. *Psychological Science,* Vol.21, No.7, (June 2010), pp. 931-937, ISSN 1467-9280.

[33] DeWall, C.N., Masten, C.L., Kashdan, T.B., Pond, Jr., R.S., Powell, C., Combs, D., & Schurtz, D.R. (2011). *Who is Most Vulnerable to Social Rejection? The Toxic Combination*

of Low Self-Esteem and Lack of Emotional Differentiation on Neural Responses to Rejection. Unpublished manuscript, University of Kentucky.

[34] DeWall, C.N., Masten, C.L., Powell, C., Combs, D., Schurtz, D.R., & Eisenberger, N.I. (2011). Does the pain of rejection depend on attachment style? An fMRI study. *Social Cognitive & Affective Neuroscience,* Advance Access, (April 2011), ISSN 1749-5016.

[35] DeWall, C.N. & Richman, S.B. (2011). Social exclusion and the desire to reconnect. *Social and Personality Psychology Compass,* Vol.5, No.11, (November 2011), pp. 919-932, ISSN1751-9004.

[36] DeWall, C.N., Twenge, J.M., Gitter, S.A., & Baumeister, R.F. (2009). It's the thought that counts: The role of hostile cognition in shaping aggressive responses to social exclusion. *Journal of Personality and Social Psychology,* Vol.96, No.1, (January 2009), pp. 45-59, ISSN 1939-1315.

[37] DeWall, C.N., Twenge, J.M., Bushman, B.J., Im, C., & Williams, K.D. (2010). Acceptance by one differs from acceptance by none: Applying social impact theory to the rejection-aggression link. *Social Psychological and Personality Science,* Vol.1, No.2, (April 2010), pp. 168-174, ISSN 1948-5506.

[38] Downey, G., Freitas, A.L., Michaelis, B., & Khouri, H. (1998). The self-fulfilling prophecy in close relationships: Rejection sensitivity and rejection by romantic partners. *Journal of Personality and Social Psychology,* Vol.75, No.2, (August 1998), pp. 545-560, 1939-1315.

[39] Eisenberger, N.I., & Lieberman, M.D. (2004). Why rejection hurts: A common neural alarm system for physical and social pain. *Trends in Cognitive Sciences,* Vol.8, No.7, (July 2004), pp.294-300, ISSN 1364-6613.

[40] Eisenberger, N.I., Lieberman, M.D., & Williams, K.D. (2003). Does rejection hurt? An fMRI study of social exclusion. *Science,* Vol.302, No.5643, (October 2003), pp. 290-292, ISSN 1095-9203.

[41] Eisenberger, N.I., Taylor, S.E., Gable, S.L., Hilmert, C.J., & Lieberman, M.D. (2007). Neural pathways link social support to attenuated neuroendocrine stress responses. *Neuroimage,* Vol.35, No.4, (May 2007), pp. 1601-1612, ISSN 1053-8119.

[42] Epker, J., & Gatchel, R.J. (2000). Coping profile differences in the biopsychosocial functioning of patients with temporomandibular disorder. *Psychosomatic Medicine,* Vol.62, No.1, (January 2000), pp. 69–75, ISSN 1534-7796.

[43] Feeney, J.A., Noller, P., & Hanrahan, M. (1994). Assessing adult attachment. In: *Attachment in adults: Clinical and developmental perspectives,* M.B. Sperling, and W.H. Berman (Eds.), pp. (128–152). Guilford Press, ISBN 0898625475, New York.

[44] Felmingham, K., Kemp, A., Williams, L., Das, P., Hughes, G., Peduto, A., and Bryant, R. (2007). Changes in anterior cingulate and amygdala after cognitive behavior thera-

py of posttraumatic stress disorder. *Psychological Science,* Vol.18, No.2, (February 2007), pp. 127–129, ISSN 1467-9280.

[45] Ferrando, M., Andreu, Y., Galdon, M.J., Dura, E., Poveda, R., & Bagan, J.V. (2004) Psychological variables and temporomandibular disorders: Distress, coping, and personality. *Oral Surgery, Oral Medicine, Oral Pathology, Oral Radiology, and Endodontics,* Vol.98, No.2, (August 2004), pp. 153–160, ISSN 1531-6541.

[46] Foltz, E.L., & White, L.E. (1968). The role of rostral cingulotomy in "pain" relief. *International Journal of Neurology,* Vol.6, pp. 353–373, ISSN 0020-7446.

[47] Fraley, R.C., & Shaver, P.R. (2000). Adult romantic attachment: theoretical developments, emerging controversies, and unanswered questions. *Review of General Psychology,* Vol.4, No.2, (June 2000), pp.132–154, ISSN 1089-2680.

[48] Gaertner, L., JIuzzini, J., & O'Mara, E.M. (2008). When rejection by one fosters aggression against many: Multiple-victim aggression as a consequence of social rejection and perceived groupness. *Journal of Experimental Social Psychology,* Vol.44, No.4 (July, 2008), pp. 958-970, ISSN 0022-1031.

[49] Gardner, W.L., Pickett, C.L., & Brewer. M.B. (2000). Social exclusion and selective memory: How the need to belong influences memory for social events. *Personality and Social Psychology Bulletin,* Vol.26, No.4, (April 2000), pp. 486–496, ISSN 1552-7433.

[50] Harmon-Jones, E., & Winkielman, P. (Eds.). (2007). *Social Neuroscience: Integrating Biological and Psychological Explanations of Social Behavior,* Guilford Publications, ISBN 978-159-3854-04-1, New York, NY.

[51] Harris, M.J. (2009). *Bullying, Rejection, and Peer Victimization: A Social Cognitive Neuroscience Perspective.* Springer Publishing Company, ISBN 978-082-6103-78-9, New York, NY.

[52] Herman, B.H., & Panksepp, J. (1978). Effects of morphine and naloxone on separation distress and approach attachment: Evidence for opiate mediation of social affect. *Pharmacology, Biochemistry, & Behavior,* Vol.9, No.2, (August 1978), pp. 213–220, 0091-3057.

[53] Hogan, R., Jones, W.H., & Cheek, J.M. (1985). Socioanalytic theory: An alternative to armadillo psychology. In: *The self and social life,* B.R. Schlenker (Ed.), pp. (175-198). McGraw-Hill, ISBN 0070553076, New York.

[54] House, J.S., Landis, K.R., & Umberson, D. (1988). Social relationships and health. *Science,* Vol.241, No.4865, (July 1988), pp. 540-545, ISSN 0036-8075.

[55] Insel, T.R., & Winslow, J.T. (1991). Central administration of oxytocin modulates the infant rat's response to social isolation. *European Journal of Pharmacology,* Vol.203, No. 1, (October 1991), pp. 149–152, ISSN 1665788.

[56] Izuma, K., Saito, D.N., & Sadato, N. (2008). Processing of social and monetary rewards in the human striatum. *Neuron*, Vol.58, No.2, (April 2008), pp. 284-294, ISSN 0896-6273.

[57] Janda, L.H., Fulk, J., Janda, M., & Wallace, J. (1995). *The development of a test of General Mental Abilities.* Unpublished manuscript, Old Dominion University.

[58] Janda, L. (1996). *The psychologists' book of self-tests.* Berkley, ISBN 0399522115, New York.

[59] Kashdan, T.B., Ferssizidis, P., Collins, R.L., & Muraven, M. (2010). Emotion differentiation as resilience against excessive alcohol use: An ecological momentary assessment in underage social drinkers. *Psychological Science,* Vol.21, No.9, (September 2010), pp.1341-1347, ISSN 1467-9280.

[60] Keefe, F.J., Brown, G.K., Wallston, K.A., Caldwell, D.S. (1989). Coping with rheumatoid arthritis pain: Catastrophizing as a maladaptive strategy. *Pain,* Vol.37, No.1, (April 1989), pp. 51–56, ISSN 0304-3959.

[61] Kehoe, P., & Blass, E. M. (1986). Opioid-mediation of separation distress in 10-day-old rats: Reversal of stress with maternal stimuli. *Developmental Psychobiology*, Vol.19, No.4, (July 1986), pp. 385–398, ISSN 1098-2302.

[62] Kirschbaum, C., Pirke, K.M., & Hellhammer, D.H. (1993). The 'Trier Social Stress Test'—a tool for investigating psychobiological stress responses in a laboratory setting. *Neuropsychobiology*, Vol.28, No.1–2, (June 1993), pp. 76–81, ISSN 0302-282X.

[63] Klaus, M., Kennel, J., Robertson, S., & Rosa, R. (1986). Effects of social support during parturition on maternal and infant mortality. *British Medical Journal,* Vol.293, No. 6547, (September 1986), pp. 585–597, ISSN 0267-0623.

[64] Konecka, A.M., & Sroczynska, I. (1990). Stressors and pain sensitivity in CFW mice: Role of opioid peptides. *Archives Internationales de Physiologie et de Biochimie,* Vol.98, No.5, pp. 245–252, ISSN 0003-9799.

[65] Kong, J., Gollub, R. L., Rosman, I. S., Webb, J. M., Vangel, M. G., Kirsch, I., and Kaptchuk, T.J. (2006). Brain activity associated with expectancy-enhanced placebo analgesia as measured by functional magnetic resonance imaging. *The Journal of Neuroscience,* Vol.26, No.2, (January 2006), pp. 381–388, ISSN 0270-6474.

[66] Kross, E., Berman, M.G., Mischel, W., Smith, E.E., & Wager, T.D. (2011). Social rejection shares somatosensory representations with physical pain. *Proceedings of the* National Academy of Sciences, Vol.108, No.15, (April 2011), pp. 6270-6275, ISSN 1091-6490.

[67] Kross, E., Egner, T., Ochsner, K., Hirsch, J., & Downey, G. (2007). Neural dynamics of rejection sensitivity. *Journal of Cognitive Neuroscience*, Vol.19, No.6, (June 2007), pp. 945–956, ISSN 1530-8898.

[68] Landro, N.I., Stiles, T.C., & Sletvold, H. (1997). Memory functioning in patients with primary fibromyalgia and major depression and healthy controls. *Journal of Psychosomatic Research,* Vol.42, No.3, (March 1997), pp. 297-306, ISSN 0022-3999.

[69] Lane, R.D., Reiman, E.M., Ahern, G.L., Schwartz, G.E., & Davidson, R.J. (1997). Neuroanatomical correlates of happiness, sadness, and disgust. *American Journal of Psychiatry,* Vol.154, No.7, (July 1997), pp. 926-933, ISSN 0002-953X.

[70] Leary, M.R. & Springer, C.A. (2001). Hurt feelings: The neglected emotion. In: *Aversive behaviors and interpersonal transgression,* R. Kowalski (Ed.), pp. (151-175). American Psychological Association, ISBN 1-55798-716-5, Washington DC.

[71] Leary, M.R., Kowalski, R.M., Smith, L., & Phillips, S. (2003). Teasing, rejection, and violence: Case studies of the school shootings. *Aggressive Behavior,* Vol.29, No.3, (June 2003), pp. 202-214, ISSN 1098-2337.

[72] Leary, M.R., Tambor, E.S., Terdal, S.J., & Downs, D.L. (1995). Self-esteem as an interpersonal monitor: The sociometer hypothesis. *Journal of Personality and Social Psychology,* Vol.68, No.3, (March 1995), pp.518-530, ISSN 0022-3514.

[73] Leary, M.R., Twenge, J.M., & Quinlivan, E. (2006). Interpersonal Rejection as a Determinant of Anger. *Personality and Social Psychology Review,* Vol.10, No.2, (May, 2006), pp. 111-132, ISSN 1532-7957.

[74] Lidderdale, J.M., & Walsh, J.J. (1998). The effects of social support on cardiovascular reactivity and perinatal outcome. *Psychology and Health,* Vol.13, No.6, (May 1998), pp. 1061–1070, ISSN 1476-8321.

[75] Lieberman, M.D. (2007). Social cognitive neuroscience, In: *Encyclopedia of Social Psychology Vol. 2,* Baumeister, R.F., & Vohs, K.D. (Eds.), pp. 885-887, Sage Press, ISBN 978-141-2916-70-7, Thousand Oaks, CA.

[76] Lieberman, M.D. (2010). Social cognitive neuroscience, In: *Handbook of Social Psychology 5th ed.,* Fiske, S.T., Gilbert, D.T., & Lindzey, G. (Eds.), pp. 143-193, McGraw-Hill, ISBN 047-013-7479, New York, NY.

[77] Lieberman, M.D., Eisenberger, N.I., Crockett, M.J., Tom, S.M., Pfeifer, J.H., & Way, B.M. (2007). Putting feelings into words: Affect labeling disrupts amygdala activity in response to affective stimuli. *Psychological Science,* Vol.18, No.5, (May 2007), pp. 421–428, ISSN 0956-7976.

[78] MacDonald, G., & Leary, M.R. (2005). Why does social exclusion hurt? The relationship between social and physical pain. *Psychological Bulletin,* Vol.131, No.2, (March 2005), pp. 202–223, ISSN 1939-1455.

[79] MacLean, P.D., & Newman, J.D. (1988). Role of midline frontolimbic cortex in production of the isolation call of squirrel monkeys. *Brain Research,* Vol.450, No.1-2, (May 1988), pp. 111-123, ISSN 0006-8993.

[80] Maner, J.K., DeWall, C.N., Baumeister, R.F., & Schaller, M. (2007). Does social exclusion motivate withdrawal or reconnection? Resolving the "porcupine problem." *Journal of Personality and Social Psychology,* Vol.92, No.1, (January 2007), pp. 42-55, ISSN 1939-1315.

[81] Masten, C.L., Telzer, E.H., & Eisenberger, N.I. (2011). An fMRI investigation of attributing negative social treatment to racial discrimination. *Journal of Cognitive Neuroscience,* Vol.23, No.5, (May 2011), pp. 1042-1051, ISSN 1530-8898.

[82] Melzack, R. & Casey, K. (1968). Sensory, motivational and central control determinants of pain: A new conceptual model. In: *The Skin Senses,* D. Kenshado (Ed.), (pp. 423-439). C. C. Thomas, ISBN BWB5857742, Springfield, IL.

[83] Mitchell, J.P., Cloutier, J., Banaji, M.R., & Macrae, C.N. (2006). Medial prefrontal dissociations during processing of trait diagnostic and nondiagnostic person information. *Social Cognitive and Affective Neuroscience,* Vol.1, No.1, (May 2006), pp. 49-55, ISSN 0006-8993.

[84] Moll, J., de Oliveira-Souza, R., Eslinger, P.J., Bramati, I.E., Mourão-Miranda, J., Andreiuolo, P.A., & Pessoa, L. (2002). The neural correlates of moral sensitivity: A functional magnetic resonance imaging investigation of basic and moral emotions. *The Journal of Neuroscience,* Vol.22, No.7, (April 2002), pp. 2730 -2736, ISSN 1529-2401.

[85] Moreland, R.L. (1987). The formation of small groups. In: *Group processes: Review of personality and social psychology Vol 8,* C. Hendrick (Ed.), pp. (80-110). Sage, ISBN 0803930720, Newbury Park, CA.

[86] Murphy, M.R., MacLean, P.D., & Hamilton, S.C. (1981). Species-typical behavior of hamsters deprived from birth of the neocortex. *Science,* Vol.213, No.4506, (July 1981), pp.459–461, ISSN 0036-8075.

[87] Naranjo, J.R., & Fuentes, J.A. (1985). Association between hypoalgesia and hypertension in rats after short-term isolation. *Neuropharmacology,* Vol.24, No.2, (February 1985), pp. 167–171, ISSN 0028-3908.

[88] Nichols, D.S., Glenn, T.M. (1994). Effect of aerobic exercise on pain perception, affect, and level of disability in individuals with fibromyalgia. *Physical Therapy,* Vol.74, No. 4, (April 1994), pp. 327–332, ISSN 8140145.

[89] Niven, C. (1985). How helpful is the presence of the husband at childbirth? *Journal of Reproductive and Infant Psychology,* Vol.3, No.2, pp. 45–53, ISSN 1469-672X.

[90] Onoda, K., Okamoto, Y., Nakashim, K., Nittono, H., Yoshimura, S., Yamawaki, S., Yamaguchi, S., & Ura M. (2010). Does low self-esteem enhance social pain? The relationship between trait self-esteem and anterior cingulate cortex activation induced by ostracism. *Social Cognitive and Affective Neuroscience,* Vol.5, No.4, (December 2010), pp.385-391, ISSN 1749-5016.

[91] Panksepp, J. (1998). *Affective neuroscience: The foundations of human and animal emotions*. Oxford University Press, ISBN 0-19-509673-8, London.

[92] Panksepp, J., Herman, B.H., Conner, R., Bishop, P., & Scott, J.P. (1978). The biology of social attachments: Opiates alleviate separation distress. *Biological Psychiatry*, Vol.13, No.5, (October 1978), pp. 607–618, ISSN 0006-3223.

[93] Panksepp, J., Vilberg, T., Bean, N.J., Coy, D.H., & Kastin, A.J. (1978). Reduction of distress vocalization in chicks by opiate-like peptides. *Brain Research Bulletin*, Vol.3, No.6,(November 1978), pp. 663-667, ISSN 0361-9230.

[94] Park, D.C., Glass, J.M., Minear, M., & Crofford, L.J. (2001). Cognitive function in fibromyalgia patients. *Arthritis and Rheumatism,* Vol.44, No.9, (September 2001), pp. 2125-2133, ISSN 0004-3591.

[95] Petrovic, P., Dietrich, T., Fransson, P., Andersson, J., Carlsson, K., & Igvar, M. (2005). Placebo in emotion processing—induced expectations of anxiety relief activate a generalized modulatory network. *Neuron*, Vol.46, No.6, (June 2005), pp. 957–969, ISSN 0896-6273.

[96] Petrovic, P., & Ingvar, M. (2002). Imaging cognitive modulation of pain processing. *Pain*, Vol.95, No.1-2, (January 2002), pp. 1-5, ISSN 0304-3959.

[97] Phillips, J.M., & Gatchel, R.J. (2000). Extraversion–introversion and chronic pain: Locus of control in the patient with chronic pain. In: *Personality characteristics of patients with pain,* R. Gatchel & J. Weisberg (Eds.), pp. (181–202). American Psychological Association, ISBN 1-55798-646-0, Washington, DC.

[98] Price, D.D. (2000). Psychological and neural mechanisms of the affective dimension of pain. *Science,* Vol.288, No.5472, (June 2000), pp. 1769–1772, ISSN 0036-8075.

[99] Rainville, P., Duncan, G.H., Price, D.D., Carrier, B., & Bushnell, M.C. (1997). Pain affect encoded in human anterior cingulated but not somatosensory cortex. *Science,* Vol.277, No.5328, (August 1997), pp. 968–971, ISSN 0036-8075.

[100] Rushen, J., Boissy, A., Terlouw, E.M., & de Passille, A.M. (1999). Opioid peptides and behavioral and physiological responses of dairy cows to social isolation in unfamiliar surroundings. *Journal of Animal Science,* Vol.77, No.11, (November 1999), pp. 2918-2924, ISSN 1525-3163.

[101] Sebastian, C.L., Roiser, J.P., Tan, G.C., Viding, E., Wood, N.W., & Blakemore, S.J. (2010). Effects of age and MAOA genotype on the neural processing of social rejection. *Genes, Brain & Behavior*, Vol.9, No.6, (June 2010), pp. 628-637, ISSN 1601-183X.

[102] Sharp, T.J., & Harvey, A.G. (2001). Chronic pain and post-traumatic stress disorder: Mutual maintenance? *Clinical Psychology Review,* Vol.21, No.6, (August 2001), pp. 857– 877, ISSN 0272-7358.

[103] Simmons, A., Matthews, S.C., Feinstein, J.S., Hitchcock, C., Paulus, M.P., & Stein, M.B. (2008). Anxiety vulnerability is associated with altered anterior cingulate re-

sponse to an affective appraisal task. *NeuroReport*, Vol.19, No.10, (July 2008), pp. 1033–1037, ISSN 0959-4965.

[104] Slavich, G.M., Way, B.M., Eisenberger, N.I., & Taylor, S.E. Neural sensitivity to social rejection is associated with inflammatory responses to social stress. *Proceedings of the National Academy of Sciences*, Vol.107, No.33, (August 2010), pp. 14817-14822, ISSN 0027-8424.

[105] Smith, A.K., White, P.D., Aslakson, E., Vollmer-Conna, U., & Rajeevan, M.S. (2006). Polymorphisms in genes regulating the HPA axis associated with empirically delineated classes of unexplained chronic fatigue. *Pharmacogenomics*, Vol.7, No.3, (April 2006), pp 387–394, ISSN 16610949.

[106] Snow-Turek, A.L., Norris, M.P., Tan, G. (1996) Active and passive coping strategies in chronic pain patients. *Pain*, Vol.64, No.3, (March 1996). pp. 455–462, ISSN 0304-3959.

[107] Solberg Nes, L., Roach, A.R., & Sergerstrom, S.C. (2009). Executive functions, self-regulation, and chronic pain: a review. *Annals of Behavioral Medicine*, Vol.37, No.2, (April 2009), pp. 173-183, ISSN 0883-6612, 1532-4796.

[108] Somerville, L.H., Kelley, W.M., & Heatherton, T.F. (2010). Self-esteem modulates medial prefrontal cortical responses to evaluative social feedback. *Cerebral Cortex*, Vol. 20, No.12, (December 2010), pp.3005-3013, ISSN 1047-3211.

[109] Spear, L.P., Enters, E.K., Aswad, M.A., & Louzan, M. (1985). Drug and environmentally induced manipulations of the opiate and serotonergic systems alter nociception in neonatal rat pups. *Behavioral and Neural Biology*, Vol.44, No.1, (July 1985), pp. 1–22, ISSN 0163-1047.

[110] Sufka, K.J., & Hughes, R.A. (1990). Dose and temporal parameters of morphine-induced hyperalgesia in domestic fowl. *Physiology & Behavior*, Vol.47, No.2, (February 1990), pp. 385–387, ISSN 0031-9384.

[111] Sufka, K.J., & Weed, N.C. (1994). Construct validation of behavioral indices of isolation stress and inflammatory nociception in young domestic fowl. *Physiology & Behavior*, Vol.55, No4., (April 1994), pp. 741–746, ISSN 0031-9384.

[112] Tugade, M.M., Fredrickson, B.L., & Barrett, L.F. (2004). Psychological resilience and emotional granularity: Examining the benefits of positive emotions on coping and health. *Journal of Personality*, Vol.72, No.6, (December 2004), pp.1161-1190, ISSN 1467-6494.

[113] Twenge, J.M., Baumeister, R.F., Tice, D.M., & Stucke, T.S. (2001). If you can't join them, beat them: Effects of social exclusion on aggressive behavior. *Personality and Social Psychology*, Vol.81, No.6, (December, 2001), pp.1058-1069, ISSN 1939-1315.

[114] Twenge, J.W., Cantanese, K.R., & Baumeister, R.F. (2002). Social exclusion causes self-defeating behavior. *Journal of Personality and Social Psychology*, Vol.83, No.3, (September, 2002), pp. 606-615, ISSN 1939-1315.

[115] Twenge, J.M., Catanese, K.R., & Baumeister, R.F. (2003). Social exclusion and the deconstructed state: Time perception, meaninglessness, lethargy, lack of emotion, and self-awareness. *Journal of Personality and Social Psychology*, Vol.85, No.3, (September 2003), pp. 409–423, ISSN 1939-1315.

[116] Way, B.M., Taylor, S.E., & Eisenberger, N.I. (2009). Variation in the mu-opioid receptor gene (OPRM1) is associated with dispositional and neural sensitivity to social rejection.*Proceedings of the National Academy of Sciences*, Vol.106, No.35, (September 2011), pp. 15079-15084, 0027-8424.

[117] Wiedenmayer, C.P., Goodwin, G.A., & Barr, G.A. (2000). The effect of periaqueductal gray lesions on responses to age-specific threats in infant rats. *Developmental Brain Research*, Vol.120, No.2, (April 2000), pp. 191–198, ISSN 0165-3806.

[118] Williams, K.D., Cheung, C.K.T., & Choi, W. (2000). CyberOstracism: Effects of being ignored over the Internet. *Journal of Personality and Social Psychology*, Vol.79, No.5, (November 2000), pp.748–762, ISSN 00223514.

[119] Zadro, L., Williams, K.D., & Richardson, R. (2004). How low can you go? Ostracism by a computer is sufficient to lower self-reported levels of belonging, control, self-esteem, and meaningful existence. *Journal of Experimental Social Psychology*, Vol.40, No. 4, (July 2004), pp. 560-567, ISSN 0022-1031.

[120] Zautra, A.J., Hamilton, N.A., & Burke, H.M. (1999). Comparison of stress responses in Women with two types of chronic pain: Fibromyalgia and osteoarthritis. *Cognitive Therapy and Research*, Vol.23, No.2, (April 1999), pp. 209–230, ISSN 1573-2819.

Chapter 25

Social Neuroscience Tasks: Employing fMRI to Understand the Social Mind

D.Y. Phua and G.I. Christopoulos

Additional information is available at the end of the chapter

http://dx.doi.org/10.5772/56630

1. Introduction

The purpose of the present chapter is to introduce and describe some fMRI-compatible tasks that social scientists can employ to study various social phenomena. fMRI (functional Magnetic Resonance Imaging) is one of the principal methods for studying *in vivo* human brain-related responses associated with cognitive functions. Given the increasing availability of brain imaging scanners, the less complicated analysis methods, and the lower costs of using fMRI, more social scientists (from e.g., social psychology, political science, business, economics, philosophy and culture science) have become interested in integrating neuroscience into their own research agenda. It is expected that this trend will continue as multidisciplinary research becomes more popular.

Crossing or mixing disciplines has always been both challenging and very rewarding. In the case of neuroscience, the unwillingness to cross disciplines is often implicated with a lack of appreciation for the potential of fMRI and neuroscience methodologies. In addition, social scientists are often overwhelmed by the complexity of neuroscience and are unsure whether it is worthwhile to invest time and energy in enriching their methodological tools with fMRI. Therefore, there is an exigency for an introductory chapter aiding non-neuroscientists (or, generally, researchers with less exposure to experimental methods of neuroscience) to understand how cognitive neuroscientists have employed fMRI in elucidating social phenomena.

As such, the general aims of this chapter are to:

i. Offer inspiration on novel approaches for studying social phenomena using neuroscience methods

ii. Help readers form and answer proper questions using an appropriate methodology

iii. Understand how neuroscience epistemologically approaches various phenomena

To achieve these aims, we here present an overview of fMRI tasks that have been successfully used to study social phenomena. Note here that we focus on the *tasks* employed and not in the social neuroscience theories or neurobiology implicated in the phenomena. We believe that this is a pedagogically more efficient way for the non-expert scientist or student to be introduced to the methodology of social neuroscience, since it approaches social neuroscience from an applied, hands-on perspective. To that end, in the following paragraphs we present tasks that are related to:

- Facial expression recognition
- Perception of biological motion
- Mimicry
- Mirror neurons system
- Mentalizing and understanding mental states of others
- Empathy
- Distinguishing self from others
- Morality
- Kinship, romantic and maternal love
- Attitudes and evaluative processes
- Race and stereotypes
- Culture
- Social interactions

We note that what follows is a brief introduction of different tasks. Most, if not all, of the tasks can be used for pure behavioral research. What makes these tasks special is that they abide to some general rules that allow for the specificities imposed by the fMRI methodology:

- The need for many repetitions (trials). Because of the large measurement error, it is advisable that each condition has a good number of trials. In other words, the design should be able to measure the same variable many times. Of course, here one needs to balance design efficiency with psychological factors such as boredom, learning and memory effects.
- The need for relatively quick presentation of stimuli. For instance, it is rather rare to have a long scenario as a stimulus, especially if one is an inexperienced researcher. Because of the temporal properties of the fMRI signal, it is also advisable, with maybe some exceptions, to present stimuli quickly, whereas the participant should be able to provide behavioral responses (if any) easily without a lot of motor or attention demands.

http://dx.doi.org/10.5772/56630

- Isomorphic stimuli. This is quite critical, as the brain is sensitive to a wide space of stimulus differences that could override our main targeted dimension. See below at the paragraphs about face recognition for a demonstration.
- Other needs such as good randomization and general or trial-specific time constraints.

Of course, the present text cannot cover all aspects of fMRI research and methods. We therefore strongly suggest that readers do get familiar with the nuances of an fMRI experiment (e.g., fMRI experiments typically require a large number of repetitions of the same condition). A good starting point would be Huettel, Song and McCarthy (2009). In addition, readers should be familiar with the basic tenets of experimental methodology (such as employing proper controls, separating dependent and independent variables, and avoiding confounding factors) which are typically more crucial in fMRI experiments.

Yet, we would like to emphasize three important points that are often misinterpreted:

1. **Ask 'How' and not (only) 'Where'.** It is important to form research questions that are related to specific cognitive mechanisms and, ideally, aim to test alternative/competing theoretical models of behavior. "Fishing expeditions" without concrete anatomical, function (and/or computational) hypotheses and models should be avoided. It is critical to understand how modern neuroscience aids in understanding *mechanisms* (i.e., *how* the brain processes information to produce behavioral response). Therefore, cognitive and / or computational models are essential before proceeding to fMRI experiments.
2. **Avoid reverse inference.** This is a very common error, which is elegantly described in Poldrack (2006). Imagine we find a brain area X (e.g., amygdala) that demonstrates a higher response to a stimulus A (e.g., a face with fear expression) than B (a control face with a neutral expression). The, to a great extent correct, inference is that amygdala responds to face stimuli depicting expressions of fear (A → X). Assume that in another study, a researcher employs stimulus Y (e.g., the face of an outgroup member) and finds that area X (i.e., amygdala) is activated. Often the conclusion drawn (and many times published) is that Y promotes the mental state associated with A (i.e., faces of outgroup members promote fear responses). Of course, this is a *reverse inference,* and is problematic. What we established was that A →X and that Y → X but not X → A or X→ Y. Amygdala responds to many mental states in addition to fear. Therefore, researchers should be very careful when attempting to infer the *mental state* based on *brain responses*. Proper experimental design with appropriate controls can help in alleviating these issues. Moreover, instead of approaching the variables categorically, many times it might be more appropriate to seek brain-related responses as a function of different levels of input stimulus intensity. For instance, instead of examining how the brain responds to positive and negative monetary rewards (e.g. receiving $10 vs. losing $10), one could allow for larger variation (e.g. outcome -$40, -$20, -$10, $10, $20, $30) of parameters.
3. **Deal with complexity**. Another common mistake made by researchers new to neuroscience is that they often try to solve very complex problems that introduce multiple variables and/or do not have concrete, measurable responses. Neuroscience methodology is most appropriate for straightforward questions that have few variables and very

specific responses. Remember that the neurofunctional data collected will be very complex. On average, a brain image is captured every two seconds by the scanner, equating to over 100,000 data points in 20 to 30 minutes. These time-series data points are then combined across participants, thus increasing the complexity of dataset. Fortunately, there are specialized software packages that can do such processing and analysis. However, this does not negate the need for careful experimental designs that take into consideration the complexity of the data. Lack of attention to the design may make subsequent interpretation difficult, if not meaningless.

Other neurobiological methods such as psychophysiology, EEG, fNIRS, genetics, psychopharmacology will not be covered. Readers are encouraged to further explore possibilities from other sources. Yet, many tasks and principles described here might be much related to these methods.

2. Facial recognition

Facial recognition goes beyond face processing; the latter refers to the mere ability to differentiate individuals. Facial recognition includes human's ability to decipher social information embedded in facial expressions. Some studies have explored the neurobiological basis for often-unconscious responses to specific facial expressions, such as emotional expressions (e.g., Harrison, Singer, Rotshtein, Dolan, & Critchley, 2006), untrustworthiness (e.g., Engell, Haxby & Todorov, 2007) or averted eye gaze (e.g., Hoffman & Haxby, 2000). In such studies, the principal methodology has been to contrast brain responses between conditions. For example, judgment studies of another's trustworthiness obtain behavioral ratings of the trustworthiness associated with different faces. These ratings were then used to regress against neural responses in order to compare brain signals when viewing faces rated as trustworthy versus viewing those rated as untrustworthy (e.g., Engell, et al., 2007; Winston, Strange, O'Doherty, & Dolan, 2002). Using a similar methodology, Hoffman and Haxby (2000) investigated the differential blood-oxygen-level-dependent (BOLD) responses [the principal signal derived from fMRI] when participants viewed photographs of faces looking directly versus away from the camera (i.e., direct versus averted eye gaze). This was to examine whether regions associated with perception of the face identity are also involved in perception of the changeable aspects of the face. Averted eye gaze redirect one's spatial attention in social communication, and thus different neural mechanisms should be involved. To better understand the different processes involved in facial processing, Narumoto and colleagues (2001) examined the differential brain responses towards faces as a function of attention focus (i.e., focus on contours, identity or emotional expressions).

Researchers have also investigated how people recognize speech from visual perception (i.e., lip-reading). Typical methodologies used include the comparison of brain responses when participants viewed videos of facial gurning (meaningless opening and closing of mouth) versus silent mouthing of random numbers (e.g., Calvert, et al., 1997; Campbell, et al., 2001). Buccino and colleagues (2004) also examined if the brain response observed when viewing

http://dx.doi.org/10.5772/56630

human verbal communicative actions are generalizable to viewing of other species' communicative actions. Their stimuli included silent video sequences of a man speaking, a monkey lip-smacking and a dog barking.

3. Perceptions of biological motion

Human beings are naturally sensitive to motion. Studies on perception of motion often concentrated on biological motion. These include facial movements such as mouth or eye movements (e.g., Puce, Allison, Bentin, Gore, & McCarthy, 1998) or bodily movements (e.g., Pelphrey, Morris, Michelich, Allison, & McCarthy, 2005). Some research has also expanded into the investigation of the neuronal mechanisms involved in perception of non-human facial movements (Buccino, et al., 2004) and complex actions that are not in one's motor repertoire (Calvo-Merino, Glaser, Grèzes, Passingham, & Haggard, 2005). Another stream of research examines biological motion depicted as point-light displays; in these tasks, subjetcs have to either judge the biological plausibility of the implied motion or to discriminate between different types of biological motion (for instance moving left or right ;McKay, et al., 2012). However, more complex discrimination tasks such as agent and expression intensity recognition (Sevdalis & Keller, 2011) have not yet been adequately examined in fMRI environments.

Individuals are also capable of perceiving implied motion. Implied motions are static images that suggest there is motion involved when the image was captured. As such, while studies on perception of actual motion often use video clips as stimuli, studies on implied motion relied on static photographs (e.g., Kourtzi & Kanwisher, 2000; Senior, et al., 2000). In these studies, participants were shown photographs of an agent in action (human, animal or a natural agent such as wave). Motion was not actually present or viewed, just implied by the static images. This allows for a deeper understanding of how humans perceive motion. Results have suggested that the extrastriate cortex is not merely involved in passive perception of motion, but also critical for visual cognition that allows for interpretation of static images with implied motion.

Furthermore, Iacoboni and colleagues (1999) investigated whether the same neural substrates are responsible for perception of motion versus actual execution of the action. This is known as *direct matching hypothesis*, which suggests that observing an action stimulates the same brain regions as actually executing the action. In addition, they also looked at the neural responses of imitation, which is learning and executing an action after observing it being performed. In most of their studies, they used actions that were familiar to participants, such as simple fingers movements. Buccino and colleagues (2004) expanded this line of research into the imitation of complex hand movements that were not in participants' motor repertoire. For instance, participants who were non-guitarists observed and later executed finger movements of guitar chords.

4. Mimicry

While imitation involves a conscious 'copying' of actions, mimicry is defined as an unconscious copying of gestures, postures or mannerisms of another person. Mimicry often takes place in social interactions and has been found to facilitate social exchanges and enhance feelings of affiliation and liking. Some mimicry actions studied include contagious yawning (Nahab, Hattori, Saad, & Hallett, 2009), smiling (Wild, Erb, Eyb, Bartels, & Grodd, 2003) and even subtle social stimuli such as pupil dilation (Harrison, et al., 2006). Typically, participants either passively viewed images (e.g., smiling or yawning) or videos (e.g., social interaction or hand movements) or instructed to perform similar actions as shown in the videos.

The effects of mimicry in social affections have been widely reported in behavioral studies. Recently, Kühn and colleagues (2010) sought to better understand this social phenomenon. Using videos of dyadic interactions, participants were instructed to take the perspective of the person filmed from first-person perspective. When the interaction partner was mimicking, stronger activation were shown in the neural reward mechanisms, which researchers infer as partially supporting the liking effect during mimicry (notice that this interpretation could be a demonstration of reverse inference – see introduction). On another note, while mimicry is unconscious, it can be enhanced or inhibited by eye contact with the partner (Wang, Ramsey, & de C. Hamilton, 2011). Wang and colleagues tested this by instructing participants to imitate the hand actions of a person in the video who is either looking at or away from the camera.

5. Mirror neurons

Mirror neurons are neurons that respond to a particular action and will respond regardless if the individual is passively observing or actively executing the action (Chong, Cunnington, Williams, Kanwisher, & Mattingley, 2008; Kilner, Neal, Weiskopf, Friston, & Frith, 2009). Studies on mirror neurons often use the paradigm of *fMRI adaptation* or *repetition suppression*. fMRI adaptation is the effect of reduced BOLD responses as a result of repeated presentation of a sensory stimulus (Chong, et al., 2008). Extending this idea to mirror neurons, if mirror neurons do exist, there should be reduced neural responses when observing an action that was previously executed relative to a novel action. This effect is also known as cross-modality adaptation, as the repetition suppression effect is observed across different 'modes' of a particular action (executive versus observation).

There have been conflicting reports on the existence of brain areas that resemble the behavior of mirror neurons in humans. Using pantomimed hand actions as stimuli, Chong and colleagues (2008) found effects of cross-modality adaptation in the right inferior parietal lobe, thus suggesting the existing of the human mirror neuron system in that region. However, their studies were refuted by Lingnau and colleagues (2009) who did not find the adaptation effect when simple right-handed motor actions were first executed and subsequently observed. This null result could be due to the use of meaningless actions, as the mirror neurons may only be activated by goal-direction actions (Kilner, et al., 2009).

6. Mentalizing and understanding mental states of others (Theory of Mind)

The term Theory of Mind (ToM) refers to the ability to attribute mental states (e.g., attitudes, perceptions, attitudes, intentions) to others in order to predict and explain behavior of another person (Premack & Woodruff, 1978). In such studies, participants are often instructed to infer the mental states of a target, which can be a person (Mitchell, Banaji, & Macrae, 2005b), a dog (Mitchell, Banaji, & Macrae, 2005a) or a character in a fictional story (Vogeley, et al., 2001). Cartoons or comic strips have also been used as stimuli in some studies (Gallagher, et al., 2000; Walter, et al., 2004). Mason and collegues (2004) found that ToM mechanisms are also involved when participants evaluate whether certain motor actions can be performed by another individual.

6.1. Empathy

There are two aspects of empathy: (i) the cognitive ability to understand and predict another person's mental states, as well as (ii) the affective ability to experience an appropriate emotion in response to another person's mental state (Baron-Cohen & Wheelwright, 2004).

Studies on empathy often use emotional facial expressions as stimuli. Typically, the tasks involve either passive observation of another's emotional expression, imitate the expressions observed or to independently generate specific emotional expression (e.g., Carr, Iacoboni, Dubeau, Mazziotta, & Lenzi, 2003; Wicker, et al., 2003). Findings from these studies often suggest the involvement of ToM and mimicry mechanisms in empathy. Merely instructing participants to describe the emotion of a facial expression was sufficient to engage brain areas commonly associated with ToM, thus suggesting the involvement of unconscious mimicry in empathy (Schulte-Rüther, Markowitsch, Fink, & Piekfe, 2007). Furthermore, BOLD responses of these regions have been positively correlated with self-ratings on the Balanced Emotional Empathy Scale (Lawrence, et al., 2006; Schulte-Rüther, et al., 2007).

Many studies have also looked into the neural responses associated with empathetic responses towards others' pain. In a study by Singer and colleagues (2004), they recruited couples and examined the BOLD response of the female partners when they saw their male partners being administered painful electric stimulation. Their findings suggest that only the affective region, but not the sensory regions, of the pain matrix (areas typically activated when a participant is experiencing pain) is involved in empathetic pain. Jackson and colleagues (2006) also demonstrated both common and distinct neural activations between imagining oneself versus imagining others in painful situations. Participants were shown pictures of feet and hands in some common painful situations (e.g., toes being caught under a heavy object), and tasked to evaluate the level of pain if they were in that situation (self condition) or another individual were in that situation (other condition).

Biases in empathy between ingroup versus outgroup targets are also common research directions. Cheon and colleagues (2011) examined how culture can affect empathy (for details, see below under the section 'Culture'). Cikara and Fiske (2011) hypothesized that individuals show differential empathy responses toward different outgroup members, depending on the stereotype of the specific outgroup. To examine this phenomenon, participants saw different types of events (e.g., "had stomachache after lunch") paired with a photograph of an outgroup member (e.g., elderly person, homeless person, business person). Furthermore, they also examined empathy for positive events (e.g., "found $5 bill") instead of merely focusing on negative or emotionally painful events.

6.2. Distinguishing self from others

Being able to distinguish the self from others is fundamental for many higher-order abilities, such as directing goal-specific actions and self-reflection of one's mental or emotional states. One way we distinguish our actions from others' is through our intentions to move. This is also known as *self-agency*, which is the knowledge of being the source of causality. A common paradigm used to investigate this phenomenon is through the mismatch of sensory-action consequence. Typically, participants performed a simple task with periodic "interference" from the experimenter. Such tasks may involve pressing a button to produce a certain color (Spengler, von Cramon, & Brass, 2009) or directing a target down a path (Farrer & Frith, 2002). During the "interference", the action produced would be the opposite of what the participants had intended, thus producing incongruence (or mismatch) between their actions and the actual consequences.

Some tasks may also use temporal mismatch rather than consequential mismatch. In other words, there will be a time lag between their actions and the perception of their actions. For example, participants performed some simple hand movements while the same actions on a screen. They may be viewing either a visual feedback of their own actions with varying degree of temporal delay (Leube, et al., 2003), or similar actions performed by an actor (Macuga & Frey, 2011). For the latter paradigm, it is important that participants know whether they are watching a pre-recorded action or live feedback of their own action.

Another aspect of distinguishing the self is our ability to recognize our own faces and own mental or emotional states. Many neuroscience studies on self-face recognition showed participants photographs of either themselves, a close or familiar other and a stranger. The stimuli were digital morphs between these pictures (Uddin, Kaplan, Molnar-Szakacs, Zaidel, & Iacoboni, 2005) or simply the original facial photographs (e.g., Platek, Keenan, Gallup Jr, & Mohamed, 2004; Platek et al., 2006). In a study on self's emotional reflection, participants were shown photographs of people displaying different emotions and asked to evaluate their own emotional response to each photograph. This revealed a neural network in the prefrontal cortex that is specifically recruited for evaluation of own emotions (Ochsner et al., 2004). In another study on high-functioning individuals with autism, this ability to separate self from others has also been demonstrated; these individuals were impaired in representing their own social intention but not in distinguishing others' social actions from the self (Chiu et al., 2008).

6.3. Morality

The main focus of experimental studies of morality is the evaluation of moral judgments. Thus, participants are commonly asked to either evaluate the morality of an action (for instance,

http://dx.doi.org/10.5772/56630

driving away after bumping another parked car) or to decide how they themselves would respond in a moral dilemma (e.g., smothering a loud crying baby in order to save oneself and fellows from enemy soldiers). Moll and colleagues (2002) compared emotionally charged pictures that either conveyed moral violations (e.g., physical assaults) or not (e.g., accidents). In a later study (Moll et al., 2005), in order to examine neurobehavioral differences between emotion with and without moral aspect, they also contrasted pictures of pure disgust with and without indignation.

Besides identifying the neural correlates, the principal goal is often to uncover the dimensions that can account for moral judgments. Often, these are in comparison with basic emotions. For instance, recent research has compared 'utilitarian' and 'non-utilitarian' choices. A classic example is the choice of whether to adopt a new vaccine that will ensure immunity to the vast majority of people but will also introduce the disease to a small minority. Adopting the vaccine is the utilitarian choice as the decision criterion is overall welfare; non-utilitarian option is not to adopt the vaccine in order to avoid harm to innocent people. Other dimensions include personal versus impersonal dilemmas (Greene, Sommerville, Nystrom, Darley, & Cohen, 2001); consequences, action (versus inaction) and intentionality (Borg, Hynes, Van Horn, Grafton, & Sinnott-Armstrong, 2006); and harm versus dishonesty versus disgust (Parkison et al., 2011; Haidt, 2002; Rozin, Lowery, Imada, & Haidt, 1999). A useful resource of condensed moral vignettes can be found at Knutson et al. (2010).

We would like to add two important words of caution with regards to the morality studies. Firstly, it is unlikely that participants have encountered the situations used in such experiments. This raises reasonable criticisms as to whether the moral judgments made during the experimental sessions are realistic or credible. Secondly, the experimental moral situations versus either (i) other moral situations or (ii) non-moral situations might be different on more than one dimension. For instance, the contrasted scenarios often also differ in the emotional reaction elicited, levels of social desirability or involvement of bodily harm (Heekeren, Wartenburger, Schmidt, Schwintowski & Williringer, 2003). Therefore, great care should be given in the development of proper controls.

6.4. Kinship, romantic and maternal love

The aim of this group of studies is often to examine the neuronal responses associated with positive feelings towards close others (such as romantic partners or children). The universality of both romantic and maternal love (Jankowiak & Fischer, 1992) has been associated with the involvement of a strong, evolutionary mechanism. Typically, in these studies, participants engage in passive viewing of pictures of a person they 'feel love' towards, often a romantic partner or an infant (Aron et al., 2005; Strathearn, Fonagy, Amico & Montague, 2009; Strathearn, Li, Fonagy, & Montague, 2008; Zeki, 2007). Mashek, Aron and Fisher (2000) suggested that a photograph is more effective in eliciting a positive response compared to either touch or voice of the person.

In such studies, the experimenter often has to control for familiarity. Some common controls used include pictures of familiar but emotionally neutral acquaintances, such as photographs of babies whom they have known for the same amount of time. Another important methodological issue to note is with regard to one's emotional response. Intensity of emotional response towards a pictorial stimulus often diminishes after 30 seconds of exposure (Mashek, Aron & Fisher, 2000).

7. Attitudes and evaluative processes

Two fundamental concepts in social psychology are attitudes (i.e., 'relatively stable ideas about whether something is good or bad'; Cunningham & Zelazo, 2007, pp. 97) and evaluation (i.e., temporally narrower appraisals). Theoretically, a major categorization of processes refers to implicit (mostly automatic) versus explicit (mostly controlled) processes (also known as Dual Attitude Theory; see Greenwald & Banaji, 1995). Implicit, automatic attitudes towards specific social constructs are often measured by the Implicit Association Test (IAT; Greenwald, McGhee, & Schwartz, 1998). Information gleaned from IAT is valuable in an fMRI context as it can be correlated with brain responses.

One common method used to study 'automatic/implicit' versus 'controlled/explicit' attitude-related responses is the manipulation of the duration of presentation of a stimulus. Stimuli can be presented to participants subliminally by carefully manipulating the duration of presentation such that presentation is so short that stimulus cannot be consciously perceived by participants but yet long enough for unconscious perception to take place. Cunningham and colleagues (2004) found higher amygdala response in participants when Afro-American faces, relative to Caucasian faces, were presented subliminally. This effect was not found when the stimuli were presented supraliminally. Another method of inducing automatic responses is to frame the task in an analogous way. For instance, Cunningham and colleagues (2003, 2004) asked participants to either evaluate a word (e.g., "murder") as a 'good' or 'bad' word, or to evaluate whether it is an abstract or concrete term. The former evaluation is more likely to elicit controlled (and thus socially desirable) responses from the participants.

8. Race and stereotype

The effect of controlled versus automatic responses is a major question in studies on racial stereotyping as well. Physiological measures, and in particular, neuronal responses have an advantage over traditional behavioral measures, particularly self-report, as the responses are less likely to be 'controlled' by the participants. This aids in reducing social desirability issues as well as uncovering 'hidden' temporal patterns to responses; in addition, behavioral responses can be more easily influenced by extraneous factors, such as fatigue and order effects. Thus, neuroscience is a good supplement to the more traditional behavioral methodologies. The prototypical fMRI task in such studies involves the presentation of same- and other-race face stimuli; the amygdala is often activated when out-group faces are presented (Hart et al., 2000; but see next paragraph & Phelps et al., 2000).

It is important to point out some important issues that should be taken into consideration when designing a study related to race. Firstly, neural responses to race-based stimuli might depend, to a great extent, on the duration of the presentation itself. Differences in behavioral and brain responses are likely to be related to whether the stimulus is presented subliminally or supraliminally. Differences in neural responses to ingroup and outgroup faces disappear when the stimuli are presented for a longer duration, which allows for conscious elaboration (see Phelps et al., 2000).

Secondly, the mere exposure to outgroup faces might induce cognitive depletion effect for some participants. Richeson and colleagues (2003) found that performance in a Stroop color-naming task was more impaired after interaction with an Afro-American person; however, this effect is limited to Caucasian participants who exhibited anti-Afro-American evaluative bias.

Thirdly, researchers should take extra care in the interpretation of differences in neural responses to the different faces. Differences may also result from mere increased familiarity (even to low level characteristics such as color patterns), perception exposure or social knowledge (in contrast to biases or stereotypes; Eberhardt, 2005). For example, Golarai and colleagues (2004) found that the difference in fusiform activation when perceiving Caucasian versus Afro-American faces extend to perception of scrambled faces as well.

One should also note that it might be easier to uncover the neural responses implicated in stereotyping when implicit (rather than explicit) measurements are used. For instance, in Phelps et al. (2000), no differences were found in amygdala response when Caucasian versus Afro-American faces were presented explicitly; however, differential amygdala response emerged when implicit attitudes measures (e.g., IAT or startle responses) were used instead. Furthermore, event-related responses employing EEG have been very prolific in showing implicit racial biases and their successful or unsuccessful regulation (see Amodio & Bartholow, 2011, for a review).

9. Culture

Behavioral evidence has demonstrated that cultural experiences can influence responses at a variety of tasks. Tasks in such cultural studies look at basic mechanisms (e.g., visual processing) as well as more complex behaviors (e.g., social cognition). One of the fundamental goals of culture neuroscience is to evaluate differences in neural responses between people with different cultural backgrounds. Below are some central phenomena that have been studied.

Holistic versus analytic processing of visual information. In such studies, common stimuli used are pictures with a foreground object against a background. It has been hypothesized that Caucasians will fixate to the central object while Asians also pay equal attention to the background (Chua, Boland, & Nisbett, 2005). This is likely to be related to a more general holistic (versus analytic) processing style characterizing Asian societies (as compared with North American societies; Chiao et al., 2009). Another task that has been used to study this phenomena is the "framed-line test" (Kitayama, Kawamura, & Larsen, 2003), which sustains that perceptual comparisons mechanisms are different between Japanese and US students.

Though widely used and examined, both the framed-line task and the cultural basis of the holistic/analytic dipole in perceptual terms have been challenged as their results either contradict basic visual science findings, which offers simpler explanations (see Zhou, Gotch, Zhou & Liu, 2008) or they seem to be produced by the natural (and not the cultural environment; Ueda & Komiya, 2012).

Emotional faces recognition. Similar to the 'racial stereotypes' experiments described in the previous section, participants in these studies are presented with emotional faces from either the same or different cultures as themselves to investigate the differences in neural activations and possible mechanisms to behaviors towards members of same versus different cultures (Chiao, et al., 2008). Adams Jr and colleagues (2009) had also used another task to study the same phenomenon. They used the 'Reading the mind in the Eyes' test (Baron-Cohen, Wheelwright, Hill, Raste, & Plumb, 2001), where participants described the mental state of a person based only on the picture of his/her eyes.

Social elements, such as responses to status and hierarchy. Freeman and colleagues (2009) examined how culture shapes individual's neural reward mechanisms. In their study, participants passively viewed pictures of body displays displaying dominate body posture or subordinate body posture. Cheon and colleagues (2011) also examined how cultural preferences for social hierarchy influence empathy for outgroup members, while further examining how these preferences interact with different cultural groups (Japanese / Korean vs. European / Americans). Participants viewed photographs of scenes that were emotionally painful or neutral. They also indicated their empathy level toward the person in the photograph through a button-press controller in the scanner.

Recent developments in culture science suggest that a valuable method will be to use individuals who have been acculturated into two different cultures (i.e., biculturals). This allows researchers to compare differential cultural responses within the same individuals, instead of comparing between groups of participants. Such differential responses can be elicited by priming these bicultural participants with different cultural images, hence activating the respective cultural frame (for more details see Hong, Morris, Chiu, & Benet-Martinez, 2000; Oyserman & Lee, 2008). This line of research demonstrated that priming biculturals with different cultural images is associated with differential behavioral and neuronal responses (for instance, priming with either Asian or Western images is associated with different responses in culturally sensitive tasks). This suggests that cultural experience influences brain function, which is at odds with the static and essentialistic interpretation of inherent cultural differences in neural responses.

It is important for researchers to be mindful of some major issues when engaging in cross-cultural research. In addition to the issues that are also present in cross-cultural behavioural research (e.g., matched samples across cultures, properly translated instructions), cross-cultural neuroimaging has additional caveats that researchers need to be aware of:

1. **Scanner compatibility**. Typically, neuroimaging data from different cultures are captured by different scanners. It is thus important that researchers should strive to collect data that are captured by the same, or not very highly similar, MRI machines with the same

http://dx.doi.org/10.5772/56630

sequences. Furthermore, interscanner reliability tests have to be carried out to ensure that any differences found that not due to scanner variability (see Sutton et al., 2008).

2. **Task selection**. Standardised, non-verbal visuospatial tasks are preferable to numerical tasks, as the potential for language-mediated cultural bias is higher in the latter. Hedden et al. (2002) have a list of useful neuropsychological tests that researchers may want to refer to.

3. **Brain templates**. Most brain templates used by major neuroimaging software are based on data collected from Caucasian samples. However, there is a demonstrated variability in brain structural anatomy across populations (Tang et al., 2010), which may make the existing templates inappropriate. Some possible solutions include either finding a more appropriate template or to use templates derived from the sampled population itself.

10. Social interactions

A revolutionary development in social neuroscience is the introduction of game-theoretic methods in the study of social interactions. In such experiments, participants typically play particular roles in highly-structured social exchanges (i.e., games) to stimulate real social interactions.

There are multiple advantages to this methodology. Firstly, the interactions involve real monetary outcomes, thus making available a metric of phenomena that are otherwise difficult to capture (e.g., the dilemmas involved in trust or reciprocity). In addition, incentivizing the choices participants make allows researchers to study certain social phenomena that are highly sensitive to social desirability problems. For instance, participants might self-report that they are very trusting towards other people; however, they might behave otherwise when their choices or actions have financial consequences.

Another advantage is that there is a wealth of behavioral, simulation and theoretical studies related to such social-economic games. Optimal or 'rational' behaviors (e.g., Nash equilibrium) are well-characterized and studied. The mathematical descriptions of the various choices one can make are also readily available. Furthermore, behavioral explanations and formal computational models of the different choices have also been described in sufficient details. As such, it is possible to model the underlying, hidden computations or processes that change parametrically on a trial-by-trial basis. Moreover, one can study how dynamic social interactions evolve over time. This allows social neuroscientists to move beyond mere brain mapping to the examination of brain mechanisms involved, as described in the introduction of this chapter. In other words, the research focus can thus move from social neuroscience to *social computational neuroscience*.

Social computational neuroscience should include the following steps:

1. Adopt or develop a simulated interaction (game), based on the phenomenon being studied. Ensure that the behavioral choices actually reflect the phenomenon in question.

2. Find one or more algorithms that can describe the choices available. In dynamic environments, the focus can be on how previous interactions update the representations of the self, partner and their relationship.

3. Fit the algorithm/s to actual behaviors in order to identify the optimal values for the parameters of each model.

4. Identify whether the computations implied by the models are actually taking place in the brain. This will ensure the biological plausibility of the model.

An excellent example of the social computational neuroscience approach is the study by Behrens, Hunt, Woolrich and Rushworth (2008). The authors wanted to study the behavioral and neuronal responses related to advice from a partner. In their game, participants had to choose between different cards offering rewards associated with different and unknown probabilities (step 1). At the same time, they received advice from a partner who may or may not be trustworthy. Thus, participants needed to update, on each trial, the probability that the advisor is reliable (or not). The authors hypothesized that the probability was updated using basic reinforcement learning algorithms (step 2). They then fitted their model to the behavior in order to identify the (otherwise hidden) probability associated with their partner's perceived reliability. The outcome was a time-series of updated values, on a trial-by-trial basis (step 3). Next, they fitted the time-series values to the BOLD responses, which thus allowed them to identify the brain areas that appeared to subserve the relevant calculations (step 4).

As illustrated, this approach allows the identification of individual or group differences of very subtle and hidden computations. Another example of its application is the differentiation of participants according to whether they update the value of social gestures in social interactions. This is well demonstrated in a series of studies by King-Casas and colleagues (2005, 2008). In these studies, the authors used a paradigm known as the Trust Game (see figure 1).

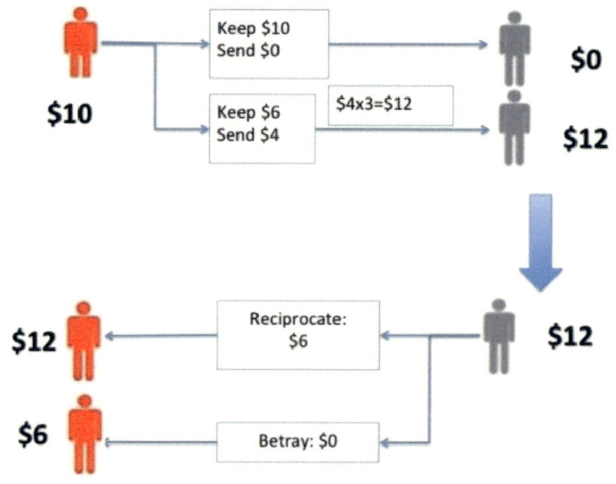

Figure 1. The Trust Game

http://dx.doi.org/10.5772/56630

In a typical trust game, player A (red in figure 1) is given an amount (e.g., $10). Player A has the choice to keep the amount, or to send ('invest') a portion of it to player B (gray in figure 1). The amount sent to the partner will then be tripled by the experimenter (as shown in our example, $4 was invested, and thus player B receives $12). Next, player B has to decide how much of the received amount (i.e., $12 in this case) to send back to player A as an indication reciprocity.

King-Casas and colleagues measured the BOLD responses as participants played the trust game. In the first study, they identified a neuronal response reflecting the computational learning signals that represented the intentions of the social partner. In the following study, they included participants who were diagnosed with Borderline Personality Disorder (BDP). This disorder is strongly characterized by inadequate and brief social relationships (see American Psychiatric Association, 2000, for more details). By analyzing behavior in the trust game, they found that participants with BDP cooperated less as the game progressed, thus indexing a breach of trust in the dyad. This was reflected in these participants' insula brain responses, which failed to represent the magnitude of offers received. This series of studies elegantly demonstrated how a simple simulated social interaction can be used to uncover social deficits.

Many similar games have been used to describe or simulate various social interactions. It is beyond the scope of the current chapter to elaborate on each game. Nevertheless, we present a list of relevant papers together with the major constructs they measured in Table 1.

Game / Task	**Concept**	**References (fMRI or behavioural studies)**
Prisoner's Dilemma; ultimatum game	Cooperation / Competition; unfairness	Rilling et al., (2004)
Social allocations / social value orientation / fairness	Social predisposition	Hsu, Anen & Quartz (2008) Haruno & Frith (2010) Tricomi, et al. (2010)
Ball-tossing task (Cyberball)	Social Rejection	Williams, Cheung & Choi (2000)
Trust game	Trust, reciprocation	King-Casas, et al. (2005, 2008)
Observational learning	Learning from others	Burke, et al. (2010)
Altruistic punishment	Altruistic punishment	de Quervain, et al. (2004)
Work-or-shirk	Mentalising	Hampton, et al. (2008)
Beauty contest	Depth of strategic reasoning	Coricelli & Nagel (2009) See also Bhatt & Camerer (2005)
Social comparisons, envy and Schadenfreude	Schadenfreude occurs when envied persons fall from grace	Takahashi, et al. (2009); Fliessbach et al. (2007)
Punishment games	Parochial altruism; norm enforcement	Baumgartner, et al. (2011)
Promise game	Promise keeping	Baumgartner, et al. (2009)
Bargaining Games	Bargaining/Suspicion/Strategic Deception	Bhatt, et al. (2010)

Table 1. A limited list of games employed in the study of social interactions

11. Criticisms, issues and the future

A common criticism to social neuroscience in general (and games in particular) is the lack of ecological validity as the tasks seem to fail to represent the complexity of actual social interactions. The same argument has been directed towards non-social decision-making studies in neuroeconomics (Volz & Gigerenzer, Chapter 22, this Book). There have been plenty of responses to this criticism.

Firstly, participants do appear to be sensitive to the experimental manipulations and thus display the behaviors that truly represent their social preference or even social pathology (see above for example of participants with BDP). This is illustrated in many neuroeconomic studies where participants demonstrated consistent risk aversion (Christopoulos, Tobler, Bossaerts, Dolan, & Schultz, 2009), even with small monetary rewards. In addition, these games can be the basis or building block of more realistic and sophisticated simulations of social interactions (see Wolf, Dziobej, & Heekeren, 2010). Furthermore, this approach allows for an understanding of basic or fundamental cognitions which can contribute to subsequent models of more complex behaviors.

Nevertheless, this criticism is not unfounded and has been constructive in bringing improvements to the field of social neuroscience. A number of recent studies have tried to relate neuroimaging results with real life behaviors. Kanai, Feilden, Firth and Rees (2011) used brain volumetric techniques to correlate neuroanatomical structural differences (i.e., size of specific brain structures) with the size of one's social networks (e.g., Facebook). In another study, Falk, Berkman, Whalen and Lieberman (2011) demonstrated that neural responses to health messages can predict whether the participant will reduce unhealthy behaviors (i.e., smoking in their study) over-and-above self-reported intentions.

In conclusion, social neuroscience is in its infancy. It can be expected that studies trying to understand basic mechanisms will eventually corroborate with studies that more closely investigate real life, out-of-the-lab behaviors.

Acknowledgements

We thank Bobby Cheon and Mark Khei for their invaluable comments on an earlier draft of the present paper. The preparation of this article was partially supported by the Academic Research Fund (AcRF) Tier 1 (RG 1/11 M4010946.010) of the Ministry of Education, Singapore awarded to the second author.

Author details

D.Y. Phua[1,2] and G.I. Christopoulos[1,2,3]

1 Nanyang Business School, Nanyang Technological University, Singapore

2 Culture Science Institute, Nanyang Tchnological University, Singapore

3 Virginia Tech Carilion Research Institute, VA, USA

References

[1] Adams Jr, R. B., Rule, N. O., Franklin Jr, R. G., Wang, E., Stevenson, M. T., Yoshikawa, S., ... & Ambady, N. (2010). Cross-cultural reading the mind in the eyes: An fMRI investigation. *Journal of Cognitive Neuroscience*, 22(1), 97-108.

[2] American Psychiatric Association (2000). *Diagnostic and statistical manual of mental disorders* (4th ed., Text Revision). Washington, DC: Author.

[3] Amodio, D. M., & Bartholow, B. D. (2011). Event-related potential methods in social cognition. In C. Klauer, A. Voss, & C. Stahl (Eds.), *Cognitive methods in social psychology* (pp. 303–339). New York : Guilford Press.

[4] Aron, A., Fisher, H., Mashek, D. J., Strong, G., Li, H., & Brown, L. L. (2005). Reward, motivation, and emotion systems associated with early-stage intense romantic love. *Journal of Neurophysiology*, 94(1), 327–337.

[5] Baron-Cohen, S., & Wheelwright, S. (2004). The empathy quotient: An investigation of adults with Asperger syndrome or high functioning autism, and normal sex differences. *Journal of Autism and Developmental Disorders*, 34(2), 163–175.

[6] Baron-Cohen, S., Wheelwright, S., Hill, J., Raste, Y., & Plumb, I. (2001). The "Reading the Mind in the Eyes" Test Revised Version: A study with normal adults, and adults with Asperger Syndrome or high-functioning Autism. *Journal of Child Psychology and Psychiatry*, 42(2), 241–251.

[7] Baumgartner, T., Fischbacher, U., Feierabend, A., Lutz, K., & Fehr, E. (2009). The neural circuitry of a broken promise. *Neuron*, 64(5), 756–770.

[8] Baumgartner, T., Knoch, D., Hotz, P., Eisenegger, C., & Fehr, E. (2011). Dorsolateral and ventromedial prefrontal cortex orchestrate normative choice. *Nature Neuroscience*, 14(11), 1468–1474.

[9] Behrens, T. E, Hunt, L. T, Woolrich, M. W, Rushworth, M. F (2008). Associative learning of social value. *Nature*, 456(7219), 245–249.

[10] Bhatt, M., & Camerer, C. F. (2005). Self-referential thinking and equilibrium as states of mind in games: fMRI evidence. *Games and Economic Behavior*, 52(2), 424–459.

[11] Bhatt, M. A., Lohrenz, T., Camerer, C. F., & Montague, P. R. (2010). Neural signatures of strategic types in a two-person bargaining game. *Proceedings of the National Academy of Sciences*, 107(46), 19720-19725.

[12] Blanke, O., Mohr, C., Michel, C. M., Pascual-Leone, A., Brugger, P., Seeck, M., et al. (2005). Linking out-of-body experience and self processing to mental own-body imagery at the temporoparietal junction. *The Journal of Neuroscience*, 25(3), 550–557.

[13] Borg, S. J., Hynes, C., Van Horn, J., Grafton, S., & Sinnott-Armstrong, W. (2006). Consequences, action, and intention as factors in moral judgments: An fMRI Investigation. *Journal of Cognitive Neuroscience, 18*(5), 803–817.

[14] Buccino, G., Lui, F., Canessa, N., Pastteri, I., Lagravinese, G., Benuzzi, F., et al. (2004). Neural circuits involved in the recognition of actions performed by nonconspecifics: An fMRI study. *Journal of Cognitive Neuroscience*, 16(1), 114–126.

[15] Buccino, G., Vogt, S., Ritzl, A., Fink, G. R., Zilles, K., Freund, H.-J., et al. (2004). Neural circuits underlying imitation learning of hand actions: An event-related fMRI study. *Neuron*, 42(2), 323–334.

[16] Burke, C. J., Tobler, P. N., Baddeley, M., and Schultz, W. (2010). Neural mechanisms of observational learning. *Proceedings of the National Academy of Sciences*, 107(32), 14431–14436.

[17] Calvert, G. A., Bullmore, E. T., Brammer, M. J., Campbell, R., Williams, S. C. R., McGuire, P. K., et al. (1997). Activation of auditory cortex during silent lipreading. *Science*, 276(5312), 593–596.

[18] Calvo-Merino, B., Glaser, D. E., Grèzes, J., Passingham, R. E., & Haggard, P. (2005). Action observation and acquired motor skills: An fMRI study with expert dancers. *Cerebral Cortex*, 15(8), 1243–1249.

[19] Campbell, R., MacSweeney, M., Surguladze, S., Calvert, G., McGuire, P., Suckling, J., et al. (2001). Cortical substrates for the perception of face actions: An fMRI study of the specificity of activation for seen speech and for meaningless lower-face acts (gurning). *Cognitive Brain Research*, 12(2), 233–243.

[20] Carr, L., Iacoboni, M., Dubeau, M.-C., Mazziotta, J. C., & Lenzi, G. L. (2003). Neural mechanisms of empathy in humans: A relay from neural systems for imitation to limbic areas. *Proceedings of the National Academy of Sciences*, 100(9), 5497–5502.

[21] Cheon, B. K., Im, D.-m., Harada, T., Kim, J.-S., Mathur, V. A., Scimeca, J. M., ..., Chiao, J. Y. (2011). Cultural influences on neural basis of intergroup empathy. *NeuroImage*, 57(2), 642-650.

[22] Chiao, J. Y., Harada, T., Komeda, H., Li, Z., Mano, Y., Saito, D. N., ..., Iidaka, T. (2009). Dynamic cultural influences on neural representations of self. *Journal of Cognitive Neuroscience*, 22 (1), 1–11.

[23] Chiao, J. Y., Iidaka, T., Gordon, H. L., Nogawa, J., Bar, M., Aminoff, E., . . ., Ambady, N. (2008). Cultural specificity in amygdala response to fear faces. *Journal of Cognitive Neuroscience*, 20, 2167–2174.

[24] Cikara, M., & Fiske, S. T. (2011). Bounded empathy: Neural responses to outgroup targets' (mis)fortunes. *Journal of Cognitive Neuroscience*, 23(12), 3791-3803.

[25] Chiu, P. H., Kayali, M. A., Kishida, K. T., Tomlin, D., Klinger, L. G., Klinger, M. R., & Montague, P. R. (2008). Self responses along cingulate cortex reveal quantitative neural phenotype for high-functioning autism. *Neuron*, 57(3), 463-473.

[26] Chong, T. T. J., Cunnington, R., Williams, M. A., Kanwisher, N., & Mattingley, J. B. (2008). fMRI adaptation reveals mirror neurons in human inferior parietal cortex. *Current Biology*, 18(20), 1576–1580.

[27] Christopoulos, G. I., Tobler, P. N., Bossaerts, P., Dolan, R. J., Schultz, W. (2009). Neural correlates of value, risk, and risk aversion contributing to decision making under risk. *Journal of Neuroscience*, 29(40), 12574–12583.

[28] Chua, H. F., Boland, J. E., & Nisbett, R. E. (2005). Cultural variation in eye movements during scene perception. *Proceedings of the National Academy of Science*, 102 (35), 12629–12633.

[29] Coricelli, G., & Nagel, R. (2009). Neural correlates of depth of strategic reasoning in medial prefrontal cortex. *Proceedings of the National Academy of Sciences*, 106(23), 9163–9168.

[30] Cunningham, W. A., Johnson, M. K., Gatenby, J. C., Gore, J. C., & Banaji, M. R. (2003). Neural components of social evaluation. *Journal of Personality and Social Psychology*, 85(4), 639–649.

[31] Cunningham, W. A., Johnson, M. K., Raye, C. L., Gatenby, C. J., Gore, J.C., & Banaji, M.R. (2004). Separable neural components in the processing of Black and White faces. *Psychological Science*, 15(12), 806–813.

[32] Cunningham, W. A., Raye, C. L., & Johnson, M. K. (2004). Implicit and explicit evaluation: fMRI correlates of valence, emotional intensity, and control in the processing of attitudes. *Journal of Cognitive Neuroscience*, 16(10), 1717–1729.

[33] Cunningham, W. A., & Zelazo, P. D. (2007). Attitudes and evaluations: A social cognitive neuroscience perspective. *Trends in Cognitive Sciences*, 11(3), 97–104.

[34] Engell, A. D., Haxby, J. V., & Todorov, A. (2007). Implicit trustworthiness decisions: Automatic coding of face properties in the human amygdala. *Journal of Cognitive Neuroscience*, 19(9), 1508–1519.

[35] Eberhardt, J. L. (2005). Imaging race. *American Psychologist*, 60(2), 181–190.

[36] Falk, E. B., Berkman, E. T., Whalen, D., & Lieberman, M. D. (2011). Neural activity during health messaging predicts reductions in smoking above and beyond self-report. *Health Psychology*, 30(2), 177–185.

[37] Farrer, C., & Frith, C. D. (2002). Experiencing oneself vs another person as being the cause of an action: The neural correlates of the experience of agency. *NeuroImage*, 15(3), 596-603.

[38] Fliessbach, K., Weber, B., Trautner, P., Dohmen, T., Sunde, U., Elger, C. E., & Falk A. (2007). Social comparison affects reward-related brain activity in the human ventral striatum. *Science*. 318(5854), 1305–8.

[39] Freeman, J. B., Rule, N. O., Adams, R. B., & Ambaby, N. (2009). Culture shapes a mesolimbic response to signals of dominance and subordination that associates with behaviour. *NeuroImage*, 47(1), 353–359.

[40] Gallagher, H. L., Happé, F., Brunswick, N., Fletcher, P. C., Frith, U., & Frith, C. D. (2000). Reading the mind in cartoons and stories: An fMRI study of 'theory of mind' in verbal and nonverbal tasks. *Neuropsychologia*, 38(1), 11–21.

[41] German, T. P., Niehaus, J. L., Roarty, M. P., Giesbrecht, B., & Miller, M. B. (2004). Neural correlates of detecting pretense: Automatic engagement of the intentional stance under covert conditions. *Journal of Cognitive Neuroscience*, 16(10), 1805–1817.

[42] Golarai, G., Ghahremani, D. G., Whitfield-Gabrieli, S., Reiss, A., Eberhardt, J. L., Gabrieli, J. D., & Grill-Spector, K. (2007). Differential development of high-level visual cortex correlates with category-specific recognition memory. *Nature Neuroscience*, 10(4), 512–522.

[43] Greenwald, A. G., & Banaji, M. R. (1995). Implicit social cognition: Attitudes, self- esteem, and stereotypes. *Psychological Review*, 102(1), 4–27.

[44] Greenwald, A. G., McGhee, D. E., & Schwartz, J. K. L. (1998). Measuring individual differences in implicit cognition: The implicit association test. *Journal of Personality and Social Psychology*, 74(6), 1464–1480.

[45] Greene, J. D., Sommerville, R. B., , Nystrom, L. E., Darley, J. M., & Cohen, J. D. (2001). An fMRI investigation of emotional engagement in moral judgment. *Science*, 293(5537), 2105-2108.

[46] Haidt, J. (2002). Dialogue between my head and my heart: Affective influences on moral judgment. *Psychological Inquiry*, 13(1), 54–56.

[47] Hampton, A. N., Bossaerts, P., & O'Doherty, J. P. (2008). Neural correlates of mentalizing-related computations during strategic interactions in humans. *Proceedings of the National Academy of Sciences*, 105(18), 6741–6746.

[48] Harenski, C. L., Kim, S., & Hamann, S. (2009). Neuroticism and psychopathy predict brain activation during moral and nonmoral emotion regulation. *Cognitive, Affective, and Behavioral Neuroscience*, 9(1), 1–15.

[49] Harrison, N. A., Singer, T., Rotshtein, P., Dolan, R. J., & Critchley, H. D. (2006). Pupillary contagion: Central mechanisms engaged in sadness processing. *Social Cognitive and Affective Neuroscience*, 1(1), 5–17.

[50] Hart, A.J., Whalen, P.J., Shine, L.M., McInerney, S.C., Fischer, H., & Rauch, S.L. (2000). Differential response in the human amygdala to racial outgroup vs ingroup face stimuli. *Neuroreport*, 11(11), 2351– 2355.

[51] Haruno, M., & Frith, C. (2010). Activity in the amygdala elicited by unfair divisions predicts social value orientation. *Nature Neuroscience*, 13(2), 160–161.

[52] Hedden, T., Park, D.C., Nisbett, R., Ji, L.-J., Jing, Q., & Jiao, S. (2002). Cultural variation in verbal versus spatial neuro- psychological function across the life span. *Neuropsychology*, 16(1), 65–73.

[53] Heekeren, H.R., Wartenburger, I., Schmidt, H., Schwintowski, H.P., & Villringer, A. (2003) An fMRI study of simple ethical decision-making. *Neuroreport*, 14(9), 1215–1219.

[54] Hoffman, E. A., & Haxby, J. V. (2000). Distinct representations of eye gaze and identity in the distributed human neural system for face perception. *Nature Neuroscience, 3*(1), 80-84.

[55] Hong, Y., Morris, M.W., Chiu, C., & Benet-Martinez, V. (2000). Multicultural minds: A dynamic constructivist approach to culture and cognition. *American Psychology*, 55(7), 709–720.

[56] Hsu, M., Anen, C., & Quartz, S. R. (2008). The right and the good: Distributive justice and neural encoding of equity and efficiency. *Science*, 320(5879), 1092–1095.

[57] Huettel, S. A., Song, A. W., & McCarthy, G. (2009). *Functional Magnetic Resonance Imaging* (2nd ed.). Massachusetts: Sinauer Associates.

[58] Iacoboni, M., Woods, R. P., Brass, M., Bekkering, H., Mazziotta, J. C., & Rizzolatti, G. (1999). Cortical mechanisms of human imitation. *Science, 286*(5449), 2526–2528.

[59] Jackson, P. L., Brunet, E., Meltzoff, A. N., & Decety, J. (2006). Empathy examined through the neural mechanisms involved in imagining how I feel versus how you feel pain. *Neuropsychologia*, 44(5), 752–761.

[60] Jankowiak, W. R., & Fischer, E. F., (1992). A cross-cultural perspective on romantic love. *Ethnology*, 31(2), 149-155.

[61] Kanai, R., Feilden, T., Firth, C., & Rees, G. (2011). Political orientations are correlated with brain structure in young adults. *Current Biology*, 21(8), 1–4.

[62] Kilner, J. M., Neal, A., Weiskopf, N., Friston, K. J., & Frith, C. D. (2009). Evidence of mirror neurons in human inferior frontal gyrus. *The Journal of Neuroscience*, 29(32), 10153–10159.

[63] King-Casas, B., Sharp, C., Lomax-Bream, L., Lohrenz, T., Fonagy, P., & Montague, P. R. (2008). The rupture and repair of cooperation in borderline personality disorder. *Science,* 321(5890), 806–810.

[64] King-Casas, B., Tomlin, D., Anen, C, Camerer, C. F., Quartz, S. R., Montague, P. R. (2005). Getting to know you: Reputation and trust in a two-person economic exchange. *Science,* 308(5718),78-83.

[65] Kitayama, S., Kawamura, T., & Larsen, J. T. (2003). Perceiving an object and its context in different cultures: A cultural look at new look. *Psychological Science,* 14(3), 201–206.

[66] Knutson, K. M., Krueger, F., Koenigs, M., Hawley, A., Escobedo, J. R., Vasudeva, V., Adolphs, R., & Grafman, J. (2010). Behavioral norms for condensed moral vignettes. *Social Cognitive and Affective Neuroscience,* 5(4), 378–384.

[67] Kourtzi, Z., & Kanwisher, N. (2000). Activation in human MT/MST by static images with implied motion. *Journal of Cognitive Neuroscience,* 12(1), 48–55.

[68] Kühn, S., Müller, B. C. N., van Baaren, R. B., Wietzker, A., Dijksterhuis, A., & Brass, M. (2010). Why do I like you when you behave like me? Neural mechanisms mediating positive consequences of observing someone being imitated. *Social Neuroscience,* 5(4), 384–392.

[69] Lawrence, E. J., Shaw, P., Giampietro, V. P., Surguladze, S., Brammer, M. J., & David, A. S. (2006). The role of 'shared representations' in social perception and empathy: An fMRI study. *NeuroImage,* 29(4), 1173–1184.

[70] Leube, D. T., Knoblich, G., Erb, M., Grodd, W., Bartels, M., & Kircher, T. T. J. (2003). The neural correlates of perceiving one's own movements. *NeuroImage,* 20(4), 2084–2090.

[71] Lingnau, A., Gesierich, B., & Caramazza, A. (2009). Asymmetric fMRI adaptation reveals no evidence for mirror neurons in humans. *Proceedings of the National Academy of Sciences,* 106(24), 9925–9930.

[72] Macuga, K. L., & Frey, S. H. (2011). Selective responses in right inferior frontal and supramarginal gyri differentiate between observed movements of oneself vs. another. *Neuropsychologia,* 49(5), 1202–1207.

[73] Mason, M. F., Banfield, J. F., & Macrae, C. N. (2004). Thinking about actions: The neural substrates of person knowledge. *Cerebral Cortex,* 14(2), 209–214.

[74] Mashek, D., Aron, A., & Fisher, H. E. (2000). Identifying, evoking, and measuring intense feelings of romantic love. *Representative Research in Social Psychology,* 24, 48–55.

[75] McKay, L. S., Simmons, D. R., McAleer, P., Marjoram, D., Piggot, J., & Pollick, F. E. (2012). Do distinct atypical cortical networks process biological motion information in adults with Autism Spectrum Disorders? *NeuroImage,* 59(2), 1524-1533.

[76] Mitchell, J. P., Banaji, M. R., & Macrae, C. N. (2005a). General and specific contributions of the medial prefrontal cortex to knowledge about mental states. *NeuroImage*, 28(4), 757–762.

[77] Mitchell, J. P., Banaji, M. R., & Macrae, C. N. (2005b). The link between social cognition and self-referential thought in the medial prefrontal cortex. *Journal of Cognitive Neuroscience*, 17(8), 1306-1315.

[78] Moll, J., de Oliveira-Souza, R., Eslinger P. J., Bramati, I. E., Mourão-Miranda, J., Andreiuolo, P. A., & Pessoa, L. (2002). The neural correlates of moral sensitivity: A functional magnetic resonance imaging investigation of basic and moral emotions. *Journal of Neuroscience*, 22(7), 2730–2736.

[79] Moll, J., de Oliveira-Souza, R., Moll, F.T., Ignacio,F.A., Bramati, I.E., Caparelli-Daquer, E. M., Eslinger, P. J. (2005). The moral affiliations of disgust: A functional MRI study. *Cognitive Behavioural Neurology*, 18(1), 68–78.

[80] Nahab, F. B., Hattori, N., Saad, Z. S., & Hallett, M. (2009). Contagious yawning and the frontal lobe: An fMRI study. *Human Brain Mapping*, 30(5), 1744–1751.

[81] Narumoto, J., Okada, T., Sadato, N., Fukui, K., & Yonekura, Y. (2001). Attention to emotion modulates fMRI activity in human right superior temporal sulcus. *Cognitive Brain Research*, 12(2), 225–231.

[82] Ochsner, K. N., Knierim, K., Ludlow, D. H., Hanelin, J., Ramachandran, T., Glover, G., & Mackey, S. C. (2004). Reflecting upon feelings: An fMRI study of neural systems supporting the attribution of emotion to self and other. *Journal of cognitive neuroscience*, 16(10), 1746-1772.

[83] Oyserman, D., & Lee, S. W. (2008). Priming 'culture': Culture as situated cognition. In Shinobu Kitayama & Dov Cohen (Eds.), *Handbook of Cultural Psychology* (pp. 255–276). New York: Guilford Press.

[84] Parkinson, C., Sinnott-Armstrong, W., Koralus, P., Mendelovici, A., McGeer, V., & Wheatley, T. (2011). Is morality unified? Evidence that distinct neural systems underlie judgments of harm, dishonesty, and disgust. *Journal of Cognitive Neuroscience*, 23(10), 3162–3180.

[85] Pelphrey, K. A., Morris, J. P., Michelich, C. R., Allison, T., & McCarthy, G. (2005). Functional anatomy of biological motion perception in posterior temporal cortex: An fMRI study of eye, mouth and hand movements. *Cerebral Cortex*, 15(12), 1866–1876.

[86] Phelps, E. A., O'Connor, K. J., Cunningham, W. A., Funayama, E. S., Gatenby, J. C., Gore, J. C., & Banaji, M. R. (2000). Performance on indirect measures of race evaluation predicts amygdala activation. *Journal of Cognitive Neuroscience*, 12(5), 729–738.

[87] Platek, S. M., Keenan, J. P., Gallup Jr, G. G., & Mohamed, F. B. (2004). Where am I? The neurological correlates of self and other. *Cognitive Brain Research*, 19(2), 114–122.

[88] Platek, S. M., Loughead, J. W., Gur, R. C., Busch, S., Ruparel, K., Phend, N., et al. (2006). Neural substrates for functionally discriminating self-face from personally familiar faces. *Human Brain Mapping, 27*(2), 91–98.

[89] Premack, D., & Woodruff, G. (1978). Does the chimpanzee have a theory of mind? *Behavioral and Brain Sciences*, 1(04), 515-526.

[90] Poldrack, R. (2006). Can cognitive processes be inferred from neuroimaging data? Trends in Cognitive Sciences. 10(2), 59–63.

[91] Puce, A., Allison, T., Bentin, S., Gore, J. C., & McCarthy, G. (1998). Temporal cortex activation in humans viewing eye and mouth movements. *The Journal of Neuroscience*, 18(6), 2188–2199.

[92] de Quervain, D. J., Fischbacher, U., Treyer, V., Schellhammer, M., Schnyder, U., Buck, A., Fehr, E. (2004). The neural basis of altruistic punishment. *Science*, 305(5688), 1254–1258.

[93] Richeson, J. A., Baird, A. A., Gordon, H. L., Heatherton, T. F, Wyland, C. L., Trawalter, S., & Shelton, J. N. (2003). An fMRI investigation of the impact of interracial contact on executive function. *Nature Neuroscience*, 6(12), 1323–1328.

[94] Rilling, J. K., Sanfey, A. G., Aronson, J. A., Nystrom, L. E., Cohen, J. D. (2004). The neural correlates of theory of mind within interpersonal interaction. *Neuroimage*, 22, 1694–1703.

[95] Rozin, P., Lowery, L., Imada, S., & Haidt, J. (1999). The CAD triad hypothesis: A mapping between three moral emotions (contempt, anger, disgust) and three moral codes (community, autonomy, divinity). *Journal of Personality and Social Psychology*, 76(4), 574–586.

[96] Schulte-Rüther, M., Markowitsch, H. J., Fink, G. R., & Piefke, M. (2007). Mirror neuron and theory of mind mechanisms involved in face-to-face interactions: A functional magnetic resonance imaging approach to empathy. *Journal of Cognitive Neuroscience*, 19(8), 1354–1372.

[97] Sevdalis, V., & Keller, P. E. (2012). Perceiving bodies in motion: expression intensity, empathy, and experience. *Experimental Brain Research*, 222(4), 447-453.

[98] Senior, C., Barnes, J., Giampietroc, V., Simmons, A., Bullmore, E. T., Brammer, M., et al. (2000). The functional neuroanatomy of implicit-motion perception or 'representational momentum'. *Current Biology*, 10(1), 16–22.

[99] Singer, T., Seymour, B., O'Doherty, J., Kaube, H., Dolan, R. J., & Frith, C. D. (2004). Empathy for pain involves the affective but not sensory components of pain. *Science*, 303(5661), 1157–1162.

[100] Spengler, S., von Cramon, D. Y., & Brass, M. (2009). Was it me or was it you? How the sense of agency originates from ideomotor learning revealed by fMRI. *Neuro-Image*, 46(1), 290–298.

[101] Strathearn, L., Li, J., Fonagy, P., & Montague, P. R. (2008). What's in a smile? Maternal brain responses to infant facial cues. *Pediatrics*, 122(1), 40–51.

[102] Strathearn, L., Fonagy, P., Amico, J., & Montague, P. R. (2009). Adult Attachment Predicts Maternal Brain and Oxytocin Response to Infant Cues. *Neuropsychopharmacology*, 34(13), 2655–2666.

[103] Sutton, B. P., Goh, J., Hebrank, A., Welsh, R. C., Chee, M. W., & Park, D. C. (2008). Investigation and validation of intersite fMRI studies using the same imaging hardware. *Journal of Magnetic Resonance Imaging*, 28(1), 21-28.

[104] Takahashi, H., Kato, M., Matsuura, M., Mobbs, D., Suhara, T., & Okubo, Y. (2009). When your gain is my pain and your pain is my gain: Neural correlates of envy and schadenfreude. *Science*, 323(5916), 937–939.

[105] Tang, Y., Hojatkashani, C., Dinov, I. D., Sun, B., Fan, L., Lin, X., et al. (2010). The construction of a Chinese MRI brain atlas: A morphometric comparison study between Chinese and Caucasian cohorts. *NeuroImage*, 51(1), 33–41.

[106] Tricomi, E., Rangel, A., Camerer, C. F., & O'Doherty, J. P. (2010). Neural evidence for inequality-averse social preferences. *Nature*, 463(7284), 1089–91.

[107] Uddin, L. Q., Kaplan, J. T., Molnar-Szakacs, I., Zaidel, E., & Iacoboni, M. (2005). Self-face recognition activates a frontoparietal "mirror" network in the right hemisphere: An event-related fMRI study. *NeuroImage*, 25(3), 926–935.

[108] Ueda, Y., & Komiya, A. (2012). Cultural adaptation of visual attention: Calibration of the oculomotor control system in accordance with cultural scenes. *PloS one*, 7(11), e50282.

[109] Vogeley, K., Bussfeld, P., Newen, A., Herrmann, S., Happé, F., Falkai, P., et al. (2001). Mind reading: Neural mechanisms of theory of mind and self-perspective. *Neuro-Image*, 14(1), 170-181.

[110] Volz, K.G. & Gigerenzer, G. (2013) The brain is not "as-if" –Taking stock of the neuroscientific approach on decision making. In T.D. Papageorgiou, G. I. Christopoulos & S. M. Smirnakis (Eds.) Advanced Brain Neuroimaging Topics in Health and Disease – Methods and Applications. InTech (Rijeka, Croatia)

[111] Walter, H., Adenzato, M., Ciaramidaro, A., Enrici, I., Pia, L., & Bara, B. G. (2004). Understanding intentions in social interaction: The role of the anterior paracingulate cortex. *Journal of Cognitive Neuroscience*, 16(10), 1854-1863.

[112] Wang, Y., Ramsey, R., & de C. Hamilton, A. F. (2011). The control of mimicry by eye contact is mediated by medial prefrontal cortex. *The Journal of Neuroscience*, 31(33), 12001-12010.

[113] Wicker, B., Keysers, C., Plailly, J., Royet, J.-P., Gallese, V., & Rizzolatti, G. (2003). Both of us disgusted in my insula: The common neural basis of seeing and feeling disgust. *Neuron*, 40(3), 655-664.

[114] Wild, B., Erb, M., Eyb, M., Bartels, M., & Grodd, W. (2003). Why are smiles contagious? An fMRI study of the interaction between perception of facial affect and facial movements. *Psychiatry Research: Neuroimaging*, 123(1), 17-36.

[115] Williams, K. D., Cheung, C. K. T., Choi, W. (2000). Cyberostracism: Effects of being ignored over the internet. *Journal of Personality and Social Psychology*, 79(5), 748–762.

[116] Winston, J. S., Strange, B. A., O'Doherty, J., & Dolan, R. J. (2002). Automatic and intentional brain responses during evaluation of trustworthiness of faces. *Nature Neuroscience*, 5(3), 277–283.

[117] Wolf, I., Dziobek, I., & Heekeren, H. R. (2010). Neural correlates of social cognition in naturalistic settings: A model-free analysis approach. *NeuroImage*, 49(1), 894–904.

[118] Zeki, S. (2007). The neurobiology of love. *Federation of European Biochemical Societies*, 581(14*), 2575–2579.*

[119] Zhou, J., Gotch, C., Zhou, Y., & Liu, Z. (2008). Perceiving an object in its context—is the context cultural or perceptual? *Journal of Vision*, 8(12), 1-5.